PHOTOIONIZATION AND PHOTODETACHMENT

Part II

Advanced Series in Physical Chemistry

Published

Vol. 1: Physical Chemistry of Solids: Basic Principles of Symmetry and Stability of Crystalline Solids
H. F. Franzen

Vol. 2: Modern Electronic Structure Theory
ed. D. R. Yarkony

Vol. 3: Progress and Problems in Atmospheric Chemistry
ed. J. R. Barker

Vol. 4: Molecular Dynamics and Spectroscopy by Stimulated Emission Pumping
eds. H.-L. Dai and R. W. Field

Vol. 5: Laser Spectroscopy and Photochemistry on Metal Surfaces
eds. H.-L. Dai and W. Ho

Vol. 6: The Chemical Dynamics and Kinetics of Small Radicals
eds. K. Liu and A. Wagner

Vol. 7: Recent Developments in Theoretical Studies of Proteins
ed. R. Elber

Vol. 8: Charge Sensitivity Approach to Electronic Structure and Chemical Reactivity
R. F. Nolewajski and J. Korchowiec

Vol. 9: Vibration-Rotational Spectroscopy and Molecular Dynamics
ed. D. Papoušek

Advanced Series in Physical Chemistry – Vol. 10B

PHOTOIONIZATION AND PHOTODETACHMENT

Part II

Editor

Cheuk-Yiu Ng

Iowa State University, USA

World Scientific
Singapore • New Jersey • London • Hong Kong

Published by

World Scientific Publishing Co. Pte. Ltd.

P O Box 128, Farrer Road, Singapore 912805

USA office: Suite 1B, 1060 Main Street, River Edge, NJ 07661

UK office: 57 Shelton Street, Covent Garden, London WC2H 9HE

Library of Congress Cataloging-in-Publication Data

Photoionization and photodetachment / editor, Cheuk-Yiu Ng.

p. cm. -- (Advanced series in physical chemistry ; vol. 10)

Includes bibliographical references and index.

ISBN 9810238924 (set) -- ISBN 9810239513 (pt. 1) -- ISBN 9810239521 (pt. 2)

1. Photoionization 2. Photodissociation I. Ng, C. Y. (Cheuk-Yiu), 1947– II. Series.

QD716.P48.P46 1999

541.3'5--dc21 99-057907

British Library Cataloguing-in-Publication Data

A catalogue record for this book is available from the British Library.

Printed in Singapore by Uto-Print

ADVANCED SERIES IN PHYSICAL CHEMISTRY

INTRODUCTION

Many of us who are involved in teaching a special-topic graduate course may have the experience that it is difficult to find suitable references, especially reference materials put together in a suitable text format. Presently, several excellent book series exist and they have served the scientific community well in reviewing new developments in physical chemistry and chemical physics. However, these existing series publish mostly monographs consisting of review chapters of unrelated subjects. The modern development of theoretical and experimental research has become highly specialized. Even in a small subfield, experimental or theoretical, few reviewers are capable of giving an in-depth review with good balance in various new developments. A thorough and more useful review should consist of chapters written by specialists covering all aspects of the field. This book series is established with these needs in mind. That is, the goal of this series is to publish selected graduate texts and stand-alone review monographs with specific themes, focusing on modern topics and new developments in experimental and theoretical physical chemistry. In review chapters, the authors are encouraged to provide a section on future developments and needs. We hope that the texts and review monographs of this series will be more useful to new researchers about to enter the field. In order to serve a wider graduate student body, the publisher is committed to making available the monographs of the series in a paperbound version as well as the normal hardcover copy.

Cheuk-Yiu Ng

PREFACE

Photoionization and photodetachment processes are the simplest electron transfer reactions induced by photoabsorption. Thus, the understanding of these processes is of fundamental importance to physics, chemistry and biology. Although the physics and chemical applications of photoionization and photodetachment have much in common, these areas have been arbitrarily viewed as two different research fields in the past. Considering the similarity in experimental techniques used for the study of photoionization and photodetachment processes, the lack of communication between researchers in these research fields is surprising. Due to the low electron affinities of atomic and molecular species, photodetachment studies can be made using commercially available high-resolution visible and ultraviolet (UV) lasers. The high ionization energies (IEs) for atomic and molecular species require the use of vacuum ultraviolet (VUV) light sources for single-photon ionization studies. The recent development of VUV lasers and third generation synchrotron sources, together with the introduction of the pulsed field ionization photoelectron (PFI-PE) and ion detection schemes, has significantly broadened the scope of VUV photoionization and photoelectron studies. With these technical advances, it is now possible to measure rovibronically state-resolved transitions in photoionization of diatomic and simple polyatomic molecules. Thus, the resolution gap between photoionization and photodetachment studies has essentially been eliminated. The reverse processes of photoionization and photodetachment are electron–ion recombination and electron attachment, respectively. Significant experimental progress has also been made in the studies of these processes. This two-volume review represents the first attempt to present the topics of photoionization, photodetachment, electron–ion recombination, and electron attachment in a unified manner, covering some of the most important experimental and theoretical developments in recent years.

Volume 10A consists of twelve chapters. In Chapter 1, Parker gives a detailed account of the velocity-imaging method, a new variant of the ion imaging technique, for the investigation of photodissociation and dissociative photoionization processes. It is shown that an immersion lens can be used to map all ions with the same velocity to the same point on a two-dimensional detector, irrespective of their positions of origin. This velocity-imaging scheme is expected to make its way into experimental studies employing linear time-of-flight ion detection schemes, especially in experiments involving the use of a position sensitive detector. Gordon and Zhu describe in Chapter 2 the first experimental demonstration on the control of dissociative photoionization channels by employing the coherent phase control scheme originally proposed by Brumer and Shapiro. The focus of Chapters 3 and 4 is on the development and application of femtosecond laser time-resolved photoelectron and photoionization techniques for studies of radiationless relaxation processes, dissociation dynamics, and photochemical reactions in molecules and molecular clusters. The femtosecond laser pump-probe approach together with photoelectron detection schemes promises to add a new dimension for molecular dynamics studies. High-resolution spectroscopic experiments demand detailed theoretical interpretations. Rosmus and Chambaud review in Chapter 5 some highly accurate *ab initio* theoretical calculations, focusing on the interpretation of spectroscopic data for selected molecular ion systems in terms of Renner–Teller interactions and anharmonic resonances. The role of degenerate electronic states in the stability, protonation, charge transfer, and photodissociation of small molecular ions are also discussed from the theoretical prospective. In Chapter 6 Kimura summarizes selected PFI-PE studies of aromatic-Ar van der Waals (vdW) complexes using two-color laser excitation schemes. These high-resolution photoelectron measurements have yielded accurate vibrational and structural information for these complexes. The detection of mass-analyzed threshold ions (MATI) formed in PFI is first introduced by Johnson. In addition to reviewing the MATI scheme, Johnson also presents in Chapter 7 an ion spectroscopic technique, photoinduced Rydberg ionization (PIRI) spectroscopy, involving the detection of ions produced by photoexcitation of high-n Rydberg species. The availability of modern synchrotron radiation sources has been a key factor for the recent progress in VUV photoionization and photoelectron studies. Chapter 8 by Morioka reviews high-resolution synchrotron based threshold photoelectron (TPE) and TPE-photoion coincidence (TPEPICO) measurements on diatomics and rare-gas dimers using the penetrating field scheme. Ng provides in Chapter 9 a detailed account of experiments performed using the newly developed synchrotron based PFI-PE detection scheme at the Chemical Dynamics Beamline of the Advanced Light Source (ALS). Selected UV/VUV laser PFI-PE experiments on simple polyatomic molecules are also discussed with the emphasis on developing a semi-empirical simulation scheme for accurate IE determinations. Chapter 10 by Weitzel is devoted to unimolecular kinetics and dynamics studies using synchrotron based TPEPICO methods and two-color laser resonant enhanced multiphoton ionization schemes. The use of a reflection time-of-flight mass spectrometer in TPEPICO studies has made possible the measurement of dissociation rates in the range of 10^{-3}–10^{-7} s^{-1} for energy-selected molecular ions. Photoionization and photoabsorption studies of liquids represent an unexplored research area. The review of Chapter 11 by Ogawa concerns the development of two-photon laser ionization schemes as sensitive molecular detection methods based on conductivity measurements in solution and on surfaces in ambient air. Faubel pioneers the technique for the formation of liquid jets in high vacuum, allowing photoelectron studies of the liquid surfaces in collision free environments. He presents in Chapter 12 the experimental consideration for the formation of liquid jets in vacuum and the essence of selected photoelectron experiments involving the study of solutions in the form of liquid jets.

Volume 10B comprises of Chapters 13–22. The central theme of Chapter 13 by Larsson is on the study of dissociative electron–ion recombination processes using ion synchrotrons. The cross-section measurements for electron–ion recombination reactions are among the most difficult experiments. The recent ion synchrotron experiments can be considered a technical breakthrough and promise to provide fundamental insight into these complicated processes. Continetti reviews in Chapter 14 the recent development in combining the multi-coincidence and dissociative photodetachment methods for studying the energetics and dissociation dynamics of transient neutral species. This multi-coincidence scheme requires the use of a time and position sensitive detector and is directly applicable for studies of dissociative photoionization processes. Similar to PFI-PE spectroscopy on cations, anion-zero kinetic energy (ZEKE) photoelectron spectroscopy is a special development of photoelectron spectroscopy on anions. The resolution achieved in anion-ZEKE studies are usually more than an order of magnitude better than conventional anion-photoelectron spectroscopy. In Chapter 15, Boesl *et al.* reviews the results of recent anion-ZEKE studies. By using laser ablation for the preparation of non-volatile metal oxides, Wang has performed a series of anion-photoelectron studies of transition metal oxides, yielding valuable energetics and structural information for these oxides. The results of these studies are summarized in Chapter 16. In Chapter 17 Simons presents the theory for underlying electron detachment mechanisms in molecular anions, including detachment induced by photoabsorption, collisions, and intramolecular energy transfers. In accordance with the detachment mechanisms, various kinds of stability for anions are classified. The focus of Chapter 18 by Pratt is on the decay processes of molecular Rydberg states at energies higher than the first IE of the molecule. Since such Rydberg resonances may decay by autoionization and/or neutral dissociation, the study of such decay processes is relevant to dissociative recombination. Illenberger reviews in Chapter 19 molecular decay processes induced by electron capture at electron kinetic energies below the IE of the molecule. This chapter gives special emphasis on electron attachment reactions in binary vdW complexes and on dissociation processes of molecules adsorbed on metallic substrates and condensed molecules. The study of electron correlations and many-body interactions from first principle calculations is a main concern of theoretical research in atomic and molecular physics. Lin and Morishita present in Chapter 20 first principle calculations on doubly and triply excited states of atoms using hyperspherical coordinates. These theoretical studies reveal interesting symmetry properties for the joint motions of two and three electrons in doubly and triply excited atomic states, respectively. Chapter 21 by Berrah reports some recent state-of-the-art photoionization and photoelectron experiments conducted at the high-resolution atomic and molecular beamline of the ALS, covering the photon energy range of 25–350 eV. These experiments have yielded new observations and insights into many-body interactions in excited atomic and molecular states. The interpretation of these experimental results has also presented new challenges to the development of atomic and molecular theories. In this Chapter 22, Southworth *et al.* discuss the basic physics of X-ray interactions with atoms and molecules, together with illuminating examples of elastic and inelastic X-ray scattering experiments using X-ray beamlines at the National Synchrotron Light Source and the Advanced Photon Source.

In addition to presenting the recent technical and scientific progress, individual chapters also contain an expert assessment of the future research directions. It is my hope that this review will become a useful reference for graduate students and new researchers interested in the subject of photoionization and photodetachment. Finally, I would like to thank all contributors of Vols. 10A and 10B for their energy and enthusiasm devoted to this project. The opportunity for me to read and appreciate their state-of-the-art scientific writings was my best Christmas present.

Cheuk-Yiu Ng
Ames, Iowa
December, 1999

CONTENTS

Part I

Part II

Index

PART II

CHAPTER 13

DISSOCIATIVE ELECTRON–ION RECOMBINATION STUDIES USING ION SYNCHROTRONS

Mats Larsson
Department of Physics, Stockholm University
P.O. Box 6730, S-113 85 Stockholm, Sweden
E-mail: mats.larsson@physto.se

Contents

1. Introduction

Many chapters in this book deal with photoionization, which is a process where an electron bound to an atom or molecule is transferred to the electronic continuum by means of photon absorption. For an atomic ion A^{q+}, radiative recombination of $A^{(q+1)+}$ can be viewed as the inverse process of photoionization of A^{q+}. The free electron attaches to $A^{(q+1)+}$ to form A^{q+}, and the excess energy is carried away by a photon. If the ion $A^{(q+1)+}$ contains bound electrons, the free electron can excite one of these electrons while entering a bound atomic orbital. This is called dielectronic capture and results in the formation of a doubly excited ion $[A^{q+}]^{**}$, which is located in the ionization continuum of $A^{(q+1)+}$ and therefore inclined to autoionize. If the capture process is stabilized by photon emission from the doubly excited ion, the process dielectronic recombination has been completed.

When a molecular ion captures an electron, the molecular analogues of radiative and dielectronic recombination are in principle possible, however, they are usually completely overshadowed by a process which is far more effective than any of the atomic processes: dissociative recombination. It can be described as

$$\mathrm{AB}^+ + e^- \rightarrow \mathrm{A} + \mathrm{B}\,, \tag{1}$$

where A and B are atoms or molecules.

The reason for this process being so much more efficient than the atomic processes is that a molecular ion can stabilize the capture of an electron by making use of its internal structure, which allows for a radiationless rearrangement of the nuclei on a time scale (< 1 ps) much shorter than that for radiative transitions (> 1 ns). Figure 1 illustrates the process. The free electron can deposit its excess energy to the molecular ion either by exciting a bound electron or by exciting a vibrational mode. In either case, the neutral molecule now resides in a 'superexcited' state, which is located in the ionization continuum. In order to prevent autoionization, the molecule must lower its electronic energy. This is achieved by rapid molecular dissociation along the repulsive potential energy curve of the doubly excited state (the direct mechanism), hence the name 'dissociative recombination'. If the free electron loses its energy by vibrationally exciting the ion core, a radiationless transition (predissociation) from the vibrationally excited Rydberg state to the repulsive doubly excited state is required before dissociative stabilization of the electron capture can occur (the indirect

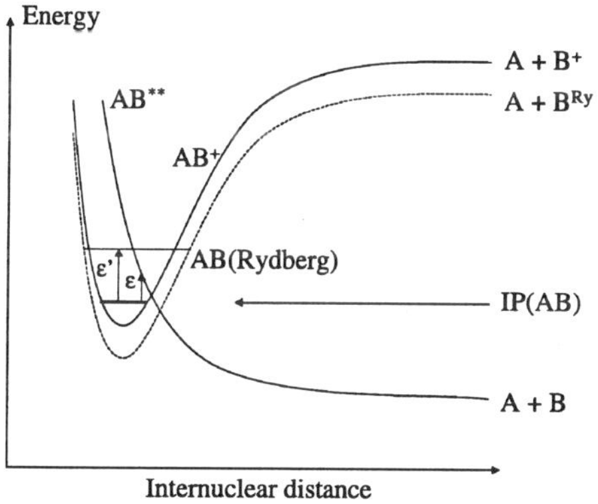

Fig. 1. A schematic illustration of dissociative recombination. The direct mechanism is initiated when the molecular ion AB^+ captures an electron with kinetic energy ε. The potential energy function for the neutralized and electronically *doubly* excited molecule AB^{**} is strongly repulsive, which renders the two atoms A and B to rapidly depart from each other. When the potential energy has dropped below the ionization potential, denoted IP(AB), the electron capture process has been stabilized. The direct dissociative recombination process has been completed. The indirect mechanism involves the capture of an electron with kinetic energy ε', so that a vibrationally and electronically *singly* excited molecule AB(Rydberg) is formed. A radiationless transition (predissociation) AB(Rydberg) $\rightarrow$ AB^{**} followed by the rapid departure of A and B from each other stabilizes the electron capture process and completes the indirect dissociative recombination process.

mechanism). In contrast with the atomic case, the first step in the direct mechanism, dielectronic capture, is not sharply resonant in the molecular case. This is because the vibrational motion relaxes the resonant constraint so that recombination can occur over a wide range of electron energies.

Dissociative recombination plays a very important role in many circumstances. It is the recombination mechanism that controls the charged-particle balance in any plasma cold enough to contain molecular ions. It occurs in planetary ionospheres, in comas of comets, in interstellar molecular clouds, in fusion devices, and in laboratory plasmas used in plasma processing technologies.

Dissociative recombination involves radiationless transition(s) from a molecular ion and a continuum electron to a nuclear continuum, and it could involve a large number of intermediate Rydberg states. This renders the process very difficult to treat theoretically. A further problem for theoretical calculations is the sparsity of relevant molecular data, such as potential energy surfaces, electronic capture widths, coupling elements, etc.

A number of factors also conspire to make the process difficult to study experimentally. Molecular ions are difficult to prepare and maintain at high concentrations, and they are particularly difficult to prepare with a well-defined vibrational distribution. The ions should further be brought to interact with electrons with a *known* velocity distribution. Depending on the experimental technique, the production of neutral particles (beam techniques) or the removal of charge (afterglow techniques, ion traps) is monitored. In both cases there are other processes that give rise to the same effect, and these processes must be completely understood so that they can be corrected for.

The technique relying on the monitoring of neutral particles is based on accelerated charged-particle beams. In order to have access to very-low-energy collisions ($\sim$ 1 meV), the ion and electron beams are merged at zero-degree angle over a known distance. Ion storage rings represent the most recent advancement of the merged-beams technique. In a storage ring, the molecular ions continuously circulate at very high speed in an extremely high vacuum. The ions all have nearly the same kinetic energy, and they occupy, for infrared active modes, the lowest vibrational quantum level. This very clean sample of ions serves as a well-defined target for a cold beam of electrons which interact with the molecular ion beam over a distance of about one meter. The application of ion storage ring technology to the process of dissociative recombination, and related processes, has led to a breakthrough in our understanding of this complicated process. This article reviews this development.

2. History and Applications of Dissociative Recombination

The mechanisms of dissociative recombination are shown schematically in Fig. 1. The direct mechanism was first suggested by Bates.[1] It is deceptively simple, but was long in coming, as described in some detail in Bates' final article on the subject.[2] It was not until the microwave experiments by Biondi and Brown[3,4] that Bates was finally convinced that molecular ions effectively recombines with electrons. The direct dissociative recombination cross-section is inversely proportional to the electron energy, ε, and proportional to the electronic capture width, Γ, which determines the likelihood for the initial electron capture process, the Franck–Condon factor, FC, which is the square of the vibrational overlap between the ion state and the resonant state, and the survival factor, S, which also depends on the electronic capture width and regulates the degree of autoionization before the electronic energy has been lowered below the ionization limit. Thus, the direct DR cross-section σ_{DR} can simplistically be expressed as $\sigma_{DR} = (\Gamma/\varepsilon)(FC)(S)$.[5]

The indirect mechanism was proposed much later by Bardsley[6] in order to account for differences between the electron-temperature dependences derived from the direct mechanism and those found experimentally. The indirect mechanism involves a sequence of two radiationless transitions, where the electron first is captured into a vibrationally excited Rydberg state, which is then subsequently predissociated by the resonant state. Whereas the direct mechanism gives a cross-section which is proportional to $1/\varepsilon$, the interference between the direct and the indirect mechanism often leads to a structured cross-section, as first demonstrated experimentally by McGowan *et al.*[7] for H_2^+ and later explained theoretically by Giusti[8] when she developed the multichannel quantum-defect theory (MQDT) to include also the dissociation of a molecule into two atoms (an important step towards a unified MQDT treatment of the direct and indirect mechanisms was taken a few years earlier by Lee[9]).

There are several reviews of dissociative recombination that give a comprehensive coverage of the subject;[5,10,11] more references can be found in the recent review of recombination of complex molecular ions.[12] The published proceedings from the three dissociative recombination conferences are also very useful sources of information.[13–15] The development of theoretical methods has been reviewed,[2,16–23] and theoretical discussions of issues closely related to dissociative recombination are given in Refs. 24 and 25. The application of ion storage rings to the study of dissociative recombination started about six years ago, with the first results being reported in a single issue of the *Physical Review Letters*.[26–28] The development since the first experiments has been reviewed in Refs. 29–31, and 32. Ion storage rings can also be used to study related processes, such as dissociative excitation. These processes fall outside the scope of the present review, but are comprehensively covered in a recent review by Dunn and Djuric.[33]

In a plasma where the gas temperature is of the order of a few thousand degrees or less, and the electron temperature is lower than a few eV, dissociative recombination is the dominant neutralizing agent. The historical example is the ionized layers of the Earth's upper atmosphere, for which Bates and Massey[2,34,35] sought suitable recombination mechanisms that could account for a total rate coefficient larger than 10^{-8} cm^3s^{-1}. Bates seminal paper in 1950[1] together with the early experiments by Biondi and Brown[3,4] led to the insight that dissociative recombination is the dominant recombination mechanism throughout much of the upper atmosphere. In the E and F regions (90–250 km) the most important molecular ions are O_2^+, NO^+ and N_2^+, whereas in the D region (80–90 km) also ions such as H_3O^+ and $H_3O^+{\cdot}H_2O$ have significant concentrations. Likewise, dissociative recombination plays an important role in the planetary ionospheres of Mars, Venus, Jupiter, Saturn and Uranus, and in the satellite of Saturn, Titan. The exothermicity of dissociative recombination leads to the production of suprathermal atoms in the atmosphere. At altitudes above the exobase (200–250 km for Mars and Venus), dissociative recombination is an important nonthermal source of translationally excited oxygen and nitrogen, with speeds exceeding the escape speed. Thus, dissociative recombination contributes to the loss of some species of planetary atmospheres.[36]

In the inner coma of comets, photoionization of the gas released from the comet nucleus produces ions and electrons, while dissociative recombination acts as the sink term for ions and also leads to a cooling of the electron gas.[37–39]

The discoveries of a large number of polyatomic molecules in dark molecular clouds led to the suggestion that they are formed by gas phase reactions.[40] Dissociative recombination was invoked as the final step in a chain of ion-molecule reactions that build up increasingly complex polyatomic molecular ions. Today, model calculations of molecular clouds involve hundreds of species and several thousand reactions.[41] The rate coefficients for these reactions are available at the University of Manchester Astrophysics Group homepage (http://saturn.ma.umist.ac.uk:8000/). About 10% of the almost 4000 reactions contained in the data base concerns dissociative recombination reactions. A major problem with the listed rate coefficients for polyatomic molecular ions is the uncertainties concerning the product branching ratios. In many cases they are no more than guesses.

The experimental study of dielectronic recombination of atomic ions, which started about 15 years ago,[42–44] was motivated partly because the process plays an important role in fusion plasmas. Low-charge impurities sputtered from the walls of Tokamak fusion devices are rapidly converted to highly charged ions by electron impact ionization. Dielectronic recombination leads to energy losses by the photon emission process which stabilizes the electron capture. The process is now fairly well understood,[45] and the recombination issues which will be of concern in the next generation fusion device, the International Thermonuclear Experimental Reactor (ITER), are related to the operation of a divertor to exhaust impurities from the plasma, and the fusion edge plasma. The colder part of these regions will contain a molecular component, and there is a demand for an understanding of electron impact related processes such as dissociative recombination and dissociative excitation.[46]

Finally, it is interesting to note that dissociative recombination has been suggested to play an important part in the fragmentation of molecules in intense laser fields.[47,48] The idea is that when the laser field has ionized the molecule AB to form AB^+, the free electron can be recaptured by AB^+ so that a doubly excited molecule AB^{**} is formed. This is the initial phase of dissociative recombination. If the laser field is still present during the dissociation phase, the doubly excited dissociating molecule can be ionized again, now at a larger internuclear distance. This would cause a bond stretching effect which has been observed but proved difficult to explain.

3. Ion Synchrotrons and Storage Rings

Storage rings have been used since 1988 for atomic physics experiments, and since 1992 for experiments involving molecular ions. They combine a source for ion production, an acceleration device, and a ring system in which a beam of charged particles are confined by magnetic fields. Some of the rings presently used for atomic physics operate also as synchrotrons, that is, the particles are further accelerated in the ring while the magnets are ramped synchronously to keep the beam orbit constant. Figure 2 shows the CRYRING facility at the Manne Siegbahn Laboratory at Stockholm University in Sweden.

The ring itself consists of a periodic structure of bending and focusing magnets, and devices for beam diagnostics. In order to allow long storage times, the vacuum system in which the ion beam is confined by means of the magnets is kept at a very low pressure ($\sim 10^{-11}$ Torr). The storage lifetimes exceed typical vibrational relaxation times of heteronuclear molecules by several orders of magnitude. This makes experiments with vibrationally cool molecular ions possible even if they contain substantial vibrational excitations when injected in the ring. It is worth noting that even a respectable vaccum of 10^{-9} Torr would be detrimental to any

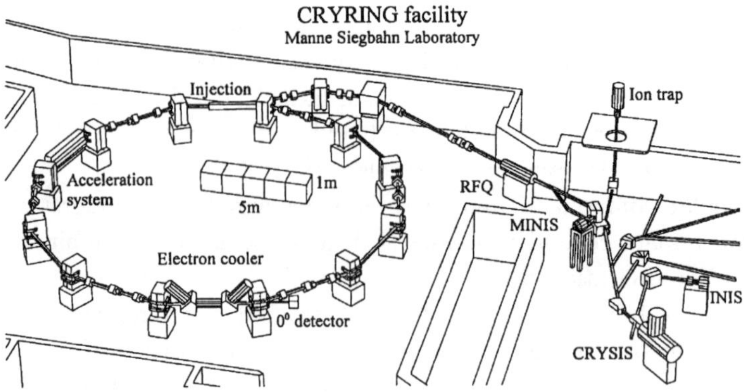

Fig. 2. The CRYRING facility at the Manne Siegbahn Laboratory, Stockholm University.

molecular physics ion storage ring experiment, since the ion beam would be destroyed in a fraction of a second by breakup on ambient gas particles.

Table 1 lists the ion storage rings used for molecular physics experiments, and some relevant parameters for these rings.[49–52] CRYRING has twelve bending magnets and six-fold symmetry. Six of the straight sections are left without focusing and correction magnets. The six other straight sections contain quadrupole magnets (four poles) as focusing elements and sextupole magnets (six poles) for second-order corrections to the beam orbit. Present days synchrotrons build on the strong focusing principle,[53,54] with alternating focusing and defocusing magnets.

A further advantage of long storage times is that the beam can be manipulated prior to the experiment. The manipulation usually involves a repeated passage of the ion beam through a beam of velocity-matched electrons in one section of the storage ring. Since the electron beam can be made translationally much colder than the ion beam at the time of injection, the passage of the ion beam through the electron beam transfers heat from the warm ions to the cold electrons. The procedure is called electron cooling and leads to a reduction of the movement of the ions with respect to each other and an increase of the ion density in phase space; a stored beam of small divergence, small cross-sectional area and narrow momentum spread is obtained. The cooling time scales as Z^2/M, where Z is the charge state and M is the atomic mass. For molecular ions, Z is usually equal to one, which implies that the cooling time becomes prohibitively long for systems that do not contain hydrogen and helium.

Table 1. Ion storage rings for molecular physics.

Ring	Location	Circumference (m)	Magnetic rigidity (Tm)	Max energy for H_2^+ beam (MeV/u)	Reference
Test Storage Ring (TSR)	Max-Planck-Institut für Kernphysik, Heidelberg, Germany	55.4	1.6	32	49
Test Accumulation Ring for the Numatron Accelerator Facility (TARN II)	Institute for Nuclear Study (INS), Tokyo, Japan	78.0	6.1	465	50
Aarhus Storage Ring Denmark (ASTRID)	University of Aarhus, Denmark	40.0	1.6	32	51
Cryogenic Ion Source Ring (CRYRING)	Manne Siegbahn Laboratory, Stockholm University, Sweden	51.6	1.4	24	52

4. Experimental Procedure

Ion storage technology allows very 'clean' experimental conditions to be obtained. The stored ion beam passes through an electron target in one section of the ring, which so far has been identical with the electron cooler beam. (There are plans at some rings to install a second, separate, electron target.) The target is shaped as a cylinder of about one-meter in length. It is approximately homogeneous and it completely overlaps the stored ion beam even when the ions are not effectively cooled translationally. This means that the beam overlap factor need not be measured. During the translational cooling phase, which can last for $\sim$ 5–30 s, the acceleration voltage of the electron cooler is set so that the electrons move with the same velocity as the ions. For light ions coasting in the ring, this criterion will be automatically fulfilled; the frictional force between the ion and electron beams will accelerate or decelerate the ions so that the ion beam velocity becomes equal to the electron beam velocity. This is the condition that maximizes the production of neutral particles by dissociative recombination, since the cross-section usually scales as the inverse of the collision energy. Likewise, for heavier molecular ions, for which one cannot rely on the frictional force to 'lock' the two beams to each other, the production of neutral particles can be used as a criterion that the velocity of the two beams is equal.

The velocity difference between the two beams is usually referred to as the detuning velocity:

$$\nu_d = |\bar{\nu}_e - \bar{\nu}_i| , \tag{2}$$

where $\bar{\nu}_i$ and $\bar{\nu}_e$ are the average ion and electron laboratory velocities, respectively. The detuning velocity is, nonrelativistically, related to the detuning energy through $E_d = m_e\nu_d^2/2$. When the ion beam is electron cooled, $E_d = \nu_d = 0$ and the laboratory electron energy is equal to E_{cool}. When data are taken at a desired detuning energy, the electron cooler cathode voltage is changed so that electron beam energy becomes E_{meas}. The detuning energy and the two laboratory electron beam energies are related through

$$E_d = (\sqrt{E_{meas}} - \sqrt{E_{cool}})^2 , \tag{3}$$

which means that the detuning velocity is fully controlled by means of the acceleration voltage of the electron cooler. It is worth noting that that the electron beam energy is not equal to the cathode voltage times the electron charge, eU_c. The cylindrical electron beam consists of a uniform distribution of negatively charged particles, which modifies the electron energy in the center of the beam to[55]

$$E_e = eU_c - \frac{I_e r_e m_e c^2 (1 - \xi(E_e))}{e\nu_e} \left[1 + 2\ln(b/a)\right] , \tag{4}$$

where E_e is E_{cool} or E_{meas}, ν_e the corresponding electron velocity, I_e the electron current, r_e the classical electron radius, b the radius of the beam tube, and a the radius of the electron beam. The potential of the cathode U_c is essentially the applied voltage V_c minus a contact potential ϕ_c. The term $\xi(E_e)$ is included in order to account for the compensation of the electron space charge potential caused by slow ions trapped in the electron beam. These ions derive from electron-impact ionization of ambient gas particles.

In the first generation of electron coolers/targets the transverse electron-velocity spread was essentially determined by the cathode temperature of about 1000 K, and only the longitudinal electron-velocity distribution was compressed by the acceleration of the electrons. These coolers were operated with a transverse electron temperature of $kT_{e\perp} \approx 100$ meV. The development of the adiabatic expansion technique by Danared *et al.*[56] has radically changed this condition, and the coldest electron beams are now produced by expanding the electron beam starting from a superconducting magnet. Thus, the electron coolers at both CRYRING and TARN II are now capable of delivering electron beams having $kT_{e\perp} = 1$ meV.

The electron-velocity distribution in the beam is given by the following two-component Maxwell velocity distribution:

$$f(\boldsymbol{\nu}_e) = \frac{m_e}{2\pi kT_{e\perp}}\left(\frac{m_e}{2\pi kT_{e\parallel}}\right)^{1/2} \times \exp\left(-\frac{m_e\nu_{e\perp}^2}{2kT_{e\perp}} - \frac{m_e\nu_{e\parallel}^2}{2kT_{e\parallel}}\right), \tag{5}$$

where the longitudinal electron temperature, $kT_{e\parallel}$, is of the order of 0.1 meV. The interaction via a cross-section $\sigma(\nu)$ of an electron beam characterized by the distribution given in Eq. (5) and an ion beam completely overlapped by the electron beam results in a rate of interaction per unit time equal to

$$R = n_e\langle\nu\sigma\rangle Nl/C\,, \tag{6}$$

where n_e is the electron density, N the number of stored ions, l the interaction length, C the circumference of the storage ring, and $\langle\nu\sigma\rangle$ the rate coefficient in the electron cooler, which is given by

$$\alpha(\nu_d) = \langle\nu\sigma\rangle = \int \sigma(\nu)\nu f(\boldsymbol{\nu}_e)d^3\boldsymbol{\nu}_e\,. \tag{7}$$

Assuming that the rest frame of the electrons is used implies that $f(\nu_e)$ is given by Eq. (5), the velocity difference ν between an electron and an ion is given by $\nu^2 = (\nu_{e\perp})^2 + (\nu_d + \nu_{e\parallel})^2$, and $d^3\nu_e = 2\pi\nu_{e\perp}d\nu_{e\perp}\nu_{e\parallel}$. Equation (7) implies that the velocity spread in the ion beam can be neglected. This is usually a valid assumption owing to the large mass difference between an ion and an electron, which results in an $(m_e/M)^{0.5}$ smaller velocity spread in the ion beam if the two beams have the same temperature. The effect of a finite electron velocity distribution is shown for model cross-sections in Refs. 29 and 30. Equation (7) is suitable for comparisions of experiment and theory. The theoretical cross-section is folded with the electron-velocity distribution used in the experiment, which gives a theoretical electron-velocity dependent rate coefficient $\alpha(\nu_d)$ that can be directly compared with the experimental one.

The detuning energy equals the energy in the electron–ion center-of-mass frame (essentially given by $m_e\nu^2/2$) as long as it is larger than the transverse thermal energy of the electron beam (i.e. $kT_{e\perp}$). Whereas E_d can be smaller than $kT_{e\perp}$, the center-of-mass energy approaches $kT_{e\perp}$ when $E_d \rightarrow 0$.

Dissociative recombination of molecular ions in the electron cooler results in the production of neutral particles. The first dipole (bending) magnet after the electron cooler filters the neutral products from the ion beam. For singly charged molecules, neutral particles are usually the dominant DR products; they continue through the dipole magnet and impinge on a detector, as shown in Fig. 2. The recombination products are easily separated from the neutral products occurring from ion breakup on rest-gas particles, $AB^+ + Rg \rightarrow A^+ + B$ or $A + B^+$, by means of an ion-implanted-silicon detector. At the high energies used in storage rings ($\sim$ MeV), the background process $AB^+ + Rg \rightarrow A + B$ is often negligible and does not interfere with the recombination signal.

Different schemes can be used to measure $\langle\nu\sigma\rangle$. The electron cooler can be continuously ramped a number of times from E_{cool} to an E_{meas}(max) while the number of neutral products are counted. It can also be switched back and forth between E_{cool} and E_{meas} during each injection cycle. In either case, a relative value of $\langle\nu\sigma\rangle$ is obtained. If the number of stored ions is measured, the absolute value of $\langle\nu\sigma\rangle$ can be obtained. The number of stored ions is usually measured with a current transformer, which measures the magnetic field created by the circulating ions. The experimental procedures are described in more detail in Refs. 30, 57 and 58.

When an energy-dependent cross-section $\sigma(E)$ has been determined, the thermal DR rate coefficient $\alpha(T_e)$ for an isotropic Maxwellian electron-velocity distribution at an electron temperature T_e is obtained by integrating the cross-section over the electron energy:

$$\alpha(T_e) = \int_0^\infty \frac{8pm_e E}{(2pm_e kT)^{3/2}}\, \sigma(E) \times \exp(-E/kT)\, dE\,. \tag{8}$$

In order to calculate the thermal rate coefficient, $\sigma(E)$ must first be extracted from the measured $\langle \nu\sigma \rangle$. When the detuning energy is larger than $kT_{e\perp}$, a good approximation is to use $\langle \nu\sigma \rangle / \nu_d$ as the cross-section. When the detuning is smaller than $kT_{e\perp}$, a deconvolution method must be used, as described by Mowat *et al.*[59]

The chemical modeling of plasmas such as the interstellar medium or planetary ionospheres requires the knowledge also of recombination products and their state of excitation. For diatomic molecular ions, the state of excitation is directly determined by the amount of kinetic energy which is released in the recombination process. The kinetic energy can be measured by recording the image of the neutral dissociation products at some distance from the electron cooler. The imaging device can consist of one or several multichannel plates (MCPs) linked to a phosphor screen and a charged-coupled-device (CCD) camera,[60,61] but also a less expensive solution combining an ion-implanted-silicon detector with a wedge shaped foil has been used[62] (at the expense of lower resolution). The transverse separation, D (i.e. the projected distance on the surface of the imaging detector), between the dissociation products can be expressed in terms of the kinetic energy release, E_{KER}; the ion beam energy, E_b; the masses of the neutral fragments, m_{A} and m_{B}, respectively; the distance between the location of the recombination event and the imaging detector, L; and the angle θ between the molecular axis and the electron beam during dissociation:

$$D = \sqrt{\frac{E_{\mathrm{KER}}}{E_b}}\, \frac{(m_{\mathrm{A}} + m_{\mathrm{B}})}{\sqrt{m_{\mathrm{A}} m_{\mathrm{B}}}}\, L \sin\theta\,. \tag{9}$$

The extended electron–ion interaction region, which leads to an uncertainty regarding the location of the recombination event, and the unknown angle θ, leads to a distribution, $P(D)$, of projected distances. The most recent imaging detectors used at ion storage rings include the possibility to measure also the time difference between the arrival of particles A and B on the detector.[63,64]

For dissociative recombination of polyatomic molecular ions it has so far been possible in storage rings to measure only the branching ratios into neutral products, but not their state of excitation. The technique used for doing these measurements will be described in Sec. 7.

5. Dissociative Recombination through a Crossing of Potential Curves

The mechanism for dissociative recombination shown in Fig. 1 has been given the descriptive name the *crossing mode.*[2] It was for a long time believed that a favorable crossing was a prerequisite for the occurrence of dissociative recombination. As we will see in the next section, a second mode of recombination, the *tunneling mode*, has recently been recognized as important. The recognition of this second mode came almost half a century after Bates' seminal paper.[1] Thus, it is natural that the majority of molecular ions for which theoretical and experimental dissociative recombination data are available belongs to the crossing mode category.

Table 2 lists all ion storage ring experiments known to the author concerning dissociative recombination that have resulted in a published paper, a paper that is in press, or a paper that has been submitted for publication (by August 1998), and the type of results that have been obtained in these experiments. Experimental results that have appeared exclusively in doctorial theses are not included.

5.1. *Diatomic Hydrogen Ions*

Being the simplest diatomic molecule, it is natural that H_2^+ and its isotopomers represent benchmark systems for dissociative recombination, in particular with respect to the comparison of experiment and theory. The experimental and theoretical studies of DR of H_2^+ and its isotopomers in their zeroth vibrational levels are described in some detail in two recent reviews,[2,32] and only a brief summary of the present situation will be given here.

The development of a single-pass merged electron-molecular ion beams technique[65] played an important role for the understanding of DR of H_2^+. Because of the rapid conversion of H_2^+ to H_3^+ by ion–molecule reactions, H_2^+ is inaccessible to the afterglow techniques. The merged beams technique also allows the measurement of DR cross-sections or rate coefficients $\langle \nu\sigma \rangle$, which are more suitable quantities for comparison with theory than is the thermal rate coefficient measured in the afterglow techniques. Whereas the

Table 2. Listing of ion storage ring measurements of dissociative recombination.

Ion	Cross sections	Thermal rate coefficient at 300 K	Branching ratios	Reference (Ring)	Comments
H_2^+	Yes		Yes	62 (CRYRING)	Vibrationally excited
H_2^+	Yes			233 (CRYRING)	Vibrationally excited
H_2^+	Yes			73 (ASTRID)	Vibrationally excited
H_2^+	Yes		Yes	75 (ASTRID)	Vibrationally excited
HD^+	Yes (relative)			27, 237 (TSR)	$\nu = 0$
HD^+	Yes			72 (CRYRING)	$\nu = 0$; includes theory
HD^+	Yes (relative)			74 (TARN II)	$\nu = 0$, includes theory
HD^+	Yes			76 (CRYRING)	$\nu = 0$; includes theory
HD^+	No		Yes	61 (TSR)	$\nu = 0$
HD^+	No		Yes	81 (TSR)	$\nu = 0$; includes theory
HD^+	Yes		Yes	82 (TSR)	$\nu < 7$; includes theory
D_2^+	Yes			234 (CRYRING)	Vibrationally excited
$^3HeH^+$	Yes (relative)			26 (TARN II)	$\nu = 0$
$^4HeH^+$, $^4HeD^+$	Yes (relative)		Yes	142 (TARN II)	$\nu = 0$
$^3HeH^+$	Yes			143 (CRYRING)	$\nu = 0$
$^3HeH^+$	Yes			59 (CRYRING)	$\nu = 0$
$^4HeH^+$	Yes	3×10^{-8} cm^3s^{-1}	Yes	145 (CRYRING)	$\nu = 0$
$^3HeD^+$, $^4HeH^+$	No		Yes	146 (CRYRING, TSR)	$\nu = 0$

Table 2. (*Continued*)

Ion	Cross sections	Thermal rate coefficient at 300 K	Branching ratios	Reference (Ring)	Comments
$^{3,4}HeH^+$, $^{3,4}HeD^+$	Yes (relative)			148 (TARN II)	$\nu = 0$; includes theory
$^{3}HeH^+$	Yes			149 (CRYRING)	$\nu = 0$
CH^+	Yes (relative)		Yes	58, 229 (TSR)	$\nu = 0$
CD^+	Yes (relative)			235 (TSR)	$\nu = 0$
OH^+	Yes (relative)			236 (TSR)	$\nu = 0$; includes theory
OH^+	No		Yes	84 (CRYRING)	$\nu = 0$; includes theory
N_2^+	No		Yes	116 (ASTRID)	Vibrationally excited
N_2^+	Yes	$1.75 \times 10^{-7}\,cm^3s^{-1}$	Yes	110 (CRYRING)	$\nu = 0$ (46%), $\nu = 1$ (27%)
NO^+	Yes	$4 \times 10^{-7}\,cm^3s^{-1}$	Yes	126 (ASTRID)	$\nu = 0$
O_2^+	No		Yes	106 (ASTRID)	Vibrationally excited
O_2^+	Yes	$2 \times 10^{-7}\,cm^3s^{-1}$	Yes	86 (CRYRING)	$\nu = 0$
CO^+	Yes	$2.75 \times 10^{-7}\,cm^3s^{-1}$	Yes	133 (CRYRING)	$\nu = 0$
CN^+	Yes	$3.4 \times 10^{-7}\,cm^3s^{-1}$	Yes	136 (CRYRING)	$\nu = 0$ of two states
H_3^+	Yes	$1.15 \times 10^{-7}\,cm^3s^{-1}$		28, 168 (CRYRING)	$\nu = 0$
H_3^+	No		Yes	181 (CRYRING)	$\nu = 0$
H_2D^+	Yes	$0.6 \times 10^{-7}\,cm^3s^{-1}$	Yes	175, 176, 204 (CRYRING)	$\nu = 0$
HD_2^+, D_3^+	Yes (relative)		Yes	177 (TARN II)	$\nu = 0$ and $\nu > 0$ (see text)

Table 2. (*Continued*)

Ion	Cross sections	Thermal rate coefficient at 300 K	Branching ratios	Reference (Ring)	Comments
D_3^+	Yes	$0.27 \times 10^{-7}\ cm^3s^{-1}$		173, 174 (CRYRING)	$\nu = 0$; electric field added
H_3O^+	Yes (relative)		Yes	205, 207 (ASTRID)	$\nu = 0$
CH_3^+	Yes (relative)		Yes	207 (ASTRID)	$\nu = 0$
H_2O^+	No		Yes	207 (ASTRID)	$\nu = 0$
CH_2^+	Yes	$6.4 \times 10^{-7}\ cm^3s^{-1}$	Yes	209 (CRYRING)	$\nu = 0$
CH_5^+	Yes	$2.8 \times 10^{-7}\ cm^3s^{-1}$	Yes	220 (CRYRING)	$\nu = 0$
NH_2^+	Yes (relative)		Yes	208 (CRYRING)	$\nu = 0$
NH_4^+	Yes (relative)		Yes	208 (CRYRING)	Vibrationally excited
HCN^+/HNC^+	No		Yes	218 (CRYRING)	$\nu = 0$

early experimental and theoretical papers[6,7] provided the first qualitatively correct insight into the DR of H_2^+, it was not until more than ten years later that a complete quantitative comparison of experiment and theory finally was accomplished by the research groups of Giusti-Suzor and Mitchell.[66–69] The comparison revealed a good agreement both in terms of the absolute values of the cross-section and the positions of several 'window' resonances caused by the interference between the direct and indirect mechanisms. The comparison concerned only DR at electron energies well below the ion dissociation energy, where the $(2p\sigma_u)^2\,{}^1\Sigma_g^+$ resonant state completely dominates the recombination. It was not possible, however, to unambiguously assign the experimental results to H_2^+ populating only its zeroth vibrational level. Neglected in this comparison was the rotation of the molecule. The effect of internal rotation of the recombining ion was included in order to provide useful thermal rate coefficients for a wide range of electron temperatures.[70] Still neglected in this calculation was the effect of couplings introduced by the H_2^+ rotation. When rotational coupling finally was included in MQDT calculations,[71] it turned out that the 'window' resonances were markedly affected. A new comparison of theory and experiment was obviously called for.

Because of its dipole moment, the isotopomer HD^+ is preferred in storage ring experiments since it will vibrationally relax by emission of infrared radiation. It was also the molecule chosen for one of the very first experiments.[27] Although this experiment did not reveal any new information about the $(2p\sigma_u)^2\,{}^1\Sigma_g^+$ resonant state low-energy recombination, it showed for the first time new modes of DR taking place at 'high' electron energies (~ 9 eV), which occur through Rydberg states converging to HD^+ $(2p\sigma_u)$. The mechanism leading to this broad high-energy resonance is now well understood,[72] although there remain some discrepancies between the results from different storage ring experiments.[27,72–75] The invention of the adiabatically expanded electron beam[56] allowed the low-energy resonances to be observed, something which was exploited in two different comparisons of theory and experiment. The first was based on relative cross-sections measured in TARN II[74] by means of a $kT_{e\perp} = 10$ meV electron beam, and the apparent agreement between experiment and theory was only qualitative. In the second comparison, the absolute cross-sections from CRYRING, also obtained by means of a 10 meV electron beam, were compared with the results of an MQDT calculation which also included rotational coupling.[76] In the low-energy region, up to 0.3 eV, the theoretical results are on average 50% larger than the experimental results. The positions and shapes of the resonances, however, compare well with the experimental ones. The theoretical results of Takagi[71,74] are even larger. The agreement between the two storage ring experiments is remarkable; every low-energy resonance appears with identical shape and position (an absolute comparison is unfortunately not possible). The absolute cross-sections from CRYRING are also in good agreement in the thermal energy region with the results from ASTRID.[75] Very recently the DR cross-section of HD^+ has been measured in CRYRING with a $kT_{e\perp} = 1$ meV electron beam, which has allowed resonances below 10 meV to be observed. A detailed comparison of the experimental cross-sections from ASTRID, CRYRING and TSR is also being planned.

The rotation of HD^+ affects mainly the indirect process through bound Rydberg states.[71,76] A question is naturally whether rotational effects are expected to play an important role also for other molecules. The answer here seems, in general, to be no. Calculations by the MQDT approach show that rotational effects play a role only for light ions.[77]

So far, we have said nothing about the state of excitation of hydrogen and deuterium when the DR process is completed. In other words, what is n and n' in the reaction $HD^+ + e^- \to H(n) + D(n')$? An early attempt to measure this (for the isotopomer D_2^+) was made by Dunn and coworkers[78,79] using an optical method to detect the state of excitation of one of the D atoms. At storage rings, the issue has been addressed at TSR,[61] CRYRING (for H_2^+)[62] and ASTRID[75] by means of the two-dimensional imaging techniques described in Sec. 4. It was discovered that when the electron collision energy was 0 eV (i.e. $E_d = 0$ eV), one atom was found in $n = 1$ and the other in $n = 2$, Ref. 61; no events could be related to both atoms being in $n = 1$.[61,62,75] This is an understandable result if we consider the DR mechanism of HD^+ and the energetics shown in Fig. 3. At 0 eV collision energy, the DR process is totally dominated by the $(2p\sigma_u)^2\,{}^1\Sigma_g^+$ resonant state, which goes to $H^+ + D^-$ in the diabatic limit. In order for the dissociation flux to be redirected to the $H(1s) + D(1s)$ limit, the resonant state

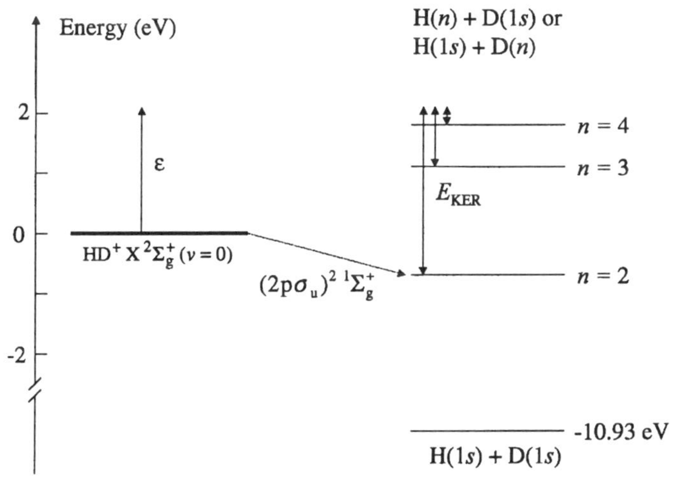

Fig. 3. Energy level diagram for HD^+ and $H + D$ atomic limits. The electron energy is given by the length of the arrow labeled ε.

must interact with a state that correlates with this limit; no such state exists. The complete dominance of the H(1*s*) + D(2*s*) limit (or vice versa) at 0 eV shows that the assumption of the $(2p\sigma_u)^2\,{}^1\Sigma_g^+$ state adiabatically dissociating to H(1*s*) + D(2*s*)[80] is physically correct. When the collision energy is sufficient to make the H(1*s*) + D($n \geq 3$) limits energetically allowed, a redistribution of the flux occurs, as demonstrated in an experiment at the TSR which made use of a three-dimensional imaging technique.[81] The branching ratio results from this experiment are shown in Fig. 4. The results have been normalized so that the branching into $n = 2$ is unity when the electron energy is smaller than or equal to 1.1 eV. It is striking that all

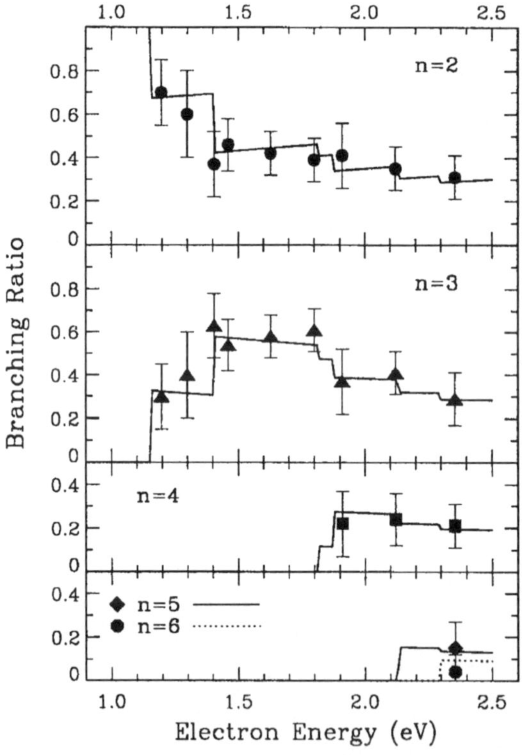

Fig. 4. Branching ratios into H($n = 2, 3, 4, 5, 6$) + D(1*s*) (or vice versa) for dissociative recombination of HD^+ as a function of electron energy. The full drawn lines (dotted for $n = 6$) show the results of theoretical calculations based on a Landau–Zener formulation of the crossing process, as described briefly in the text. (Reproduced with permission from Ref. 81.)

dissociation limits acquire finite branching ratios as soon as they become energetically allowed. When the electron energy is 2.35 eV, all channels up to $n = 6$ are open. At this energy, no single channel dominates and no channel has diminished. The results can be explained by a relatively simple Landau–Zener treatment (full drawn lines in Fig. 4) of the crossings of the $(2p\sigma_u)^2\,{}^1\Sigma_g^+$ resonant state with a series of $1s\sigma_g nl\sigma_g$ Rydberg states, where l is 0 or 2.[81] These crossings takes place in a narrow range of internuclear distances (2.6–3.0 a_0), and leads to a redistribution of flux from the $n = 2$ limit to more highly excited limits.

The presence of vibrationally excited levels has often been a problem in DR experiments, and ion storage rings have proved an effective remedy for this. But there is also an interest in determining DR properties of vibrationally excited levels, and this requires a technique which allows the population of vibrationally excited levels to be controlled or monitored. The experimenter is here faced with a formidable challenge. An elegant technique that allows the simultaneous measurement of the DR cross-section and the vibrational population has recently been developed at the TSR and applied to HD^+.[82] While the recombination fragment imaging spectrum of HD^+ was measured in the electron cooler/target section of the ring by means of a two-dimensional imaging technique,[61] a fraction of the stored HD^+ beam was extracted from the ring and sent through a thin foil. The foil serves to strip the electron off the HD^+ ion, which leaves two positively charged atomic ions that suffer a strong Coulomb repulsion. The ions were detected downstream from the foil by means of a three-dimensional imaging detector which gives a 'fingerprint' of the two nuclei when they enter the foil. The foil/imaging technique is known as Coulomb explosion imaging (CEI),[83] and has been used extensively to determine molecular geometric structures. When applied at the TSR, it allowed the determination of the vibrational population of the HD^+ ions. Combined with the two-dimensional imaging spectra of DR events, the individual rate coefficients (see Eq. (7)) of excited vibrational levels could be determined with respect to the that for the zeroth vibrational levels. The results were compared with those of theoretical calculations and the agreement was found to be good or fairly good for all levels except $\nu = 5$.[82] For this level, the result of the MQDT calculations by Schneider and Suzor-Weiner[82] is a factor 250 larger than the experimental result. The reason for this is not yet understood.

5.2. *Monohydride Ions*

Dissociative recombination of monohydride ions has been reviewed recently.[2,32] The process which involves an electronic excitation of the ionic core has not been further investigated since the last review.[32] An imaging experiment with OH^+ in CRYRING led to discovery of a very small kinetic energy release when the electron energy was sufficient for the $O(^3P)+H(n = 2)$ dissociation limit to become energetically allowed.[84] Recombination of OH^+ is known to proceed through the $2\,^2\Pi$ resonant state,[85] which dissociates diabatically to $O(^1D) + H(n = 1)$. The imaging result shows that part of the dissociating flux is redirected, probably at avoided crossings between the $2\,^2\Pi$ state and the $3\,^2\Pi$, $4\,^2\Pi$ and $5\,^2\Pi$ states, which all correlate diabatically to the $O(^3P) + H(n = 2)$ limit.

5.3. *Nonhydride Diatomic Ions*

This category is dominated by the three atmospherically important ions O_2^+, N_2^+, and NO^+, but also CO^+ and CN^+ will be discussed.

5.3.1. *Oxygen*

The thermal rate coefficient for O_2^+ occupies an unusual role in the respect that different experiments over a time span of thirty years almost unanimously agree[2] that $\alpha(O_2^+) \approx 2 \times 10^{-7}$ cm^{3}s^{-1}, the very recent measurement based on cross-section data from CRYRING being no exception.[86] With few exceptions,[87] most theoretical work has been devoted to the understanding of the formation of $O(^1S)$ in dissociative recombination of O_2^+. It is the $O(^1S) \rightarrow O(^1D)$ transition in neutral atomic oxygen which gives rise to the green airglow at 5577 Å in the terrestrial atmosphere. It is worth noting that the earliest mention of dissociative recombination was made in an attempt to identify a source of the green oxygen line.[88] Guberman showed[89,90] that DR of $O_2^+(\nu < 10)$ leading to the production of $O(^1S)$ must occur through the $^1\Sigma_u^+$ state of O_2 that dissociates to $O(^1S)+O(^1D)$, and proceeded to calculate, using the MQDT approach, the partial rate coefficient for DR of O_2^+ through the $^1\Sigma_u^+$ state.[91] By dividing the partial rate coefficient with the (total) thermal rate coefficient measured in the laboratory, Guberman arrived at a quantum yield of 0.0016 for $O(^1S)$ ($O_2^+(\nu = 0)$; $T_e = 300$ K).[91]

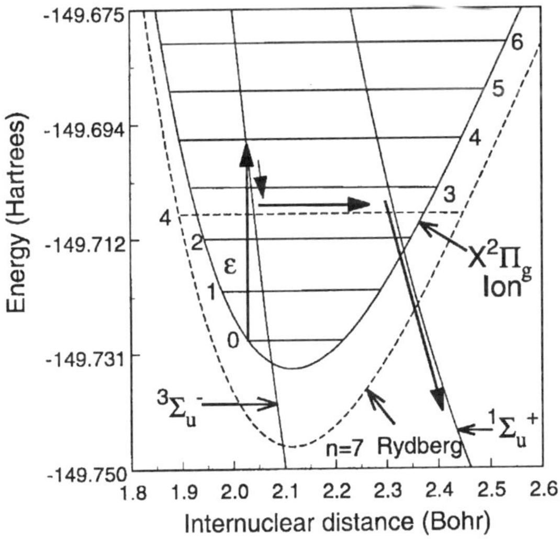

Fig. 5. Potential energy curves for O_2^+ and O_2. The ${}^1\Sigma_u^+$ and ${}^3\Sigma_u^-$ states are repulsive states of O_2 and the electronic ground state of O_2^+ is labeled $X\,{}^2\Pi_g$. The vertical arrow shows direct capture into the dissociative ${}^3\Sigma_u^-$ state. The Rydberg state is of mixed ${}^1\Sigma_u^+$ and ${}^3\Sigma_u^-$ character. Dissociation to $O({}^1S)$ + $O({}^1D)$ occurs along the repulsive ${}^1\Sigma_u^+$ state. (Reproduced with permission from Ref. 107.)

Figure 5 shows the relevant potential curves of O_2^+ and O_2. The ${}^1\Sigma_u^+$ state crosses the ion ground state close to the right turning point of $\nu = 2$. It is immediately obvious from the potential curves that $\nu = 0$ is expected to have a much lower quantum yield than are $\nu = 1$ and 2 (neglecting for a moment the new mechanism shown in Fig. 5 and discussed below). Quantitative calculations by Guberman confirm that direct DR for the ${}^1\Sigma_u^+$ channel leads to quantum yields for $\nu = 1$ and 2 that are factors 23 and 78, respectively, higher than the quantum yield for $\nu = 0$.[92] These numbers make the problem apparent; any experiment or observation of the $O({}^1S)$ quantum yield will be sensitively dependent on the vibrational excitation in the recombining O_2^+ ions. Rocket and ground based observations of the Earth's ionosphere have given quantum yields in the range 0.01 to 0.23,[93–98] and indirect evidence from laboratory experiments have suggested a $\nu = 0$ quantum yield larger than the calculated one.[99–102] However, none of these results concerns a direct measure of the $O({}^1S)$ quantum yield for DR of $O_2^+(\nu = 0)$. Bates suggested[97] that inclusion of the indirect mechanism could increase the production of $O(^1S)$ from DR of O_2^+ in the zeroth vibrational level, but MQDT calculations by Guberman and Suzor-Weiner[103] conclusively showed that this route offered no solution. In fact, the $\nu = 0$ quantum yield even dropped to 0.0012 when the indirect mechanism was taken into account.[103] After having analyzed results from rocket measurements of the F region,[96,104] which provided data on the electron density and the quantum yields of $O(^1S)$ and $O(^1D)$ generated from O_2^+ recombination, Bates[105] was quite convinced that the *ab initio* calculations nevertheless underestimated the quantum yield of $O(^1S)$. One piece of indirect evidence[2] was that the discrepancy was large at low altitudes, where $O_2^+(\nu)$ ions are most effectively vibrationally deactivated. However, Bates was never able to propose a mechanism that could allow for a sizeable $O(^1S)$ quantum yield of $\nu = 0$.

The breakthrough came in 1997, both experimentally and theoretically. Using the ion storage ring ASTRID and a two-dimensional imaging detector, Andersen and coworkers found the $O(^1S)$ quantum yield for DR of $O_2^+(\nu = 0)$ to be 0.05 ± 0.02,[106] thus providing the first *direct* measurement of this quantity. Figure 6 shows the projected distances, $P(D)$, measured in ASTRID from dissociative recombination of O_2^+.[106] A complicating factor in the experiment was that the O_2^+ beam was not vibrationally cold; thus, the results shown in Fig. 6 pertain to O_2^+ populating at least six vibrational levels. Despite this complication, it was possible to isolate the result for the zeroth vibrational level, as shown by the small peak at $R \approx 9$ mm in Fig. 6. Guberman presented a new mechanism at the American Physical Society Meeting in April 1997 (given as Ref. 27,[106]), later followed by the full account.[107] Figure 5 shows how the new mechanism works. The $^3\Sigma_u^-$ resonant state crosses close to the minimum of the ion ground state, which, over a wide range of electron energies ε, renders a favourable vibrational overlap $\langle O_2; {}^3\Sigma_u^-; \chi_d | O_2^+; X\ {}^2\Pi_g; \chi_{\nu=0}\rangle$, where the continuum vibrational wavefunction of the resonant state is given by $|O_2; {}^3\Sigma_u^-; \chi_d\rangle$ and the vibrational wavefunction of the zeroth vibrational level of the ion ground state is given by $|O_2^+; X\ {}^2\Pi_g; \chi_{\nu=0}\rangle$, respectively. Electron capture into the $^3\Sigma_u^-$ state leads to dissociation to the $O(^3P) + O(^1D)$ asymptotic limit, as calculated by Guberman.[92] The ASTRID experiment also reveals that 42% of all recombination events involving $O_2^+(\nu = 0)$ leads to the generation of $O(^3P) + O(^1D)$.[106] In the new calculations,[107] spin-orbit coupling between Rydberg states of $^1\Sigma_u^+$ and $^3\Sigma_u^-$ symmetry have been included. These

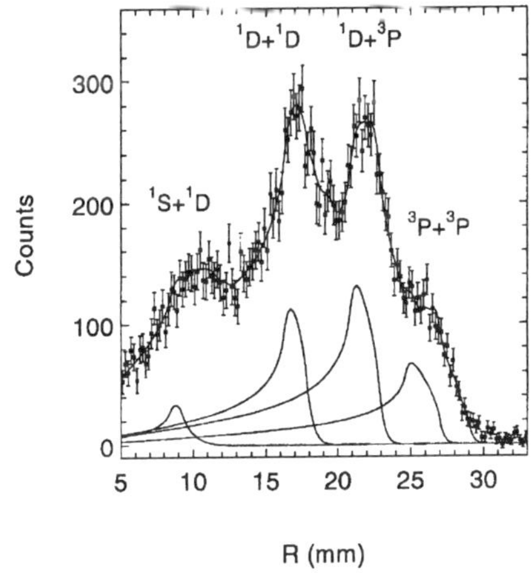

Fig. 6. The distribution of projected distances, $P(D)$, for dissociative recombination of O_2^+ (populating an unknown distribution of vibrational levels) measured in the ASTRID ion storage ring. The data points were recorded with a two-dimensional imaging detector consisting of a 77-mm-diameter MCP detector with a phosphor anode and a CCD camera. The detector was located about 10 m from the midpoint of the electron cooler/target in ASTRID. The figure also shows the fit to the data and the contribution from $\nu = 0$ for each final dissociation limit. The small peak at $R \approx 9$ mm gives the $O(^1S)$ quantum yield for $\nu = 0$. (Reproduced with permission from Ref. 106.)

Rydberg states have, in the single configuration picture, a common ion core described by $1\sigma_g^2 1\sigma_u^2 2\sigma_g^2 2\sigma_u^2 3\sigma_g^2 1\pi_u^4 1\pi_g$ and a diffuse $n\pi_g$ Rydberg orbital. Since the spin-orbit coupling is dominated by the valence $1\pi_g$ orbital, the coupling is essentially independent of the principal quantum number n. As shown with the arrows in Fig. 5, the spin-orbit coupling redirects a fraction of the dissociation flux from the initial electron capture state $^3\Sigma_u^-$ to a Rydberg state of mixed $^3\Sigma_u^-/^1\Sigma_u^+$ symmetry. This Rydberg state is predissociated by the $^1\Sigma_u^+$ resonant state that leads to the generation of $O(^1S) + O(^1D)$. Thus, even a small part of the flux being redirected is sufficient to increase the quantum yield for $O(^1S)$ from 0.0012 to 0.020 at $T_e = 300$ K.[107]

The experimental result from the ASTRID storage ring[106] and the theoretical result of Guberman[107] are not exactly comparable quantities. The experimental result concerns a measurement carried out only at an electron energy of nominally 0 eV ($E_d = 0$ eV), whereas the calculated result derives from an electron-energy dependent cross-section integrated over a thermal energy distribution according to Eq. (8). The theoretical cross-section integrated over the electron-velocity distribution in the ASTRID electron cooler according to Eq. (7) would provide a quantity directly comparable with the experimental result. It would most probably increase the theoretical value and make the agreement between experiment and theory even better.

It is finally interesting to note that, according to Guberman,[107] the amount of $^3\Sigma_u^-$(Rydberg) $\sim^1 \Sigma_u^+$(Rydberg) mixing varies with the principal quantum number n. This is countered by an increasing predissociation width as n decreases. Since Rydberg states of different n values are involved at different electron energies ε, one would anticipate that the partial cross-section for $O(^1S)$ + $O(^1D)$ generation should vary as a function of ε. CRYRING has recently been supplied with a cooled hollow cathode ion source that delivers O_2^+ ions populating primarily the zeroth vibrational levels,[86] and a three-dimensional imaging detector. A variation in the experimental $O(^1S)$ quantum yield is indeed observed[86] and seems to be in good agreement with theoretical predictions for electron energies below 40 meV. A detailed comparison with the theoretical partial cross-section would provide a very stringent test of the new mechanism. Further experiments with O_2^+ are scheduled at CRYRING in order to improve and extend the experimental data.

5.3.2. *Nitrogen*

The thermal rate coefficient for DR of $N_2^+(\nu = 0)$ has been calculated by means of the MQDT approach and found to be 1.6×10^{-7} cm^{3}s^{-1} when the indirect mechanism was included.[108] This is slightly lower than results from several afterglow experiments, as reviewed in Ref. 2, but significantly higher than the rate coefficient extracted from a single-pass merged beams experiment.[109] This raises the interesting question whether there could be some fundamental difference between the afterglow experiments on the one hand and the merged beams technique on the other. Judging from the recent data from CRYRING, the answer is no. A thermal rate coefficient of $\alpha(T_e) = 1.75 \pm 0.09 \times 10^{-7}(300/T_e)^{-0.30\pm0.09}$ cm^{3}s^{-1} was derived from the cross-section data measured in CRYRING,[110] which is in very good agreement with the result of Guberman.[108] Some caution must be exercised in this comparison since the theoretical result[108] pertains to $\nu = 0$ and the experimental result was obtained for ions with an estimated population of 46% in $\nu = 0$, 27% in $\nu = 1$, and 26% in $\nu = 2$ and 3.[110] The internal state distribution in the CRYRING experiment[110] was measured by means of a three-dimensional imaging technique[64] and with the added assumption that the DR cross-section is only weakly dependent on the vibrational excitation of N_2^+. The rationale for making this assumption is based on the fact that essentially the same cross-section was measured in CRYRING when the N_2^+ ions were extracted from a hot filament ion source as when they were extracted from a hollow cathode source, the latter being able to produce significantly vibrationally colder ions than the hot filament source.[110] There is also evidence from a laser-induced-fluorescence study of an afterglow plasma that the thermal rate coefficient varies weakly for the lowest vibrational levels.[111]

N_2^+ ions in the zeroth vibrational level recombining with 0 eV electrons give rise to the following atomic fragments:

$$N_2^+ + e^- \rightarrow N(^4S) + N(^4S) + 5.82\ \text{eV} \tag{10.1}$$

$$\rightarrow N(^4S) + N(^2D) + 3.44\ \text{eV} \tag{10.2}$$

$$\rightarrow N(^4S) + N(^2P) + 2.25\ \text{eV} \tag{10.3}$$

$$\rightarrow N(^2D) + (^2D) + 1.06\ \text{eV} \tag{10.4}$$

$$\rightarrow N(^2D) + N(^2P) - 0.13\ \text{eV (accessible for } \nu > 0\text{)}\,. \tag{10.5}$$

Guberman calculated[112] that DR of $N_2^+(\nu = 0)$ in 88% of the recombination events would lead to Eq. (10.2). In doing this calculation, he inferred from MQDT calculations[113] that DR of the lowest vibrational level is totally dominated by the C′ $^3\Pi_u$ state (sometimes also labeled 2 $^3\Pi_u$), which dissociates to $N(^4S) + N(^2P)$ in the adiabatic limit, and used a Landau–Zener approach to show that the separating atoms have a chance of 88% of traversing the avoided crossing between the C′ $^3\Pi_u$ and C $^3\Pi_u$ states at 1.3 Å and reach the limit, Eq. (10.2). The remaining 12% of the separating fragments would then traverse the avoided crossing between the C′ $^3\Pi_u$ and 3 $^3\Pi_u$ states and reach the limit, Eq. (10.4).

Apart from being an interesting problem in molecular dynamics, the question of the branching ratios in DR of N_2^+ is of importance for understanding the isotopic enhancement in the Martian atmosphere of the heavier ^{15}N as compared with ^{14}N.[36,114,115] Dissociative recombination creates translationally energetic nitrogen atoms, but the energy is just sufficient to

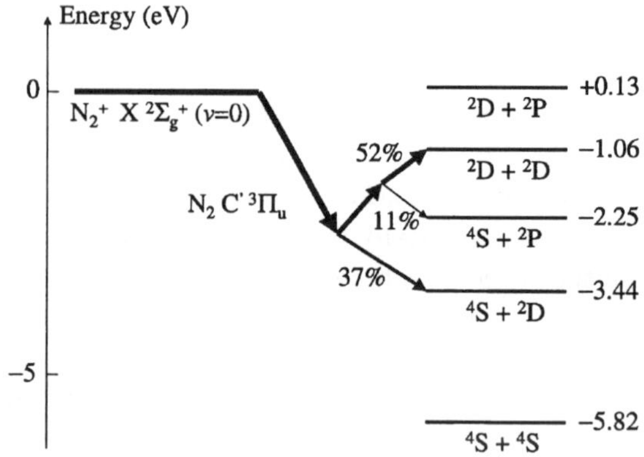

Fig. 7. Energy level diagram which shows the location of $N_2^+(\nu = 0)$ with respect to the relevant asymptotic limits of N_2 and the experimental results from the CRYRING experiment.[110]

remove ^{14}N whereas ^{15}N remains trapped in the Martian atmosphere by the gravitational field. The $^{15}N/^{14}N$ ratio in the Martian atmosphere has been measured to be 1.62 times the terrestrial value by the Viking landers.[36]

The branching ratios into the Eqs. (10.1) to (10.4) limits have been measured in storage rings by means of imaging techniques, first in ASTRID[116] and somewhat later in CRYRING.[110] The experimental results are in quite good agreement with each other, but in variance with Guberman's theoretical results.[112] Figure 7 shows the energetics of the N_2^+ electronic ground state and the neutral asymptotic limits, and the results from the CRYRING experiment.[110] The figure shows how DR of $N_2^+(\nu = 0)$ proceeds through the C′ $^3\Pi_u$ valence state of N_2, with 52% of the dissociation flux going to the N(^2D) + (^2D) limit, 11% to the N(^4S) + N(^2P) limit, and 37% to the N(^4S)+N(^2D) limit. The corresponding numbers from the ASTRID experiment are 46%, 8%, and 46%, respectively.[116] The product state distribution has also been measured using a glow discharge time-of-flight technique,[117] and the results are in agreement with those from the storage rings. Theory and experiment agree completely that DR of N_2^+ does not lead to the formation of N(^4S) + N(^4S). The major disagreement between experiment and theory concerns the formation of N(^4S) + N(^2D), which is the limit that gives the translationally most energetic nitrogen atoms. Table 3 shows the experimental and theoretical results discussed in this section, including recent yet unpublished results by Guberman.[118]

Table 3. Experimental and theoretical branching ratios for dissociative recombination of N_2^+. All results except those of Ref. 117 concern $\nu = 0$.

Asymptotic limit	Theory[112]	Theory[118]	Exp.[116]	Exp.[110]	Exp.[117]
$N(^4S) + N(^2D)$	88%	70%	46%	37%	45%
$N(^4S) + N(^2P)$	0%	3%	8%	11%	10%
$N(^2D) + (^2D)$	12%	27%	46%	52%	45%

5.3.3. *Nitric Oxide*

Dissociative recombination of NO^+ and O_2^+ and photodissociation of O_2 are believed to be the origin of the oxygen geocorona, a layer of suprathermal oxygen atoms surrounding the Earth at high altitude.[36,119] DR of NO^+ has been theoretically estimated to be a fast process,[120] and several independent experiments have corroborated this.[121–125] A recent experiment[126] at ASTRID gave a thermal rate coefficient of 4×10^{-7} cm^{3}s^{-1}, which is in very good agreement with previous results. By comparison, the MQDT calculation by Sun and Nakamura[127] gave a result which is more than a factor of two smaller.

The imaging technique was used in the ASTRID experiment to show that DR of NO^+ with an 85% probability leads to $N(^2D) + O(^3P)$. The result is in good agreement with a measurement of the $N(^2D)$ quantum yield.[128]

5.3.4. *Other Ions*

There are no nonhydride diatomic ions that have occupied nearly as much interest as the three atmospheric molecules O_2^+, N_2^+, and NO^+. Even a molecule like CO^+ has not been much studied with respect to dissociative recombination, neither theoretically nor experimentally. One problem for the afterglow technique seems to have been the formation of $(CO)_nCO^+$ clusters, which have a large thermal rate coefficient.[129] A single-pass merged beams experiment[130] gave a rate coefficient of 10^{-7} cm^{3}s^{-1}, which is significantly lower than the results for the cluster ions in the afterglow experiment.[129] The clustering problem was apparently avoided in a flowing afterglow — Langmuir probe (FALP) experiment[131] which gave a result that $\alpha(CO^+)$ is 1.6×10^{-7} cm^{3}s^{-1} at 300 K. Thus, just as in the nitrogen case, the merged beams experiment gave a rate coefficient smaller than that of an afterglow experiment. This may indicate that the particle detector used in the single-pass merged beams experiment had a less than 100% efficiency.[132] The detector efficiency is not likely to pose a problem at the high energies used in storage rings, and the $\alpha(CO^+) = 2.75 \times 10^{-7}$ cm^{3}s^{-1} measured in CRYRING[133] is even larger than the FALP result.

Dissociative recombination of CO^+ is different from the atmospheric ions in that the formation of ground state atomic fragments dominates (76%) at 0 eV collision energy. This suggests that the neutral $D'\,^1\Sigma^+$ state, which diabatically dissociates to $C(^3P) + O(^3P)$, is the main contributor to DR. The potential energy curve for this state has been constructed on the basis of spectroscopic data,[134] according to which the $D'\,^1\Sigma^+$ state crosses the electronic ground state CO^+ just above the right turning point of $\nu = 0$. This should account for a favorable vibrational overlap, but the electronic capture width is as yet unknown. According to *ab initio* calculations, the $X\,^2\Sigma^+$ state of CO^+ is also crossed by a neutral $^1\Pi$ state. It is likely that also this state contributes to DR of CO^+, but to what extent cannot be assessed before more detailed calculations of the electronic capture width have been carried out. At 0 eV, 24% of the dissociation flux goes to excited atomic products. Already at 0.4 eV this number has increased to 47%, and at 1.0 eV the production of excited atoms dominates.[133]

The CN^+ ion was a white spot on the DR map until very recently. The ion has an $X\,^1\Sigma^+$ electronic ground state and an $a\,^3\Pi$ state lying only 0.45 eV above.[135] Not even storage in CRYRING for 6 s should be sufficient to remove the $a\,^3\Pi$ state from the CN^+ beam, and the deduced rate coefficient, $\alpha(CN^+) = 3.4 \times 10^{-7}$ cm^{3}s^{-1}, must be regarded as that characteristic of ions in a mixture of the lowest electronic states.[136] The cross-sections measured in CRYRING are larger than those measured in the single pass experiment,[132] probably for the same detector efficiency problem as mentioned above.

6. Tunneling Dissociative Recombination

Convinced by overwhelming experimental evidence that HCO^+ recombines with a rate coefficient larger than 10^{-7} cm^{3}s^{-1}, and by theoretical predictions that its electronic ground state does not possess a favorable neutral state crossing, Bates introduced a multistep indirect mechanism which does not require a curve crossing.[137] He later gave this mechanism the decriptive name 'tunneling dissociative recombination'.[2] Figure 8 shows how Bates conceived of the tunneling mode of recombination.

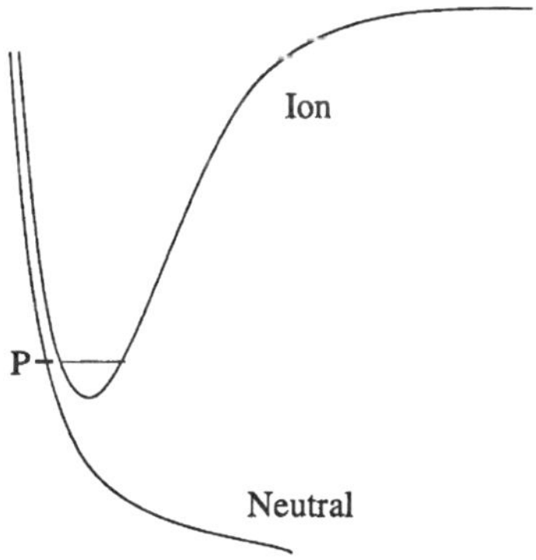

Fig. 8. A schematic illustration of tunneling dissociative recombination. The ionic curve includes the energy of the electron. The equi-energy point (P) on the neutral curve is marked. (Based on Fig. 11 in Ref. 2.) A more detailed and realistic view of recombination without a curve crossing is given for HeH^+ in Ref. 140.

Two ions for which it is known with certainty that the crossing mode recombination with thermal energy electrons is not operative for the zeroth vibrational level are HeH^+ and H_3^+. The former ion lacks a neutral state crossing of its electronic ground state altogether, and not even vibrationally excited levels can recombine by the crossing mode. By contrast, vibrationally excited H_3^+ recombines effectively by the crossing mode, and the issue at stake concerns the DR properties of its lowest vibrational levels. Both ions have been studied extensively in storage rings, as described below.

6.1. *HeH*$^+$

The FALP measurement[138] of a thermal rate coefficient less than 1×10^{-10} cm^{3}s^{-1} for HeH^+ was seemingly in perfect harmony with the absence of a curve crossing. The sizeable cross-section measured in a merged beams experiment[139] therefore came as a surprise. The findings inspired theoretical work,[140,141] which finally identified coupling by the nuclear kinetic-energy operator of the HeH^+ ground state to close lying (but not crossing) Rydberg states of HeH as the driving mechanism. Quantitative calculations were performed on $^3HeH^+$[140] and $^4HeH^+$[141] and found to be in fairly good agreement with the original merged beams cross-section data[139] and results coming from the storage rings TARN II[26,142] and CRYRING.[59,143–145] The comparison of experiment and theory for HeH^+

firmly established that tunneling recombination, driven by nonadiabatic coupling between the electron-molecular ion system and neutral Rydberg states, can be much more effective than had been previously thought. A signature of the mechanism is that indirect DR through Rydberg states has a larger cross-section than direct DR along the repulsive ground state of HeH. Further evidence are provided by imaging experiments at TARN II, CRYRING and TSR,[142,145,146] which show that DR of HeH^+ with thermal energy electrons leads to the formation of $He(^1S) + H(n = 2)$, in agreement with Guberman's MQDT calculations.[140,147]

Experiments have now been performed on HeH^+ using the superconducting electron coolers at TARN II[148] and CRYRING,[149] both having transverse electron temperatures of about 1 meV. The low-temperature electron beams allowed new resonances to be observed. At the same time, Takagi performed new MQDT calculations which included rotational effects and higher-order effects,[148] and obtained better agreement with theory than did the earlier calculations. The experimental results from TARN II and Takagi's calculated results are shown in Fig. 9. It should be noted that the experimental rate coefficients are relative, so it is only relevant to compare the positions and shapes of the resonances. However, even with this reservation, it is obvious that the agreement is good for all isotopomers save $^3HeD^+$. Most of the resonances were identified as deriving from an initial electron capture into a dissociativelly closed channel followed by a radiationless transition to a dissociative channel. Takagi's theoretical rate coefficients[148] for $^3HeH^+$ can be compared with the absolute results from CRYRING,[149] a comparison which shows that Takagi's results are higher than the experimental ones by at least a factor of two.

Energetic electrons can be captured into doubly excited states of $^2\Sigma$ and $^2\Pi$ symmetry with electronic configurations $(1\sigma 2\sigma)n\sigma$ and $(1\sigma 2\sigma)n\pi$, respectively. These resonant states can then either autoionize or dissociate. In the former case, the autoionization process can lead to the formation of HeH^+ in the vibrational continuum of its electronic ground state, that is, $He + H^+$.[150] This resonant-enhanced dissociative excitation (RDE) process was first observed experimentally for HeH^+ in CRYRING.[145] The dissociation process gives rise to a peak in the DR cross-section around 20 eV.[26,143,145,151] Once the electron has been captured into a resonant state, the RDE and DR processes obviously compete with each other. The DR process has a cross-section about three times larger than that of

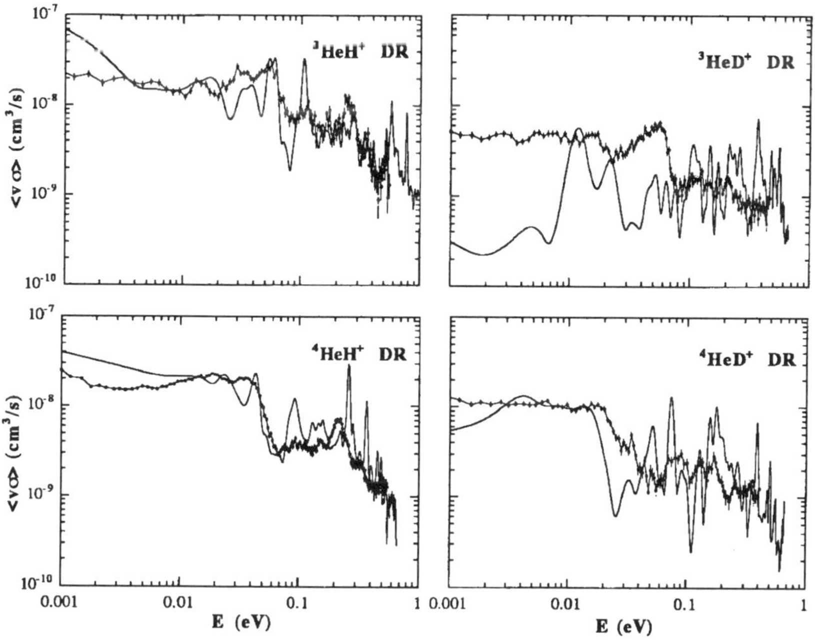

Fig. 9. Comparison of the relative DR rate coefficients measured in TARN II (dots) of four isotopomers of HeH$^+$ and the theoretical results of Takagi (absolute values). (Reproduced with permission from Ref. 148.)

RDE.[145] The RDE process is important because it can produce fragments that are different from those arising from direct dissociative excitation and this at an electron energy that may be much lower than the threshold for the direct process.[33,150]

The thermal rate coefficient at 300 K derived for ^{4}HeH$^+$ based on the cross-section data from CRYRING is 3×10^{-8} cm^{3}s-1,[145] which is significantly larger than the 1×10^{-10} cm^{3}s^{-1} quoted in Ref. 138 but also smaller than the crossing mode rate coefficients discussed in the previous section. KrH$^+$, which also is expected to recombine by the tunneling mode, has a rate coefficient similar to that of HeH$^+$ according to two different FALP experiments.[131,152] In contrast, XeH$^+$ is found to recombine much faster (8.3×10^{-8} cm^{3}s^{-1}),[152] the reason being that the electronic ground state of XeH$^+$ is crossed by the lowest $^2\Sigma$ state. The crossing mode becomes operative, at least for vibrationally excited XeH$^+$ that are present in the afterglow, and the DR rate increases.

6.2. H_3^+

It is appropriate to start this section by citing T. Oka:[153] "My love affair with H_3^+ over many years, has led me to believe that some beautiful subtleties of physics do not appear until one faces a three-particle system". The subtleties associated with the spectroscopy of H_3^+ and the reasons why this simplest polyatomic molecule plays such an important role in astrophysics and astrochemistry can be found in several review articles.[154–157] The recent discoveries of H_3^+ in interstellar molecular clouds[158,159] represent milestones in molecular astrophysics. In diffuse molecular clouds, H_3^+ is expected to be destroyed primarily by dissociative recombination. It is formed by the ion-molecule reaction $H_2^+ + H_2 \rightarrow H_3^+ + H$. Thus, the number density of H_3^+ in diffuse clouds is proportional to the molecular hydrogen density, $[H_2]$, and the ionizing flux, and inversely proportional to the electron density and the DR rate coefficient.[160] This, in combination with the recent indications that chemical models of interstellar molecular clouds may have bistable solutions,[161] brings again the question of the DR rate coefficient of $H_3^+(\nu = 0)$ into focus, a controversial topic indeed.

The different twists and turns in the history of the DR rate coefficient for H_3^+ have been reviewed many times[2,5,32,155] and will not be repeated here. Instead an attempt to summarize the present position will be made.

The electronic ground state of H_3^+ is crossed by the neutral resonant state $^2A_1$ close to the outer turning point of $\nu = 3$.[162] This crossing state provides an effective route for low-energy DR for vibrational levels $\nu \geq 3$, and for high-energy DR (9.5 eV) for $\nu \geq 0$,[2,28,163] but it cannot account for a 300 K thermal DR rate coefficient of the order of 10^{-8} cm^{3}s^{-1} for H_3^+ occupying only its lowest vibrational level. Thus, there is presently no claim, with the possible exception of Ref. 164 (see also footnote in Ref. 165), for a rate coefficient sufficiently low to motivate the crossing mode recombination to be invoked. Another mechanism must be sought, and one such mechanism could be the multistep indirect mechanism.[2,137,165] However, even if Bates *et al.*[165] assumed that the mechanism could lead to a rate coefficient as large as 1.5×10^{-7} cm^{3}s^{-1}, they did not provide any quantitative evidence. There is presently no *ab initio* calculation of the rate, although efforts in this direction are being taken.[166]

Whereas the two most recent merged beams experiments agree that the rate coefficient is slightly above 10^{-7} cm^{3}s^{-1} (1.2×10^{-7} cm^{3}s^{-1}) from a single-pass experiment[167] and 1.15×10^{-7} cm^{3}s^{-1} from CRYRING,[168] the position with respect to the three most recent afterglow experiments is less clear. Using a flowing afterglow apparatus supplied with a mass spectrometer (FALP-MS), Laubé *et al.*[169] found $\alpha(H_3^+, \nu = 0)$ to be $(0.78 \pm 2.3) \times 10^{-7}$ cm^{3}s^{-1}, which is slightly lower but still consistent with the results of the beam experiments. Smith and Španel derived a rate coefficient of $(0.1 - 0.2) \times 10^{-7}$ cm^{3}s^{-1} based on their FALP data[170] and concluded from the decay of the afterglow plasma that vibrationally excited H_3^+ recombines much faster than $H_3^+(\nu = 0)$. A similar nonlinear decay was found by Gougousi *et al.*,[164] but their interpretation is different. On the basis of a detailed consideration of the formation and quenching of H_3^+, they concluded that H_3^+ ions in $\nu > 1$ are not present in the plasma. They then inferred, from the observation that the deionization of the plasma proceeded slower when the electron and hydrogen density was lower, that the deionization of the plasma is due to three-body recombination. Following the initial electron capture leading to the formation of H_3^{**} Rydberg molecules, Gougousi *et al.* argues,[164] collisions and electric microfields could stabilize the capture. A consequence of this hypothesis would be that H_3^+ is removed fast in a plasma where collisions and electric microfields are present but very slowly in a cold low-density plasma such as an interstellar molecular cloud. It is noteworthy that neither the nonlinear decay of the afterglow plasma nor the hydrogen density effect was observed in the FALP-MS experiment.[169] The behavior of H_3^+ in an afterglow plasma is apparently not yet well understood.

Tangible evidence for the hypothesis of Gougousi *et al.*[164] being plausible would be that the DR process is enhanced by the presence of an electric field in the interaction region. This is known to occur for dielectronic recombination of atomic ions.[171] In the CRYRING electron cooler, a well-controlled electric field can be applied in the electron–ion interaction region by making use of the guiding magnetic field produced by the cooler solenoid. This magnetic field is normally parallel to the electron beam and, because of the merged beams geometry, also to the ion beam. If the ion beam is slightly misaligned with respect to the electron beam and the solenoid magnetic field, a motional electric field is induced when the ions crosses the perpendicular magnetic field component which is introduced by the misalignment. The technique was used in CRYRING to study electric field effects for dielectronic recombination of Si^{11+},[172] and dissociative recombination of D_3^+.[173,174] No effect on the DR rate of D_3^+ was observed when an electric field of 30 V/cm was added to the interaction region. Possible interpretations are discussed in Ref. 173 where it is argued that the most likely explanation is that an electric field of 30 V/cm is too small to affect the recombination process. This does not rule out the hypothesis of Gougousi *et al.*,[164] but it certainly does not strengthen it either.

Whereas the FALP-MS experiment[169] found the thermal rate coefficient of D_3^+ to be only slightly lower than that of H_3^+, larger differences were found for the CRYRING results for H_3^+ ($\alpha(H_3^+) = 1.15 \times 10^{-7}$ cm^3s^{-1}),[168] H_2D^+ ($\alpha(H_2D^+) = 0.6 \times 10^{-7}$ cm^3s^{-1}),[175,176] and D_3 ($\alpha(D_3^+) = 0.27 \times 10^{-7}$ cm^3s^{-1}).[173,174] The results are all combined in Fig. 10; it should be noted here that the rate coefficients on the y-axis are not the thermal ones but those obtained from Eq. (6). The high-energy peak for D_3^+ is narrower than that for H_3^+, which reflects the fact that the zeroth vibrational level in D_3^+ lies deeper in the potential surface than the corresponding level in H_3^+. Results for HD_2^+ and D_3^+ from TARN II[177] are in agreement with the CRYRING results in that the low-energy cross-section for D_3^+ is lower than that for HD_2^+. Finally, it is noted that the CRYRING result for D_3^+ is in good agreement with the result from Mitchell's group using a single-pass merged beams apparatus.[178] Takagi's theoretical results[148] for HeH^+

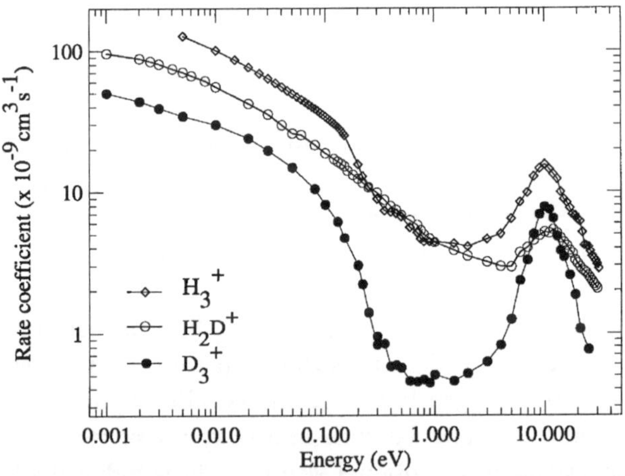

Fig. 10. The rate coefficients, $\langle v\sigma \rangle$, measured in the CRYRING electron cooler as a function of detuning energy. The data below 0.2 eV have been modified to account for the fact that the results for H_2D^+[175] and D_3^+[173,174] were obtained with an electron beam having a transverse energy spread of 10 meV, whereas the data for H_3^+ were obtained with a 100 meV electron beam.[28,168] (Reproduced with permission from Ref. 174.)

show that an isotope effect is expected when the tunneling mode of recombination is operative. The storage ring results for H_3^+ and its isotopomers are not in contradiction with this.

The breakup pattern into atomic and molecular fragments in DR of H_3^+ provides additional information about the DR mechanisms. For $H_3^+(\nu = 0)$ recombining with 0 eV electrons the following channels are open (neglecting the formation of a metastable H_3 Rydberg molecule):

$$H_3^+(\nu = 0) + e^- \rightarrow H(n = 1) + H(n = 1) + H(n = 1) \qquad (11.1)$$

$$\rightarrow H_2(^1\Sigma_g^+) + H(n = 1)\,. \qquad (11.2)$$

Figure 11, a modified version of Fig. 1 in Ref. 179, shows the energetics of H_3^+ and the asymptotic limits of H_3. The branching ratios into (11.1) and (11.2) were first measured for vibrationally hot ions in a single-pass experiment,[180] and for vibrationally cold ions in CRYRING for both H_3^+,[181] and H_2D^+.[175] In the CRYRING experiments, the electron energy was varied from 0 to 25 eV. The results show that the three-body breakup (11.1) dominates over the two-body breakup (11.2) by a 3 : 1 ratio for thermal energy electrons. The two-body breakup becomes important at electron energies for which the crossing resonant state $^2A_1$ is important, but for some reason the peak in the cross-section at 9.5 eV does not coincide with the peak in the two-body breakup, which occurs at 5 eV for both H_3^+ and H_2D^+. Additional information about the dynamics of the particle breakup in DR of H_3^+ and its isotopomers was provided by the TARN II group.[177] Using a 1 mm bar in front of a position sensitive detector, they were able to measure, as a function of time after injection of ions into TARN II, the dissociation energies involved in DR of D_3^+. The results showed that for the earliest time window [0–10 ms], when plenty of vibrationally excited D_3^+ were present in the ion beam, the data were best fitted assuming a breakup into $D_2(^1\Sigma_g^+) + D(n = 2)$. This is in agreement with DR along the $^2A_1$ resonant state, although this state in the diabatic limit correlates with the $D_2^+ + D^-$ ion-pair limit. Calculations (for H_3^+) by Kulander and Guest[182] show that the strong coupling between states dissociating to $D_2(^1\Sigma_g^+) + D(n = 2)$ and the resonant state will lead to a preferable production of $D_2(^1\Sigma_g^+) + D(n = 2)$ instead of the ion pairs. Further support of this scenario is provided by the high-energy DR cross-section measurements of the neutral channel[28] and of the ion-pair channel.[183,184] A comparison

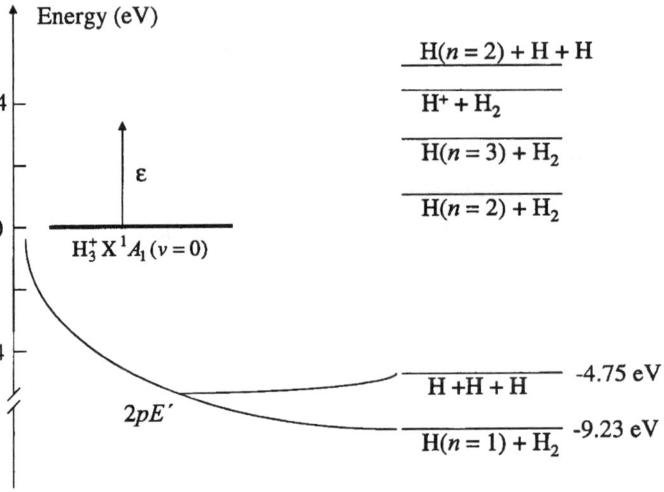

Fig. 11. Schematic energy level diagram of H_3^+ and the lowest asympotic limits of H_3. (Based on Fig. 1 in Ref. 179.)

shows that despite the rather high energy which goes into the nuclear coordinates at 9.5 eV, less than 2% of the DR events lead to the formation of ion pairs. When the TARN II measurements were performed in the [0.5–1.0 s] time window, the data were best reproduced when a three-body breakup was assumed.[177] Thus, the experiments at the storage rings show that the crossing mode recombination can be accessed at will, either by increasing the electron energy when the ions are cold, or by using vibrationally excited ions. They also show that the branching ratios change radically when DR no longer can occur through the $^2A_1$ resonant state.

Dissociative recombination of H_3^+ in its zeroth vibrational level with low-energy electrons must go via the H_3 Rydberg states. The dissociation dynamics of Rydberg states *below* the ionozation limit, where predissocion and radiative dissociation is responsible for the breakup, have been extensively investigated both experimentally[179,185,186] and theoretically.[187,188] The predissociation is driven by nonadiabatic coupling to two states that correlate asymptotically to $H_2(^1\Sigma_g^+)+H(n=1)$ and $H(n=1)+H(n=1)+H(n=1)$.[187] It seems likely that the same mechanism is operative above the ionization limit. This would mean that dissociative recombination of H_3^+ is driven by the same nonadiabatic mechanism which drives DR of HeH^+.

7. Products of Dissociative Recombination of Polyatomic Ions

Polyatomic neutral molecules in dark interstellar clouds are believed to be synthesized in a sequence of gas-phase reactions with dissociative recombination of polyatomic ions with electrons as the final step.[40,189] Thus, knowledge of product branching fractions in DR of polyatomic ions is of vital importance for the modeling of dark clouds. Attempts to theoretically predict the DR breakup patterns[190–195] have met with only moderate success. Spectroscopic techniques have been applied to DR of thermalized polyatomic ions in flowing afterglow plasmas[196–200] in order to determine the branching into specific decay channels, such as OH and H, but they have not given the complete set of branching fractions into all decay channels. A totally different technique, to be discussed further below, was used by Mitchell *et al.*[180] in a single-pass merged beams experiment to measure the branching ratios for DR of H_3^+. The results, however, concerned vibrationally hot ions; attempts to measure the branching for vibrationally cold H_3^+ was made, but the low available beam current made the experiment impracticable.[201]

Whereas Herbst[190] used a statistical phase space theory to make predictions of DR branching fractions, Bates[191] used arguments based on the chemical bonding in the molecule. For example, in CH_2^+ the two hydrogen atoms are connected to the C^+ ion by paired-electron valence bonds. In dissociative recombination, a free electron enters an antibonding orbital simultaneously with the conversion of one of the bonding electrons to an antibonding electron. Thus, the argument goes, the DR process should lead to ejection of one hydrogen atom, i.e. $CH_2^+ + e^- \rightarrow CH + H$.[191] Inspired by the flowing afterglow experiments,[196–200] Bates proceeded to develop the valence bond theory,[194] but soon after, when he had realized the importance of the tunneling mode,[137] he modified the theory to accommodate both modes of recombination. This, however, did not increase the predictive power of the method.

The ion storage ring experiments to determine branching fractions of DR of polyatomic ions build on the technique used by Mitchell *et al.*,[180] which in turn derives from an idea by Berkner *et al.*[202] The grid technique is described in some detail in Refs. 30–32, 175, 181, 201, and 203. Briefly, a grid of known transmission, t, is placed in front of an energy dispersive particle detector (for example, an ion-implanted-silicon detector). The probability for one particle passing the grid is t, for two particles passing

t^2, three particles passing t^3, etc. Let us assume that the molecular ion ABC^+ breaks up as follows in recombination with electrons:

$$ABC^+ + e^- \rightarrow \begin{cases} A + B + C & (a) \\ AB + C & (b) \quad (a + b + c = 1) \; . \\ A + BC & (c) \end{cases} \tag{12}$$

without the grid, the three decay channels will all give rise to a peak in the pulse height spectrum corresponding to the full beam energy. When the grid is inserted, there is a probability $(1 - t)$ that one of the particles, say A, is blocked by the grid and hence lost, something which results in a peak at $(m_B + m_C/m_{ABC})$ of the full beam energy. The pulse height spectrum is therefore different with and without grid, and it is this redistribution of particles from the full beam energy peak to other peaks which gives information about the branching fractions a, b and c. A particularly elegant version of the grid technique was developed by Datz *et al.*[175,181,204] It makes use of the fact that the beam energy in a storage ring can be made sufficiently high to allow the grid to be translucent to the recombination products of light molecules such as H_3^+ and H_2D^+.

The first heavy-atom containing polyatomic ion ever to be investigated with respect to branching ratios in a storage ring, H_3O^+,[205] immediately led to a controversy, the reason being that the result was in disagreement with the result from a flowing afterglow experiment that for the first time yielded a complete set of branching fractions.[206] It is the distribution into the $H_2O + H$ and $O + H + H_2$ which is the main source of discrepancy. The experiment at ASTRID[205] has given a branching ratio of 0.33 for the $H_2O + H$ channel and only 0.01 for the $O + H + H_2$ channel, whereas the flowing afterglow spectroscopy experiment[206] has given a branching ratio of 0.05 for water production and 0.30 for the $O + H + H_2$ channel. A simple observation in the ASTRID experiment, which completely bypasses the analysis of the grid technique, favors the storage ring results. When no grid was inserted in front of the detector, a secondary peak at 18/19 of the full beam energy peak was observed. This secondary peak represents DR events leading to $H_2O + H$ and with the hydrogen atom missing the detector because of its considerable kinetic energy. No other channel can produce H atoms with sufficient energy to cause a similar particle loss. Thus, a lower limit to the branching fraction for the $H_2O + H$ channel of 10% is given by the relative size of the secondary peak.[205] Although impressive, it is inescapable that the flowing afterglow experiment[206] leaves an impression of being very complicated.

The triatomic ions that have been studied in storage rings with respect to branching ratios show similar dissociation dynamics. The three-body channel dominates in all cases, with branching fractions 0.75 (H_3^+),[181] 0.73 (H_2D^+),[175] 0.68 (H_2O^+),[207] 0.66 (NH_2^+)[208] and 0.63 (CH_2^+).[209] This is significantly more fragmentation than has been assumed in the standard set of reaction rates.[41,210] The mechanisms for the three-body breakup are presently not well understood. In addition to a simultaneous breakup into three particles, one can also conceive of a sequential breakup which could occur if sufficient reaction energy is converted into internal energy of the diatomic part of the two-body channel.[211] Experimental schemes to explore the breakup dynamics are presently being considered at CRYRING.[212]

The triatomic systems listed above all contain at least two hydrogen atoms. No information on the excitation state of the dissociation products has been obtained. Vibrational populations in diatomic recombination products of HCO^+, N_2H^+,[213] CO_2^+,[196,214–216] and H_3O^+,[217] have been probed by combining spectroscopic methods with the flowing afterglow technique. These experiments provide no information on whether three-body breakups occur. Recently, the grid technique was applied in CRYRING in order to measure the branching ratios for DR of HCN^+.[218] A preliminary analysis of the data shows that the breakup to H + CN accounts for about 50% of the events.

The additional three polyatomic ions that have been studied so far in storage rings contain three, four and five hydrogen atoms, respectively. The recombination breakup of CH_3^+ is dominated (branching fraction 0.40) by the CH_2 + H channel,[205] which is in quite good agreement with Bates' prediction[191,194] but in disagreement with the result of the phase space theory calculation.[190] For NH_4^+ the single H atom release was found to be even more dominating. In a recent CRYRING experiment,[208] the branching fraction for the $NH_4^+ + e^- \rightarrow NH_3 + H$ reaction was determined to be 0.67. It is interesting to note the similarities with the photophysics of the isoelectronic CH_4 molecule. Photodissociation at 121.6 nm results in a simple C–H bond fission leading to CH_3 + H in about 50% of the events.[219] In contrast to DR of CH_3^+ and NH_4^+, only 5% of recombining CH_5^+ ions dissociates to the single H atom channel.[220] This is in complete disagreement with Bates' prediction, based on experimental results,[200] of a branching ratio as large as 0.82.[194] A possible explanation for this large discrepancy could be that a substantial part of the CH_3 + H + H channel, with a measured branching ratio of 0.70,[220] is formed from secondary decomposition of highly excited CH_4 molecules.

Experimental data on products of dissociative recombination of polyatomic ions are given in Table 4. Only measurements of complete product distributions are included.

Table 4. Results from complete branching ratio measurements for dissociative recombination of polyatomic ions with thermal energy electrons.

Ion	Branching fractions		Technique	Reference
H_3^+	H + H + H	0.75	Storage ring (CRYRING)	181
	H_2 + H	0.25		
H_3^+	H + H + H	0.52	Single-pass merged beams apparatus	quoted in Ref. 5
	H_2 + H	0.40		
	H_3(Rydberg)	0.08		
H_2D^+	H + H + D	0.73	Storage ring (CRYRING)	175
	HD + H	0.19		
	H_2 + D	0.09		
CH_2^+	C + H + H	0.63	Storage ring (CRYRING)	209
	CH + H	0.25		
	C + H_2	0.12		
H_2O^+	O + H + H	0.68	Storage ring (ASTRID)	207
	OH + H	0.22		
	O + H_2	0.10		
NH_2^+	N + H + H	0.66	Storage ring (CRYRING)	208
	NH + H	0.34		
H_3O^+	OH + H + H	0.48	Storage ring (ASTRID)	205, 207
	O + H_2 + H	0.01		
	H_2O + H	0.33		
	OH + H_2	0.18		
H_3O^+	OH + H + H	0.29	Flowing afterglow	206
	O + H_2 + H	0.30		
	H_2O + H	0.05		
	OH + H_2	0.36		

Table 4. (*Continued*)

Ion	Branching fractions		Technique	Reference
CH_3^+	CH + H + H	0.16	Storage ring (ASTRID)	207
	C + H_2 + H	0.30		
	CH_2 + H	0.40		
	CH + H_2	0.14		
NH_4^+	NH_2 + H + H	0.24	Storage ring (CRYRING)	208
	NH_3 + H	0.69		
	NH_2 + H_2	0.09		
CH_5^+	CH_3 + H + H	0.70	Storage ring (CRYRING)	220
	CH_2 + H_2 + H	0.17		
	CH + H_2 + H_2	0.03		
	CH_4 + H	0.05		
	CH_3 + H_2	0.05		

Herbst and Lee[221] have found from model calculations on dark interstellar clouds that although the breakup into neutral reaction products in DR of polyatomic ions is more substantial than previously believed, a general branching fraction of 0.30 for the single H atom channel is still sufficient to give molecular abundances in excellent agreement with observations. However, if this branching fraction is lowered to 0.05, a larger change in predicted abundances occurs. Such low branching was indeed measured for CH_5^+, but so far this is the exception rather than the rule, at least as far as the storage ring results are concerned.

8. Conclusions and Outlook

This article has focussed almost exclusively on the use of ion storage rings for the study of dissociative recombination of positively charged molecular ions with electrons. It is pertinent at this point to recall that ion storage rings by no means are confined to the study of this particular molecular process. Dissociative excitation has already been mentioned in passing, with a reference given to a recent review of the topic.[33] But there are also other aspects of molecular physics that can be addressed by means of the storage ring technology, as discussed in a recent article by Zajfman,[222] and as witnessed by publications on electron impact detachment of negative molecular ions,[223,224] laser photodetachment spectroscopy of fullerenes,[225] and laser photofragment spectroscopy.[226] An interesting point with the latter experiment, which was performed at TSR, is that it shows that no heating processes prevents a complete rovibrational thermalization of a stored CH^+ beam.[226]

Dissociative recombination of the simplest ion, H_2^+ and its isotopomers, is beginning to be well understood. Yet, discrepancies between experiment and theory still remain. They will most probably be cleared up within the next couple of years. A good understanding of DR of HeH^+ is also emerging, although there are still large discrepancies between experiment and theory for some of the isotopomers. The importance of HeH^+ lies in fact that it is the simplest system to recombine by means of the nonadiabatic radial coupling (tunneling mode).

Recombination of heavier diatomics usually proceed by means of the curve crossing mechanism, and here the challenge is, in addition to the the DR cross-section and rate coefficient, the experimental and theoretical determination of the recombination products. The dissociation dynamics of an important system such as N_2^+ is presently not well understood. The development of better imaging techniques at storage rings will surely lead to a deeper understanding of how diatomic ions break up after recombining with electrons.

Angular distributions in DR of diatomic ions is another topic that is likely to be further explored in the future. It has been approached experimentally by means of the same imaging techniques as used to determine reaction products, but a clear understanding has not yet emerged. Theoretical work on anisotropies in electron–molecule collisions started already a long time ago,[227,228] but has proved difficult to reconcile with some results from storage ring experiments.[58,146,229]

When one leaves the domain of diatomic systems, guidance from theory basically cease to exist. There is still no first principle calculation of thermal energy electron DR with a polyatomic ion, although it seems likely that we will see such calculations within the next couple of years. The most urgent problem is of course low-energy DR of H_3^+. Not only is H_3^+ the simplest polyatomic molecule (although this 'simplicity' has proved elusive to many researchers over the years), it is also the most important molecular ion in astrophysics and the ion for which no experimental consensus has been reached as to how fast it recombines with electrons. The important problem of branching fractions in DR of polyatomic ions will probably have to rely solely on experiment for some time to come. The discrepancy between the storage ring result and the flowing afterglow result for H_3O^+ need to be resolved. Since no clear trend has emerged as to how polyatomic ions break up when neutralized by thermal energy electrons, more experiments are needed in order to gain a better understanding of the breakup process and to provide data for astrochemical modeling. Investigations of isotope effects are likely to shed more light on the mechanisms.

Ion storage rings are yet another example of how development of new technology leads to development of fundamental science. The technical development of storage rings and associated experiments will continue. Better vacuum will allow longer storage times; a second electron cooler will radically increase the efficiency of experiments since one-electron beam could serve as cooler and the other one as target; the combination of laser spectroscopy and DR will open new possibilities; etc. One must, however, also be aware of the limitations with the storage ring technique. Interactions of a stored ion beam with a collinear electron beam of equal velocity puts an upper limit to the ion mass; for example, a DR experiment in CRYRING with C_{60}^+ would require a 0.1 eV electron beam, which is obviously impossible. A larger range of ions would be accessible if the magnetic rigidity (see Table 1) was increased, but this is an expensive solution. An alternative development is represented by electrostatic storage devices.[230–232] Since these devices concern storage of keV ion beams, the merged beams concept must to be abandoned altogether. On the other hand, an electrostatic storage ring like ELISA[230] will be more suitable for storage of heavy ions than a magnetic ring.

The study of dissociative recombination in ion storage rings has led to a wealth of new information and insight. The rapid pace of the development is well illustrated by the fact that much material discussed in this review is taken from papers published after the most recent review appeared in 1997.[32] It is a safe bet that the rapid pace will continue.

Acknowledgments

This work was supported by the Swedish Natural Science Research Council, the Göran Gustafsson Foundation, and by the Human Capital and Mobility programme of EU. I would like to thank those who provided preprints or unpublished material that was used in this article, and A. Suzor-Weiner,

S. L. Guberman, G. H. Dunn, and W. Shi for valuable comments on early drafts of the manuscript.

References

1. D. R. Bates, *Phys. Rev.* **78**, 492 (1950).
2. D. R. Bates, *Adv. At. Mol. Opt. Phys.* **34**, 427 (1994).
3. M. A. Biondi and S. C. Brown, *Phys. Rev.* **75**, 1700 (1949).
4. M. A. Biondi and S. C. Brown, *Phys. Rev.* **76**, 1697 (1949).
5. J. B. A. Mitchell, *Phys. Rep.* **186**, 215 (1990).
6. J. N. Bardsley, *J. Phys.* **B1**, 365 (1968).
7. J. Wm. McGowan, R. Caudano and J. Keyser, *Phys. Rev. Lett.* **36**, 1447 (1976).
8. A. Giusti, *J. Phys.* **B13**, 3867 (1980).
9. C. M. Lee, *Phys. Rev.* **A16**, 109 (1977).
10. M. A. Biondi and J. N. Bardsley, *Adv. At. Mol. Phys.* **6**, 1 (1970).
11. A. V. Eletskii and B. M. Smirnov, *Sov. Phys. Usp.* **25**, 12 (1982).
12. J. B. A. Mitchell and C. Rebrion-Rowe, *Int. Rev. Phys. Chem.* **16**, 201 (1997).
13. J. B. A. Mitchell and S. L. Guberman, Eds., *Dissociative Recombination: Theory, Experiment and Applications* (World Scientific, Singapore, 1989).
14. B. R. Rowe, J. B. A. Mitchell and A. Canosa, Eds., *Dissociative Recombination: Theory, Experiment, and Applications*, NATO ASI Ser. B Phys., Vol. 313 (Plenum, New York, 1993).
15. D. Zajfman, J. B. A. Mitchell, D. Schwalm and B. R. Rowe, Eds., *Dissociative Recombination: Theory, Experiment and Applications III* (World Scientific, Singapore, 1996).
16. A. Giusti-Suzor, I. F. Schneider and O. Dulieu, in Ref. 14, p. 11.
17. A. Suzor-Weiner and I. F. Schneider, in Ref. 15, p. 1.
18. A. Giusti-Suzor, in *Atomic Processes in Electron–Ion and Ion–Ion Collisions*, Ed. F. Brouillard (Plenum, New York, 1986), p. 223.
19. S. L. Guberman, in *Thermophysical Aspects of Re-entry Flows, Prog. Astronaut. Aeronaut Ser. Vol. 103*, Ed. J. N. Moss and C. D. Scott, (Am. Inst. Aeronaut. Astronaut, New York, 1986), p. 225.
20. M. R. Flannery, *Adv. At. Mol. Opt. Phys.* **32**, 117 (1994).
21. S. L. Guberman, in *Atomic Collisions; A Symposium in Honour of Christopher Bottcher* (1945–1993), AIP Conference Proceeding Vol. 347 (American Institute of Physics, New York, 1995), p. 88.
22. L. Carata, I. F. Schneider and A. Suzor-Weiner, *Phil. Trans. Royal Soc. London* **355**, 1677 (1997).
23. G. V. Golubkov, M. G. Golubkov, S. V. Drygin and G. K. Ivanov, *Russ. Chem. Bull.* **45**, 1265 (1996).
24. T. F. O'Malley, *Adv. At. Mol. Phys.* **7**, 223 (1971).
25. H. Nakamura, *Annu. Rev. Phys. Chem.* **48**, 299 (1997).

26. T. Tanabe, I. Katayama, N. Inoue, K. Chida, Y. Arakaki, T. Watanabe, M. Yoshizawa, S. Ohtani and K. Noda, *Phys. Rev. Lett.* **70**, 422 (1993).
27. P. Forck, M. Grieser, D. Habs, A. Lampert, P. Repnow, D. Schwalm, A. Wolf and D. Zajfman, *Phys. Rev. Lett.* **70**, 426 (1993).
28. M. Larsson, H. Danared, J. R. Mowat, P. Sigray, G. Sundström, L. Broström, A. Filevich, A. Källberg, S. Mannervik, K.-G. Rensfelt and S. Datz, *Phys. Rev. Lett.* **70**, 430 (1993).
29. M. Larsson, *Phys. Scripta* **T59**, 361 (1995).
30. M. Larsson, *Int. J. Mass Spectrom. Ion Proc.* **149/150**, 403 (1995).
31. M. Larsson, *Rep. Prog. Phys.* **58**, 1267 (1995).
32. M. Larsson, *Annu. Rev. Phys. Chem.* **48**, 151 (1997).
33. G. H. Dunn and N. Djuric, in *Novel Aspects of Electron–Molecular Scattering*, Ed. K. Becker (World Scientific, Singapore, 1998), p. 241.
34. D. R. Bates and H. S. W. Massey, *Proc. R. Soc. London Ser.* **A192**, 1 (1947).
35. D. R. Bates, *J. Geophys. Res.* **99**, A10, 19101 (1994).
36. B. D. Shizgal and G. G. Arkos, *Rev. Geophys.* **34**, 483 (1996).
37. P. Puhl, T. E. Cravens and J. Lindgren, *Astrophys. J.* **418**, 899 (1993).
38. P. Eberhardt and D. Krankowsky, *Astron. Astrophys.* **295**, 795 (1995).
39. R. H. Häberli, K. Altwegg, H. Balsiger and J. Geiss, *J. Geophys. Res.* **101**, A7, 15579 (1996).
40. E. Herbst and W. Klemperer, *Astrophys. J.* **185**, 505 (1973).
41. T. J. Millar, P. R. Farquhar and K. Willacy, *Astron. Astrophys. Suppl.* **121**, 139 (1997).
42. J. B. A. Mitchell, C. T. Ng, J. L. Forand, D. P. Levac, R. E. Mitchell, A. Sen, D. B. Miko and J. Wm. McGowan, *Phys. Rev. Lett.* **50**, 335 (1983).
43. D. S. Belić, G. H. Dunn, T. J. Morgan, D. W. Mueller and C. Timmer, *Phys. Rev. Lett.* **50**, 339 (1983).
44. P. F. Dittner, S. Datz, P. D. Miller, C. D. Moak, P. H. Stelson, C. Bottcher, W. B. Dress, G. D. Alton, N. Nešković and C. M. Fou, *Phys. Rev. Lett.* **51**, 31 (1983).
45. Y. Hahn, *Rep. Prog. Phys.* **60**, 691 (1997).
46. J. B. A. Mitchell, in *Atomic and Molecular Physics in Fusion Edge Plasmas*, Ed. R. K. Janev (Plenum Press, New York, 1995), p. 225.
47. A. Talebpour, C.-Y. Chien and S. L. Chin, *J. Phys.* **B29**, L677 (1996).
48. A. Talebpour, S. Larochelle and S. L. Chin, *J. Phys.* **B30**, 1927 (1997).
49. D. Krämer, G. Bisoffi, M. Blum, A. Friedrich, Ch. Geyer, B. Holzer, H. W. Heyng, D. Habs, E. Jaeschke, M. Jung, W. Ott, R. E. Pollock, R. Repnow and F. Schmitt, *Nucl. Instr. Meth. Phys. Res.* **A287**, 268 (1990).
50. T. Tanabe, K. Noda, T. Honma, M. Kodaira, K. Chida, T. Watanabe, A. Noda, S. Watanabe, M. Mizobuchi, M. Yoshizawa, T. Katayama, H. Muto and A. Ando, *Nucl. Instr. Meth. Phys. Res.* **A307**, 7 (1991).
51. S. P. Mϕller, in *Conference Record of the 1991 IEEE Particle Accelerator Conference*, Ed. K. Berkner (IEEE, New York, 1991), p. 2811.

52. K. Abrahamsson, G. Andler, L. Bagge, E. Beebe, P. Carlé, H. Danared, S. Egnel, K. Ehrnstén, M. Engström, C. J. Herrlander, J. Hilke, J. Jeansson, A. Källberg, S. Leontein, L. Liljeby, A. Nilsson, A. Paál, K.-G. Rensfelt, U. Rosengård, A. Simonsson, A. Soltan, J. Starker, M. af Ugglas and A. Filevich, *Nucl. Instr. Meth. Phys. Res.* **B79**, 269 (1993).
53. E. Courant, M. S. Livingstone and H. Snyder, *Phys. Rev.* **88**, 1190 (1952).
54. E. Courant and H. Snyder, *Ann. Phys.* **3**, 1 (1958).
55. G. Kilgus, D. Habs, D. Schwalm, A. Wolf, N.R. Badnell and A. Müller, *Phys. Rev.* **A46**, 5730 (1992).
56. H. Danared, G. Andler, L. Bagge, C. J. Herrlander, J. Hilke, J. Jeansson, A. Källberg, A. Nilsson, A. Paál, K.-G. Rensfelt, U. Rosengård, J. Starker, and M. af Ugglas, *Phys. Rev. Lett.* **72**, 3775 (1994).
57. C. Strömholm, J. Semaniak, S. Rosén, H. Danared, S. Datz, W. J. van der Zande and M. Larsson, *Phys. Rev.* **A54**, 3086 (1996).
58. Z. Amitay, D. Zajfman, P. Forck, U. Hechtfischer, B. Seidel, M. Grieser, D. Habs, R. Repnow, D. Schwalm and A. Wolf, *Phys. Rev.* **A54**, 4032 (1996).
59. J. R. Mowat, H. Danared, G. Sundström, M. Carlson, L. H. Andersen, L. Vejby-Christensen, M. af Ugglas and M. Larsson, *Phys. Rev. Lett.* **74**, 50 (1995).
60. A. J. R. Heck and D. W. Chandler, *Annu. Rev. Phys. Chem.* **46**, 335 (1995).
61. D. Zajfman, Z. Amitay, C. Broude, P. Forck, B. Seidel, M. Grieser, D. Habs, D. Schwalm and A. Wolf, *Phys. Rev. Lett.* **75**, 814 (1995).
62. W. J. van der Zande, J. Semaniak, V. Zengin, G. Sundström, S. Rosén, C. Strömholm, S. Datz, H. Danared and M. Larsson, *Phys. Rev.* **A54**, 5010 (1996).
63. Z. Amitay and D. Zajfman, *Rev. Sci. Instrum.* **68**, 1387 (1997).
64. S. Rosén, R. Peverall, J. ter Horst, G. Sundström, J. Semaniak, O. Sundqvist, M. Larsson, M. de Wilde and W. J. van der Zande, *Hyperfine Interact.* **115**, 201 (1998).
65. D. J. Auerbach, R. Cacak, R. Caudano, T. D. Gaily, C. J. Keyser, J. Wm. McGowan, J. B. A. Mitchell and S. F. J. Wilk, *J. Phys.* **B10**, 3797 (1977).
66. P. Van der Donk, F. B. Yousif, J. B. A. Mitchell and A. P. Hickman, *Phys. Rev. Lett.* **67**, 42 (1991).
67. I. F. Schneider, O. Dulieu and A. Giusti-Suzor, *J. Phys.* **B24**, L289 (1991).
68. I. F. Schneider, O. Dulieu and A. Giusti-Suzor, *Phys. Rev. Lett.* **68**, 2251 (1992).
69. P. Van der Donk, F. B. Yousif, J. B. A. Mitchell and A. P. Hickman, *Phys. Rev. Lett.* **68**, 2252 (1992).
70. I. F. Schneider, O. Dulieu, A. Giusti-Suzor and E. Roueff, *Astrophys. J.* **424**, 983 (1994).
71. H. Takagi, *J. Phys.* **B26**, 4815 (1993).

72. C. Strömholm, I. F. Schneider, G. Sundström, L. Carata, H. Danared, S. Datz, O. Dulieu, A. Källberg, M. af Ugglas, X. Urbain, V. Zengin, A. Suzor-Weiner and M. Larsson, *Phys. Rev.* **A52**, R4320 (1995).
73. H. T. Schmidt, L. Vejby-Christensen, H. B. Pedersen, D. Kella, N. Bjerre and L. H. Andersen, *J. Phys.* **B29**, 2485 (1996).
74. T. Tanabe, I. Katayama, H. Kamegaya, K. Chida, Y. Arakaki, T. Watanabe, M. Yoshizawa, M. Saito, Y. Haruyama, K. Hosono, K. Hatanaka, T. Honma, K. Noda, S. Ohtani and H. Takagi, *Phys. Rev. Lett.* **75**, 1066 (1995).
75. L. H. Andersen, P. J. Johnson, D. Kella, H. B. Pedersen and L. Vejby-Christensen, *Phys. Rev.* **A55**, 2799 (1997).
76. I. F. Schneider, C. Strömholm, L. Carata, X. Urbain, M. Larsson and A. Suzor-Weiner, *J. Phys.* **B30**, 2687 (1997).
77. B. Vâlcu, I. F. Schneider, M. Raoult, C. Strömholm, M. Larsson and A. Suzor-Weiner, *Eur. Phys. J.* **D1**, 71 (1998).
78. R. A. Phaneuf, D. H. Crandall and G. H. Dunn, *Phys. Rev.* **A11**, 528 (1975).
79. M. Vogler and G. H. Dunn, *Phys. Rev.* **A11**, 1983 (1975).
80. A. U. Hazi, C. Derkits and J. N. Bardsley, *Phys. Rev.* **A27**, 1751 (1983).
81. D. Zajfman, Z. Amitay, M. Lange, U. Hechtfischer, L. Knoll, D. Schwalm, R. Wester, A. Wolf and X. Urbain, *Phys. Rev. Lett.* **79**, 1829 (1997).
82. Z. Amitay, A. Baer, M. Dahan, L. Knoll, M. Lange, J. Levin, I. F. Schneider, D. Schwalm, A. Suzor-Weiner, Z. Vager, R. Wester, A. Wolf and D. Zajfman, *Science* **281**, 75 (1998).
83. Z. Vager, R. Naaman and E. P. Kanter, *Science* **244**, 426 (1989).
84. C. Strömholm, H. Danared, Å. Larson, M. Larsson, C. Marian, S. Rosén, B. Schimmelpfennig, I. F. Schneider, J. Semaniak, A. Suzor-Weiner, U. Wahlgren and M. Larsson, *J. Phys.* **B30**, 4919 (1997).
85. S. L. Guberman, *J. Chem. Phys.* **102**, 1699 (1995).
86. R. Peverall, S.Rosén, J. R. Peterson, M. Larsson, A. Al-Khalili, L. Vikor, J. Semaniak, A. Le Padelled and W. J. van der Zande, *J. Chem. Phys.* (in preparation).
87. J. H. Seong and H. S. Sun, *Bull. Korean Chem. Soc.* **17**, 1065 (1996).
88. J. Kaplan, *Phys. Rev.* **38**, 1048 (1931).
89. S. L. Guberman, in *Physics of Ion–Ion and Electron–Ion Collisions*, Ed. F. Brouillard and J. Wm. McGowan (Plenum Press, New York, 1983), p. 167.
90. S. L. Guberman, *Int. J. Quantum Chem., Symp.* **13**, 531 (1979).
91. S. L. Guberman, *Nature* **327**, 408 (1987).
92. S. L. Guberman, in Ref. 13, p. 45.
93. D. R. Bates, in *Applied Atomic Collision Physics*, Vol. 1, Eds. H. S. W. Massey and D. R. Bates (Academic Press, New York, 1982), p. 149.
94. J. E. Frederick, D. W. Rusch, G. A. Victor, W. E. Sharp, P. B. Hays and H. C. Brinton, *J. Geophys. Res.* **81**, 3923 (1976).
95. P. B. Hays and W. E. Sharp, *J. Geophys. Res.* **78**, 1153 (1973).
96. H. Takashi, B. R. Clemesha, P. P. Batista, Y. Sahai, M. A. Abdu and P. Muralikrishna, *Planet. Space Sci.* **38**, 547 (1990).

97. D. R. Bates, *Planet. Space Sci.* **38**, 889 (1990).
98. V. J. Abreu, S. C. Solomon, W. E. Sharp and P. B. Hays, *J. Geophys. Res.* **88**, 4140 (1983).
99. E. C. Zipf, *Bull. Am. Phys. Soc.* **15**, 418 (1970).
100. E. C. Zipf, *J. Geophys. Res.* **85**, 4232 (1980).
101. E. C. Zipf, *Planet. Space Sci.* **36**, 621 (1988).
102. J. L. Queffelec, B. R. Rowe, F. Vallée, J. C. Gomet and M. Morlais, *J. Chem. Phys.* **91**, 5335 (1989).
103. S. L. Guberman and A. Giusti-Suzor, *J. Chem. Phys.* **95**, 2602 (1991).
104. J. H. Sobral, H. Takahashi, M. A. Abdu, P. Muralikrishna, Y. Sahai and C. J. Zamlutti, *Planet. Space Sci.* **40**, 607 (1992).
105. D. R. Bates, *Planet. Space Sci.* **40**, 893 (1992).
106. D. Kella, L. Vejby-Christensen, P. J. Johnson, H. B. Pedersen and L. H. Andersen, *Science* **276**, 1530 (1997); *ibid.* **277**, 167 (1997).
107. S. L. Guberman, *Science* **278**, 1276 (1997).
108. S. L. Guberman, *Geophys. Res. Lett.* **18**, 1051 (1991).
109. C. Noren, F. B. Yousif and J. B. A. Mitchell, *J. Chem Soc. Faraday Trans.* **85**, 1697 (1989).
110. J. R. Peterson, A. Le Padellec, H. Danared, G. H. Dunn, M. Larsson, Å. Larson, R. Peverall, C. Strömholm, S. Rosén, M. af Ugglas and W. J. van der Zande, *J. Chem. Phys.* **108**, 1978 (1998).
111. E. C. Zipf, *Geophys. Res. Lett.* **7**, 645 (1980).
112. S. L. Guberman, *Geophys. Res. Lett* **18**, 1051 (1991).
113. S. L. Guberman, in Ref. 15, p. 64.
114. J. L. Fox and A. Dalgarno, *J. Geophys. Res.* **88**, 9027 (1983).
115. J. L. Fox, *J. Geophys. Res.* **98**, 3297 (1993).
116. D. Kella, P. J. Johnson, H. B. Pedersen, L. Vejby-Christensen and L. H. Andersen, *Phys. Rev. Lett.* **77**, 2432 (1996).
117. S. Oddone, J. W. Sheldon, K. A. Hardy and J. R. Peterson, *Phys. Rev.* **A56**, 4737 (1997).
118. S. L. Guberman, Institute of Scientific Research, Lexington, Ma. (personal communication).
119. V. I. Shematovitch, D. V. Bisikalo and J. C. Gérard, *J. Geophys. Res.* **99**, 23217 (1994).
120. J. N. Bardsley, *Planet. Space Sci.* **31**, 667 (1983).
121. F. L. Walls and G. H. Dunn, *J. Geophys. Res.* **79**, 1911 (1974).
122. J. L. Dulaney, M. A. Biondi and R. Johnsen, *Phys. Rev.* **A36**, 1342 (1987).
123. E. Alge, N. G. Adams and D. Smith, *J. Phys.* **B16**, 1433 (1983).
124. P. Španěl, L. Dittrichová and D. Smith, *Int. J. Mass Spectrom. Ion Proc.* **129**, 183 (1993).
125. P. M. Mul and J. Wm. McGowan, *J. Phys.* **B12**, 1591 (1979).
126. L. Vejby-Christensen, D. Kella, H. B. Pedersen and L. H. Andersen, *Phys. Rev.* **A57**, 3627 (1998).
127. H. Sun and H. Nakamura, *J. Chem. Phys.* **93**, 6491 (1990).

128. D. Kley, G. M. Lawrence and E. J. Stone, *J. Chem. Phys.* **66**, 4157 (1977).
129. M. Whitaker, M. A. Biondi and R. Johnsen, *Phys. Rev.* **A23**, 1481 (1981).
130. J. B. A. Mitchell and H. Hus, *J. Phys.* **B18**, 547 (1985).
131. M. Geoghegan, N. G. Adams and D. Smith, *J. Phys.* **B24**, 2589 (1991).
132. A. Le Padellec, C. Sheehan and J. B. A. Mitchell, *J. Phys.* **B31**, 1725 (1998).
133. S. Rosén, R. Peverall, M. Larsson, A. Le Padellec, J. Semaniak, Å. Larson, C. Strömholm, W. J. van der Zande, H. Danared and G.H. Dunn, *Phys. Rev.* **A57**, 4462 (1998).
134. W.-Ü. L Tchang-Brillet, P. S. Julienne, J. M. Robbe, C. Letzelter and F. Rostas, *J. Chem. Phys.* **96**, 6735 (1992).
135. D. Hirst, *Mol. Phys.* **82**, 359 (1994).
136. A. Le Padellec, J. B. A. Mitchell, A. Al-Khalili, H. Danared, A. Källberg, Å. Larson, S. Rosén, M. af Ugglas, L. Vikor and M. Larsson, *J. Chem. Phys.* **110**, 890 (1999).
137. D. R. Bates, *J. Phys.* **B25**, 5479 (1992).
138. N. G. Adams and D. Smith, in *Rate Coefficients in Astrochemisry*, Eds. T. J. Millar and D. A. Williams (Kluwer Academic, Dordrecht, 1988), p. 173.
139. F. B. Yousif and J. B. A. Mitchell, *Phys. Rev.* **A40**, 4318 (1989).
140. S. L. Guberman, *Phys. Rev.* **A49**, R4277 (1994).
141. B. K. Sarpal, J. Tennyson and L. A. Morgan, *J. Phys.* **B27**, 5943 (1994).
142. T. Tanabe, I. Katayama, N. Inoue, K. Chida, Y. Arakaki, T. Watanabe, M. Yoshizawa, M. Saito, Y. Haruyama, K. Hosono, T. Honma, K. Noda, S. Ohtani and H. Takagi, *Phys. Rev.* **A49**, R1531 (1994).
143. G. Sundström, S. Datz, J. R. Mowat, S. Mannervik, L. Broström, M. Carlson, H. Danared and M. Larsson, *Phys. Rev.* **A50**, R2806 (1994).
144. M. Larsson and C. Strömholm, in Ref. 15, p. 74.
145. C. Strömholm, J. Semaniak, S. Rosén, H. Danared, S. Datz, W. van der Zande and M. Larsson, *Phys. Rev.* **A54**, 3086 (1996).
146. J. Semaniak, S. Rosén, G. Sundström, C. Strömholm, S. Datz, H. Danared, M. af Ugglas, M. Larsson, W.J. van der Zande, Z. Amitay, U. Hechtfischer, M. Grieser, R. Repnow, M. Schmidt, D. Schwalm, R. Wester, A. Wolf and D. Zajfman, *Phys. Rev.* **A54**, R4617 (1996).
147. S. L. Guberman, in *The Physics of Electronic and Atomic Collisions, 19th Int. Conf.* (Whistler, Can.), Eds. L. J. Dubé, J. B. A. Mitchell, J. W. McConkey and C. E. Brion, AIP Conference Proceeding, Vol. 360 (American Institute of Physics, New York, 1995), p. 307.
148. T. Tanabe, I. Katayama, S. Ono, K. Chida, T. Watanabe, Y. Arakaki, Y. Haruyama, M. Saito, T. Odagiri, K. Hosono, K. Noda, T. Honma and H. Takagi, *J. Phys.* **B31**, L297 (1998).
149. A. Al-Khalili, H. Danared, M. Larsson, A. Le Padellec, R. Peverall, S. Rosén, J. Semaniak, M. af Ugglas, L. Vikor and W. J. van der Zande, *Hyperfine Interact.* **114**, 281 (1998).
150. A. E. Orel and K. C. Kulander, *Phys. Rev.* **A54**, 4092 (1996).

151. A. E. Orel, K. C. Kulander and T.N. Rescigno, *Phys. Rev. Lett.* **74**, 4807 (1995).
152. A. Le Padellec, S. Laubé, O. Sidko, C. Rebrion-Rowe, B. R. Rowe, B. Sarpal and J. B. A. Mitchell, *J. Phys.* **B30**, 963 (1997).
153. T. Oka, *Physics Today* **50**, No. 7, 90 (1997).
154. T. Oka, *Rev. Mod. Phys.* **64**, 1141 (1992).
155. A. Dalgarno, *Adv. At. Mol. Opt. Phys.* **32**, 57 (1994).
156. I. R. McNab, *Adv. Chem. Phys.* **89**, 1 (1995).
157. J. Tennyson, *Rep. Prog. Phys.* **58**, 421 (1995).
158. T. R. Geballe and T. Oka, *Nature* **384**, 334 (1996).
159. B. J. McCall, T. R. Geballe, K. H. Hinkle and T. Oka, *Science* **279**, 1910 (1998).
160. B. J. McCall, K. H. Hinkle, T. R. Geballe and T. Oka, *Faraday Disc.* **109**, 267 (1998).
161. H.-H. Lee, E. Roueff, G. Pineau des Forêts, O. M. Shalabiea, R. Terzieva and E. Herbst, *Astron. Astrophys.* **334**, 1047 (1998).
162. H. H. Michels and R. H. Hobbs, *Astrophys. J.* **286**, L27 (1984).
163. A. E. Orel and K. C. Kulander, *Phys. Rev. Lett.* **71**, 4315 (1993).
164. T. Gougousi, R. Johnsen and M. F. Golde, *Int. J. Mass Spectrom. Ion Proc.* **149/150**, 131 (1995).
165. D. R. Bates, M. F. Guest and R. A. Kendall, *Planet. Space Sci.* **41**, 9 (1993).
166. I. F. Schneider, Université Paris-Sud (personal communication).
167. F. B. Yousif, M. Rogelstad and J. B. A. Mitchell, in *Atomic and Molecular Physics: 4th US/Mexico Symp.*, Eds. I. Alvarez, C. Cisneros and T. J. Morgan (World Scientific, Singapore, 1995), p. 343.
168. G. Sundström, J. R. Mowat, H. Danared, S. Datz, L. Broström, A. Filevich, A. Källberg, S. Mannervik, K.-G. Rensfelt, P. Sigray, M. af Ugglas and M. Larsson, *Science* **263**, 785 (1994).
169. S. Laubé, A. Le Padellec, O. Sidko, C. Rebrion-Rowe, J. B. A. Mitchell and B. R. Rowe, *J. Phys.* **B31**, 2111 (1998).
170. D. Smith and P. Španěl, *Int. J. Mass Spectrom. Ion Proc.* **129**, 163 (1993).
171. A. Müller, D. Belić, B. DePaola, N. Djuric, G. H. Dunn, D. W. Mueller and C. Timmer, *Phys. Rev. Lett.* **56**, 127 (1986).
172. T. Bartsch, A. Müller, W. Spies, J. Linkemann, H. Danared, D. R. DeWitt, H. Gao, W. Zong, R. Schuch, A. Wolf, G. H. Dunn, M. S. Pindzola and D. C. Griffin, *Phys. Rev. Lett.* **79**, 2233 (1997).
173. M. Larsson, H. Danared, Å. Larson, A. Le Padellec, J. R. Peterson, S. Rosén, J. Semaniak and C. Strömholm, *Phys. Rev. Lett.* **79**, 395 (1997).
174. A. Le Padellec, M. Larsson, H. Danared, Å. Larson, J. R. Peterson, S. Rosén, J. Semaniak and C. Strömholm, *Phys. Scripta* **57**, 215 (1998).
175. S. Datz, M. Larsson, C. Strömholm, G. Sundström, V. Zengin, H. Danared, A. Källberg and M. af Ugglas, *Phys. Rev.* **A52**, 2901 (1995).
176. M. Larsson, S. Lepp, A. Dalgarno, C. Strömholm, G. Sundström, V. Zengin, H. Danared, A. Källberg, M. af Ugglas and S. Datz, *Astron. Astrophys.* **309**, L1 (1996).

177. T. Tanabe, I. Katayama, H. Kamegaya, K. Chida, Y. Arakaki, T. Watanabe, M. Yoshizawa, M. Saito, Y. Haruyama, K. Hosono, K. Hatanaka, T. Honma, K. Noda, S. Ohtani and H. Takagi, in Ref. 15, p. 84.
178. P. Van der Donk, F. B. Yousif and J. B. A. Mitchell, *Phys. Rev.* **A43**, 5971 (1991).
179. J. R. Peterson, P. Devynck, Ch. Hertzler and W. G. Graham, *J. Chem. Phys.* **96**, 8128 (1992).
180. J. B. A. Mitchell, J. L. Forand, C. T. Ng, D. P. Levac, R. E. Mitchell, P. M. Mul, W. Clayes, A. Sen and J. Wm. McGowan, *Phys. Rev. Lett.* **51**, 885 (1983).
181. S. Datz, G. Sundström, Ch. Biedermann, L. Broström, H. Danared, S. Mannervik, J. R. Mowat and M. Larsson, *Phys. Rev. Lett.* **74**, 896 (1995).
182. K. C. Kulander and M. F. Guest, *J. Phys.* **B12**, L501 (1979).
183. B. Peart, R. A. Foster and K. Dolder, *J. Phys.* **B12**, 3441 (1979).
184. F. B. Yousif, P. J. T. Van der Donk and J. B. A. Mitchell, *J. Phys.* **B26**, 4249 (1993).
185. H. Helm, in Ref. 14, p. 145.
186. U. Müller and P. C. Cosby, *J. Chem. Phys.* **105**, 3532 (1996).
187. A. E. Orel and K. C. Kulander, *J. Chem. Phys.* **91**, 6086 (1989).
188. J. L. Krause, A. E. Orel, B. H. Lengsfield and K. C. Kulander, in *Time-Dependent Quantum Molecular Dynamics*, Eds. J. Broeckhove and L. Lathouwers (Plenum Press, New York, 1992), p. 131.
189. A. Dalgarno, *J. Chem. Soc. Faraday Trans.* **89**, 2111 (1993).
190. E. Herbst, *Astrophys. J.* **222**, 508 (1978).
191. D. R. Bates, *Astrophys. J.* **306**, L45 (1986).
192. D. R. Bates, *Astrophys. J.* **344**, 531 (1989).
193. E. T. Galloway and E. Herbst, *Astrophys. J.* **376**, 531 (1991).
194. D. R. Bates, *J. Phys.* **B24**, 3267 (1991).
195. D. R. Bates, *Proc. R. Soc. Lond.* **A443**, 257 (1993).
196. F. Vallée, B. R. Rowe, J. C. Comet, J. L. Queffelec and M. Morlais, *Chem. Phys. Lett.* **124**, 317 (1986).
197. B. R. Rowe, F. Vallée, J. L. Queffelec, J. C. Gomet and M. Morlais, *J. Chem. Phys.* **88**, 845 (1988).
198. N. G. Adams, C. R. Herd and D. Smith, *J. Chem. Phys.* **91**, 963 (1989).
199. C. R. Herd, N. G. Adams and D. Smith, *Astrophys. J.* **349**, 388 (1990).
200. N. G. Adams, C. R. Herd, M. Geoghegan, D. Smith, A. Canosa, B. R. Rowe, J. L. Queffelec and M. Morlais, *J. Chem. Phys.* **94**, 4852 (1991).
201. J. B. A. Mitchell and F. B. Yousif, in Ref. 13, p. 109.
202. K. H. Berkner, T. J. Morgan, R. V. Pyle and J. W. Stearns, in *The Physics of Electronic and Atomic Collisions, 7th Int. Conf.* (Amsterdam, The Netherlands, 1971), p. 422.
203. J. L. Forand, J. B. A. Mitchell and J. Wm. McGowan, *J. Phys.* **E18**, 623 (1985).
204. S. Datz, M. Larsson, C. Strömholm, G. Sundström, V. Zengin, H. Danared, A. Källberg and M. af Ugglas, in Ref. 15, p. 151.

205. L. H. Andersen, O. Heber, D. Kella, H. B. Pedersen, L. Vejby-Christensen and D. Zajfman, *Phys. Rev. Lett.* **77**, 4891 (1996).
206. T. L. Williams, N. G. Adams, L. M. Babcock, C. R. Herd and M. Geoghegan, *Mon. Not. R. Astron. Soc.* **282**, 413 (1996).
207. L. Vejby-Christensen, L. H. Andersen, O. Heber, D. Kella, H. B. Pedersen, H. T. Schmidt and D. Zajfman, *Astrophys. J.* **483**, 531 (1997).
208. L. Vikor, A. Al-Khalili, H. Danared, N. Djuric, G. H. Dunn, M. Larsson, A. Le Padellec, S. Rosén and M. af Ugglas, *Astron. Astrophys.* **344**, 1027 (1999).
209. Å. Larson, A. Le Padellec, J. Semaniak, C. Strömholm, M. Larsson, S. Rosén, R. Peverall, H. Danared, N. Djuric, G.H. Dunn and S. Datz, *Astrophys. J.* **505**, 459 (1998) .
210. T. J. Millar, A. Bennet, J. M. C. Rawlings, P. D. Brown and S. B. Charnley, *Astron. Astrophys. Suppl.* **87**, 585 (1991).
211. R. P. A. Bettens and E. Herbst, *Int. J. Mass Spectrom. Ion Proc.* **149/150**, 321 (1995).
212. S. Datz, Oak Ridge National Laboratory (personal communication).
213. J. M. Butler, L. M. Babcock and N. G. Adams, *Mol. Phys.* **91**, 81 (1997).
214. R. A. Gutcheck and E. C. Zipf, *J. Geophys. Res.* **78**, 5429 (1973).
215. M. Tsuji, M. Nakamura, Y. Nishimura and H. Obase, *J. Chem. Phys.* **103**, 1413 (1995).
216. M. Tsuji, M. Nakamura, Y. Nishimura and H. Obase, *J. Chem. Phys.* **108**, 8031 (1998).
217. T. Gougousi, R. Johnsen and M. F. Golde, *J. Chem. Phys.* **107**, 2430 (1997).
218. A. Al-Khalili, L. Vikor, M. af Ugglas, Å. Larson, A. Le Padellec, S. Rosén and M. Larsson (unpublished).
219. D. H. Mordaunt, I. R. Lambert, G. P. Morley, M. N. R. Ashfold, R. N. Dixon, C. M. Western, L. Schnieder and K. H. Welge, *J. Chem. Phys.* **98**, 2054 (1993).
220. J. Semaniak, Å. Larson, A. Le Padellec, C. Strömholm, M. Larsson, S. Rosén, R. Peverall, H. Danared, N. Djuric, G. H. Dunn and S. Datz, *Astrophys. J.* **498**, 886 (1998).
221. E. Herbst and H.-H. Lee, *Astrophys. J.* **485**, 689 (1997).
222. D. Zajfman, in *Photonic, Electronic and Atomic Collisions, 20th Int. Conf. on the Physics of Electronic and Atomic Collisions* (Vienna, Austria), Eds. F. Aumayr and HP. Winter (World Scientific, Singapore, 1998), p. 279.
223. L. H. Andersen, P. Hvelplund, D. Kella, P. H. Mokler, H. B. Pedersen, H. T. Schmidt and L. Vejby-Christensen, *J. Phys.* **B29**, L643 (1996).
224. L. H. Andersen, T. Andersen and P. Hvelplund, *Adv. At. Mol. Opt. Phys.* **38**, 155 (1998).
225. C. Brink, L. H. Andersen, P. Hvelplund, D. Mathur and J. D. Voldstad, *Chem. Phys. Lett.* **233**, 52 (1995).
226. U. Hechtfischer, Z. Amitay, P. Forck, M. Lange, J. Linkemann, M. Schmitt, U. Schramm, D. Schwalm, R. Wester, D. Zajfman and A. Wolf, *Phys. Rev. Lett.* **80**, 2809 (1998).

227. G. H. Dunn, *Phys. Rev. Lett.* **8**, 62 (1962).
228. T. F. O'Malley and H. S. Taylor, *Phys. Rev.* **176**, 207 (1968).
229. D. Zajfman, Z. Amitay, P. Forck, D. Habs, D. Schwalm and A. Wolf, in Ref. 15, p. 114.
230. S. P. Mϕller, *Nucl. Instr. Meth. Phys. Res.* **A394**, 281 (1997).
231. D. Zajfman, O. Heber, L. Vejby-Christensen, I. Ben-Itzhak, M. Rappaport, R. Fishman and M. Dahan, *Phys. Rev.* **A55**, R1577 (1997).
232. M. Dahan, R. Fishman, O. Heber, M. Rappaport, N. Altstein, D. Zajfman, and W. J. van der Zande, *Rev. Sci. Instrum.* **69**, 76 (1998).
233. M. Larsson, L. Broström, M. Carlson, H. Danared, S. Datz, S. Mannervik and G. Sundström, *Phys. Scripta* **51**, 354 (1995).
234. M. Larsson, M. Carlson, H. Danared, L. Broström, S. Mannervik and G. Sundström, *J. Phys.* **B27**, 1397 (1994).
235. P. Forck, C. Broude, M. Grieser, D. Habs, J. Kenntner, J. Liebmann, R. Repnow, D. Schwalm, A. Wolf, Z. Amitay and D. Zajfman, *Phys. Rev. Lett.* **72**, 2002 (1994).
236. Z. Amitay, D. Zajfman, P. Forck, T. Heupel, M. Grieser, D. Habs, R. Repnow, A. Wolf and S. L. Guberman, *Phys. Rev.* **A53**, R644 (1996).
237. P. Forck, M. Grieser, D. Habs, A. Lampert, R. Repnow, D. Schwalm, A. Wolf and D. Zajfman, *Nucl. Instr. Meth. Phys. Res.* **B79**, 273 (1993).

CHAPTER 14

DISSOCIATIVE PHOTODETACHMENT STUDIES OF TRANSIENT MOLECULES BY COINCIDENCE TECHNIQUES

R. E. Continetti

Department of Chemistry and Biochemistry, 0314
University of California, San Diego
9500 Gilman Drive La Jolla, California 92093-0314

Contents

1. Introduction

The dynamics and energetics of transient species are important to understanding complex chemical systems, and remain an important area of research in gas-phase physical chemistry. To apply new optical and molecular-beam techniques developed in the study of dissociation dynamics and energetics, it is necessary to have well-characterized methods of producing transient species. In recent years, the use of pyrolysis[1] and photodissociation,[2] in addition to more traditional *in situ* chemical[3] and discharge sources[4] of reactive species has continued to broaden the range of systems amenable to study. As a result of this, the energetics and reaction dynamics of reactive species have been studied using a variety of techniques including photoionization,[5,6] photodissociation[7] and reactive scattering.[4] Another important route to the preparation of transient neutral species involves the photodetachment of negative ions. This approach has been taken to produce stable free radicals[8] and mass selected neutral clusters.[9]

In the experiments discussed in this review, photodetachment and coincident energy analysis of the photoelectron is used to create transient neutral species with a known internal energy. Following dissociation, the neutral photofragments are detected in coincidence. These experiments combine the traditional measurements of photoelectron[10] and photofragment translational spectroscopies[11] to provide a unique insight into the energetics and dynamics of transient species.

Photoelectron spectra of negative ions have been obtained for a variety of systems, ranging from atoms and atomic clusters to molecular clusters and polyatomic molecules.[10] When an anion has Franck–Condon overlap with a bound state of a neutral molecule, photoelectron spectra reveal vertical and adiabatic electron affinities, and can often be used to determine spectroscopic constants of both the neutral and precursor anion. Photodetachment can be particularly valuable in the characterization of excited states that are not optically accessible from the ground state of the neutral. Electron affinities, coupled with gas-phase acidities determined in studies of ion–molecule reactions, can furthermore reveal important thermochemical properties such as heats of formation and bond dissociation energies.[12] With the continued development of sources of molecular and cluster anions, these measurements are being made on an ever-expanding set of systems. An example of this is provided by the recent study of the photodetachment of the citric acid dianion by Wang and coworkers.[13]

There is another important class of photodetachment processes — those that occur when photodetachment of the anion produces an unstable neutral molecule or complex. Analysis of the photoelectron spectra observed from such processes has provided an approach to a spectroscopy of the transition state, since the photoelectron spectra provide detailed information on the Franck–Condon overlap between the parent anion and the unstable region of the neutral potential energy surface. Lineberger and coworkers have used this technique to probe the isomerization dynamics of organic molecules, such as the acetylene-vinylidene isomerization[14] and conformational changes in cyclooctatetraene.[15] Neumark and coworkers have carried out pioneering studies of photodetachment to dissociative states of neutral complexes, providing important insights into the potential energy surfaces governing fundamental bimolecular reactions such as the hydrogen exchange reactions X + HY, where X, Y = F, Cl, Br, I,[16] and the fundamental $F + H_2$ system.[17] These experiments have been extended to polyatomic systems as well.[18,19] The studies of both isomerization dynamics and the transition state region of reactive potential energy surfaces rely on the overlap of the Franck–Condon region for photodetachment of the ground state of the anions with the relevant portions of the neutral potential energy surface. The photodetachment studies of the transition state region for bimolecular reactions have provided inspiration for theoretical studies of the relevant potential energy surfaces and reaction dynamics, providing a connection between the observed photoelectron spectra, scattering dynamics and the potential energy surfaces for these fundamental reactions.[20]

The study of neutral dissociation reactions by dissociative photodetachment (DPD) of molecular or cluster anion precursors represents an extension of the ideas of anion photodetachment transition state spectroscopy. In addition to the measurement of the photoelectron kinetic energy, which determines the energy remaining in the neutral complex, a kinematically complete experiment can be carried out by measuring the photofragment translational energy release *in coincidence.* This type of experiment can reveal the asymptotic correlation between the dissociation dynamics and the Franck–Condon region accessed by photodetachment, providing qualitatively new information on the dynamics of the dissociation processes that occur on the neutral potential energy surface upon photodetachment.

1.1. *Dissociative Photodetachment Processes*

Two limiting types of DPD processes can be considered — direct and sequential. In the direct case, dissociation and photodetachment both occur on a timescale less than a rotational period. This occurs when photodetachment of a stable anion accesses a repulsive region of the corresponding neutral potential energy surface. In such cases, the neutral will survive for less than a vibrational period and the photoelectron spectrum will be expected to have no discernable structure, providing little insight into the structure of the anion or neutral. In the sequential case, departure of the photoelectron and the subsequent dissociation are uncoupled. The photoelectron spectrum alone will provide information on the excited state of the neutral, but will yield no insights into the subsequent dissociation process. In either case, to kinematically characterize DPD processes it is necessary to carry out a photoelectron-photofragment coincidence (PPC) experiment involving detection of *all* the photoproducts.

PPC experiments can reveal both angular and energetic correlations between the products of a DPD event. The bulk of the work reviewed here will focus on the correlations of the photoelectron kinetic energy and the photofragment translational energy release. It is useful to consider at this point the types of energetic correlations that will be observed in direct and sequential DPD processes. Photoelectron-photofragment angular correlations observed in the DPD of O_4^- provide important insights into the electronic structure and photodetachment dynamics of O_4^- and will be discussed further in Sec. 3.1.4.

Figure 1(a) shows a schematic of the potential energy surfaces involved in the hypothetical sequential DPD of a diatomic anion. Photodetachment to the bound excited state ($AB^* + e^-$) yields a structured photoelectron spectrum, with intensities determined by the Franck–Condon factors for photodetachment from AB^- to AB^*.[6,10] The photoelectron kinetic energy (eKE) spectrum is shown to the left of the potential energy curves. The total energy available for the DPD process is fixed by the photon energy, $E_{h\nu}$, the bond dissociation energy of AB^-, and in the case shown here, the electron affinity of atom A. In the absence of internal structure

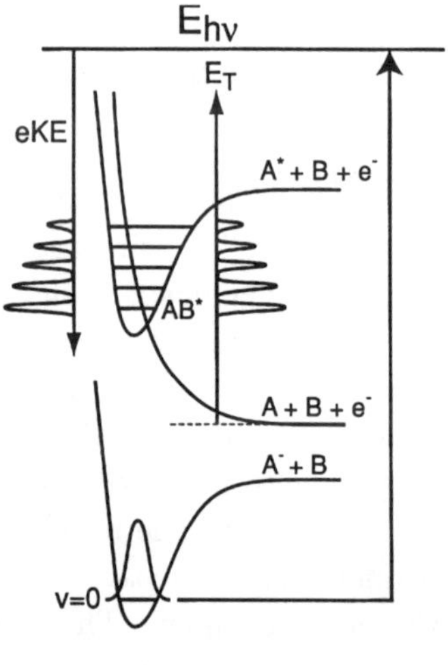

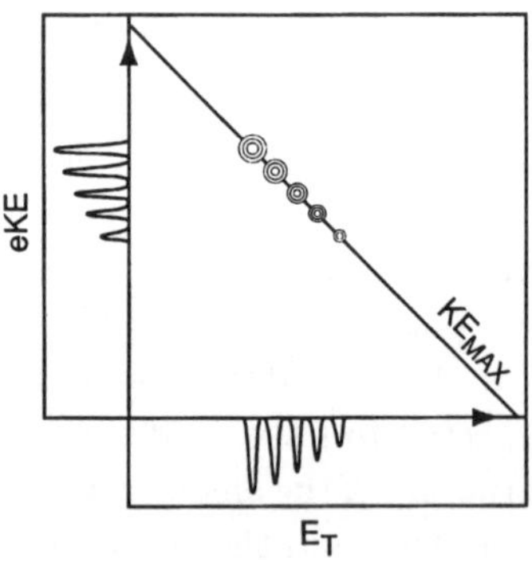

Fig. 1. (a) Schematic illustrating the energetics of a sequential DPD process of AB^-. (b) N(E_T, eKE) correlation spectrum expected for the sequential DPD shown in A. The E_T and eKE spectra measurable in a noncoincidence experiment are displayed along the x- and y-axes, respectively.

in the products A or B, the kinetic energy release is fixed and the center-of-mass translational energy, E_T, released between A and B is the 'inverse' of the photoelectron spectrum. The E_T spectrum provides information complementary to the photoelectron spectrum, revealing directly the excited state energies relative to the dissociation asymptote. This is more clearly shown in the schematic photoelectron-photofragment correlation spectrum in Fig. 1(b). This contour plot shows the number of events with a given translational energy and photoelectron kinetic energy, N(E_T, eKE). The diagonal line marked KE_{MAX} represents the total available energy that, in this case, is only partitioned between eKE and E_T. Thus, the photoelectron-photofragment correlation spectrum is a series of spots along KE_{MAX}. Integrating over the complementary variable will yield either the eKE spectrum shown along the y-axis or the E_T spectrum shown along the x-axis.

Figure 2(a) shows the case of direct DPD. The bound anion wavefunction is projected onto a repulsive neutral curve correlating to two ground state atoms and a free electron. Here, measurement of the photoelectron spectrum still provides a measure of the vertical electron affinity of the

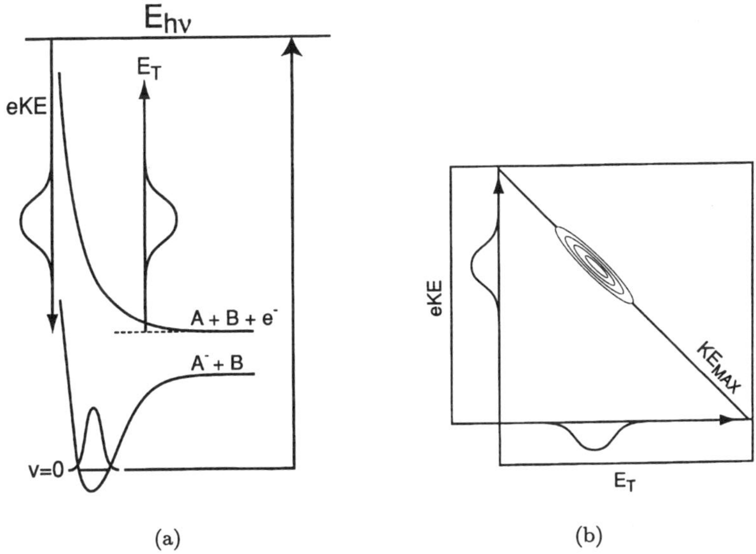

Fig. 2. (a) Schematic illustrating the energetics of a direct DPD process of AB^-. (b) N(E_T, eKE) correlation spectrum expected for the direct DPD shown in A.

neutral excited state and the distribution of photoelectron kinetic energies is dictated by the Franck–Condon principle. In this case, however, one must consider the bound-free Franck–Condon factors.[21] Again, as in the case of sequential DPD, E_T directly shows the repulsion between the neutral products.

As shown in Fig. 2(b), direct DPD will produce a diagonal feature in the N(E_T, eKE) correlation spectrum, resulting from projection of the wavefunction describing the bound diatom onto the repulsive nuclear potential in a bound-to-free Franck–Condon absorption. The width of the projection of this feature unto either the eKE or E_T axes are identical, and the length of the projection onto these kinetic energy axes is a function of both the equilibrium distance along the bound coordinate and the slope of the repulsive nuclear potential reached by photodetachment. The steeper the slope of the repulsive potential, the longer the ridge in the N(E_T, eKE) correlation spectrum. An analogous statement is true for all bound-free spectroscopies, including photon absorption, emission, and processes closely related to DPD, such as dissociative electron attachment and dissociative photoionization. Bound-free Franck–Condon processes for dissociative electron attachment, for example, are discussed at some length by Illenberger and Momigny.[22]

No example of DPD of a diatomic anion has been studied to date, but in Secs. 3.1 and 3.2, examples of direct and sequential DPD processes of small molecules, O_4^- and O_3^-, respectively, will be discussed. In these cases, the products have electronic, vibrational and rotational degrees of freedom available for energy partitioning, leading to more complicated correlation spectra. However, the correlation spectra for these two systems clearly illustrate features corresponding to the limiting cases discussed here.

1.2. *Three-Body Dissociative Photodetachment*

The PPC technique has recently been extended to the study of dissociation processes yielding three atomic or molecular neutral products. This is an important development since three-body chemical reactions among neutral species play an important role in a number of atmospheric and combustion chemistry systems.[23] In the past, however, the detailed dynamics of these reactions have proven difficult to study due to difficulties in determining the correlation between the distribution of internal and kinetic energies among the products. Significant experimental progress along these lines has been made in studies of the dissociation of multiply-charged cations and other high energy systems.[24–27] Detailed experimental studies of the dynamics of three-body reactions of *neutral* species, however, have been limited to excited state photodissociation reactions leading to either prompt or sequential dissociation processes.[28] In these studies, measurements of the uncorrelated asymptotic properties of the photofragments such as momentum and quantum state have been used to infer the three-body dissociation dynamics.[28] While significant insights into both concerted and sequential dissociation reactions have been gained, a general technique capable of performing the coincidence measurements required to measure the partitioning of energy and recoil angle correlations among the photofragments was not previously available for neutral dissociation processes. The PPC technique can be applied to these problems using a multiparticle time- and position-sensitive detector to record the velocities and recoil angles of *all* the fragments in a three-body dissociation. In the discussion of experimental techniques below, the focus will be on the second generation photoelectron-photofragment spectrometer, equipped with a multiparticle detector capable of recording all the photofragments produced in a three-body DPD. The first experiments using this apparatus, concerning the DPD of O_6^-, are reviewed in Sec. 3.3.

2. Coincidence Techniques for the Study of Dissociative Photodetachment

As discussed above, DPD processes by nature involve two continua — the photoelectron and photofragment continuum. Thus, to carry out a kinematically well-defined study of such processes, it is necessary to measure these photoproducts in coincidence. A measurement of this type can be carried out by performing translational spectroscopy on both the photoelectron and atomic or molecular photofragments in a fast ion beam. The principle of the experiments is straightforward: measure the kinetic energies and recoil angles of the photoelectron and all the photofragments produced in a single DPD event.

Use of a fast ion beam has particular advantages for carrying out translational spectroscopy measurements on atomic and molecular products. The high laboratory velocity of the precursor negative ion permits efficient, universal detection of the products using microchannel-plate-based detectors. Furthermore, dissociation kinematics in the fast beam constrain the photofragments to recoil in a small cone about the beam velocity vector, allowing collection of all the photofragments with a small-area detector. This consequence of momentum conservation, coupled with high detection efficiencies for fast ions and neutrals, allows the dynamics of photodissociation processes to be studied using a time- and position-sensitive detection scheme to record the recoil velocities of the photofragments. This variant of photofragment translational spectroscopy was pioneered in charge-exchange studies of the highly excited states of diatomic molecules by Los and coworkers,[29,30] and was later used by Helm and Cosby to study both diatomic and polyatomic molecules.[31] Neumark and coworkers more recently extended this technique to study the photodissociation of neutral free radicals produced by photodetachment of a fast negative ion beam.[32]

Carrying out photoelectron spectroscopy in a fast ion beam, however, poses special difficulties, particularly when it is desired to collect a large fraction of the photodetached electrons in a coincidence experiment. The ion beam velocity is no longer negligible when a low energy photoelectron is produced, leading to a large Doppler shift in the laboratory velocity of the photoelectron as a function of recoil angle from the ion beam. Thus, it is essential to measure the recoil angle of each electron to accurately determine the center-of-mass (CM) electron kinetic energy.[33,34]

Carrying out photoelectron and photofragment translational spectroscopy in coincidence makes these experiments analogous to a number of studies of dissociative photoionization processes.[24,35] In the case of dissociative photoionization, product detection is simpler since only charged products are detected, and the mass of the ionic product can be determined with great accuracy using standard techniques. An important difficulty in photoionization processes of transient species, however, is determining the identity of the parent molecule. Indeed, photoelectron-photoion coincidence experiments can be used to resolve this question.[36,37] Angle-resolved photoelectron-photoion coincidence experiments in which the product translational energies are fully resolved provide a level of information similar to that obtained in a PPC experiment.[38,39]

The basic principle of a PPC experiment is presented here. Following photodetachment, photoelectrons of reasonable kinetic energies ($>$ 100 meV) recoil out of the beam and can be detected. The neutral fragments from the dissociation event also recoil out of the beam during the flight time to the detector. Residual ions are deflected out of the beam electrostatically after the interaction region, allowing only neutral fragments to reach the detector. A detector then records the time and position of arrival of the fragments. For both photoelectrons and photofragments, given the parent ion mass and velocity, measuring the time and position of arrival of the particle allows determination of the center-of-mass recoil angles and the kinetic energy release. In the case of the photofragments, the product mass ratio can also be determined.[29]

In our laboratory, we have built two PPC spectrometers based on the fast negative ion beam photodetachment technique. The first spectrometer is capable of detecting a photoelectron and two neutral photofragments in coincidence, and has been previously described in some detail.[40,41] When the nascent neutral dissociates into two fragments, kinematics provide a separation of the fragments in space and/or time once they reach the detector. This separation allows the photofragment detector to be segmented into two halves with only a small loss of efficiency at the seam between the two detector sections. The second generation apparatus extends the PPC technique by allowing the detection of a photoelectron and three (up to eight) photofragments in coincidence. When dissociation occurs into three or more fragments, the kinematics no longer compel the fragments to be well separated in space or time upon arrival at the detector. This makes detection considerably more complicated, as it is necessary to build an ideal detector capable of resolving several particles regardless of position and time of arrival. This second generation apparatus is described below and in detail elsewhere.[42]

2.1. *Photoelectron-Photofragment Coincidence Spectrometer*

A schematic of the second generation PPC spectrometer is shown in Fig. 3. The apparatus is composed of an ion source, an acceleration region, a time-of-flight region, and an ultrahigh vacuum region containing the laser/ion beam interaction volume and particle detectors. The apparatus is differentially pumped, with operating pressures of $1\text{–}5 \times 10^{-4}$ in the source region, $1\text{–}4 \times 10^{-5}$ torr in the acceleration region, 1×10^{-6} torr in the TOF region and $\approx 1 \times 10^{-9}$ torr in the detection region. The ion source and TOF mass spectrometer are discussed in more detail in the next section, followed by a description of the detection techniques used in these experiments.

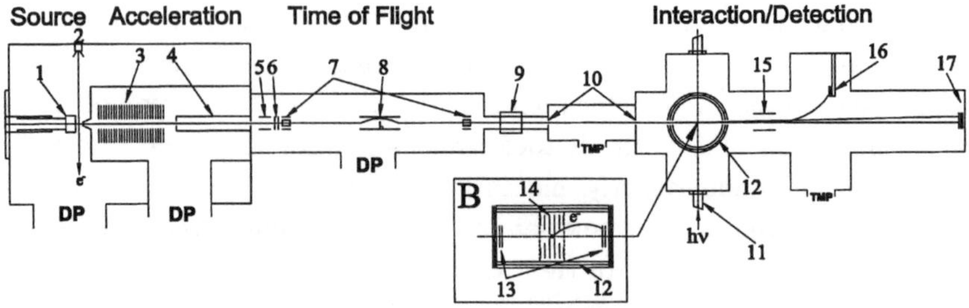

Fig. 3. Schematic of the apparatus, showing the ion source, acceleration, mass-selection and interaction/detection regions. The different regions are pumped by diffusion and turbomolecular pumps, DP and TMP, respectively. See text for detailed description.

The anions discussed in this review were produced by electron beam irradiation (2) of a pulsed free-jet expansion (1) at repetition rates of 0.6–1 kHz. This has proven to be a versatile method of creating both positive and negative molecular and cluster ions.[43] A 1 keV electron beam is focused on the expansion, producing low energy secondary electrons that attach to molecules via three-body collisional processes or cluster evaporation.[44] The nascent anions, produced in the continuum flow region of the pulsed molecular beam, are subsequently cooled in the supersonic expansion. Vibrational temperatures of 200–500 K are typical for small molecular anions (O_3^-, CCO^-, and N_3^-), with rotational temperatures < 50 K expected.[8] O_4^- and O_6^- were produced by expansion of pure O_2 at a backing pressure of 1.5 atm through a 0.25 mm piezoelectric pulsed nozzle operating at 600 Hz–1 kHz.[45] O_3^- was generated by electron impact on a mixture of O_3 and O_2.

The anions pass through a skimmer, enter a differentially pumped chamber, and are accelerated to an energy of 2.5–8 keV (3). A long-focal-length einzel lens embedded in the acceleration stack is used to optimize ion transport into the dissociation region. After acceleration, the ion beam is re-referenced to ground potential using a high-voltage switch (4). This device is a stainless steel tube with small apertures at each end. Ions enter the tube while it is at the full acceleration potential, and it is then switched to ground before the ions exit so the rest of the apparatus does not need to be floated at the acceleration potential.

Ions that pass through the switch are mass selected using a TOF method due to Bakker.[46] A rapidly switched potential perpendicular to the beam axis (5) is used to permit only a small packet of ions to pass undeflected through 1 mm entrance apertures to the interaction region (10). Over the 1.9 m flight path from these plates to the laser interaction region, the ions separate into mass-selected packets by TOF. This approach was chosen as it induces minimal axial energy spread in the ion beam. The beam energy spread is estimated to be $\leq$ 5 V, giving $\Delta E_{beam}/E_{beam} \leq 0.2\%$, allowing the accurate determination of the photofragment recoil energy and angles. To optimize overlap with the pulsed laser, an axial pulsed field ($\approx$ 35 V/cm) between a set of disks (6) can be used to focus the ions at the interaction region at the expense of some kinetic energy resolution. Electrostatic deflectors (7) are used to transport the ion beam into the detection region. The ions are also guided over a beam-block situated on the beam axis (8) that removes any fast neutral particles from the beam.[47]

Anions at the mass of interest are intersected at a right angle by a laser pulse (11) from a Nd:YAG laser (mode-locked, *Q*-switched, cavity dumped Quantronix 116). This laser produces short pulses ($\approx$ 100 ps) that are required for high-resolution photoelectron TOF measurements given the short (5-cm) nominal photoelectron flight path. The laser is focused to a $\approx$ 0.5-mm diameter spot at the point of ion laser interaction. In the experiments described here, the Nd:YAG laser produced 100–140 μJ/pulse at 532 nm, and 20–60 μJ/pulse at 355 and 266 nm.

Photoelectrons traverse a nominal field-free flight-path of 50-mm and impinge on a 100-mm diameter detector entrance grid centered perpendicular to the ion beam and laser propagation directions. In Fig. 3 the photoelectron detectors are centered above and below the plane of the figure. The arrangement of the photoelectron detectors in this perpendicular plane is shown in inset B in the figure. A triple layer of magnetic shielding (12) surrounds the entire region. The defining apertures and entrance-grid to the photoelectron detectors are shown (14), along with the two opposed time- and position-sensitive detectors (13). The laboratory kinetic energy of the electron is measured by time-of-flight and the recoil angle by detected position. Measurement of the electron recoil angle is essential to allow correction for both the Doppler broadening due to the fast ion beam and the actual flight path in the large solid angle detector. With this correction, the electron kinetic energy in the center-of-mass frame (eKE) is determined. The photoelectron spectrometer has an effective angular acceptance of $\approx$ 4% of $4\pi\, sr$. The resolution in eKE is $\approx$ 4% $\Delta E/E$ as determined by measurements of O_2^- and I^- photoelectron spectra.

Photofragments recoil out of the beam over a 104-cm flight path and impinge on the multiparticle time- and position-sensitive detector. The ion beam and any charged photofragments are electrostatically deflected off the beam axis (15) into a detector to allow continuous monitoring of the ion beam (16). Only neutral photofragments are detected at the multiparticle detector (17). This detector, discussed in more detail below, encodes the time- and position-of-arrival of up to eight photofragments after the laser shot. The time and position information for the coincident photofragments is used to calculate the photofragment mass ratio, center-of-mass translational energy (E_T) and recoil angle for each event.[29,32] Once the mass of the photofragments is determined, the kinetic energy and recoil angle of the photofragments may be calculated. The results are then histogrammed in a raw translational energy and angular distribution, $N(E_T, \theta)$. The translational energy resolution is $\leq 8\%$ $\Delta E/E$ at $E_T = 0.55$ eV, with a precision of $\approx \pm 0.02$ eV.[69] Photofragment mass resolution $m/\Delta m \approx 15$ has been shown to be routinely achievable. If the dissociation lifetime is significant compared to the several μs flight time from the interaction region to the detector, it is still possible to determine that the complex dissociates, and an assessment of the dissociation lifetime can be made by varying either the incident beam energy or the neutral flight path.

Depending on the dissociation processes that may occur, each event is required to consist of two or three neutral photofragments and a photoelectron. Using the time- and position-of-arrival information, the conservation of linear momentum among the detected photofragments can be analyzed to determine if the photofragments originate in a single dissociation event by checking each of the chemically possible dissociation pathways. Detection efficiency and the count rate determine correlation of the photoelectron and the neutral photofragments. To ensure valid coincidences, the dissociation rate must be $\ll 1$ event per laser pulse, as discussed below.

2.2. *Particle Detection*

There are several viable approaches to recording the kinetic energy and recoil angle of photoelectrons and photofragments. To carry out coincidence experiments of the type discussed in this review, it is necessary that the detection scheme be fully multiplexed, e.g. after any event particles of any energy and recoil angle may be detected. Furthermore, it is necessary to operate these experiments in the low-event probability, high repetition rate limit to avoid false coincidences. This requires the detection scheme to allow rapid acquisition and storage of data on an event-by-event basis.

For both neutral and charged particles at beam energies of several keV, it is essential to amplify the signal by production of secondary electrons. Microchannel plates (MCPs) provide the optimum means for doing this over a large detector area (4-cm diameter field-of-view), providing gain on the order of 10^7. When a fast neutral or a photoelectron strikes the microchannel plate, secondary electrons are ejected and amplified by the MCPs. High gains (on the order of 2×10^7) and narrow charge distributions can be achieved using three-MCP 'Z' stacks.[48] The electron cloud emerges from the last MCP in the stack and is collected on the anode and analyzed for time- and position-of-arrival. Thus, the question rests on how to encode the time- and position-of-arrival of the electron clouds emerging from the MCP. We have used two solutions to this problem. On our first apparatus, and for the photoelectron detectors on the second apparatus, a charge-division technique based on the wedge-and-strip anode[49] is used to encode time and position for a single particle per laser shot. The multiparticle detector in the second generation spectrometer uses a time-division technique, in the form of a crossed-delay line anode, to encode the time- and position-of-arrival of up to eight particles per laser shot. Both of these detection schemes are briefly reviewed here.

2.2.1. *Wedge-and-Strip Anode*

The wedge-and-strip is a type of charge-division anode, with another important example being the resistive anodes.[50] On a wedge-and-strip anode, the charge cloud from the MCP is divided among three conductors. The charge division occurs between a series of wedges tapering linearly in one coordinate, a series of strips with widths decreasing linearly along an orthogonal coordinate, and the Z or zigzag conductor, which interleaves between the wedge and strip to provide a charge normalization. The x, y positions are then proportional to the ratio of the wedge and strip charges to the total charge, respectively.[49] The pattern period on a typical wedge-and-strip anode is much larger (800 μm) than the resolution desired (70 μm is easily achievable). In practice, the charge cloud from the MCP is allowed to spread over several periods of the anode to allow for accurate interpolation of the centroid.[51]

The primary limitation of the wedge-and-strip anode for experiments involving the production of more than one detectable particle per shot involves the measurement of the charge. First, accurate charge measurement requires the use of charge-sensitive preamplifiers and shaping amplifiers prior to digitization, leading to several μs dead-times. This requires the use of a split wedge-and-strip anode to record two photofragments in the case of a two-body breakup.[32] Even in this case, however, the size of the charge cloud required for accurate interpolation of the centroid requires a 4–6 mm wide dead area across the detector at the boundary of the two anodes to prevent charge spill-over and the resulting distortion of the detected positions. This makes use of the wedge-and-strip anode problematic for any more than two particles per detector.

2.2.2. *Multiparticle Detection with Crossed Delay Lines*

To carry out detailed measurements of three-body dissociation processes a detector capable of recording the time- and position-of-arrival of all three particles in coincidence is needed. While measuring two out of three particles suffices to determine the momentum conservation if the fragmentation pattern is known, detection of all three particles is necessary to reduce the possibility of false coincidences in the data and to reveal the fragmentation pattern in unknown cases. In these experiments, all the fragments arrive at the detector in typically < 500 ns. To avoid bias in the data set it is thus necessary to minimize dead time and dead area on the detector face. Another important consideration is that the data acquisition system is required to operate at a high repetition rate (≈ 1 kHz). These considerations impose significant constraints on the chosen detection scheme. Nonetheless, there are several possible approaches to solving this problem.

Although it is beyond the scope of this review to discuss in detail the detection of multiple particles at laboratory energies of several keV,[52] a brief review is in order. A 'pixellated' anode, composed of discrete detector elements of the desired resolution, provides an ideal multiparticle anode. Charge-coupled or charge-injection devices (CCDs or CIDs, respectively) provide one approach to this limit. These devices generally lack the data throughput rates required in a coincidence experiment. Vager and coworkers have recently developed a CCD-based device that can detect the time- and position-of-arrival for several particles in coincidence, however, by using alternate means of encoding the time-of-arrival.[53] The primary limitation of this technique is the low data readout rates currently achievable with CCDs (50 Hz). X-ray astronomy[54] and more recently, real-time X-ray diffraction experiments[55] have pushed adaptation of true time- and position-sensitive pixellated detectors based on application-specific-integrated-circuits (ASICs) developed for high-energy physics detectors. In these devices, each pixel is 'fully instrumented', capable of recording charge and time information. However, at the desired spatial resolution (< 100 μm over a 4 cm field-of-view), the number of pixels required and the complexities of ASIC development and data acquisition make this an expensive solution at the present time.

An alternate approach involves the use of a number of parallel wires forming a wire plane. Two orthogonal wire planes can be used to give (x, y) information for particles striking the detector. Position measurement with wire planes has been accomplished in several ways. One approach is to instrument each wire: in atomic and molecular physics this approach has been taken in a number of Coulomb explosion experiments.[56–58] The time and charge of the signal on each wire is measured, resulting in a large number of electronic channels. This type of detector has the advantage of being able to measure a large number of particles arriving virtually simultaneously, but often has resolution limited by the number of wires. A second technique, charge division, joins neighboring wires with resistors.[59] Each end of the resistor chain is coupled to charge sensitive electronics. By measuring the fraction of charge at each end of the resistor chain, the position is inferred. This technique requires charge integration and is capable of determining the position of only one particle per resistor chain. A third practical approach couples each wire to a delay line.[60] By measuring the relative time of arrival of the signal at each end of the delay line, the position is determined. Multiple hits in one delay line result in a number of signals from each end of the delay line, however, which may be difficult to disentangle.[61] The two latter techniques have the advantage that a small number of electronics channels are required for the position encoding (two channels per resistor chain or delay line).

The multiparticle detector we have built uses the last approach: determining the position by time-division on a delay line.[62,63] This represents a compromise with respect to the number of particles detectable per shot, but maintains high spatial and temporal resolution. In Fig. 4, the layout of the anode is shown schematically. There are four independent quadrants, each

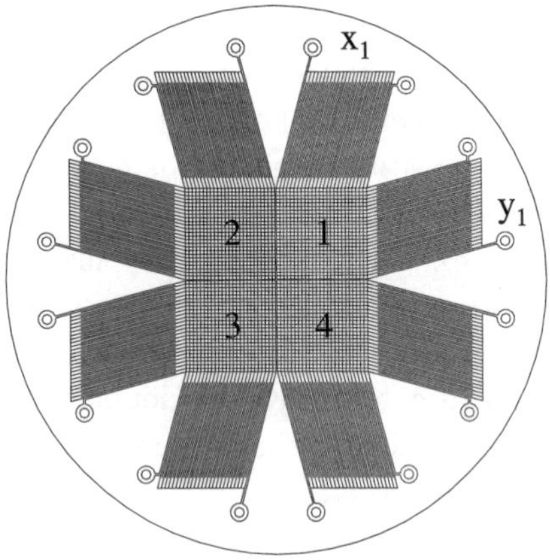

Fig. 4. Schematic layout of the quadrant crossed-delay-line (QXDL) anode used for multiparticle detection. The four quadrants are labeled 1–4, with the x, y delay-lines for quadrant 1 indicated. The time difference for the signals is measured at the circular pads on the end of each delay line. (Adapted with permission from Ref. 42.)

with an active area of 20×20 mm. The serpentine delay lines are positioned outside of the field of view, and charge is collected and divided among the $x-y$ delay lines for a given quadrant by a set of crossed conductive fingers in the active area. This is achieved using a multilayer circuit board and laser ablation to remove the insulating material between the upper and lower sets of conductors. The charge cloud that hits the anode is divided between the two delay lines, and propagates along the delay line to each end, producing pulses of ≈ 3 ns FWHM. The time difference relates linearly to position, and has been shown in the laboratory to give the position of the centroid of the electron cloud to ≈ 70 microns. Given the 20–25 ns propagation time of the signal across the serpentine delay lines, this resolution requires timing resolution on the order of 50 ps in each coordinate.

The four quadrants are arranged next to each other with no dead space between anodes. This allows the detection of any two particles except when both particles arrive in the center of the detector. In that case, the first particle triggers all the electronics and the second particle would be lost. However, an advantage of carrying out high-resolution timing on short delay-lines is that the dead time for the detector is small. The minimum deadtime imposed by the electronics is ≈ 10 ns. We have taken advantage of this by instrumenting each delay line with a purpose built two-hit time-to-amplitude converter, allowing the measurement of two particles per quadrant. Thus, the multiparticle detector can record up to eight photofragments per laser shot in a case of ideal kinematics (two particles per quadrant, with a 10 ns temporal separation). This multiplicity is required to enable detection of three particles arriving in nearly any order and orientation.

2.3. *Data Acquisition*

The data acquisition system for the experiment is shown in Fig. 5. The central clock for the experiment is the pulsed photodetachment laser, with

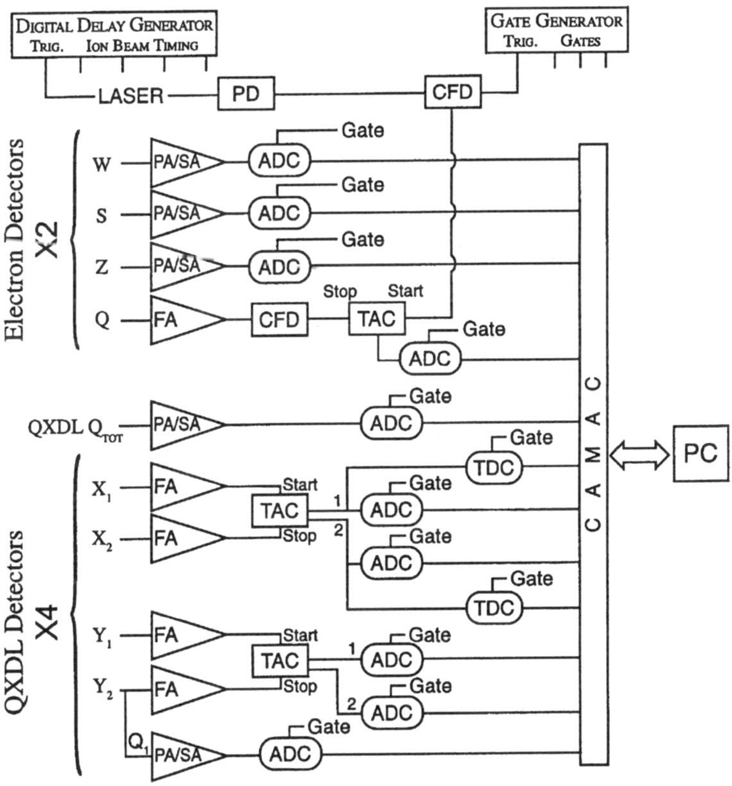

Fig. 5. Data acquisition system for the photoelectron-multiple-photofragment coincidence spectrometer. See the description in the text. (Adapted with permission from Ref. 42.)

the previous laser shot providing the start time for production of the pulsed ion beam. A digital delay generator (SRS DG535) controls the pulsed ion beam switching. Photoelectron and photofragment data acquisition and timing is initiated by the photodetachment laser pulse as recorded by a fast-photodiode triggering one channel of a quad constant-fraction-discriminator (CFD, Tennelec TC454). This provides the logic signal for measuring the time-of-flight of the photoelectron and the photofragments and generating the gates required for the eight-channel peak-sensing analog-to-digital (ADC, LeCroy 3351) and time-to-digital converters (TDC, LeCroy 3371). Custom charge-sensitive preamplifier/shaping amplifier combinations (PA/SA) and fast (300 MHz) timing amplifiers (FA) are used to amplify the signal from the detector anodes. Using custom-built CFD/double-hit TAC modules, the signals from the delay line are processed prior to digitization. All the data are transferred to a PC via a CAMAC interface after each laser shot. The detector has a total of 16 delay-line signal outputs instrumented with 16 two-hit TACs followed by 16 ADC channels to encode the position-of-arrival of up to two particles per quadrant. In addition, there are 8 TDC channels for recording the time-of-arrival of two particles per quadrant, 4 ADCs for the charge in each quadrant and 1 ADC for the total charge for that event. This is a total of 29 channels. Coupled with the two wedge-and-strip-anode photoelectron detectors, each of which records three charges and one-electron time-of-flight per event, the overall experiment has 37 data acquisition channels.

2.4. *Data Analysis*

2.4.1. *Fast Beam Photoelectron Spectroscopy*

The principles of fast beam photoelectron spectroscopy have been previously discussed,[40] and are straightforward. The detected position and time-of-flight of the photoelectron yields the laboratory kinetic energy and the recoil angle of the photoelectron in the center-of-mass frame. This requires a one-to-one mapping of the position of the photoelectron at the detector entrance grid. The short flight path and large entrance apertures used in these electron spectrometers requires that the electron recoil angle be determined in order to obtain the flight path for each electron. In addition, due to the fast ion beam, the electron recoil angle is necessary for conversion of the laboratory energy of the photoelectron to the center-of-mass frame. Given the incident beam velocity and the laboratory velocity of the photoelectron, the center-of-mass electron energy (eKE) is determined by vector subtraction, removing the large Doppler broadening due to the ion beam when photoelectrons are detected over a wide range of angles.[64]

2.4.2. *Two-Body Dissociation Kinematics*

Given the velocity and mass of the parent ion beam and the position-of-arrival and the arrival time difference of two coincident photofragments, it is straightforward to determine the recoil velocities, angles and mass ratio of the photofragments. The experimental observables required are the positions of impact of the two particles and $\tau = t_1 - t_2$, the difference in time-of-arrival for the two photofragments. The kinematic analysis of such data has been previously discussed.[29,65] Knowing the beam velocity, v_0, and the photofragment flight path L, the photofragment mass ratio can be found using

$$\frac{m_1}{m_2} = \frac{r_2}{r_1}\left(1 - \frac{v_0\tau}{L}\right) . \tag{1}$$

In this equation, the mass resolution is determined by the recoil radii r_1, r_2 relative to the ion beam spot-size, and is modest, $m/\Delta m \approx 15$.[30] Once the dissociation channel is determined, the center-of-mass translational energy (E_T) and polar recoil angle (θ) of the photofragments are found using the following equations,[29,32]

$$E_T = E_0 \frac{m_1 m_2}{M^2} \frac{[(v_0\tau)^2 + R^2]}{L^2} \left[1 + 2\left(\frac{m_2 - m_1}{M}\right)\frac{v_0\tau}{L}\right] \tag{2}$$

$$\theta = \tan^{-1}\left(\frac{R}{v_0\tau}\right) . \tag{3}$$

In these equations, $R = r_1 + r_2$ is the distance between the photofragments at the detector, the parent mass is M, and E_0 is the energy of the ion beam. With a two-dimensional detector, the azimuthal recoil angle can also be recorded.

2.4.3. *Three-Body Dissociation Kinematics*

In the case of three-body dissociation, where all the photofragments are recorded in coincidence, a similar analysis can be carried out. Gardner *et al.* discussed fast-beam three-body dissociation experiments, and presented

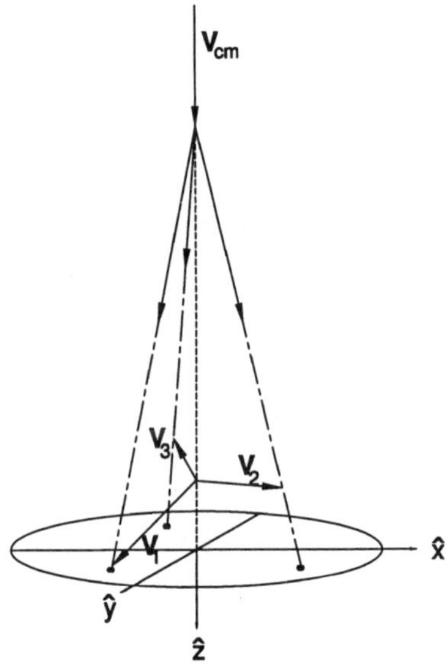

Fig. 6. Schematic view of a detected three-body dissociation event. The velocity of the incident ion beam is shown as V_{cm}, and the center-of-mass velocities of the three fragments are shown as V_1, V_2, and V_3. The x, y positions of impact at the detector are shown as well. (Adapted with permission from Ref. 42.)

equations for determining the photofragment mass from the two-dimensional projection of the center-of-mass velocities onto a photofragment detector.[47] With the detection scheme employed here, the three-dimensional recoil velocities can be recorded. The kinematic analysis of a three-body dissociation relies on conservation of linear momentum in the center-of-mass frame, as discussed previously in the literature.[25,28,47] The technique described here is specific to an experiment recording the time- and position-of-arrival of *all* particles in a three-body dissociation in coincidence.

Figure 6 illustrates a three-body dissociation originating in the interaction region and propagating to a time- and position-sensitive detector, yielding x, y, t data for each particle. Each three-body dissociation event is checked for conservation of linear momentum in a center-of-mass reference frame using the largest particle recoil velocity vector as the positive z-axis. The principles of conservation of mass and momentum are then applied:

$$\begin{aligned} m_1 + m_2 + m_3 &= M \\ p_1 + p_2 + p_3 &= 0\,. \end{aligned} \tag{4}$$

The variables m_i are the product masses, M is the parent mass, and p_i are the product momenta in the center-of-mass frame. The momentum of each particle can be decomposed into components along each of the Cartesian axes, and a system of three equations in the three unknowns m_1, m_2 and m_3 can be solved to yield, for example,

$$m_3 = M\left[\frac{\nu_{1z}\nu_{2y}}{\nu_{3y}\nu_{2z} - \nu_{3z}\nu_{2y} - \nu_{3y}\nu_{1z} + \nu_{1z}\nu_{2y}}\right]. \tag{5}$$

The masses of the other particles can then be determined. Mass spectra of the three O_2 photofragments produced in the three-body DPD of O_6^- show a nominal mass resolution $m/\Delta m \approx 2.5$. In the dissociation of O_6^-, two fragments are fast and one is slow, barely recoiling out of the beam. This limits the mass resolution, due to the distribution of velocities of the center-of-mass. Improvements in the photofragment mass resolution will require reducing this distribution, or, in other words, the size of the ion beam at the multiparticle detector. This may be achieved in the future using appropriate collimating ion optics.

Once the photofragment masses are assigned, the uncertainty due to the finite size of the ion beam can be removed since the true center-of-mass for the event is known. Given the photofragment masses and center-of-mass velocities, the translational energy released in the three-body dissociation can be calculated by simply adding up the contributions from each particle:

$$E_T = \frac{1}{2}m_1\nu_1^2 + \frac{1}{2}m_2\nu_2^2 + \frac{1}{2}m_3\nu_3^2\,. \tag{6}$$

The product angular distributions can provide important insights into the dissociation mechanism, and can be extracted from this data as well.

There is considerably more detail to be considered in the study of three-body dissociation dynamics. Two primary three-body dissociation mechanisms can be defined: concerted and sequential processes.[28] Maul and Gericke[28] further classify concerted processes into two subcategories, synchronous and asynchronous concerted reactions, according to whether or not multiple bond-breaking processes occur simultaneously. These distinctions are not made in the data analysis described here, which only considers conservation of linear momentum. In the case when the break-up is sequential, a more complex treatment is needed to fully describe the kinematics of the three-body dissociation process.[66] In the future, as the product angular distributions are studied, it is anticipated that insights into the detailed mechanisms of three-body dissociation will be obtained.

2.4.4. *Photofragment Center-of-Mass Differential Cross-sections*

The experimental translational energy and angular distributions of the photofragments are binned into a raw CM energy and angular distribution, $N(E_T, \theta)$. If the detector is not ideal, i.e. it has some dead time or a limited angular acceptance for the photofragments, the raw $N(E_T, \theta)$ will differ from the true center-of-mass differential cross-sections, $P(E_T, \theta)$. This is the case for our earlier experiments using a split wedge-and-strip anode. In that case, however, a straightforward numerical direct-inversion procedure to transform the experimental distributions to $P(E_T, \theta)$ has been previously described.[32] The new multiparticle detector, described in Sec. 2.2.2, more closely approximates the ideal detector. The work described in this review primarily focuses on the energetic correlations of the photoelectron and photofragments in a DPD process, so the numerical procedures to extract the center-of-mass differential cross-sections will not be discussed further.

2.5. *Coincidence Considerations*

The calculation of the rate of detection of true coincidences is an essential consideration in the design of a multiparticle coincidence experiment. Using pulsed laser sources it is frequently possible to obtain in excess of one detachment event per laser shot given reasonable photodetachment cross-sections ($\sigma = 10^{-17}$ cm^2) even with target number densities on the order of 10^5 cm^{-3}. To ensure that photoelectron-photofragment coincidences are valid, however, lower event rates are required. The coincidence experiments discussed here involve the detection of one photoelectron and two or three photofragments in coincidence. The number of true coincidences recorded per laser shot can be written

$$n_{tco} = P(1) \cdot (f_n \cdot \Delta\Omega_n)^p \cdot f_e \cdot \Delta\Omega_e \cdot T_e\,. \qquad (7)$$

In this equation, $P(1) = 0.09$ is the Poisson probability for a single event when the average number of events per shot is 0.1, $f_n \cdot \Delta\Omega_n$ is the product of the detection and collection efficiency for each of the p neutral fragments to be detected in coincidence and $f_e \cdot \Delta\Omega_e \cdot T_e$ is the product of the detection, collection and transmission efficiencies for the measurement of the photoelectron. For fast neutral particles and photoelectrons accelerated to several hundred volts, the detection efficiencies f_n and f_e are $\approx 50\%$ with the microchannel-plate detectors.[67] For DPD yielding a photoelectron and two photofragments, if we assume that the photofragment detector geometry allows 100% of the photodissociation events to be collected in coincidence, $(\Delta\Omega_n)$, and 20% of the electron solid angle $(\Delta\Omega_e)$ is collected with a transmission efficiency (T_e) of 81%, the average number of successful triple-coincidences per laser shot will be 1.8×10^{-3}. This yields a signal count rate of 1.8 Hz when the experiment is run at 1 kHz. For three-body DPD yielding a photoelectron and three photofragments, $(p = 3)$, the signal count rate expected is 0.9 Hz.

Background and false coincidences are an important consideration in the design of an experiment of this type. Multiparticle false coincidences can be eliminated by hardware and software charge-discrimination techniques. The most important source of random noise is correlated with the photoelectron signal in time, resulting from stray photoelectrons produced by laser light striking surfaces in the analyzer. The laser-correlated background has a common signature of being an increasing signal at low eKE, and only becomes a serious problem at wavelengths less than 300 nm.

An intrinsic source of noise is false coincidences arising from two dissociation events followed by detection of the photoelectron from one and the coincident photofragments from the other. The neutral particle detectors allow an independent determination that the photofragments are from the same dissociation event by the conservation of momentum, allowing the elimination of many false coincidences. An estimate of the number of false coincidences per laser shot involving an uncorrelated photoelectron and p correlated photofragments is given by

$$\begin{aligned} n_{fco} = P(2) \cdot 2 \cdot (f_n \cdot \Delta\Omega_n)^p \cdot (1 - f_n \cdot \Delta\Omega_n)^p \cdot (f_e \cdot \Delta\Omega_e \cdot T_e) \\ \cdot (1 - f_e \cdot \Delta\Omega_e \cdot T_e) = 1.8 \times 10^{-4} \,. \end{aligned} \tag{8}$$

where $P(2) = 0.004$ is the probability for two detachment events occurring per laser shot when the average number of events per shot is 0.1. The term $2(f_n \cdot \Delta\Omega_n)^p$ accounts for the probability for detecting p correlated fragments out of the $2p$ fragments produced in two dissociations. The term $f_e \cdot \Delta\Omega_e \cdot T_e$ accounts for the probability of detecting only one out of the two photoelectrons. The terms $(1 - f_n \cdot \Delta\Omega_n)^p$ and $1 - f_e \cdot \Delta\Omega_e \cdot T_e$ account for the fact that the remaining photofragments and photoelectron are not detected. This expression gives a false coincidence count rate for $p = 2$ of 0.05 Hz when the experiment is run at 1 kHz, and for $p = 3$, 0.01 Hz. For the two-body dissociation, this is approximately 2% of the coincidence count rate and for the three-body dissociation it is $\approx 1\%$.

3. Experimental Examples

In this section, three examples of coincidence studies of the DPD of molecular and cluster anions will be reviewed. The first two examples, the direct DPD of O_4^- and the sequential DPD of O_3^-, were carried out on the first spectrometer in the laboratory and involve the detection of one photoelectron and two atomic or molecular photofragments per laser shot. The final example, concerning the three-body DPD of O_6^- was carried out on the second generation apparatus, discussed in some detail in Sec. 2.

3.1. *Direct Dissociative Photodetachment: O_4^-*

The most dramatic illustration of a 'direct' DPD process found to date is provided by the O_4^- system.[41,68] The dynamics of DPD in this system provide both important structural insights into O_4^- and a characterization of the low-lying repulsive states of neutral O_4 accessed in photodetachment. The results discussed below illustrate the utility of coincidence measurements, which reveal both vibrationally resolved translational energy distributions and evidence of an anisotropic molecular-frame photoelectron angular distribution in the DPD of O_4^-.

The dimer anion of oxygen, O_4^-, undergoes photodissociation and DPD from 1064 nm to 213 nm. A thorough review of earlier work on this system is given by Sherwood *et al.*[69] in a paper concerning the ionic photodissociation of O_4^-: $O_4^- + h\nu \rightarrow O_2 + O_2^-$. Ionic photodissociation competes with DPD at all wavelengths studied so far and is observed to proceed via a strong parallel transition, in other words, the products are strongly peaked along the electric vector of the laser. The studies of ionic photodissociation have confirmed the O_2–O_2^- bond energy $D_0 \approx 0.46$ eV, in accord with mass spectrometric measurements by Hiraoka.[70] They have also provided evidence that O_4^- is a doublet radical anion from the observation that the lowest observed photodissociation channel at 532 nm: $O_4^- + h\nu \rightarrow O_2(a^1\Delta_g) + O_2^-(X^2\Pi_g)$ is spin-forbidden from a quartet ground state.

Other previous studies of O_4^- most relevant to the current DPD experiments include photoelectron spectroscopy and photodestruction measurements by Johnson and coworkers.[71] Photoelectron spectra recorded at 355 and 532 nm revealed no resolved vibrational structure. In our laboratory, low-resolution PPC experiments at 523 and 349 nm[72] showed that DPD of O_4^- produces O_2 photofragments with wavelength dependent angular distributions, and provided evidence for a weak non-Franck–Condon process at 349 nm. This is an exception to the direct Franck–Condon DPD processes that are the focus of this section.

Previous theoretical studies of O_4^- were limited to semi-empirical or small basis set *ab initio* calculations due to the complexity of the electronic structure of this anion.[73–75] Recently, however, an extensive set of high-level *ab initio* calculations have been completed by Taylor and coworkers.[76] Using the complete-active-space + second-order perturbation theory method of Bjorn and Roos,[77] they determined that the ground state of O_4^- is a doublet and that the molecule is a symmetric species of D_{2h} symmetry, as suggested by the dynamics of DPD at 532 nm[68] and recent density functional theory (DFT) calculations by Chertihin and Andrews.[78] The calculations predict a stability with respect to $O_2 + O_2^-$ of ≈ 0.5 eV, in reasonable accord with the experimental quantities.[69]

DPD or photodissociation of O_4^- at 532–266 nm has been observed to lead to only equal-mass photofragment pairs. Analysis of data taken under a variety of beam energies and detector positions showed no evidence for other products such as O_3 + O or stable O_4 after dissociation. In the following sections, we first present the N(E_T, eKE) correlation spectra, introduce a bound-free Franck–Condon model for the partitioning of translational energy and discuss the implications for the repulsive states of O_4 accessed by photodetachment. This is followed by a discussion of the total translational energy release spectra, and a bound-bound Franck–Condon model for the partitioning of vibrational energy among the O_2 products. The angular correlation of the photoelectron and photofragments are then presented along with a brief discussion of the electronic structure of O_4^-.

3.1.1. *Photoelectron-Photofragment Kinetic Energy Correlations: N(E_T, eKE) Correlation Spectra*

The N(E_T, eKE) correlation spectra recorded for O_4^- at 532, 355 and 266 nm, (2.33, 3.49 and 4.66 eV, respectively), are shown in Figs. 7, 8 and

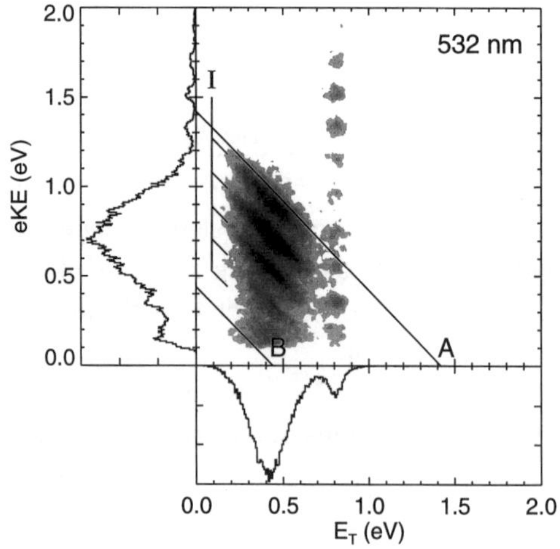

Fig. 7. Photoelectron-photofragment energy correlation spectrum $N(E_T, \text{eKE})$ for $O_4^- + h\nu \rightarrow O_2 + O_2 + e^-$ at 532 nm. $N(E_T, \text{eKE})$ is represented as a two-dimensional gray-scale histogram. The histogram along the y-axis shows the photoelectron spectrum, and the x-axis shows the photofragment translational energy distribution. The features and energetic limits shown are discussed in the text. (Adapted from Ref. 41, with permission from Elsevier Science.)

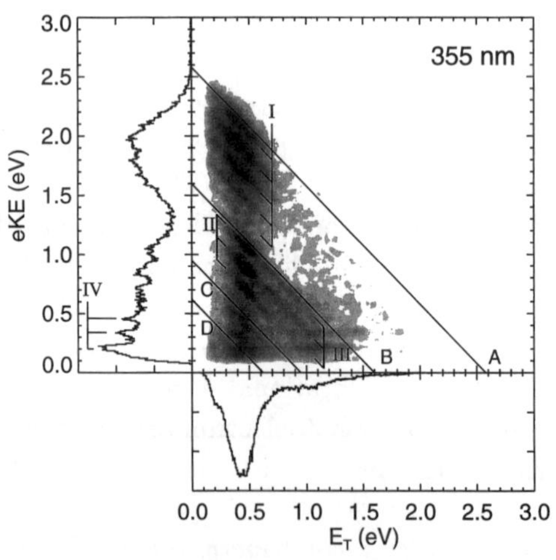

Fig. 8. Photoelectron-photofragment energy correlation spectrum $N(E_T, \text{eKE})$ for $O_4^- + h\nu \rightarrow O_2 + O_2 + e^-$ at 355 nm. See the text for discussion of the features and energetic limits shown. (Adapted from Ref. 41, with permission from Elsevier Science.)

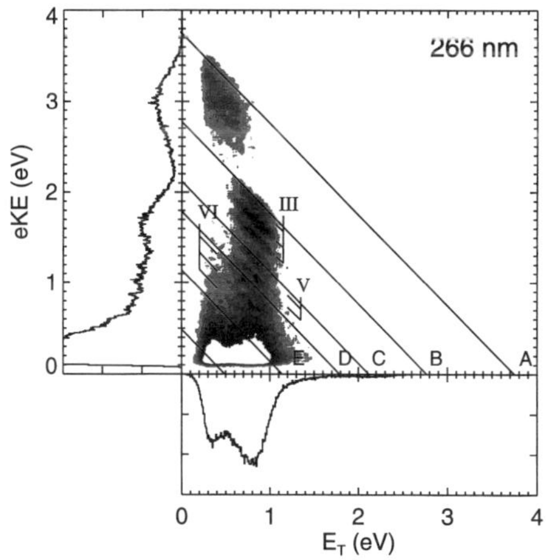

Fig. 9. Photoelectron-photofragment energy correlation spectrum N(E_T, eKE)) for $O_4^- + h\nu \rightarrow O_2 + O_2 + e^-$ at 266 nm. See the text for discussion of the features and energetic limits shown. (Adapted from Ref. 41, with permission from Elsevier Science.)

9, respectively. N(E_T, eKE) is shown in these figures as a two-dimensional histogram of the correlation between electron kinetic energy (eKE) along the y-axis and the photofragment translational energy release (E_T) along the x-axis. The one-dimensional spectra shown along each axis are obtained by integrating the correlation spectrum over the complementary variable, and represent the photoelectron kinetic energy and photofragment translational energy release spectra measurable in a noncoincidence experiment. The correlation spectra reveal considerably more information than the one-dimensional spectra. On the spectra, several maximum translational energy release limits are shown as the diagonal lines marked A, B, C, etc. These show the maximum kinetic energy that can be partitioned between photoelectron kinetic energy and photofragment translational energy release for a specific product channel. The limits shown correspond to DPD to the ground and low-lying excited states of O_2, with the energetics collected in Table 1.

At 532 nm, the correlation of the broad peaks in the eKE and E_T spectra reveals a series of five diagonal ridges, (marked I in Fig. 7), centered at $E_T = 0.4$ eV. There is also a vertical row of spots at $E_T = 0.8$ eV. The eKE spectrum of the spots in the vertical row at $E_T = 0.8$ eV are consistent with the known peaks in the O_2^- photoelectron spectrum,[79] as

Table 1. Energetics of the low-lying DPD pathways for $O_4^- + h\nu \rightarrow O_2 + O_2 + e^-$. These energetics are determined by the bond dissociation energy of $O_4^- \rightarrow O_2^- + O_2$,[70] the electron affinity of O_2,[79] and the excited state energies of O_2.[89] (Ref. 41)

Label	Products	Energy Relative to O_4^- (eV)
A	$O_2(X^3\Sigma_g^-) + O_2(X^3\Sigma_g^-) + e^-$	0.91
B	$O_2(a^1\Delta_g) + O_2(X^3\Sigma_g^-) + e^-$	1.89
C	$O_2(b^1\Sigma_g^+) + O_2(X^3\Sigma_g^-) + e^-$	2.54
D	$O_2(a^1\Delta_g) + O_2(a^1\Delta_g) + e^-$	2.87
E	$O_2(b^1\Sigma_g^+) + O_2(a^1\Delta_g) + e^-$	3.52

noted in earlier photoelectron spectroscopy measurements by Johnson and coworkers.[71] These features arise from a second photon detaching the electron from an $O_2^-(X^2\Pi_g)$ photofragment produced in the photodissociation of O_4^- ($O_4^- + h\nu \rightarrow O_2(a^1\Delta_g) + O_2^-(X^2\Pi_g)$).

The diagonal ridges observed as the dominant features in the correlation spectrum are a signature of a direct DPD process. Energy conservation dictates that all events that lie within a single diagonal ridge have a well-defined total kinetic energy, ($E_{TOT} = E_T +$ eKE). Within a given ridge, however, there is a range of kinetic energy partitioning between the three products. The ridges are separated by 0.19 eV, which is identified as the vibrational spacing in the O_2 products. The simplest explanation of these features is that they correspond to DPD on vibrationally adiabatic repulsive curves correlating with the different vibrational states of two ground electronic state O_2 products, as discussed further below.

The 355 nm correlation spectrum is shown in Fig. 8. The channel A observed at 532 nm is observed at this wavelength as well, with an increase in eKE given by the difference of the photon energies at 532 and 355 nm. This shows that Franck–Condon DPD produces $O_2(X^3\Sigma_g^-) + O_2(X^3\Sigma_g^-) + e^-$ at this wavelength. Evidence for a non-Franck–Condon feature previously observed at 349 nm producing ground state O_2 products is seen in the low-energy shoulder in the E_T spectra at 0.2 eV.[72] This feature is much less significant at 355 nm, implying that it is related to a relatively narrow resonance embedded in the DPD continuum. At this photon energy, new product channels are energetically accessible: the limits B and C correspond to the KE_{MAX} values for production of $O_2(a^1\Delta_g)+O_2(X^3\Sigma_g^-)+e^-$ and $O_2(b^1\Sigma_g^+) + O_2(X^3\Sigma_g^-) + e^-$, respectively. A prominent set of short diagonal ridges is observed beginning below limit B (feature II). This is consistent with DPD on vibrationally adiabatic curves corresponding to the production of $O_2(a^1\Delta_g) + O_2(X^3\Sigma_g^-) + e^-$, with the repulsive energy between the fragments appearing as $E_T = 0.47$ eV. There is also a set of fainter diagonal ridges below maximum B, peaking at $E_T = 0.8$ eV (feature III). This implies that there is a second, more repulsive curve producing $O_2(a^1\Delta_g)+O_2(X^3\Sigma_g^-)+e^-$. Evidence for signal below maximum C, corresponding to $O_2(b^1\Sigma_g^+)+O_2(X^3\Sigma_g^-)+e^-$, with $E_T \approx 0.4$ eV is also observed. The final observation to be made on the 355-nm spectrum is the appearance of some faint lines parallel to the E_T axis with $E_T \approx 1$ eV (feature IV). These features arise from photodissociation of O_4^- producing vibrationally excited O_2^- that undergoes rapid vibrational autodetachment[80] yielding electrons of a well-defined kinetic energy. This process appears in the photoelectron spectrum as small narrow peaks superimposed on the broad DPD continuum. The three peaks observed here correspond to autodetachment transitions from $O_2^-(v = 5, 6, 7)$ to $O_2(v = 0)$. Allan and coworkers have recently observed similar phenomena in the dissociative electron attachment to O_3 which yields O + vibrationally excited O_2^-.[81]

The 266 nm correlation spectrum is shown in Fig. 9. At 266 nm, pathway A is observed again, with an increase in eKE corresponding to the increase in the photon energy. The diagonal features in the data for pathway A are not as clearly visible due to the loss of resolution in the TOF measurement for the fast photoelectrons. Pathway B is observed to occur only *via* the more repulsive state observed weakly at 355 nm (feature III). The kinetic energy release in this channel peaks at $E_T = 0.8$ eV, as observed at 355 nm. There is also evidence for signal appearing below limits C (feature V) and D (feature VI), indicating that these dissociation pathways are also open. These repulsive curves also yield O_2 products with repulsion on the order of 0.8 eV appearing in translation between the photofragments. It is noted that pathway C has also moved to a higher kinetic energy release ($E_T = 0.8$ eV) compared to 355 nm, implying two distinct repulsive curves yielding products in pathway C as well as B.

3.1.2. *Franck–Condon Model for Product Translational and Vibrational Energy Partitioning*

In this case of direct DPD, the bound O_4^- anion is projected unto a neutral O_4 potential energy surface that is unbound along at least one coordinate.

As mentioned in the last section, the overall wavelength independence of the individual features in the correlation spectra suggests application of a Franck–Condon model for both the product translational and vibrational energy partitioning. The translational energy distribution should be governed by the bound-free Franck–Condon factors for DPD, and the product O_2 vibrational energy distribution is governed by Franck–Condon overlap between O_4^- and two free O_2. The partitioning of internal energy between vibrational and rotational degrees of freedom of the products provides information on the short O–O bonds in O_4^- and the relative configuration of the nuclei. The model makes significant approximations in neglecting vibrational degrees of freedom other than the ones that map onto the reaction coordinate and the 'local mode' vibrations of the individual O_2 moieties in O_4^-. In spite of this, it contains the essential elements required to reproduce the product energy partitioning.

Extraction of quantitative information on the repulsive states of O_4 requires consideration of the bound-free Franck–Condon factors governing DPD. LeRoy and coworkers[82] have developed a formalism for calculating bound-free Franck–Condon factors (FCFs) for diatomic molecules, originally to model continuum absorption coefficients. They define the bound-free FCF as the photodissociation cross-section expressed in terms of the radial matrix elements between the bound and continuum states of a diatomic molecule as follows:

$$\sigma_v(v, J, E) \propto CvG_{el}|\langle\Psi_{E,J}|M(R)|\Psi_{v,J}\rangle|^2 . \tag{9}$$

In this equation $\Psi_{v,J}$ is the bound-state wavefunction for a given vibrational (v) and rotational (J) state and $\Psi_{E,J}$ is an appropriately normalized continuum wavefunction for the dissociation into two atoms with kinetic energy E. $M(R)$ is the transition dipole function, which in general is radially dependent, but was assumed constant in these calculations. C is a constant, v is the frequency of the excitation, and G_{el} is the ratio of final state to initial state electronic degeneracies. An important assumption made in applying this formalism to the interpretation of DPD is that the continuum wavefunction should rigorously be treated as a product of the electron and nuclear continuum wavefunctions. Detailed theoretical treatments of the bound-continuum Franck–Condon factors in the case of DPD have recently been reported for the $ClHCl^- \rightarrow Cl + HCl + e^-$ system, as discussed by Takatsuka[83] and Balint-Kurti and Schatz.[84] In the future, generalization and application of these formalisms to DPD of larger polyatomic systems will provide a more accurate description of the dynamics of this bound-free process. For present purposes, however, only an approximate measure of the slope of the repulsive neutral surfaces accessed in this experiment is sought.

In this calculation, O_4^- is approximated as a pseudo-diatomic molecule along the coordinate connecting the O_2 centers of mass using a Morse potential for $[O_2\text{–}O_2]^-$ with a harmonic frequency given by the totally symmetric 371 cm^{-1} vibration obtained from the *ab initio* calculations of Taylor and coworkers.[76] Modeling this vibration as a Morse potential,

$$V(r) = \varepsilon(e^{-\beta(r-r_e)} - 1)^2 , \tag{10}$$

with classical well depth $\varepsilon = 3896\ \text{cm}^{-1}$, $r_e = 2.108$ Å and $\beta = 4.08$ Å^{-1}, yields a potential with classical turning points for $v = 0$ at 2.05 Å and 2.16 Å. The extent of this ground vibrational state wavefunction determines the region of the excited state potential energy surface sampled in photodetachment. Using the BCONT program developed by LeRoy[82] to model photodetachment from the ground vibrational state of this one-dimensional Morse potential to repulsive excited state curves, we can fit the kinetic energy distributions given by the projection of the diagonal features in the correlation spectra onto either the eKE or E_T axes. The fits were performed for the $(v = 0, v = 0)$ eKE distributions from the correlation spectra, in feature I at 532 nm, feature II at 355 nm and feature III at 266 nm.

The results are shown as the solid line fits to the eKE data on the potential energy diagram in Fig. 10, with the best-fit excited state Morse parameters shown in Table 2. In this figure, the vertical energy spectrum of the repulsive excited states is shown relative to the O_4^- ground state at 0 eV. These curves are repulsive in the region of Franck–Condon overlap, and have a small attractive well at large distances of 75 cm^{-1}. This long-range well is qualitatively consistent with the 87 cm^{-1} van der Waals well depth for the ground state of $(O_2)_2$ reported by Long and Ewing.[85] As noted in the discussion of the N(E_T, eKE) spectra above, evidence is also observed for at least two repulsive states correlating to $O_2(b^1\Sigma_g^+)+O_2(X^3\Sigma_g^-)+e^-$ with similar slopes and mean repulsion as the $O_2(a^1\Delta_g)+O_2(X^3\Sigma_g^-)+e^-$ curves. The quality of the data for these higher energy repulsive states is degraded by laser-related background in the 266 nm spectrum, so a simulation of that data was not attempted. These results show how the study of DPD

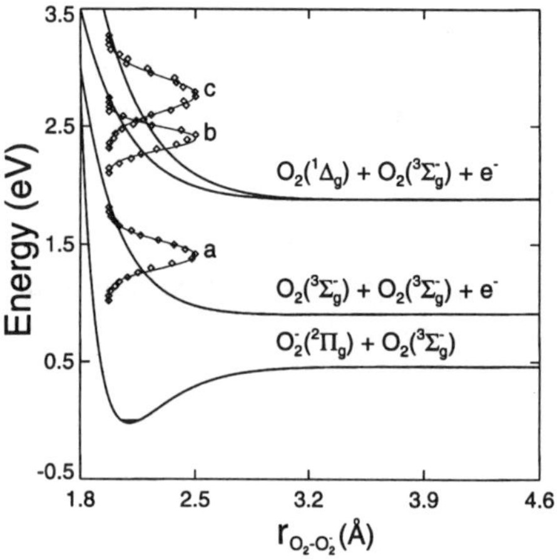

Fig. 10. Schematic of the Morse potentials used to represent the ground state of O_4^- and the low-lying states of O_4 as a pseudo-diatomic molecule in the bound-free Franck–Condon calculation discussed in the text. The photoelectron spectra for the $O_2(v = 0) + O_2(v = 0)$ peak of the features I (532 nm), II (355 nm) and III (266 nm) in the correlation spectra are shown as the diamonds along the vertical scale, with the fits shown as the thin solid curves. (Reprinted from Ref. 41, with permission from Elsevier Science.)

Table 2. Best fit Morse parameters for the bound-free Franck–Condon simulation of the DPD of O_4^-, treating O_4^- and neutral O_4 as pseudo-diatomic molecules (O_2–O_2) with a reduced mass of 16 amu. Potential energy curves and fits to the data are shown in Fig. 10. (Ref. 41)

Dissociation Asymptote	$\varepsilon(\text{cm}^{-1})$	$r_e(\text{Å})$	$\beta(\text{Å}^{-1})$
$O_2^-(X^2\Pi_g) + O_2(X^3\Sigma_g^-)$	3896	2.10	4.08
$O_2(X^3\Sigma_g^-) + O_2(X^3\Sigma_g^-) + e^-$	75	3.15	2.05
$O_2(a^1\Delta_g) + O_2(X^3\Sigma_g^-) + e^-$	75	3.40	1.66
$O_2(a^1\Delta_g) + O_2(X^3\Sigma_g^-) + e^-$	75	3.59	1.61

in O_4^- provides an experimental characterization of the repulsive curves of O_4. In the future, it is hoped that this data can be used to test of the neutral O_2–O_2 potential energy surface. Preliminary calculations by Taylor and coworkers are consistent with these experimental observations, showing that the repulsion between $O_2(^3\Sigma_g^-)$ and $O_2(^3\Sigma_g^-)$ at the geometry of the O_4^- ground state is ≈ 0.4 eV. It will be of interest to compare these experimental results to global O_2–O_2 potential energy surfaces[86,87] when a more complete theoretical treatment of DPD is available and applied to O_4^-.

3.1.3. *Internal Energy Spectra: Franck–Condon Model for Product Vibrational Energy Partitioning*

Another informative way to view the N(E_T, eKE) correlation spectra is by examination of the total translational energy spectra generated by the sum of the photoelectron kinetic energy and the translational energy release for each event: $E_{TOT} = \text{eKE} + E_T$. Given data at several wavelengths, it is more convenient to plot these spectra against the product internal energy, $E_{INT} = E_{h\nu} - E_{TOT} - \Delta D_0$, where ΔD_0 is the bond dissociation energy. The E_{INT} spectra at the three wavelengths are shown in Fig. 11. Now the diagonal features in the N(E_T, eKE) spectra appear as a resolved spectrum of the correlated product O_2 vibrational distribution. In the 532 nm data, for example, the peak at $E_{INT} = 0.08$ eV corresponds to two $O_2(v = 0)$, the peak at 0.29 eV corresponds to $O_2(v = 0) + O_2(v = 1)$

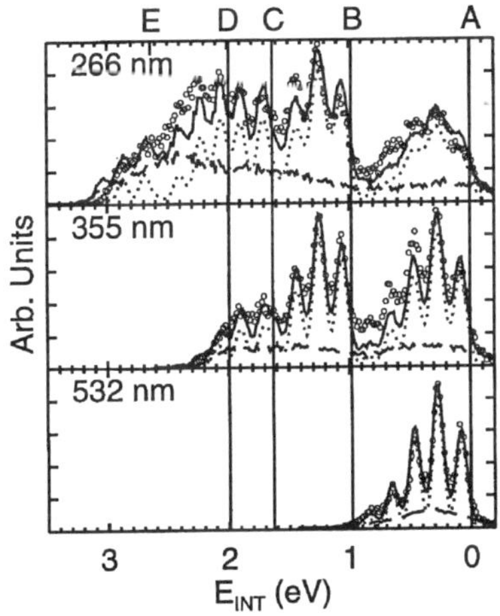

Fig. 11. Vibrationally resolved product internal energy (E_{INT}) spectra for the DPD of O_4^- at 532, 355 and 266 nm as labeled. The minimum product internal energies for the DPD pathways listed in Table 1 are marked as A–E. The open circles correspond to the data, the dotted curves correspond to the local mode Franck–Condon simulation discussed in the text and the long-dashed curves correspond to an upper-limit estimate on the noncoincident background in the data. The solid line fit corresponds to the sum of the simulation components. (Adapted from Ref. 41, with permission from Elsevier Science.)

and the peak at 0.48 eV corresponds to unresolved contributions from both $O_2(v = 1) + O_2(v = 1)$ and $O_2(v = 0) + O_2(v = 2)$ and so forth. The 532 nm spectrum shows the best resolution, since the photoelectron resolution degrades significantly with eKE.

The $E_{\rm INT}$ spectra at 355 and 266 nm are similar, with the exception that the 266 nm spectrum shows a rising signal at large internal energies due to laser-correlated background in the photoelectron spectrum. The similarity between the spectra is remarkable since pathways B and C predominantly occur *via* new, more repulsive O_4 potential energy surfaces at 266 nm. At 266 nm at least two identifiable peaks are observed above the E_{INT} threshold for pathway D, confirming the presence of this dissociation pathway. While signal is also observed above threshold E, it is dominated by laser-related photoelectron background, so the evidence for occurrence of this dissociation pathway is weak. The observed independence of the product vibrational state distribution on photon energy suggests that a Franck–Condon transition is depositing neutral O_4 on a repulsive potential energy surface.

The fits shown with the spectra consist of two components — one from a local mode Franck–Condon simulation[68] discussed in further detail below, and a second component, a background feature given by the random correlation of the N(eKE) and N(E_T) spectra. This background feature is scaled to the low (negative) internal-energy tail of the spectra and appears to account for this quite well. With this scaling, the noncoincident background contributes 26%, 28% and 35% of the total signal at 532, 355 and 266 nm, respectively. This background contribution to the E_{TOT} spectra is significantly larger than the $\approx 5\%$ predicted by the statistical calculation as discussed in the experimental section, and is likely to represent an upper limit. The discrepancy can be attributed to several sources, including overestimation of detection efficiencies and the fact that these large contributions are an overestimate since some of the low internal-energy tail must originate from internally excited parent anions.

Matrix isolation vibrational spectra are consistent with two identical O_2 moieties in the O_4^- anion.[88] Taking this as a starting point, the approximation is made that O_4^- has two important coordinates — the distance between the O_2 moieties and the O–O bond length in each subunit. This local mode model neglects interaction with the other vibrational degrees of freedom and the detailed form of the normal coordinates for O_4^-. Given this model, the Franck–Condon overlap of the vibrational wavefunctions of the two 'independent' O_2 moieties in O_4^- with two free $O_2(X^3\Sigma_g^-)$, (equilibrium bond length $r_e = 1.207$ Å[89]) can be calculated. The overall Franck–Condon factors for each energetically allowed combination of product states are calculated as products of the individual diatom Franck–Condon factors,[90] including weighting for the different product state degeneracies. The calculations assumed a local mode O_2 vibrational frequency of 1335 cm^{-1} with an anharmonicity of 8.1 cm^{-1}. The equilibrium bond length, r_e, was then varied to obtain the best fit, with the results plotted as the solid lines in Fig. 11. The agreement between the simulation assuming equal bond lengths of 1.272 Å and the data is excellent at 532 nm. The predicted (uncorrelated) O_2 product vibrational-state distributions for DPD of O_4^- to $O_2 + O_2 + e^-$ are reported in Ref. 41. The sensitivity of the simulation to changes in the O_2 moiety bond lengths has been previously shown: for example, a reduction of the bond lengths to 1.26 Å doubles the (0, 0) peak.[68] Simulations of the 355 and 266 nm data are also shown in Fig. 11, illustrating contributions from the higher-lying dissociation asymptotes, such as $O_2(a^1\Delta_g) + O_2(X^3\Sigma_g^-) + e^-$. While the agreement of the simulation with the data at these wavelengths is not as good as at 532 nm, it is reasonable given this simple model. The overall success of this simple Franck–Condon model in predicting the correlated O_2 product vibrational distributions provides strong evidence that in the gas-phase, the excess electron in O_4^- is delocalized over two identical O_2 subunits, with a short O–O bond of 1.272 Å, nearly half way between the ground state bond lengths of $O_2(r_e = 1.207$ Å$)$[89] and $O_2^-(r_e = 1.35$ Å$)$.[89]

Examination of the offset of the vibrational peaks from the internal energy origin in the 532 nm spectrum gives a peak rotational energy summed over the two photofragments of 80 ± 30 meV. The error bars are determined by taking the errors in eKE and E_T in quadrature, along with a 20-meV allowance for the precision of the E_T measurement.[69] If it is assumed that this rotational energy is shared equally between the two diatom products, a peak value of $J \approx 14$, corresponding to 40 meV in rotation of each O_2, is predicted. The narrow product rotational distribution inferred from the sharp vibrational features implies that dissociation of O_4 occurs from a structure with a small exit-channel impact parameter. This is consistent with the interpretation of the matrix spectra in terms of a planar D_{2h} structure,[78] and recent *ab initio*[76] and density functional[78] calculations.

3.1.4. *Photoelectron-photofragment Angular Correlation*

The experiment measures photoelectron and photofragment recoil angles in coincidence, so any correlation between them will be observable in the data. As our first studies of the DPD of O_4^- at 523 nm showed,[72] there is a strong alignment of the O_2 products induced by photodetachment, implying dissociation of the molecular framework on a timescale rapid relative to rotation. Given a rapid dissociation, measurement of the asymptotic product recoil angles thus provides a measure of the instantaneous alignment of the molecule at the time of photon absorption. If the photoelectrons are distributed anisotropically in the molecular frame, evidence of this distribution will be observable in the data. In the case of O_4^-, such a correlation is observed, as discussed here.

Photodetachment or photodissociation of an isotropic ensemble of molecules, within the electric-dipole approximation, is known to yield product angular distributions given by the equation[91]

$$I(\theta) = \frac{d\sigma}{d\Omega} = \frac{\sigma}{4\pi}(1 + \beta P_2(\cos\theta)), \tag{11}$$

where σ is the total cross-section and $P_2(\cos\theta)$ is a Legendre polynomial. The polar angle between the photofragment recoil direction and the electric vector of the laser is denoted as θ. The anisotropy parameter β ranges between $-1[I(\theta) \propto \sin^2\theta]$ and $+2$ $[I(\theta) \propto \cos^2\theta]$. At 532 nm, both photoelectrons and O_2 photofragments produced in the DPD of O_4^- are preferentially scattered in the plane perpendicular to the E vector ($\beta \approx -0.8$ and -0.5 for photofragments[72] and photoelectrons[92], respectively).

A striking angular correlation is observed between the photoelectron and the photofragments at all wavelengths. This angular correlation is most easily visualized by considering the 532 nm data shown in Fig. 12. The data shown here are surface density plots of the O_2 photofragment angular distribution. Frame B shows the relative geometry of the photoelectron recoil direction and the electric vector of the laser, E. In these experiments, the ion beam direction was also parallel to the E vector. The E_T-integrated photofragment angular distribution, $N_n(\theta, \phi)$, is extracted by integrating the data over $0 < E_T < 0.6$ eV to give the number of events as a function of θ and ϕ. Here, n denotes the number of particles detected in coincidence; $n = 2$ for $O_2 + O_2$, $n = 3$ for $O_2 + O_2 + e^-$ and ϕ is the azimuthal angle around the E vector. In frame A of Fig. 12 the surface-density plot

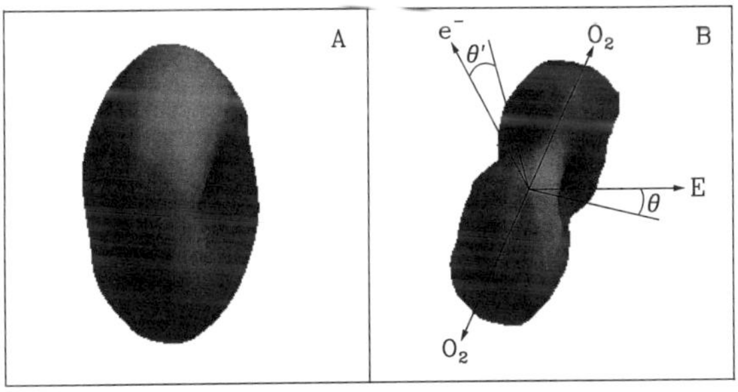

Fig. 12. Surface density plots of the coincident O_2–O_2 photofragment angular distribution illustrating the photoelectron-photofragment angular correlation observed at 532 nm. Frame A shows the O_2-product laboratory angular distribution observed without coincident detection of the photoelectron, $N_2(\theta, \phi)$. This is a $\sin^2\theta$ distribution relative to the laser E vector. Frame B shows the O_2 product angular distribution observed with coincident detection of the photoelectron, $N_3(\theta, \phi)$, showing that the cylindrical symmetry about the E vector is broken. Frame B also shows the E-vector and photoelectron-recoil-vector axes.

of $N_2(\theta, \phi)$ is shown. Since the O_2 photofragment angular distribution is nearly $\sin^2\theta$, this distribution is a torus. Owing to the restricted range of electron recoil angles that can be detected, the angular correlation between the photoelectron and photofragments breaks the cylindrical symmetry of $N_2(\theta, \phi)$ about E, as shown in $N_3(\theta, \phi)$ plotted in frame B of Fig. 12. The pronounced reduction of $O_2 + O_2$ photofragment recoils detected parallel to the direction of photoelectron detection shows that photodetachment and the subsequent dissociation of O_4 all occurs on a timescale fast relative to molecular rotation ($\approx$ 1–10 ps). Furthermore, the data indicates that the distribution of photoelectron recoil angles peaks *perpendicular* to the dissociation coordinate in O_4.

The photoelectron-photofragment angular correlation can be further understood by considering the photofragment angular distribution in a laboratory frame of reference using the photoelectron-recoil direction as the z-axis, as opposed to the direction of the E vector. This distribution is generated by dividing $N_3(\theta, \phi)$ by $N_2(\theta, \phi)$, effectively removing from $N_3(\theta, \phi)$ weighting due to both the photofragment angular distribution in $N_2(\theta, \phi)$ and the finite angular acceptance of the photofragment detector. Figure 13 shows fits to the photoelectron-recoil-frame product angular distributions. At 532 nm, the fit shows that the angular distribution should peak at $\theta' = 90°$, where θ' is now defined as the polar angle of photofragment

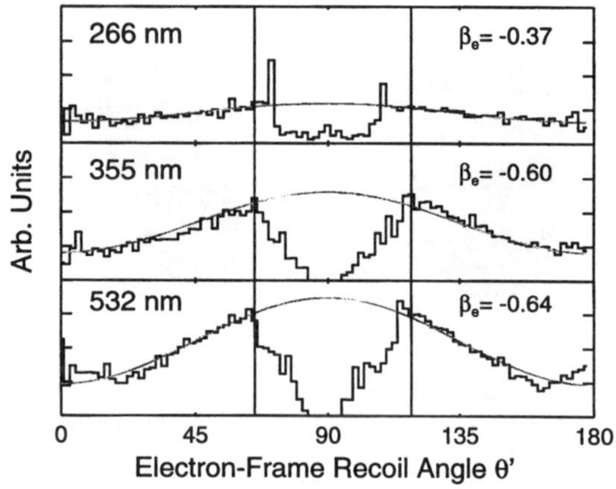

Fig. 13. Photoelectron-recoil-frame O_2 product angular distributions at 532, 355 and 266 nm, generated as in Fig. 12 by dividing $N_3(\theta, \phi)$ by $N_2(\theta, \phi)$. The histogram represents the data, the solid-line curve the fit using Eq. (1), and the vertical bars show the range of angles included in the fit, due to limited detector acceptance. The anisotropy parameters β_e indicate that photofragments preferentially recoil perpendicular to the photoelectron at all wavelengths. (Adapted from Ref. 41, with permission from Elsevier Science.)

recoil relative to the nominal photoelectron-recoil direction. A fit to Eq. (1) shows that this distribution is characterized by $\beta_e \approx -0.64$ at 532 nm, so this distribution is nearly $\sin^2\theta'$. The anisotropy parameter β_e refers to the photofragment angular distribution in the photoelectron-recoil-frame. The value of β_e is expected to be an upper bound on the anisotropy parameter since this treatment of the data assumes that the photoelectron recoils at a single nominal angle in the laboratory frame. Figure 13 shows that at 355 and 266 nm anisotropic angular distributions with $\beta_e < 0$ are also observed, consistent with photofragment recoil predominantly perpendicular to the recoil direction of the photoelectron.

3.1.5. *The Electronic Structure of O_4^- and Angular Distributions in Dissociative Photodetachment*

The electronic structure of O_4^- has been studied in some detail by Aquino *et al.*[76] Their high-level calculations indicate that the ground state of O_4^- is rectangular, with D_{2h} symmetry and a^2A_u electronic ground state. The dominant configuration of the ^2A_u ground state can be understood by considering overlap of the π_g anti-bonding orbitals of O_2 and O_2^- as shown

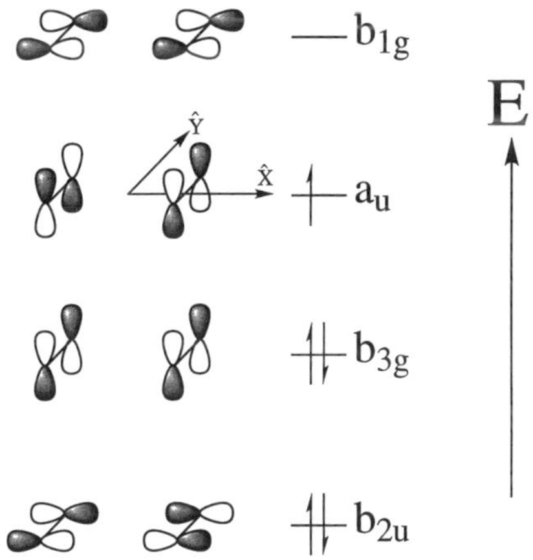

Fig. 14. A schematic diagram of the valence orbitals for O_4^-, formed by interaction of the π_g anti-bonding orbitals of O_2 and O_2^-. (Reprinted from Ref. 41, with permission from Elsevier Science.)

in Fig. 14. Using this MO configuration the lowest repulsive curve of O_4 accessed by photodetachment is expected to be a singlet state produced by removal of the lone electron in the a_u highest occupied molecular orbital (HOMO) of the anion. The other curves, correlating to singlet + triplet O_2 are low-lying triplet states of O_4 resulting from photodetachment of electrons in the b_{3g} and b_{2u} orbitals. This single configuration picture, however, is only approximate due to the highly multiconfigurational nature of the electronic structure of O_4^-.

A consideration of the dipole selection rules for photodetachment of molecular anions allows the implications of the electronic structure of O_4^- for the photoelectron and photofragment angular distributions to be examined. The electronic orbital angular momentum, l, is a good quantum number in a central potential, so the dipole selection rules for atomic anions are $\Delta l = \pm 1$.[93] However, in a polyatomic molecular anion, l is no longer well-defined, and a detailed treatment of the photodetachment dynamics entails a high-level electron–molecule scattering calculation. In lieu of this detailed treatment, the formalism of Brauman and coworkers[94] for evaluation of photodetachment cross-sections and continuum dipole selection rules for polyatomic molecules can be applied to O_4^-.

If the photodetached electron and the resulting molecular core can be treated in a one-electron approximation, the allowed continuum states for photodetachment are those for which the direct product of the symmetry of the HOMO, the transition dipole μ_x, μ_y, or μ_z, and the final state is the totally symmetric irreducible representation.[94] The overall symmetry of the final state reached in photodetachment is then determined by the direct product of the symmetry of the electron continuum and the neutral core. In the DPD of O_4^- at 532 nm, the transition moment for DPD must lie along either the y or z axes, since the photofragments are preferentially scattered in the plane perpendicular to the electric vector of the laser. Thus, given that the symmetry of the molecular core after photodetachment of the a_u electron is ^1A_g, the photoelectron continua must be of either $B_{2g}(y)$ or $B_{1g}(z)$ symmetry. These irreducible representations transform like d_{xz} and d_{xy} orbitals, respectively. This is consistent with the observation at 532 nm that the photoelectron angular distribution peaks are perpendicular to the laser electric vector, indicating that a strong preference for d-wave photodetachment is observed, even at a photon energy 1.42 eV above threshold.

The conclusion that d-wave photodetachment dominates near threshold for a system with the electronic structure of O_4^- can also be directly reached by qualitative consideration of the a_u HOMO for O_4^- shown in Fig. 14. This HOMO has three angular nodes, analogous to an atomic f orbital, and is expected to lead to both $d(l = 2)$ and g-wave $(l = 4)$ photodetachment. This situation was discussed in a study by Gygax *et al.*[95] of the photodetachment cross-section for the cyclooctatetraene anion. They showed that the photodetachment cross-section from the f-like HOMO of this anion is consistent near threshold with a d-wave photoelectron continuum. Above threshold, however, higher angular momentum states of the photoelectron continuum are accessible and are expected to result in an energy-dependent photoelectron angular distribution. This, and the contributions of new product electronic states, is likely to be the cause of the energy dependence observed in the photoelectron-recoil frame angular distributions in Fig. 13.

The strong photofragment-photoelectron angular correlation observed indicates a highly anisotropic molecular-frame photoelectron angular distribution (MF-PAD). Since the dissociation of O_4 is prompt, the recoil direction of the O_2–O_2 photofragments is determined by the alignment of the dissociation coordinate at the time of photon absorption. Dill showed that photoelectron angular distribution for a space-fixed molecule, in contrast to Eq. (1) is given by:

$$\frac{d\sigma(E)}{d\Omega} = \sum_{l=0}^{2l_{max}} \sum_{m} A_{lm}(E) Y_{lm}(\vartheta, \varphi) \,. \qquad (12).^{96}$$

In this equation, l_{max} is the maximum value for the photodetached electron orbital angular momentum and m corresponds to the azimuthal quantum number for the photodetached electron. The spherical harmonics $Y_{lm}(\vartheta, \varphi)$ are referenced to the molecular frame by polar and azimuthal angles ϑ, φ. The magnitudes of the coefficients $A_{lm}(E)$ are in general energy dependent and are determined by electric-dipole selection rules and interference between the degenerate photoelectron continuum channels. In addition to determining the relative contributions of the spherical harmonics Y_{lm}, these coefficients contain detailed information on both the photodetachment dynamics and the orbital from which the electron is ejected.[96–99] This has been exploited experimentally in studies of dissociative photoionization by Dehmer and Dill,[100] Golovin *et al.*,[101] Powis and coworkers,[102] Kaesdorf *et al.*,[103] and Shigemasa *et al.*[39] Zare and coworkers have also studied the MF-PAD by vibrational and rotational state-selective photoionization of dipole aligned NO, providing a detailed partial-wave decomposition of the photoelectron continuum.[99] Due to the multiconfigurational character of the ground electronic state of O_4^-, a detailed analysis of the photoelectron continuum will not be attempted here. Nonetheless, the DPD of O_4^- provides the first observation of an anisotropic MF-PAD in photodetachment of a negative ion. With future improvements in electronic structure and electron-molecule scattering calculations, it is hoped that a more complete interpretation of the anisotropic MF-PADs observed in the DPD of O_4^- will be possible.

3.1.6. *Conclusions — Direct DPD of O_4^-*

The data presented here provide experimental insights into the dynamics of DPD, the repulsive states of O_4, and the structure of O_4^-. No stable states of O_4 were observed from the photodetachment of O_4^-, consistent with the observation of no vibrational structure in the O_4^- photoelectron spectrum. These data provide firm evidence that the DPD of O_4^- is a direct process at these wavelengths, resulting in the observation of diagonal structure in the N(E_T, eKE) spectra. The strong photoelectron-photofragment angular correlation also shows that the dissociation of the molecular framework is much faster than molecular rotation.

Direct DPD alone does not guarantee that diagonal structure will be observed in the correlation spectra. If O_4^- were a diatomic molecule, this would by necessity be true, due to the absence of a broad range of product states in atomic products. Given that two O_2 molecules, with electronic, vibrational, and rotational degrees of freedom are formed, however, it is surprising that the partitioning of energy among the products is so constrained. This requires formation of products with low rotational excitation. These dynamics are a consequence of the symmetric structure of O_4^-, leading to a restricted range of exit-channel impact parameters and rotational excitation.

On a general level, these experiments provide a fine example of the utility of PPC measurements in the study of a direct DPD process. Direct DPD is expected to be common in systems with a weakly bound neutral ground state and a strongly bound anionic ground state. The DPD of O_4^- provides the best example observed to date of such direct DPD processes, wherein dissociation occurs much faster than molecular rotation. Without a coincidence measurement, determination of the partitioning of kinetic energy between the photoelectron and photofragments is not possible in such DPD processes. Furthermore, the coincidence measurement of photofragment and photoelectron angular distributions provides a practical means of observing anisotropic molecular-frame photoelectron angular distributions in anions that undergo direct DPD.

3.2. *Sequential Dissociative Photodetachment: O_3^-*

The DPD of the low-lying excited states of O_3 provides the best example of a sequential process in which electron departure and dissociation of the molecular frame are distinct, uncoupled events.[104] O_3 provides a particularly inviting target for study by PPC spectroscopy due to the existence of several low-lying excited states. An energy level diagram for O_3^- and O_3 is shown in Fig. 15, providing an illustration of the advantage that negative ion photodetachment provides for the study of optically forbidden excited states. Optical excitation of the $^1A_2$ state is dipole-forbidden from the $^1A_1$ ground state, and the three low-lying triplet states are also optically forbidden. Although these states have been studied by high-sensitivity absorption techniques,[105,106] the small optical cross-sections have prevented

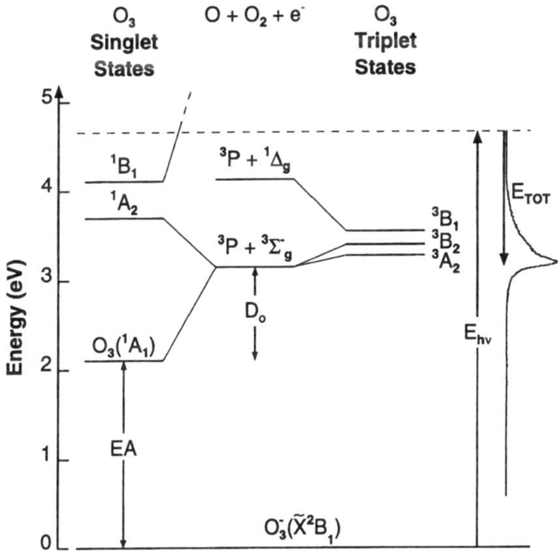

Fig. 15. The energetics diagram for O_3^- and O_3, showing the total translational energy release spectrum ($E_{TOT} = E_T$ + eKE). The electronic states are labeled in C_{2v} symmetry and the energetics are defined in the text. This diagram uses the energetics of Arnold *et al.*,[113] and shows the adiabatic correlations between the excited states and the dissociation channels.[115] (Reprinted with permission from Ref. 104.)

measurements of the dissociation dynamics. Photodetachment of the $^2B_1$ ground state of O_3^-, on the other hand, can produce all of these states by fully allowed one-electron transitions. This has led to several previous studies of the energetics and structure of the excited states of O_3 by electron scattering[107–110] and photodetachment[111,112] experiments. Recently, Arnold *et al.*[113] reported high-resolution photoelectron spectra that reveal features assigned to the five lowest excited states of O_3. Thus, O_3^-/O_3 provides an ideal system for study by the PPC technique.

The energetics and dynamics of the low-lying excited states of O_3 are of considerable interest due to the important role played by O_3 in atmospheric chemistry. In particular, there has been interest in the possible existence of bound excited states of O_3. Kinetic studies have suggested that a bound low-lying excited state may explain discrepancies observed in measured rates of ozone production through the important three-body recombination reaction $O + O_2 + M \rightarrow O_3 + M$.[114] A yet-unobserved metastable singlet state of D_{3h} symmetry has been predicted to exist[115,116] and may play a role in this reaction. There are, however, the three known low-lying triplet states ($^3A_2$, $^3B_2$, and $^3B_1$) which are poorly characterized. Experimental[106,113,117] and recent theoretical[118,119] studies have indicated that the triplet states all lie above the energy required for dissociation of O_3 into ground state O + O_2 products as shown in Fig. 15.

It has been suggested that the low-lying triplet states may explain vibronic structure in the lowest energy electronic excitation known as the Wulf band.[120–124] High-sensitivity absorption spectroscopy[106,117,125] has been used to study these very weak transitions. High-resolution spectra measured by Abel *et al.*[125] showed rotational structure superimposed on the origin and first bending vibration of the $^3A_2$ state. Considerable line broadening was observed for these transitions, indicating that the lifetime for the $^3A_2$ electronic state is less than 100 ps. Thus, the triplet states of O_3 provide an ideal case study of sequential DPD, where there is a significant time delay between photoelectron ejection and dissociation of the molecule.

3.2.1. *Photoelectron Spectra*

To assess the stability of the excited states reached by photodetachment, a search was made for stable O_3 in coincidence with photoelectrons from photodetachment to the excited states of O_3. In this measurement the split-anode particle detector was translated so that stable O_3 could be detected on the lower half of the detector. These photoelectron spectra recorded at 266 nm (4.66 eV) showed the expected features in the range 1.7 eV < eKE < 3.0 eV corresponding to production of ground state $O_3(\tilde{X}\,^1A_1)$ and at eKE < 1.5 eV, the excited triplet and low-lying singlets of O_3. Any contributions to the spectra from stable states of O_3 can be identified in these experiments by selecting only photoelectrons coincident with detection of a single particle at the time and position-of-arrival of the incident ion beam. No evidence outside of a flat false-coincidence background was observed for stable O_3 in coincidence with photoelectrons originating from the excited states of O_3. This shows that the low-lying excited states of O_3 dissociate faster than the 6 μs flight time to the detector at a beam energy of 6 keV, consistent with the recent observation by Abel and coworkers the lifetimes of the $^3A_2$ state are < 100 ps.[125]

3.2.2. $N(E_T$, *eKE*) *Correlation Spectra*

The correlation between the photoelectron kinetic energy and the photofragment center-of-mass translational energy release is shown as

N(E_T, eKE) in Fig. 16. The observation of resolved structure in the photoelectron spectra allows a study of the predissociation dynamics of internal-energy-selected O_3 excited states by the PPC measurement. Spectra for laser polarizations parallel and perpendicular to the electron detection direction are shown in Fig. 16A and Fig. 16B, respectively. The two diagonal lines represent the maximum translational energy limits for the production of ground state photofragments $[O(^3P) + O_2(^3\Sigma_g^-, v = 0) + e^-]$ and photofragments with one quantum of vibration in O_2. These thresholds are determined from the photon energy ($E_{h\nu}$ = 4.66 eV at 266 nm), the electron affinity of O_3 (EA = 2.103 eV),[111] and the O–O_2 bond dissociation energy (D_o = 1.05 eV).[126] The contours extend over the maximum translational

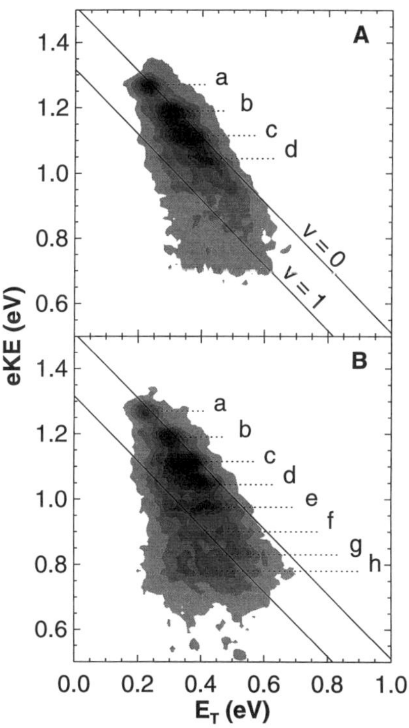

Fig. 16. The N(E_T, eKE) correlation spectrum for the DPD of O_3^-. Energetic thresholds for $O(^3P) + O_2(^3\Sigma_g^-, v = 0, 1) + e^-$ are also shown. Frames A and B show parallel and perpendicular laser polarizations, as defined in the text, respectively. The photoelectron kinetic energy axis is truncated at 0.5 eV as no significant features were observed in the correlation spectrum below that energy. (Reprinted with permission from Ref. 104.)

energy limit for $v = 0$, indicating the production of some internally excited O_3^- in the ion source.[127]

There are four prominent peaks evident in the contours of the correlation spectra (Fig. 16). In the parallel polarization (Fig. 16A) the three peaks (a–c) are arrayed along the translational energy limit corresponding to ground state products. Thus, nearly all the available energy appears in translation with very little vibrational or rotational excitation. In the perpendicular polarization (Fig. 16B) there are four peaks (a–d). In addition, bands are observed at lower eKE (e–h). This shows that photofragments produced in coincidence with the $^3B_1$ and $^1A_2$ states are characterized by a broader distribution of rotational and/or vibrational energy.

As seen in the case of O_4^-, the total translational energy release spectrum, E_{TOT} = eKE + E_T, also provides an informative way of examining these results. The E_{TOT} spectrum is shown vertically on the right side of Fig. 15. Unlike the O_4^- case, there is no significant resolved structure in the E_{TOT} spectrum for O_4^- due to the lack of diagonal features in the correlation spectra. The E_{TOT} spectrum shows, however, that DPD of O_3^- yields only ground electronic state products $[O(^3P) + O_2(^3\Sigma_g^-) + e^-]$. As the adiabatic correlations in Fig. 15 show, the $^1A_2$, $^3A_2$ and $^3B_2$ states adiabatically correlate with the observed ground state products.[115] The $^3B_1$ state, however, correlates to $[O(^3P) + O_2(^1\Delta_g) + e^-]$ and is well below this dissociation asymptote. Dissociation of the $^3B_1$ state to ground state products must occur by a nonadiabatic interaction.

3.2.3. *Dissociation Dynamics of the Low-lying Excited States of O_3*

These studies of the DPD of O_3^- were the first to probe the dissociation dynamics of the low-lying excited states of O_3. Since the measurement allows simultaneous observation of all energetically accessible dissociative states, some general observations on the dissociation dynamics are easily made. The N(E_T, eKE) correlation spectra show that dissociation of the two lowest triplet states ($^3A_2$ and $^3B_2$, features a–c) leads to virtually all the available energy appearing in translation. For the $^3B_1$ and $^1A_2$ states, (features d–h) considerably more energy appears in rotation and/or vibration of the O_2 product. Given the adiabatic correlations shown in Fig. 15, the results also indicate that the $^3B_1$ state dissociates nonadiabatically to ground electronic state products.

Dissociation of the low-lying excited states of O_3 likely proceeds by distortion from C_{2v} to C_s symmetry along the asymmetric stretch coordinate. An alternative mechanism, involving elimination of the central O atom, preserving C_{2v} symmetry, is also plausible.[111] Elimination of the central O atom would be expected to produce significant vibrational excitation of the nascent O_2 fragment. The asymmetric dissociation mechanism, on the other hand, is expected to yield a large fraction of energy in translation along with some rotational and vibrational excitation, consistent with the observed dynamics. In the low levels of the $^3A_2$ and $^3B_2$ states, there is no evidence at all of vibrational excitation. As dissociation of the $^3B_1$ and $^1A_2$ states occurs, more energy appears in rotation and perhaps vibration as well. The absence of resolved O_2 vibrational states in the translational energy distribution indicates, however, that rotational excitation in excess of the vibrational spacing of O_2 occurs in this energy range. These results are most consistent with the asymmetric dissociation pathway.

Levene and Valentini have discussed the photodissociation dynamics of the higher-lying $^1B_1$ state in terms of a modified impulse model, in which the partitioning of energy between translation and rotation is determined by the impulse along the breaking bond.[128] The modified impulse model ignores impulsive vibrational excitation, with the product vibrational distribution determined by Franck–Condon overlap between the excited state of O_3 and the product O_2. This model provides a qualitative explanation of the broader product translational energy distribution observed in dissociation of the $^1A_2$ and $^3B_1$ states (peaks d–h in the N(E_T, eKE) correlation spectra).

The impulse model apparently fails to describe dissociation of the low levels of the $^3A_2$ and $^3B_2$ states. Peaks a–c in Fig. 16, assigned to these states, exhibit no product vibrational and little rotational excitation. The impulse model predicts 0.06 eV in rotation for dissociation of the $^3A_2$ state, using the theoretically predicted equilibrium bond angle of 99.2°,[118] significantly larger than the observed value of 0.01 eV. For the impulse model to hold, it must be arbitrarily assumed that the excited molecule distorts into a linear configuration before the bond breaks impulsively. A factor contributing to the breakdown of the impulse approximation is the small available energy in dissociation of the $^3A_2$ and $^3B_2$ states. This provides greater opportunity for exit-channel interactions, contrary to the fundamental assumption of a sudden impulse. As Schinke has pointed out, the details of the bending potential and exit-channel interactions can play a crucial role in rotational excitation.[129] As more detailed potential energy surfaces for the low-lying states of O_3 become available, it should be possible to carry out realistic dynamics calculations to compare with the present experiments.

The nonadiabatic dissociation of the $^3B_1$ state further confirms that the excited state dynamics of O_3 are complicated. Theoretical studies have shown that both singlet and triplet conical intersections occur in C_s symmetry involving the $^1B_1$, $^1A_2$ and $^3B_1$, $^3A_2$ states, respectively.[120,124] Thus, it is most likely that dissociation of the $^3B_1$ state to ground state products occurs through this interaction with the lower $^3A_2$ state. In C_{2v} symmetry the $^3A_2$ and $^3B_1$ states cross, however, they become the two lowest $^3A''$ states in C_s symmetry, forming the conical intersection.[118] As dissociation begins along the asymmetric stretch coordinate, the molecular symmetry drops to C_s and the molecule may find itself on the lower dissociative surface. Tsuneda *et al.* have determined that the conical intersection between the $^3A_2$ and $^3B_1$ states should lie near the $^3A_2$ equilibrium geometry.[118] Since $O_3^-(\tilde{X}^2B_1)$ has a bond angle of 115.5° while the $^3A_2$ and $^3B_1$ states of O_3 have bond angles of 99.2° and 123.7°, respectively,[118] photodetachment is expected to yield significant bending excitation. This bending excitation may promote sampling of the conical intersection leading to facile dissociation of the $^3B_1$ state to ground state photoproducts via the intermediate $^3A_2$ state. It is interesting to note that peak c in the correlation spectra, assigned to the $^3B_1$ state by Arnold *et al.*,[113] marks the beginning of the trend to greater rotational excitation in the O_2 products. This may be a direct signature of the role played by bending excitation and the nonadiabatic interaction in the dissociation.

3.2.4. *Conclusions — Sequential DPD of O_3^-*

The DPD of O_3^- provides an excellent example of a sequential process, in which the photoelectron leaves behind a metastable state of the neutral that subsequently dissociates. This provides an example of the ability of PPC experiments to provide insight into the dissociation dynamics of a wide range of excited state levels, with spectroscopic identification of the excited states made from the structured photoelectron spectrum. These experiments also constituted the first study of the dissociation dynamics of the triplet states of ozone, illustrating the strength of photodetachment methods for probing optically forbidden excited states of neutral molecules.

In spite of the large body of information now available on O_3, there are some important questions that remain to be answered. These include isotopic fractionation effects observed in the stratosphere[130] and in laboratory experiments[131] and the possible existence of the metastable D_{3h} cyclic isomer of ozone.[115] With respect to isotopic fractionation effects there have been suggestions that both the ground state of ozone[132] and possible low-lying excited states may play an important role.[133] The possibility has also been raised that highly vibrationally excited O_2 may provide a previously unknown source of ozone.[134] PPC measurements should be able to help resolve questions about isotope effects in the low-lying excited states of O_3, but higher mass resolution ($m/\Delta m > 20$) is required to cleanly resolve $^{16}O/^{18}O$ photofragments. Photodissociation experiments in our laboratory have indicated that it may be possible to prepare an excited state of O_3^- with a different geometry from the ground state of O_3^-,[135,136] providing a Franck–Condon route to probing different regions of the O_3 neutral potential energy surface. In the experiments carried out to date, however, evidence has only been found for photodissociation of this excited state.

3.3. *Three-Body Dissociative Photodetachment: O_6^-*

In this section, the application of the PPC technique to the study of three-body dissociation reactions is discussed. These experiments were carried out with the new spectrometer reviewed in Sec. 2. As discussed in the introduction, photodetachment of anions can produce a neutral complex far from the equilibrium geometry, and the repulsive energy in the cluster can lead to a rapid dissociation. To determine the energetics and dissociation dynamics of such short lived neutral clusters, a measurement of the photoelectron kinetic energy alone is insufficient, it is necessary to measure the photoelectron kinetic energy *in coincidence* with the translational energy release. In the case of three-body dissociation, in particular when the identity of the products is unknown, it is essential to measure all the fragments in coincidence to fully describe the dissociation kinematics.

The first system we have studied with this new technique is the three-body dissociation reaction initiated by DPD of the O_6^- cluster anion.[137] The homogeneous cluster anions of O_2 are known to exhibit a rich photochemistry, undergoing both photodissociation and photodetachment with visible photons.[69,71] Mass spectrometric studies have shown that O_6^- is stabilized relative to O_4^- by 0.11 eV,[70] thus DPD of O_6^- at 532 nm will yield a free electron and three ground state O_2 molecules. The experiments discussed below reveal a vibrationally resolved translational energy distribution as well as new photodissociation channels for O_6^- that were unobserved in O_4^-.

3.3.1. *Three-Body N(E_T, eKE) Correlation Spectra*

The O_6^- data at 532 nm has been analyzed for both two- and three-body dissociation pathways, and the only dissociation pathway observed is $O_6^- + h\nu \rightarrow 3O_2(^3\Sigma_g^-) + e^-$. This is in accord with our studies of O_4^-, which show that no stable O_4 is produced. To examine the partitioning of translational energy among the photoelectron and the three photofragments produced in the DPD of O_6^-, it is useful to examine the photoelectron-photofragment correlation spectra, N(E_T, eKE), as shown in Fig. 17. The limit A corresponds to the maximum kinetic energy among the four particles $(3O_2 + e^-)$, given the 0.11 eV stability of O_6^- relative to $O_4^- + O_2$.

The dominant feature observed in Fig. 17 arises from the three-body DPD process $(O_6^- + h\nu \rightarrow 3O_2(X^3\Sigma_g^-) + e^-)$. This pathway appears as the series of diagonal ridges marked by I. These ridges are very similar to the features previously observed in the DPD of O_4^- at 532 nm as discussed in

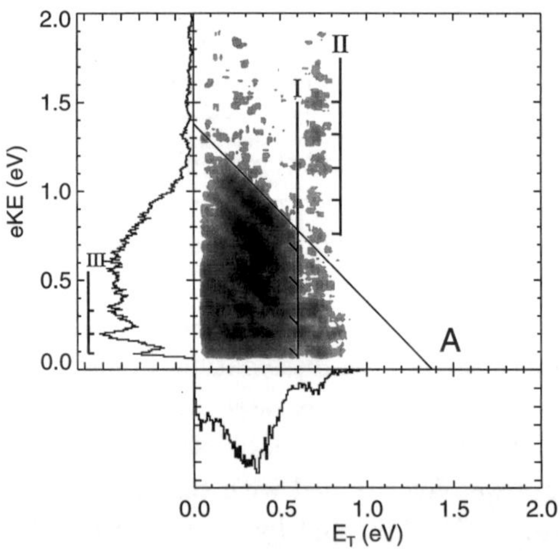

Fig. 17. N(E_T, eKE) correlation spectrum for O_6^- at 532 nm. The energetic limit A for the process $O_6^- + h\nu \rightarrow O_2(X^3\Sigma_g^-) + O_2(X^3\Sigma_g^-) + O_2(X^3\Sigma_g^-) + e^-$ is shown. The contours are linearly scaled and the marked features are discussed in the text. (Adapted with permission from Ref. 137.)

Sec. 3.1.1.[41,68] In the DPD of O_4^-, a series of 5 diagonal ridges was observed in the photoelectron-photofragment correlation spectrum, corresponding to direct DPD on a series of vibrationally adiabatic repulsive curves correlating to ground electronic state $O_2(v=0)+O_2(v=0)$, $O_2(v=0)+O_2(v=1)$, and so forth. In the O_6^- spectrum, these diagonal bands are shifted to a lower eKE and E_T, due to the 0.11 eV stability of O_6^- relative to $O_4^-+O_2$. Internal energy of the photofragments appears as a displacement along the diagonal between the origin and the limit A. The separation of the ridges along the diagonal is ≈ 0.19 eV, consistent with vibrational excitation of the O_2 photofragments.

There are two other important features in Fig. 17. Feature II, the vertical row of spots at $E_T = 0.7$ eV, results from the photodissociation process $O_6^- + h\nu \rightarrow O_2(^1\Delta_g, v=0) + O_2^-(^2\Pi_g, v=0) + O_2(^3\Sigma_g^-)$. The analogous process $O_4^- + h\nu \rightarrow O_2(^1\Delta_g, v=0) + O_2^-(^2\Pi_g, v=0)$ was previously observed, at an $E_T = 0.8$ eV.[69] The lower E_T observed in the photodissociation of O_6^- once again arises due to the 0.11 eV stability of O_6^- relative to $O_4^-+O_2$. As in the case of O_4^-, these features are observable due to photodetachment of the nascent $O_2^-(v=0)$ photofragments by a second photon, yielding the well-known photoelectron spectrum of $O_2^-(v=0)$[71,138] as the vertical row of spots. Finally, the feature III is a second photodissociation process, not observed in O_4^- at 532 nm; $O_6^- + h\nu \rightarrow O_2^-(^2\Pi_g, v>3) + O_2(^3\Sigma_g^-) + O_2(^3\Sigma_g^-)$. The peaks visible in the photoelectron spectrum appear at energies of 0.09, 0.19 and 0.30 eV, consistent with autodetachment of $O_2^-(v=4,5,6) \rightarrow O_2(v=0)+e^-$. This process was also observed in the photodissociation of O_4^- at 355 nm, as discussed in Sec. 3.1.1. Higher-resolution photoelectron spectra from this laboratory have shown that the peak for $O_2^-(v=7)$ autodetachment is also observed at 0.41 eV.[139] A consideration of the energetics indicates that this process can only occur if no $O_2(^1\Delta_g)$ is produced in this photodissociation channel.

3.3.2. *Three-Body E_{TOT} Spectra*

In Fig. 18 the E_{TOT} spectra for O_4^- (from Ref. 68) and O_6^- are shown in the lower and upper frames, respectively. These spectra are generated from the correlation spectra by histogramming $E_{TOT} = \mathrm{eKE} + E_T$ for each event. The shift of the threshold A in the O_6^- spectrum is due to the stability of O_6^- relative to $O_4^-+O_2$. The diagonal ridges of Fig. 17 now appear as resolved

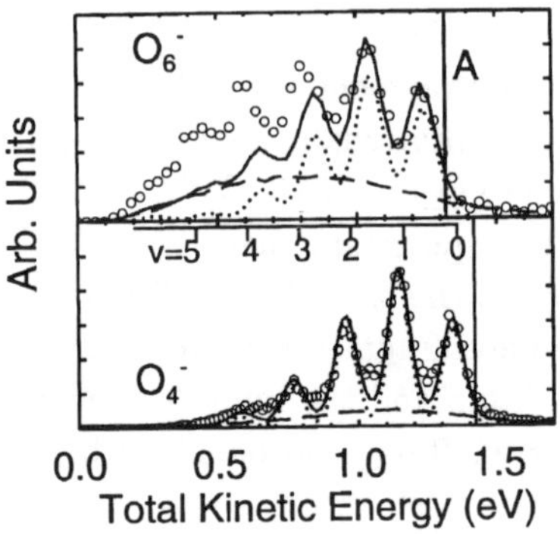

Fig. 18. Total kinetic energy (E_{TOT}) spectra for DPD of O_4^- and O_6^- are shown in the lower and upper frames, respectively. The spacing of the first five vibrational levels of O_2 are shown in the lower frame for reference. The circles represent the data, and the solid line represents the sum of the contributions from the local mode Franck–Condon fit (dotted-line curve) and a background contribution generated from the random correlation of the eKE and E_T spectra (dashed-line curve). (Adapted with permission from Ref. 137.)

peaks in the O_6^- data. The differences between these spectra shows that the DPD of O_6^- is characterized by an increase in the relative intensities of the peaks corresponding to vibrationally excited O_2. There is also evidence of increased rotational excitation at higher vibrational energies, as shown by the shift of the peaks and reduction in resolution at lower E_{TOT}. The fact that these peaks are resolved indicates that the rotational distribution in the photofragments, even in the case of O_6^-, is narrow and generally less than a vibrational quantum.

The low observed rotational excitation indicates that, like O_4^-, there is a restricted range of product angular momenta produced in the dissociation of the nascent O_6 neutral. This result is consistent either with a weakly bound ion-dipole complex, in which the third O_2 is merely a spectator, or a molecular O_6^- anion with a reasonably high symmetry. Within the framework of the simple local mode Franck–Condon simulation previously used in the analysis of O_4^-, the observed product vibrational distribution is not consistent with either extreme limit. As discussed in Sec. 3.1.3, the fit to the $O_4^- E_{TOT}$ spectrum was generated assuming two O_2 moieties with equal bond lengths midway between O_2 and O_2^-. In the limit of a weak interaction between O_2 and O_4^-, a similar distribution would be expected for O_6^-. An examination of Fig. 18 shows that the O_4^- distribution predicts lower vibrational excitation than observed in O_6^-. An alternative possibility for O_6^- is that all three short O–O bonds are equal, with $r_e = 1.255$ Å, one third of the difference between O_2 and O_2^-. This fit, shown on the O_6^- spectrum, does not reproduce the high vibrational excitation observed in the correlated product vibrational distribution. A better fit to the data will require more information on the structure of O_6^- and the dynamics of the three-body dissociation of O_6.

3.3.3. *Conclusions — Three-Body DPD of O_6^-*

This study represented the first complete kinematic description of a three-body DPD process. The results show that the DPD dynamics of O_6^- are remarkably similar to O_4^-, but a more marked change is observed in the photodissociation dynamics. The low rotational excitation in the products permits the observation of vibrationally resolved translational energy distributions and argues for either an O_4^- core bound electrostatically to a third O_2 or perhaps, a highly symmetric structure for O_6^-. Studies of the photodestruction dynamics at other wavelengths and theoretical calculations of the O_6^- structure are underway to resolve this question. It is expected that analysis of the three-body photofragment angular distributions will also be of assistance.

4. Future Directions and Prospects

The experimental techniques of PPC spectroscopy as discussed in this review provide a unique method of preparing transient neutrals with a known internal energy and studying the subsequent dissociation dynamics. The basic requirement for carrying out a PPC experiment is that the precursor anion have Franck–Condon overlap with a dissociative ground or excited state at low enough energies to be reached by photodetachment. Using existing anion sources, extension of these experiments to some of the important transition state spectroscopy systems mentioned in the introduction, nitrogen oxides,[140] anion-water cluster complexes[141,142] and small organic radicals will be straightforward. To broaden the range of systems that can be studied, it will be necessary to apply some of the new techniques for ion production, such as matrix-assisted laser-desorption ionization (MALDI)[143] and electrospray ionization (ESI).[144] As mentioned in the introduction, Wang and coworkers have already used the ESI technique interfaced to a photoelectron spectrometer to obtain the photoelectron spectrum of a doubly charged anion.[13]

Current photoelectron detection techniques have sufficient resolution to give vibrational state-specific information for small molecules, but as extensions are made to larger molecules this will be lost. Significant improvements in photoelectron resolution, such as that afforded by zero electron kinetic energy (ZEKE) techniques,[145] will be difficult to achieve. ZEKE spectroscopy can only be applied to the lowest dissociative states. At photoelectron thresholds for higher-lying states, photodetachment to dissociative states at lower energies will still produce photofragments rendering correlation of photoelectrons and photofragments impossible. An additional consideration as the PPC technique is pushed to studying larger complexes is the requirement for ns-timescale dissociation processes due to the use of a fast ion beam and the reduced detection efficiency of MCPs for higher-mass ions. In principle much higher beam energies can be achieved by use of linear acceleration techniques,[146] however that will only serve to make the dissociation lifetime constraint more stringent.

In future studies, the time-scale for three-body dissociation will be addressed by examination of the photoelectron-photofragment and photofragment-photofragment angular correlations, and this technique will be extended to important systems including the three-body association reaction responsible for the formation of ozone: $O + O_2 + M \rightarrow O_3 + M$. It is hoped that these experiments will determine what happens when a dissociative state of a molecular species within a complex is created by photodetachment. Is there a 'collisional stabilization' by a third body, in the half-collision version of a termolecular association process? This will result in two fragments, the stabilized species and the third body, with a characteristic translational energy distribution. If the third body does *not* stabilize the molecular species, three fragments will be produced. A third possibility involves dissociation of the initially excited species followed by an intra-cluster reaction with the third body. Studies of the internal energy and third-body dependence of these processes should provide a qualitatively new view into the dynamics of three-body reactions of neutral species.

Acknowledgments

I would like to acknowledge the contributions of my research group, including Dr. K. A. Hanold, M. C. Garner, Dr. M. S. Resat, A. K. Luong and T. G. Clements, to these studies. R. E. C. is a Camille Dreyfus Teacher-Scholar, A. P. Sloan Research Fellow and a Packard Fellow in

Science and Engineering. This work was supported by the NSF under grant CHE-9700142 and the U.S. Air Force Office of Scientific Research, AFOSR-F49620-96-1-0220.

References

1. D. W. Kohn, H. Clauberg and P. Chen, *Rev. Sci. Instrum.* **63**, 4003 (1992).
2. P. J. Dagdigian, *J. Chem. Phys.* **90**, 2617 (1989).
3. M. Lock, R. Barnes and A. Sinha, *J. Chem. Phys.* **104**, 1350 (1996).
4. M. Alagia, N. Balucani, P. Casavecchia, D. Stranges and G. G. Volpi, *J. Chem. Phys.* **98**, 2459 (1993).
5. J. Berkowitz and B. Ruscic, in *VUV Photoionization and Photodissociation of Molecules and Clusters*, Ed. C. Y. Ng (World Scientific, Singapore, 1991), pp. 1–41.
6. P. Chen, in *Unimolecular and Bimolecular Ion-Molecule Reaction Dynamics*, Eds. C. Y. Ng, T. Baer and I. Powis (Wiley, New York, 1994), pp. 371–425.
7. L. J. Butler and D. M. Neumark, *J. Phys. Chem.* **100**, 12801 (1996).
8. R. E. Continetti, D. R. Cyr, R. B. Metz, D. M. Neumark, *Chem. Phys. Lett.* **182**, 406 (1991).
9. S. Suzuki, T. Wakabayashi, H. Matsuura, H. Shiromaru, C. Kittaka and Y. Achiba, *Chem. Phys. Lett.* **182**, 12 (1991); O. Heber, D. Kella, D. Zajfman, H. Feldman, D. Majer, Z. Vager and R. Naaman, *Nuc. Inst. Meth.* **B79**, 227 (1993).
10. K. M. Ervin and W. C. Lineberger, in *Advances in Gas Phase Ion Chemistry*, Vol. 1, Ed. N. G. Adams and L. M. Babcock (JAI Press, Greenwich, CT, USA, 1992), pp. 121–166.
11. J. M. Farrar, in *Cluster Ions*, Eds. T. Baer, C. Y. Ng and I. Powis (Wiley, New York, 1993), pp. 243–317.
12. M. S. Robinson, M. L. Polak, V. M. Bierbaum, C. H. DePuy and W. C. Lineberger, *J. Am. Chem. Soc.* **117**, 6766 (1995).
13. X. B. Wang, C. F. Ding and L. S. Wang, *Phys. Rev. Lett.* **81**, 3351 (1998).
14. K. M. Ervin, J. Ho and W. C. Lineberger, *J. Chem. Phys.* **91**, 5974 (1989).
15. P. G. Wenthold, D. A. Hrovat, W. T. Borden and W. C. Lineberger, *Science* **272**, 1456 (1996).
16. R. B. Metz, S. E. Bradforth and D. M. Neumark, *Adv. Chem. Phys.* **81**, 1 (1992).
17. D. E. Manolopoulos, K. Stark, H.-J. Werner, D. W. Arnold, S. E. Bradforth and D. M. Neumark, *Science* **262**, 1852 (1993).
18. E. de Beer, E. H. Kim, D. M. Neumark, R. F. Gunion and W. C. Lineberger, *J. Phys. Chem.* **99**, 13627 (1995).
19. D. W. Arnold, C. Xu and D. M. Neumark, *J. Chem. Phys.* **102**, 6088 (1995).
20. W. H. Thompson and W. H. Miller, *J. Chem. Phys.* **101**, 8620 (1994); D. C. Clary, J. K. Gregory, M. J. T. Jordan and E. Kauppi, *J. Chem. Soc. Farad. Trans.* **93**, 747 (1997).

21. J. Tellinghuisen, in *Photodissociation and Photoionization*, Ed. K. P. Lawley (Wiley, New York, 1985), pp. 299–369.
22. E. Illenberger and J. Momigny, *Gaseous Molecular Ions: An Introduction to Elementary Processes Induced by Ionization* (Springer-Verlag, New York, 1992), pp. 254–260.
23. W. H. Miller, *J. Phys. Chem.* **99**, 12387 (1995); R. J. Duchovic, J. D. Pettigrew, B. Welling and T. Shipchandler, *J. Chem. Phys.* **105**, 10367 (1996) ; A. Gross, G. D. Billing, *Chem. Phys.* **217**, 1 (1997); R. T. Pack, R. B. Walker and B. Kendrick, *J. Chem. Phys.* **109**, 6701 (1998).
24. J. H. D. Eland, in *VUV Photoionization and Photodissociation of Molecules and Clusters*, Ed. C. Y. Ng (World Scientific, Singapore, 1991), pp. 297–343.
25. S. Hsieh and J. H. D. Eland, *J. Phys.* **B30**, 4515 (1997).
26. U. Werner, K. Beckord, J. Becker and H. O. Lutz, *Phys. Rev. Lett.* **74**, 1962 (1995).
27. T. Masuoka, *J. Chem. Phys.* **98**, 6989 (1993).
28. Recent experiments on three-body dissociation are extensively reviewed in: C. Maul and K. H. Gericke, *Int. Rev. Phys. Chem.* **16**, 1 (1997).
29. D. P. de Bruijn and J. Los, *Rev. Sci. Instrum.* **53**, 1020 (1982).
30. W. J. van der Zande, W. Koot, J. R. Peterson and J. Los., *Chem. Phys.* **126**, 169 (1988); J. H. M. Beijersbergen, S. Kornig, W. J. van der Zande, P. G. Kistemaker and J. Los, *J. Phys. Chem.* **95**, 9059 (1991).
31. H. Helm and P. C. Cosby, *J. Chem. Phys.* **86**, 6813 (1987); H. Helm and C. W. Walter, *J. Chem. Phys.* **98**, 5444 (1993).
32. R. E. Continetti, D. R. Cyr, D. L. Osborn, D. J. Leahy and D. M. Neumark, *J. Chem. Phys.* **99**, 2616 (1993).
33. H. Hotop and W. C. Lineberger, *J. Phys. Chem. Ref. Data* **14**, 731 (1985).
34. D. Hanstorp, *Nuc. Inst. Meth.* **B100**, 165 (1995).
35. J. H. D. Eland, in *Electron Spectroscopy: Theory, Techniques and Applications*, Vol. 3, Eds. C. R. Brundle and A. D. Baker (New York, Academic Press, 1979), pp. 231–303.
36. K. Rademann, T. Rech, B. Kaiser, U. Even and F. Hensel, *Rev. Sci. Instrum.* **62**, 1932 (1991).
37. D. Van Pijkeren, J. Van Eck and A. Niehaus, *Chem. Phys. Lett.* **96**, 20 (1983).
38. I. Powis, *Chem. Phys. Lett.* **189**, 473 (1992).
39. E. Shigemasa, J. Adachi, M. Oura and A. Yagishita, *Phys. Rev. Lett.* **74**, 359 (1995).
40. K. A. Hanold, C. R. Sherwood, M. C. Garner and R. E. Continetti, *Rev. Sci. Instrum.* **66**, 5507–5511 (1995).
41. K. A. Hanold and R. E. Continetti, *Chem. Phys.* **239**, 493 (1998).
42. K. A. Hanold, A. K. Luong, T. G. Clements and R. E. Continetti, *Rev. Sci. Instrum.* **70**, 2268 (1999).
43. M. A. Johnson, M. L. Alexander and W. C. Lineberger, *Chem. Phys. Lett.* **112**, 285 (1984).

44. M. A. Johnson and W. C. Lineberger, in *Techniques for the Study of Ion-Molecule Reactions*, Eds. J. M. Farrar and W. H. Saunders (New York, Wiley, 1988), pp. 591–635.
45. D. Proch and T. Trickl, *Rev. Sci. Instrum.* **60**, 713 (1989).
46. J. M. B. Bakker, *J. Phys.* **E6**, 785 (1973); *ibid.* **7**, 364 (1974).
47. L. D. Gardner, M. M. Graff and J. L. Kohl, *Rev. Sci Instrum.* **57**, 177 (1986).
48. O. H. W. Siegmund, K. Coburn and R. F. Malina, *IEEE Trans. Nuc. Sci.* **32**, 443 (1985).
49. C. Martin, P. Jelinsky, M. Lampton, R. F. Malina and H. O. Anger, *Rev. Sci. Instrum.* **52**, 1067 (1981); O. H. W. Siegmund, M. Lampton, J. Bixler, S. Chakrabarti, J. Vallerga, S. Bowyer and R. F. Malina, *J. Opt. Soc. Am.* **A3**, 2139 (1986).
50. M. Lampton and C. W. Carlson, *Rev. Sci. Instrum.* **50**, 1093 (1979).
51. A. Smith, R. Kessel, J. S. Lapington and D. M. Walton, *Rev. Sci. Instrum.* **60**, 3509 (1989).
52. J.-C. Brenot and M. Durup-Ferguson, *Adv. Chem. Phys.* **82**, 309 (1992).
53. Z. Amitay and D. Zajfman, *Rev. Sci. Instrum.* **68**, 1387 (1997).
54. J. G. Timothy, *Opt. Engineering* **24**, 1066 (1985).
55. P. Datte, E. Beuville, J. F. Beche, C. Cork, T. Ernest, J. Millaud, D. Nygren, H. Padmore, B. Turko and N-H. Xuong, *Nuc. Inst. Meth. Phys. Res.* **A391**, 471 (1997).
56. A. Belkacem, A. Faibis, E. P. Kanter, W. Koenig, R. E. Mitchell, Z. Vager and B. J. Zabransky, *Rev. Sci. Instrum.* **61**, 945 (1990).
57. W. Koenig, A. Faibis, E. P. Kanter, Z. Vager and B. J. Zabransky, *Nuc. Inst. Meth.* **B10/11**, 259 (1985).
58. J. Becker, K. Beckord, U. Werner and H. O. Lutz, *Nuc. Inst. Meth.* **A337**, 409 (1994).
59. G. Knapp, *Rev. Sci. Instrum.* **49**, 982 (1978).
60. M. B. Williams and S. E. Sobottka, *IEEE Trans. Nuc. Sci.* **36**, 227 (1989).
61. J. H. D. Eland and A. H. Pearson, *Meas. Sci. Technol.* **1**, 36 (1990).
62. M. Lampton, O. Siegmund and R. Raffanti, *Rev. Sci. Instrum.* **58**, 2298 (1987).
63. P. G. Friedman, R. A. Cuza, J. R. Fleischman, C. Martin, D. Schiminovich and D. J. Doyle, *Rev. Sci. Instrum.* **67**, 596 (1996).
64. H. Hotop and W. C. Lineberger, *J. Phys. Chem. Ref. Data* **14**, 731 (1985).
65. D. L. Osborn, H. Choi, D. H. Mordaunt, R. T. Bise, D. M. Neumark and C. M. Rohlfing, *J. Chem. Phys.* **106**, 3049 (1997).
66. M. Lavollee and H. Bergeron, *J. Phys. B: At. Mol. Opt. Phys.* **25**, 3101 (1992).
67. A. Muller, N. Djuric, G. H. Dunn and D. S. Belic, *Rev. Sci. Instrum.* **57**, 349 (1986).
68. K. A. Hanold, M. C. Garner and R. E. Continetti, *Phys Rev. Lett.* **77**, 3335 (1996).

69. C. R. Sherwood, K. A. Hanold, M. C. Garner, K. M. Strong and R. E. Continetti, *J. Chem. Phys.* **105**, 10803 (1996).
70. K. Hiraoka, *J. Phys. Chem.* **89**, 3190 (1988).
71. L. A. Posey, M. J. Deluca and M. A. Johnson, *Chem. Phys. Lett.* **131**, 170 (1986); M. J. DeLuca, C. C. Han and M. A. Johnson, *J. Chem. Phys.* **93**, 268 (1990); C. C. Han and M. A. Johnson, *Chem. Phys. Lett.* **189**, 460 (1992).
72. C. R. Sherwood, M. C. Garner, K. A. Hanold, K. M. Strong and R. E. Continetti, *J. Chem. Phys.* **102**, 6949 (1995); K. A. Hanold, C. R. Sherwood and R. E. Continetti, *J. Chem. Phys.* **103**, 9876 (1995).
73. D. C. Conway, *J. Chem. Phys.* **50**, 3864 (1969).
74. K. Ohta and K. Morokuma, *J. Phys. Chem.* **91**, 401 (1987).
75. X. Y. Li, A. M. Tian, F. C. He and G. S. Yan, *Chem. Phys. Lett.* **233**, 227 (1995).
76. A. Aquino, S. P. Walch and P. R. Taylor, (to be published).
77. K. Andersson, P. A. Malmqvist, B. O. Roos, A. J. Sadlej and K. Wolinski, *J. Phys. Chem.* **94**, 5483 (1990).
78. G. V. Chertihin and L. Andrews, *J. Chem. Phys.* **108**, 6404 (1998).
79. M. J. Travers, D. C. Cowles and G. B. Ellison, *Chem. Phys. Lett.* **164**, 449 (1989).
80. G. J. Schulz, *Rev. Mod. Phys.* **45**, 423 (1973).
81. M. Allan, *J. Phys. B: At. Mol. Opt. Phys.* **28**, 5163 (1995).
82. R. J. Le Roy, R. G. Macdonald and G. Burns, *J. Chem. Phys.* **65**, 1485 (1976); R. J. Le Roy, *Computer Phys. Comm.* **52**, 383 (1989).
83. K. Takatsuka, *Phys. Rev.* **A55**, 347 (1997).
84. G. G. Balint-Kurti and G. C. Schatz, *J. Chem. Soc. Farad. Trans.* **93**, 755 (1997).
85. C. A. Long and G. E. Ewing, *J. Chem. Phys.* **58**, 4824 (1973).
86. P. E. S. Wormer and A. van der Avoird, *J. Chem. Phys.* **81**, 1929 (1984).
87. D. Lauvergnat and D. C. Clary, *J. Chem. Phys.* **108**, 3566 (1998).
88. L. Manceron, A-M. Le Quere and J-P. Perchard, *J. Phys. Chem.* **93**, 2960 (1989).
89. K. P. Huber and G. Herzberg, *Molecular Spectra and Molecular Structure IV* (Van Nostrand, New York, 1979), p. 490.
90. K. M. Ervin, J. Ho and W. C. Lineberger, *J. Phys. Chem.* **92**, 5405 (1988).
91. R. N. Zare, *Mol. Photochem.* **4**, 1 (1972).
92. C. R. Sherwood, Ph.D. Thesis, University of California, San Diego, 1995.
93. J. Cooper and R. N. Zare, *J. Chem. Phys.* **48**, 942 (1968).
94. K. J. Reed, A. H. Zimmerman, H. C. Andersen and J. I. Brauman, *J. Chem. Phys.* **64**, 1368 (1976).
95. R. Gygax, H. L. McPeters and J. I. Brauman, *J. Am. Chem. Soc.* **101**, 2567 (1979).
96. D. Dill, *J. Chem. Phys.* **65**, 1130 (1976).
97. N. Chandra, *Chem. Phys.* **108**, 301 (1986).

98. H. Rudolph and V. McKoy, *J. Chem. Phys.* **91**, 2235 (1989).
99. H. K. Park and R. N. Zare, *J. Chem. Phys.* **104**, 4554 (1996), *ibid.* **104**, 4568 (1996).
100. J. L. Dehmer and D. Dill, *Phys. Rev.* **A18**, 164 (1978).
101. A. V. Golovin, N. A. Cherepkov and V. V. Kuznetsov, *Z. Phys.* **D24**, 371 (1992).
102. I. Powis, *Chem. Phys. Lett.* **189**, 473 (1992); K. L. Reid and I. Powis, *J. Chem. Phys.* **100**, 1066 (1994).
103. S. Kaesdorf, G. Schönhense and U. Heinzmann, *Phys. Rev. Lett.* **54**, 885 (1985).
104. M. C. Garner, K. A. Hanold, M. Sowa Resat and R. E. Continetti, *J. Phys. Chem.* **101**, 6577 (1997).
105. J. I. Steinfeld, S. M. Adler-Golden and J. W. Gallagher, *J. Phys. Chem. Ref. Data* **16**, 911 (1987).
106. S. M. Anderson and K. Mauersberger, *J. Geophys. Res.* **100**, 3033 (1995).
107. N. Swanson and R. J. Celotta, *Phys. Rev. Lett.* **35**, 783 (1975).
108. M. Allan, N. J. Mason and J. A. Davies, *J. Chem. Phys.* **105**, 5665 (1996); N. J. Mason, J. M. Gingell, J. A. Davies, H. Zhao, I. C. Walker, M. R. F. Siggel, *J. Phys.* **B29**, 3075 (1996).
109. M. Allan, K. R. Asmis, D. B. Popovic, M. Stepanovic, N. J. Mason and J. A. Davies, *J. Phys.* **B29**, 4727 (1996); I. C. Walker, J. M. Gingell, N. J. Mason and G. Marston, *J. Phys.* **B29**, 4749 (1996).
110. J. D. Skalny, S. Matejcik, A. Kiendler, A. Stamatovic and T. D. Märk, *Chcm. Phys. Lctt.* **255**, 112 (1996).
111. S. E. Novick, P. A. Engelking, P. L. Jones, J. H. Futrell and W. C. Lineberger, *J. Chem. Phys.* **70**, 2652 (1979).
112. L. J. Wang, S. B. Woo and E. M. Helmy, *Phys. Rev.* **A35**, 759 (1987).
113. D. W. Arnold, C. Xu, E. H. Kim and D. M. Neumark, *J. Chem. Phys.* **101**, 912 (1994).
114. J. R. Locker, J. A. Joens and E. J. Blair, *J. Photochem.* **36**, 235 (1987).
115. P. J. Hay and T. H. Dunning, Jr., *J. Chem. Phys.* **67**, 2290 (1977).
116. T. J. Lee, *Chem. Phys. Lett.* **169**, 529 (1990); S. S. Xantheas, G. J. Atchity, S. T. Elbert and K. Ruedenberg, *J. Chem. Phys.* **94**, 805 (1991); A. Banichevich and S. D. Peyerimhoff, *Chem. Phys.* **174**, 93 (1993).
117. A. J. Bouvier, R. Bacis, B. Bussery, S. Churassy, D. Inard, M. Nota, J. Brion, J. Malicet and S. M. Anderson, *Chem. Phys. Lett.* **255**, 263 (1996).
118. T. Tsuneda, H. Nakano and K. Hirao, *J. Chem. Phys.* **103**, 6520 (1995).
119. P. Borowski, M. Fulscher, P. Malmqvist and B. O. Roos, *Chem. Phys. Lett.* **237**, 195 (1995).
120. M. Braunstein, P. J. Hay, R. L. Martin and R. T. Pack, *J. Chem. Phys.* **95**, 8239 (1991).
121. M. Braunstein and R. T. Pack, *J. Chem. Phys.* **96**, 6378 (1992).
122. S. M. Anderson, P. Hupalo and K. Mauersberger, *J. Chem. Phys.* **99**, 737 (1993).

123. B. Minaev and H. Ågren, *Chem. Phys. Lett.* **217**, 531 (1994).
124. M. Braunstein, R. L. Martin and P. J. Hay, *J. Chem. Phys.* **102**, 3662 (1995).
125. B. Abel, A. Charvat and S. Deppe, *Chem. Phys. Lett.* **277**, 347 (1997).
126. J. L. Gole and R. N. Zare, *J. Chem. Phys.* **57**, 5331 (1972).
127. M. C. Garner, C. R. Sherwood, K. A. Hanold, R. E. Continetti, *Chem. Phys. Lett.* **248**, 20 (1996).
128. H. B. Levene and J. J. Valentini, *J. Chem. Phys.* **87**, 2594 (1987).
129. R. Schinke, *Photodissociation Dynamics* (Cambridge Press, Cambridge, UK, 1993), pp. 251–255.
130. P. L. Houston, A. G. Suits and R. Toumi, *J. Geophys. Res.* **101**, 18829 (1996).
131. M. H. Thiemens and J. E. Heidenreich, *Science* **219**, 1073 (1983); J. Wen and M. H. Thiemens, *J. Geophys. Res.* **96**, 10911 (1991).
132. G. I. Gellene, *Science* **274**, 1344 (1996).
133. S. M. Anderson, D. Hulsebusch and K. Mauersberger, *J. Chem. Phys.* **107**, 5385 (1997).
134. R. L. Miller, A. G. Suits, P. L. Houston, R. Toumi, J. A. Mack and A. M. Wodtke, *Science* **265**, 1831 (1994).
135. M. C. Garner, C. R. Sherwood, K. A. Hanold and R. E. Continetti, *Chem. Phys. Lett.* **248**, 20 (1996).
136. Q. Cui and K. Morokuma, *J. Chem. Phys.* **108**, 7684 (1998).
137. K. A. Hanold, A. K. Luong and R. E. Continetti, *J. Chem. Phys.* **109**, 9215 (1998).
138. M. J. Travers, D. C. Cowles and G. B. Ellison, *Chem. Phys. Lett.* **164**, 449 (1989).
139. R. Li, K. A. Hanold, M. C. Garner, A. K. Luong and R. E. Continetti, *Faraday Discuss.* **108**, 115 (1997).
140. M. S. Resat, V. Zengin, M. C. Garner and R. E. Continetti, *J. Phys. Chem.* **A102**, 1719 (1998).
141. C. R. Sherwood and R. E. Continetti, *Chem. Phys. Lett.* **258**, 171 (1996).
142. J. B. Kim, P. G. Wenthold and W. C. Lineberger, *J. Chem. Phys.* **108**, 830 (1998).
143. M. Karas, D. Bachmann, U. Bahr and F. Hillenkamp, *Int. J. Mass Spec. Ion Proc.* **78**, 53 (1987).
144. J. B. Fenn, M. Mann, C. K. Meng, S. F. Wong and C. M. Whitehouse, *Science* **246**, 64 (1989).
145. T. N. Kitsopoulos, I. M. Waller, J. G. Loeser and D. M. Neumark, *Chem. Phys. Lett.* **159**, 300 (1989); G. Drechsler, U. Boesl, C. Bassmann and E. W. Schlag, *J. Chem. Phys.* **107**, 2284 (1997).
146. E. Hendell and U. Even, *Rev. Sci. Instrum.* **66**, 3901 (1995).

CHAPTER 15

MASS SELECTED ANION-ZERO KINETIC ENERGY PHOTOELECTRON SPECTROSCOPY

Ulrich Boesl,* Carsten Bäßmann and Edward W. Schlag

Institut für Physikalische und Theoretische Chemie
Technische Universität München
Lichtenbergstr. 4, 85748 Garching
Germany

Contents

---

*Corresponding author, E-mail: ulrich.boesl@ch.tum.de

1. Introduction

Mass selective photoelectron spectroscopy of negatively charged atoms, molecules and clusters open many technical and experimental possibilities to analytical chemistry, research on gas-phase reactions, molecular structure studies, and investigations of vibrational and electronic states and thus on chemical bonds. Like cations, negatively charged species may easily be mass selected prior to subsequent experiments. In opposite to those, these end up in neutrals after emission of an electron. The combination of both mass selection and subsequent emission of an electron therefore allows access to many neutral species independent on their stability and availability as pure substances. If the photoelectron emission is induced by light of a discrete wavelength, photodetachment photoelectron spectroscopy (PD-PES), in the literature sometimes also termed anion photoelectron spectroscopy (anion-PES) may be performed, and that for mass selected anions. Anion-ZEKE spectroscopy is a special development of photoelectron spectroscopy on anions, exhibiting an enhancement of spectroscopic resolution by more than one order of magnitude. Concerning selection rules, anion-ZEKE is subject to much stricter conditions than photodetachment-PES and therefore, in some sense, complementary to it.

In Fig. 1, three concepts to perform mass selective spectroscopy following neutral ← anion transitions by recording photoelectron current, namely photodetachment photoelectron spectroscopy, photodetachment spectroscopy, and anion-ZEKE photoelectron spectroscopy, are compared in a schematic way. All three of them allow us to study the electronic and vibronic states of neutral atoms, molecules or clusters in the gas-phase and their anionic states, which show up as hot bands in the spectra. These methods differ in the application of fixed or tunable wavelength light sources and in recording kinetic energy selective or total integrated electron currents.

Photodetachment photoelectron spectroscopy uses a fixed-frequency light source and measures the energy of the detached electrons. The energy spectra of electrons photodetached from molecular negative ions will, in

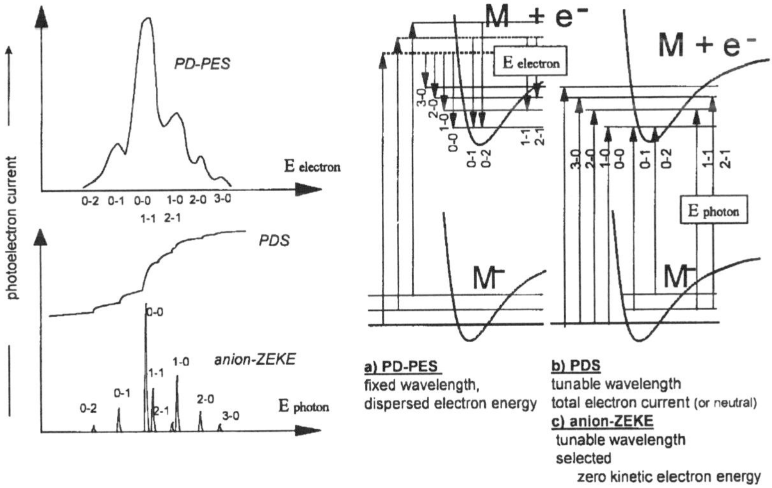

Fig. 1. Transition scheme and exemplary spectra of photodetachment photoelectron spectroscopy (PD-PES), photodetachment spectroscopy (PDS), and anion zero kinetic energy photoelectron spectroscopy (anion-ZEKE).

general, exhibit multiple peaks (similar to conventional PES) whose bandwidth reduces the spectral resolution considerably. Each of these features corresponds to neutral $\leftarrow$ anion transitions as indicated in Fig. 1. The 0–0 peak is of particular importance as it represents the adiabatic detachment threshold, commonly known as electron affinity EA.

A different, resonance-type spectrum is obtained, if the wavelength of the light source is tuned while recording the total electron current. These photodetachment spectra (PDS) resemble photoionization spectra and show a step function characteristic since the total photodetachment cross-section and thus the total electron current increase whenever a neutral state is reached by the photon energy during the wavelength scan. However, rather than a rectangular staircase shape of these steps as seen in the case of photoionization, photodetachment spectra exhibit slowly increasing thresholds, as described by the threshold laws (see Sec. 1.4) and indicated in Fig. 1. The accuracy of photodetachment spectroscopy, at least in principle, only depends on the accuracy and bandwidth of the wavelength of the used light source. In practice, however, the signal-to-noise ratio tends to be poor due to the gradual onset of the steps. Individual transitions often can no longer be discerned and are buried in a monotonously increasing, total electron signal. Instead of recording the total electron current, the neutral yield can also be measured, which however is subject to even more noise due to detachment processes on surfaces and during gas-phase collisions.

At anion-ZEKE spectroscopy, one takes advantage of the fact that during a scan of the wavelength, electrons with zero kinetic energy (ZEKE) are only released when the exciting light is in resonance with one of the final neutral states. This gives rise to well resolved spectral lines, rather than to the step function which characterized the photodetachment spectra (see Fig. 1) or to low-resolution bands as in PD-PES. Similar to PDS and in contrast to PD-PES, where fixed photon energy is used and energy accuracy is determined by the photoelectron energy spectrometer, the accuracy of absolute energies in anion-ZEKE spectroscopy is tied to the absolute wavelength calibration of the light source.

1.1. *Photodetachment Photoelectron Spectroscopy*

To understand the roots of anion-ZEKE spectroscopy some fundamental and historical features of photodetachment-PES will shortly be illuminated before introducing anion-ZEKE spectroscopy itself. One outstanding virtue of mass selective PD-PES already mentioned above is its capability to yield spectroscopic information about neutral molecular systems even if they are present in highly complex mixtures or are short lived due to their reactivity or instability. Examples are homogeneous and heterogeneous carbon, metal or semiconductor complexes, or the much weaker bound molecular clusters such as homogeneous and heterogeneous water clusters. Another characteristic of mass selective PD-PES is that, owing to different or less strict selection rules or due to favorable geometrical changes during transitions from anion to neutral, more vibrational and even excited electronic neutral states are accessible than by neutral-neutral transitions. Anion-photoelectron spectroscopy may also serve as a guide to map the multitude of neutral electronic states even of different multiplicity. Chemically important parameters such as the electron affinity or (following the simple molecular picture of Koopmans' theorem) approximate energies of molecular orbits (HOMO-LUMO gap) may be extracted from PD-PE spectra. Also stable excited anionic states and dipole bound states may be studied by mass selective photodetachment experiments. Both of them are special states which only exist under certain circumstances such as high photodetachment thresholds and large neutral dipole moments, respectively. Even charge transfer processes in heterogeneous clusters of an anionic solute and solvent molecules have been investigated by this method. To summarize, the typical features of the combination of negative ion sources, mass separation, and photoelectron detection are (i) the possibility of mass separation prior to spectroscopy, (ii) access to neutral systems and neutral states which are not accessible by neutral $\leftarrow$ neutral transitions, and (iii) accurate and unambiguous determination of electron affinities.

Historically, photoelectron spectroscopy of negative ions has been invented more than 30 years ago,[1] and has been developed as high-resolution technique for the determination of atomic and molecular electron affinities.[2–9] More than one decade later, other groups incorporated this technique into their experiments. Additional concepts, such as time-of-flight mass separators, supersonic molecular-beams, and laser desorption[10–15] opened new options to study systems that were hardly accessible before, such as larger molecules, atomic and molecular clusters.

1.2. *Anion*-ZEKE *Spectroscopy*

However, photoelectron spectroscopy of negatively charged species suffers from the same problem of low-resolution as conventional photoelectron spectroscopy involving cation $\leftarrow$ neutral transitions. There, this problem is overcome by resonance photoelectron spectroscopy and ZEKE spectroscopy. In both schemes photoelectrons with nearly vanishing kinetic energy are discriminated against kinetic electrons. In the former case, this discrimination is achieved in steradiancy spectrometers[16–17] with a resolution of some 25 cm^{-1}. ZEKE- (zero kinetic energy electron) spectroscopy[18] is an elegant combination of several concepts: (i) excitation of long lived, very high Rydberg states (so called ZEKE states), (ii) field ionization and (iii) discrimination against kinetic electrons by delayed extraction between excitation and field ionization and by time-of-flight differences. The best resolution of 0.15 cm^{-1} has been achieved for NO.[19] Unfortunately, this method is not applicable to negatively charged molecules, for which no Rydberg-like states exist, and which cannot be ionized by small electrical fields.

Nevertheless, parts of the ZEKE concept can still be carried over to the resonance photoelectron spectroscopy on anions giving rise to the so-called 'anion-ZEKE spectroscopy'. Photodetached zero kinetic energy electrons are discriminated from kinetic ones by differences in steradiancy and flight time in field free drift tube. These differences are substantially enhanced by introducing a temporal delay between photodetachment and electron extraction. During this time lag, electrons with kinetic energy spread out in space while those without energy remain. After the delay (typically some hundred nanoseconds) an extraction pulse is applied which draws all electrons which are still within a small detection volume into the field free time-of-flight tube where they are detected within a small preset time window.

During a scan of the wavelength, the ZEKE electrons are released only while the exciting light is in resonance with one of the final neutral states. Therefore, narrow spectral lines are observed at anion-ZEKE spectroscopy in opposite to step functions in photodetachment spectroscopy. To emphasize once again, the accuracy of absolute energies in anion-ZEKE spectroscopy is determined by the absolute wavelength calibration of the light source, which is similar to PDS, and not by the energy accuracy of the photoelectron energy spectrometer, as at PD-PES, where the laser wavelength is fixed.

As for the short history of anion-ZEKE spectroscopy, its development[20–22] bases on the cation $\leftarrow$ neutral ZEKE method[18] and has been applied to negative ions by Neumark *et al.* for the first time.[20] One of the early systems investigated by this new method is a text book example of chemical reaction kinetics, namely I H I.[23] This experiment demonstrates one of the outstanding features of photodetachment-PES and in particular of anion-ZEKE: the possibility to study very short lived transition states which usually are not accessible for spectroscopy and even hardly for femtosecond experiments. A second class of systems which have been studied by anion-ZEKE are homogeneous complexes of metals (e.g. Au, Ag[21]), semiconductors (e.g. Si[24] or Ge[25]) or carbon (e.g. C_{60},[26] C_5,[27] C_6[28]) and heterogeneous complexes of metal and carbon (e.g. FeC_2[22,29]) or oxygen (FeO[30]). Also the third class of systems, weakly bound molecular clusters, represent examples of otherwise hardly or not at all accessible neutral species. In particular halogen containing molecular clusters have been investigated, such as $I(CO_2)$,[31] Ar_nBr, Ar_nI, KrI,[32,33] or $I(H_2O)$.[34] In the meantime, the energy resolution, which is determined by the quality of zero kinetic energy electron discrimination, could be pushed near to the magic limit of 1 cm^{-1}.[30,35]

1.3. *Photodetachment Threshold Laws*

There exists a further fundamental difference between photoionization and photodetachment processes which effects anion-ZEKE spectroscopy. While being emitted from a neutral molecule (photoionization), the electron experiences the Coulomb field of the positively charged core left behind. An electron ejected from an anion is subjected to angular momentum interaction with the residual neutral core. As a consequence we observe a steep staircase threshold for photoionization, but a much less sharp onset of the energy dependent photodetachment cross-section, which additionally depends on the angular momentum of the emitted electron. As a further consequence, anions do not possess Rydberg-analogous states which converge to neutral states (comparable to neutral Rydberg states converging to cationic states). If the permanent dipole moment of the neutral core is sufficiently large, so called dipole bound stable anion states may exist which have no analog in neutral systems.

The interaction between the detached electron and the neutral which it leaves behind follows a r^{-4} potential. The long-range interaction will then be governed by the centrifugal term in the effective potential

$$V(r) = \frac{\hbar l(l+1)}{2mr^2}\,, \tag{1}$$

($\hbar l$ = the angular momentum of the detached electron). According to Wigner[36] the leading term in the energy dependence of the photodetachment cross-section near-threshold is given by

$$\sigma_l = a(\nu - \nu_{thr})^{l+1/2} = bE^{l+1/2}\,, \tag{2}$$

with the laser frequency n; the threshold energy $h\nu_{thr}$; the center of mass electron energy E; and constants a and b. The Wigner law (Eq. (2)) has been found to yield a faithful description of the experimental cross-section behavior within a few meV above threshold for several cases of s-wave (l=0)[4,8,37] and p-wave detachment ($l = 1$).[4,37–39] In some cases, the detachment threshold may be steeper than suggested by the above equation. Since electron-dipole interaction also shows a $1/r^2$ dependence, the presence of a permanent molecular dipole moment calls for a modification of Wigner's threshold law. The OH^- data, as an example, follow a $\sigma_l = bE^{1/4}$ ($l = 0$) dependence.[40] Detailed discussions of the wavelength dependence of photodetachment cross-section for atomic negative ions can be found in the literature.[4,8,37]

Anion-ZEKE spectroscopy is strongly affected by photodetachment thresholds not exhibiting a staircase step function behavior anymore (as mentioned above) but a slowly increasing $E^{1/2}$ function (for emitted electrons with *s*-wave character) or, even worse, a $E^{3/2}$ behavior if *p*-wave electrons are emitted. The *l*-wave character of the emitted electron is determined by the rule: $l = l_{bound} \pm 1$ (with the angular momentum l_{bound} of the atomic or molecular orbital from which the electron is ejected.[41] In practice, this threshold behavior is the reason, that only detachment processes with *s*-wave emission are accessible to anion-ZEKE spectroscopy; for $l > 0$, the intensity at the threshold is vanishing. This is a restriction for anion-ZEKE on the one hand. On the other hand, these more strict selection rules allow additional information on the symmetry of involved neutral and anionic electronic states. In this sense, the comparison of anion-ZEKE and PD-PES is a very valuable tool for the interpretation of photodetachment photoelectron and anion-ZEKE spectra.

2. Experimental Principles and Details

2.1. *Experimental Principles*

In Fig. 2 the scheme of an exemplary apparatus is shown which allows performance of photodetachment photoelectron spectroscopy, photodetachment spectroscopy and anion-ZEKE spectroscopy. Conventionally, such apparatus consist of four essential components: (i) a negative ion source, (ii) a mass separator, (iii) a light source for photodetachment, and (iv) an electron energy analyzer or discriminator, respectively. While in this chapter general aspects of these four components are shortly discussed, particular details, as used in our experiment[22,29,30,42–44] are presented in the following sections.

(i) The formation of anions in the gas-phase may be achieved by a number of different processes. The most important are the combination of atomic/molecular neutral source with (a) electron attachment, (b) dissociative electron attachment and (c) ion/molecule reactions. One way to realize this is adiabatic expansion of a mixture of neutral molecules diluted in a carrier gas (typically a noble gas) and the subsequent attachment of a free low-energy electron to these molecules. A variety of techniques is known which provides low-energy electrons: (α) by focusing a laser pulse onto different kinds of metal surfaces inducing

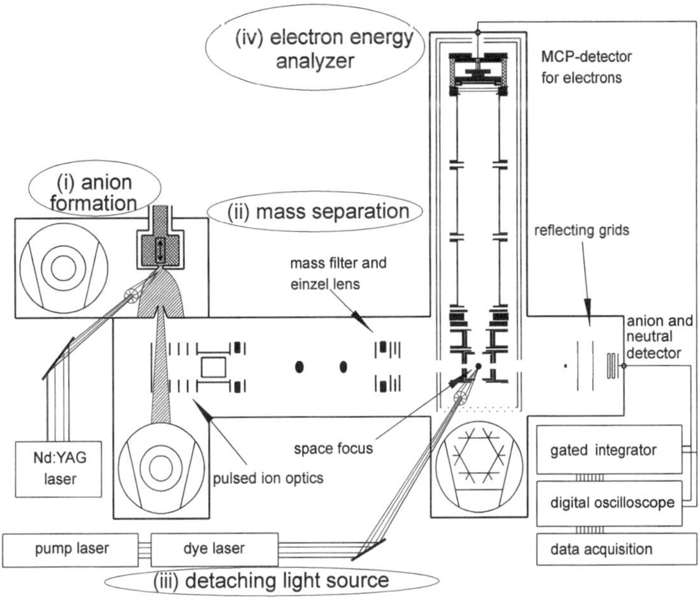

Fig. 2. Typical experimental set-up for anion-ZEKE spectroscopy consisting of (i) an anion formation chamber, (ii) a time-of-flight mass separator, (iii) an arrangement for photodetachment including a detaching light source (laser), and (iv) a time-of-flight electron energy analyzer with arrangement for zero kinetic energy electron selection.

electron emission; (β) by electric discharge sources; (γ) by negatively biased hot metal filaments which act as a thermionic electron emitter; (δ) by crossing a high energy electron beam (typically 1 keV) with a gas jet. In the latter one, the formation of negatively charged compounds results from attachment of slow secondary electrons generated by ionization of the carrier gas. Another anion source, which is particularly well suited for metal containing compounds, are sputter sources working on the basis of laser ablation combined with a gas jet. By focusing high intensity laser pulses onto a metal target, metal atoms are sputtered simultaneously with photo emission of electrons. After subsequent negative ion–molecule reaction a variety of negatively charged metal containing complexes is usually available in the molecular-beam.

(ii) Concerning the second component, various types of mass spectrometers (MS) on the basis of magnetic fields, electrostatic fields, or both have been applied for mass separation. Most commonly, linear Wiley–McLaren type time-of-flight mass analyzers[45] are used, where the accelerated ions are separated in time according to their mass-dependent flight times in a field free drift region. The so called 'space focus' is a particular point in this drift region, where the ion pulses are compressed in the direction of ion trajectories and which, therefore, represents a position of high mass separation. This is the point of choice to record mass spectra or perform photodetachment at high mass resolution. In some experiments also a space-saving quadrupole mass spectrometer or a Wien velocity filter is incorporated for mass separation. The latter has been used in the very early photodetachment photoelectron experiments[6,7] and still is used by Lineberger and coworkers.[46] That latter two means of mass separation are well adapted to continuous anion sources, while the former is optimally combined with pulsed anion sources. But this correlation is not exclusive and other combinations have been realized such as pulsed anion sources and quadrupole mass analyzers.

(iii) As third component, the photodetaching light source, lasers are used nearly exclusively. Their excellent spectral, temporal and spatial features made photodetachment photoelectron spectroscopy in general, and anion-ZEKE spectroscopy in particular possible in most cases and were crucial for the development of the whole field of research. The availability of high photon densities in narrow wavelength intervals is absolutely necessary for efficient photodetachment. One should consider that one mass separated anion pulse usually contains 10^3 to 10^4 ions and therefore only few electrons are photodetached and recorded per light pulse. Similar are the conditions for continuous anion beams. Furthermore, in time-of-flight mass separators the ion pulses pass the space focus within few nanoseconds. Only fast and very accurately synchronized light pulses allow efficient photodetachment. On the other hand, in the case of continuous ion beams optimal overlapping of ion beam and light beam is only possible for light sources with very small divergence. In addition, lasers offer a large spectrum of features, such as a wide range of pulse lengths from nanoseconds to femtoseconds and a wide range of wavelengths from the infrared to the ultraviolet. The variety of laser types applied includes solid state lasers like pulsed Nd:YAG lasers (1064 nm) and their second (532 nm), third (355 nm) or fourth (266 nm) harmonics, continuous wave argon ion lasers in the green and blue, excimer lasers in the UV and VUV, as well as wavelength tunable dye-lasers and their frequency doubled emission, Stokes Raman lines of laser light shifted in high-pressure H_2-cells, tunable Titan-Saphir lasers in the red and optical parametric oscillators (OPO)-lasers supplying wavelengths over the whole visible range and down into the IR (i.e. 4000 cm^{-1} and further).

There are two further aspects of the light source to be source to be considered. Firstly, tunable lasers are needed for PD and anion-ZEKE spectroscopy. But also for PD-PES tunable lasers are useful, since they allow to study the same neutral ← anion transition by PD-PES at different excess energies. Small excess energies result in slow electrons for which a higher energy resolution is achieved in time-of-flight analyzers. Secondly, since the polarization of the exciting light fixes the orientation of the transition dipole moment in the laboratory frame one may learn something from the angular distribution of the emitted electrons. This can be studied by simply rotating the laser light polarization and supplies additional valuable information about the molecular system under investigation. But this information is also of fundamental interest for anion-ZEKE spectroscopy, since only photodetachment transitions which lead to the emission of an *s*-wave electron (spherical emission, therefore no dependence on polarization) are detectable by anion-ZEKE.

(iv) The fourth component of a photodetachment photoelectron or anion-ZEKE spectrometer is the energy analyzer for the photodetached electrons. There exist three widely spread types of instruments. The simplest one is a time-of-flight photoelectron spectrometer consisting of a μ-metal shielded, field-free electron flight tube. Its advantage is a relatively good energy resolution. For a review of laser induced photoelectron analyzers see Ref. 47. A great disadvantage is the small acceptance angle and thus a small collection efficiency for photoelectrons. This problem is avoided in the second type of analyzer, the magnetic bottle electron spectrometer.[48] Its high collection efficiency of nearly 100% is achieved by using magnetic fields guiding the emitted electrons. However, collecting electrons from all directions has to be put up with large differences in flight paths and thus in flight times resulting in a reduced energy resolution. For detection of ZEKE electrons, the discrimination against kinetic electrons by their different flight time is not sufficient. Additional means like discrimination due to steradiancy enhanced by delayed electron extraction are necessary.

2.2. *Anion Formation of Atomic and Molecular Systems*

In Fig. 2 on the left side, the anion source is displayed. Figure 3 shows three variations of this anion source in more detail. All of them consist of a pulsed valve for formation of a supersonic beam. The cooling effect of the supersonic expansion is of fundamental importance for a satisfying high-resolution anion-ZEKE spectroscopy. In particular for molecular clusters, excitation of rotational levels with large quantum numbers and of low-frequency vibrational modes would lead to congestion of spectral lines which is not resolvable by anion-ZEKE spectroscopy. About 20-mm after the gas expansion nozzle, a skimmer is used to achieve a well defined molecular-beam as well as to allow differential pumping of anion formation chamber and ion optics chamber. Very near to the nozzle to the pulsed valve, a metal surface (e.g. wire, rod, disc) is placed and irradiated by a laser beam. This results in a pulsed electron emission. Since a laser wavelength is usually chosen which is not too far above the work function of the metal, the kinetic energy of most of the emitted electrons (space charge effects may increase the electron energy) is small enhancing direct electron attachment, dissociative attachment or formation of anionic molecular cluster which cool down and stabilize by evaporation of clustering molecules or atoms. Also chemical reactions with anionic products may occur in the dense region of the supersonic beam giving rise to a mixture of carrier gas, neutral molecules and diverse anion species.

Variant A in Fig. 3 is particularly adopted to the formation of molecular cluster anions. Here target molecules (e.g. CO_2, O_2) or precursor molecules (e.g. iodobenzene as source of iodine atoms) are seeded in the carrier gas (mostly noble gases). The whole gas mixture passes the vapor phase or even the fluid phase of the cluster forming molecular species (e.g. H_2O) and is then expanded into the vacuum of the anion formation chamber. A laser pulse which induces photoemission of electrons from a metal needle is synchronized with the gas pulse and gives rise to rich mixture of anionic molecular cluster.

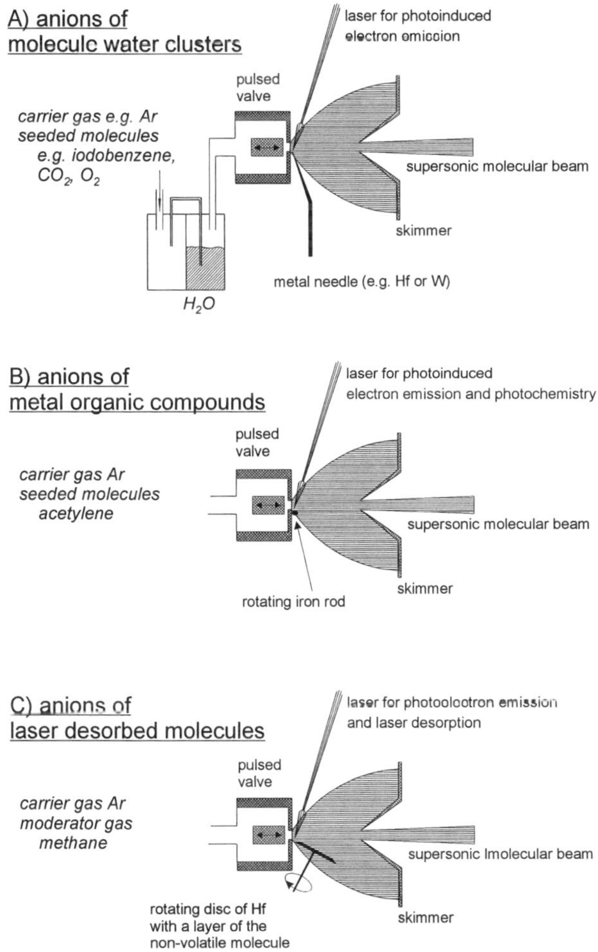

Fig. 3. Three variations of an anion source which consisits of a supersonic molecular-beam and laser-induced photoemission of electrons from a metal surface near the beam orifice.

Variant B in Fig. 3 represents the development of a cold anion source for metal organic complexes. In this case, carrier gas and seeded organic precursor molecules, mostly acetylene, are expanded through the nozzle of the pulsed valve. Metal atoms are supplied after the expansion but still in the high density region of the supersonic beam by laser ablation from a thin metal rod (e.g. iron, nickel, brass). This rod is continuously rotated and translated to avoid deterioration of the metal surface which would induce strong fluctuation of metal atom concentration in the gas-phase. Complex, mostly unknown chemical and photochemical reactions, which may take place on the metal surface or in the gas-phase, in the presence of photoemitted electrons lead to diverse anionic metal organic compounds. These may form a series of molecular clusters e.g. with the seeded organic precursor molecules or even with the carrier gas.

Variant C in Fig. 3 has been developed for investigation of molecules with low vapour pressure, e.g. polycyclic or substituted aromatics or biomolecules. Carrier gas is expanded into the anion formation chamber. In front of the nozzle of the pulsed valve a rotational metal disc has been mounted with a layer of the adsorbed nonvolatile molecule on its edge. A laser pulse is slightly focused onto this edge and induces desorption of the nonvolatile molecules. Due to the use of a disc out of hafnium or tungsten, with a laser wavelength of 266 nm from a Nd-YAG-laser, electrons are photoemitted synchronously. The so-formed molecular anions are cooled and transported into the ion source optics by the supersonic beam.

Figure 4 represents an exemplary mass spectrum of metal-carbonhydride complexes obtained by anion source B in Fig. 3. For conventional, nonmass selective spectroscopy a source of such high complexity would be

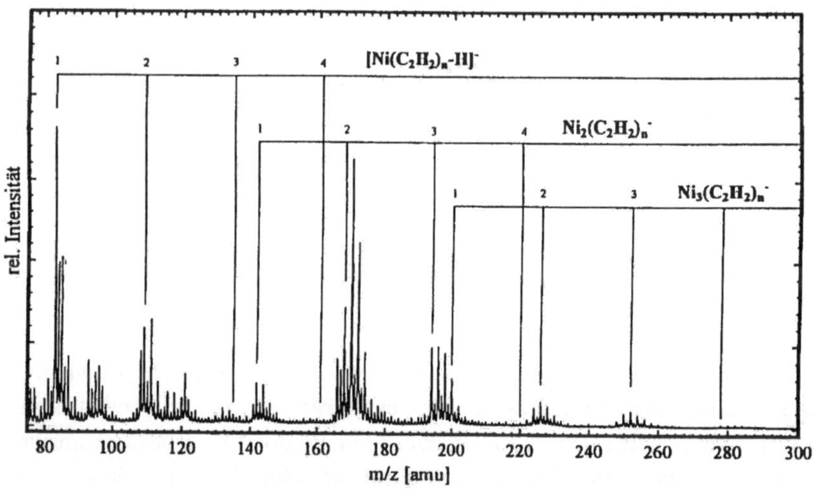

Fig. 4. Example of metal carbon hydride formation by anion source (B) in Fig. 3, illustrating the large variety of anionic species formed by this kind of anion source.

useless. Thanks to the possibility of mass separation prior to spectroscopy, photodetachment photoelectron, photodetachment and anion-ZEKE spectroscopy may benefit from this wide variety of anionic species.

2.3. *Mass Separation prior to Spectroscopy*

After anion formation the whole mixture of carrier gas, neutral and anionic species drifts down into the ion source optics of a linear time-of-flight mass analyzer (see Fig. 2). When this gas pulse arrives between the electrodes of this ion optics, a negative high voltage pulse is applied to the repelling electrode and the manifold of different anionic species is extracted into the field free drift region. After being collimated by electrodes of the ion optics itself as well as by a separate einzel lens, the bunches of anions cross the space focus of the pulsed ion source. A typical distance between both is about 1 m. This drift length is enough to separate anion bunches of different mass efficiently due to their different flight times and due to the ion bunch compression in the space focus. If the ion mass of interest is preceded or followed by very intense ion pulses of slightly different mass, these unwanted ions may cause a considerable background due to electron detachment at collisions with solid or residual gas. To avoid that problem an additional pulsed mass filter (typical mass resolution of $R = 100$) is often used in front of the point of photodetachment.

In particular for heavier anions with masses above 100 u, a good mass separations is essential for unambiguous spectroscopy. This is achieved by a careful design of the time-of-flight spectrometer. Figure 5 shows a computer simulation of the anion trajectories in an existent spectrometer[22,29,30,34,35] as well as mass peaks obtained by this same instrument. This simulation exhibits a radial focusing of the anion beam half way between pulsed ion optics and space focus at 950 mm. The longitudinal focusing (space focus) is displayed in the inset of Fig. 5 on the right on a strongly expanded scale. While the beam width is 7 mm, the distance of neighbored ion clouds ($M = 101$ to $M = 100$) is $dx_{\Delta M=1} = 10$ mm and the width of one individual ion cloud $\Delta x = 1.2$ mm. This gives a theoretical mass resolution of $R = M\, dx/\Delta x \approx 850$, which is fairly well reproduced by the experimental resolution as demonstrated by the time-of-flight spectrum in Fig. 5 on the left. The inset in the spectrum shows a group of peaks around the mass $M = 100$ u, which are due to $^{64}ZnO_2$ (96 u), $^{66}ZnO_2$ (98 u), and $^{68}ZnO_2$ (100 u),

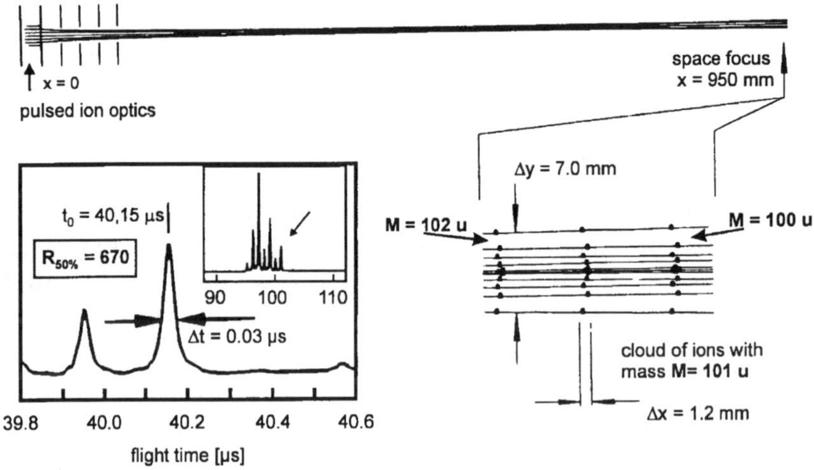

Fig. 5. Computer simulated ion trajectories starting at the ion source optics of a time-of-flight mass spectrometer. At the right: a snap shot of the ion clouds near the space focus at 950 mm after a time of about 40 μs. Ion clouds of masses 102 u, 101 u and 100 u are well separated (estimated mass resolution R $\approx$ 850). At the left: measured time-of-flight spectrum of masses 101 and 100 (measured mass resolution $R = 670$). The converted mass spectrum over a larger mass range 90 u to 110 u is shown in the inset.

as well as ^{64}ZnO(OH) (97 u), ^{66}ZnO(OH) (99 u), and ^{68}ZnO(OH) (101 u). The mass peaks $M = 100$ u and $M = 101$ u are displayed in the time-of-flight spectrum and appear at 39.95 μs and 40.15 μs and exhibit a width of $\Delta t = 0.03$ μs. An experimental mass resolution of $R_{50\%} = 670$ is deduced either from the formula $R_{50\%} = 1/2t/\Delta t$ or $R_{50\%} = M\, dt_{\Delta M=1}/\Delta t$.

2.4. *Laser-Induced Photodetachment*

Since the space focus is the position of optimum mass separation, this is the position of choice for mass selective photodetachment. This is performed by overlapping the detaching laser beam, which crosses the nion trajectories in the space focus and the ion bunch of one individual mass. To achieve that, the laser is exactly synchronized with the transit time of the anion bunch. Laser pulses of less than 10 ns with a jitter of triggering of less than 1 ns are necessary, which both are standard conditions for Nd:YAG lasers or Nd:YAG laser pumped dye lasers. The anions passing the point of detachment as well as the formed neutrals are detected by a multichannel plates detector (MCP) which is placed as near as possible to the space

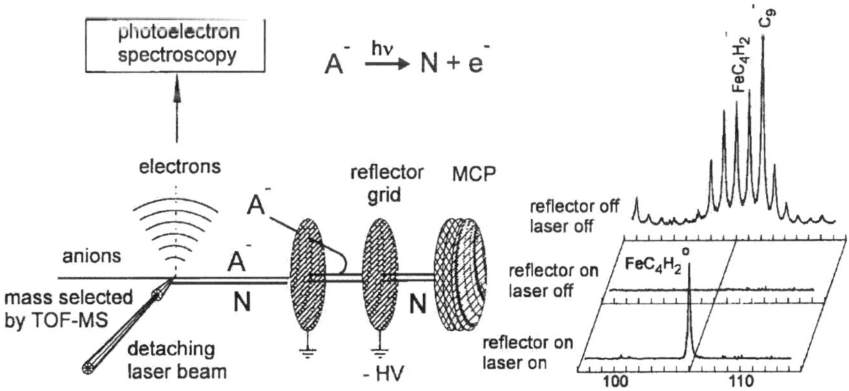

Fig. 6. Mass selective photodetachment experiment in three steps [(i) reflector off/laser off, (ii) reflector on/laser off, (iii) reflector on/laser on] illustrating the high mass selectively which is possible by the combination of laser-induced photodetachment and time-of-flight mass separation.

focus to achieve a satisfying mass resolution for anion and neutral mass spectra. To discriminate neutrals from anions, a negative high voltage can be applied to a reflecting grid so that no anions may reach the MCP and neutrals are detected solely.

In Fig. 6 an experiment is displayed which demonstrates effective mass selected photodetachment. On the left side, the scheme of a three-step experiment (reflector off/laser off; reflector on/laser off; reflector on/laser on) is presented, while on the right side the three corresponding measured mass spectra are shown. At the first experimental step (reflector off/laser off) the variety of anions A^- reach the MCP detector without any change. The corresponding mass spectrum exhibits anions of carbon and iron carbon hydrogen complexes in the range between 100 u and 110 u. At the second experimental step the reflecting grid is switched on. No anions A^- can reach the MCP detector anymore and the corresponding mass range is represented by the base line. At the third experimental step, the detaching laser is switched on additionally. In Fig. 6 its timing was synchronized with the transit time of the $FeC_4H_2^-$ anion through the space focus. Thus neutral FeC_4H_2 complexes which are formed are not affected by the reflecting field and reach the MCP detector at the same time as corresponding anions at the first experiment.

Besides the illustration of how mass selective the photodetachment process is, this experiment serves to the PD-PES or anion-ZEKE experimenter as a guarantee that the electrons formed during the photodetachment process are only due to the one preselected individual mass. These electrons now can be analyzed for their kinetic energy, giving rise to photoelectron spectra; or their collected intensity can be recorded as a function of the detaching laser wavelength for photodetachment spectroscopy; or zero kinetic energy electrons can be recorded selectively as a function of laser wavelength (anion-ZEKE spectroscopy).

2.5. *Electron Energy Analyzer/Zero Kinetic Energy Electron Discriminator*

After photodetachment the kinetic energy of the photoelectrons is analyzed or descriminated by their flight time in a drift region which is kept free of magnetic and electric fields as far as possible. Magnetic fields are shielded by a double wall metal tube as indicated in Fig. 2. Electric residual fields are often due to surface effects and are minimized by special surface treatments for which various 'philosophies' exist, e.g. graphite surfaces or clean gold surfaces.

PD-PES is performed by applying fixed detachment wavelengths and analyzing the electron energy. In this mode the place of detachment is kept free too, so that the flight time for a fixed distance (e.g. 40 cm) depends only on the square root of the electron kinetic energy. For PD spectroscopy the total electron current is measured as a function of the tuned laser wavelength. To measure the total ion current, two electrodes above and below the space focus are mounted to extract the electrons by a static electric field. For anion-ZEKE spectroscopy, in addition to a tunable wavelength several means for discrimination between kinetic and zero kinetic energy electrons is necessary. This is achieved by a delay time between photodetachment and electron collection by a small pulsed field. The idea behind it is that the kinetic electrons will spread out in a field free region after the photodetachment process. The opening in the attracting electrode will act as a steradiancy filter. The longer the delay time is the further the kinetic electrons will spread out and the better the discrimination of zero kinetic energy electrons (which in the ideal case stay at the place of photodetachment) against kinetic electrons will be. Only kinetic electrons emitted in or opposite to the direction of electron extraction are not discriminated by this measure. But they will show up in the electron signal at different flight times, since their displacement during the delay time is

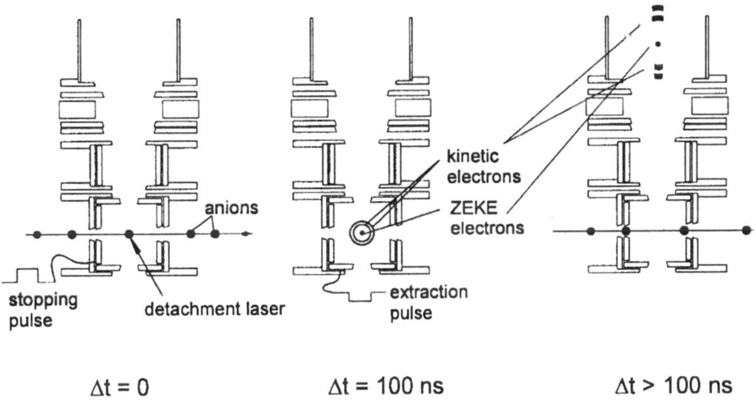

Fig. 7. Electrode arrangement for zero kinetic energy electron selection at three different times: at $\Delta t = 0$: Laser-induced photodetachment takes part followed immediately by the stopping pulse for the detached electrons. $\Delta t = 100$ ns: After a delay time of 100 ns (the kinetic electrons have spread out, but not the ZEKE electrons) ZEKE electrons are preferentially extracted by a small electrical pulse. $\Delta t > 100$ ns: After about 1 μs the ZEKE electrons have passed the field free drift tube and are separated in flight time from kinetic electrons which have spread out in or opposite to the direction of extraction.

transformed in a significant difference of potential energy in the pulsed electric extraction field.

In Fig. 7 different situations during and after photodetachment and electron extraction for anion-ZEKE experiment are displayed. The area of photodetachment is surrounded by electrodes. The electrodes on the left and the right have openings for the entering (from left) and exiting anions with different mass passing the space focus at different times. Through the electrode on the top those electrons are extracted which enter the field free drift region and are recorded by the MCP detector (see Fig. 2). Two further electrodes above and below the drawing plane, which are not shown here for clearness, have openings for the photodetaching laser beam. The different situations represented in Fig. 7 are snapshots at $\Delta t = 0$; $\Delta t = 100$ ns and $\Delta t > 100$ ns. $\Delta t = 0$ is the moment where the preselected anion bulk of one individual mass passes the space focus and the detaching laser pulse induces ejection of an electron. The electric field within the detachment region is zero. After a delay time of $\Delta t = 100$ ns an electron extraction is applied to the lower electrode. During this delay time the kinetic electrons spread out on spherical shells if *s*-wave electrons are emitted. The situation several 100 ns after accelerating the whole assembly of electrons is shown on the right side of Fig. 7. The anions as well as the neutrals formed during the detachment have moved to the right towards the anion and neutral MCP detector. The extracted electrons have reached the field free drift region. Most of the kinetic electrons have been retained by the upper electrode representing an electron optical diaphragm. Only the kinetic electrons spread out in or opposite to the direction of electron extraction pass this diaphragm but show up at different flight times than the ZEKE electrons due to their different potential in the pulsed electric field.

However, there exists one problem which has not yet been considered in this hypothetical anion-ZEKE experiment. Detached ZEKE-electrons may possess zero kinetic electron energy in the molecular frame, but are subjected to the original anion velocity in the laboratory frame. Similar arguments are valid for kinetic electrons. The 'true' ZEKE-electrons therefore move during the delay time by some tenth of a millimetre towards the anion and neutral MCP detector (to the right in Fig. 7) and cannot pass the above mentioned diaphragm anymore. The PD-PES is subject to similar problems. To solve this problem for PD-PES, the anion bunch as a whole may be decelerated from some 1000 eV to below 100 eV. In anion-ZEKE experiments the 'inherited' translational momentum of the ZEKE electrons can be compensated by a kind of stopping electrical pulse. This pulse is applied directly after the photodetachment (see Fig. 7 on the left) and produces an electric field of typically some V/cm for few nanoseconds.

In Fig. 8 the realistic situation (photodetachment of the FeC_2-anion) as observed at the oscilloscope screen is shown. Several time-resolved signals of electrons reaching the electron MCP detector are represented for different wavelengths of the photodetaching laser. If the laser is tuned from a photon energy (600 nm) above the neutral $\leftarrow$ anion resonance of FeC_2 at 605.5 nm to a photon energy (610 nm) below this resonance the following changes of the time-resolved signal will occur. At high excess energy (600 nm, no resonance) only kinetic electrons are formed which give rise to a widely separated double peak due to the electrons spread out in two directions during the delay time. During tuning to longer wavelengths the excess energy and thus the kinetic energy of the emitted electrons become smaller and smaller resulting in a decreasing gap of the two-electron signals in Fig. 7. Finally these two peaks pile up to one single peak, the anion-ZEKE peak for resonant photon energies. This signal decreases and finally vanishes below the resonance (no double peak since no related kinetic electrons are formed).

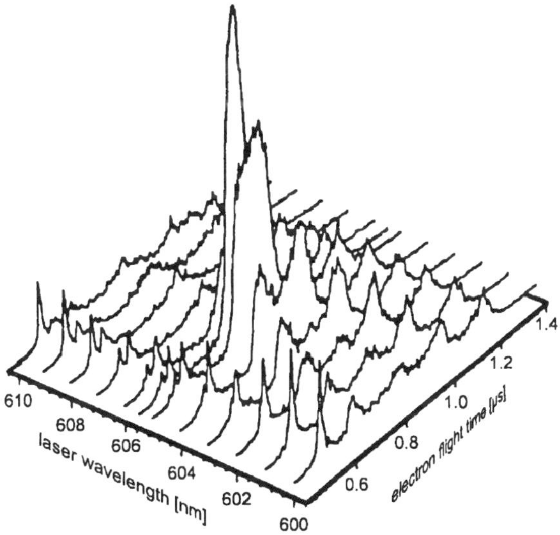

Fig. 8. Photoelectron time-of-flight spectra for different wavelengths of the detaching laser. This wavelength is tuned over a resonance of $FeC_2 \leftarrow FeC_2^-$. Starting at 600 nm and tuning to 610 nm, the electron spectra reveal a pair of kinetic electrons (due to photon excess energy) which collapse to the ZEKE peak at the resonance. (For reasons of presentation, a spectral resolution has been chosen which is not optimum.)

In addition, another pair of electron signals (at 0.6 μs and 1.3 μs) appear in the spectra which seem to be not affected by the tuning of the photon energy. The considerably wider spread in flight time is caused by a considerably larger kinetic energy due to a higher excess energy. The corresponding neutral $\leftarrow$ anion resonance of FeC_2, where this pair of peaks will collapse to an anion-ZEKE peak, lies below 625 nm. It is obvious but should be mentioned nevertheless, that the electron flight time is not related to the kinetic energy by a simple square root function, anymore. This is due to the electric extraction field and due to the delayed application of this field.

3. Examples of Mass Selected Anion-ZEKE Spectra

3.1. *Resolution of Anion-*ZEKE *Spectroscopy*

In Fig. 9 one single rotational line of the OH^--anion-ZEKE spectrum is displayed as a function of the delay time Δt. The larger the delay time is the larger will be the discrimination of kinetic and zero kinetic energy

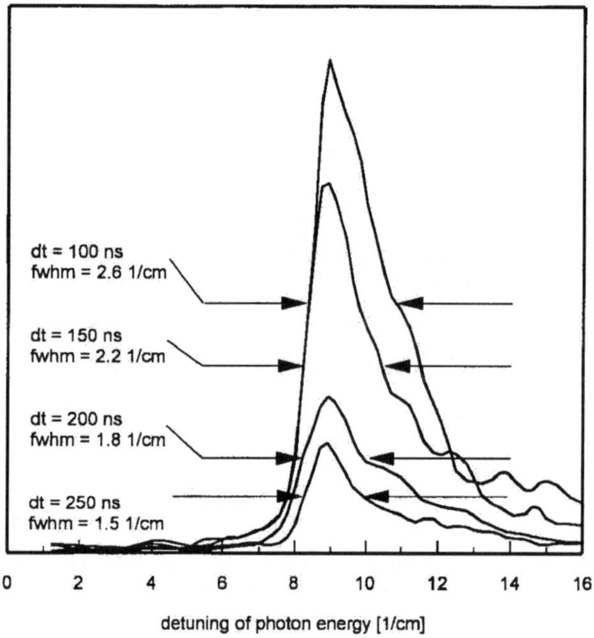

Fig. 9. The dependence of the spectral resolution of anion-ZEKE lines on the delay time between photodetachment and anion extraction (see Sec. 2.5) illustrated at the $Q3(1)$ line of the OH $\leftarrow$ OH^- anion-ZEKE spectrum (see Sec. 3.2) at 14 703 cm^{-1}.

electrons and thus the spectroscopic resolution. For a $\Delta t = 100$ ns, in this experiment a full linewidth at half maximum of 2.6 cm^{-1} has been received, while for $\Delta t = 250$ ns it decreased to 1.5 cm^{-1}. Unfortunately, this improvement of resolution has been received at the expense of intensity. This decrease is considerably stronger than proportional which is a consequence of the detachment threshold laws. While for a staircase shaped threshold a linear dependence of the intensity on the resolution may be expected (at least in principle), the $I \propto E^{l+1/2}$ dependence of detachment is responsible for this unfavorable situation. Nevertheless, a resolution in photoelectron spectroscopy which approaches the 1 cm^{-1} limit is sufficient for many new spectroscopic experiments with mass selected neutrals as final states.

In Fig. 10, two further examples of anion-ZEKE spectra are shown. In Fig. 10(a), the $^3P_2$(neutral) $\leftarrow$ $^2P_{3/2}$(anion) transition of the sulfur atom is displayed where a resolution of 1.6 cm^{-1} has been achieved. In addition, the neutral signal as a function of photon energy is shown exhibiting the

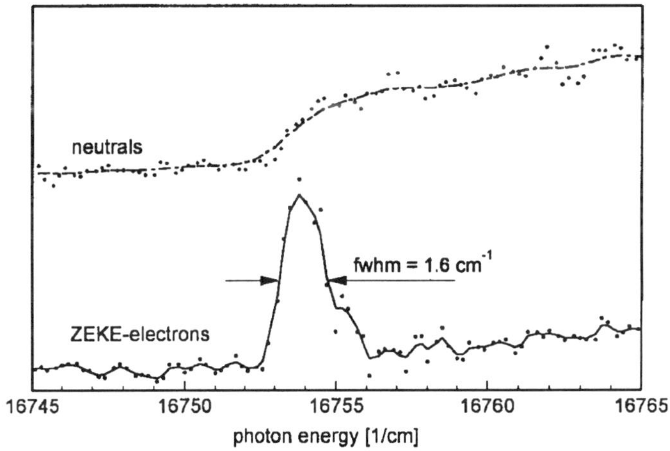

(a)

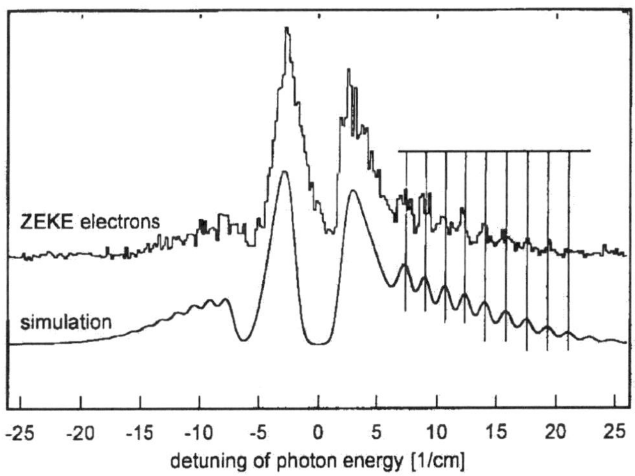

(b)

Fig. 10. Illustration of anion-ZEKE resolution: 10(a) The $S(^3P_2) \leftarrow S^-(^2P_{3/2})$ transition of the atomic sulfur anion-ZEKE spectrum exhibiting a linewidth of 1.6 cm^{-1}. Above, the photodetachment spectrum of this single atom–anion line. 10(b) The vibrationless origin of the FeO $\leftarrow$ FeO$^-$ anion ZEKE spectrum (see also Sec. 3.4) exhibiting a rotational contour with four branches $\Delta J = -3/2$, $-1/2$, $1/2$, and $3/2$. Below, the simulated band shape, which showed optimum agreement for a rotational temperature of 35 K and a convolution with a 1.1 cm^{-1} Gaussian linewidth. Please notice the indication of single rotational lines on the high energy side ($+3/2$ branch) of the experimental spectrum.

$E^{1/2}$ dependence of the stepwise increase of the intensity typical for the threshold of emitted *s*-wave electrons. However, it is easily conceivable that at higher line densities the overlapping photodetachment steps may congest the spectrum considerably and small step may be hidden by signal fluctuation of lower energetic intense steps. In principle, the absolute accuracy of the energy scale only depends on the calibration of the laser wavelength which may be two or more orders of magnitude better than that of conventional photoelectron spectrometers.

Figure 10(b) presents the rotational envelope of the electronic origin of an iron monoxide anion-ZEKE spectrum (FeO $(X^5\Delta_4) \leftarrow$ FeO$^-$ $(X^4\Delta_{7/2})$ transition). There exist four rotational branches $\Delta J = -3/2$, $-1/2$, $1/2$, and $3/2$ due to the four possibilities to combine rotational transition rules with spin up or down of the emitting *s*-wave electron. While the branches $\Delta J = \pm 1/2$ give rise to the narrow bands in the center of the envelope, the branches $\Delta J = \pm 3/2$ are represented as wings on the low and high energy side of the structure. The high energy $\Delta J = +3/2$ even shows indications of single rotational lines. A simulation with a Gaussian linewidth of 1.1 cm^{-1} and a rotational temperature of 35 K gave the best fit of the experimental curve. This demonstrates that anion-ZEKE spectroscopy is able to resolve rotational envelopes which have been used for long time in spectroscopy to determine rotational constants of small and medium sized molecules with high accuracy.

3.2. *Anion*-ZEKE *Spectrum of the OH Radical*

The OH radical is involved in many chemical reactions and one of the most important radicals in the upper and lower atmosphere as well as in combustion processes. It is one of the spectroscopically best investigated molecules.[49] Thus it is the first radical to be characterized by microwave spectroscopy in the gas-phase.[50] Photodetachment experiments on the OH anion has also a long history starting with the pioneering work of Smith and Branscomb.[51] They studied photodetachment cross-sections and observed an increase at 1.78 eV (14 355 cm^{-1}). In 1966 Branscomb could improve the resolution up to 200 cm^{-1} by using a monochromator instead of frequency filters. The missing of a vibrational progression and the very small isotope shift already indicated a very similar bond length of the anion and the neutral. This has been confirmed by two groups[52,53] using photodetachment photoelectron spectroscopy. A resolution of 5 to 7 meV has been achieved by Hotop *et al.*[53] revealing the P, $Q_{3/2}$, $Q_{1/2}$, and R rotational branch of the OH ← OH^- transition. They even found the first vibrational band however with intensity smaller by a factor of 150 indicating a change in bond length by 0.0011 nm and of the rotational constant B_0 by 0.4 cm^{-1}.

Concerning resolution, the development of photodetachment spectroscopy has reached a top position under spectroscopic methods involving photodetachment processes. As for the hydroxy radical, in a first step,[54] a resolution of 25 cm^{-1} has been reached already indicating the appearance of new photodetachment channels. In 1982, an improvement by two orders of magnitude was obtained.[40] This allowed for the first time to resolve single rotational levels as steps in a photodetachment spectrum. Even the shape of the steps could be analyzed and a cross-section $\sigma \propto (E - E_{thr})^{1/4}$ dependence has been found, which is a steeper threshold than predicted by Wigner's threshold laws. An explanation is the coupling of the angular momentum interaction and the large dipole moment of OH. Recently, photodetachment spectra near the threshold of single rotational lines of the OH ← OH^- transition have been published with an excellent resolution below 0.01 cm^{-1}.[55] Now, even L-doubling, the splitting of rotational lines due to interaction of rotational and electronic angular momentum, could be resolved with a very interesting and, at the first glance astonishing threshold behavior. Since both components of one rotational level differ by parity, only one of these components is accessible by photodetachment due to selection rules. It turned out, that the threshold behavior differ significantly depending on whether the lower or higher energetic component is excited. For the lower energetic one nearly a $(E - E_{thr})^{1/2}$ was found, in other words the outgoing electron does not feel the permanent dipole moment of the neutral OH core. The theorical explanation is that any state having a definite parity cannot possess a dipole moment.[55] The more visual explanation is that a dipole which rotates around an axis perpendicular to its orientation (no rotational component around the dipole axis as is the case for single states of a L doublet) does not exhibit a permanent dipole moment in the laboratory frame anymore.[55] This is an illustrative example how selection rules and threshold behavior (which also affect anion-ZEKE spectra by the way) deliver additional information about fundamental physical questions.

With view of all this highly sophisticated spectroscopic data on the hydroxide anion and its neutral radical counterpart it is a challenge to test the possibilities of anion-ZEKE spectroscopy of this molecule. Before

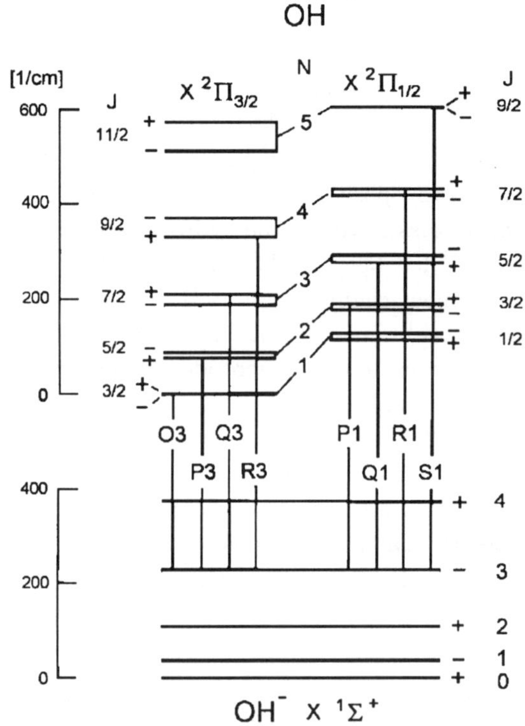

Fig. 11. Level scheme of $OH(X^1\Sigma^+)$ and $OH^-(X^2\Pi)$[40] with total angular momentum J, mechanical angular momentum N and parity + or −. The $OH^-(X^2\Pi)$ levels show Λ-doubling (splitting is expanded by a factor of 50). All possible photodetachment transitions starting at the $J'' = 3$ level of $OH(X^1\Sigma^+)$ are marked.

presenting this spectrum, energy levels and possible transitions of neutral and anionic OH will be discussed. In Fig. 11, a corresponding scheme is displayed. The rotational levels of the $X^1\Sigma^+$ ground state of OH^- are characterized by the quantum numbers $J = 0, 1, 2, 3, 4 \ldots$ of the total angular momentum with the corresponding parity + or − for even and uneven quantum numbers, respectively. Much more complicated are the rotational levels of the neutral radical. The $X^2\Pi$ ground state splits into the two spin-orbit components $X^2\Pi_{3/2}$ and $X^2\Pi_{1/2}$ with the mechanical rotational quantum numbers $N = 1, 2, 3, 4, \ldots$ and the total angular momentum $J = 3/2, 5/2, 7/2 \ldots$ and $J = 1/2, 3/2, 5/2 \ldots$, respectively. In addition, single rotational lines are subject to Λ-doubling with alternating parity of the sublevels and increasing (for $X^2\Pi_{3/2}$) respectively decreasing (for $J < 9/2$ of $X^2\Pi_{1/2}$) splitting of $OH^- \leftarrow OH$. The splitting is represented in Fig. 11 with 50-fold expansion. Transitions are only possible between levels of equal parity. All possible transitions starting from the $J''(OH^-) = 3$ level are marked in Fig. 11. Due to $\Delta N = -2, -1, 0, +1 +2$ O, P, Q, R, and S branches are observable. Transitions to the $X^2\Pi_{3/2}$ are characterized by the number 3, those to the $X^2\Pi_{1/2}$ by the number 1. The same notation is used in Fig. 12; in addition, the J quantum number of the initial state is given in brackets. The notation $R3(0)$ is therefore used to mark transition $X^2\Pi_{3/2}(J = 3/2, \text{parity}+)(OH) \leftarrow X^1\Sigma^+(J = 0)(OH^-)$. This is the transition between the energetically lowest levels of OH^- and OH and thus corresponds to the adiabatic electron affinity (EA) of OH.

The anion-ZEKE spectrum of the $X^2\Pi_{3/2}(OH) \leftarrow X^1\Sigma^+(OH^-)$ transition is shown in Fig. 12 together with its assignment. Single rotational lines appear in the spectrum with a line width (FWHM) of 1.5 cm^{-1} (due to a delay time of 250 ns; see Secs. 2.5 and 3.1). From intensity distribution, a rotational temperature of 30 K is estimated. The energy scale (laser wavelength) has been calibrated by a simultaneously measured spectrum of molecular iodine. The measured value of EA = 14 740.3 cm^{-1} is in very

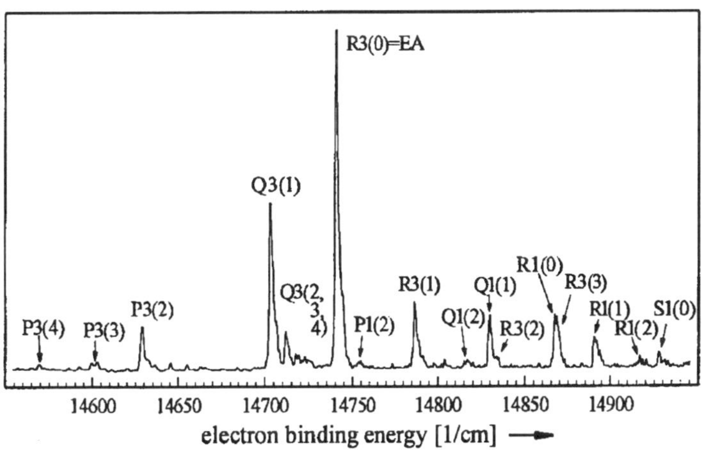

Fig. 12. Rotationally resolved anion-ZEKE spectrum of OH^-: the $OH(X^1\Sigma^+) \leftarrow OH^-(X^2\Pi)$ transition. The wavelength of the photodetaching laser has been tuned by steps of 0.3 cm^{-1} and calibrated by an iodine spectrum recorded simultaneously. The rotational temperature is 30K. Assignments of the single rotational lines are given. See also Table 1.

Table 1. Transitions in the anion-ZEKE spectrum of OH^-.

Assignment	J′	Energy [cm^{-1}] experimental	Energy [cm^{-1}] $-R3(0)$ [cm^{-1}]	$E0_{J'}$(OH) $E0_{J''}$(OH^-)	Rel. Intensity $I/I(R3(0))$
$P3(4)$	7/2	(14569,3)	−171	−172, 1	0,018
$P3(3)$	5/2	14599,8	−140, 5	−140, 9	0,026
$P3(2)$	3/2	14628,8	−111, 5	−112, 4	0,130
$Q3(1)$	3/2	14703,0	−37, 5	−37, 4	0,490
$P1(3)$	3/2	(14703,5)	−36, 8	−36, 8	–
$Q3(2)$	5/2	14711,5	−28, 8	−28, 5	0,109
$Q3(3)$	7/2	14718,6	−21, 7	−22, 2	–
$Q3(4)$	9/2	(14722,6)	−17, 7	−18, 1	0,031
$Q3(5)$	11/2	–	–	−16, 9	–
$R3(0)$ = **EA**	3/2	14740,3	0	0	1,000
$P1(2)$	1/2	(14754,8)	14,5	14,1	0,023
$R3(1)$	5/2	14786,7	46,4	46,2	0,193
$Q1(5)$	9/2	–	–	47,7	–
$Q1(4)$	7/2	–	–	55,3	–
$Q1(3)$	5/2	(14803,9)	63,6	64,2	0,029
$Q1(2)$	3/2	(14815,1)	74,8	75,1	0,021
$Q1(1)$[a]	1/2	14829,2[a]	88,9	88,8	0,113[a]
$R3(2)$[a]	7/2	14829,9[a]	89,6	89,5	0,156[a]
$R1(0)$[b]	1/2	14867,1[b]	126,8	126,5	0,159[b]
$R3(3)$[b]	9/2	14870,0[b]	129,7	130,5	0,095[b]
$R1(1)$	3/2	14890,4	150,1	150,3	0,093
$R3(4)$	11/2	–	–	169,6	–
$R1(2)$	5/2	14917,0	176,7	176,6	0,037
$S1(0)$	3/2	14927,6	187,3	187,5	0,047
$R1(3)$	7/2	–	–	204,5	–
$S1(1)$	5/2	–	–	251,3	–

[a], [b]: pairs of overlapping lines.

good (with respect of our resolution) agreement with the high-resolution value of 14741.02(3) cm^{-1}.[40] The specific benefit of anion-ZEKE spectroscopy is obvious from this spectrum. Every single rotational transition gives rise to a distinct line and not a stepwise increase. In particular, on the high energy side of Fig. 12, this is crucial for the unambiguous identification of single lines with relatively small intensities. In the photodetachment spectrum, these transitions are represented as small steps of the photodetachment cross-sections on the intense background of all transitions with lower energy. Small steps will be buried under fluctuations of this intense background signal and will then not be identified anymore. In addition to detectability of these lines, anion-ZEKE spectroscopy also will supply more reliable values on their relative intensity.

This is not only important for the spectroscopy of the molecular system under investigation, but also for principal understanding of the photodetachment process itself. For photodetachment of atomic anions one assumed the formation of an *LS* coupled, excited anionic complex which decays into a *jj*-coupled electron-neutral molecule state.[56,57] Similarly for a diatomic molecule the photon absorption has been dealt as an $^1\Sigma - {}^1\Pi$ transition with four open channels of electron emission from the final $^1\Pi$ state. For the photon absorption, then, the well-known Hönl–London factors combined with the Boltzmann distribution describe the rotational transition moments.[58,59] However, these transition probabilities together with those of the further transition to the final neutral state do not represent the intensity distribution of the anion-ZEKE spectrum in Fig. 12. Obviously, the photodetachment process has to be described by a more general formalism of angular momentum coupling[60,61] and/or further arguments (e.g. different threshold behavior) have to be considered.

A check for the reliability of anion-ZEKE spectroscopy is the comparison of measured anion-ZEKE line positions with theorectical energy levels. In Table 1 this comparison is presented. In the first three columns, assignment, final J quantum number, and measured frequency (correspond to electron binding energy) of the single lines in Fig. 12 are given. Column four gives the line shifts relative to the measured adiabatic electron affinity [$R(3)0$ line]. In the next column, the theoretical shifts determined from highly accurate parameters of the neutral[49] and the anion[40] are displayed, while the last column shows the relative intensities of the anion-ZEKE lines. In most cases the deviation of the experimental value is less than half a wavenumber which is reasonable in respect of the resolution and the very low intensity of some lines.

3.3. *Anion-*ZEKE *Spectrum of Metal Carbon Complexes:* FeC_2

Metal carbon complexes, as well as their hydrides, play an important role for heterogeneous catalytic reactions of hydrocarbons on metal surfaces (e.g. acetylene on iron).[62–64] In order to understand these processes it is necessary to study those intermediate molecular species. As an example, acetylene reacting on an iron surface gives rise to a multitude of intermediate $Fe_xC_yH_z$ complexes, which are precursors of the final catalytic products, e.g. polymers of the primary molecule. The unequivocal study of these molecular complexes on surfaces is mostly not possible due to the many different sites and geometric arrangements a primary molecule may find on the metal surface. For some small primary molecules, such as CO, surface spectroscopic techniques (e.g. high-resolution electron energy loss spectroscopy (HREELS))[65] help to characterize the different possible intermediates. As soon as more complex intermediates attach to the surface it becomes impossible to discriminate single ones. For a better insight into catalytic mechanisms it is therefore valuable to isolate possible single intermediate complexes in the gas-phase and to study them separately. Several spectroscopic and spectrometric techniques have already been applied to isolated iron carbon complexes such as mass spectrometry,[66] FTIR-spectroscopy in noble gas matrices,[67–69] or ion mobility spectroscopy.[70]

Photodetachment photoelectron spectroscopy is one of the most promising approaches to this problem, which is two-fold: (a) isolation of single intermediate species and (b) appropriate spectroscopy to study them. It has also been applied to isolated $Fe_xC_yH_n$ complexes.[22,29,71,72] In particular, anion-ZEKE spectroscopy is here outstanding due to its high-resolution and its different selection rules. We have studied one of the smallest complexes of the iron/acetylene system: FeC_2 by anion-ZEKE spectroscopy. FeC_2 is formed in a gas-phase or near-surface reaction of iron atoms, which were due to laser ablation from an iron wire, and acetylene seeded in the supersonic beam. Anions are formed due to additional laser induced electron emission from the iron wire. For more detail see Sec. 2.2. A large variety of different negatively charged species is thus obtained, containing all kinds of intermediates composed of iron, carbon and hydrogen.

In Fig. 13, anion-ZEKE spectra of the FeC_2 anion are represented delivering spectroscopic information about the neutral FeC_2. In addition, Fig. 13 shows a well-resolved photodetachment photoelectron spectrum of this species. Due to the necessity of changing laser dyes for different wavelength ranges and of keeping the anion source constant for considerably longer times than for the photodetachment PE-spectrum, the single ZEKE spectra are much shorter than a single PD-PE spectrum. In addition, the

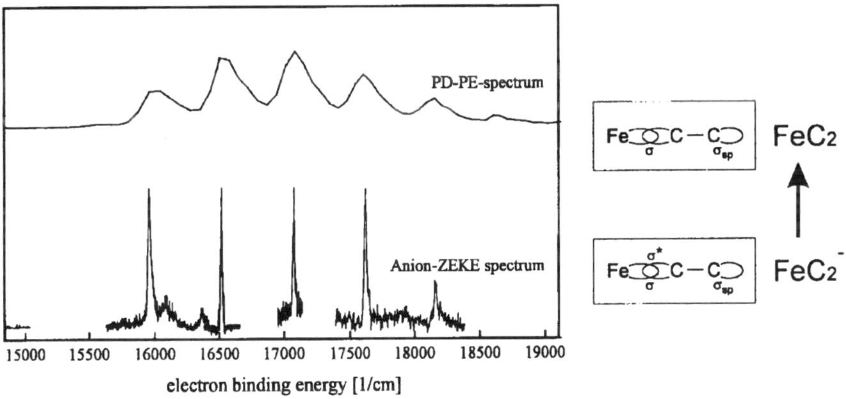

Fig. 13. Photodetachment photoelectron (PD-PE) and anion-ZEKE spectrum of FeC_2^-. The PD-PE spectrum has been taken with a laser wavelength of 355 nm. The anion-ZEKE spectra consist of several single spectra. (Due to insufficient overlap the relative intensities are unknown and therefore the single spectra are normalized for the same maximum intensity.) On the right side, the electron configurations of FeC_2^- and FeC_2 are presented. The additional antibonding σ^* electron in FeC_2^- results in a less strong F–C–bond and a larger F–C–distance. This gives rise to the broad Franck–Condon envelope of the progression in the $FeC_2 \leftarrow FeC_2^-$ PE spectrum.

complexity and therefore long-term reproducibility of the anion source did not allow to reliably compare the anion-ZEKE spectra of FeC_2 taken at different time. Therefore, the single ZEKE spectra are presented on a normalized intensity scale. The PD-PE spectrum has been taken at a laser wavelength of 355 nm, for the ZEKE spectra the laser wavelength has been tuned between 650 and 540 nm. The electron binding energies (energy scale in Fig. 13) are determined from the flight times (representing the kinetic energies) of the detached electrons (PD-PES) or the detaching laser wavelength (PD-ZEKE). The latter gives much more accurate values, being nearly independent of disturbing fields (e.g. of charged surfaces, space charges etc.). Therefore, one valuable advantage of ZEKE spectroscopy is the possibility to calibrate photoelectron data precisely. This is of particular importance for FeC_2 and similar compounds since for them no other relevant spectroscopic data (i.e. from photodetachment spectroscopy) are available up to now. Of course, the main advantage of high-resolution ZEKE is to reveal or discriminate substructures (e.g. spin-orbit components) which may either shift the centroids of PES-peaks resulting in wrong molecular constants or congest the PES-spectra totally (e.g. for larger members of the iron/acetylene system).

The spectra in Fig. 13 are due to transitions to the neutral X exhibiting a broad Franck–Condon envelope of the Fe–C stretching frequency. This envelope may be interpreted as a strong change of the Fe–C bond length during the neutral $\leftarrow$ anion transition. Considering possible electron configurations of Fe and C_2, this strongly suggests a $3d^6\sigma^2\sigma_{sp}^2$ electronic configuration of the neutral FeC_2 complex in its electronic ground state with the extra electron of the anion in the σ^* orbital of the Fe–C bond (see Fig. 13). The spectroscopic results deduced from the anion-ZEKE spectra in Fig. 13 are (i) the adiabatic electron affinity EA = 15 955 cm^{-1}, (ii) the iron carbon stretching frequency with $\omega_e = 561.6$ cm^{-1} in the neutral X ground state, (iii) a small anharmonicity constant $\omega_e x_e$ of 2.1 cm^{-1} (deduced from four quanta of this vibration), and (iv) several small peaks which are tentatively assigned to spin-orbit components. From the Fe–C stretching frequency and the anharmonicity of this vibration a dissociation energy of the iron carbon bond of 37 500 cm^{-1} or 450 kJ/mol could be estimated.

3.4. *Anion*-ZEKE *Spectrum of Metal Oxides:* FeO

Discussing metal carbon complexes, one should also mention other small metal containing complexes where anion-PES contributed important information, namely metal oxides. It is well-known for transition metal oxides that many low-energetic electronic states with high multiplicity as well as numerous perturbations make interpretation of their spectra very difficult. FeO is a typical case for these problems. Many types of spectroscopy have been applied to it such as ESR-spectroscopy,[73] chemiluminescence,[74] matrix-IR spectroscopy,[75], IR-emission spectroscopy,[76] fluorescence spectroscopy,[77] photodetachment photoelectron spectroscopy (PD-PES),[78,79] autodetachment spectroscopy[80] or high-resolution FT-fluorescence emission spectroscopy,[81] but nevertheless questions even about the electronic ground state were still open. Recent anion-ZEKE spectroscopy[30] revealed new details on the neutral ground and low electronically excited states of FeO. As for other metal oxides (e.g. CrO[82]), mass selected photoelectron spectroscopy in general turns out to be a very valuable and unequivocal spectroscopic method for such problematic systems.

So far it is certain that the neutral ground state of FeO is a ${}^5\Delta_i$ state with the electron configuration $(4s\sigma)^1(3d\delta)^3(3d\pi)^2$. But there was a discrepancy concerning the vibrational frequency of this electronic state. From photodetachment photoelectron spectra (PD-PES) a frequency of 970 cm^{-1} was found,[78,79] while high-resolution IR-spectroscopy supplied a value of 871 cm^{-1}.[83] In addition, a neutral $A^5\Sigma^+$-state with the electron configuration $(4s\sigma)^2(3d\delta)^2(3d\pi)^2$ has been found in the PD-PES spectra[78,79] with an energy of about 3990 cm^{-1} above the X-ground state; this A-state is not accessible in optical spectra and therefore has not been observed there. Finally, the anionic ground state $X^4\Delta_{7/2}((4s\sigma)^2(3d\delta)^3(3d\pi)^2)$ is involved in photodetachment processes. One advantage of photodetachment experiments is that selection rules are not as strict as for neutral to neutral state transitions and therefore may reveal more electronic states. The main problem of PD-PES is that resolution is often not sufficient. It turned out that this was the main reason for the discrepancy of FeO vibrational frequencies in the neutral ground state.

In Fig. 14 at the top, a well-resolved photodetachment PE spectrum is shown for which a detachment laser wavelength of 532 nm has been used.

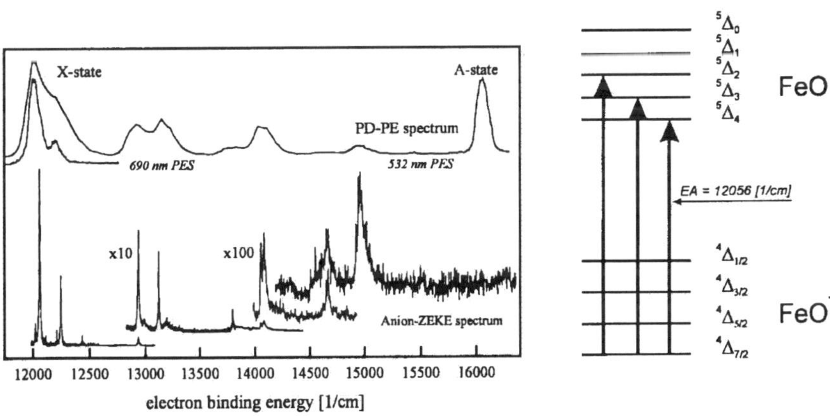

Fig. 14. Photodetachment photoelectron (PD-PE) and anion-ZEKE spectrum of FeO^- revealing progressions of the transitions $FeO(X^5\Delta) \leftarrow FeO^-(X^4\Delta_{7/2})$ and $FeO(a^7\Sigma^+) \leftarrow FeO^-(X^4\Delta_{7/2})$(see text), as well as a PE-band due to $FeO(A^5\Sigma^+) \leftarrow FeO^-(X^4\Delta_{7/2})$. The PD-PE spectra have been taken with laser wavelengths of 532 nm and 690 nm and shifted by 100 cm^{-1} for best overlap of the X-bands of PE- and anion-ZEKE spectrum. On the right side, the ground states of FeO^- and FeO with their spin-orbit sublends are shown. All possible FeO $\leftarrow$ FeO^- transitions starting at the anionic $X^4\Delta_{7/2}$ ground state are indicated.

Two dominant bands at about 12 000 and 16 000 cm^{-1} appear in this PE spectrum which appeared as strong bands in other photodetachment PE spectra.[78,79] They have been assigned to FeO($X^5\Sigma^+$) $\leftarrow$ FeO$^-$($X^4\Delta_{7/2}$) and FeO($A^5\Delta_4$) $\leftarrow$ FeO$^-$($X^4\Delta_{7/2}$) transitions. Their energies correspond to the electron binding energy of the neutral ground state (adiabatic EA) and the neutral excited electronic state A. While the latter is accompanied by a very weak vibrational band at about 16 900 cm^{-1} (not shown here) the former origin shows an intense vibrational progression revealing a splitting of the four bands between 12 000 and 15 000 cm^{-1} which is not resolved in earlier photodetachment PE spectra.[78,79]

For higher resolution, one actually expects a splitting due to the transitions from the anionic $^4\Delta_{7/2}$ ground state to the three accessible spin-orbit components $^5\Delta_4$, $^5\Delta_3$, $^5\Delta_2$ the neutral ground state. A first indication of these spin-orbit components may be detected on the high energy side of the origin at 12 000 cm^{-1}. To increase the resolution, photodetachment with a laser wavelength of 690 nm has been performed (see Fig. 14). Now, at least the $^5\Delta_4$ and $^5\Delta_3$ components are revealed. Their intensity distribution, however, do not represent the true ratios since they were subject to energy dependent photoelectron detection cross-sections as known for time-of-flight energy analyzers. These cross-sections decrease strongly for slow electrons (as in the case of the 690 nm spectrum). Similar intensity patterns are expected for the vibrational progression bands.

However, the splitting of these progression bands in the 532 nm PE spectrum does not show the behavior expected for the spin-orbit components. Instead, the ratios of the intensities of the two band maxima change along the progression with the more energetic component gaining more and more relative intensity. It is obvious that the double headed band structure in Fig. 14 is not due to spin-orbit components of single vibrational bands of one progression but most likely to a second progression band system with a different origin due to a new neutral excited electronic state. Such a state has already been supposed by Merer *et al.*[81] because of perturbations of single rotational lines in the infrared spectrum due to dark states (not accessible to IR spectroscopy), as well as by theoreticians who expected a low energetic state $a^7\Sigma^+$(FeO).[84]

In Fig. 14 in addition to the photodetachment PE spectrum, high-resolution anion-ZEKE spectra are shown.[30] Due to overlapping energy ranges these spectra could be calibrated relatively to each other. Now, all three accessible spin-orbit components of the (0,0)-band are fully resolved and confirm that the PES-band at 12 000 cm^{-1} is due to the $^5\Delta_4$, $^5\Delta_3$ and $^5\Delta_2$ spin-orbit components of the neutral X-state. In addition, weak lines appear in the anion-ZEKE spectrum (e.g. at 12 018, $\approx$ 12 110, $\approx$ 2160, and 12 208 cm^{-1}). These bands can be explained by transitions between the initial $\Omega = 5/2$, $3/2$, and $1/2$ spin-orbit components of the anionic ground state (weakly populated) and the final $\Omega = 3$, 2, 1, and 0 spin-orbit components of the neutral ground state. Together with values from autodetachment work[80] all spin-orbit splittings of the anionic and the neutral ground states can be determined. The values for the neutral spin-orbit splittings show good agreement with high-resolution IR spectroscopy[81] although the assignment differs by one quantum number.[30]

In contrast to the photodetachment PE spectrum, the anion-ZEKE spectra exhibit the expected spin-orbit components not only for the (0,0)-band, but also for the (1,0)-, and the (2,0)-bands. The proposed second progression system seems to be missing in the anion-ZEKE spectrum. A more thorough study of the spectrum reveals a small peak at 13 198 cm^{-1} and a correlated series of following peaks which correspond the second head of the (1,0)-photodetachment PE-band and its progression. Obviously, this new progression appears with a considerably smaller intensity in the ZEKE than in the PE-spectrum. This is a further indication for a new electronic state with a symmetry different from that of the X ground state giving rise to different intensities due to different selection rules.

To summarize the results, we found two progressions, one with ω_e = 882 cm^{-1} and an anharmonicity $\omega_e x_e$ = 4 cm^{-1} and one with ω_e = 887 cm^{-1} and $\omega_e x_e$ = 2.5 cm^{-1}. This explains the discrepancy between photodetachment PES and IR spectroscopy in the literature and confirms the IR value of 880.4 cm^{-1}. In addition, we assign the new origin at 13 198 cm^{-1} and its progression to the $a^7\Sigma^+(FeO) \leftarrow X^5\Delta_{7/2}(FeO^-)$ transition. Thus, a binding energy of the $a^7\Sigma^+$ of 1144 cm^{-1} has been found. There is a major problem with the $a^7\Sigma^+$ state, however. A band due to $a^7\Sigma^+(FeO) \leftarrow X^5\Delta_{7/2}(FeO^-)$ should not appear in any photoelectron spectrum since this is a two-electron transition: the electron configuration of the FeO^- ground state is $(4s\sigma)^2(3d\delta)^3(3d\pi)^2$, while that of the neutral ground state is $(4s\sigma)^1(3d\delta)^2(3d\pi)^2(3d\sigma)^1$. A possible explanation is extraordinary strong configuration interaction resulting in strong state mixing. Very strong state mixing is predicted for higher electronic states. But obviously, even the low energy region is subjected to it. This may, furthermore, influence the character of the outgoing electron behaving neither as a pure s-, nor as a pure $l = 1$ or $l > 1$ wave and thus induce the strong difference of intensity for photodetachment PE and anion-ZEKE spectroscopy.

There is another line in Fig. 14 where the ZEKE and PES spectra differ significantly. The strong A-band in the PES-spectrum has totally vanished in the ZEKE spectrum. This is in agreement with the dependence of the PES-band on the polarization of the detaching laser light which is not shown here. While changing the polarization of the detaching laser light from parallel (with respect to the direction of the observed emitted photoelectrons) to vertical the intensity of the A-band collapses by a factor of 5 relative to the intensity change of the X-band system (origin and progression). Thus, the polarization ratios $\perp$ to $\|$ are found to lie between 0.33 and 0.25 for the bands $X(v,0)$ and 0.07 for the $A(0,0)$ band. This difference in polarization may be expected since the A-state has a different symmetry from the X-state. Excitation of the A-band causes the ejection of a $(3d\delta)$ electron and an emitted electron wave with $l \geq 1$ ($l = l_{\text{bound}} \pm 1$/see Sec. 1.3). As discussed above, for $l > 0$ one expects a very slowly increasing cross-section (e.g. $\sigma \propto E^{3/2}$) and therefore a vanishing ZEKE intensity. This is the case for the A-band in our anion-ZEKE spectrum as mentioned above and is an illustrative example for the complementing selection rules of photodetachment-PE and anion-ZEKE spectroscopy.

3.5. *Anion-ZEKE Spectrum of Heterogeneous Water Clusters:* $I^-(H_2O)$

Halogen water clusters are of particular interest for environmental chemistry (e.g. in the stratosphere) as well as for fundamental research on van der Waals, atom dipole and ion dipole interaction and thus for effects of solvation. Three regimes of cluster size may be distinguished: large cluster sizes with totally solvated halogen atoms and ions, small cluster sizes with $n = 2$ to about $n = 15$, and clusters of molecules or atoms with one single water molecule. The spectroscopy of halogen water clusters, even of the smallest one, is difficult for several reasons. It is not possible to produce selected neutral water clusters $M(H_2O)_n$ with discrete n-number and (for $n > 1$) with discrete isomeric structure; the available concentrations of specific $M(H_2O)_n$ systems are very small, and the study of the primarily interesting ground state would only be possible by IR spectroscopy in a wavelength region where no appropriate tunable lasers are available. One solution of this problem is to start the experiment with anions instead of neutrals and to apply photodetachment PE and anion-ZEKE spectroscopy to it. Low-resolution photoelectron spectra have been published about water clusters of all three size regimes containing iodine, chlorine or bromine.[85] The only high-resolution anion-ZEKE spectrum of molecule water clusters (i.e. of $I^-(H_2O)$)[34] will be discussed in the following after referring to theoretical and experimental work related to the iodine water cluster.

There is some theoretical work on the smallest halogen-water clusters (with $n = 1$) Kaldor[86] and Jortner[87] gave structure values and ionization energies for $I^-(H_2O)$ and Jortner *et al.*[87] some information about $I^-(H_2O)$, such as the charge distribution or the angle between the H–O–H-and the I–O–H bonds. Nearly no information exists about neutral $Cl{-}H_2O$ and $I{-}H_2O$ clusters. Valuable insight may therefore be obtained by considering the spectroscopy of other I^--clusters such as $I^-(CH_3I)$,[88] $I^-(CO_2)$,[31] I^-Ar, and I^-Kr.[32] For all these systems, anion-ZEKE spectra have been published. The spectra of all four systems are dominated by the energy gap of the first $(X_{1/2})$ and the third electronic state $(II_{1/2})$ of the neutral cluster, which approaches the spin-orbit splitting of iodine atoms by few per cent. In addition, the degeneracy of the lower of the two components $(^2P_{3/2})$ in iodine is lifted for the clusters, resulting in a $X_{1/2}$ and $I_{3/2}$ state (for nomenclature see Ref. 28). This splitting is smaller than the spin-orbit splitting by a factor of 50 and is a hint for the strength of the interaction of the clustering molecule with the iodine atom. Finally, the blue shift of the lowest anion-neutral transition in the cluster (compared with that in the iodine) delivers information about the binding energy of the anionic and the neutral cluster.

As an example, for $I^-(CO_2)$,[31] the splitting of the $I(^2P_{3/2})(CO_2)$-state was found to be 225 cm^{-1}, a mean energy gap $\{1/2(I_{3/2} + X_{1/2}) - II_{1/2}\}$ corresponding to the spin-orbit splitting of iodine of 7588 cm^{-1}. The shift of the lowest anion-neutral transition in comparison with iodine is 0.181 eV (+1460 cm^{-1}). Two low-frequency modes have been detected with 30 cm^{-1} and 60 cm^{-1}. The former has been assigned to a C–I stretching mode of the neutral cluster, the latter of the anionic cluster. In addition, a dissociation energy of 44 meV has been determined for the neutral cluster. A similar picture is supplied by the most simple iodine containing clusters, namely

I^-Ar and I^-Kr.[32] A spin-orbit splitting of 7605 cm^{-1} (IAr and IKr) has been found (to be compared with 7603 of iodine itself). The splitting of the degenerate $P_{3/2}$ state of iodine is 37 cm^{-1} and 53 cm^{-1}, respectively, and vibrational frequencies of 27.7 and 21.8 cm^{-1} ($X_{1/2}/I_{3/2}$) for IAr and 21.9, 18.9, and 21.4 cm^{-1} ($X_{1/2}/I_{3/2}/II_{1/2}$) for IKr have been measured.

The only experimental information for iodine water clusters besides anion-ZEKE spectra (one is shown in Fig. 15) have been obtained by Cheshnovsky *et al.*[85] They used a magnetic bottle TOF-PES spectrometer and a large photon energy of 7.5 eV and obtained a width of the $P_{3/2}$-PES-band of > 1500 cm^{-1} and maximum of peak energy at 28 312 cm^{-1} (3.51 eV). Even considerably better resolved PES-spectra would only deliver very rough energies, e.g. about the electronic origin. That is due to the broad envelope of these bands as well as to uncertainties of the absolute energy scale of PES-spectra. On the other hand, the energy scaling of ZEKE spectra is determined by laser wavelength and not by electron energy (as discussed in Sec. 1.2).

In Fig. 15 an anion-ZEKE spectrum of $I^-(H_2O)$ is shown in the energy region of the $^2P_{3/2}$ spin-orbit component of the iodine atom and covering an energy range of about 900 cm^{-1}. Two spectral regions can be distinguished. Region A between 27 600 cm^{-1} and 28 060 cm^{-1} exhibits sharp structure on the low energy side and ends in a congested region or a broad vibronic band on the high energy side. Region B between 28 060 cm^{-1} and 28 500 cm^{-1} consists of few fairly broad bands sitting on top of a broad unresolved spectral background whose intensity decreases for higher energies. The narrowest band in region A has an energy of 27 800 cm^{-1}, its width is only 3 cm^{-1}, while the bands in region B nearly are 20 cm^{-1} broad. In region A five sharp peaks appear. The first two bands $A0$ and $A1$ differ in energy by 45 cm^{-1}, while the gaps between $a2$, $a3$ and $a4$ are some 35 cm^{-1}. In addition, the maximum of the broad unstructured band $A2$ is shifted by 200 cm^{-1} from the band $A0$. In region B the first two bands $B0$ and $B1$ show a spacing of 45 cm^{-1}. Also, the energy difference between bands $B0$ and $B2$ of 210 cm^{-1} is similar to that between bands $A0$ and $A2$. Region B mainly differs from region A by two facts: missing peaks comparable to bands $a2$ to $a4$ and a considerably larger peak bandwidth.

Following the finding that for several molecular clusters containing one iodine atom the spin-orbit splitting of the iodine nearly is conserved, but the energetically lower $^2P_{3/2}$ component is split into two states ($X_{1/2}$ and $I_{3/2}$),

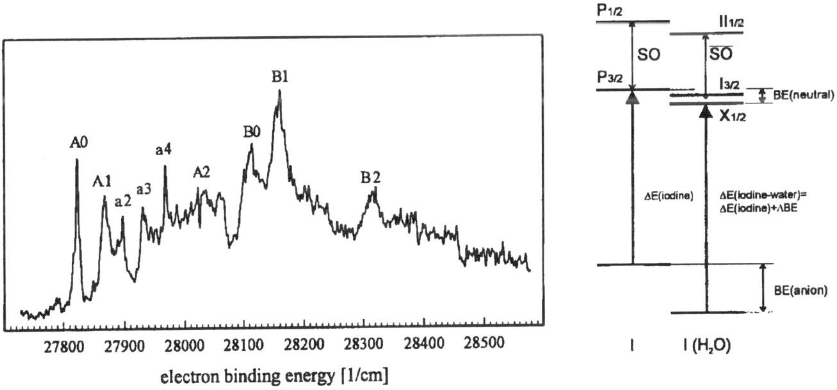

Fig. 15. Anion-ZEKE spectrum of I^- (H_2O). The band systems A and B correspond to the $I(^2P_{3/2}) \leftarrow I^-$ anion-ZEKE transition of iodide which exhibits a splitting in the anion-ZEKE transition of the iodide water cluster. On the right side, level schemes of I^-/I and $I^-(H_2O)/I(H_2O)$, illustrate the effect of the binding energies BE(anion) and BE(neutral) onto the transition, the nearly constant spin-orbit splitting in iodine and in its water cluster as well as the splitting of the degenerate $P_{3/2}$ iodine ground state into the $X_{1/2}$ and $I_{3/2}$ cluser states. The indicated $I(H_2O) \leftarrow I^-(H_2O)$ transition corresponnds to the $A0$ band in the spectrum.

we assign the band system A to the $X_{1/2}$, the band system B to the $I_{3/2}$ and a third band system starting at 35,700 cm^{-1} (not shown here) to the $II_{1/2}$ states. The bands $A0$ and $B0$ then are the vibrationless electronic origins of the corresponding anion neutral transitions. A $X_{1/2} - I_{3/2}$ splitting of 280 cm^{-1} and a $X_{1/2} - II_{1/2}$ splitting of 7778 cm^{-1} are found corresponding to a spin-orbit splitting of 7638 cm^{-1}. Finally, the anion neutral transition to the neutral $X_{1/2}$ cluster state is shifted by $\Delta E(\text{EA}) = +3150$ cm^{-1} (0.391 eV) in comparison with the anion neutral transition to the $^2P_{3/2}$ ground state of atomic iodine. These are the largest deviations in energy (for splitting and shift) from atomic iodine found for iodine clusters studies up to now (with the exception of $I^-(CH_3CN)$[89]) and can be explained by the interaction of the iodine anion with the large dipole moment of water as well as by the inclusion of a hydrogen bond. Jortner *et al.*[87] give an estimation for the binding energy of the anionic cluster between 0.422 eV and 0.400 eV. In combination with our value DE(EA) of 0.391 eV this gives a binding energy of the neutral cluster between 0.031 and 0.01 eV. This is smaller than the value found for $I(CO_2)$ (0.044 eV[31]) which is unexpected. Probably, the real binding energy of the anionic cluster is somewhat larger than the theoretical value of 0.422 eV.

Assuming the assignment proposed above, the bands $A1$ and $B1$ might correspond to a very low frequency vibrational mode with some 50 cm^{-1}. Since the iodine-hydrogen bond length becomes significantly longer during the neutral $\leftarrow$ anion transition one may expect the appearance of the iodine-water stretching mode. Therefore, a first tentative assignment of the bands $A1$ and $B1$ is to this van der Waals stretching mode of the neutral cluster. The 200 cm^{-1} band marked in band systems A and B as $A2$ and $B2$ is too small for an internal vibration of the water molecule and, therefore, has to be due to an intermolecular bending I–H_2O mode, vibrating in the plane or out of the plane of the water molecule. Concerning the additional sharp structure in band system A, one explanation might be a system of progression bands due to the I–H_2O stretching mode. The constant, but smaller energy gaps between bands $A1$, $a2$, $a3$ and $a4$ comparison to the spacing between bands $A0$ and $A1$ contradict this argument. A second explanation is the excitation of a third van der Waals mode with a frequency of 35 cm^{-1} which seems fairly small. Unpublished first results from our laboratory on isotope substituted spectra indicate a not yet explained irregularity of the whole band system A of the $I(H_2O)$ cluster.

The bandwidths of system B in Fig. 15 are considerably larger than that of A-bands. Keeping in mind that the dissociation energy found for $I(CO_2)$ is 355 cm^{-1},[31] that the dissociation energy of $I(H_2O)$ should be at least as large, and that the $A0 - B0$ energy gap is only 280 cm^{-1}, predissociation is unreasonable to be the reason for the large bandwidths of the B-band system. Fast H atom tunneling effects or fast isomerization processes with thresholds below 280 cm^{-1} could be responsible for these bandwidths. Furthermore, rotational constants B and C could differ between band system A and B and thus strongly influence the width of vibronic bands whose rotational substructure is not resolved.

To summarize, a mass selected and high-resolution anion-ZEKE spectrum of $I^-(H_2O)$ has been presented. Linewidths as narrow as 3 cm^{-1} have been observed for a resolution of the apparatus of 2 cm^{-1}. Unresolved rotational substructure may be responsible for the observed line width. From a first tentative assignment we deduce electronic and vibrational information about the neutral ground and the lowest electronic excited states of the iodine water cluster. The spectra exhibit the typical iodine spin-orbit splitting, two low frequency vibrational modes near 50 cm^{-1} and 200 cm^{-1} and a shift of the whole transition system by 3150 cm^{-1} which is mainly due to the strong interaction of the iodine anion with the water molecule. Further experiments on deuterated iodine water clusters which are in progress promise detailed information about vibrational frequencies, potentials and binding energies of the three electronic states and last but not least about the anionic cluster.

4. Conclusion

Mass selective photodetachment PE spectroscopy is a well established spectroscopic method used by many research groups all over the world to get access to isolated atomic and molecular systems which otherwise are hardly or not at all to be investigated. Anion-ZEKE spectroscopy represents the highly sophisticated outcome of further development of this promising technique. Of course, the instrumental and experimental effort is considerably larger. This may be the reason that only few research groups apply anion-ZEKE spectroscopy despite its intriguing benefits such as the enhancement of spectral resolution by more than one order of magnitude and its selection rules which are more strict than for photodetachment PES and allow the study of different neutral states than e.g. IR spectroscopy. Another reason for the small acceptance of anion-ZEKE spectroscopy may be the fact, that it has only been invented nine years ago[20] and it is still in a developing state while photodetachment spectroscopy is more than 30 years old.[1,8,9]

The large variety of atomic and molecular systems already studied by photodetachment PE and anion-ZEKE spectroscopy up to now (for a review see Ref. 90) promises further detailed information on a microscopic scale for several fields of fundamental research such as

— semiconductors
— catalytic reactions
— solute/solvent interaction
— kinetics of chemical reactions.

In this context, information on a microscopic scale means atomic and molecular parameters (e.g. energy and character of electronic states, vibrational modes, eventually even rotational motion) and thus structure and bonding of isolated atomic and molecular systems (e.g. Ge-, Se-, C-clusters or metal hydrocarbon complexes, molecular water clusters or transition states such as IHI). In particular, anion-ZEKE spectroscopy is able to contribute such details due to its high-resolution power. Therefore, one may hope that more research groups will deal with this spectroscopic method and its development in spite of its considerable instrumental effort.

Acknowledgment

The authors are grateful to Dr. Gerhard Drechsler and to Rainer Käsmeier for their important contributions to the work presented here, as well as to Volker Distelrath and Franz Schlicht for recent results on the iodine water cluster. The authors thank the Deutsche Forschungsgemeinschaft (grant DFG Bo718/5) as well as the Fonds der Chemie for financial support.

References

1. B. Brehm, M. A. Gusinov and J. L. Hall, *Phys. Rev. Lett.* **19**, 737 (1967).
2. M. W. Siegel, R. J. Cellotta, J. L. Hall, J. Levine and R. A. Bennett, *Phys. Rev.* **A6**, 607 (1972).
3. R. J. Cellotta, R. A. Bennett, J. L. Hall, M. W. Siegel and J. Levine, *Phys. Rev.* **A6**, 631 (1972).
4. H. Hotop, R. A. Bennett and W. C. Lineberger, *J. Chem. Phys.* **58**, 2373 (1973).
5. R. J. Cellotta, R. A. Bennett and J. L. Hall, *J. Chem. Phys.* **60**, 1740 (1974).
6. A. Kasdan, E. Herbst and W. C. Lineberger, *Chem. Phys. Lett.* **31**, 78 (1975).
7. A Kasdan, E. Herbst and W. C. Lineberger, *J. Chem. Phys.* **62**, 541 (1975).
8. W. C. Lineberger and B. W. Woodward, *Phys. Rev. Lett.* **25**, 424 (1970).
9. H. Hotop, W. C. Lineberger, *J. Chem. Phys.* **58**, 2379 (1973).
10. L. A. Posey, M. J. Delucca and M. A. Johnson, *Chem. Phys. Lett.* **131**, 170 (1986).
11. G. Ganteför, M. Gausa, K. H. Meiwes-Broer and H. O. Lutz, *Faraday Discuss. Chem. Soc.* **86**, 1 (1998).
12. K. M. Hugh, J. G. Eaton, G. H. Lee, H. W. Sarkas, L. H. Kidder, J. T. Snodgrass, M. R. Manaa and K. H. Bowen, *J. Chem. Phys.* **91**, 3792 (1989).
13. J. Hoe, K. M. Ervin and W. C. Lineberger, *J. Chem. Phys.* **93**, 6987 (1990).
14. K. J. Taylor, C. L. Pettiette-Hall, O. Cheshnovsky and R. E. Smalley, *J. Chem. Phys.* **96**, 3319 (1992).
15. C.-Y. Cha, G. Ganteför and W. Eberhardt, *Rev. Sci. Instrum.* **63**, 5661 (1992).
16. T. Baer, W. B. Peatman and E. W. Schlag, *Chem. Phys. Lett.* **4**, 243 (1969).
17. W. B. Peatman, T. Baer and E. W. Schlag, *Chem. Phys. Lett.* **3**, 492 (1969).
18. K. Müller-Dethlefs, M. Sandner and E. W. Schalg, *Chem. Phys. Lett.* **12**, 291 (1984); *Z. Nat.forsch. A: Phys. Sci.* **39**, 1089 (1984).
19. H.-J. Dietrich, K. Müller-Dethlefs and L. Y. Baranov, *Phys. Rev. Lett.* **76**, 3530 (1996).
20. T. N. Kitsopoulos, I. M. Waller, J. G. Loeser and D. M. Neumark, *Chem. Phys. Lett.* **159**, 300 (1989).

21. G. Ganteför, D. M. Cox and A. Kaldor, *J. Chem. Phys.* **93**, 8395 (1990).
22. G. Drechsler, C. Bäßmann, U. Boesl and E. W. Schlag, *Z. Nat.forsh. A: Phys. Sci.* **49**, 1256 (1994).
23. I. M. Waller, T. M. Kitsopoulos and D. M. Neumark, *J. Phys. Chem.* **94**, 2240 (1990).
24. C. C. Arnold and D. M. Neumark, *J. Chem. Phys.* **99**, 3353 (1993).
25. C. C. Arnold, C. Xu, G. R. Burton and D. M. Neumark, *J. Chem. Phys.* **102**, 6982 (1995).
26. L. S. Wang, J. Conceicao and R. E. Smalley, *Chem. Phys. Lett.* **182**, 5 (1991).
27. T. N. Kitsopoulos and C. J. Chick, Y. Zhao, D. M. Neumark, *J. Chem. Phys.* **95**, 5479 (1991).
28. C. C. Arnold, Y. X. Zhao, T. N. Kitsopoulos and D. M. Neumark, *J. Chem. Phys.* **97**, 6121 (1992).
29. G. Drechsler, C. Bäßmann, U. Boesl and E. W. Schlag, *J. Mol. Struct.* **348**, 337 (1995).
30. G. Drechsler, U. Boesl, C. Bäßmann and E. W. Schlag, *J. Chem. Phys.* **107**, 2284 (1997).
31. Y. Zhao, C. C. Arnold and D. M. Neumark, *J. Chem. Soc. Faraday Trans.* **89**, 1449 (1993); D. W. Arnold, S. E. Bradforth, E. H. Kim and D. M. Neumark, *J. Chem. Phys.* **102**, 3493 (1995).
32. Y. Zhao, L. Yourshaw, G. Reiser, C. C. Arnold and D. M. Neumark, *J. Chem. Phys.* **101**, 6538 (1994).
33. I. Yourshaw, Y. Zhao and D. M. Neumark, *J. Chem. Phys.* **105**, 351 (1996).
34. C. Bäßmann, U. Boesl, D. Yang, G. Drechsler and E. W. Schlag, E. W., *Int. J. Mass Spectrom. Ion Process.* **159**, 153 (1996).
35. C. Bäßmann, G. Drechsler, R. Käsmeier and U. Boesl, in *Resonance Ionization Spectroscopy 1996*, Ed. N. Winograd and J. E. Parks (AIP Press, Woodbury, 1997), p. 403.
36. E. P. Wigner, *Phys. Rev.* **73**, 1002 (1948).
37. H. Hotop, T. A. Patterson and W. C. Lineberger, *Phys. Rev.* **A8**, 762 (1973).
38. D. Feldmann, *Phys. Lett.* **A53**, 82 (1975).
39. D. Feldmann, *Z. Phys.* **A277**, 19 (1976).
40. P. A. Schulz, R. D. Mead, P. L. Jones and W. C. Lineberger, *J. Chem. Phys.* **77**, 1153 (1982).
41. J. Cooper and R. N. Zare, *J. Chem. Phys.* **48**, 347 (1979).
42. U. Boesl, R. Weinkauf, C. Weickhardt and E. W. Schlag, *Int. J. Mass Spectrom. Ion Proc.* **131**, 87 (1994).
43. G. Drechsler, C. Bäßmann, W.-D. V. Fraunberg, U. Boesl and E. W. Schlag, *Rev. Sci. Instrum.* **65**, 3172 (1994).
44. U. Boesl, R. Weinkauf and E. W. Schlag, *Int. J. Mass Spectrum. Ion Proc.* **112**, 121 (1992).
45. W. C. Wiley and I. H. McLaren, *Rev. Sci. Instrum.* **41**, 741 (1970).
46. P. G. Wenthold, J. Hu, R. R. Squires and W. C. Lineberger, *J. Am. Chem. Soc.* **118**, 475 (1998).

47. K. Kimura, *Rev. Phys. Chem.* **6**, 195 (1978).
48. P. Kruit and F. H. Read, *J. Phys. E: Sci. Instrum.* **16**, 313 (1983).
49. J. A. Coxon, *Can. J. Phys.* **85**, 933 (1980).
50. G. C. Dousmanis, T. M. Saunders and C. H. Townes, *Phys. Rev.* **100**, 1735 (1955).
51. S. J. Smith and L. M. Branscomb, *Phys. Rev.* **99**, 1657 (1955).
52. R. J. Celotta, R. A. Bennett and J. L. Hall, *J. Chem. Phys.* **60**, 1806 (1974).
53. F. Breyer, P. Frey and H. Hotop, *Z. Phys.* **A300**, 7 (1981).
54. H. Hotop, T. A. Patterson and W. C. Lineberger, *J. Chem. Phys.* **60**, 1806 (1974).
55. J. R. Smith, J. B. Kim and W. C. Lineberger, *Phys. Rev.* **A55**, 2036 (1977).
56. W. C. Lineberger and B. W. Woodward, *Phys. Ref. Lett.* **25**, 424 (1970).
57. A. R. P. Rau and U. Fano, *Phys. Rev.* **A4**, 1751 (1971).
58. T. E. Walker, *Chem. Phys. Lett.* **19**, 493 (1973); T. E. Walker, P. M. Dehmer and J. Berkowitz, *J. Chem. Phys.* **59**, 4292 (1973).
59. B. K. Janousek and J. I. Braumann, *Phys. Rev.* **A23**, 1673 (1981).
60. J. Xie and R. N. Zare, *J. Chem. Phys.* **93**, 3033 (1990); **97**, 2891 (1992).
61. R. N. Zare, *Angular Momentum* (Wiley, New York, 1988).
62. K. Ziegler, E. Holzkampf, H. Breil and H. Martin, *Angew. Chem.* **67**, 541 (1955).
63. G. Natt, *J. Polym. Sci.* **16**, 143 (1955).
64. R. P. Quirk, *Transition Metal Catalyzed Polymerizations* (Cambridge University Press, Cambridge, 1988).
65. St. L. Bernasek, *Ann. Rev. Phys. Chem.* **44**, 265 (1993).
66. B. C. Guo and A. W. Castleman, *J. Am. Chem. Soc.* **114**, 6152 (1992).
67. Z. H. Kafafi, R. H. Hauge and J. L. Margrave, *J. Am. Chem. Soc.* **107**, 7550 (1985).
68. E. S. Kline, Z. H. Kafafi, J. L. Hauge and J. L. Margrave, *J. Am. Chem. Soc.* **107**, 7559 (1985).
69. D. W. Ball, R. G. S. Pong and Z. H. Kafafi, *J. Am. Chem. Soc.* **115**, 2864 (1993).
70. G. von Helden, N. G. Gotts, P. Maitre and M. T. Bowers, *Chem. Phys. Lett.* **227**, 601 (1994).
71. J. Fan, L. Lou and L.-S. Wang, *J. Chem. Phys.* **102**, 2701 (1995).
72. J. Fan and L.-S. Wang, *J. Phys. Chem.* **98**, 11814 (1994).
73. W. Weltner, *Ber. Bunsenges.* **82**, 80 (1978).
74. J. B. West and H. P. Broida, *J. Chem. Phys.* **62**, 2566 (1975).
75. D. W. Green, G. T. Reedy and J. G. Key, *J. Mol. Spectrosc.* **78**, 257 (1979).
76. S. M. Harris and R. F. Barrow, *J. Mol. Spectrosc.* **84**, 334 (1980).
77. A. S. C. Cheung, R. M. Gordon and A. J. Merer, *J. Mol. Spectrosc.* **87**, 289 (1981); A. S. C. Cheung, N. Lee, A. M. Lyrra, A. J. Merer and A. W. Taylor, *J. Mol. Spectrosc.* **95**, 213 (1982) and **102**, 224 (1983).
78. J. Fan and L.-S. Wang, *J. Chem. Phys.* **102**, 8714 (1995).
79. P. C. Engelking, W. C. Lineberger, *J. Phys. Chem.* **66**, 5054 (1977).

80. T. Anderson, R. Lykke, D. M. Neumark and W. C. Lineberger, *J. Chem. Phys.* **86**, 1858 (1987).
81. A. W. Taylor, A. S. C. Cheung and A. J. Merer, *J. Mol. Spectrosc.* **113**, 487 (1985).
82. P. G. Wenthold, R. F. Gunion and W. C. Lineberger, *Chem. Phys. Lett.* **258**, 101 (1996); P. G. Wenthold, K.-L. Jonas and W. C. Lineberger, *J. Chem. Phys.* **106**, 9961 (1997).
83. A. J. Merer, *Annu. Rev. Phys. Chem.* **40**, 407 (1989).
84. M. Krauss and W. J. Stevens, *J. Chem. Phys.* **82**, 5584 (1985).
85. G. Markovich, R.Giniger, M. Levin and O. Cheshnovsky, *J. Chem. Phys.* **95**, 9416 (1991); G. Markovich, S. Pollack, R. Giniger and O. Cheshnovsky, *J. Chem. Phys.* **101**, 9344 (1994).
86. U. Kaldor, *Z. Phys.* **D31**, 279 (1994).
87. J. E. Combariza, N. R. Kestner and J. Jortner, *J. Chem. Phys.* **100**, 2851 (1994).
88. C. C. Arnold, D. M. Neumark, D. M. Cyr and M. A. Johnson, *J. Chem. Phys.* **99**, 1633 (1995).
89. C. E. H. Dessent, C. G. Bailey and M. A. Johnson, *J. Chem. Phys.* **103**, 2006 (1995).
90. U. Boesl and W. Knott, Mass Spectrom. Reviews **17**, 275 (1998) and references cited herein.

CHAPTER 16

PHOTODETACHMENT PHOTOELECTRON SPECTROSCOPY OF TRANSITION METAL OXIDE SPECIES

Lai-Sheng Wang

Department of Physics, Washington State University
Richland, WA 99352 USA
and
Environmental Molecular Sciences Laboratory
Pacific Northwest National Laboratory
MS K8-88, P. O. Box 999, Richland, WA 99352 USA

Contents

1. Introduction

Transition metal-containing species are important because of their interesting chemical and physical properties and their broad applications in catalysis and materials sciences.[1] Thus, understanding the chemical bonding and structure of molecules containing transition metals are of both basic and applied interest. Not surprisingly, extensive experimental and theoretical efforts have been directed at elucidating the chemical bonding properties of the *d*-orbitals. However, the complicated electronic structure of the transition metal species still poses a considerable challenge to both experimental and theoretical studies. The bonding of the *d*-orbitals and many transition metal species are still not well understood.[2] Theoretical studies of transition metal species have been reviewed in an excellent chapter in a previous volume of this series by Bauschlicher Jr *et al.*[3] Earlier, Morse reviewed experimental and theoretical investigations of transition metal clusters.[4] Additionally, Merer reviewed diatomic oxide species of 3*d* transition metals.[5] In the past few years, we have engaged in a comprehensive research program to investigate the electronic structures of transition metal clusters and molecules using photodetachment photoelectron spectroscopy (PES) of anionic species. In a recent book chapter, we reviewed our efforts on the transition metal clusters.[6] In the present chapter, we will focus on our work on transition metal oxide species. We have also investigated several transition metal carbide species,[7–15] which will not be included in this chapter. Researches on the transition metal oxide molecules and clusters are still being actively pursued in our laboratory and other groups. This review should be read as a progress report.

Gas phase spectroscopic studies of transition metal oxide molecules have been rather scarce, except perhaps for the diatomic monoxides[2] because a number of these molecules are important in the interstellar medium, as reviewed by Merer.[5] Several metal ion reactions with O_2 have been studied in the gas phase.[16–19] However, matrix isolation experiments[20–26] have been mainly used to study the interaction of neutral metal atoms and O_2 along with a few gas phase studies.[27] Most of the reaction products involve the monoxides or the dioxide (MO_2). The spectroscopy of some MO_2 species has been probed in the low temperature matrix experiments.[20–26] However, there are few gas phase studies of these species and their structures and bonding are not well understood. For the higher oxides, MO_3 and MO_4, or generally M_xO_y, even less is known.

The difficulty in studying these species in the gas phase is how to generate them with sufficient number density. We have found that PES of anionic species using a magnetic-bottle time-of-flight (TOF) photoelectron analyzer and a laser vaporization source provides an ideal technique to investigate transition metal oxide species in the gas phase.[6,28] PES is a powerful technique to obtain size-selected electronic and spectroscopic information of the gas phase oxide species, providing direct measurement of their electron affinities (EAs), low-lying electronic states, and vibrational information.[29–39] The laser vaporization technique can produce a beam of nearly any oxide species; the high sensitivity of the magnetic-bottle photoelectron analyzer allows PES spectra to be taken even for relatively low abundant species. Lineberger's group has pioneered the photodetachment PES of transition metal species, starting from the transition metal atoms.[40–43] They have investigated several transition metal oxide systems,[44–48] in addition to their work on transition metal hydrides,[49,50] carbonyl compounds,[51] and small clusters.[52–54]

However, the versatility and sensitivity of the magnetic-bottle-TOF-PES technique has allowed us to investigate a broad range of oxide species systematically,[29–39,55–63] particularly for species with extremely high EAs. We have studied MO_y-types of species for all the 3*d* transition metals from Sc to Cu,[29–39] except Cr, Mn, and Co. We have performed more extensive studies for Cu and Fe oxide clusters. For the former, we studied the Cu_2O_y clusters as well as density functional theory (DFT) calculations.[34] Our most extensive studies involved the Fe oxide systems,[29–32] for which we studied Fe_xO_y clusters for $x = 1$–4 and $y = 1$–6, as well as Fe_xO clusters with $x = 1$–16 and a preliminary DFT study for $x = 2$–6. Among the notable observations, we have found that the EA of transition metal oxide species increases monotonically with the oxygen content in the clusters until the maximum oxidation states of the metals are reached. This is a phenomenon that we called sequential oxidation, indicating the ionic bonding characters between the metals and oxygen. For oxides with oxygen content exceeding the maximum oxidation states of the metal, we began to observe O_2-complex formation, i.e. superoxide or peroxide species. We have also observed isomers in many oxide species, due to the O_2-complex formation, especially for the late 3*d* metals. For the monoxides, many low-lying excited states, which were previously unknown, have been observed. In particular, we have observed strong electron correlation effects in TiO, ScO, and YO, evidenced by strong two-electron transitions in the PES spectra.

This chapter is organized as follows. In Sec. 2, we describe in detail about the experimental techniques and procedures. In Sec. 3, we review our investigations on the metal oxide systems starting from the copper oxides and moving to the left of the periodic table to scandium oxides. Results and discussion will be presented for each oxide system including the relevant literature. Our conclusions are given in Sec. 4.

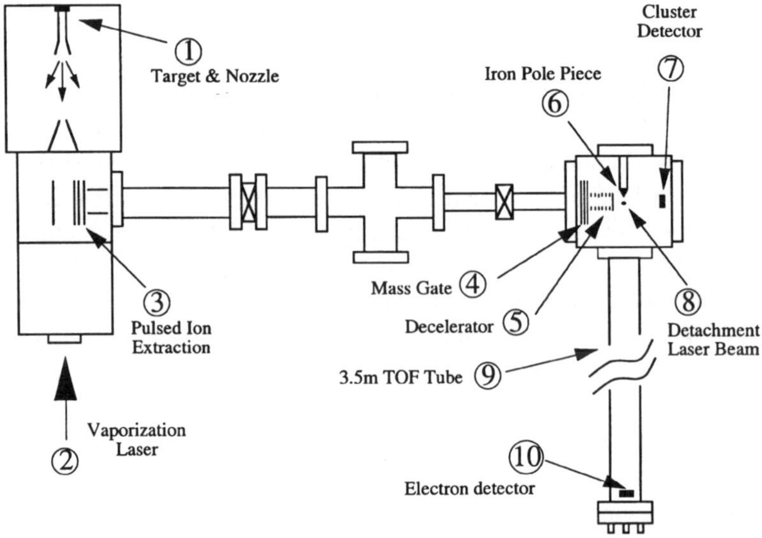

Fig. 1. Schematic view of the magnetic-bottle time-of-flight photoelectron apparatus coupled with a laser vaporization supersonic cluster source and a time-of-flight mass spectrometer.

2. Experimental Technique

The details of our apparatus and its performance have been well documented in previous publications,[6–15,28–39,55–63] particularly, in our recent review concerning the transition metal clusters.[6] A summary is provided in the section. Figure 1 shows a schematic of our experimental apparatus, which consists of a pulsed laser vaporization cluster source, a modified Wiley–McLaren TOF mass spectrometer,[64,65] a mass gate, a momentum decelerator,[6,28,66] and a magnetic-bottle TOF photoelectron analyzer.[67,68]

2.1. *Pulsed Laser Vaporization Cluster Source, Mass Selection and Ion Deceleration*

We produce the transition metal oxide anions using a laser vaporization supersonic cluster beam source. As shown in Fig. 1, the vaporization laser beam is directed collinearly with the cluster beam and hits a metal target disk of interest after passing a 2 mm nozzle orifice. The laser beam, typically 10–20 mJ from the second harmonic of a Nd : YAG laser, is focused down to a 1 mm diameter spot onto the target. Two pulsed molecular beam valves (R. M. Jordan Co., CA), symmetrically mounted, are used to deliver a short and intense helium carrier gas pulse (10 atm stagnation pressure). To produce oxide species, a small amount of O_2 is usually seeded in the He carrier gas. This intense carrier gas pulse mixes with and cools the laser-induced plasma. The O_2 in the carrier gas reacts with the vaporized metal species to produce the desired oxide molecules although monoxide species can usually be produced with enough abundance even with a pure He carrier gas due to the surface contamination of most metal targets. The cluster/He mixture undergoes a supersonic expansion and is skimmed to form a collimated cluster beam into the ion extraction chamber.

Negatively charged species are extracted perpendicularly from the beam by a 1 kV high voltage pulse and are subjected to a TOF mass analysis. Our mass spectrometer is a modified Wiley–McLaren-type for large volume ion extraction and simultaneous high mass resolution.[65] The major modification is an addition of a short free-flight zone in between the two acceleration stages of the original Wiley–McLaren design. This modification allows us to achieve a mass resolution ($M/\Delta M$) of more than 500 at low masses with a large extraction volume. The resolution deteriorates slightly at higher masses, mainly limited by the fringe field effect due to the ion steering optics to compensate for the transverse velocity of the clusters.

During PES experiments, only clusters of interest are allowed to enter the interaction zone. The clusters are selected by a mass gate and decelerated by a momentum decelerator.[6,28,66] A three-grid mass gate is used for mass selection as shown in Fig. 2. The first and third grids are grounded, and the middle grid is at a negative high voltage (−1 kV) so that no negative clusters are able to pass. Once the desired cluster arrives at the first grid, the high voltage is pulsed to ground for a short period allowing the cluster to pass unaffected.

Following the mass gate, the selected cluster beam enters the momentum decelerator as shown in Fig. 2. Once the cluster packet passes the third grid of the mass gate, a positive square high voltage pulse is applied to this grid for the momentum deceleration. The high voltage is pulsed to ground before the ion packet leaves the deceleration stack, which consists of ten guarded rings to ensure a uniform deceleration field. Both the pulse amplitude and the pulse width can be varied to achieve the best deceleration effect. All ions experience the same decelerating force within the same time period, thus will be decelerated by the same amount of linear momentum.

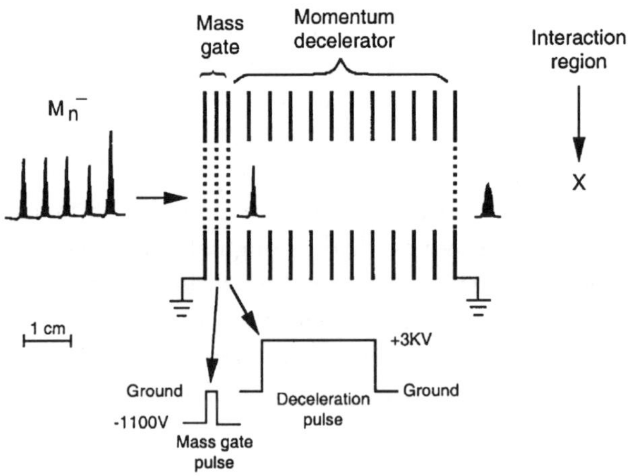

Fig. 2. The mass gate and the momentum decelerator.

In this deceleration method, ions are decelerated in the momentum space and the initial ion energy spread is reduced after the deceleration, in contrast to the conventional retarding field deceleration in which the initial ion energy spread remains the same after the deceleration. The deceleration step is crucial to improve the MTOF electron energy resolution due to the minimization of the Doppler-broadening caused by the anion beam velocity. The momentum decelerator allows us to decelerate a given cluster ion packet down to such low kinetic energies that the Doppler-broadening is no longer a factor.

Figure 3 shows the operation of the mass selection and deceleration. The top panel displays the two isotopes of Cu^- at a mass resolution of about 550. The middle panel shows the mass-selected peak of the $^{65}Cu^-$ isotope. The bottom panel is the same $^{65}Cu^-$ peak after a deceleration by using a 2 kV and 550 ns high voltage pulse. The ion energy is reduced to about 30 eV from the original 1000 eV. Further deceleration to even lower energy is routinely used in our PES experiments as will be shown below.

2.2. *Magnetic-Bottle Time-of-Flight Photoelectron Analyzer*

The magnetic-bottle type TOF photoelectron spectrometer first described by Kruit and Read is ideal for the gas phase study of atomic clusters and the oxide species due to its high collecting efficiency (2π solid angle),[67] which

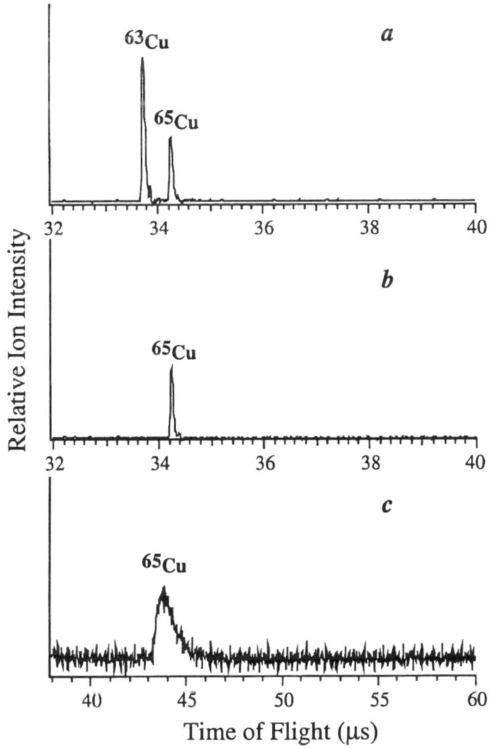

Fig. 3. Mass selection and deceleration. (a) Mass spectrum of Cu^-, showing the two well resolved isotopes of Cu (^{63}Cu, 69%, and ^{65}Cu, 31%). The mass resolution here was $M/\Delta M = 550$. (b) Mass-selected $^{65}Cu^-$ isotope peak. (c) Decelerated $^{65}Cu^-$ isotope peak with a high voltage pulse (2kV pulse height, 550 ns pulse width).

compensates for the low number density typically found with the laser vaporization cluster source. The original design has been implemented for negative cluster studies and configured to collect all 4π solid angles.[68,69] Our apparatus is a modified 4π version of the MTOF. As shown in Fig. 1, the MTOF is located at the end of the TOF mass spectrometer. The high B-field is generated by a permanent magnet mounted outside of the UHV interaction chamber. A soft iron cone is used as a pole piece to focus the magnetic field to a spot. The tip of the soft iron pole is 8 mm away from the center of the detachment zone, which is 2.5 cm down stream from the momentum decelerator. A set of in-line micro-channel plates, about 12 cm down stream from the detachment zone, are used as the cluster detector.

The MTOF tube is 3.5 m long, at the end of which is located a fast Z-stack electron detector (R. M. Jordan Co., CA). Not shown in the figure are the low B-field solenoid and the double μ-metal shielding along the 3.5 m MTOF tube.

Our detachment lasers include the four harmonics of a Nd : YAG laser (1064 nm–1.16 eV, 532 nm–2.33 eV, 355 nm–3.49 eV, 266 nm–4.66 eV) and an ArF Excimer laser (193 nm–6.42 eV). Significant background electrons can arise at high photon energies due to photoemission from surfaces near the interaction zone. At 3.49 eV or lower photon energies which do not produce background electrons, photoelectron TOF spectra are measured at 10 Hz repetition rate. At 4.66 and 6.42 eV, spectra are taken at 20 Hz with the anion beam on and off at alternating laser shot for background subtraction. For relatively weak signals, the 4.66 and 6.42 eV spectra can still exhibit significant noise at low electron energies due to imperfect background subtraction. The electron TOF spectra are converted to kinetic energy distributions, calibrated with the known spectra of Cu^-, and smoothed with a 3 to 10 meV square window function. The binding energy (BE) spectra are obtained by subtracting the kinetic energy spectra from the corresponding photon energies.

2.3. *Performance: PES of* Cu^- *at* 3.49, 4.66 *and* 6.42 *eV*

Figure 4 shows the effect of parent ion deceleration on the electron energy resolution for the Cu^- PES spectrum at 3.49 eV photon energy. Figure 4c shows the best resolution achieved with our current apparatus. The peak widths (FWHM) for the $^2S_{1/2}$, $^2D_{5/2}$, and $^2D_{3/2}$ states are 47, 22, and 19 meV, respectively, indicating the energy resolution dependence on the electron kinetic energies. We can decelerate the Cu^- ion beams down to such low energies that the Doppler broadening is no longer the dominating effect. For the spectrum in Fig. 4c, the Cu^- anion beam was decelerated to < 10 eV mean kinetic energy. The PES spectrum of Cu^- is used routinely for the spectrometer calibration in the TOF-to-kinetic energy conversions. Figure 5 shows the Cu^- PES spectra in BE at three-photon energies 3.49, 4.66 and 6.42 eV. The increasing spectral widths with photon energies reflect the resolution dependence on the electron kinetic energies. The band width of the excimer laser ($\sim$20 meV) also contributes to the broadening of the 193 nm spectrum. From Fig. 5, it can be seen clearly that the noise begins to show up at the high BE side of the 266 nm spectrum

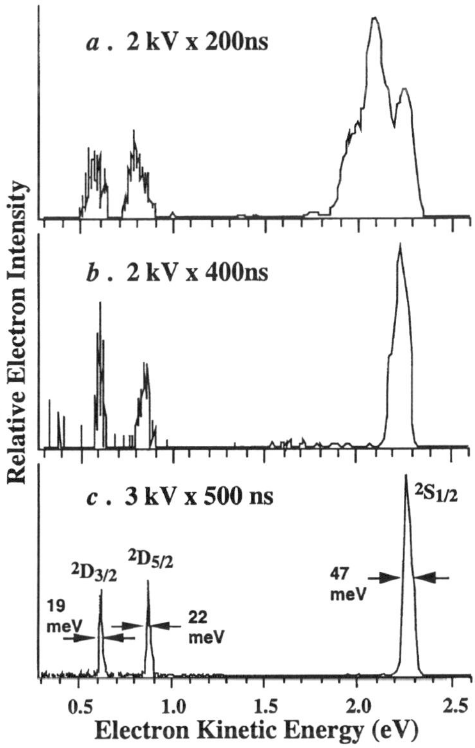

Fig. 4. Photoelectron spectra of Cu^- at 3.49 eV photon energy, showing the effect of ion deceleration on the photoelectron energy resolution. The deceleration pulses and the best resolution are indicated.

above ~3.5 eV BE. The 193 nm spectrum indicates that the noise becomes significant above 4.6 eV BE and often low photon flux has to be used at the higher photon energies to reduce the noise problem.

We also observed two other new features in Fig. 5: (1) the photon energy dependence of detachment cross sections for different detachment transitions, and (2) a new peak near 5 eV BE at 193 nm. Since our spectrometer collects nearly 100% of the photoelectrons, the relative peak intensities represent the relative total detachment cross-sections. We observed that the cross section for the ^2D states increased with photon energies relative to that for the ^2S state. This is consistent with the general observation in PES that cross sections for emitting electrons from higher angular momentum states increase with photon energies.[70]

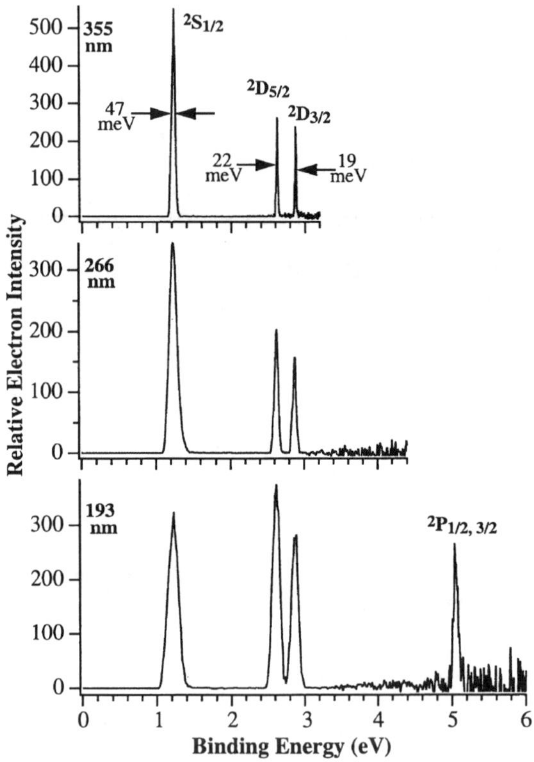

Fig. 5. Photoelectron spectra of Cu^- at 355 (3.49 eV), 266 (4.66 eV), and 193 (6.42 eV) nm.

The new peak observed at 5 eV BE is interesting since it is due to a two-electron transition. This peak can be assigned from the Cu atomic energy levels.[71] It is due to the ^2P excited state of Cu atom with a $3d^{10}4p^1$ electron configuration. There is a small spin-orbit splitting (30 meV), that is not resolved in the 193 nm spectrum. Cu^- has a $3d^{10}4s^2$ configuration. Thus, the ^2P state is resulted from detaching a $4s$ electron and at the same time exciting another $4s$ electron to the $4p$ orbital. Such multielectron transitions are due to electron correlation effects in multielectron systems. PES is the major experimental technique to study electron correlation effects in atoms and molecules.[72] Multielectron transitions are called "satellites" in PES and usually exhibit very weak intensities. It is surprising that the intensity of the ^2P state observed here are almost comparable to the main transitions. This suggests very strong electron correlation effects between the two 4*s* electrons in Cu^-. As will be discussed later, multi-electron transitions are common in the photodetachment of the transition metal oxide species,[37,39] suggesting the importance of electron correlation effects in these systems.

Figure 5 shows the advantages of performing PES experiments at various photon energies using the MTOF technique. Low photon energies allow optimum spectral resolution and are noise-free while high photon energies can reveal more highly excited states. Additionally, photon energy dependence of detachment cross sections can provide useful information in spectral assignments concerning the initial states of the photoemitted electrons. However, the MTOF technique is angular-integrated by definition and cannot provide angular-dependent information.

3. Results and Discussion

3.1. $Cu_xO_y^-$ ($x = 1, 2$; $y = 1-6$)

The chemical bonding between copper atom and oxygen is interesting both as a prototype system to understand the basic chemical bonding properties between transition metals and oxygen and due to the importance of copper in bioinorganic chemistry for dioxygen metabolism.[73–75] It is known that at room temperature Cu atom can only form mono- and bis(dioxygen) complexes, $Cu(O_2)$ and $Cu(O_2)_2$, with O_2 because the O-atom transfer reaction is endothermic.[20,27,76,77] It has been shown in low temperature matrix experiments that a copper dioxide molecule, OCuO, as well as a copper ozone complex, $Cu(O_3)$, can be formed upon UV photolysis.[20,22,78–80] Among the Cu/O species, the copper monoxide (CuO)[2,5,81–84] and $Cu(O_2)$[20–22,27,76–80,85–87] complex are perhaps the most studied and best understood molecules both experimentally and theoretically. Several matrix experiments have observed the copper dioxide molecule, OCuO,[20–22,78–80] but its structure and bonding are poorly characterized. The $Cu(O_2)$ complex has been calculated to take on a bent CuOO(C_s) structure.[20–22,76–80]

We conducted a systematic and extensive PES study of all the CuO_y (y = 1–6) species.[33,35] Several new Cu/O molecules were observed for the first time and their structure and bonding information was obtained from their PES spectra taken at various detachment photon energies. Both oxides and O_2-complexes were observed for $y > 1$. We also observed wavelength-dependent photodissociations of the anions for the O_2-complexes, yielding direct structural information.

3.1.1. *CuO and CuO⁻: Excited State of CuO⁻*

The only previous PES study on CuO^- was taken at 3.531 eV by Polak *et al.*[88] We obtained the CuO^- spectra at four-photon energies (2.33, 3.49, 4.66, and 6.42 eV), as shown in Fig. 6. Photon energy-dependent detachment cross sections of different transitions were observed. The high photon

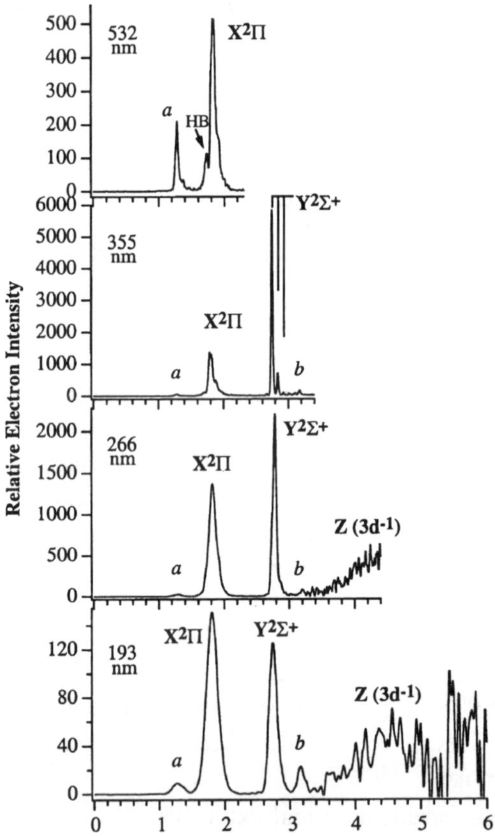

Fig. 6. Photoelectron spectra of CuO^- at 532, 355, 266, and 193 nm. HB stands for hot band transitions. From Ref. 35.

energies also allowed transitions to higher excited states of CuO to be observed. Interestingly, we also observed features due to a metastable triplet excited state of CuO^-.

The chemical bonding of CuO may be viewed as $Cu^+(3d^{10})O^-(2p^5)$ with a valence configuration of $3d\sigma^2 3d\pi^4 3d\delta^4 2p\sigma^2 2p\pi^3 4s\sigma^0 (X^2\Pi)$, where the empty $4s\sigma$ molecular orbital (MO) is mainly Cu $4s$.[81–84] In the anion, the extra electron enters the $2p\pi$ orbital, resulting in a closed-shell ground state for the CuO^- anion ($X^1\Sigma^+$), which can be roughly described as $Cu^+(3d^{10})O^{2-}(2p^6)$. Removal of a $2p\pi$ electron led to the $X^2\Pi$ ground state of CuO at a BE of 1.78 eV while detaching a $2p\sigma$ electron gave rise to the $Y^2\Sigma^+$ state at a BE of 2.75 eV (Fig. 6). The X and Y states were the only two features observed in the previous PES work by Polak *et al.*[88] The spin-orbit splitting of the $X^2\Pi$ ground state (277 cm^{-1}) had been well resolved in the previous work while it was not resolved in Fig. 6. The EA and the X–Y separations we obtained (Table 1) agreed well with the previous study. Figure 6 indicated that intensity of the X band increased with photon energies relative to the Y band. The photon energy dependence of the relative intensities of the X and Y bands was consistent with their orbital characters, an indirect support for their assignments. The observations of the a and b features were surprising since they were not observed in the previous work.[88] They were due to an electronically excited state of CuO^-. The feature a was observed to be very strong in the 532 nm spectrum. Both features a and b should have the same origin because their

Table 1. Observed binding energies (BE) and spectroscopic constants for CuO^- and CuO.

	State	BE (eV)^a	Term value (eV)^b	Vib. freq. (cm^{-1})^b
CuO^-	X ($^1\Sigma^+$)	0	0	640 (80)
	a'' ($^3\Pi_2$)	1.36	0.42	
	a ($^3\Pi_1$)	1.27	0.51 (2)	
	a' ($^3\Pi_0$)	1.03	0.75 (3)	
CuO	X ($^2\Pi$)	1.78	0	640 (80)
	Y ($^2\Sigma^+$)	2.75	0.97 (2)	
	b ($^4\Sigma^-$)	3.18	1.91 (2)	
	Z ($3d^1$)	~4	~2.2	

^{a}Uncertainty ± 0.04 eV.
^{b}Relative energies can be determined more accurately.

intensities depended on the source conditions in the same fashion. As will be shown in the following when CuO_3^- is discussed, very hot CuO^-, both vibrationally and electronically, was produced from photodissociation of CuO_3^- at 355 nm, giving more intense signals for the *a* and *b* features.

Observation of excited states for anions is rare. Recently, calculations have predicted excited triplet states for several cluster anions.[89] The lowest excited state for CuO^- would be to excite a $2p\pi$ electron to the empty $4s\sigma$ orbital, giving a triplet excited state $3d\delta^4 3d\pi^4 3d\sigma^2 2p\sigma^2 2p\pi^3 4s\sigma^2 (^3\Pi)$. This suggested that when the CuO^- anion was formed, there was a finite probability for the extra electron to enter the $4s\sigma$ orbital to give the metastable $^3\Pi$ excited state, which could be roughly viewed as $Cu(3d^{10}4s^1)$-$O^-(2p^5)$, i.e. Cu^0O^-. This is equivalent to a charge transfer transition from the ground state Cu^+O^{2-} anion to a Cu^0O^- anion (O $2p \rightarrow$ Cu $4s$). From the BE of feature *a*, this excitation energy was 0.51 eV. The feature *a* was then formed by a transition from the excited CuO^- anion to the ground state of CuO, i.e. by removing the $4s\sigma$ electron. Detachment of a $2p\pi$ electron from the excited CuO^- anion would lead to a quartet excited state for $CuO(3d\delta^4 3d\pi^4 3d\sigma^2 2p\sigma^2 2p\pi^2 4s\sigma^1, {}^4\Sigma^-)$. We assigned the feature *b* to this transition, yielding an excitation energy of 1.91 eV above the CuO ground state. A quartet state had been observed and assigned to CuO in previous optical experiments with a similar excitation energy. However, a different electron configuration was assigned.[2,5] In light of the current observation, this previous assignment needs to be reevaluated. An energy diagram is presented in Fig. 7, showing the transitions resulting in the features X, Y, *a* and *b*, as well as the features Z. The transitions a'' and a' were observed from hot CuO^- from photodissociation of CuO_3^- which will be discussed below.

The 266 nm spectrum of CuO^- showed a tail of broad band at high BE (~4 eV). This band was seen to extend to about 5 eV in the 193 nm spectrum. The broad nature of this band was unusual and suggested that there was a significant geometry change from the anion to the neutral. From the ground state configuration of CuO^-, this band was likely to be due to detachment of the $3d$ electrons ($3d\delta^4 3d\pi^4 3d\sigma^2$). This would result in excited states which might be viewed as $Cu^{2+}O^{2-}$, i.e. $3d$(Cu)$\rightarrow 2p$(O) charge transfer states. Both Cu and O atoms are in valence two in these excited states compared to valence one in the ground state. This change of bonding character was expected to lead to shorter bond length in the

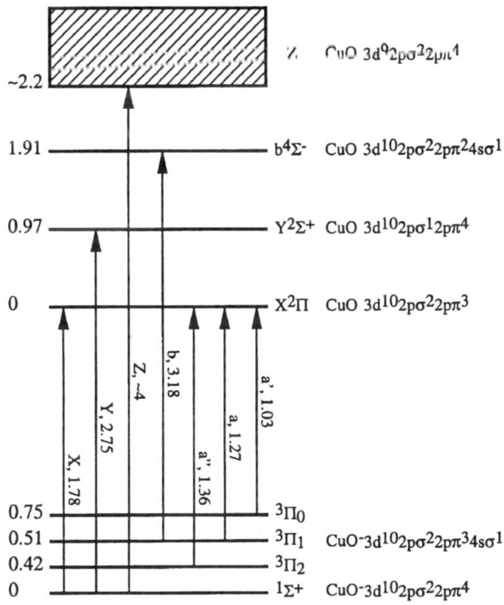

Fig. 7. Schematic diagram of the energy levels of CuO^- and CuO, showing the photodetachment transitions. Energies are in eV. From Ref. 35.

excited states, resulting in the broad PES spectrum. Previous calculations predicted that such charge transfer states should occur 10 000–30 000 cm^{-1} above the ground state,[81–84] consistent with our observation. Table 1 summarizes all the spectroscopic constants and BEs obtained from the current work for CuO and CuO^-. Figure 7 shows a schematic diagram of all the observed energy levels for CuO and CuO^-.

3.1.2. *CuO_2 and CuO_2^-: Two Isomers — $Cu(O_2)$ and OCuO*

Figure 8a shows the spectrum of CuO_2^- obtained at 355 nm detachment wavelength (fluence: 30 mJ/cm^2). Surprisingly, the very familiar peaks due to Cu^- used in our routine calibration were observed in the spectrum as the strongest signals. When the detachment laser fluence was reduced to 7 mJ/cm^2 the ratio of the electron signals due to Cu^- and CuO_2^- was substantially decreased (Fig. 8b). This suggested that Cu^- anion came from photodissociation of the parent CuO_2^- ion within the same detachment laser pulse (pulse width 8 ns FWHM). A vibrationally resolved band near

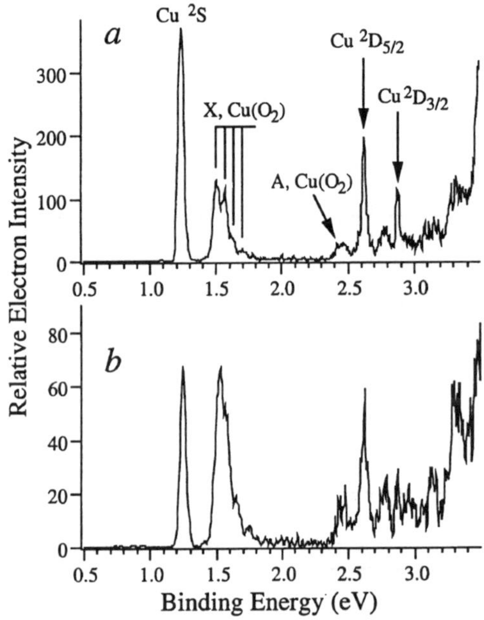

Fig. 8. Photoelectron spectra of CuO_2^- at 355 nm, showing signals from Cu^- due to photodissociation and that from a $Cu(O_2)^-$ complex. (a) High laser fluence, 30 mJ/cm^2; (b) lower laser fluence, 7 mJ/cm^2. From Ref. 33.

1.5 eV was observed due to CuO_2^- with a frequency of about 530 cm^{-1}. Transitions to excited states of CuO_2 were also observed starting at about 2.4 eV. The electron signals coming from the CuO_2^- anion were rather weak at the 355 nm detachment wavelength even though the CuO_2^- mass signal was quite strong. However, a strong and rising signal was observed near 3.45 eV, suggesting strong transitions might be observed at higher photon energies.

Figure 9 displays PES spectra of CuO_2^- taken at 266 and 193 nm, which differ remarkably from those at 355 nm (Fig. 8). Four intense and well-resolved bands (X, A, B and C) were clearly observed at 266 nm whereas two additional high BE features (D and E) were observed at 193 nm. Vibrational structures were resolved for the A and B bands (266 nm) both with a similar frequency of about 640 cm^{-1}. Interestingly, the Cu^- peaks as seen in the 355 nm spectra (Fig. 8) were not present in either the 266 or 193 nm spectrum while the 1.5 eV band was observed as a very weak feature. This indicated that photodissociation did not take place at 266 or 193 nm for the CuO_2^- anion. It was concluded that two isomers were present. One isomer, constituting a very small portion of the CuO_2^- anion beam and dissociating at 355 nm, had a low EA of about 1.5 eV. The second isomer, with dominating abundance, yielded the intense spectral features in the 266 and 193 nm spectra (X, A, B, C, D and E in Fig. 9) and had a very high EA of about 3.46 eV. These two isomers actually corresponded to the two isomers observed in the previous matrix studies. Figure 10 shows a higher resolution spectrum taken at 532 nm for the low EA isomer. The obtained vibrational frequency of 530 (50) cm^{-1} was in

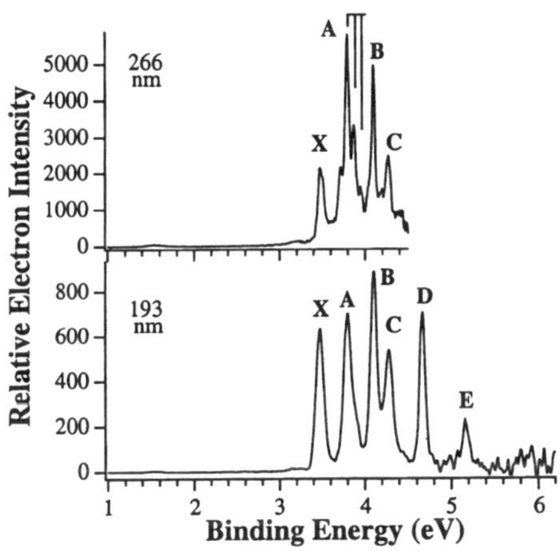

Fig. 9. Photoelectron spectra of CuO_2^- at 266 and 193 nm. From Ref. 35.

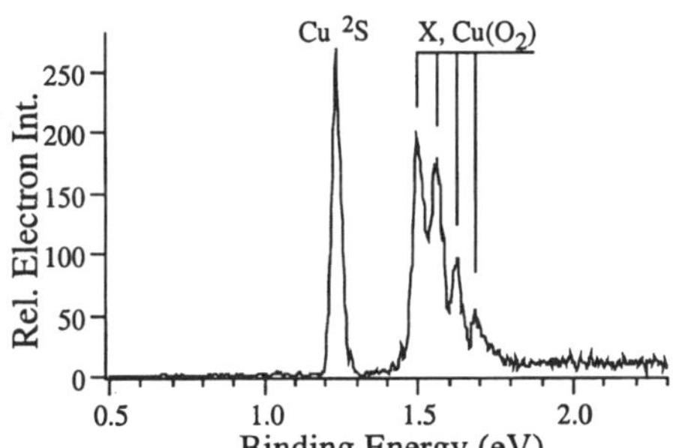

Fig. 10. Vibrationally resolved photoelectron spectrum of $Cu(O_2)^-$ complex at 532 nm and the Cu^- peak due to photodissociation. From Ref. 33.

excellent agreement with the Cu-(O_2) stretching frequency observed for the Cu(O_2) complex. The 640 cm^{-1} frequency observed in the A and B bands in Fig. 9 agreed well with the symmetric stretching frequency assigned to the copper dioxide isomer in previous matrix studies. Therefore, it was concluded that the minor species in the CuO_2^- beam corresponded to the $Cu(O_2)^-$ complex, which underwent photodissociation at 532 and 355 nm. The dominating species in the CuO_2^- beam corresponded to the copper dioxide molecule.

As shown in Fig. 8, excited state features were present for the Cu(O_2) complex between 2.42 and 3.3 eV, overlapped with the two ^2D peaks of Cu. However, detailed characterization and assignments of these excited state features were difficult due to the rather weak signals. It was only concluded that the lowest excited state of Cu(O_2) occurs near 2.42 eV, yielding a term value of about 7400 cm^{-1}. The obtained spectroscopic information for the Cu(O_2) complex is summarized in Table 2. The ground state of Cu(O_2), $^2A''$, has a bent CuOO structure with an unpaired electron.[85–87] Its main bonding character can be described as $Cu^+O_2^-$ with covalent bonding character. The bonding energy was measured to be about 15 kcal/mol.[3] The ground state of the anion was expected to be closed-shell with a $^1A''$ symmetry. The fact that we observed anion photodissociation at 532 and 355 nm suggested that the anion probably had resonant absorbtions at these wavelengths. The fact that we only observed Cu^- but not O_2^- was due to the large EA of Cu compared to that of O_2.

The PES spectra of the $OCuO^-$ isomer revealed an extensive set of low-lying excited state for OCuO. The obtained spectroscopic information for the copper dioxide molecule is gathered in Table 3. A simple MO scheme was used to give a qualitative interpretation of the PES spectrum and the electronic structure of the copper dioxide molecule, whose true ground state structure was not known. A linear OCuO structure was consistent with previous matrix studies and had been assumed for the copper dioxide isomer previously.[20–22] There was only one previous *ab initio* calculation of

Table 2. Spectroscopic constants and energies of the Cu(O_2) complex.

	Term Value (cm^{-1})	Vibrational frequency (cm^{-1})
X	0	530 (50)
A	7400 (300)	
EA (eV)	1.503 (10)	

Table 3. Observed binding energies (BE) and spectroscopic constants for CuO_2^- and CuO_2.

	State	BE (eV)[a]	Term value (eV)[b]	Vib. freq. (cm^{-1})[b]
$OCuO^-$	$X\ (^1\Sigma_g^+)$	0		600 (80)
OCuO	$X\ (^2\Pi_g)$	3.47	0	
	A	3.79	0.32 (2)	640 (80)
	B	4.10	0.63 (2)	640 (80)
	C	4.28	0.81 (2)	
	D	4.67	1.20 (2)	
	E	5.16	1.69 (2)	

[a]Uncertainty ±0.04 eV.
[b]Relative energies can be determined more accurately.

the OCuO isomer,[90] in which a linear structure was assumed. The predicted first excited state was 2 eV above the ground state, which was clearly wrong considering the current PES spectra, which yielded a term value of only 0.32 eV for the first excited state (Table 3). Ozin *et al.* proposed $O^-Cu^{2+}O^-$ to describe the bonding in the copper dioxide molecule.[20] This was a reasonable suggestion comparing to CuO, which could be viewed as Cu^+O^- with significant covalent bonding character. Bondybey and English discussed the bonding in OCuO by considering it as a 13 electron system with a linear ground state structure and disregarding the Cu $3d$.[21] The later assumption was problematic because it was expected that there should be more extensive covalent bonding contributions to the Cu–O bond between the Cu $3d$ and O $2p$ because of the $3d$ hole (d^9).

We discussed the bonding in the copper dioxide molecule by considering a $Cu^{2+}(d^9)$ ion interacting with 2 $O^-(p^5)$ ions in a linear structure. Figure 11 shows a schematic of the MOs formed from the $3d$ and $4s$ orbitals of the Cu atom with the $2p$ orbitals of the O atoms. The relative ordering and occupations of these MOs were not known. Figure 11 depicted one such possibility, which we felt was reasonable based on the MOs from CuO. According to Fig. 11, the copper dioxide ground state should be $^2\Pi_g$, with a configuration of $1\delta_g^4 1\pi_u^4 2\pi_u^4 1\sigma_g^2 2\sigma_g^2 1\pi_g^3$. Therefore, the $OCuO^-$ anion should have a closed-shell $^1\Sigma_g^+$ ground state. The $1\pi_g$ and $2\sigma_g$ orbitals are similar to the 2π and 2σ orbitals in CuO, as discussed above. Removal of an electron from these two orbitals resulted in the X and A bands that

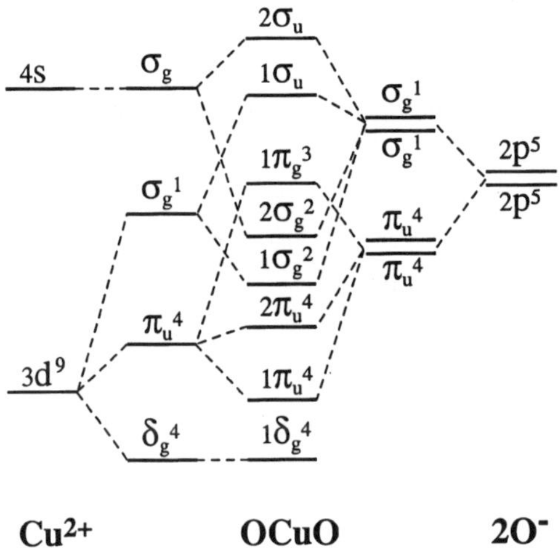

Fig. 11. A schematic molecular orbital diagram for the copper dioxide molecule.

also had similar intensity ratio compared to that of the CuO^- spectrum, resulting in a $^2\Pi_g$ and a $^2\Sigma_g^+$ state, respectively. The fact that the X band width was larger than the instrumental resolution was consistent with assignment of the $^2\Pi_g$ ground state which was expected to have a small spin-orbit splitting. The B, C, D and E bands were assigned to be due to detachment from the $1\sigma_g$, $2\pi_u$, $1\pi_u$ and $1\delta_g$ orbitals, respectively. Therefore, the 193 nm spectrum shown in Fig. 9 revealed a one-to-one correspondence between the PES features and the six valence MOs of OCuO. A more definitive assignment would require accurate *ab initio* calculations for the molecular structure and its low-lying electronic states. The OCuO copper dioxide molecule may be considered the simplest case involving an open *d* shell and should be a good test case for *ab initio* calculations.

3.1.3. *CuO_3 and CuO_3^-: Photodissociation of $CuO_3^- \to CuO^- + O_2$*

The PES spectra of CuO_3^- are shown in Fig. 12 at three-photon energies. The 355 nm spectrum showed an intense sharp peak at 3.39 eV labeled as A and a weak peak at 3.19 eV labeled as X. The rest of the features could all be attributed to CuO^-. At 266 nm, more features at higher BE were revealed, B, C and D, while the features due to CuO^- disappeared. A very weak feature (note the $\times 60$ amplification) near 1.23 eV was observed.

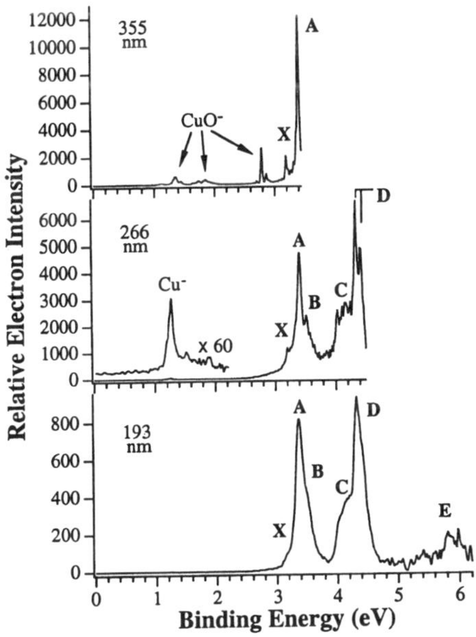

Fig. 12. Photoelectron spectra of CuO_3^- at 355, 266 and 193 nm. From Ref. 35.

Table 4. Observed binding energies (BE) and spectroscopic constants for CuO_3.

	State	BE (eV)[a]	Vib. Freq. (cm^{-1})[b]
$OCuO_2$	X	3.19	
	A	3.39	
	B	3.51	
	C	4.02	
	D	4.34	660 (90)
	E	~5.8	

[a]Uncertainty ± 0.04 eV.
[b]Relative energies can be determined more accurately.

At 193 nm, all the features observed at 266 nm (X–D) were reproduced and, additionally, there was a weak feature near 5.8 eV labeled as E. In the 266 nm spectrum, vibrational structure was clearly resolved in the D band with a frequency of 660 cm^{-1} (Table 4). The C band also contained vibrational structures, but no regular progression could be identified. The feature B was separated from A by 0.12 eV and could be due to a vibrational level above A, but no higher members could be identified even though there were clearly features above B. The feature X was rather weak, but was clearly present in all three spectra. It was assigned to be the ground state of the CuO_3^-, yielding an EA of 3.19 eV for CuO_3 neutral.

The observation of CuO^- features in the 355 nm spectrum of CuO_3^- was obviously due to photodissociation of the parent CuO_3^- anion, $CuO_3^- \rightarrow CuO^- + O_2$. This dissociation was apparently wavelength-dependent and only occurred efficiently at 355 nm, similar to that observed for the $Cu(O_2)^-$ complex. The O_2 unit was suggested to be a chromophore that absorbed at 355 nm. It followed that the CuO_3 molecule must also contain an O_2 unit bonded to a CuO unit, i.e. $OCuO_2$, which may be viewed as replacing an O atom in OCuO by an O_2 unit. This CuO_3 molecule was never observed before and its electronic structure was not known. In analogy to OCuO, the CuO_3 molecule may be regarded as $O^-Cu^{2+}(O_2)^-$. The electronic states of the CuO_3 molecule seemed to group into two regions, one including features X, A, and B and the other C and D. These two groups could be viewed as derived from the X and Y states of CuO perturbed by an O_2 molecule. Indeed, the separation between the two groups of states in CuO_3 was very similar to that between the X and Y states in CuO (Fig. 6).

The weak feature near 1.23 eV in the 266 nm spectrum was due to Cu^-, which must be from photodissociation of $CuO_3^- \rightarrow Cu^- + O_3$. However, this dissociation was not observed at 355 nm. This suggested that there was another isomer in the CuO_3^- beam that might be depicted as $Cu(O_3)^-$ or a Cu/ozone complex. Such a complex had been observed in previous matrix experiments when the Cu/O_2 matrix was photolyzed.[20] From the weak intensity of the Cu^- peak, it was inferred that the abundance of this complex in the CuO_3^- beam was very small.

The spectrum of CuO^- from photodissociation of CuO_3^- was compared in Fig. 13 to that of the direct CuO^- spectrum in the same energy and intensity scales. It was notable that the dissociation product CuO^- was vibrationally very hot as shown in the significant hot band transitions. It was also electronically hot as seen from the strong intensities of the features a and b, which were due to transitions from an excited state of CuO^- (Fig. 6). The feature b in Fig. 13b overlapped with the feature X of CuO_3^- itself. Interestingly, Fig. 13b revealed two additional features, a' and a'',

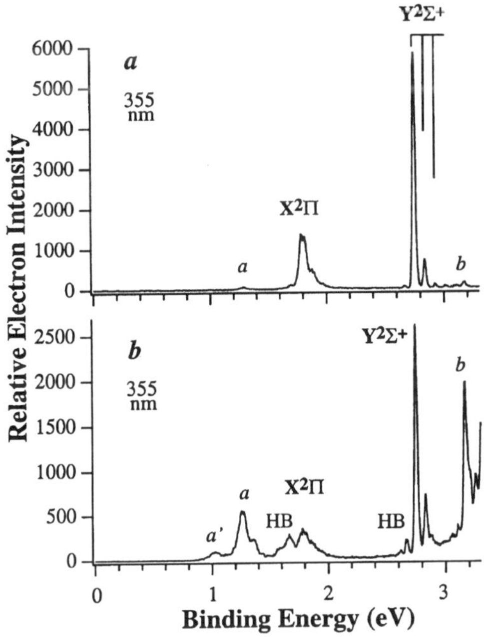

Fig. 13. (a) CuO^- spectrum at 355 nm, same as in Fig. 6; (b) Hot CuO^- spectrum at 355 nm from photodissociation: $CuO_3^- \rightarrow CuO^- + O_2$, same as in Fig. 12. HB stands for hot band transitions. From Ref. 35.

around feature *a*. These extra features were assigned to be from excited state of the CuO^- anion. This was possible because of the considerable internal energy partitioned to CuO^- from the CuO_3^- dissociation. In the CuO^- spectra (Fig. 6), the feature *a* was assigned to be due to a $^3\Pi$ excited state of CuO^-. This state should exhibit a spin-orbit splitting to three components: $^3\Pi_{2,1,0}$. Therefore, the *a* feature was assigned to the $^3\Pi_1$ component and the a' and a'' to the $^3\Pi_0$ and $^3\Pi_2$ components, respectively (Fig. 7 and Table 1). It was not clear why only the $^3\Pi_1$ spin-orbit component was populated in the CuO^- spectra whereas all three spin-orbit components seemed to be populated in CuO^- from the dissociation of CuO_3^-.

3.1.4. CuO_{4-6}^-: O_2-*Complexes*

The spectra of CuO_4^- were obtained at three-photon energies and are shown in Fig. 14. The 355 nm spectrum showed a strong band at 3.09 eV BE,

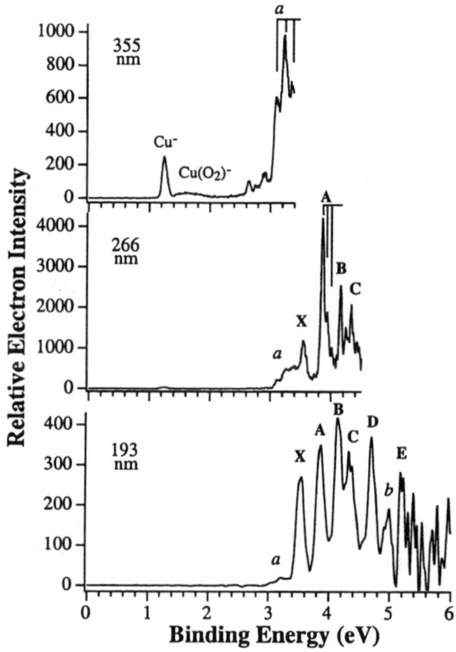

Fig. 14. Photoelectron spectra of CuO_4^- at 355, 266 and 193 nm. From Ref. 35.

labeled as *a*, and other low BE features that were clearly due to Cu^- and $Cu(O_2)^-$. The *a* band was vibrationally resolved with a frequency of about 1010 cm^{-1}. At 266 nm, several new features were observed, *X*, *A*, *B*, and *C* and the *a* band was observed not to be the dominant feature. The *A* band was vibrationally resolved with a frequency of 640 cm^{-1}. At 193 nm, three more features were further revealed, *D*, *b* and *E*, although the spectrum at high BE was quite noisy. Interestingly, the Cu^- and $Cu(O_2)^-$ signals due to photodissociations of the parent CuO_4^- were negligible in the 266 and 193 nm spectra.

The features observed in the 266 nm spectrum of CuO_4^- (*X*, *A*, *B*, *C*) were identical to those of $OCuO^-$ (Fig. 9), except that each feature was shifted to higher BE by 0.06 eV (Table 5). Even the vibrational frequency observed for the *A* band was identical to that in $OCuO^-$. The 193 nm spectrum of CuO_4^- was also identical to that of $OCuO^-$ except for the features *a* and *b*. Therefore, the major features (*X*, *A*, *B*, *C*, *D* and *E*) were assigned to be due to a copper dioxide solvated by an O_2 molecule, $(OCuO^-)O_2$.

Table 5. Observed binding energies (BE) and spectroscopic constants for CuO_4.

	State	BE (eV) [a]	Vib. freq. (cm^{-1})[b]
$(O_2)Cu(O_2)$	*a*	3.09	1010 (90)
	b	4.99	
$OCuO(O_2)$	*X*	3.53	
	A	3.85	640 (80)
	B	4.16	640 (80)
	C	4.34	
	D	4.71	
	E	5.20	

[a]Uncertainty ±0.04 eV.
[b]Relative energies can be determined more accurately.

The 0.06 eV shift in BE was then due to the O_2 solvation effect. Here the O_2 seemed to have very little perturbation to the OCuO molecule. Since it was known that the copper dioxide molecule did not dissociate under our experimental conditions, the observation of Cu^- and $Cu(O_2)^-$ at 355 nm must be due to photodissociations of another CuO_4^- isomer. The feature labeled, *a*, then represented the direct photodetachment transitions from this isomer. Again, the photodissociation was wavelength-dependent since there was little dissociation observed at the higher photon energies. This isomer was less abundant compared to the $(OCuO^-)O_2$ species. We knew, from both CuO_2^- and CuO_3^-, that isomers containing an O_2 bonded to Cu tended to absorb at 355 nm and underwent efficient dissociation. This suggested that the dissociating CuO_4^- isomer contained two O_2 units for which we designated $(O_2)Cu(O_2)^-$. The 1010 cm^{-1} vibrational frequency observed in the *a* band was likely due to an O–O stretching mode. In the 266 nm spectrum, we did not observe any other transitions that could be attributed to this isomer beside the feature *a*. In the 193 nm spectrum, an extra feature, labeled as *b*, was observed that was not due to the $(OCuO^-)O_2$ complex. We attributed this feature to the $(O_2)Cu(O_2)^-$ isomer.

The CuO_6^- mass signal was quite low. We could only take its spectra at 355 and 266 nm (Fig. 15). Interestingly, the spectra CuO_6^- were quite similar to that of CuO_4^-. The 355 nm spectrum again showed features of Cu^- that must be due to photodissociation of the parent CuO_6^- anion. The 266 nm spectrum was identical to that CuO_4^- and showed no energy shift within our experimental uncertainty relative to the features of CuO_4^-

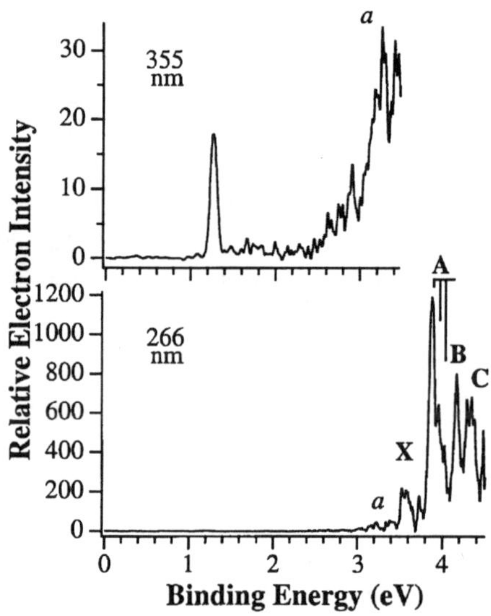

Fig. 15. Photoelectron spectra of CuO_6^- at 355 and 266 nm. From Ref. 35.

Table 6. Observed binding energies (BE) and spectroscopic constants for CuO_6.

$Cu(C_2)_3$	a	BE (eV)[a]	Vib. freq. (cm^{-1})[b]
$Cu(O_2)_3$	a	~3.2	
$OCuO(O_2)_2$	X	3.53	
	A	3.87	640 (90)
	B	4.17	
	C	4.34	

[a]Uncertainty ±0.04 eV.
[b]Relative energies can be determined more accurately.

(Table 6). In analogy to CuO_4^-, we concluded that CuO_6^- also had two isomers. The dominant one yielded the features X, A, B, and C and could be described as a copper dioxide/O_2 complex, $(OCuO^-)(O_2)_2$. The vibrational frequency observed here in the A band was also identical to that in OCuO. The other isomer which was much less abundant gave the feature a and dissociated into Cu^- at 355 nm. This isomer was due to a Cu/O_2 complex, $Cu(O_2)_3^-$. The spectra of CuO_2^-, CuO_4^-, and CuO_6^- at 266 nm are compared in Fig. 16 and their similarity is obvious.

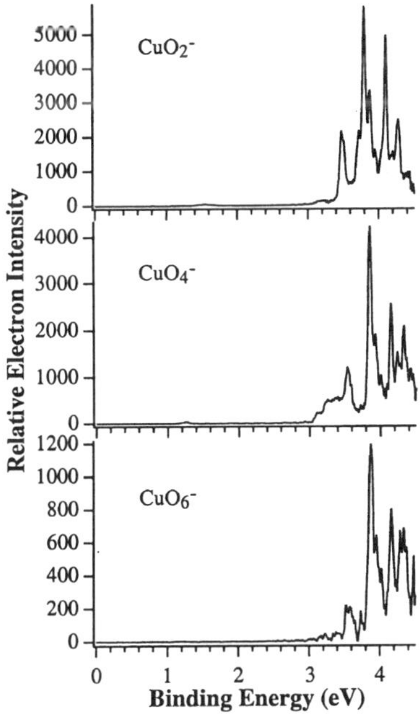

Fig. 16. Comparison of the 266 nm spectra of CuO_2^-, CuO_4^- and CuO_6^-. From Ref. 35.

The spectra of CuO_5^- shown in Fig. 17 were quite similar to that of CuO_3^-. The spectral features showed a slight shift to higher BE (Table 7). The vibrational frequency resolved in the feature *D* was also indentical to that as in CuO_3^-. This suggested that CuO_5^- was from a CuO_3^- solvated by a molecular O_2, $(OCuO_2^-)O_2$. The second O_2 apparently was much more weakly bonded than the first O_2 in CuO_3 since the second O_2 perturbed the CuO_3 spectrum very slightly. The 266 nm spectrum of CuO_5^- is compared to that of CuO_3^- and CuO^- in Fig. 18.

3.1.5. *Structure and Bonding Evolution of* CuO_y $(y = 1–6)$

Figure 19 summarizes all the isomers and the proposed structures for the CuO_y species observed as well as their EAs. Essentially, three types of structures were found, (1) oxides, CuO, OCuO, $OCuO_2$; (2) Cu/O_2 complexes, $Cu(O_2)$, $Cu(O_2)_2$, $Cu(O_2)_3$, and (3) O_2-solvated oxides, $(OCuO)(O_2)$, $(OCuO)(O_2)_2$, $(OCuO_2)(O_2)$. The $Cu(O_3)$ may be considered as a fourth structure. Many of these structures were not known and

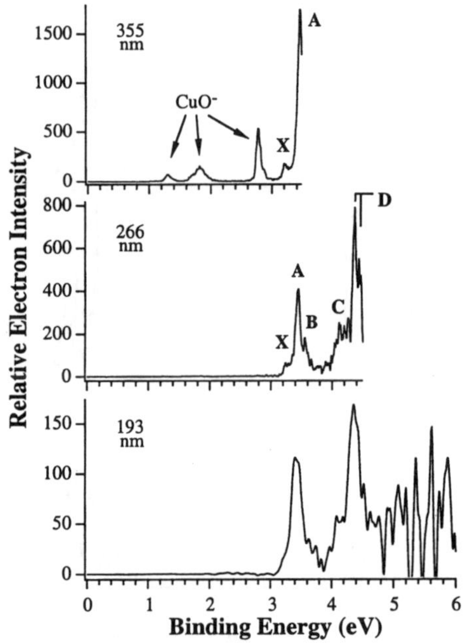

Fig. 17. Photoelectron spectra of CuO_5^- at 355, 266 and 193 nm. From Ref. 35.

Table 7. Observed binding energies (BE) and spectroscopic constants for CuO_5.

	State	BE(eV) [a]	Vib. freq. (cm^{-1})[b]
$OCuO_2(O_2)$	X	3.21	
	A	3.43	
	B	3.55	
	C	4.05	
	D	4.39	660 (80)

[a]Uncertainty ±0.04 eV.
[b]Relative energies can be determined more accurately.

those shown in Fig. 19 were only suggestions. The $Cu(O_2)$ complex had been studied extensively previously. The most recent calculation predicted that it has a C_{2v} ground state (side-on structure) as shown in Fig. 19.[85–87] Based on this, we drew all the Cu/O_2 complexes as side-on bonding. Copper plays an important role in biological dioxygen metabolism.[74,75] The $Cu(O_2)$

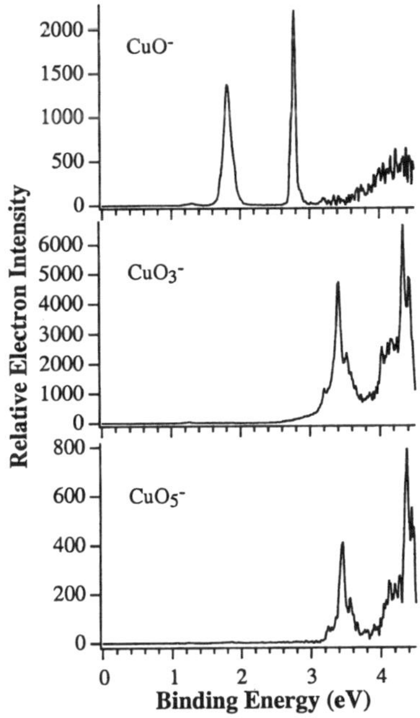

Fig. 18. Comparison of the 266 nm spectra of CuO^-, CuO_3^- and CuO_5^-. From Ref. 35.

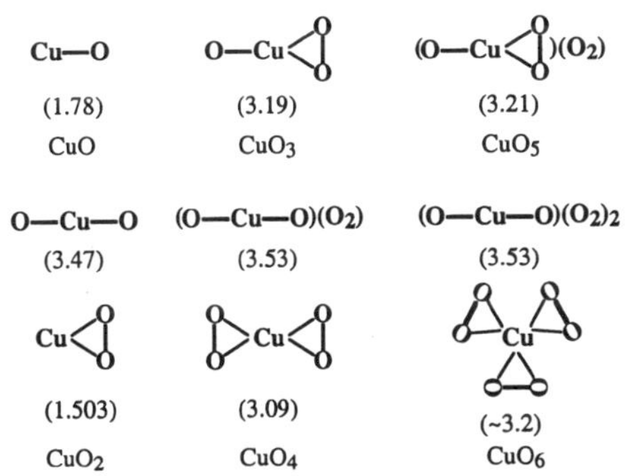

Fig. 19. Schematic diagram of the structures and isomers of the CuO_y ($y = 1$–6) species. The numbers in the parentheses are the measured electron affinities (eV). From Ref. 35.

complex was viewed as the simplest model to study the Cu-O_2 interactions. The new Cu/O_2 complexes provided additional models to study the Cu-O_2 interactions relevant to biological O_2 transport or activation of O_2 by Cu. It is also interesting to consider the oxidation states of Cu in the

CuO_y series. Although in bulk CuO oxide the Cu and O atoms both have oxidation states of two,[73] in the isolated CuO diatomic molecule each is of oxidation state one. The molecule with oxidation state two ($Cu^{2+}O^{2-}$) is significantly higher in energy (Fig. 7). The high oxidation states are stabilized in the bulk because of the Madelung energies or the multiple coordination in the crystal.[73] As y increases, we saw either O_2-complexes or O_2-solvated oxide clusters. The latter were probably formed during the supersonic expansion where the O_2 molecules in the carrier gas condensed onto the smaller CuO_y^- species. This seemed to be particularly efficient for the copper oxide systems, since we did not observe such O_2-solvated oxide clusters among the other transition metal cluster systems under similar experimental conditions, as will be discussed later.

3.1.6. $Cu_2O_y^-$ ($y=1-4$): *Sequential Oxidation and Density Functional Calculations*

We conducted a combined experimental and theoretical study for the Cu_2O_y (y = 1–4) cluster series, using photodetachment PES of the anions and density functional theory (DFT) calculations.[34] Detailed electronic and vibrational information of these clusters was revealed from the PES experiments. We determined the cluster structures by comparing the experimental observables, such as the EAs and the vibrational frequencies, with the DFT calculations. It was found that from Cu_2O to Cu_2O_3 each O atom bridge-bonds to the Cu_2 dimer, accompanied by a systematic EA increase. Cu_2O_4, at which there was no EA increase, assumed a hexagonal planner structure with two O_2 units bridge-bonded to the Cu_2 dimer. The copper atom in these systems was sequentially oxidized from an oxidation state of +1 in Cu_2O to +3 in Cu_2O_3, in which the oxidation was saturated, leading to two O–O units in Cu_2O_4. This observation was consistent with the known oxidation states for copper in the bulk oxides that can have a range of compositions due to the variable oxidation states of copper and suggested that the cluster studies could provide information relevant to the bulk.[91]

Figure 20 shows the PES spectra of $Cu_2O_y^-$ (y = 1–4) with well-resolved ground states (labeled X) and low-lying excited state features (labeled A, B, C, and so on in ascending order). The measured BEs of all the states are listed in Table 8. The adiabatic EA of the neutral clusters, derived from the ground state feature, clearly showed a step-wise increase with y up to 3 and leveled off at y = 4. Extensive vibrational structures were resolved

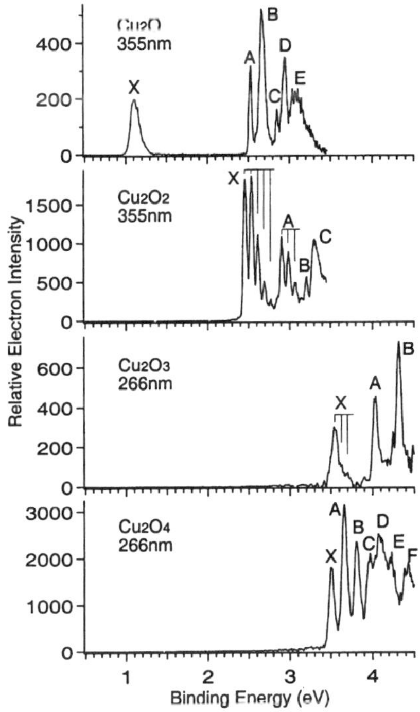

Fig. 20. Photoelectron spectra of $Cu_2O_y^-$ (y = 1–4). The ground states of the neutral clusters are labeled "X", and their low-lying electronic excited states are labeled with the alphabet. The vibrational structures for the "X" and A bands of Cu_2O_2 and the X band of Cu_2O_3 are also indicated. From Ref. 34.

for the X and A bands of Cu_2O_2. However, the spectra of all the other clusters showed relatively sharp electronic transitions with little vibrational excitation, indicating there was little geometry change from the anions to the neutrals. Discernible vibrational structures could be seen for the X band of Cu_2O_3 and the E band of Cu_2O. The peak width of the X band of Cu_2O indicated that there was vibrational excitation. But it could not be resolved even with a 25 meV (200 cm^{-1}) resolution when the 532 nm detachment light was used. The obtained vibrational frequencies are also listed in Table 8.

The PES spectra (Fig. 20) contained a wealth of electronic structure information for the clusters. The Cu_2O spectrum showed a large gap between the X and A bands, representing the energy gap between the lowest

Table 8. Observed binding energies (BE) and vibrational frequencies (ν) of Cu_2O_y ($y = 1$–4) clusters and predicted electron affinities (EA) and ground state vibrational frequencies.

	BE[a] (eV)	ν (exp)[b] (cm^{-1})	EA[c] (eV)	ν (theo)[d] (cm^{-1})
Cu_2O[e]				
X	1.10	<200	1.10	$\nu_1 = 681$[f]
A	2.53			$\nu_2 = 156$
B	2.66			$\nu_3 = 586$
C	2.85			
D	2.95			
Cu_2O_2				
X	2.46	630 (30)	2.12	182, 302, 466
A	2.91	650 (30)		493, 653, 718[f]
B	3.12			
C	3.30			
Cu_2O_3				
X	3.54	(640)	3.03	259, 259, 318
A	4.02			321, 321, 351
B	4.32			608, 608, 678[f]
Cu_2O_{2e}				
X	3.50		2.94	119, 222, 244
A	3.66			267, 277, 354
B	3.80			533, 612, 612
C	3.95			647, 912, 985[f]

[a]Uncertainty: ±0.03 eV. The BE of ground state (*X*) yields the measured adiabatic electron affinity.
[b]Measured symmetric stretching vibrational frequencies.
[c]Calculated adiabatic electron affinities.
[d]Calculated vibrational frequencies for the ground state cluster structures.
[e]More highly excited states are not listed due to their broad and overlapping nature.
[f]Totally symmetric vibrational modes.

unoccupied MO (LUMO) and the highest occupied MO (HOMO) of Cu_2O. The presence of the large HOMO–LUMO gap suggested Cu_2O neutral was electronically closed-shell. Similarities in both spectral feature and intensity ratio were seen between the *A* and *C* bands and between the *B* and *D* bands. These suggested that the origins of these bands should be due to

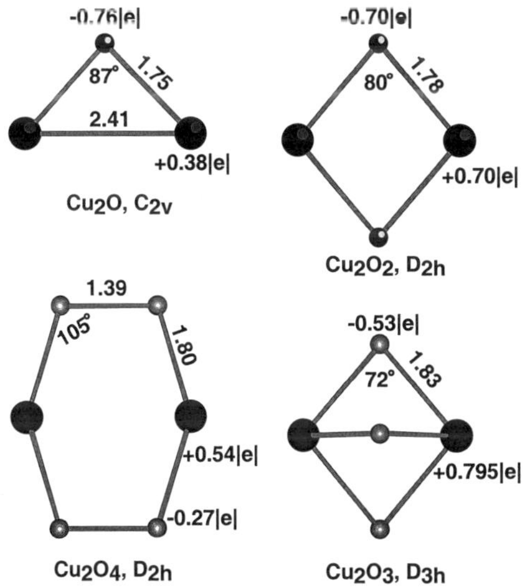

Fig. 21. The optimized structures and Mulliken charge distributions from density functional theory calculations for Cu_2O_y ($y = 1$–4). From Ref. 34.

two triplet-singlet pair states, arising from removing either a spin-down or spin-up electron from the inner filled orbitals just below the highest singly occupied orbital of Cu_2O^-. The details of the electronic structures and chemical bonding of the cluster series were better understood through the DFT calculations.

The fully optimized equilibrium structures of the Cu_2O_y ($y = 1$–4) clusters are shown in Fig. 21, including geometric parameters and atomic charges from Mulliken population analyses. The ground states of Cu_2O and Cu_2O_2 were found to be a C_{2v} triangle and a D_{2h} rhombus, respectively. For Cu_2O_3, two isomers with close energies were found: a D_{3h} bipyramid (Fig. 21) and a C_{2v} bent structure with an O–Cu–O–Cu–O atomic arrangement. The C_{2v} structure was slightly favored by about 0.5 eV, but it yielded an EA about 1 eV higher than the experimental value. The D_{3h} structure gave an EA that was in better agreement with the experiment and was responsible for the observed PES spectrum. A D_{2h} hexagonal ring with two O–O bonds was found to be the most stable structure for Cu_2O_4. The calculated EAs agreed well with the experimental values, as shown in Fig. 22. For Cu_2O_3, the D_{3h} isomer was chosen for comparison because the calculated EA for the C_{2v} bent isomer of Cu_2O_3 (4.4 eV) was out of the scale

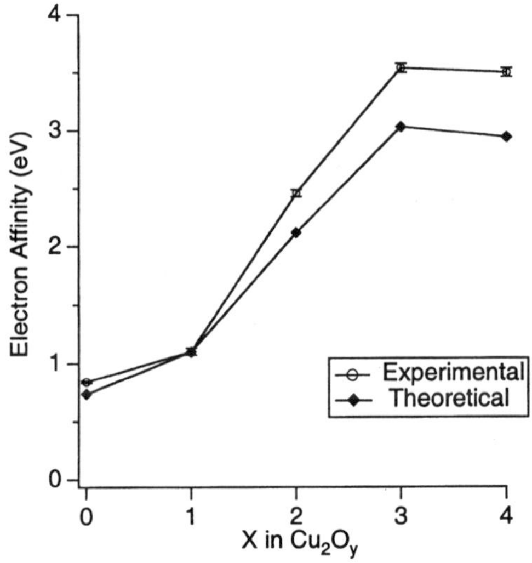

Fig. 22. Comparison of the calculated adiabatic electron affinities to that from the experiments. From Ref. 34.

of Fig. 22. The four clusters were all found to have closed-shell electronic structure in the neutral ground states. The calculated HOMO–LUMO gap for Cu_2O was 1.2 eV and in the range of 0.2–0.6 eV for $Cu_2O_{2\text{–}4}$, in good agreement with the experiments.

Vibrational structures resolved in the PES experiments provided important information for the cluster structures. The calculated vibrational frequencies (Table 8) for the Cu_2O_y clusters agreed well with the observed ones for $x = 1\text{–}3$, where vibrational information was available. The unresolved vibrational structure in the X band of Cu_2O largely involved the Cu–Cu stretching vibration since the LUMO of Cu_2O was mainly of Cu $4s\sigma$ bonding character. The experimental estimate of the upper bound of this mode was 200 cm^{-1}, which was consistent with the calculation. The 630 cm^{-1} vibrational frequency resolved for the ground state of Cu_2O_2 was due to the totally symmetric mode because only totally symmetric vibrational modes were expected to be active in PES experiments. The calculated frequency for this mode was 724 cm^{-1}, in reasonable agreement with the experiment. The estimated 640 cm^{-1} frequency for the ground state of Cu_2O_3 also compared favorably to the totally symmetric mode of 678 cm^{-1} from the calculation.

Several interesting trends were observed in the evolution of the atomic structures and chemical bonding in the Cu_2O_y clusters. First, the Cu–O bond length steadily increased as more oxygens (y = 1–3) were added to the Cu_2 dimer, all occupying the bridge positions. The increase in Cu–O bond length was accompanied by a decrease of atomic charges at the respective O sites. The atomic charge distribution suggested significant ionic bonding between Cu–O. Second, the Cu–O–Cu bond angle decreased from near 90° in Cu_2O to about 70° in Cu_2O_3. The smaller bond angle allowed the bridge-bonded oxygens to separate further away from each other and induced a shortening of the Cu–Cu distance. Third, the oxidation state of the copper atoms in the clusters as formally indicated by the number of directly connected oxygens raised from +1 in Cu_2O to +3 in Cu_2O_3, and then dropped to +2 in Cu_2O_4. The lowering of the oxidation state for Cu_2O_4 was caused by the formation of two O–O bonds.

Therefore, an interesting trend of sequential oxidation and oxidation saturation was clearly observed. The Cu atoms were oxidized formally from +1 to +3 from Cu_2O to Cu_2O_3 saturating at Cu_2O_4, where the further increase in the oxygen content did not lead to further oxidation and the coordination of copper was reduced to two. This was consistent with the known chemistry of copper, whose common oxidation states were +1 and +2. The +3 oxidation state was rare. The +4 oxidation state was extremely rare and only occurred in certain compounds of fluorine,[92] the most oxidizing chemical element. Therefore, it was not surprising that the +4 oxidation state was not realized in Cu_2O_4, which should instead be considered as a peroxide. The sequential oxidation had a strong effect on the stability of the extra charge as manifested by the large stepwise EA increase from Cu_2O to Cu_2O_3, caused by the increasing charge depletion on the Cu sites.

The Cu–O bond in the Cu_2O_y clusters was complicated involving both ionic and covalent characters. The best case for illustrating the polarized covalent bonding nature between Cu–O was the Cu_2O triatomic, that had a triangular structure with a nearly perfect Cu–O–Cu bond angle for the O $2p$-bonding. It was seen clearly from Fig. 21 that there was a large charge transfer from Cu to O. Further analysis, based on the single particle density matrix and in terms of MOs, showed that the charge transfer was mainly from the Cu $4s$ shell to the O $2p$ shell. The occupation of the O $2s$ shell changed very little ($< 0.1|e|$). On the Cu atoms, the $4s$ electron was partially promoted to the $4p$ orbitals; the sp hybrids thus formed were spatially more oriented than the spherical 4s orbitals for optimal overlaps with the O 2p. The Cu 3d orbitals also interacted with the O 2p but the net contributions to the Cu–O bonding was negligible because of the almost exact cancellation between the bonding and antibonding components. The Cu–Cu interactions in the Cu_2O_y (y = 1–4) clusters were essentially electrostatic. The covalent interaction was found to be negligible due to the cancellation between the bonding and antibonding overlap contributions. Therefore, the bonding in the Cu_2O_y clusters was dominated by the strong polarized covalent interaction between Cu–O and the intramolecular O (O_2) Coulomb repulsion.

3.2. NiO_y^- ($y = 1, 2$)

Nickel is next to copper in the periodic table and expected to have similar bonding with oxygen as copper. However, the open d-shell in Ni is expected to result in much more complicated electronic structures for Ni-containing species. There have been several calculations[93–98] and experimental studies[99–104] about the nickel monoxide molecule (NiO), whose ground state is known to be a $^3\Sigma^-$ state. But very little is known about its electronic excited states. The Ni(O_2) complex has been studied both experimentally[105,106] and theoretically.[86,107] In the CuO_2 system, both a linear dioxide OCuO and a Cu(O_2) complex are known to exist, as discussed above. However, it was not known if there existed a similar nickel dioxide molecule until very recently.[106] We reported the first photodetachment PES study of NiO^- and NiO_2^- and indeed found two isomers for NiO_2, the nickel dioxide (ONiO) and nickel-O_2 complex [Ni(O_2)], Ref. 36 similar to the CuO_2 system. Several closely spaced and previously unknown low-lying states were observed and assigned for NiO. Again, the dioxide was found to have a much higher EA than that of the Ni(O_2) complex. After our work appeared, a new PES study on NiO^- was published by Moravee and Jarrold who disputed some of our original assignments.[108] The NiO^- spectra will be discussed with the new study in mind and a new set of assignments will be proposed.

3.2.1. *NiO and NiO⁻: Complicated Low-lying Electronic States of NiO*

Figure 23 shows the PES spectra of NiO^- at three-photon energies. The 532 nm spectrum revealed two electronic states of the NiO molecule with

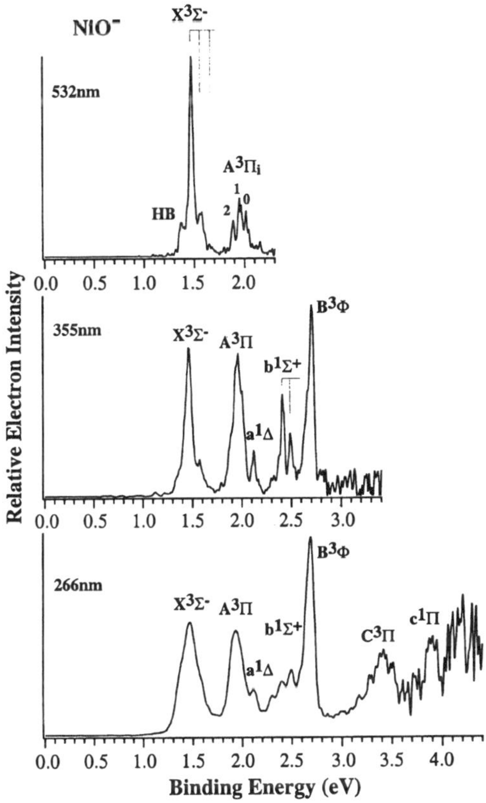

Fig. 23. Photoelectron spectra of NiO^- at 532, 355 and 266 nm. The spectra assignments and vibrational structures are indicated. HB stands for hot bands. From Ref. 36.

partially resolved vibrational structures. Three more states at 2.1, 2.4 and 2.7 eV were further revealed at 355 nm. The 2.1 eV feature exhibited a strong photon energy dependence. The 2.4 eV feature showed a vibrational progression while the 2.7 eV peak was sharp and intense. More states were revealed at 266 nm, but the signal-to-noise ratio was rather poor at the high BE side. Nevertheless, at least two states could be discerned at about 3.3 and 3.8 eV. The BEs of all these states and the obtained spectroscopic constants are listed in Table 9.

Several calculations had been performed on the electronic structure of NiO and its ground state properties were well understood.[93–104] The

Table 9. Observed binding energies (BE) and spectroscopic constants for NiO^- and NiO.

	State	BE (eV)	Term value (eV)		Vib. freq. (cm^{-1})	
			This work	Previous	This work	Previous
NiO	$X\,(^3\Sigma^-)$	1.46 (2)[a]	0		800 (50)	839[d]
	$A\,(^3\Pi_2)$	1.89 (4)	0.43 (4)			
	$(^3\Pi_1)$	1.95 (5)	0.49 (5)			
	$(^3\Pi_0)$	2.01 (6)	0.55 (6)	0.537[b]		780.8[b]
				0.532[c]		777.1[c]
	$a\,(^1\Delta)$	2.12 (5)	0.66 (5)			
	$b\,(^1\Sigma^+)$	2.41 (4)	0.94		680 (60)	
	$B\,(^3\Phi)$	2.70 (2)	1.24 (3)			
	$C\,(^3\Pi)$	3.26 (10)	1.80 (10)			
	$c\,(^1\Pi)$	3.84 (10)	2.38 (10)			

[a]Adiabatic electron affinity of NiO.
[b]From Ref. 103.
[c]From Ref. 102.
[d]From Ref. 101.

bonding between Ni and O can be viewed as Ni^+O^- with substantial covalent character. The valence MOs of NiO can be considered as derived from Ni^+ and O^-, as schematically shown in Fig. 24. The $^3\Sigma^-$ ground state of NiO was well established with a $1\sigma^2 1\pi^4 1\delta^4 2\sigma^2 2\pi^2$ configuration. The 1σ and 1π MOs are bonding orbitals between $O2p\sigma$–$3d\sigma$ and $O2p\pi$–$3d\pi$, respectively; the 2σ and 2π MOs are slightly antibonding; the 1δ is a non-bonding MO with pure $3d$ character. In the NiO^- anion, the extra electron was expected to enter the 2π orbital to give a $1\sigma^2 1\pi^4 1\delta^4 2\sigma^2 2\pi^3$ configuration, with a $^2\Pi$ ground state. As pointed out in the recent theoretical work,[98] the three MOs, 1δ, 2σ and 2π, were relatively close in energy and were well separated from the 1σ and 1π bonding MOs. Therefore, it was possible that photodetachment could take place from all these three MOs at the three-photon energies used in our experiments. The configurations and electronic states resulting from single-electron detachment are as follows:

$$1\sigma^2 1\pi^4 1\delta^4 2\sigma^2 2\pi^3 (^2\Pi) \rightarrow 1\sigma^2 1\pi^4 1\delta^4 2\sigma^2 2\pi^2 : \quad ^3\Sigma^-\,,\, ^1\Delta\,,\, ^1\Sigma^+ ; \qquad (1)$$

$$\rightarrow 1\sigma^2 1\pi^4 1\delta^4 2\sigma^1 2\pi^3 : \quad ^3\Pi\,,\, ^1\Pi ; \qquad (2)$$

$$\rightarrow 1\sigma^2 1\pi^4 1\delta^3 2\sigma^2 2\pi^3 : \quad ^3\Phi\,,\, ^3\Pi\,,\, ^1\Phi\,,\, ^1\Pi . \qquad (3)$$

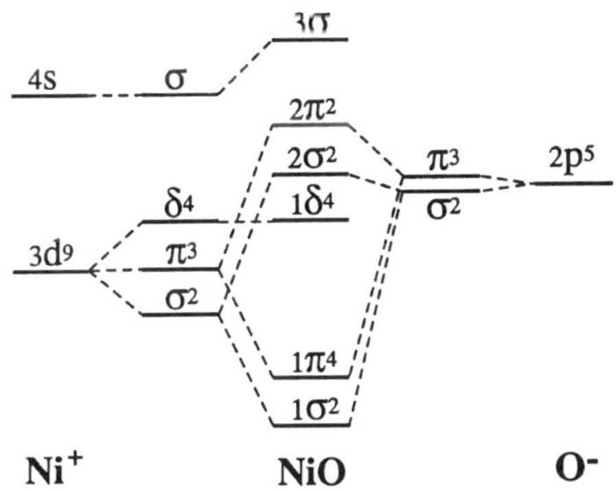

Fig. 24. A schematic molecular orbital diagram for the NiO molecule. From Ref. 36.

All of these states could be accessed in the photodetachment transitions, depending on their excitation energies relative to the ground state. However, besides the ground state ($X^3\Sigma^-$), only six excited states were observed. The assignments were indicated in Fig. 23 and the obtained spectroscopic constants are given in Table 9.

The vibrational frequency obtained for the ground state ($X^3\Sigma^-$) was 800 (50) cm^{-1} which agreed with the previous optical measurements (Table 9). The first excited state at a BE of 1.89 eV appeared to consist of a vibrational progression. However, the separation among the three features was about 60 meV (480 cm^{-1}), which seemed to be too small compared to that of the ground state. Previous optical experiments had identified a $^3\Pi_i$ excited state at about 0.53 eV above the ground state for the transition between the $^3\Sigma^-$ ground state and the $^3\Pi_0$ spin-orbit state.[102,103] This excitation energy was in excellent agreement with that corresponding to the third feature at a BE of 2.01 eV (Table 9). Thus this peak was assigned to the $^3\Pi_0$ component of the $^3\Pi$ state resulting from (2) by removing a 2σ electron. Since the MO configuration corresponding to the $^3\Pi$ state was more than half-filled, the spin-orbit components had the following order, $^3\Pi_2$, $^3\Pi_1$, $^3\Pi_0$, with increasing energy. Therefore, the peak at 1.89 eV BE was assigned to the $^3\Pi_2$ spin-orbit state, which was not observed in the previous optical experiments. The second peak at a BE of 1.95 eV was then assigned to the $^3\Pi_1$ spin-orbit state, as shown in Fig. 23. Note that the 1.89 eV feature was assigned to be from an electronically excited state of the NiO^- anion by Moravee and Jarrold.[108] Our assignment yielded a pin-orbit splitting (about 480 cm^{-1}) which was much larger than that measured in the previous optical experiments (283.37 cm^{-1} from Ref. 103 and 349.21 cm^{-1} from Ref. 102 between the $\Omega = 0$ and 1 components). The currently obtained spin-orbit splitting was reasonable compared to that found in Ni atom,[71] and clearly more accurate calculations or much higher resolution PES spectrum of NiO^- would be required to resolve this issue. In any case, there was no doubt that the first excited state of the NiO molecule is the $A\,^3\Pi$ state from (2). Each spin-orbit state was expected to have its own vibrational progression, which might overlap with states from the higher spin-orbit states. This overlap probably complicated the PES spectra in Fig. 23.

The feature near a BE of 2.12 eV, which was shown more clearly in the 355 and 266 nm spectra, was tentatively attributed to be due to the $\nu = 1$ vibrational state of the $^3\Pi_0$ spin-orbit state. Moravee and Jarrold resolved this feature slightly better and also did angular dependent studies. They assigned this feature to be the $^3\Phi$ state from (3). However, the intensity of this transition was too weak for the $^3\Phi$ state, which was expected to give an intense and sharp feature. Thus, we reassign this feature to be due to the $a\,^1\Delta$ state and subsequently reassign the 2.4 eV feature to be due to the $b\,^2\Sigma^+$ state. Both of these states are from detachment channel (1) above and have the same configuration as the ground state. The weaker intensities of these features are consistent with their being singlet states. The $b\,^1\Sigma^+$ state has a nice vibrational progression with a vibrational frequency of 680 (60) cm^{-1} and a 0.94 eV excitation energy from the ground state. The sharp and simple vibrational progression observed suggested that there was no spin-orbit splitting complication, consistent with the assignments.

The state with a BE of 2.70 eV was sharp and intense with an excitation energy of 1.24 eV above the ground state. It showed no vibrational progression, suggesting that the detached electron was from a nonbonding orbital. It was natural to assign this state to the $^3\Phi$ state, due to the detachment of a nonbonding δ electron as shown in (3) above. The width of the observed peak was about 75 meV (FWHM), much broader than experimental resolution at this energy, indicating that this peak probably contained spin-orbit components, consistent with the assignment to the $^3\Phi$ state. Moravee and Jarrold indeed resolved this feature into three components, except that they assigned them to the $^3\Pi$ state from detachment channel (3).

The higher excited states observed at the 266 nm in Fig. 23 were less well defined due to the poor signal-to-noise ratio. The state at a BE of

3.26 eV showed a vibrational progression or spin-orbit components and gave an excitation energy of about 1.8 eV. We tentatively assigned this feature to the $^1\Sigma^+$ state of (1) above previously.[36] We now reassign this feature to the $^3\Pi$ state instead due to detachment channel (3). Another state at a BE of 3.84 eV was tentatively assigned to the $^1\Pi$ state of (2), arising from the same configuration as that of the $A\,^3\Pi$ state. The two higher energy singlet states arising from detachment of a 1δ electron ($^1\Phi$ and $^1\Pi$) were expected to give sharp features similar to the $^3\Phi$ state and probably had BEs above 4 eV.

The $A\,^3\Pi$ state was the only previously characterized excited state for the NiO molecule experimentally. Theoretically, Bauschlicher *et al.* calculated an excitation energy for the $^3\Pi$ state at 0.518 eV (4180 cm^{-1}) which agreed well with the experiments.[94–96] They also calculated an excitation energy for the $^3\Phi$ state to be 0.853 eV (6880 cm^{-1}), which was apparently too small compared to the experiments. This calculated excitation energy was the main argument that Moravee and Jarrold used to assign the 2.12 eV PES feature. Walch and Goddard have calculated 20 low-lying excited states for NiO, but the ordering of the states disagreed with the experiments.[93] In particular, the first excited state was calculated to be the $^1\Delta$ state. Nevertheless, our reassignment of the 2.12 eV feature to the $^1\Delta$ state places the excitation energy of this state closer to the first excited state. Clearly, more accurate calculations for the full set of low-lying states of NiO would be needed to compare with the available experimental data and help their ultimate assignments.

3.2.2. NiO_2 *and* NiO_2^-*: Two Isomers of* NiO_2 *–* $Ni(O_2)$ *and ONiO*

Figure 25 shows the spectra of NiO_2^- at 532 and 355 nm when a large "waiting room" nozzle[6] was used to produce the NiO_2^- species. The 532 nm spectrum showed a sharp feature at 0.82 eV and the onset of a broad band following an energy gap. There were weak features around 1.2 eV which could be attributed to detachment features of Ni^-. This suggested that there was a weak dissociation channel, $NiO_2^- \rightarrow Ni^- + O_2$. As discussed earlier, similar dissociation processes were observed for the $Cu(O_2)^-$ complex, suggesting that there existed a similar $Ni(O_2)^-$ complex. The 355 nm spectrum showed a broad band at about 2 eV, in addition to the sharp peak at 0.82 eV. However, much stronger features were revealed near 3 eV. Note the $\times 10$ amplification for the features due to the $Ni(O_2)^-$ complex.

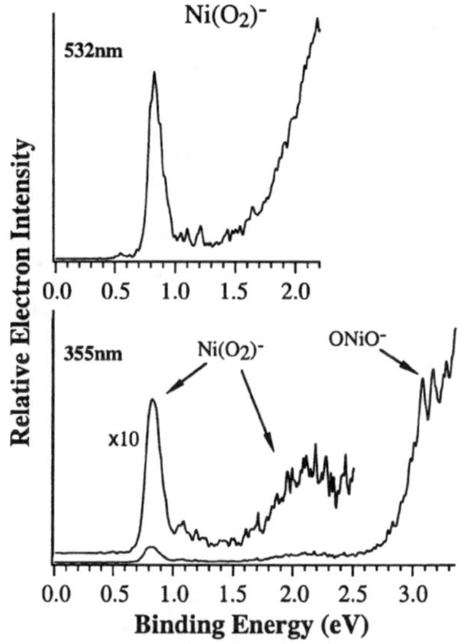

Fig. 25. Photoelectron spectra of $Ni(O_2)^-$ complex at 532 and 355 nm. The 355 nm spectrum also contains spectral features from the $ONiO^-$ dioxide molecule (see Fig. 26). Weak features around 1.2 eV are due to Ni^- as a result of photodissociation of the parent $Ni(O_2)^-$ anion. From Ref. 36.

Figure 26 shows the spectra of NiO_2^- at 355 and 266 nm when a nozzle without the large waiting room was used.[6] The 355 nm spectrum was drastically different from the one shown in Fig. 25. The weak features at lower BE disappeared and the strong features near 3 eV became a nice vibrational progression. A peak near 3.45 eV was also observed, which became an intense and sharp feature in the 266 nm spectrum (Fig. 26, lower panel). In addition, the 266 nm spectrum revealed one extra sharp feature near 3.82 eV, with indications of even higher BE features. Therefore, it was clear that this was a different isomer of NiO_2^-. The first isomer (Fig. 25) was a $Ni(O_2)$ complex. On the basis of the high EA of the second isomer and analogous to that of CuO_2, the second isomer was assigned to the $ONiO^-$ dioxide. The 355 nm spectrum shown in Fig. 25 clearly contained features from both isomers while Fig. 26 showed features only from the $ONiO^-$ isomer. The different populations of the two isomers were due to the two different nozzles used. It was also noted that the vibrational progression observed in the 355 nm spectrum of $ONiO^-$ seemed to disappear in the 266 nm spectrum (Fig. 26). The BEs and the obtained

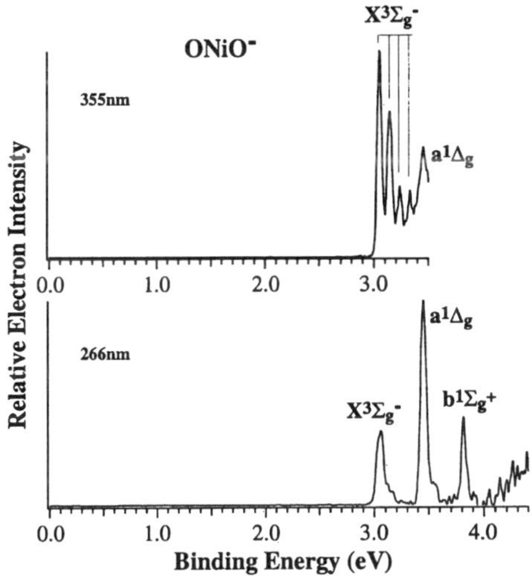

Fig. 26. Photoelectron spectra of $ONiO^-$ dioxide at 355 and 266 nm. Note the high electron affinity of the ONiO molecule and the well-resolved vibrational progression in the 355 nm spectrum. From Ref. 36.

Table 10. Observed binding energies (BE) and spectroscopic constants for $Ni(O_2)$ and ONiO.

	State	BE (eV)	Term Value (eV)	Vib. Freq. (cm^{-1})
$Ni(O_2)$	X	0.82 (3)[a]	0	
	A	~1.8	~1	
ONiO	X ($^3\Sigma_g^-$)	3.05 (1)[b]	0	750 (30)
	a ($^1\Delta_g$)	3.45 (2)	0.40 (2)	
	b ($^1\Sigma_g^+$)	3.82 (3)	0.77 (3)	

[a]Adiabatic electron affinity of $Ni(O_2)$.
[b]Adiabatic electron affinity of ONiO.

spectroscopic constants are listed in Table 10 for both the $Ni(O_2)$ complex and the ONi dioxide.

The $Ni(O_2)$ complex has a symmetric side-on C_{2v} structure, as conclude from two previous theoretical calculations and two low temperature matrix experiments.[105–107] The calculations indicated that the $Ni(O_2)$ complex had a singlet ($^1A_1$) ground state.[86,107] The interaction between Ni and O_2 involves both ionic and covalent bonding: the ionic character comes from the transfer of the Ni 4s electrons to the out of plane π_g orbital on O_2 and the covalent bonding is between the Ni 3d and the in-plane π_g orbital. The HOMO and LUMO of Ni(O_2) are derived from the bonding and antibonding interactions between the $3d_{xz}$ and the π_g orbitals, respectively. The EA for Ni(O_2) was measured to be 0.82 (3) eV, which was smaller than the EA of both Ni (1.156 eV)[40] and NiO, but higher than that of O_2 (0.451 eV).[109] The low EA for Ni(O_2) was consistent with a weakly-bonded complex and suggested that the extra electron in the Ni(O_2)$^-$ anion was probably shared between Ni and O_2, as indicated by the nature of the Ni(O_2) LUMO, which is derived from the combination of the Ni 4s and O_2 π_g orbitals. The narrow peak width of the threshold feature was in sharp contrast to the broad Franck–Condon envelope observed for the detachment of O_2^-.[110] This implied that the extra electron in Ni(O_2)$^-$ was probably more localized on the Ni atom, understandably due to the high EA of the Ni atom. The observed energy gap in the PES spectra was also consistent with the theoretical calculations that the Ni(O_2) complex was a closed-shell, singlet molecule (X^1A_1).

The detachment transition to the excited states of Ni(O_2) (Fig. 25) displayed a broad feature starting at 1.8 eV, i.e. about 1 eV above the ground state, indicating a large geometry change from the ground state of the anion. As suggested from the previous theoretical works, the O–O bond was significantly stretched in Ni(O_2). We suspected that there might be an O–O bond rupture in the excited state of Ni(O_2) to form the ONiO molecule, suggesting the reaction barrier between Ni + O_2 was less than 1 eV. Alternatively, there might be a predissociation in the excited state of Ni(O_2). This would suggest the Ni–O_2 bond energy to be less than 1 eV (23 kcal/mol). The latter suggestion was consistent with the calculation by Blomberg *et al.*, who obtained a bond energy of 18 kcal/mole,[107] but disagreed with the calculation by Bauschlicher *et al.*, who obtained a bond energy of 48 (7) kcal/mol.[86]

The only calculation for the ONiO dioxide was done by Blomberg *et al.*, who considered the O_2 chemisorption on Ni surfaces and used Ni + O_2 as the simplest case.[107] The room temperature reaction between Ni and O_2 led to the Ni(O_2) complex and the ONiO dioxide was never observed in the gas phase. The formation of the ONiO$^-$ molecule in our laser vaporization experiment was due to the high energy conditions of the laser-induced plasma. The ONiO molecule was recently observed in a low temperature matrix experiment, also produced from a laser vaporization source.[106] The PES spectra in Fig. 26 yielded a very high EA of 3.05 eV for the ONiO molecule, much higher than that of the NiO molecule. This was consistent with a dioxide molecule, suggesting that the oxidation state of Ni was increased.

The bonding of the ONiO dioxide was expected to be similar to the linear OCuO molecule. It is likely that the ONiO molecule is also linear. A similar conclusion was arrived from the matrix experiment, that yielded an antisymmetric vibrational frequency (ν_3) of 945.4 cm^{-1} for the linear ONiO molecule.[106] Our measured frequency of 750 cm^{-1} for the ground state of ONiO shound correspond to the symmetric vibrational mode (ν_1). In analogy with OCuO, the bonding in ONiO may be viewed as $O^-Ni^{2+}O^-$, including both ionic and covalent interactions. Figure 27 shows a schematic MO diagram adapted from that for OCuO (Fig. 11) The OCuO molecule has a single $1\pi_g$ hole (π_g^3) with a $^2\Pi_g$ ground state, while the $OCuO^-$ anion has a closed shell configuration with a singlet ground state $^1\Sigma_g^+$. The Ni atom has one less valence electron than the Cu atom and the ONiO molecule was expected to have two $1\pi_g$ holes (π_g^2), with a $^3\Sigma_g^-$ ground state. The $ONiO^-$ anion then was expected to have a $1\pi_g^3$ configuration similar to the OCuO neutral ground state. Detachment of one $1\pi_g$ electron would result in two excited state ($^1\Delta_g$ and $^1\Sigma_g^+$), along with the $^3\Sigma_g^-$ ground state. The observed PES features were assigned to these three states as

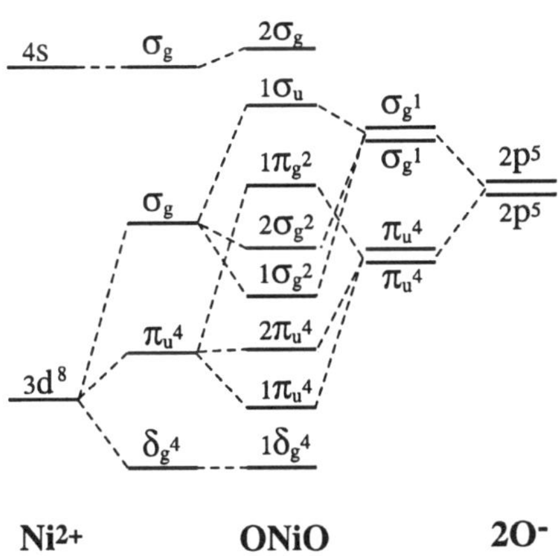

Fig. 27. A schematic molecular orbital diagram for the linear ONiO dioxide molecule. From Ref. 36.

shown in Fig. 26. The three observed states showed similar sharp spectral features and no spin-orbit splitting was apparent. The latter would broaden and complicate the spectral features. These observations were consistent with the assignments.

3.3. $Fe_xO_y^-$ ($x = 1-16$; $y = 1-6$)

Despite the wide interest in the iron-oxygen systems, few iron oxide molecules have been observed and studied. The FeO monoxide is one of the most studied and well-understood iron oxide species.[5,44,45,97,111–113] Even though many studies have been directed at the FeO_2 species[114–120] due to its relevance to oxygen transport in biological systems,[121] its true ground state structure and electronic structure are still not completely understood. There have been few observations of the higher mono-iron oxides.[115,116] and little understanding of their structure and bonding. Di-iron oxo complexes are common structural motif in a class of metalloproteins that are important as oxygen carriers and methane oxygenases.[122,123] Many inorganic complexes containing di-iron oxo groups have been synthesized and characterized.[124–126] However, few di-iron oxide molecules have been studied in the gas phase.

Iron oxides materials and thin films are some of the most important transition metal oxides. There have been extensive studied on the oxidation of single crystal iron surfaces and elementary studies on the reaction between iron and oxygen. However, a complete understanding about the interaction of oxygen with iron and the iron oxide surfaces is still lacking. Well-controlled studies on small iron oxide clusters not only provide a new avenue to obtain detailed information about the interaction between oxygen and iron but also provide models for the iron oxide materials[127] and iron oxide surfaces,[91] which are of major technical importance. Bulk iron oxides have several different forms ranging from FeO to Fe_3O_4 (magnetite) to Fe_2O_3 (hematite) each with interesting properties. Their electronic structures have been subjects of many investigations.[73,128,129] Generally, these compounds are viewed as charge transfer oxides with strong ionic interactions between the iron and oxygen atoms. From FeO to Fe_2O_3, the oxidation state of the Fe atom is increased from +2 to +3 while Fe_3O_4 is a mixed valency compound containing Fe atoms with both +2 and +3 oxidation states. All the three oxide phases can be formed during oxidation of iron surfaces from FeO to Fe_2O_3 with increasing oxygen exposure and substrate temperature.[130]

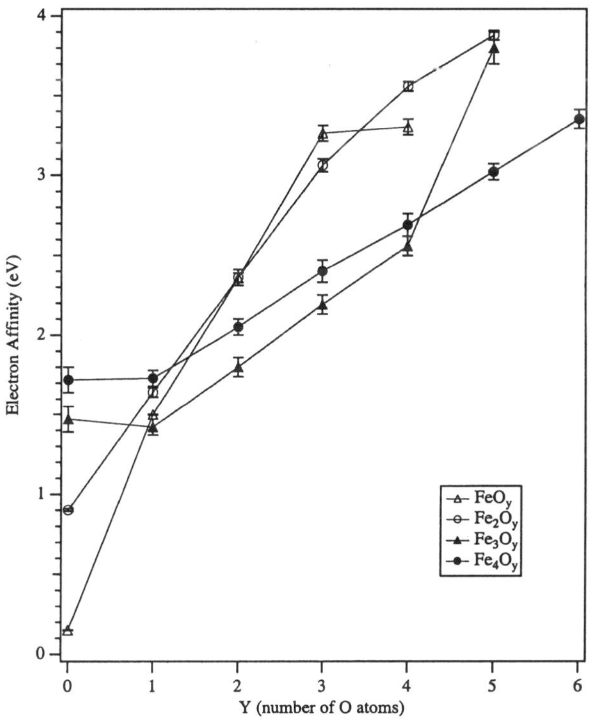

Fig. 28. The electron affinity of the Fe_yO_y clusters versus the number of O atoms. From Ref. 32.

Not surprisingly, we have focused considerable attention on the iron oxide clusters and obtained to most extensive data set concerning the $Fe_xO_y^-$ species with $x = 1$–4 and $y = 1$–6.[31,32] We have further studied clusters with a single oxygen impurity, Fe_xO^- with $x = 5$–16.[30] Our most interesting observation was the idea of sequential oxidation, i.e. for each cluster series with fixed x, we found that the EA of the oxide clusters increase with y up to the maximum oxidation states of the metals, as shown in Fig. 28. This led to the suggestion that for the Fe_3O_y and Fe_4O_y clusters, the O atoms can be viewed as sequentially adsorbed on to the cluster surfaces of Fe_3 or Fe_4. A simple electrostatic model involving a charge transfer from Fe_x to O was proposed to explain the linear dependence of EA of y. Thus, the Fe_xO_y clusters could really be viewed as microscopic models for bulk oxides or surfaces.[32]

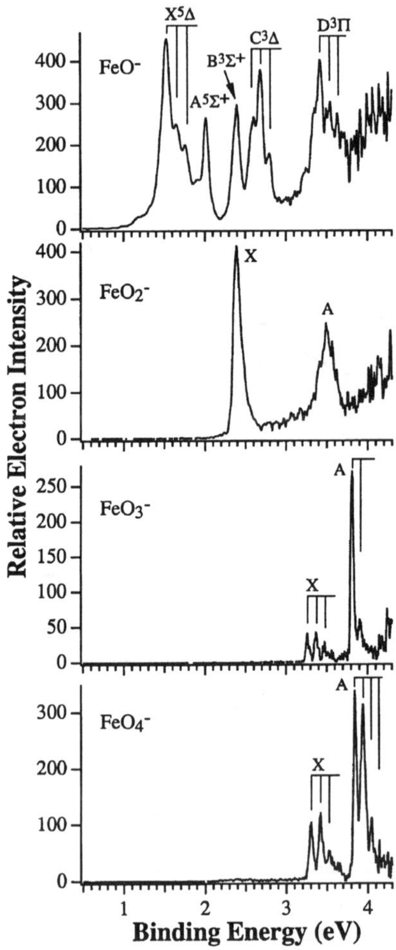

Fig. 29. Photoelecton spectra of FeO_y^- ($y = 1$–4) at 266 nm. The vertical lines indicate the resolved vibrational structures. X represents the ground state of the neutrals and the other letters represent the excited states. Note the sharp, well-resolved, and rather similar spectra for FeO_3^- and FeO_4^-. From Ref. 31.

3.3.1. FeO_y^- ($y = 1$–4): *Sequential Oxidation of Fe*

The PES spectra of FeO_y^- ($y = 1$–4) at 266 nm are shown in Fig. 29. We also measured the spectra of FeO^- and FeO_2^- at 355 nm with slightly higher resolution. Previously, Engelking and Lineberger measured the spectrum of FeO^- at 488 nm and only observed the first two transitions (X and A).

Table 11. Observed binding energies (BE) and spectroscopic constants for FeO_y (y = 1–4).

	State	BE (eV)	Term value (eV)[b]	Vib. freq. (cm^{-1})[b]
FeO	X	1.50 (4)	0	950 (80)
	A	1.98 (4)	0.48 (2)	
	B	2.36 (4)	0.86 (2)	
	C	2.56 (4)	1.06 (2)	800 (60)
	D	3.39 (6)	1.89 (3)	850 (90)
FeO_2	X	2.36 (3)	0	
	A	3.31 (6)[a]	0.95 (4)	
FeO_3	X	3.26 (4)	0	850 (50)
	A	3.81 (4)	0.55 (2)	850 (50)
FeO_4	X	3.30 (4)	0	920 (50)
	A	3.84 (4)	0.54 (2)	850 (50)

[a]Estimated adiabatic binding energy. The vertical binding energy is 3.46 eV as determined from the band maximum.
[b]Relative energies can be determined more accurately.

Four excited states of FeO were clearly observed in the 266 nm spectrum of FeO^-. The BE and term values are listed in Table 11. Two major features (X and A) were observed for FeO_2^- at 266 nm with some possible higher BE features. Unlike NiO_2^- and CuO_2^-, no $Fe(O_2)$ weakly-bonded complexes were observed, which would be expected to have low EA, although such complexes were observed in matrix IR experiments of the Fe/O_2 system.[114]

The FeO^- spectra exhibited a strong photon energy dependence. Notably, the $A\,^5\Sigma^+$ peak was the strongest feature in the 3.49 eV spectrum while its intensity was substantially reduced with respect to the other features at 266 nm. The MO diagram and ordering of FeO are shown schematically in Fig. 30. In the anion, the extra electron entered the 9σ orbital.[45,111] The X, A, B, and C states all resulted from removing an electron from the 9σ or 1δ orbitals. The D state was assigned to be from removing a 4π electron, resulting in a $^3\Pi$ state. The term value for this highly excited state was 1.89 eV. It was never observed before since the transition between this state and the ground state is spin-forbidden.

Note that all the spectral features observed for FeO were derived from the Fe $3d$ orbitals, though the 9σ orbital was really a *spd* hybrid.[111] The primarily O-derived orbitals (8σ and 3π) lied much deeper in energy and could not be accessed at the photon energies used. This was essential to

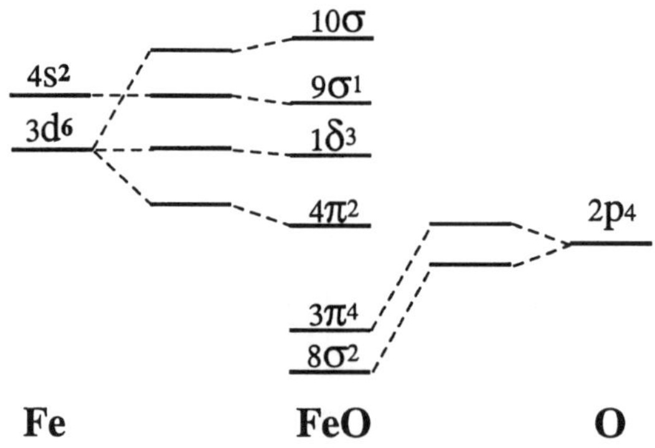

Fig. 30. The schematic ordering and occupations of the molecular orbitals of FeO, showing that the O-derived orbitals lie below those of the Fe 3*d*-derived orbitals.

understand the spectra of the higher oxide molecules, in which more 3*d* electrons were transferred to the additional O atoms, resulting in simpler electronic structures for the higher oxide molecules. This was indeed the case, as evidenced in Fig. 29 by the simpler spectra of the higher oxides. Only two states were observed in each case for FeO_2^-, FeO_3^- and FeO_4^-. Compared to the FeO^- spectrum, all the PES features of the higher oxides should be derived mainly from the Fe 3*d* orbitals. However, more detailed electronic assignments of these features were not possible without accurate theoretical calculations because the occupation and ordering of the MOs could not be determined *a priori*. Nevertheless, significant qualitative information about the structure and bonding of these oxide species could still be obtained.

From the oxidation state concept discussed above, there should be no O–O bond in FeO_2 and FeO_3. Each O atom should form a formally Fe=O double bond with the Fe atom. This saturated the oxidation state of Fe(VI) at FeO_3. Therefore it was expected that there should be an O–O bond in FeO_4, implying that the electronic structure of FeO_4 should be similar to that of FeO_3. This was evidenced by the similarity between the PES spectra of FeO_3^- and FeO_4^- and the fact that their EAs were also similar. Another piece of evidence for the existence of an O–O bond in FeO_4 came from the observation of photodissociation of the FeO_4^- anion at high laser fluences to $FeO_2^- + O_2$. A previous mass spectrometric study on the FeO_4^- anion also suggested that it contained at least one intact O–O unit.[115] However, no photodissociation of this kind were observed for FeO_2^- and FeO_3^-.

FeO_2 is a bent O=Fe=O structure with a bond angle of about 141 degree.[114] Our FeO_2^- spectrum consisted of two broad features without vibrational structure, suggesting excitation of the low frequency bending mode. This was consistent with a bent structure for FeO_2 with a low bending frequency. There were few studies on the FeO_3 and FeO_4 molecules before. A previous matrix study suggested the observation of a D_{3h} FeO_3.[116] The calculated symmetric stretching frequency for the D_{3h} FeO_3 was 792 cm^{-1}. Our PES spectrum for FeO_3^- was surprisingly simple with well-resolved vibrational structures for both the observed states (X and A). The observed vibrational frequency was about 850 cm^{-1} for both states, in reasonable agreement with the previous calculated symmetric stretching frequency for the D_{3h} FeO_3. The rather high Fe–O stretching frequency in FeO_3 suggested that the Fe–O bonds are quite strong, comparable to that in the FeO monoxide. It was concluded that the observed FeO_3 was indeed due to the D_{3h} FeO_3 with Fe=O bonds.

The similar electronic structure between FeO_4^- and FeO_3^-, as evidenced by their similar PES spectra and EAs, indicated that FeO_4 was formed simply by replacing an O atom in FeO_3 with an O–O unit while maintaining a similar bonding character. It followed that the FeO_4 molecule should contain two Fe=O bonds and an O–O unit bonded to the Fe atom. The O–O bond was most likely perpendicular to the FeO_2 plane, resulting in a C_{2v}

Table 12. Observed binding energies and spectroscopic constants for FeO_2O_y ($y = 1$–5).

	State	BE (eV)	Term value (eV)[b]	Vibrational frequency (cm^{-1})[b]
Fe_2O	X	1.63 (4)	0	
	A	2.02 (6)	0.39 (3)	
Fe_2O_2	X	2.36 (5)	0	670 (70)
	A	2.80 (7)[a]	0.44 (5)	
Fe_2O_3	X	3.06 (4)	0	810 (100)
	A	3.31 (7)[a]	0.25 (5)	650 (80)
Fe_2O_4	X	3.56 (4)	0	710 (60)
	A	3.74 (4)	0.18 (2)	720 (60)
Fe_2O_5	X	3.88 (4)	0	750 (50)
	A	4.06 (5)	0.18 (2)	790 (60)

[a]Estimated binding energy. The spectrum is complicated by the presence of another isomer.
[b]Relative energies can be determined more accurately.

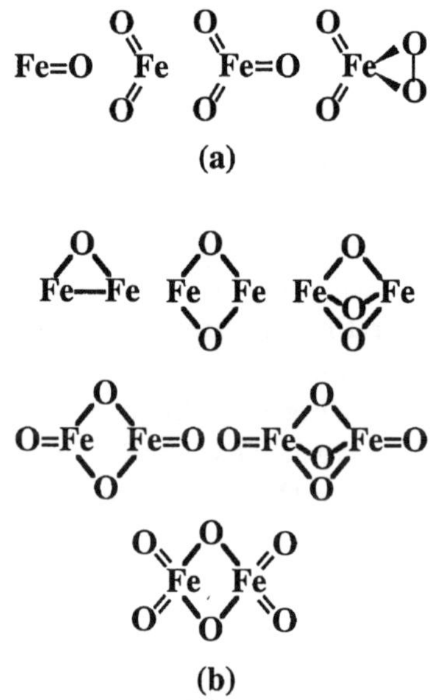

Fig. 31. Schematic drawing of the structures of the mono-iron oxides (a) and di-iron oxides (b). From Ref. 31.

FeO_4. The structures of the FeO_y ($y = 1$–4) molecules are summarized in Fig. 31, which shows clearly in the formal sense the increase of the oxidation state of Fe from +2 in FeO to +6 in FeO_3 as well as in FeO_4.

3.3.2. $Fe_2O_y^-$ ($y = 1$–5): *Sequential Oxidation of* Fe_2

The PES spectra of $Fe_2O_y^-$ are shown in Fig. 32. The most striking finding in the di-iron oxide series was the observation of the prominent ground state peak in all the molecules except $Fe_2O_2^-$ and $Fe_2O_3^-$, where broader features were observed at lower BE along with prominent features. Similar to the mono-iron series, the spectra became quite sharp and simple for the higher oxides with well-resolved vibrational structures. The broad features at the lower BE side in the spectra of $Fe_2O_2^-$ and $Fe_2O_3^-$ were due to other isomers, possibly involving intact O–O unit. Similar broad features at the lower BE side could also be discerned for the $Fe_2O_4^-$ and $Fe_2O_5^-$ spectra, but at a much lower signal level (Fig. 32). They were more clearly observed at 3.49 eV photon energy, suggesting that similar isomers also existed for $Fe_2O_4^-$ and $Fe_2O_5^-$, though at a much lower abundance.

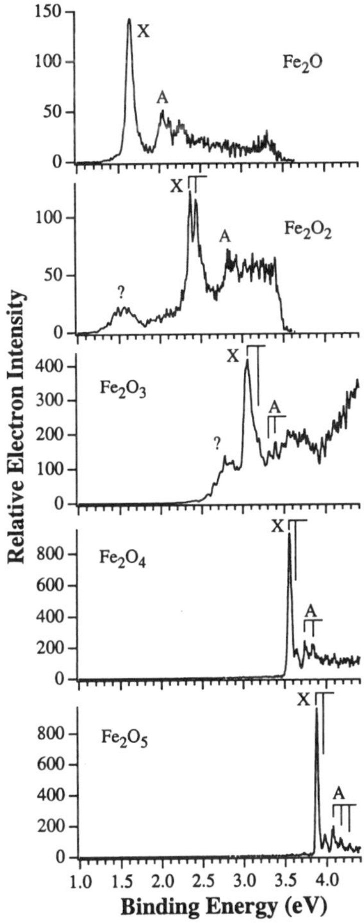

Fig. 32. Photoelectron spectra of $Fe_2O_y^-$ ($y = 1$–5). The spectra of Fe_2O^- and $Fe_2O_2^-$ were taken at 355 nm while that of the higher oxides were taken at 266 nm. The vertical lines indicate the resolved vibrational structures. "X" represents the ground state of the neutrals and "A" represents the excited states. The "?" in the spectra of $Fe_2O_2^-$ and $Fe_2O_3^-$ indicates features of possible isomers. Note the sharp and well-resolved spectra for $Fe_2O_4^-$ and $Fe_2O_5^-$. From Ref. 31.

Beside the prominent ground state peak, there were also lower intensity excited state features in the di-iron oxide spectra. The spectra of $Fe_2O_2^-$ and $Fe_2O_3^-$ were complicated by the existence of isomers that had substantial abundance. The spectra of $Fe_2O_4^-$ and $Fe_2O_5^-$ were best resolved, each with one excited state. These two spectra were also very similar, except that Fe_2O_5 had a higher EA and higher vibrational frequencies in both the ground and excited states. The EA vs. O atom trend for the di-iron series was similar to the mono-iron series, suggesting that this series could be viewed as a sequential oxidation of the Fe_2 dimer — each O atom was individually bonded to the two Fe atoms with a similar amount of charge transfer. Detailed assignments of the observed electronic states were not attempted and only qualitative structure and bonding aspects of these molecules were discussed. The best way to understand the electronic structure of these oxides was to start from the Fe_2 dimer and to compare with the mono-iron oxides discussed above.

The Fe_2 ground state configuration might be described as $(3d^{14})\ \sigma_{4s}^2$, where $3d^{14}$ represented collectively the MOs formed from the $3d$ atomic orbitals and σ_{4s}^2 represented the σ bonding orbital between the Fe atoms.[131] Analogous to the mono-iron oxide series, when Fe_2 was oxidized the O-derived orbitals lied below the Fe $3d$-derived orbitals (see Fig. 30), formally resulting in a charge transfer of two electrons from Fe_2 to the O atom. Subsequent oxidation would result in more charge transfer from the Fe_2 to the O atoms; each O addition increased formally the oxidation state of the Fe atoms by +1. The observed spectral features should all be derived from the orbitals involving the Fe atoms. Similar to the mono-iron series, the increased charge transfer from the Fe atoms to the O atom explained the fact that the electronic structure of the higher oxides appeared to be simplified, as evidenced by the seemingly simpler PES spectra. The rather high vibrational frequencies observed suggested that the interactions between the Fe and O atoms were strengthened at the higher oxides. The sharp spectral features indicated that the detached electrons were from non-bonding or weakly bonding orbitals, consistent with the nature of the Fe $3d$-derived orbitals.

On the basis of the previous discussions, possible structures of the di-iron oxides were proposed, as shown in Fig. 31 along with those of the monoxide species. All the structures were consistent with the sequential oxidation concept, and the oxidation states of the Fe atoms could be counted by the bonds they formed with O atoms and they increased sequentially from +1 in Fe_2O to +5 in Fe_2O_5. The oxidation state of Fe was still not saturated at Fe_2O_5. This was also obvious from the EA vs. O atom curve shown in Fig. 28. It was predicted that there should be a stable

Fe_2O_6 molecule, whose possible structure was also included in Fig. 31. Indeed, the $Fe_2O_6^-$ anion was observed in our mass spectra with very weak abundance, which prevented us from measuring its PES spectrum. Several other M_2O_2 oxides were known to possess the rhombus structure, including Cu_2O_2 (Fig. 21), Si_2O_2,[132] and Ge_2O_2.[56] The Fe_2O_3 molecule might have several possible structures. The one shown in Fig. 31 with three bridging O atoms was the only one with three equivalent O atoms. Cu_2O_3 was also found to have a similar structure (Fig. 21). The structures of Fe_2O_4 and Fe_2O_5 could be viewed to be built upon the structures of Fe_2O_2 and Fe_2O_3, respectively, each with two more Fe=O double bonds. The valence of the Fe atoms was better satisfied in the higher oxides, strengthening the overall Fe–O bonding as evidenced by their higher vibrational frequencies.

3.3.3. *$Fe_3O_y^-$ and $Fe_4O_y^-$: Sequential O Atom Chemisorption on Cluster Surfaces*

The PES spectra of $Fe_3O_y^-$ (y = 1–5) and $Fe_4O_y^-$ (y = 1–6) are shown in Fig. 33. The $Fe_3O_y^-$ and $Fe_4O_y^-$ spectra became increasingly more complicated and discrete electronic feature were more difficult to resolve. We also performed experiments at 355 nm and obtained slightly better resolved spectra. The most interesting observation was the EA increase as a function of the number of O atoms, as shown in Fig. 28. Though the first O addition did not increase the EAs in the Fe_3O_y and Fe_4O_y series, nearly linear increases were observed after that. From Fe_3O_4 to Fe_3O_5, there was a large EA increase, and Fe_3O_6 and higher clusters had EA values that were too high to be studied at the 266 nm used. The same was observed in the $Fe_4O_y^-$ series, where Fe_4O_7 had an Ea value larger than 4.2 eV, representing a significant jump from $Fe_4O_6^-$. The higher clusters all had larger EAs and could not be studied at 266 nm. Though detailed assignments of these spectra would require accurate *ab initio* calculations, a simple electrostatic model was used to interpret the linear trend. This simple consideration led to interesting suggestions for the structures of these oxide clusters and insight into their unique bonding characteristics.

The general trend of the EA increase with the number of O atoms was consistent with the increasing degree of oxidation. The interactions between the Fe atoms and the O atoms in the Fe_3O_y and Fe_4O_y clusters were expected to be predominantly ionic. The linear dependence of the EA on the number of O atoms in the Fe_3O_y (y = 1–4) and Fe_4O_y (y = 1–6) clusters was understood based on an electrostatic model, in which

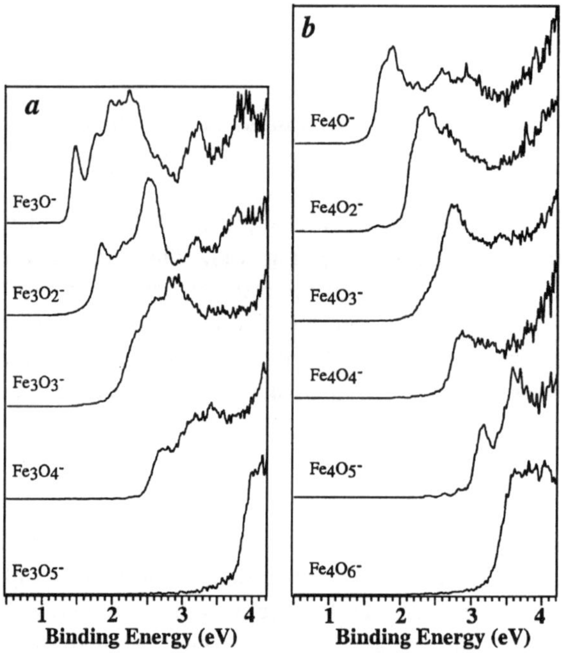

Fig. 33. Photoelectron spectra of $Fe_xO_y^-$ clusters at 266 nm. (a) $Fe_3O_y^-$ ($y = 1$–5); (b) $Fe_4O_y^-$ ($y = 1$–6). From Ref. 32.

each oxygen atom was assumed to locate on the cluster surface with a net charge transfer, δe, from the Fe_x iron cluster to each adsorbed O atoms, i.e. $(Fe_x)^{y\delta+}(O^{\delta-})_y$. The EA, by definition, was the energy required to remove the extra electron from the $Fe_xO_y^-$ clusters. Approximately, the EA was proportional to two terms: (1) the electrostatic repulsion of the extra electron with the $O^{\delta-}$ atoms, $-y\delta e^2/R$, where R was the average distance between the extra electron and the $O^{\delta-}$ atoms; and (2) the electrostatic attraction between the electron and the positively charged metal center, $y\delta e^2/r$, where $y\delta e$ was the total charge on the metal center and r was the average distance between the electron and the positive charge center. On the basis of this electrostatic model, the EA should be proportional to y, the number of O atoms, with a slope of $\delta e^2(1/r-1/R)$. This simple electrostatic consideration seemed to be able to account for the linear portions of the EA vs. y curves for Fe_3O_y and Fe_4O_y (Fig. 28) remarkably well. The underlying assumptions in this model were that the charge transfer (δe) to the O atom was the same at each step of oxidation and that the O atoms were located on the cluster surface around the metal core. However, the first O oxidation step did not produce an increase in the EA and seemed to be inconsistent with the model. One plausible interpretation was that for the Fe_3O^- and Fe_4O^- clusters, the electrostatic repulsion and attraction experienced by the extra electron canceled each other out. This would suggest that upon further oxidation the negative charges on the O atoms were screened and that on average the extra electron experienced more attraction than repulsion for the higher clusters, giving rise to the linear increase with y.

Possible structures of these oxide clusters were proposed on the basis of the experimental observations and the electrostatic model. The iron trimer was known to have an equal triangular D_{3h} structure while the tetramer has a tetrahedral T_d structure.[133] Our study on single O atom adsorption on small iron clusters found that the O atom preferred the edge site, bridging two iron atoms for the dimer, trimer and tetramer, with the face sites more favored as the iron clusters became larger (see below, Fig. 36). Studies on the O-atom adsorption on small Ni clusters also found that the O atom preferred the edge site on the tetrahedral Ni_4 cluster.[134] Nb_3O was also found to have a similar structure with the O atom bridge-bonded to two Nb atoms.[135] The bridge site adsorption in these small transition metal clusters was surprising because it was known that on bulk surfaces the four-fold sites were usually favored.[130] This was probably due to the more compact structures of the small clusters, which optimized the O $2p$ bonding to the metal atoms in the bridge sites. Therefore, it was reasonable to assume that for the higher oxide clusters the O atoms would occupy the bridge sites successively. The oxide cluster structures consistent with this picture are depicted schematically in Fig. 34.

These series of clusters were quite interesting because they covered the whole range of stoichiometries found in the bulk oxides. Assuming that the O atom had an oxidation state of -2, we could evaluate the oxidation states of the Fe atoms in each cluster, depending on the O atoms a given Fe atom was bonded to. Therefore, we found that in both Fe_3O_3 and Fe_4O_4 the Fe atoms had $+2$ oxidation state and they might be viewed as fragments of the bulk FeO oxide. In Fe_3O_4, if the fourth O atom was bridge-bonded to two Fe atoms, then we found that there were two Fe atoms with $+3$ and one with $+2$ oxidation state, similar to the bulk magnetite (Fe_3O_4).

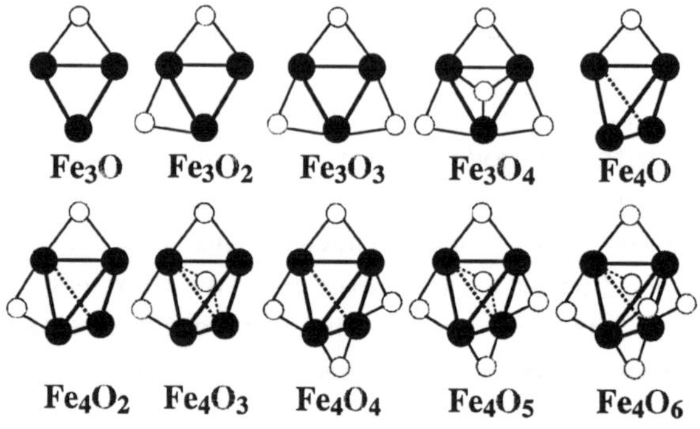

Fig. 34. Schematics of the proposed structures for the Fe_3O_y and Fe_4O_y oxide clusters. From Ref. 32.

Finally, in Fe_4O_6, each Fe atom reached the high oxidation state of +3, similar to bulk hematite. For the clusters with an O/Fe ratio less than 1, they might be compared to the initial oxidation of Fe surfaces although the four-fold sites that O atoms preferred on the bulk surfaces were not available in these small clusters. It is also important to point out that for both Fe_3O_4 and Fe_4O_6 clusters, one more O atom addition would make at least one Fe atoms to have an oxidation state beyond the normal +3. This might explain the EA jump in both cases.

All these structures are highly idealized and were suggested based on the sequential O atom adsorption on the cluster surface occupying the Fe–Fe bridge sites. This in turn was based on the electrostatic interactions between the Fe and O atoms, used to interpret the linear EA curves shown in Fig. 28 for Fe_3O_y ($y = 1$–4) and Fe_4O_y ($y = 1$–6). Final determination of the cluster structures would require accurate *ab initio* calculations and comparisons between the calculations and the experiments. Interestingly all the structures suggested in Fig. 34 have substantial Fe–Fe interactions, even in the clusters with similar stoichiometry as in the bulk, i.e. $(FeO)_n$ ($n = 3$ and 4), Fe_3O_4 and $(Fe_2O_3)_2$. These clusters may be too small to model bulk properties, but they may provide interesting insight into the nature of the bonding and magnetic properties of the bulk materials. All the structures shown in Fig. 34 appear to optimize the Fe–O bonding and allow Fe–Fe metallic bonding at the same time. This may be a major difference between the finite systems and the bulk where the metal atoms are well separated and cannot have direct metal–metal bonding.

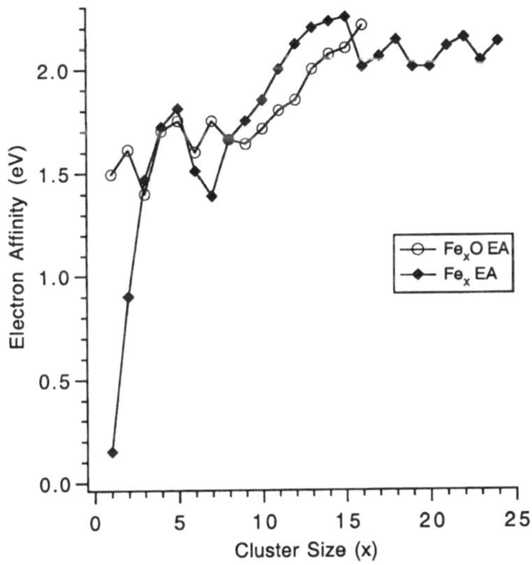

Fig. 35. Comparison of the electron affinities of the chemisorbed clusters, Fe_xO ($x =$ 1–16) to those of the bare Fe_x clusters ($x =$ 1–24). From Ref. 30.

3.3.4. *Fe_xO^- ($x = 1$–16): O Atom Chemisorption on Iron Cluster Surfaces*

We further studied iron clusters with a single O atom as an impurity to explore the idea of using clusters as models for oxygen chemisorption on iron surfaces. The PES spectra of Fe_xO^- (x = 1–4) are discussed above. We measured PES spectra for Fe_xO^- with x up to 16. The spectra were generally broad with few resolved features, revealing the complicated electronic structures expected for these system. Figure 35 shows the measured EAs of the Fe_xO species compared to that of the pure Fe clusters. It was interesting to observe that most of the Fe_xO species had lower EAs than the pure clusters. DFT calculations were further performed on Fe_xO (x = 2–6) to understand the structures of the chemisorbed species and determine the trend and adsorption sites of the O atom.

Figure 36 shows the different structures obtained for Fe_xO (x = 2–6). The ground state structure for the Fe_2O cluster was found to be an isosceles triangle with a Fe–O bond length of 1.77 Å and Fe–O–Fe bond angle of 77.2°. The spin multiplicity of this structure was S = 3. The Fe–Fe bond distance increases from 1.98 Å for the Fe_2 dimer to 2.05 Å in Fe_2O. The ground state for Fe_3O had a planar diamond-like structure

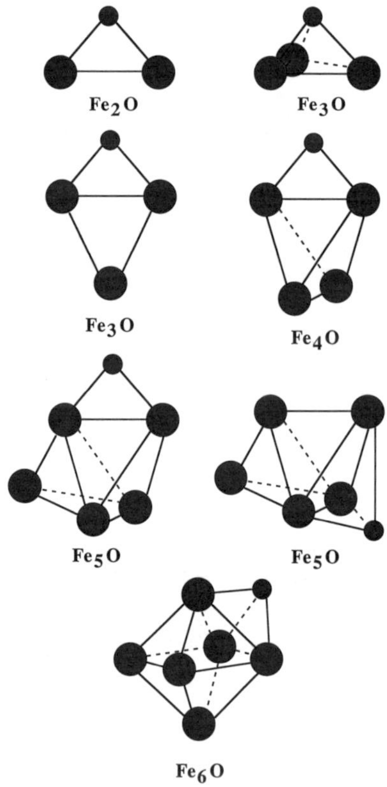

Fig. 36. The structures of the chemisorbed clusters, Fe_xO (x = 2–6), obtained from density functional theory calculations. From Ref. 30.

with the oxygen atom capped on the shorter side of the triangle formed by the three Fe atoms. The Fe–O bond length was 1.77 Å and the Fe–O–Fe bond angle was 84.8°. The ground state structure had a spin multiplicity of $S = 5$. Another structure was also found for this cluster, a trigonal pyramid with the oxygen atom residing above the plane formed by the three Fe atoms. This structure was 1.1 eV higher in energy and had a spin multiplicity of $S = 4$.

The Fe_4O cluster was found to have a three-dimensional structure with the O atom attached to an edge of the tetrahedron formed by the four Fe atoms. The Fe–O bond length was 1.78 Å and the Fe–O–Fe bond angle was 81.9°. The spin multiplicity was $S = 6$. For this cluster, capping the oxygen on a three-fold face led to an unstable structure. For Fe_5O, the 5-atom bare iron cluster Fe_5 was found to be a trigonal bipyramid. The O atom could be capped either on an edge or on a trigonal face. The edge-capping, for which the oxygen atom bridged over an apex site and an equilateral site, led to a slightly more stable structure for this cluster. The face-capping structure was less stable only by 0.06 eV. The two competing isomers had the same spin multiplicity $S = 8$. For Fe_6O, the ground state structure of the corresponding Fe_6 cluster was found to be an elongated octahedral form. The preferred chemisorption site for the O atom was found to be the 3-fold site with a spin multiplicity of $S = 10$. The bridge site was no longer stable. The trend of the chemisorption sites was quite interesting and suggested that only at larger clusters the face sites are preferred. Therefore, a sufficient large clusters would be needed if one is to use a cluster model to understand the bulk chemisorption although undoubtedly the smaller clusters would allow a more detailed analysis of the chemical bonding.

3.4. VO_y^- ($y = 1$–4)

VO is very important in astrophysics, because it is the second in abundance after TiO in the spectra of metal-rich M-type stars and has been extensively studied.[2,5,136–143] The ground state of VO is known to be a quartet state ($X^4\Sigma^-$). Doublet excited states of VO were much less well characterized because the quartet nature of the VO ground state prevented direct optical transitions between the ground state and any doublet excited states. The VO_2^- and VO_3^- anions were observed in the gas phase previously.[144] Very few studies have been carried out on the higher VO_y oxides. Recently, an electron spin resonance (ESR) study was reported on VO, VO_2 and VO_3 in neon matrix at 4K, in combination with an *ab initio* theoretical study.[145] The ESR study showed that VO_2 has a C_{2v} structure with a $^2A_1$ ground state. VO_3^- was suggested, from a previous HF-SCF calculation, to have a planar D_{3h} structure with a VO distance of 1.601 Å,[146] whereas the recent *ab initio* calculation indicated that neutral VO_3 had a C_{2v} symmetry, due to a Jahn–Teller distortion.[145] A superoxovanadium VO_4 species was suggested to be observed in photo-oxidation of $V(CO)_6$ in low temperature matrices.[147] An FTIR matrix investigation was also reported for VO, VO_2, VO_4 and V_2O_2.[148] Bulk VO_2 is a nonmagnetic insulator and a technologically important material. The VO_4 unit in condensed phase has attracted much attention because of its important role in oxovanadium catalysis.[149]

In condensed phase, the pseudotetrahedral oxovanadium group has three basal-plane oxygens and a terminal, double-bonded oxygen. Recently, well-defined absorption, emission and Raman spectra of the vanadium oxide species have revealed new electronic structure information for the pseudotetrahedral oxovanadium.[150] We reported the first systematic study on the monovanadium oxides, VO_y ($y = 1$–4), and their EAs.[38] For VO, we observed five low-lying excited states, including two new doublet states. For both VO_2 and VO_3, we observed distinct low-lying excited states from the PES spectra of VO_2^- and VO_3^-, which were interpreted using the available *ab initio* calculations. The PES spectrum of VO_4^- showed broad features, indicating a significant geometric change from the anion to the neutral.

3.4.1. *VO and VO⁻: Low-lying Electronic States of VO*

Figure 37 shows the PES spectra of VO^- at four-photon energies. The 532 nm spectrum exhibited two intense vibrational progressions (X and a) with well-resolved vibrational structures and a very weak feature at low BE (X'). Three more weak features beyond the 532 nm photon energy were revealed at the 355 nm spectrum. The feature at 3.079 eV was shown more clearly in the 266 nm spectrum, which also seemed to reveal more transitions at higher BE, but no definitive assignments could be made due to the poor signal-to-noise ratio. The vibrational progression of a band was observed to exhibit remarkable changes at 266 nm: the intensity of the $\nu = 1$ peak was decreased while that of the $\nu = 2$ peak was strongly enhanced. Such intensity changes suggested that there might be overlapping electronic states whose detachment cross sections were dependent on photon energies. The 193 nm spectrum supported this suggestion and showed even more dramatic intensity changes in the region of the a band, even though the resolution and count rate were poor at 193 nm. The weak feature at the low BE (X') was due to an electronic excited state of the VO^- anion.

The ground state of VO was well established to be a quartet state, $X^4\Sigma^-$, with an electronic configuration of $3\pi^4 8\sigma^2 9\sigma^1 1\delta^2$, where the three unpaired electrons predominantly occupy the bonding $4s(\sigma)$ and nonbonding $3d(\delta)$ MOs. Due to the quartet nature of the VO ground state, the low-lying quartet states of VO were well known, including the $A'\,{}^4\Phi$ and $A\,{}^4\Pi(9\sigma^1 1\delta^1 4\pi^1)$, $B\,{}^4\Pi(1\delta^2 4\pi^1)$, $C^4\Sigma^-(1\delta^2 10\sigma^1)$ and $D^4\Delta(9\sigma^1 1\delta^1 10\sigma^1)$ states. The many low-lying doublet states, however, were not known due to the forbidden optical transitions between the quartet and the doublet

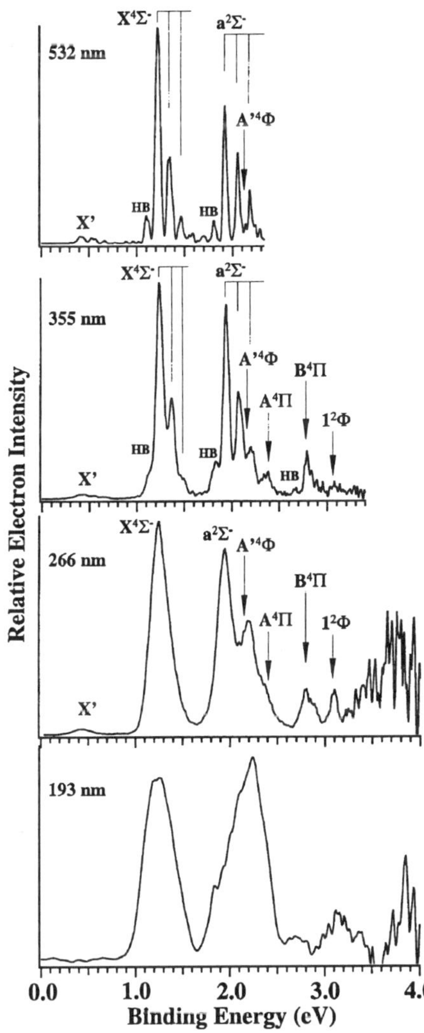

Fig. 37. Photoelectron spectra of VO^- at 532, 355, 266 and 193 nm. HB stands for hot band. From Ref. 38.

states. The lowest known excited state for VO was the $A'\,^4\Phi_{3/2}$ state at 6996.8 cm^{-1} above the ground state.[136] However, our observed lowest excited state in the VO^- PES spectrum was the progression starting at 1.927 eV, which gave an excitation energy of 5630 cm^{-1}, much lower than the known lowest quartet state ($A'\,^4\Phi_{3/2}$).

The ground state of the VO^- anion was not known. The extra electron in the anion could enter either the 4π, 9σ, or 1δ orbital to give a configuration of $3\pi^4 8\sigma^2 9\sigma^1 1\delta^2 4\pi^1$, $3\pi^4 8\sigma^2 9\sigma^2 1\delta^2$, or $3\pi^4 8\sigma^2 9\sigma^1 1\delta^3$, respectively. The third configuration could be eliminated based on two observations. First, the occupation of the 1δ orbital would lose the exchange energy which was known to be important. Second, the 1δ orbital was non-bonding and detachment of a 1δ electron should yield a feature without significant vibrational structure. The occupation of the 9σ orbital could also be eliminated. This orbital was a bonding one and its occupation was shown to enhance the VO bonding as in the $^2\Delta(9\sigma^2 1\delta^1)$ state, which had a bondlength even shorter than the VO ground state.[136] The VO^- anion was more weakly bonded than the neutral ground state, as suggested by the smaller vibrational frequency of the anion. Therefore, the only reasonable occupation of the extra electron in VO^- was in the 4π orbital, giving a $X^5\Pi(3\pi^4 8\sigma^2 9\sigma^1 1\delta^2 4\pi^1)$ state. This was also consistent with the ground state configuration of CrO, which is isoelectronic to VO^- and has a vibrational frequency similar to that of VO^-.[5]

From the ground state configuration of VO^-, three one-electron detachment channels were expected and could result in many low-lying excited states as follows:

$$3\pi^4 8\sigma^2 9\sigma^1 1\delta^2 4\pi^1 (X^5\Pi) \rightarrow 3\pi^4 8\sigma^2 9\sigma^1 1\delta^2 : \; ^4\Sigma^-, {}^2\Sigma^-, {}^2\Gamma, {}^2\Sigma^+ \quad (4)$$

$$\rightarrow 3\pi^4 8\sigma^2 9\sigma^1 1\delta^1 4\pi^1 : \; ^4\Phi, {}^4\Pi, {}^2\Phi(2), {}^2\Pi(2) \quad (5)$$

$$\rightarrow 3\pi^4 8\sigma^2 1\delta^2 4\pi^1 : \; ^4\Pi, {}^2\Pi(2), {}^2H, {}^2\Phi . \quad (6)$$

Among these states, all the quartet states were well characterized and should be easy to assign in principle. Among the doublets, only the $^2\Sigma^+$ state from (4) was known. The $^4\Sigma^-$ state, the ground state of the neutral VO, resulted from removing the single 4π electron in the anion. As pointed out above, the first excited state observed in our PES spectra at 5630 cm^{-1} did not agree with the known lowest quartet state, $A'\,^4\Phi$ from (5). We assigned this state as the $^2\Sigma^-$ state, arising from the same electron configuration as the ground state. The prominent vibrational progression was similar to that of the ground state, as expected. The $A'\,^4\Phi$ state with its four spin-orbit components should occur between the $\nu = 1$ and 2 peak of the $a\,^2\Sigma^-$ state and might overlap slightly with these two peaks, complicating the definitive observation of the $A'\,^4\Phi$ state. Further complication derived from the fact that the $A'\,^4\Phi$ state had very low intensity at the low detachment photon energies where the resolution was higher. The small feature in between the $\nu = 1$ and 2 peaks of the $a\,^2\Sigma^-$ state in the 532 nm spectrum was due to the $A'\,^4\Phi$ state. However, its appearance was more prominent at the 266 and 193 nm spectra, distorting the vibrational progression of the $a\,^2\Sigma^-$ state.

The weak feature at 2.4 eV of the 355 nm spectrum gave an excitation energy of 9400 cm^{-1}, which was consistent with the position of the $A\,^4\Pi$ state from (5). Its intensity was also enhanced at the 266 and 193 nm spectra, consistent with detaching a δ electron. The feature observed at 2.78 eV was assigned to the $B\,^4\Pi$ state due to the detachment of the single 9σ electron. The derived excitation energy of 12 500 cm^{-1} was in excellent agreement with the known value of the $B\,^4\Pi$ state. A further higher energy

Table 13. Observed binding energies (BE) and spectroscopic constants for VO^- and VO.

		Term value (eV)		Vib. freq. (cm^{-1})	
	BE (eV)	This work	Previous[b]	This work	Previous[c]
$VO^- X'$	0.42 (3)	6400 (200)			
$X\,^5\Pi(9\sigma^1 1\delta^2 4\pi^1)$		0		900 (50)	
VO $X^4\Sigma^-(9\sigma^1 1\delta^2)$	1.229 (8)[a]	0		980 (60)	1001.8
$a^2\Sigma^-(9\sigma^1 1\delta^2)$	1.927 (8)	5630 (80)		1090 (80)	
$A'^4\Phi_{3/2}(9\sigma^1 1\delta^1 4\pi^1)$	2.1		6996.8		936.5
$^4\Phi_{5/2}$			7171.4		
$^4\Phi_{7/2}$			7342.7		
$^4\Phi_{9/2}$			7509.0		
$A^4\Pi_{-1/2}(9\sigma^1 1\delta^1 4\pi^1)$	2.395 (8)	9400 (80)	9449.7		884
$^4\Pi_{1/2}$			9477.8		
$^4\Pi_{3/2}$			9512.4		
$^4\Pi_{5/2}$			9555.5		
$B^4\Pi_{-1/2}(1\delta^2 4\pi^1)$	2.780 (9)	12 500 (90)	12 518.3		901
$^4\Pi_{1/2}$			12 571.7		
$^4\Pi_{3/2}$			12 637.1		
$^4\Pi_{5/2}$			12 711.9		
$1\,^2\Phi(9\sigma^1 1\delta^1 4\pi^1)$	3.079 (15)	14 920 (90)			

[a]Adiabatic electron affinity of VO.
[b]From Ref. 136.
[c]From Ref. 5."

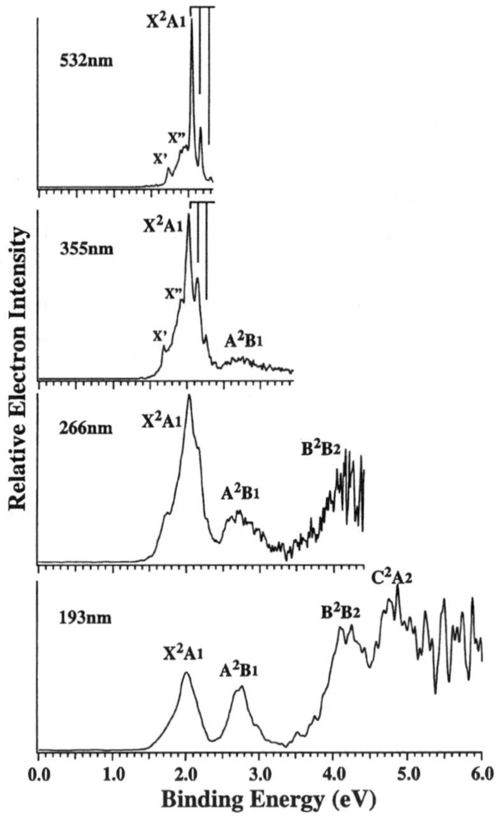

Fig. 38. Photoelectron spectra of VO_2^- at 532, 355, 266 and 193 nm. From Ref. 38.

feature, which was observed more prominently in the 266 nm spectrum was assigned to $a\,^2\Phi$ state from (5). This feature had very low intensity at the 355 nm and its intensity was enhanced with photon energy. This photon energy dependence was similar to that of the $A'\,^4\Phi$ and $A\,^4\Pi$ states, which were derived from the same configuration, and was consistent with detaching a δ electron. The observed BE and spectroscopic constants are summarized in Table 13.

3.4.2. *VO_2 and VO_2^-*

Figure 38 shows the PES spectra of VO_2^- at four-photon energies. The 532 nm spectrum displayed a simple vibrational progression (X). There were also a weak feature (X') and a broad unresolved feature (X'') at the low BE side. The 355 nm spectrum showed an extra, albeit weak and broad, feature (A) at the high BE side beyond the energy range accessible in the 532 nm detachment energy. The intensity of this feature (A) seemed to be enhanced in the 266 nm spectrum, which revealed yet another feature (B) at the high BE side although the statistics of this feature was poor due to the presence of noise in the high BE range. The intensities of the A and B feature appeared to become even stronger in the 193 nm spectrum, which revealed yet another feature (C) at the high BE side. Thus, from 532 to 193 nm each higher photon energy revealed successively a higher BE feature. The resolution for the X feature was seen to deteriorate significantly from 532 to 193 nm due to the increasing electron kinetic energies. The X' and X'' features became only shoulders in the high photon energy spectra. It was also interesting to observe that the intensity of the A feature showed such strong photon energy dependence. The X' and X'' features, which showed weak dependence on source conditions but could not be quite completely eliminated, were attributed to be due to excited states of the VO_2^- anion. The observed BE and spectroscopic constants were listed in Table 14 for VO_2 and VO_2^-. The A, B, and C bands were all quite broad without any resolved fine features and their BE could only be estimated.

Table 14. Observed binding energies (BE) and spectroscopic constants for VO_2 and VO_2^-.

		Term Value (eV)		Vib. Freq. (cm^{-1})	
	BE (eV)	This Work	Ref. 145	This Work	Ref. 148
VO_2^-					
X'	1.72 (2)	0.31			
X''	~1.9	~0.1			
$X^1A_1(1a_2^2 6b_2^2 3b_1^2 10a_1^2)$					
VO_2					
$X^2A_1(1a_2^2 6b_2^2 3b_1^2 10a_1^1)$	2.03 (1)[a]	0	0	970 (40)	946.3
$A^2B_1(1a_2^2 6b_2^2 3b_1^1 10a_1^2)$	2.6 (1)	0.6	1[b]		
$B^2B_2(1a_2^2 6b_2^1 3b_1^2 10a_1^2)$	4.0 (1)	2.0	2.0[c]		
$C^2A_2(1a_2^1 6b_2^2 3b_1^2 10a_1^2)$	4.6 (1)	2.6	3.0[c]		

[a]Adiabatic electron affinity of VO_2.
[b]Estimate from g-tensor analysis. The value from *ab initio* calculation is 0.3 eV (Ref. 145).
[c]Obtained from *ab initio* calculation (Ref. 145).

The recent ESR study by Knight *et al.* showed that VO_2 was non-linear with an unpaired electron predominantly occupying a vanadium $3d$-type orbital, giving a X^2A_1 ground state. Their calculations yielded a O–V–O bond angle of 110.5° and a V–O bondlength of 1.653 Å with a C_{2v} symmetry.[145] From the g-tensor analysis, they were able to deduce an A^2B_1 low-lying excited state for VO_2 with about 1 eV excitation energy. Their calculation gave a value of 0.3 eV for the A^2B_1 excited state. The X^2A_1 and A^2B_1 states were considered to correlate with the $^2\Delta$ state in a linear OVO case. They also predicted two more excited states correlating with the $^2\Pi$ state of the linear OVO case. These two states were calculated to be 2.0 eV (B^2B_2) and 3.0 eV (C^2A_2) above the X^2A_1 ground state. The uncertainty in the calculations was expected to be ±0.3 eV. Chertihin *et al.* in a matrix FTIR experiment on VO_2 estimated a O–V–O bond angle of $118 \pm 3°$ and measured the ν_1 and ν_3 vibrational frequencies to be 946.3 and 935.9 cm^{-1}, respectively.[148] In an earlier mass spectrometry experiment, Rudnyi *et al.* measured the enthalpy of formation of VO_2^- and estimated the EA of VO_2 to be 2.3 ± 0.2 eV.[144]

The valence MO configuration of VO_2 was $8a_1^2 5b_2^2 1a_2^2 9a_1^2 3b_1^2 6b_2^2$ $10a_1^1$ (X^2A_1).[145] The $10a_1$ orbital was primarily of V $3d$ character; all the other orbitals involved primarily the O $2p$ orbitals. The $10a_1$ and $3b_1$ orbitals correlated with the δ orbital in the linear OVO and the $1a_2$ and $5b_2$ correlated with the π orbital in the linear OVO, as mentioned above. In the anion, most likely the extra electron occupied the $10a_1$ orbital, giving a close-shell ground state for the VO_2^- anion (X^1A_1), consistent with its high enthalpy of formation and stability in the gas phase.[10] Then detachment from each filled orbital of VO_2^- would yield one band in the PES spectrum. As shown in Fig. 38, the 532 nm spectrum of VO_2^- exhibited a well resolved vibrational progression with a frequency of 970 cm^{-1}, which was in good agreement with the ν_1 frequency (946.3 cm^{-1}) measured for VO_2 in the matrix FTIR experiment.[14] This progression corresponded to the ground state (X^2A_1) of VO_2 and was due to detachment of a $10a_1$ electron. The ν_1 vibrational progression suggested that there was a slight V–O bondlength change between the anion and neutral ground state. Very little bending vibration was observed as seen from the sharp ν^1 peaks (the bending vibration would broaden the ν_1 peaks since its frequency was probably too small to be resolved), suggesting that there was little O–V–O bond angle change between the anion and neutral ground state. This was consistent with the nature of the $10a_1$ orbital, which was mainly of $3d$ character. The 0–0 transition at 2.03(1) eV defined the adiabatic EA for VO_2, suggesting that the EA estimated previously was inaccurate and too high.[144]

Beside the main vibrational progression, there were more features at the low BE side in the 532 nm spectrum. These were possibly due to hot band transitions, but could not be attributed to hot bands entirely since similar sequence bands were not observed on the high BE side. Therefore, we attributed these features to be from electronic excited states of the VO_2^- anion. The lowest BE feature at 1.72 eV was quite sharp, yielding an excitation energy of 0.31 eV for the anion excited state. The low BE features could also be due to $V(O_2)^-$ complexes. Similar $M(O_2)^-$ complexes were common for the late transition metal atoms, as discussed earlier for CuO_2 and NiO_2. However, this kind of complex seemed to be less likely for the early transition metals due to the strong M–O bonding. For example, the $V(O_2)$ complex was not observed in the previous matrix studies.[145,148] Therefore, the features that we observed were more likely to be due to electronic excited states of the OVO^- anion.

In the higher photon energy spectra of VO_2^-, three more features were revealed, which were due to detachment of deeper valence orbitals and corresponded to excited states of the VO_2 molecule. All these higher BE features showed broad bands without any resolved vibrational structure, suggesting that there were significant geometry changes between the anion ground state and the excited states of VO_2. These changes were most likely in the O–V–O bond angle such that the ν_2 mode was active, leading to the broad detachment bands. The bond angle change was consistent with the fact that all these features were due to detachment from orbitals involving the O $2p$. The second detachment band near 2.6 eV showed very strong photon energy dependence. We assigned this feature to the A^2B_1 state, as suggested in the previous ESR experiment.[145] Our estimate of 0.6 eV for the excitation energy was in reasonable agreement with that found in the ESR experiment. the A^2B_1 state was due to detachment of an electron from the $3b_1$ orbital of VO_2^-. The enhancement of detachment cross section with photon energy for this detachment channel was consistent with the fact that the $3b_1$ orbital corresponded to a component of the δ orbital in the linear OVO case.[70] We assigned the two higher energy features at 4.0 and 4.6 eV to the two excited states, B^2B_2 and C^2A_2, respectively, as predicted in the previous *ab initio* calculations.[145] The overall assignments were shown in Fig. 38. The calculated excitation energies for the B^2B_2 and C^2A_2 states were surprisingly in good agreement with our experimental estimates. The B^2B_2 and C^2A_2 states were due to detachment from the $5b_2$ and $1a_2$ orbitals, respectively, suggesting that the MO ordering, as given in the previous calculation, needed to be changed to the following: $8a_1^2 5b_2^2 9a_1^2 1a_2^2 6b_2^2 3b_1^2 10a_1^1$ if the calculated excited states were correct. However, calculations of excited states were known to be rather difficult. Our assignments of the excited states according to the existing calculations should probably be viewed as tentative, pending confirmation by more accurate calculations.

3.4.3. VO_3^- and VO_4^-

Figure 39 displays the PES spectra of VO_3^- at 266 and 193 nm, as well as the spectrum of VO_4^- at 193 nm. the VO_3^- anion exhibited such high BE that the 266 nm photon was barely enough to detach it and the 266 nm spectrum only showed an onset of the VO_3^- spectrum. At 193 nm, three

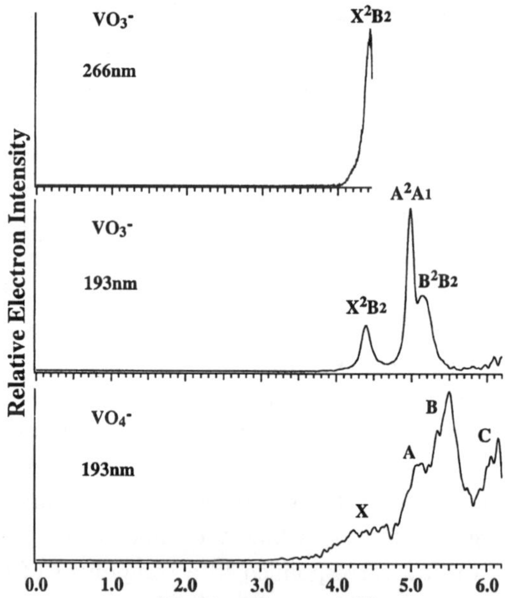

Fig. 39. Photoelectron spectra of VO_3^- at 266 and 193 nm, and that of VO_4^- at 193 nm. From Ref. 38.

Table 15. Observed binding energies (BE) and spectroscopic constants for VO_3 and VO_4.

		BE (eV)	Term Value (eV)
VO_3^-	$X^1A_1(6b_2^2 12a_1^2 7b_2^2)$		
VO_3	$X^2B_2(6b_2^2 12a_1^2 7b_2^1)$	4.36 (5)[a]	0
	$A^2A_1(6b_2^2 12a_1^1 7b_2^2)$	4.95 (5)	0.59 (5)
	$B^2B_2(6b_2^1 12a_1^2 7b_2^2)$	5.15 (5)	0.79 (5)
VO_4	X	4.0 (1)[b]	0
	A	5.0 (1)	1.0 (1)
	B	5.4 (1)	1.4 (1)
	C	6.0 (1)	2.0 (1)

[a]Adiabatic electron affinity of VO_3.
[b]Estimated adiabatic electron affinity of VO_4.

rather sharp and well separated detachment features were revealed. The VO_4^- anion could be produced but with much less abundance compared to VO_3^-. Thus the VO_4^- spectrum showed poorer signal-to-noise ration, displaying four rather broad features (X, A, B and C). The X, A, and B features of VO_4^- seemed to be similar to the three features of VO_3^-. The observed BEs for VO_3^- and VO_4^- are summarized in Table 15. The BEs for VO_4^- could only be estimated due to the broad nature of the spectrum.

The BE of VO_3^- was observed to increase dramatically compared to that of VO_2^-. We obtained an adiabatic EA of 4.36 eV for VO_3, which was more than twice as much as that of VO_2. All the three detachment transitions in VO_3^- gave fairly sharp features, suggesting that the VO_3^- anion and VO_3 neutral both had very similar geometry. The narrow features in the VO_3^- spectra were in sharp contrast to those in the VO_2^- spectra, where both stretching and bending vibrations were shown to be active. The high EA of VO_3 could be easily understood because the V atom only has five valence electrons and the additional electron in the anion would make VO_3^- a closed-shell molecule, thus stabilizing the extra electron. This suggested that VO_3 was open-shell with an unpaired electron, which was most likely on the O atoms. The consideration was consistent with the ESR experiment by Knight *et al.*[145] who observed an ESR signal which showed a ^{51}V hyperfine splitting about 10 times smaller than that in VO_2 and tentatively attributed it to VO_3. Their *ab initio* calculations supported the assignment to VO_3, yielding a planar (C_{2v}) VO_3 with an unpaired electron (a $^2B_2$ ground state).

The calculations showed that VO_3 had a short V–O bond (1.576 Å) along the C_2 symmetry axis and two equivalent longer V–O bonds (1.677 Å) off the C_2 axis with a smaller O–V–O bond angle of 110.6°. The unpaired electron in the X^2B_2 ground state of VO_3 was predicted to occupy a MO ($7b_2$) of mainly $2p_z$ character on the two off-axis O atoms. In the anion, the extra electron entered into the $7b_2$ orbital to give the closed-shell VO_3^- ($^1A_1$) anion. Therefore, only the bending vibration involving the two off-axis O atoms was expected to be active in the detachment of a $7b_2$ electron. This was consistent with our observation of the first detachment feature (X^2B_2, Fig. 39), showing no V–O stretching vibration which would have been easily resolvable. Any bond angle change between the anion and neutral was likely to be rather small since no extensive bending vibration was observed as indicated by the relatively narrow detachment peak.

The next two MOs of VO_3, as given by the *ab initio* calculations, are $12a_1$ and $6b_2$ which are both of mainly O $2p$ character. We assigned the two higher BE features in the PES spectrum of VO_3^- to the detachment of an electron from these two orbitals, respectively, giving the two low-lying excited states of VO_3: A^2A_1 and B^2B_1, as shown in Fig. 39. It was surprising that the A and B detachment features were also rather sharp with little resolved vibrational excitation, meaning that there were no significant geometry changes between the VO_3^- anion and the two excited states of VO_3. This suggested that VO_3 was a rather rigid and stable molecule. There was very little known about the VO_3 molecule except the work by Knight *et al.* VO_3^- was observed in the previous mass spectrometric work,[144] but the VO_3 molecule was not observed in the recent FTIR experiment of V + O_2.[148] Our results suggested that the VO_3^- anion was also likely to have a C_{2v} structure, the same as the neutral, in contrast to the previous HF-SCF calculation which suggested that the VO_3^- anion had a planar D_{3h} structure.[146]

Surface-dispersed vanadium oxide oxide catalysts were composed of pseudotetrahedral VO_4 oxovanadium groups ($^-O_3V{=}O$), whose electronic structure were extensively investigated and debated owing to its important role in catalysis.[150] It has a C_{3v} structure with a terminal V=O bond; the three basal O atoms are bonded to the substrates. Isolated VO_4 species was never observed in the gas phase and is unlikely to adopt the C_{3v} structure as on the surfaces. In two previous matrix IR experiments,[147,148] the VO_4 species was observed and proposed to contain a VO_2 unit and a peroxo-O_2

(η^2) unit. The IR experiments showed that the OVO bond angle of the VO_2 unit in this VO_4 species was nearly identical to that in VO_2 itself. The V atom has five valence electrons and the VO_4 species is O-excess or electron-deficient. We have discussed similar O-excess oxide species previously, such as in FeO_4 and CuO_3. In these O-excess metal oxide species, the peroxo unit is very common to accommodate the additional O atoms, and all exhibit very high EAs. We observed that the VO_4 species behaved similarly. The PES spectrum of VO_4^- was similar to that of VO_3^-, but much broader. Our estimated EA of about 4.0 eV for VO_4 was quite high, suggesting that the valence MO where the extra electron resided in VO_4^- was most likely composed of the O $2p$-type orbitals similar to that in VO_3^-. The broad features observed in the VO_4^- PES spectrum indicated that there were significant geometry changes between VO_4^- and VO_4, in particular, the VO_4 ground state which gave a very broad band ranging from about 3.9 to 4.7 eV. The VO_4^- anion had been proposed to contain a peroxo-O_2 (η^2-O_2) unit while the VO_4 neutral contained a superoxo-O_2 (σ-O_2) unit where the O_2 was tilted and only one of the O atoms was bonded to V.[147] This proposal was entirely consistent with our broad detachment feature for the X ground state of VO_4. The A, B, and C features in the VO_4^- spectrum corresponded to detachment of more tightly bound electrons. The estimated BEs for the four detachment features of VO_4^- were summarized in Table 15.

3.4.4. *Electronic and Structural Evolution from VO to VO_4*

Vanadium has four principal oxides: VO, V_2O_3, V_2O_4, and V_2O_5 with oxidation states of vanadium ranging from $+2$ to $+5$.[151] The maximum oxidation state of $+5$ originates from the five valence electrons of the V atom. The chemical bonding between V and O is extremely strong with both covalent and ionic characters. We can view the series from VO to VO_3 as a sequential oxidation in which the oxidation state of V increases from $+2$ in VO to $+5$ in VO_3. VO_3 is interesting because six electrons are needed to saturate the valences of the O atoms, making VO_3^- a closed-shell species with an extremely high BE. The numbers of unpaired electrons in the VO_y species correlate with this trend, with three in VO and one in VO_2 all on the V atom. The ESR experiment showed that VO_3 also had an unpaired electron, but mostly on the O atoms.[145] In VO_4, there are not enough valence electrons so that the molecule contains a superoxo σ-O_2 units.[147,148]

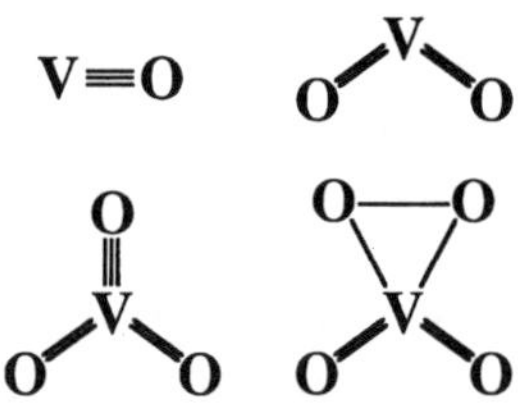

Fig. 40. Schematic structures of the VO_y ($y = 1$–4) species. From Ref. 38.

The structures of the VO_y series of species are interesting. The VO diatomic has a bondlength of 1.589 Å and is viewed theoretically as a triple bond due to strong π interactions between the O lone pair and the empty d orbitals on the V atom. As obtained from *ab initio* calculations by Knight *et al.*, the VO_2 molecule has a C_{2v} structure with a V–O bondlength of 1.653 Å and O–V–O bond angle of 110.5°. The V–O bondlength in VO_2 is longer than that in the VO diatomic and can be viewed as a V=O double bond. Most interesting are the structures of VO_3 and VO_4, both containing a VO_2 unit nearly identical with that of VO_2 itself in terms of both the V–O bondlength and O–V–O bond angle. The axial V–O bondlength in VO_3, as obtained by Knight *et al.* is 1.576 Å, which is even shorter than that in the VO diatomic, indicating the axial V–O bond can be characterized as a triple bond. The off-axis V–O bonds are shown to have a bondlength of 1.667 Å and a O–V–O bond angle of 110.6°, almost the same as in VO_2 itself. Our PES spectrum suggests that the VO_3^- anion has nearly identical geomentry to the neutral. The VO_3^- anion has been suggested to have a planar D_{3h} structure,[146] which is not consistent with our observation that there is little geometry change between the anion and the neutral. In VO_4, the O_2 unit can replace either the triple bonded O or one of the double-bonded O atoms. The matrix IR spectrum of VO_4 shows that the VO_4 molecule contains a VO_2 unit nearly identical to that of VO_2 itself,[148] suggesting that the O_2 unit in VO_4 replaces the triple-bonded O atom in VO_3. The structures of the VO_y ($y = 1$–4) species are schematically summarized in Fig. 40. This series of oxide molecules exhibits rather interesting and unusual structural and bonding properties.

3.5. $Ti_xO_y^-$ $(x = 1-4; y = 1-8)$

Titanium dioxide is one of the most technologically important oxide materials and has been widely studied as a prototypical transition metal oxide due to its rather simple electronic structure.[91] Hagfeldt *et al.* calculated the structure and energy levels of several $(TiO_2)_n$ clusters for $n = 1$–3 as models to understand the chemical and physical properties of bulk TiO_2.[152,153] Experimentally, there were two previous mass spectrometry studies on the $Ti_xO_y^+$ cluster distributions[154,155] and no spectroscopic information was known for any Ti_xO_y clusters except TiO and TiO_2, although there been considerable interest in the nanoparticles of TiO_2.[156] The TiO molecule has been extensively studied spectroscopically because of its astrophysical importance.[5,157–160] Its electronic structure is well understood.[96–98,161] Considerably less has been reported about the TiO_2 molecule. There are only two previous calculations[161,162] and three experimental studies,[163–165] which showed that TiO_2 is a bent molecule and has a singlet ground state.

We reported a comprehensive study on two series of small titanium oxide clusters, TiO_y ($y = 1$–3) and $(TiO_2)_n$ ($n = 1$–4).[37] Vibrationally resolved spectra were obtained for TiO^- and TiO_2^-. Six low-lying electronic excited states were observed for TiO and were assigned using the previous spectroscopic information available for TiO. Strong electron correlation effects were observed in TiO since five of the observed excited states were due to multielectron detachment transitions. TiO_2 was found to be a closed-shell molecule with a 2 eV HOMO–LUMO gap. The ν_1 vibrational frequency of the TiO_2 ground state was measured to be 960 (40) cm^{-1}. The EA of TiO_2 (1.59 eV) was only slightly higher than that of TiO (1.30 eV) while TiO_3 had a very high EA of about 4.2 eV. The larger $(TiO_2)_n$ clusters were all found to be closed-shell systems with increasing EA, but the HOMO–LUMO gaps of these larger clusters were about the same as that of the monomer TiO_2. This surprising observation of a smaller gap than the bulk rutile TiO_2 was interpreted on basis of the structure and bonding of the clusters, which possessed a lower coordination for the Ti atoms compared to the bulk.

3.5.1. *TiO and TiO^-: Low-lying States and Strong Electron Correlation Effects*

The PES spectra of TiO^- are shown in Fig. 41 with six vibrationally resolved bands. The observed BE and spectroscopic information are listed in Table 16. The electronic structure and many low-lying excited states of TiO were known and these made the assignment of the PES spectra quite

Table 16. Observed binding energies (BE) and spectroscopic constants for TiO^- and TiO.

	State	BE (eV)	Term value (cm^{-1})[a]	Vib. freq. (cm^{-1})[a]
TiO^-	$X^1\Sigma^+(9\sigma^1 1\delta^2)$	0	0	800 (60)
TiO	$X^3\Delta(9\sigma^1 1\delta^1)$	1.300 (0.030)	0	1000 (60)
	$a^1\Delta(9\sigma^1 1\delta^1)$	1.725 (0.030)	3430 (60)	1000 (60)
	$d^1\Sigma^+(9\sigma^2)$	2.00 (0.08)	5650 (100)	1000 (80)
	$E^3\Pi(9\sigma^1 4\pi^1)$	2.80 (0.08)	12 100 (100)	
	$A^3\Phi(1\delta^1 4\pi^1)$	3.05 (0.08)	14 110 (100)	
	$b^1\Pi(9\sigma^1 4\pi^1)$	3.13 (0.08)	14 760 (100)	
	$B^3\Pi(1\delta^1 4\pi^1)$	3.31 (0.08)	16 210 (100)	

[a]Relative energies can be determined more accurately.

straightforward, as shown in Fig. 41. However, many of these states required multielectron excitations from the photodetachment of TiO^-. These multielectron excitations appeared to be particularly strong for TiO^-. TiO has a ground state configuration of $9\sigma^1 1\delta^1 4\pi^0 (X^3\Delta)$, all the three valence orbitals are primarily of Ti $3d$ character.[96–98] In the anion, the extra electron can enter either the 9σ or the 1δ orbital. Since TiO^- is isoelectronic with VO, which has a $9\sigma^1 1\delta^2$ configuration, the extra electron in TiO was expected to enter the 1δ orbital to give a configuration for TiO^- as $9\sigma^1 1\delta^2 (^4\Sigma^-)$, similar to that of VO. The removal of one δ electron from TiO^- results in the $X^3\Delta$ ground state of TiO, as well as the $a^1\Delta$ excited state. The 532 nm spectrum revealed these two states. The spin-orbit splitting for the ground state was too small to be resolved at the experimental resolution. The observed vibrational frequencies and excitation energy for the $a^1\Delta$ state were in good agreement with the previous optical measurements.[5]

The Franck–Condon factors for the higher vibrational levels of the $a^1\Delta$ state seemed to increase from 532 nm to 355 nm, as shown in Fig. 41. In particular, the vibrational progression for the $a^1\Delta$ state became rather unusual in the 266 nm spectrum, suggesting that there was another electronic state overlapping with the $a^1\Delta$ state and whose detachment cross section increased with photon energy. This state actually was the known $d^1\Sigma^+$ state, whose vibrational frequency was similar to that of the $a^1\Delta$ state and whose excitation energy overlapped with the $\nu = 2$ level of the $a^1\Delta$ state, within our experimental accuracy. The $d^1\Sigma^+$ state, however, must be due

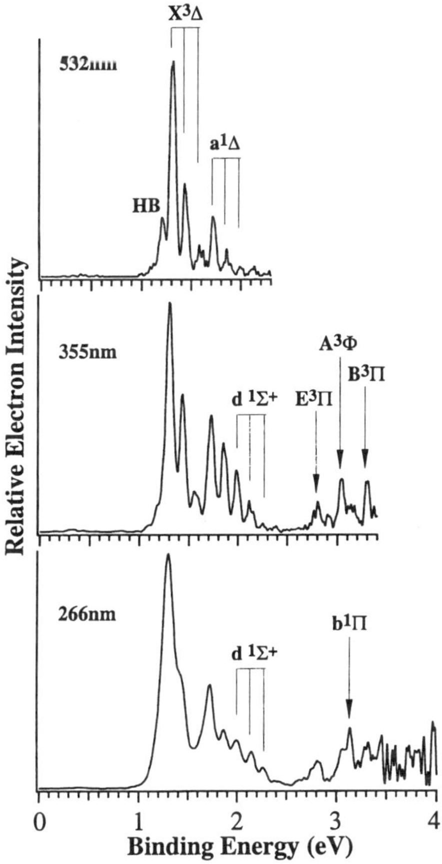

Fig. 41. Photoelectron spectra of TiO^- at 532, 355 and 266 nm. HB stands for hot band. From Ref. 37.

to a two-electron detachment process from the ground state of the anion ($9\sigma^1 1\delta^2$, $^4\Sigma^-$) because it corresponded to a $9\sigma^2$ configuration.[5] Thus, its weak intensity was understandable.

As shown in the 355 nm spectrum, at least four more states were observed at BEs of 2.80, 3.05, 3.13 and 3.31 eV. These features all had rather weak intensities and were also observed in the 266 nm spectrum. In particular, the 3.13 eV feature was observed more clearly in the 266 nm spectrum. Similar to the $d^1\Sigma^+$ state, all of these low-lying states were due to multi-electron transitions in the detachment processes, consistent with their weak intensities. The $E^3\Pi$ and $b^1\Pi$ states were derived from a $9\sigma^1 4\pi^1$ configuration, due to detachment of a 1δ electron and promotion of the other 1δ electron to the 4π orbital. Similarly, the $A^3\Phi$ and $B^3\Pi$ states were from a $1\delta^1 4\pi^1$ configuration, due to detachment of a 1δ electron and promotion of the 9σ electron to the 4π orbital. Of course, it would have been very difficult to assign these states without the knowledge of the energy levels from the previous optical experiments. The excitation energies for these states were all consistent with the known energy levels of TiO except that the spin-orbit splittings were not resolved for the $E^3\Pi$, $A^3\Phi$ and $B^3\Pi$ states in the PES experiments.

Multielectron excitations are well known in PES,[72] which is the major experimental technique to study electron correlation effects in atoms, molecules and solids. However, observation of such extensive multielectron excitation in the TiO^- system is rather unusual. Of course, it is well known that there are strong electron correlation effects in the transition metal system. Theoretically, systems involving transition metals are very difficult to handle and multireference configuration interaction techniques are required to accurately treat these systems. For the TiO^- system, the current experiment suggests that at least configurations involving transfer of electrons among the three d-type orbitals (9σ, 1δ, 4π) are important to describe the ground state of TiO^-. As will be discussed below, strong multielectron transitions are also observed in the PES spectra of ScO^- and YO^-.

3.5.2. *TiO_2 and TiO_2^-*

Compared to TiO, considerably less was reported about the TiO_2 molecule, which was known to be a closed-shell molecule with a C_{2v} structure and a bond angle of about 110°. Its vibrational frequencies were measured in a matrix IR experiment to be 946.9 and 917.7 cm^{-1} for the ν_1 and ν_3 modes, respectively.[165] An emission band at ~5300 Å (2.34 eV) was observed for TiO_2 also in a low temperature matrix.[164] Theoretically, Ramana and Philips first treated the ground ($^1A_1$) and lowest-lying excited states ($^3B_2$ and $^1B_2$) of TiO_2 in detail using SCF and CI calculations.[162] Their calculations gave an excitation energy of about 1.63 eV for the $^3B_2$ and $^1B_2$ states, which they found were nearly degenerate. Interestingly, their calculations suggested that the TiO_2 molecule was linear in the excited states.

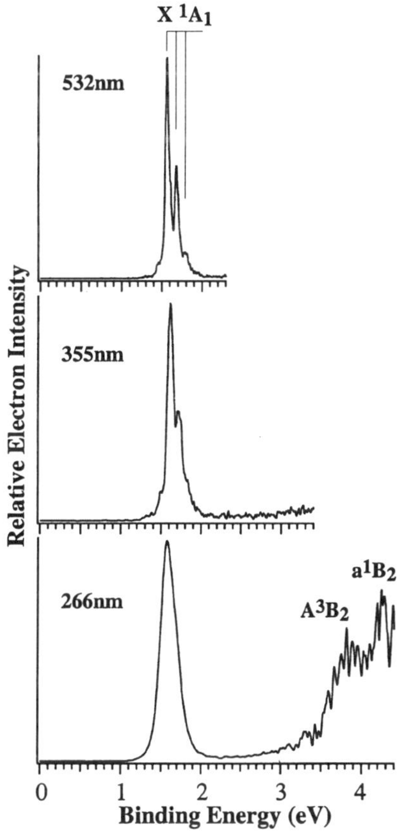

Fig. 42. Photoelectron spectra of TiO_2^- at 532, 355 and 266 nm. From Ref. 37.

The PES spectra of TiO_2^- are shown in Fig. 42 at three detachment energies. The 532 and 355 nm spectra were very simple, each giving one vibrationally-resolved band corresponding to the ground state of TiO_2. The observed vibrational mode should be due to the totally symmetric ν_1 mode. The ν_2 mode was also allowed, but its frequency was probably too low to be resolved. Our observed frequency for the ν_1 mode was 940 (40) cm^{-1}, in good agreement with the previous matrix IR experiment.[165] The HOMO of TiO_2 is $6b_2$, a weak O–O bonding orbital. The LUMO, where the extra electron enters in TiO_2^-, is $10a_1$, a nonbonding orbital of Ti $3d$ character.[162] The removal of the extra electron from the $10a_1$ orbital of TiO_2^- resulted in the X^1A_1 ground state of TiO_2. The vibrational progression of the ground state suggested that there was only a slight Ti–O bondlength change between the anion and the neutral. Therefore, the anion must also have a C_{2v} structure with a similar bond angle as the neutral TiO_2.

Two more features were observed in the 266 nm spectrum at high BE, due to the removal of a $6b_2$ electron (the HOMO of TiO_2), and were assigned to the $^3B_2$ and $^1B_2$ excited states that were studied in the previous theoretical work.[162] The broad nature of the PES spectrum indicated that there must be a large geometry change between the anion and the two B_2 states. The lack of any resolved vibrational structure suggested that the active vibrational mode involved was likely due to the low frequency ν_2 mode. Therefore, the observed broad detachment features were consistent with the theoretical prediction that two B_2 states might be linear. The BE of the $^3B_2$ state was estimated to be 3.55 eV. Due to the overlapping nature between the $^1B_2$ and the $^3B_2$ states, the BE for the $1B_2$ state was estimated to be at about 4 eV. These yielded the excitation energies for the $^3B_2$ and the $^1B_2$ to be 1.96 and 2.4 eV, respectively. The 2.4 eV excitation energy for the $^1B_2$ state was very close to the 2.34 eV emission spectrum observed for TiO_2 in the matrix experiment, suggesting that the unassigned emission band was most likely from the $^1B_2$ excited state to the $^1A_1$ ground state.[164] Apparently, the previous calculation underestimated the excitation energies for the two B_2 excited states.[162]

3.5.3. TiO_3^-

The spectrum of TiO_3^- is shown in Fig. 43, in comparison to those of TiO^- and TiO_2^-. The TiO_3 molecule was observed to have a very high EA of about 4.2 eV and the detachment spectrum could only be studied at 193 nm. Besides the ground state feature at 4.2 eV, higher energy states were also observed. But due to the poor signal-to-noise statistics, these higher energy features could not be well identified. The TiO_3 molecule has never been studied before either experimentally or theoretically. From TiO to TiO_2, the formal oxidation state of Ti is increased from +2 to +4, the maximum oxidation state of Ti. TiO_3 is oxygen-rich because Ti cannot be oxidized to +6. In this case, the TiO_3 molecule can be viewed to be formed by replacing an O atom in TiO_2 with an O_2 peroxide unit. The bonding between such peroxide unit and the TiO fragment will be interesting to consider theoretically, because the O_2 peroxide unit is believed to exist on

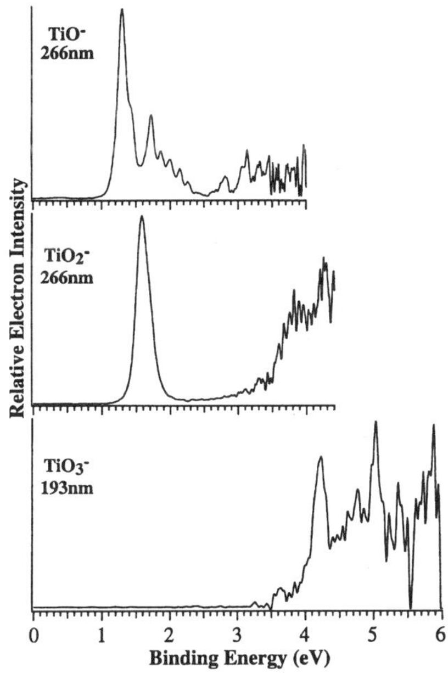

Fig. 43. Photoelectron spectra of TiO_3^- at 193 nm compared to the 266 nm spectra of TiO^- and TiO_2^-. From Ref. 37.

TiO_2 bulk surfaces.[91] The understanding of this seemingly simple molecular case may help gain insight into the properties of the bulk surfaces.

3.5.4. $(TiO_2)_n$ ($n = 2$–4): *Evolution of HOMO–LUMO Gap*

Using a simple pair-potential ionic model, Yu and Freas first calculated the minimum energy structures of several Ti_xO_y clusters including the $(TiO_2)_n$ clusters for $n = 2$–4.[154] Hagfeldt *et al.* carried out calculations on $(TiO_2)_2$ and $(TiO_2)_3$, using *ab initio* SCF methods.[152,153] They studied both the geometrical and electronic structures of these clusters and were interested in using these clusters as models to understand properties of bulk TiO_2. As expected, both of these clusters were found to be closed-shell electronic systems with large HOMO–LUMO gaps, which were overestimated in the calculations. Ti_2O_4 was shown to have a C_{2h} structure with a Ti_2O_2 rhombus and two terminal O atoms which were bent out of the Ti–Ti axis. The Ti_2O_4 structure was found to be very similar to that of Si_2O_4 which has a D_{2h} structure with the two terminal O atoms on the Si–Si axis.[57] The subtle difference between the structures of Ti_2O_4 and Si_2O_4 reflects the difference in the bonding between Ti and Si with O: *s* and *d* orbitals are involved in Ti while mostly *s* and *p* orbitals are used in Si. The structure of Ti_3O_6 found in the previous calculations was also similar to that of Si_3O_6 which has a D_{2d} structure with two perpendicular Si_2O_2 rhombuses sharing a central Si atom and two terminal O atoms on the Si–Si–Si axis.[57] The two terminal O atoms in Ti_3O_6 are again bent relative to the Ti–Ti–Ti axis, yielding a C_s structure. For Ti_4O_8, the previous simple pair-potential calculation gave a complicated three-dimensional structure.[154] There was no other more accurate calculation.

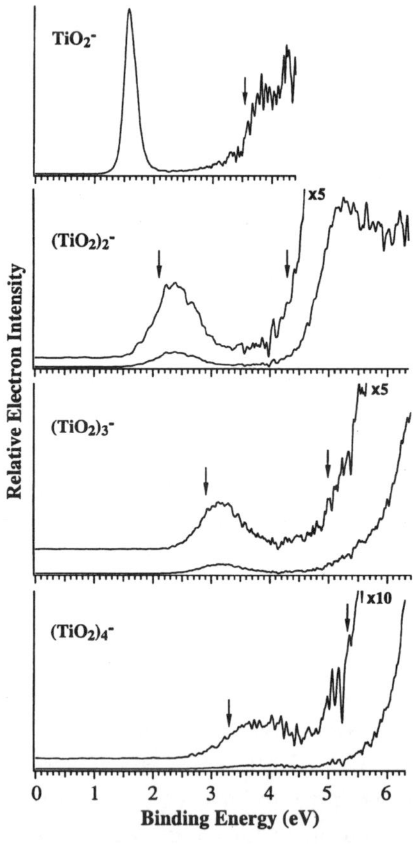

Fig. 44. Comparison of the photoelectron spectra of $(TiO_2)_n^-$ for $n = 1$–4. The arrows indicate the LUMO and HUMO levels. From Ref. 37.

Our PES spectra for $(TiO_2)_n^-$ ($n = 2$–4) are shown in Fig. 44 at 193 nm and compared to that of TiO_2^-. We also obtained data at 266 and 355 nm and measured better resolved spectra. All of these clusters showed a large HOMO–LUMO energy gap for the neutral clusters, consistent with the previous calculations. However, the calculations overestimated the gaps considerably, probably because relaxation effects were not included.[153] In the anions, the extra electron entered the LUMO of the neutral clusters, producing a doublet ground state for the anions. In the PES spectra, the first feature derived from the removal of this extra electron from the LUMO of the neutral clusters, yielding the singlet ground state for the neutrals. The second feature in the spectra corresponded to the removal of an electron from the HOMO of the neutral clusters. Thus the HOMO–LUMO gaps for the neutral clusters were directly measured in the PES spectra. The EAs for the $(TiO_2)_n$ clusters, measured from the detachment threshold, are summarized in Table 17. The HOMO and LUMO levels are indicated for each cluster in Fig. 44. The LUMO level for TiO_2 was more accurately measured and corresponded to the peak maximum. The HOMO–LUMO gaps for all the clusters were observed to be similar, with a value of about 2 eV. It was surprising that all the larger clusters showed similar HOMO–LUMO gaps to that of the TiO_2 monomer. It was well known that the bulk band gap of the rutile TiO_2 is 3.1 eV.[166] The question was why the cluster energy gaps are smaller than the bulk value.

There were two possible explanations. First, it was possible that the HOMO levels were underestimated for the $(TiO_2)_3$ and $(TiO_2)_4$ clusters because of the uncertainty in determining the HOMO levels in the $(TiO_2)_3^-$ and $(TiO_2)_4^-$ cases as seen in Fig. 44. Second and more plausible, the smaller HOMO–LUMO gaps were due to the low coordination number of the Ti atoms compared to that of bulk TiO_2. Bulk TiO_2 crystal has the rutile lattice, where each Ti is coordinated by six O atoms.[91] From our discussions above the previously calculated structures for Ti_2O_4 and Ti_3O_6, they both could be viewed as polymers of the TiO_2 monomer and the

Table 17. Measured adiabatic electron affinities (EA) of TiO_y ($y = 1$–3) and $(TiO_2)_n$ ($n = 1$–4).

	TiO	TiO_2	TiO_3	Ti_2O_4	Ti_3O_6	Ti_4O_8
EA (eV)	1.30 (3)	1.59 (3)	4.2 (1)	2.10 (8)	2.9 (1)	3.3 (2)

Ti atoms were tetra-coordinated while the terminal Ti atoms were only coordinated to three O atoms with probably a terminal Ti=O double bond. The LUMO levels (corresponding to the conduction band in the bulk) of the clusters should be mainly of Ti 3*d* character while the HOMO levels (corresponding to the valence and band in the bulk) should be mainly of O 2*p* character. The lower coordination number in the clusters implied a weaker interaction between Ti and O and were probably the cause for the smaller HOMO–LUMO energy splitting. Accordingly, the smaller HOMO–LUMO gap in the $(TiO_2)_4$ cluster might suggest that its structure was similar to the $(TiO_2)_3$ cluster with three perpendicular Ti_2O_2 rhombuses and two terminal O atoms. This was only a tentative observation. More accurate calculations would be desirable in light of these experiments and should be able to resolve the question regarding the smaller band gaps in these clusters.

3.6. ScO_y^- and YO_y^- $(y = 1–5)$

ScO is the simplest transition metal oxide with only one 3*d* electron. The ScO and YO diatomics have been well studied both experimentally[2,5,167–176] and theoretically.[96–98] However, investigation of the higher oxide molecules was scarce. Recently, Chertihin *et al.* reported an infrared study of ScO_y $(y = 1–4)$ formed from reactions of laser-ablated Sc and O_2 in a low temperature matrix.[177] Two forms of ScO_2 were observed, a bent OScO dioxide and a $Sc(O_2)$ complex. They also observed two forms of ScO_3, an $OScO_2$ molecule and a $Sc(O_3)$ ozonide species. The observed ScO_4 species was attributed to an $(O_2)Sc(O_2)$ molecule. There is no other report on the ScO_y and YO_y species. We very recently completed a systematic PES study on ScO_y^- $(y = 1–4)$ and YO_y^- $(y = 1–5)$.[39] We obtained vibrationally resolved PES spectra for both the mono- and dioxide species. The monoxide PES spectra of ScO^- and YO^-were found to be nearly identical. Strong two-electron detachment transitions were observed in the PES spectra of ScO^- and YO^-, suggesting unusually strong electron correlation effects in the anionic ground states, similar to that observed in TiO^-. Weak detachment features due to anionic excited states were also observed in the PES spectra of the monoxides. We observed four similar detachment features for the dioxide species with those of YO_2^- having lower electron BE. Isomers due to O_2-complexes were also observed for the ScO_2 and YO_2 systems. We were only able to obtain very broad and featureless spectra for the higher oxide species, MO_y^- ($y \geq 3$), for which at least two isomers were observed with low and high BEs, respectively.

3.6.1. *ScO and YO: Low-lying States and Strong Electron Correlation Effects*

Figure 45 displays the PES spectra of ScO^- at three-photon energies. The 532 nm spectrum showed a well-resolved vibrational progression, representing the ground state of ScO ($X^2\Sigma^+$) and indicating that there was bond length change between the ground states of ScO^- and ScO. Hot band

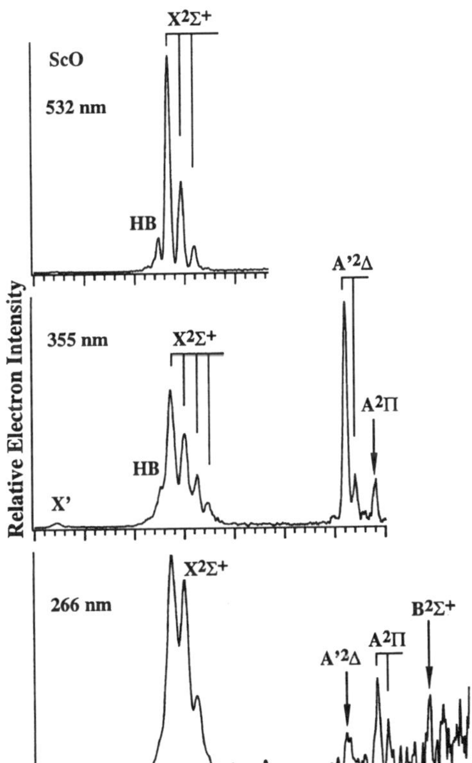

Fig. 45. Photoelectron spectra of ScO^- at 532, 355 and 266 nm. HB stands for hot band transitions. Vibrational progressions are indicated with vertical lines. From Ref. 39.

Table 18. Observed binding energies (BE) and spectroscopic constants for ScO^- and ScO.

				Term value (cm^{-1})		Vib. freq. (cm^{-1})	
	State	Configuration	BE (eV)	This work	Previous[b]	This work	Previous[b]
ScO^-	$X^3\Delta$	$8\sigma^2 3\pi^4 9\sigma^1 1\delta^1$	0	0		840 (60)	
	$X'^1\Sigma^+$	$8\sigma^2 3\pi^4 9\sigma^2$	0.23 (6)	9 000 (200)			
	$X^2\Sigma^+$	$8\sigma^2 3\pi^4 9\sigma^1$	1.35 (2)[a]	0		1000 (50)	964.95
ScO	$A'^2\Delta$	$8\sigma^2 3\pi^4 1\delta^1$	3.10 (2)	14 200 (60)	15 082.8[c]	780 (70)	845.0
	$A^2\Pi$	$8\sigma^2 3\pi^4 4\pi^1$	3.40 (3)	16 540 (80)	16 547.0[c]	830 (70)	876.0
	$B^2\Sigma^+$	$8\sigma^2 3\pi^4 10\sigma^1$	3.90 (4)	20 600 (100)	20 645.1		825.47

[a]Adiabatic electron affinity of ScO.
[b]Ref. 2.
[c]Average of the two spin-orbit components.

(HB) transitions were also observed, yielding a vibrational frequency for the ScO^- anion to be about 840 cm^{-1}, which was considerably smaller than the vibrational frequency of the neutral ScO measured to be about 1000 cm^{-1}. This suggested that the extra electron in ScO^- entered into an antibonding orbital, reducing the bond strength of ScO. The 355 nm spectrum revealed an intense peak at 3.10 eV with a weak vibrational progression and an additional weak feature at ~3.4 eV. A very weak feature at low BE(X^1) was observed in some spectra and was most likely due to an excited state of the anion. The sharp peak at 3.10 eV showed strong photon energy dependence. More features at higher BE were also observed at 266 nm, although the signal-to-noise ratio was poor. The BEs of all the observed states and obtained spectroscopic constants of ScO are listed in Table 18.

Fig. 46 shows the PES spectra of YO^- taken at three-photon energies. We observed very similar features for YO^- as for ScO^-. The 532 nm spectrum revealed a well-resolved vibrational progression for the ground state of YO. The peaks at 3.15 eV and 3.4 eV in the 355 nm spectrum exhibited doublet features, characteristic of a large spin-orbit splitting. A very weak low BE feature (X^1), possibly due to an excited state of YO^-, was also present. The feature at 3.15 eV showed a strong photon energy dependence, similar to that in ScO^-. The 266 nm spectrum of YO^- had low count rate and very poor signal-to-noise ratio and no new features could be definitively identified relative to the 355 nm spectrum. The BEs of all

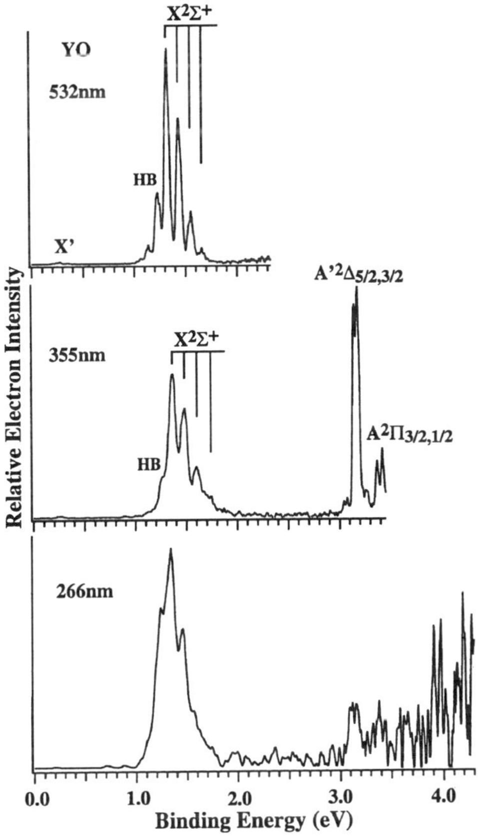

Fig. 46. Photoelectron spectra of YO^- at 532, 355 and 266 nm. HB stands for hot band transitions. Vibrational progression is indicated with vertical lines. From Ref. 39.

the assigned states and obtained spectroscopic constants of YO are listed in Table 19.

The ground electronic state of ScO was well established as $^2\Sigma^+$ with an electronic configuration of $8\sigma^2 3\pi^4 9\sigma^1$, where 8σ is from Sc $3d\sigma$ and O $2p\sigma$; 3π is from Sc $3d\pi$ and O $2p\pi$; 9σ is mainly a nonbonding $4s$ orbital. The low-lying excited states of ScO were also well established previously, including the $A'^2\Delta(8\sigma^2 3\pi^4 1\delta^1)$, $A^2\Pi(8\sigma^2 3\pi^4 4\pi^1)$, and $B^2\Sigma^+(8\sigma^2 3\pi^4 10\sigma^1)$. All of these states were observed in the PES spectra of ScO^-, as shown in Fig. 45. The obtained term values and vibrational frequencies were in good agreement with known values from the literature, as shown in Table 18.

Table 19. Observed binding energies (BE) and spectroscopic constants for YO^- and YO.

			Term value (cm^{-1})		Vib. freq. (cm^{-1})	
State	Configuration	BE (eV)	This work	Previous[b]	This work	Previous[b]
YO^- $X^3\Delta$	$11\sigma^2 5\pi^4 12\sigma^1 2\delta^1$	0	0	0	740 (60)	
$X'^1\Sigma^+$	$11\sigma^2 5\pi^4 12\sigma^2$	0.28 (5)	8 600 (300)			
YO $X^2\Sigma^+$	$11\sigma^2 5\pi^4 12\sigma^1$	1.35 (2)[a]	0	0	900 (50)	861.0
$A'^2\Delta_{3/2}$	$11\sigma^2 5\pi^4 2\delta^1$	3.14 (2)	14 500 (80)	14 531.2		794.0
$^2\Delta_{5/2}$		3.18 (2)	14 800 (80)	14 870.4		794.9
$A^2\Pi_{1/2}$	$11\sigma^2 5\pi^4 6\pi^1$	3.37 (4)	16 300 (90)	16 315.0		820.7
$^2\Pi_{3/2}$		3.42 (4)	16 700 (90)	16 742.2		822.7

[a]Adiabatic electron affinity of YO.
[b]Ref. 2.

The extra electron in the ground state of ScO^- could enter either 9σ to give a singlet anion ground state or 1δ orbital to give a triplet state. Since ScO^- is isoelectronic to TiO, which possesses a $^3\Delta(\ldots 9\sigma^1\delta^1)$ ground state, it was reasonable to assume that the extra electron in ScO^- entered the 1δ orbital. Thus the following detachment channels would be necessary to reach the neutral excited states observed in the PES spectra:

$$8\sigma^2 3\pi^4 9\sigma^1 1\delta^1(^3\Delta) \rightarrow 8\sigma^2 3\pi^4 9\sigma^1 \quad X^2\Sigma^+ \tag{7}$$

$$\rightarrow 8\sigma^2 3\pi^4 1\delta^1 \quad A'^2\Delta \tag{8}$$

$$\rightarrow 8\sigma^2 3\pi^4 4\pi^1 \quad A^2\Pi \tag{9}$$

$$\rightarrow 8\sigma^2 3\pi^4 10\sigma^1 \quad B^2\Sigma^+ . \tag{10}$$

The detachment channels (7) and (8) resulted from one-electron transition from the 1δ and 9σ orbitals, respectively. The observed photon energy dependence of the $X^2\Sigma^+$ and $A'^2\Delta$ PES features were consistent with these detachment channels, in that, the detachment cross section for removing electrons with high angular momenta increased with photon energies.[70] The spin-orbit splitting of the $A'^2\Delta$ state was known to be 106.1 cm^{-1},[2] which was too small to be resolved under the experimental resolution.

However, both detachment channels, (9) and (10) required two-electron transitions. Such multi-electron transitions, due to strong electron correlation effects, had been observed in PES of TiO^-, as discussed earlier.[37]

It was unusual to see that the intensities of the $A^2\Pi$ and $B^2\Sigma^+$ features were even stronger than that of the $A'^2\Delta$ feature in the 266 nm spectrum. This observation suggested strong electron correlation effects in the ScO^- ground state ($^3\Delta$), which must have strong configuration mixings. Our obtained term values and vibrational frequencies were in excellent agreement with those from the literature, except for that of the term value for the $A'^2\Delta$ state. Our measured value was smaller than the literature value by exactly a vibrational quantum. The literature value was derived from emmission spectra of $A'^2\Delta$ to the ground state in a chemiluminescence experiment. We suspected that the origin band was misassigned in the previous work.

The weak feature labeled as X^1 was assigned to a transition from an electronic excited state of ScO^- to the ground state of neutral ScO. We tentatively assigned this anion excited state to be $8\sigma^2 3\pi^4 9\sigma^2 (^1\Sigma^+)$, where the extra electron entered the 9σ orbital to give a singlet state. This excited state was expected to be metastable as it was spin-forbidden for it to relax down to the anion ground state. We had observed such metastable anionic excited states in several transition metal diatomic systems previously, as discussed earlier.

The PES spectra of YO^- were almost identical to that of ScO^- except a slight larger spin-orbit splitting in the $^2\Delta$ and $^2\Pi$ states, which were well within our resolution (Fig. 46). It was surprising that the EA of YO was identical to that of ScO within our experimental accuracy and even the term values of the low-lying states of YO were rather close to those of ScO, as given in Table 19. Thus the PES spectra of YO^- could be assigned similarly, as shown in Fig. 46. Analogous to ScO^-, the ground state of YO^- should be $X^3\Delta$ with a $11\sigma^2 5\pi^4 12\sigma^1 2\delta^1$ configuration. Single-electron detachment of the 2δ and 12σ electron resulted in the $X^2\Sigma^+$ and $A'^2\Delta$ states, respectively. The $A^2\Pi$ state would require a two-electron transition from the anion ground state, similar to the detachment channel (9) in ScO^-. The 266 nm spectrum of YO^- had very low count rate and poor signal-to-noise ratio, which prevented us from making definitive assignments of any higher excited states. However, the feature near 3.9 eV yielded a term value, which was very close to the literature value (20 791 cm^{-1}) for the $B^2\Sigma^+$ state. Our obtained term values and vibrational frequencies are listed in Table 19 and were in excellent agreement with those from the literature.

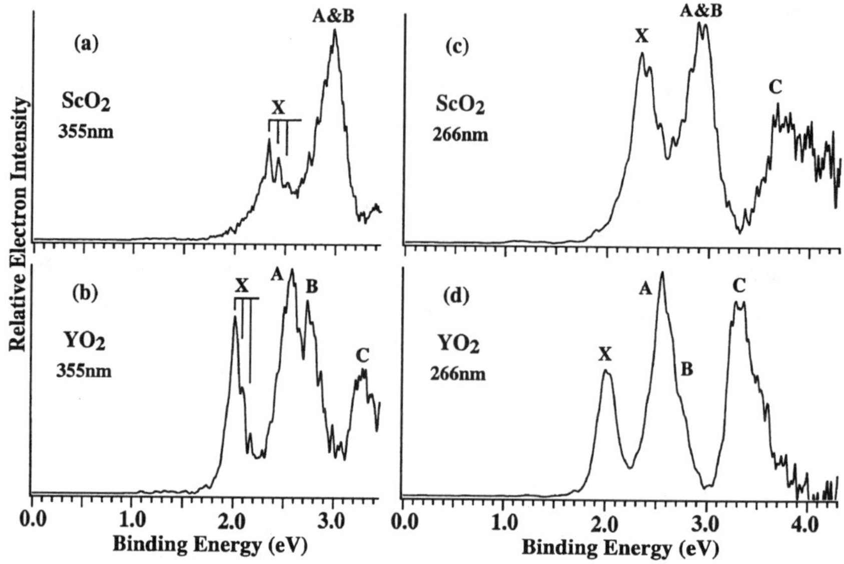

Fig. 47. Photoelectron spectra of ScO_2^- and YO_2^- at 355 and 266 nm. Vibrational progressions are indicated with vertical lines. From Ref. 39.

3.6.2. *ScO_2 and YO_2: Two Isomers — Dioxide and O_2-Complex*

Figure 47 shows the spectra of ScO_2^- and YO_2^- taken at two different photon energies. The 355 nm spectrum of ScO_2^- exhibited two major features. The features (X) around 2.3 eV revealed a resolved vibrational progression with a long tail in the low BE side. The tail might be due to either hot bands or excited states of ScO_2^- and could not be eliminated under different experimental conditions. The 266 nm spectrum of ScO_2^- revealed an additional broad feature at higher BEs. The 355 nm spectrum of YO^- exhibited four resolved features with the ground state feature (X) showing a partially resolved vibrational progression. The 266 nm spectrum of YO_2^- revealed no more new features, except for a photon energy dependence of features B and C. At 266 nm, the relative intensity of feature B was decreased whereas that of feature C was enhanced. The broad low energy tail present in the ScO_2^- spectra did not seem to be present in the YO_2^- spectra.

The 266 nm spectra of ScO_2^- and YO_2^- were similar except that the features of the YO_2^- spectrum had lower BEs with slightly narrower band

Table 20. Observed binding energies (BE) and spectroscopic constants for ScO_2 and YO_2.

	State	BE(eV)	Vib. freq. $(cm^{-1})^c$
ScO_2	X^2B_2, $(8a_1)^2(5b_2)^2(1a_2)^2(9a_1)^2(3b_1)^2(6b_2)^1$	2.32 (2)[a]	740 (80)
	A^2B_1, $(8a_1)^2(5b_2)^2(1a_2)^2(9a_1)^2(3b_1)^1(6b_2)^2$	(2.7)	
	B^2A_1, $(8a_1)^2(5b_2)^2(1a_2)^2(9a_1)^1(3b_1)^2(6b_2)^2$	(2.8)	
	C^2A_2, $(8a_1)^2(5b_2)^2(1a_2)^1(9a_1)^2(3b_1)^2(6b_2)^2$	3.6 (1)	
YO_2	X^2B_2, $(13a_1)^2(6b_2)^2(2a_2)^2(14a_1)^2(5b_1)^2(7b_2)^1$	2.00 (3)[b]	640 (80)
	A^2B_1, $(13a_1)^2(6b_2)^2(2a_2)^2(14a_1)^2(5b_1)^1(7b_2)^2$	2.46 (9)	
	B^2A_1, $(13a_1)^2(6b_2)^2(2a_2)^2(14a_1)^1(5b_1)^2(7b_2)^2$	2.75 (5)	
	C^2A_2, $(13a_1)^2(6b_2)^2(2a_2)^1(14a_1)^2(5b_1)^2(7b_2)^2$	3.28 (6)	

[a]Adiabatic electron affinity of ScO_2.
[b]Adiabatic electron affinity of YO_2.
[c]ν_1 mode.

width. Comparing the two spectra, we concluded that the intense and broad feature at 2.9 eV in the ScO_2^- spectrum should contain two overlapping features (A and B), similar to the A and B features in the YO_2^- spectrum where they were resolved at 355 nm (Fig. 47b). Comparing the 355 and 266 nm spectra of ScO_2^-, we could also see that the intensities of the X and A features were enhanced at 266 nm. The similarity between the dioxides of Sc and Y was analogous to that between their monoxides (Figs. 45 and 46). The measured BEs for ScO_2^- and YO_2^- and the obtained neutral ground state vibrational frequencies are summarized in Table 20.

There was no theoretical work concerning the electronic structure of either ScO_2 or YO_2. In the recent low temperature matrix infrared study,[177] Chertihin *et al.* used density functional theory to calculate the structure and vibrational frequencies of ScO_2. They found that ScO_2 was a bent molecule with a bond angle of 128° and a $^2B_2$ ground state. Their calculated ν_1 vibrational frequency was 742 cm^{-1}, in good agreement with our measured ν_1 frequency of 740(80) cm^{-1}. To understand the PES spectra and electronic structure of ScO_2, we could use the previous results of TiO_2 and VO_2, as discussed earlier. Both are similar to ScO_2 with a bent structure except that they have one and two extra $3d$ electrons compared to ScO_2.

The chemical bonding between Sc and O is mainly from the interactions between the Sc $3d$ and $4s$ and the O $2p$. Using the result of VO_2 by subtracting two valence electrons, one obtained the following valence configuration for ScO_2: $(8a_1)^2(5b_2)^2(1a_2)^2(9a_1)^2(3b_1)^2(6b_2)^1$ with a $^2B_2$ ground state, consistent with the result of Chertihin *et al.*[177] In ScO_2^-, the extra electron most likely entered the $6b_2$ orbital to give a closed-shell ground state for the anion. Therefore, the PES features of ScO_2^- were interpreted in the single-particle picture to be due to detachment of an electron from the top four valence MOs, respectively. The PES spectra of YO_2^- were similarly interpreted, except that the valence MOs of YO_2 were formed from the Y $4d$ and $5s$ and O $2p$. The PES spectra of ScO_2^- and YO_2^- were relatively simple because of the closed-shell nature of the anion ground states. The low energy tail present in the spectra of ScO_2^- was probably due to excited states of ScO_2^- because they could not be eliminated by better vibrational cooling. Similar low energy tails were also observed in the PES spectra of VO_2^- previously. Interestingly, the BEs of all the YO_2^- PES features were found to be lower than those of ScO_2^- whereas the PES features of the monoxides were almost identical. We suspected that the destabilization of the YO_2 valence MOs relative to those of ScO_2 might be due to the fact that they have different bond angles. The $4d$ and $5s$ orbitals of Y are more spatially diffuse compared to the $3d$ and $4s$ orbital of Sc and might lead to a larger bond angle for YO_2.

As shown in Fig. 48, sometimes a weak and low energy feature (X') could be observed around 1.1 eV in the spectra of both ScO_2^- and YO_2^-. This weak feature was most likely due to a $M(O_2)^-$ complex. This isomer possessed a much lower EA, comparable to those of the monoxides. The $Sc(O_2)$ complex was observed in the previous matrix study.[177] Transition metal-dioxygen complexes had been observed in our previous PES studies of transition metal oxides, such as $Cu(O_2)$ and $Ni(O_2)$. We found that the $M(O_2)$ complexes always have lower EAs than the OMO dioxides and similar EAs to the respective monoxides. The O_2 in the $Sc(O_2)$ and $Y(O_2)$ complexes was likely side-bonded to the metal atom and could be viewed as a peroxide (O_2^{2-}), somewhat analogous to the monoxide in term of the formal oxidation state of the metal, explaining their similar EAs.

3.6.3. *ScO_y and YO_y ($y \geq 3$) and Isomers*

The PES spectra of the higher oxide molecules are compared to the mono- and dioxides in Figs. 49 and 50 for ScO_y^- ($y = 1$–4) and YO_y^- ($y = 1$–5), respectively. The spectra of the higher oxide species were quite broad and featureless and only a few qualitative remarks could be made. It was seen

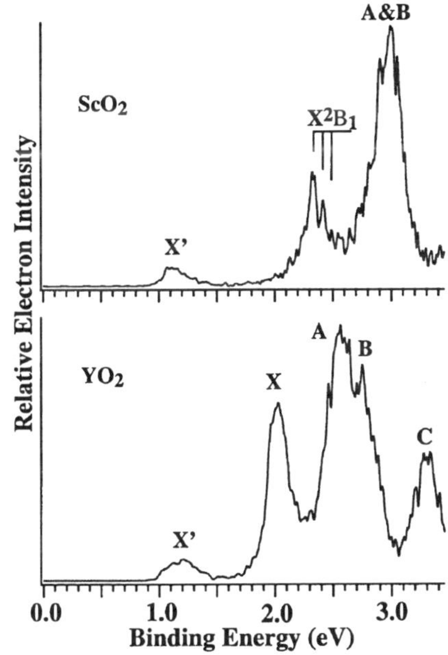

Fig. 48. Photoelectron spectra of ScO_2^- and YO_2^- at 355 nm, showing the low binding feature (X') due to an $M(O_2)^-$ complex. From Ref. 39.

that the BEs increased from MO^- to MO_3^-. The BEs of MO_3^- were so high that only a tail was observed in the 355 nm spectra. The 266 nm spectrum of ScO_3^- showed a broad feature and long low energy tail whereas that of YO_3^- appeared to have a lower (~3 eV) and a higher (~4 eV) BE feature, both very broad. In the previous matrix study of Sc/O_2, two ScO_3 species were identified, an ozonide $Sc(O_3)$ and an $(O_2)ScO$ species.[177] In the latter, the O_2 was suggested to be side-bonded to Sc (C_{2v}). We assigned the observed PES spectra of MO_3^- to the C_{2v} isomer. In the 355 nm spectrum of YO_3^-, very weak signals were discernible between 1 and 1.5 eV, which was probably due to the ozonide isomer of YO_3. However, we did not seem to observe a similar ozonide isomer of ScO_3^- with any substantial abundance. The ScO_3 and YO_3 species were similar to CuO_3, for which we observed both the ozonide and C_{2v} isomers previously, as discussed earlier.

The 355 nm spectra of ScO_4^- and YO_4^- were similar, both with broad low energy features, which disappeared in the 266 nm spectra, where only high BE features were observed. The low energy features were in fact still

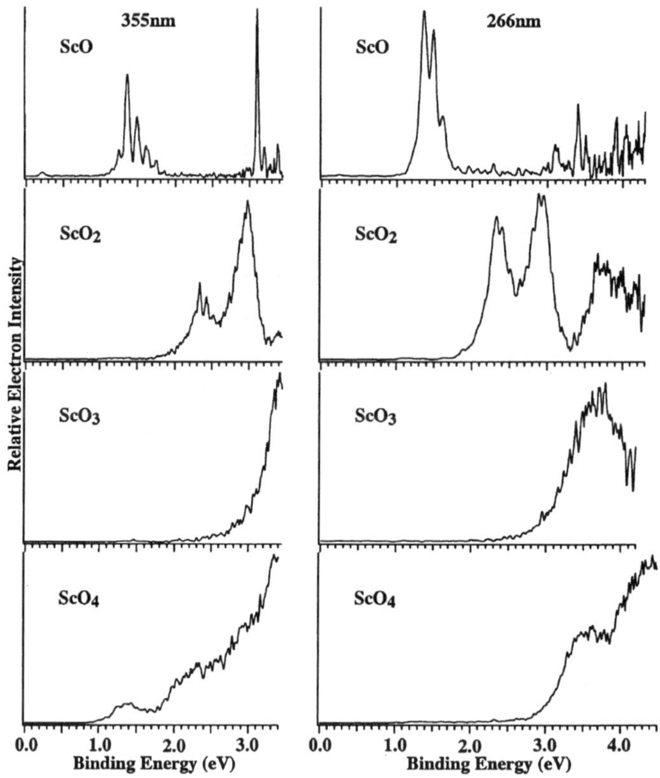

Fig. 49. Photoelectron spectra of ScO_y^- ($y = 1$–4) at 355 and 266 nm. From Ref. 39.

present in the 266 nm spectra, but with much weaker relative intensities, suggesting that the lower BE features were due to a minor isomer. An $(O_2)Sc(O_2)$ complex was identified in the previous matrix study.[177] The EA of the minor isomer was quite low, similar to that of the $M(O_2)$ complexes. Therefore, the minor isomer of MO_4 was assigned to be an $(O_2)M(O_2)$ species, where O_2 was head-bonded to the metal (superoxide-like). We assigned the major MO_4^- isomer to one in which the O_2 molecules were side-bonded to the metal atom (peroxide-like). Hence the formal oxidation state of the metal was higher in the major isomer, consistent with its high EA.

The 355 nm spectrum of YO_5^- appeared similar to that of YO_3^-, with some very weak signals between 1 and 2 eV and a tail at high BE. The

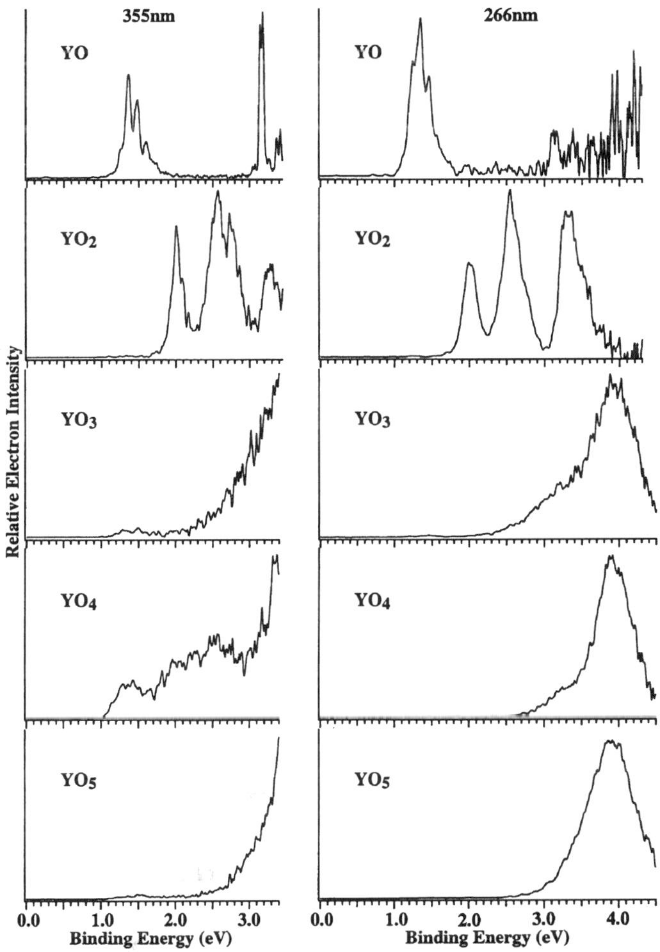

Fig. 50. Photoelectron spectra of YO_y^- ($y = 1$–5) at 355 and 266 nm. From Ref. 39.

266 nm of YO_5^- showed a broad peak, which seemed to be similar to that in both YO_3^- and YO_4^-. Again the very weak low energy signals present in the 355 nm spectrum was probably due to a minor isomer. We tentatively assigned the major isomer to be $(O_2)_2YO$. The broad nature of the PES spectra of all the higher oxides suggested that these species were probably rather floppy with significant geometry changes between the anions and the neutrals.

4. Conclusions

Our recent effort on the photodetachment photoelectron spectroscopy of the 3*d* transition metal oxide species is reviewed and summarized. It is shown that the laser-vaporization cluster source and the magnetic-bottle TOF PES technique is quite powerful to study these novel gas phase species, which are otherwise difficult to study and many of which were not known in the gas phase previously. We demonstrated that photodetachment PES is a versatile and convenient, albeit less accurate, means to unravel the complicated electronic structure and spectroscopy of these transition metal species. The gas phase studies provide information on size-selected species and complement the low temperature matrix technique which has been extensively used to investigate the reactive transition metal species.

Several conclusions can be drawn from the completed studies. For the monoxides, we observed that their electronic structure and spectroscopy become increasingly more complicated from ScO and NiO and then became relatively simple for CuO, due to the unpaired *d* electrons. Many new low-lying excited states, particularly, those with spin multiplicities different from the ground states, have been observed. The most important findings are the strong multi-electron transitions observed for ScO^-, YO^-, and TiO^-, providing the most vivid demonstration of the importance of electron correlation effects in these systems. In addition, excited states have been observed for a number of diatomic anions. For the MO_2 systems, we observed two isomers for M = Sc, Y, Ni, and Cu, an OMO dioxide molecule and an $M(O_2)$ complex. For TiO_2, VO_2, and FeO_2, only the dioxide molecules were observed, probably due to the stronger M–O bonding in these systems.

For higher oxides, we observed stable trioxide molecules and molecules containing O–O units, depending on the oxidation states of the metal. In general, we found that the EA of the oxide molecules increases with the O content up to the maximum oxidation states of the metal, beyond which oxide species containing O–O units begin to exist. For Cu, Ni, and Fe systems, we found that anions containing the O–O units undergo efficient photofragmentation to lose O_2 at 532 and 355 nm, but not at 266 or 193 nm. The fragmentation suggests that the O_2 units may serve as chromophores for the initial photoabsorption and also provides structural information. For most of the oxide species, only qualitative interpretation was possible due to lack of theoretical calculations, which are highly desirable for a more detailed understanding of these complicated electronic systems. The new photodetachment data, plus advances in *ab initio* theoretical methods and computational power, may provide enough impetus to motivate further theoretical efforts for these transition metal species.

Acknowledgments

I would like to thank major experimental contributions to this work by Dr. H. Wu. Drs. J. Fan and S. R. Desai also contributed to the early experimental work. I am indebted to Dr. L. Lou for his theoretical collaborations on the copper and iron oxide clusters. I would also like to thank Drs. J. B. Nicholas and S. D. Colson for valuable discussions on many occasions. Support of our research by the National Science foundation and the U.S. Department of Energy, Office of Basic Energy Sciences, Chemical Science Division, are gratefully acknowledged. Pacific Northwest National Laboratory is operated by Battelle for the DOE under Contract No. DE-AC06-76RLO 1830. The author is an Alfred P. Sloan Research Fellow.

References

1. *Homogeneous Transition Metal Catalyzed Reactions*, Eds. W. R. Moser and D. W. Slocum (American Chemical Society, Washington, DC, 1992).
2. K. P. Huber and G. Herzberg, *Molecular Spectra and Molecular Structure IV: Constants of Diatomic Molecules* (Van Nostrand Reinhold, New York, 1979).
3. C. W. Bauschlicher Jr., S. R. Langhoff and H. Partridge, in *Advanced Series in Physical Chemistry, Vol. 2: Modern Electronic Structure Theory*, Ed. D. R. Yarkony (World Scientific, London, 1995), p. 1280.
4. M. D. Morse, *Chem. Rev.* **86**, 1049 (1986).
5. A. J. Merer, *Annu. Rev. Phys. Chem.* **40**, 407 (1989).
6. L. S. Wang and H. Wu, in *Advances in Metal and Semiconductor Clusters, Vol. 4. Cluster Materials*, Ed. M. A. Duncan (JAI Press, Greenwich, 1998), p. 299.
7. J. Fan and L. S. Wang, *J. Phys. Chem.* **98**, 11814 (1994).
8. J. Fan, L. Lou and L. S. Wang, *J. Chem. Phys.* **102**, 2701 (1995).
9. L. S. Wang, *Surf. Rev. Lett.* **3**, 423 (1996).
10. L. S. Wang, S. Li and H. Wu, *J. Phys. Chem.* **100**, 19211 (1996).
11. L. S. Wang and H. Cheng, *Phys. Rev. Lett.* **78**, 2983 (1997).
12. S. Li, H. Wu and L. S. Wang, *J. Am. Chem. Soc.* **119**, 7417 (1997).
13. X. B. Wang, C. F. Ding and L. S. Wang, *J. Phys. Chem.* **A101**, 7699 (1997).
14. L. S. Wang, X. B. Wang, H. Wu and H. C. Cheng, *J. Am. Chem. Soc.* **120**, 6556 (1998).

15. X. Li and L. S. Wang, *J. Chem. Phys.* **109**, 5264 (1998).
16. E. R. Fisher, J. L. Elkind, D. E. Clemmer, R. Georgiadis, S. K. Loh, N. Aristov, L. S. Sunderlin and P. B. Armentrout, *J. Chem. Phys.* **93**, 2676 (1990).
17. Y. Matsuo, H. Maeda and M. Takami, *Chem. Phys. Lett.* **201**, 341 (1993).
18. Y. M. Chen and P. B. Armentrout, *J. Chem. Phys.* **103**, 618 (1995).
19. D. B. Jacobson and B. S. Freiser, *J. Am. Chem. Soc.* **108**, 27 (1986).
20. G. A. Ozin, S. A. Mitchell and J. G. Prieto, *J. Am. Chem. Soc.* **105**, 6399 (1983).
21. V. E. Bondybey and J. H. English, *J. Phys. Chem.* **88**, 2247 (1984).
22. D. E. Tevault, *J. Chem. Phys.* **76**, 2859 (1982).
23. G. V. Chertihin and L. Andrews, *J. Phys. Chem.* **99**, 6356 (1995).
24. L. Andrews, J. T. Yustein, C. A. Thompson and R. D. Hunt, *J. Phys. Chem.* **98**, 6514 (1994).
25. M. Fanfarillo, H. E. Cribb, A. J. Downs, T. M. Greene and M. J. Almond, *Inorg. Chem.* **31**, 2962 (1992).
26. S. Chang, G. Blyholder and J. Fernandez, *Inorg. Chem.* **20**, 2813 (1981).
27. C. E. Brown, S. A. Mitchell and P. A. Hackett, *J. Phys. Chem.* **95**, 1062 (1991).
28. L. S. Wang, H. S. Cheng and J. Fan, *J. Chem. Phys.* **102**, 9480 (1995).
29. J. Fan and L. S. Wang, *J. Chem. Phys.* **102**, 8714 (1995).
30. L. S. Wang, J. Fan and L. Lou, *Surf. Rev. Lett.* **3**, 695 (1996).
31. H. Wu, S. R. Desai and L. S. Wang, *J. Am. Chem. Soc.* **118**, 5296 (1996).
32. L. S. Wang, H. Wu and S. R. Desai, *Phys. Rev. Lett.* **76**, 4853 (1996).
33. H. Wu, S. R. Desai and L. S. Wang, *J. Chem. Phys.* **103**, 4363 (1995).
34. L. S. Wang, H. Wu, S. R. Desai and L. Lou, *Phys. Rev.* **B53**, 8028 (1996).
35. H. Wu, S. R. Desai and L. S. Wang, *J. Phys. Chem.* **A101**, 2103 (1997).
36. H. Wu and L. S. Wang. *J. Chem. Phys.* **107**, 16 (1997).
37. H. Wu and L. S. Wang, *J. Chem. Phys.* **107**, 8221 (1997).
38. H. Wu and L. S. Wang, *J. Chem. Phys.* **108**, 5310 (1998).
39. H. Wu and L. S. Wang, *J. Phys. Chem.* **A102**, 9129 (1998).
40. H. Hotop and W. C. Lineberger, *J. Phys. Chem. Ref. Data* **14**, 731 (1985).
41. C. S. Feigerle, Z. Herman and W. C. Lineberger, *J. Electron Specotrosc. Rel. Phenom.* **23**, 441 (1981).
42. P. C. Engelking and W. C. Lineberger, *Phys. Rev.* **A19**, 149 (1979).
43. H. Hotop and W. C. Lineberger, *J. Chem. Phys.* **58**, 2379 (1973).
44. P. C. Engelking and W. C. Lineberger, *J. Chem. Phys.* **66**, 5054 (1977).
45. T. Andersen, K. R. Lykke, D. M. Neumark and W. C. Lineberger, *J. Chem. Phys.* **86**, 1858 (1987).
46. R. F. Gunion, S. J. Dixon-Warren, W. C. Lineberger and M. D. Morse, *J. Chem. Phys.* **104**, 1765 (1996).
47. P. G. Wenthold, R. F. Gunion and W. C. Lineberger, *Chem. Phys. Lett.* **258**, 101 (1996).

48. P. G. Wenthold, K. L. Jonas and W. C. Lineberger, *J. Chem. Phys.* **106**, 9961 (1997).
49. A. E. Stevens, C. S. Feigerle and W. C. Lineberger, *J. Chem. Phys.* **78**, 5420 (1983).
50. A. E. S. Miller, C. S. Feigerle and W. C. Lineberger, *J. Chem. Phys.* **87**, 1549 (1987).
51. A. E. Stevens, C. S. Feigerle and W. C. Lineberger, *J. Am. Chem. Soc.* **104**, 5026 (1982).
52. D. G. Leopold and W. C. Lineberger, *J. Chem. Phys.* **85**, 51 (1986).
53. K. M. Ervin, H. Ho and W. C. Lineberger, *J. Chem. Phys.* **89**, 4514 (1988).
54. S. J. Dixon-Warren, R. F. Gunion and W. C. Lineberger, *J. Chem. Phys.* **104**, 4902 (1996).
55. J. Fan, J. B. Nicholas, J. M. Price, S. D. Colson and L. S. Wang, *J. Am. Chem. Soc.* **117**, 5417 (1995).
56. J. B. Nicholas, J. Fan, H. Wu, S. D. Colson and L. S. Wang, *J. Chem. Phys.* **102**, 8277 (1995).
57. L. S. Wang, H. Wu, S. R. Desai, J. Fan and S. D. Colson, *J. Phys. Chem.* **100**, 8697 (1996).
58. S. R. Desai, H. Wu and L. S. Wang, *Int. J. Mass Spectrom. Ion Processes* **159**, 75 (1996).
59. S. R. Desai, H. Wu, C. Rohfling and L. S. Wang, *J. Chem. Phys.* **106**, 1309 (1997).
60. L. S. Wang, S. R. Desai, H. Wu and J. B. Nicholas, *Z. Phys. D - Atoms, Molecules and Clusters* **40**, 36 (1997).
61. L. S. Wang, J. B. Nicholas, M. Dupuis, H. Wu and S. D. Colson, *Phys. Rev. Lett.* **78**, 4450 (1997).
62. H. Wu, X. Li, X. B. Wang, C. F. Ding and L. S. Wang, *J. Chem. Phys.* **109**, 449 (1998).
63. H. Wu, Ph.D. Thesis, Washington State University, Pullman, WA, 1997.
64. W. C. Wiley and I. H. McLaren, *Rev. Sci. Instrum.* **26**, 1150 (1955).
65. W. A. de Heer and P. Milani, *Rev. Sci. Instrum.* **62**, 670 (1991).
66. G. Markovich, R. Giniger, M. Levin and O. Cheshnovsky, *J. Chem. Phys.* **95**, 9416 (1991).
67. P. Kruit and F. H. Read, *J. Phys. E: Sci. Instrum.* **16**, 313 (1983).
68. O. Chesnovsky, S. H. Yang, C. L. Pettiette, M. J. Craycraft and R. E. Smalley, *Rev. Sci. Instrum.* **58**, 2131 (1987).
69. G. Gantefor, K. H. Meiwes-Broer and H. O. Lutz, *Phys. Rev.* **A37**, 2716 (1988).
70. S. Hufner, *Photoelectron Spectroscopy — Principles and Applications* (Springer-Verlag, New York, 1995).
71. C. E. Moore, *Atomic Energy Levels* (Natl. Bur. Stand. Circ., Washington DC, 1971) Vol. II.
72. S. Suzer, S. T. Lee and D. A. Shirley, *Phys, Rev.* **A13**, 1842 (1976).
73. P. A. Cox, *Transition Metal Oxides* (Oxford, New York, 1995).

74. Z. Tyeklar and K. D. Karlin, *Acc. Chem. Res.* **22**, 241 (1989).
75. J. A. Halfen *et al., Science* **271**, 1397 (1996).
76. C. Vinckier, J. Corthouts and S. De Jaegere, *J. Chem. Soc. Faraday Trans. II* **84**, 1951 (1988).
77. D. Sulzle, H. Schwarz, K. H. Moock and J. K. Terlouw, *Int. J. Mass Spectrom. Ion Processes* **108**, 269 (1991).
78. J. A. Howard, R. Sutcliffe and B. Mile, *J. Phys. Chem.* **88**, 4351 (1984).
79. P. H. Kasai and P. M. Jones, *J. Phys. Chem.* **90**, 4239 (1986).
80. S. M. Mattar and G. Ozin, *J. Phys. Chem.* **92**, 3511 (1988).
81. S. R. Langhoff and C. W. Bauschlicher Jr., *Chem. Phys. Lett.* **124**, 241 (1986).
82. P. V. Madhaven and M. D. Newton, *J. Chem. Phys.* **83**, 2337 (1985).
83. G. Igel *et al., J. Chem. Phys.* **81**, 2737 (1984).
84. P. S. Bagus, C. J. Nelin and C. W. Bauschlicher Jr., *J. Chem. Phys.* **79**, 2975 (1983).
85. J. Hrusak, W. Koch and H. Schwarz, *J. Chem. Phys.* **101**, 3898 (1994).
86. C. W. Bauschlicher Jr., S. R. Langhoff, H. Patridge and M. Sodupe, *J. Phys. Chem.* **97**, 856 (1993).
87. Y. Mochizuki, U. Nagashima, S. Yamamoto and H. Kashiwagi, *Chem. Phys. Lett.* **164**, 225 (1989).
88. M. L. Polak, M. K. Gilles, J. Ho and W. C. Lineberger, *J. Phys. Chem.* **95**, 3460 (1991).
89. M. Gutowski and J. Simons, *J. Chem. Phys.* **101**, 4867 (1994); **100**, 1308 (1994); *J. Phys. Chem.* **98**, 8326 (1994).
90. T. K. Ha and M. T. Nguyen, *J. Phys. Chem.* **89**, 5569 (1985).
91. V. E. Henrich and P. A. Cox, "The Surface Science of Metal Oxides" (Cambridge, New York, 1994).
92. A. F. Well, *Structural Inorganic Chemistry*, 5th Ed. (Oxford, New York, 1987).
93. S. P. Walch and W. A. Goddard III, *J. Am. Chem. Soc.* **100**, 1338 (1978).
94. C. W. Bauschlicher Jr., *Chem. Phys.* **93**, 399 (1985).
95. C. W. Bauschlicher Jr., C. J. Nelin and P. S. Bagus, *J. Chem. Phys.* **82**, 3265 (1985).
96. C. W. Bauschlicher Jr. and P. Maitre, *Theor. Chim. Acta.* **90**, 189 (1995).
97. M. Dolg, U. Wedig, H. Stoll and H. Preuss, *J. Chem. Phys.* **86**, 2123 (1987).
98. E. G. Bakalbassis, M. D. Stiakaki, A. C. Tsipis and C. A. Tsipis, *Chem. Phys.* **205**, 389 (1996).
99. B. Rosen, *Nature* (London) **156**, 570 (1945).
100. D. W. Green and G. T. Reedy, *J. Mol. Spectrosc.* **78**, 257 (1979).
101. V. I. Srdanov and D. O. Harris, *J. Chem. Phys.* **89**, 2748 (1988).
102. E. J. Friedman-Hall and R. W. Field, *J. Mol. Spectrosc.* **155**, 259 (1992).
103. R. S. Ram and P. F. Bernath, *J. Mol. Spectrosc.* **155**, 315 (1992).
104. K. Namiki and S. Saito, *Chem. Phys. Lett.* **252**, 343 (1996).
105. H. Huber, W. Klotzbucher, G. A. Ozin and A. V. Voet, *Can. J. Chem.* **51**, 2722 (1973).

106. A. Citra, G. V. Chertihin, L. Andrews and M. Neurock, *J. Phys. Chem.* **101**, 3109 (1997).
107. M. R. A. Blomberg, P. E. M. Siegbahn and A. Strich, *Chem. Phys.* **97**, 287 (1985).
108. V. D. Moravec and C. C. Jarrold, *J. Chem. Phys.* **108**, 1804 (1998).
109. R. J. Celotta, R. A. Bennett, J. L. Hall, M. W. Siegel and J. Levine, *Phys. Rev.* **A6**, 631 (1972).
110. L. A. Posey, M. J. Deluca and M. A. Johnson, *Chem. Phys. Lett* **131**, 170 (1986).
111. M. Krauss and W. J. Stevens, *J. Chem. Phys.* **82**, 5584 (1985).
112. A. W. Taylor, A. S. C. Cheung and A. J. Merer, *J. Mol. Spectrosc.* **113**, 487 (1985).
113. P. S. Bagus and H. J. T. Preston, *J. Chem. Phys.* **59**, 2986 (1973).
114. L. Andrews, G. V. Chertihin, A. Ricca and C. W. Bauschlicher, *J. Am. Chem. Soc.* **118**, 467 (1996).
115. D. Schroder, A. Fiedler, J. Schwarz and H. Schwarz, *Inorg. Chem.* **33**, 5094 (1994).
116. M. Fanfarillo, H. E. Cribb, A. J. Downs, T. M. Greene and M. J. Almond, *Inorg. Chem.* **31**, 2962 (1992).
117. J. E. Newton and M. B. Hall, *Inorg. Chem.* **23**, 4627 (1984).
118. G. Blyholder, J. Head and F. Ruette, *Inorg. Chem.* **21**, 1539 (1982).
119. C. A. Reed and S. K. Cheung, *Proc. Natl. Acad. Sci.* USA **74**, 1780 (1977).
120. D. L. Hilderbrand, *Chem. Phys. Lett.* **34**, 352 (1975).
121. Special issue on metal-dioxygen complexes: *Chem. Rev.* **94**, 567–856 (1994).
122. B. P. Murch, P. D. Boyle and L. Que, *J. Am. Chem. Soc.* **107**, 6728 (1985).
123. A. C. Rosenzweig and S. Lippard, *Acc. Chem. Res.* **27**, 229 (1994).
124. Y. Zang, Y. Dong, L. Que, K. Kauffmann and E. Munck, *J. Am. Chem. Soc.* **117**, 1169 (1995).
125. A. K. Powell, S. L. Heath, D. Gatteschi, L. Pardi, R. Sessoli, G. Spina, F. Del Giallo and F. Pierralli, *J. Am. Chem. Soc.* **117**, 2491 (1995).
126. K. L. Taft, G. C. Papaefthymiou and S. J. Lippard, *Science* **259**, 1302 (1993).
127. J. S. Corneille, J. W. He and D. W. Goodman, *Surf. Sci.* **338**, 211 (1995).
128. J. Zaanen, G. A. Sawatzky and J. W. Allen, *Phys. Rev. Lett.* **55**, 418 (1985).
129. R. J. Lad and V. E. Henrich, *Phys. Rev.* **B39**, 13478 (1989).
130. V. S. Smentkowski and J. T. Yates, Jr., *Surf. Sci.* **232**, 113 (1990).
131. T. Noro, C. Ballard, M. H. Palmer and H. Tatewaki, *J. Chem. Phys.* **100**, 452 (1994).
132. L. C. Snyder and K. Raghavachari, *J. Chem. Phys.* **80**, 5076 (1984).
133. J. L. Chen *et al., Phys. Rev.* **B44**, 6558 (1991); M. Castro and D. R. Salahub, *ibid.* **47**, 10955 (1993).
134. R. Fournier and D. R. Salahub, *Surf. Sci.* **245**, 263 (1991).
135. D. S. Yang *et al., J. Chem. Phys.* **103**, 5335 (1995).

136. A. J. Merer, G. Huang, A. S.-C. Cheung and A. W. Taylor, *J. Mol. Spectrosc.* **125**, 465 (1987).
137. G. Huang, A. J. Merer and D. J. Clouthier, *J. Mol. Spectrosc.* **153**, 32 (1992).
138. A. G. Adam, M. Barnes, B. Berno, R. D. Bower and A. J. Merer, *J. Mol. Spectrosc.* **170**, 94 (1995).
139. A. S.-C. Cheung, P. G. Hajigeorgiou, G. Huang, S.-Z. Huang and A. J. Merer, *J. Mol. Spectrosc.* **163**, 443 (1994).
140. L. Karlsson, B. Lindgren, C. Lundevall and U. Sassenberg, *J. Mol. Spectrosc.* **181**, 274 (1997).
141. J. M. Dyke, B. W. J. Gravenor, M. P. Hastings and A. Morris, *J. Phys. Chem.* **89**, 4613 (1985).
142. E. A. Carter and W. A. Goddard III, *J. Phys. Chem.* **92**, 2109 (1988).
143. E. Broclawik, *Int. J. Quant. Chem.* **56**, 779 (1995).
144. E. B. Rudbyi, E. A. Kaibicheva and L. N. Sidorov, *J. Chem. Therm.* **25**, 929 (1993).
145. L. B. Knight Jr., R. Babb, M. Ray, T. J. Banisaukas III, L. Russon, R. S. Dailey and E. R. Davidson, *J. Chem. Phys.* **105**, 10237 (1996).
146. F. Ramondo, L. Bencivenni, N. Sanna and S. N. Cesaro, *J. Mol. Struct. (Theochem)* **253**, 121 (1992).
147. M. J. Almond and R. W. Atkins, *J. Chem. Soc. Dalton Trans.* 835 (1994).
148. G. V. Chertihin, W. D. Bare and L. Andrew, *J. Phys. Chem.* **A101**, 5090 (1997).
149. G. C. Bond and S. F. Tahir, *Appl. Catal.* **71**, 1 (1991).
150. K. Tran, M. A. Hanning-Lee, A. Biswas, A. E. Stiegman and G. W. Scott, *J. Am. Chem. Soc.* **117**, 2618 (1995).
151. F. A. Cotton and G. Wilkinson, *Advanced Inorganic Chemistry*, Fifth Edition (Wiley, New York, 1988).
152. A. Hagfeldt, R. Bergstrom, H. O. G. Siegbahn and S. Lunell, *J. Phys. Chem.* **97**, 12725 (1993).
153. A. Hagfeldt, S. Lunell and H. O. G. Siegbahn, *Int. J. Quantum Chem.* **49**, 97 (1994).
154. W. Yu and R. B. Freas, *J. Am. Chem. Soc.* **112**, 7126 (1990).
155. B. C. Guo, K. P. Kerns and A. W. Castleman, *Int. J. Mass Spectrom. Ion Proc.* **117**, 129 (1992).
156. M. S. El-Shall, W. Slack, W. Vann, D. Kane and D. Hanley, *J. Phys. Chem.* **98**, 3067 (1994); J. C. Parker and R. W. Siegel, *J, Matter. Res.* **5**, 1246 (1990); C. A. Melendres, A. Narayanasamy, V. A. Maroni and R. W. Siedel, *J. Mater. Res.* **4**, 1246 (1989).
157. M. Barnes, A. J. Merer and G. F. Metha, *J. Mol. Spectrosc.* **181**, 180 (1997).
158. L. A. Kaledin, J. E. McCord and M. C. Heaven, *J. Mol. Spectrosc.* **173**, 499 (1995).
159. A. D. Sappey, G. Eiden, J. E. Harrington and J. C. Weisshaar, *J. Chem. Phys.* **90**, 1415 (1989).

160. G. M. Daly and M. S. El-Shall, *J. Chem. Phys.* **100**, 1771 (1994).
161. R. Bergstrom, S. Lunell and L. A. Eriksson, *Int. J. Quantum Chem.* **59**, 427 (1996).
162. M. V. Ramana and D. H. Phillips, *J. Chem. Phys.* **88**, 2637 (1988).
163. M. Kaufman, J. Muenter and W. Klemperer, *J. Chem. Phys.* **47**, 3365 (1967).
164. N. S. McIntyre, K. R. Thompson and W. Welter, *J. Phys. Chem.* **75**, 3243 (1971).
165. G. V. Chertihin and L. Andrews, *J. Phys. Chem.* **99**, 6356 (1995).
166. J. Pascual, J. Camassel and H. Mathieu, *Phys. Rev. Lett.* **39**, 1490 (1977); *Phys. Rev.* **B18**, 5606 (1978).
167. C. L. Chalek and J. L. Gole, *J. Chem. Phys.* **65**, 2845 (1976); *Chem. Phys.* **19**, 59 (1977).
168. S. F. Rice and R. W. Field, *J. Mol. Spectrosc.* **119**, 331 (1986).
169. S. F. Rice, W. J. Childs and R. W. Field, *J. Mol. Spectrosc.* **133**, 22 (1989).
170. J. Hoeft and T. Torring, *Chem. Phys. Lett.* **215**, 367 (1993).
171. B. Simard, A. M. James and P. A. Hacket, *J. Mol. Spectrosc.* **154**, 455 (1992).
172. T. C. Steimle and J. E. Shirley, *J. Chem. Phys.* **92**, 3298 (1990).
173. R. D. Suenram, F. J. Lovas, G. T. Fraser and K. Matsumura, *J. Chem. Phys.* **92**, 4724 (1990).
174. W. J. Childs, O. Poulsen and T. C. Steimle, *J. Chem. Phys.* **88**, 598 (1988).
175. T. C. Steimle and Y. Alramadin, *J. Mol. Spectrosc.* **122**, 103 (1987).
176. C. Linton, *J. Mol. Spectrosc.* **69**, 351 (1978).
177. G. V. Chertihin, L. Andrews, M. Rosi and C. W. Bauschlicher Jr., *J. Phys. Chem.* **101**, 9085 (1997).

CHAPTER 17

DETACHMENT PROCESSES FOR MOLECULAR ANIONS

Jack Simons*

Chemistry Department University of Utah
Salt Lake City, Utah 84112

Contents

*Henry Eyring Center for Theoretical Chemistry.

1. Introduction

1.1. *Scope of this Chapter*

In this chapter, I will attempt to describe the theory underlying electron detachment in molecular anions, including detachment induced by photon absorption (photodetachment) and by energy transfer from rotational, vibrational, or collisional motion. I will not attempt to review or even summarize all of the calculations of detachment rates that have been carried out on particular molecular anions, but I will discuss some of the more recent areas of research that I believe offer fruitful lines of pursuit. For the reader who is interested in gaining an overview of what is known experimentally and theoretically about molecular anions, several review articles, books, and other sources[1–9] can be recommended. Most of these sources focus on the nature of the electronic structure of anions rather than on how such species undergo electron detachment; therefore, I believe the present chapter is timely and hopefully informative.

1.2. *Various Stabilities of Molecular Anions*

Let me begin by making it clear what it means for a molecular anion to be electronically stable, locally geometrically stable, and thermodynamically

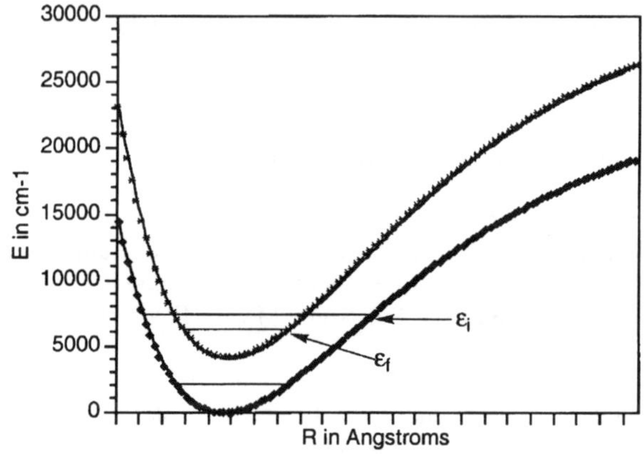

Fig. 1. Anion (lower) and neutral (upper) potential energy surfaces as functions of internuclear distance R illustrative of the ground states of NH^- and NH.

stable. A clear understanding of these terms is important for the discussion provided in this chapter.

1.2.1. *Electronically Stable*

If, at a fixed molecular geometry (whose internal coordinates are denoted $\mathbf{Q} = Q_1, Q_2, \ldots, Q_{3N-6}$ for a nonlinear molecule with N atoms), the anion's electronic energy is lower than that of the neutral molecule at this same geometry, the anion is said to be *electronically stable* at this geometry. Figure 1 shown below illustrates the electronic energies of the ground states of the NH^- anion[10] and its neutral counterpart NH as functions of the internuclear distance R. This anion is electronically stable over all internuclear distances displayed.

Figure 2 shows the ground state anion and neutral energies typical of enolate anions[11] as functions of the rotation about the C–C bond axis. In this case, the anion is electronically stable for the range of angles shown.

Figure 3 shows anion and neutral energy curves characteristic of an anion that is electronically stable over geometries accessed by its low-energy vibrational levels but is electronically unstable at other geometries.

1.2.2. *Locally Geometrically Stable*

If, at a specific geometry $\mathbf{Q}^0$, (1) the anion is electronically stable and (2) the anion's electronic energy $E(Q_i)$ as a function of displacements δQ_i away

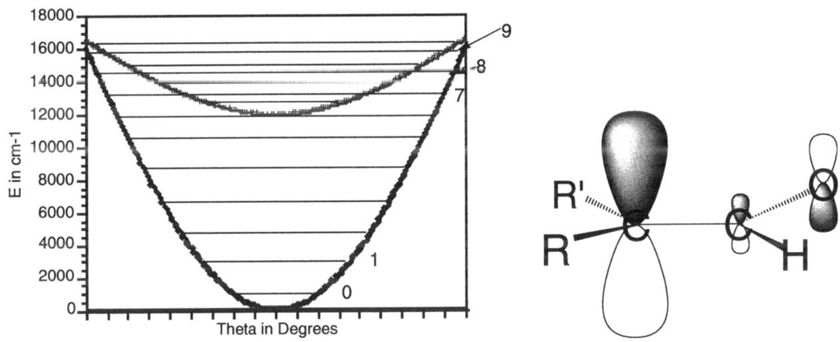

Fig. 2. Anion (lower) and neutral (upper) potential energy surfaces illustrative of enolate anions (shown at right) as functions of the torsion angle about the C–C bond axis.

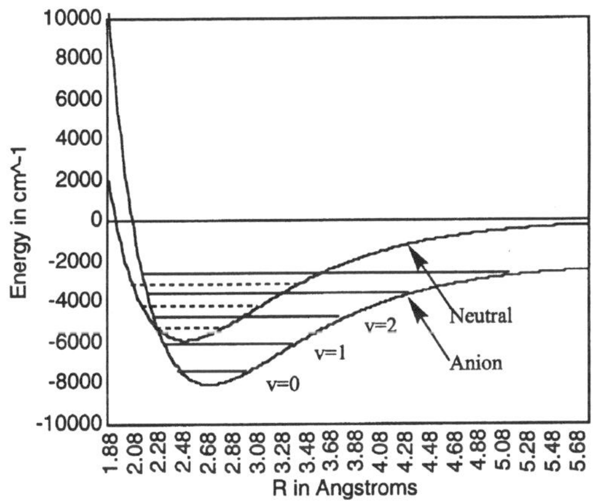

Fig. 3. Anion and neutral curves as functions of bond length for a case where the anion is electronically stable over a limited range of geometries.

from Q_i^0 obeys

$$\partial E/\partial Q_i = 0, \text{ for all } i \text{ and}$$

the square Hermitian–Hessian matrix $\partial^2 E/\partial Q_i \partial Q_j$ has all $3N - 6$ eigenvalues positive, then the geometry $\mathbf{Q}^0$ corresponds to a local minimum on the anion's energy surface. In such situations, we say that this geometry of the anion is locally (i.e. near $\mathbf{Q}^0$) geometrically stable. The NH^- anion shown in Fig. 1 and the enolate anion shown in Fig. 2 are geometrically stable at their equilibrium bond length and equilibrium angle, respectively.

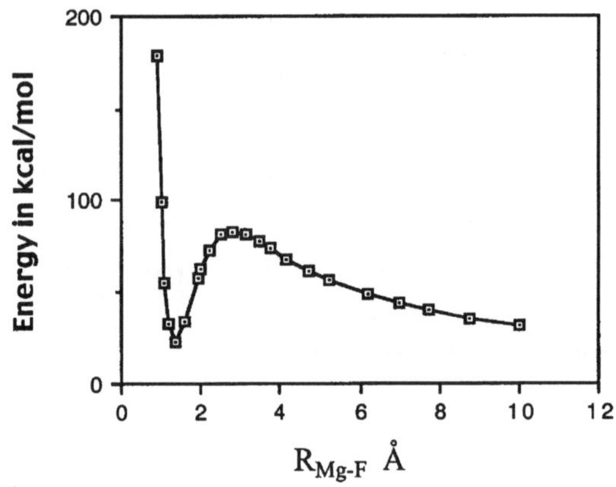

Fig. 4. Energy of MgF_4^{2-} as a function of one Mg–F bond length with all other degrees of freedom optimized (taken from Ref. 12). The energy of $MgF_3^- + F^-$ (at $R \to \infty$) is defined as zero.

1.2.3. *Thermodynamically Stable*

An anion is thermodynamically stable if it is electronically and geometrically stable and if its energy at its lowest-energy structure lies below that of any molecular or atomic fragments into which it might dissociate.

The NH^- anion is thermodynamically stable because, at its lowest-energy structure, its energy is lower than the energy of $H^- + N$ or $N^- + H$. Figure 4 shows the electronic energy of the MgF_4^{-2} dianion[12] as a function of one Mg–F bond length (with the remaining three Mg–F bond lengths and all internal angles "optimized" to minimize the energy).

This dianion is geometrically stable near $R = 1.7$ Å, and is electronically stable (i.e. MgF_4^{2-} lies below MgF_4^- plus a free electron). However, this dianion is not thermodynamically stable because its energy near $R = 1.7$ Å lies above the energy of $MgF_3^- + F^-$ by ca. 20 kcal/mol.

1.3. *Various Mechanisms for Molecular Anions to Eject an Electron*

Let us consider a molecular anion that is vibrating (and rotating) in the neighborhood of one of the local minima on its energy surface and that is electronically stable over the range of geometries accessed by these motions (including tunneling). This anion has no way to eject an electron without the input of energy from an external sources. Hence, it can exist in this state indefinitely. The lowest vibrational level of the NH^- anion shown in Fig. 1 as well as the $v = 0$ through $v = 6$ torsional levels of the enolate of Fig. 2 present examples of molecular anion states that cannot undergo detachment in the absence of some external perturbation.

Given an anion in a state that is stable to detachment as discussed above, there are two primary tools that are commonly employed to cause the anion to eject an electron:

1. One or more *photons* can be used to induce an *electronic excitation* of the anion to a final state in which an electron exists in a continuum orbital rather than in a bound orbital. Alternatively, the photon(s) may cause *excitation* of the anion's *vibrational/rotational modes* to an extent that the anion's total (electronic plus vibrational/rotational) energy lies above one or more levels of the neutral molecule. The upper vibrational levels of NH^- shown in Fig. 1 and enolate torsional levels higher than $v = 6$ shown in Fig. 2 offer examples of this situation.
2. *Collision* with an atom, molecule, or other particle projectile (e.g. cation, anion, or electron) can induce either an *electronic excitation* in the anion to a final continuum state or an *excitation of* the anion's *vibrational and/or rotational modes.*

1.3.1. *Direct Photon-Initiated Detachment*

In the photon-induced electronic transition process, if the final electronic state of the anion is electronically unstable relative to one or more states of the neutral molecule, one achieves what is termed a *direct photodetachment* event. This process can be viewed as an electronic transition connecting an electronically bound state (the anion) to an electronically unbound state (the neutral plus ejected electron system). The theory behind this family of detachment processes is treated in Sec. 2 and is very similar to the theory of molecular photoionization treated by Berrah elsewhere in this book.[13] An example of such transitions is given in Fig. 5 where $v = 0 \to v = 0, 1, 2$ transitions in $NH^- \to NH + e^-$ are illustrated.[10]

1.3.2. *Photon-Initiated Detachment Involving Non-Born–Oppenheimer Coupling*

Alternatively, one or not more photons may induce vibration/rotation excitation of the anion to a final level that lies above one or more levels of

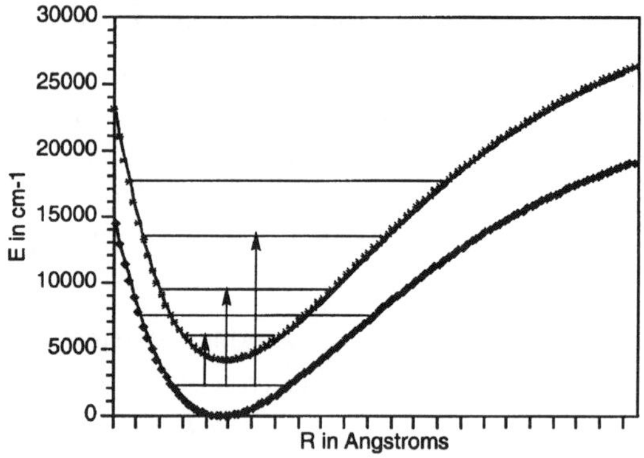

Fig. 5. Direct photodetachment from $v = 0$ NH^- to $v = 0, 1, 2$ of NH + e^-.

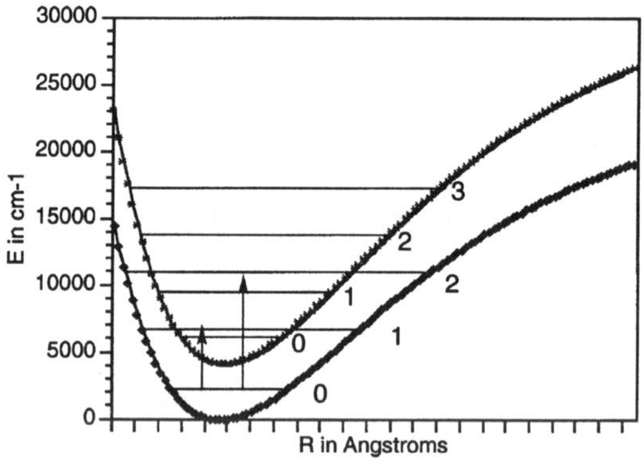

Fig. 6. Vibrational excitation of NH^- from $v = 0$ to $v = 1, 2$.

the neutral molecule. In Fig. 6, a case in which NH^- is excited from the $v = 0$ of its ground electronic state to $v = 1$ or $v = 2$ of this same electronic state is illustrated. In this situation, the $v = 1$ anion state may undergo subsequent electron detachment to the $v = 0$ level of NH by ejecting an electron with kinetic energy equal to the $v = 1$ NH^- to $v = 0$ NH energy gap. The $v = 2$ level of NH^- can decay to either $v = 1$ NH or $v = 0$ NH ejecting an electron of the appropriate kinetic energy.

The theory underlying this kind of electron detachment is covered in Sec. 3. Briefly, the photon absorption step is described by the conventional theory of infrared (when vibrational excitation is involved) absorption as discussed in a variety of spectroscopy text books.[14,15] Once the anion is vibrationally and/or rotationally excited, it must convert some or all of its excess vibrational/rotational energy into electronic energy to eject an electron. The origin of this energy transfer lies in the dynamical couplings that are ignored when the Born–Oppenheimer (BO) approximation[16] is made. Such non-BO induced electron ejection rates are treated in Sec. 3.2.

1.3.3. *Photon-Initiated Detachment Involving Shape and Feshbach Resonances*

In addition to the case discussed above in which vibration/rotation energy must be converted to electronic energy, excitation to higher vibrational levels of the anion can induce electron ejection by causing the anion to sample geometries where the anion's electronic energy lies above that of the neutral. Such a case is illustrated in Fig. 3. For this species, the lower anion vibrational levels do not sample geometries where the anion is electronically unstable (at least in the absence of great tunneling). In contrast, from $v = 2$ onward, the vibrational motion samples regions of electronic instability for the anion; within such regions, an electron can be ejected spontaneously (i.e. without any energy input from photons or via coupling to vibrational/rotational energy). The rate at which electrons are detached in such cases is determined by two factors: (1) the frequency with which vibrational motion causes such regions of spontaneous instability to be accessed, and (2) the rate at which an electron is detached within these regions. The former rate is simply the rate of vibrations along coordinates that move the system into the regions of instability. The latter rate depends strongly upon the nature of the electronic transition[17] that occurs when the anion spontaneously loses its "extra" electron.

1.3.3.1. *Shape Resonances*

Specifically, if the occupancy of spin-orbitals in the anion (which we denote $\phi_1\phi_2\cdots\phi_L\phi_K\phi_N$) differs from that in the neutral molecule produced in the detachment by the *loss of a single spin-orbital* (e.g. ϕ_N), and if the spin-orbital ϕ_N has no spherical character (i.e. if it contains only *p*, *d*, or higher angular momentum character, but no *s*-character), then the ejection rate is governed by the rate at which the electron tunnels through an angular momentum barrier. States that decay through such single-spin-orbital electron tunneling detachments are called *shape resonance* states, and they usually have decay rates in the range of 10^{14}–10^{16} sec^{-1}.

The N_2^- anion in which the "extra" electron has attached to the π_g^* antibonding orbital of N_2 offers an example of a shape resonance, as does the ethylene anion $H_2C{=}CH_2^-$ in which the electron occupies the π^* C–C orbital. In both cases, the fact that the orbital occupied by the extra electron possesses (at least) one nodal plane precludes its having any spherical character. At long range from the underlying neutral's nuclei and other electrons, the electron in the anion's highest occupied molecular orbital (HOMO) experiences an attraction to the neutral determined by a potential $V(\mathbf{r})$. This potential can be expanded in terms of spherical harmonics to display its angular and radial character (the angles θ and ϕ and the radial coordinate r describe the position of the electron $\mathbf{r}$ in a coordinate system fixed to the molecule)

$$V(\mathbf{r}) = \Sigma_{L,M} Y_{L,M}(\theta, \phi) V_{L,M}(r) .$$

This expansion posesses no $L = 0$ character because the anion's HOMO has a nodal plane. For N_2^-, its largest term would involve $L = 2$ because the π_g^* orbital, as shown below, clearly has strong d-character when expanded about the center of the N–N bond.

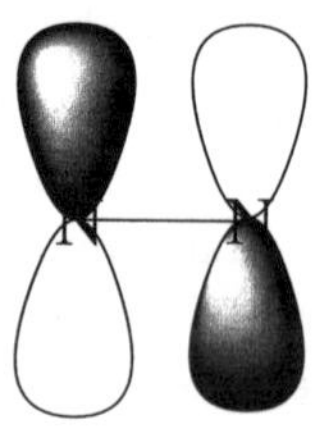

Fig. 7. The anti-bonding π^* orbital of N_2^-.

For CO^-, the anti-bonding π^* orbital clearly possesses both $L = 1$ and $L = 2$ character as shown in Fig. 8.

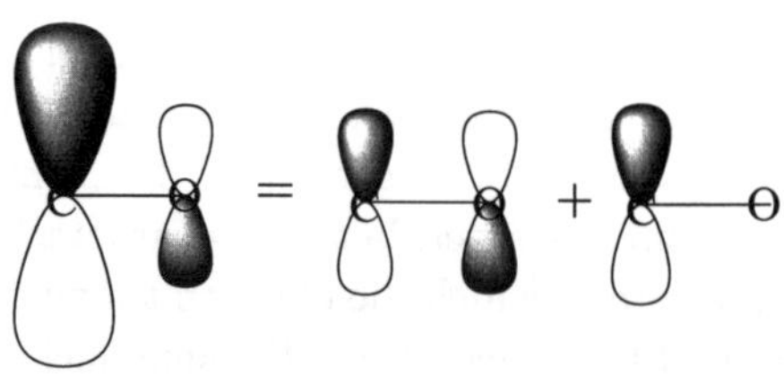

Fig. 8. The anti-bonding π^* orbital of CO^- expressed as a sum of $L = 2$ and $L = 1$ components.

When the expansion for $V(\mathbf{r})$ and a similar expansion of the wavefunction $\psi(r,\theta,\phi)$

$$\psi(\mathbf{r}) = \Sigma_{L,M} Y_{L,M}(\theta,\phi)\psi_{L,M}(r)$$

are used in the Schrödinger equation,

$$-\hbar^2/2m_e \partial/\partial r(r^2 \partial\psi/\partial r) + L^2/2m_e r^2 \psi + V(\mathbf{r})\psi = E\psi$$

a set of coupled equations is obtained for the wavefunction amplitudes $\psi_{L,M}(r)$. When there exists a single dominant $V_{L,M}(r)$ as in the case of N_2^- (where $L = 2$ dominates), the equation for the corresponding wavefunction amplitude is

$$-\hbar^2/2m_e \partial/\partial r(r^2 \partial\psi_{L,M}/\partial r) + L(L+1)/2m_e r^2 \psi_{L,M} + V_{L,M}(r)\psi_{L,M} = E\psi_{L,M}\,.$$

Depicted in Fig. 9 are the effective radial potentials $V_{\text{eff}}(r) = V(r) + L(L+1)/2m_e r^2$ for hypothetical cases with different L values.

The centrifugal potential $L(L+1)2m_e r^2$ has two effects on the decay rate of a shape resonance state. First, it moves the energy of the state to a higher value than the energy in the Schrödinger equation with the same $V(r)$ but with no centrifugal potential. This tends to destabilize the anion

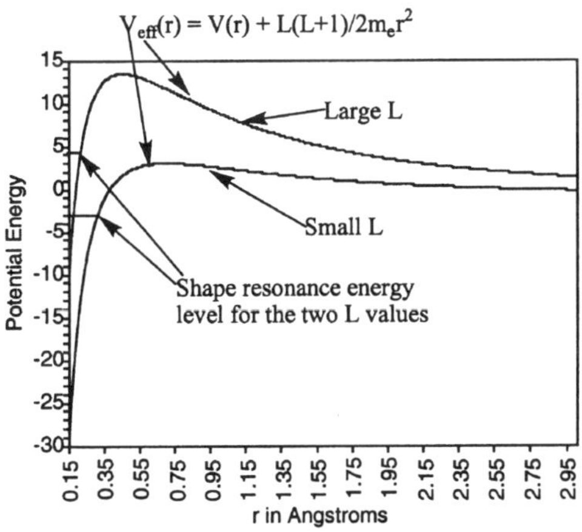

Fig. 9. The effective radial potential experienced by an electron in an orbital having angular momentum L and attracted to the underlying molecule by a potential $V(r)$.

and thus increase its decay rate. Second, the centrifugal potential provides the "barrier" through which the electron must tunnel to detach, this, of course, tends to decrease the decay rate as L increases. The net result of both effects is illustrated above in Fig. 9. Some molecular anions display a large number of shape resonance states having different L values; the pattern of decay rates of these states as functions of L is often complicated by these two competing effects of the centrifugal potential.

1.3.3.2. *Feshbach Resonances*

When the anion's spin-orbital occupancy differs from that of the daughter neutral by *two spin–orbitals* (e.g. if ϕ_N were lost to produce the ejected electron in the continuum orbital ϕ_C and ϕ_L were excited to a spin-orbital ϕ_X not present in the anion), the ejection rate is governed by the strength of coupling between the two configurations $\phi_1\phi_2\cdots\phi_L\phi_K\phi_N$ and $\phi_1\phi_2\cdots\phi_X\phi_K\phi_C$. The coupling V, induced by the electronic Hamiltonian H, between these two configurations can be expressed in terms of the two-electron integrals:

$$V = \int \mathbf{dr} \int \mathbf{dr}'\phi_C(\mathbf{r})\phi_X(\mathbf{r}')e^2/|\mathbf{r}-\mathbf{r}'|\{\phi_N(\mathbf{r})\phi_L(\mathbf{r}') - \phi_L(\mathbf{r})\phi_N(\mathbf{r}')\}\,.$$

States that decay in this manner are called *Feshbach resonance* states, and they usually have smaller electron ejection rates (ca. 10^{13}–10^{15} sec^{-1}) than do shape resonance states. The state of N_2^- in which an electron is attached to the anti-bonding π_g^* orbital and a second electron is excited from a non-bonding σ_g orbital into the π_g^* orbital provides an example of a Feshbach resonance state. To eject an electron, this state must undergo a two-electron rearrangement whose rate is determined by the square of the two-electron integral

$$V = \int \mathbf{dr} \int \mathbf{dr}'\phi_C(\mathbf{r})\sigma_g(\mathbf{r}')e^2/|\mathbf{r}-\mathbf{r}'|\pi_g^*(\mathbf{r})\pi_g^*(\mathbf{r}')$$

multiplied by the density of translational states ρ for the ejected electron.

1.3.4. *Collision-Induced Detachment*

Collision of an electronically stable anion with another species can cause electron ejection by most of the mechanisms enumerated above when photons are involved. That is, the collision can simply replace the photon in exciting the stable anion to an excited electronic state or to an excited vibration/rotation state which, subsequently, can detach an electron via non-BO coupling or by accessing geometries where the anion becomes electronically unstable to decay (via shape or Feshbach mechanism). The theory underlying collision-induced detachment is similar to that governing the photon-initiated events except for the rates and selection rules that deal with the initial energy transfer process (i.e. the photon absorption or the collisional excitation).

Before closing this Introduction, I should mention that there are molecular anions that do not lie within the scope of this chapter. Specifically, some molecular anions do not possess any range of geometries over which (1) the anion is electronically stable and (2) the anion potential energy surface has a local minimum deep enough to support one or more bound vibrational levels. In this chapter, attention is restricted to those anions that can exist as electronically and locally geometrically stable units which thus are amenable to identification and isolation (e.g. by mass selection) and which must subsequently be excited by photons or collisions to induce electron detachment. The detachment can take place by (1) direct photon excitation of the electronic degrees of freedom, (2) excitation of vibration/rotation modes followed by non-BO excitation of the electrons, or (3) excitation of vibration/rotation modes causing structural deformation into geometries where the anion is electronically unstable and thus may decay as a shape or Feshbach resonance. An example of an anion that does not have an electronically and geometrically stable state is shown in Fig. 10.

I have chosen not to examine such anions (i.e. those that exist only as resonances and have no electronically and geometrically stable states) in this chapter because (1) their experimental study involves entirely different tools than are characteristic of this Book, (2) their theoretical study is highly specialized and (3) both the experimental and theoretical aspects of their study have been reviewed in the not too distant past.[1,3]

2. Direct Photodetachment

An electronically and locally geometrically stable molecular anion interacts with the electromagnetic field of a photon source much as a neutral molecule does. The primary difference between photoionization of a neutral and photodetachment of an anion lies in the form of the wavefunction describing the ejected electron. These differences are detailed further below.

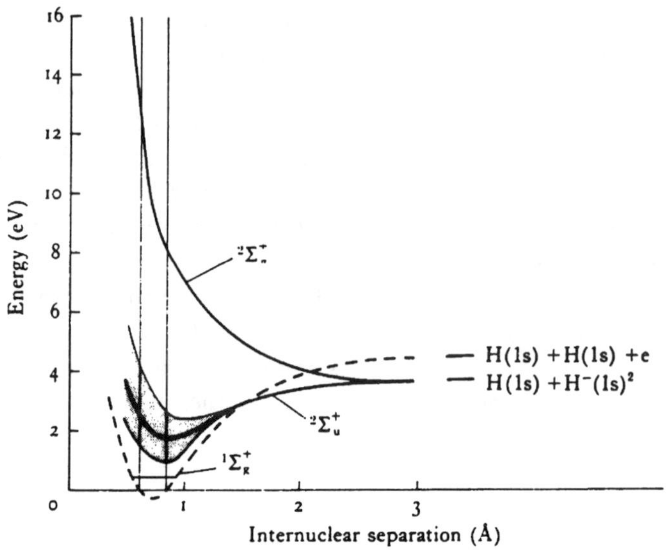

Fig. 10. Energy curves similar to those of ground state $X^1\Sigma_g^+$ H_2 and the lowest H_2^- state.

2.1. *Time Dependent Perturbation Rate Expression*

The theory used to obtain expressions for rates of electron detachment proceeds as follows.[18] First, the interaction of anion's electrons (labeled j) and nuclei (labeled a, having charges Z_a) with the vector potential $\mathbf{A}$

$$H_{\text{int}} = \Sigma_j\{(ie\hbar/m_e c)\mathbf{A}(r_j,t) \bullet \nabla_j + (e^2/2m_e c^2)|\mathbf{A}(r_j,t)|^2\}$$
$$+ \Sigma_a\{(iZ_a e\hbar/m_a c)\mathbf{A}(R_a,t) \bullet \nabla_a + (Z_a^2 e^2/2m_a c^2)|\mathbf{A}(R_a,t)|^2\}$$

is used as a perturbation in the time-dependent Schrödinger equation

$$i\hbar\partial\Psi/\partial t = (H^0 + H_{\text{int}})\Psi\,.$$

An order-by-order expansion of the wavefunction

$$\Psi = \Psi^0 + \Psi^1 + \Psi^2 + \Psi^3 + \cdots,$$

is combined with a realization that H_{int} has both first- and second- order components

$$H_{\text{int}}^1 = \Sigma_j\{(ie\hbar/m_e c)\mathbf{A}(r_j,t) \bullet \nabla_j\} + \Sigma_a\{(iZ_a e\hbar/m_a c)\mathbf{A}(R_a,t) \bullet \nabla_a\}$$
$$H_{\text{int}}^2 = \Sigma_j\{(e^2/2m_e c^2)|\mathbf{A}(r_j,t)|^2\} + \Sigma_a\{(Z_a^2 e^2/2m_a c^2)|\mathbf{A}(R_a,t)|^2\}\,.$$

The first-order wavefunction Ψ^1 is expanded in terms of the zeroth-order wavefunctions, which are the eigenstates $\{\Phi_f\}$ of the anion including those states lying in the continuum (i.e. corresponding to a neutral molecule and a free electron).

$$\Psi^1 = \Sigma_f D_f(t)\Phi_f \exp(-iE_f^0 t/\hbar)\,.$$

When used in the first-order Schrödinger equation and assuming that at $t = 0$ (just prior to when the photon field is turned on) the anion is in the electronically and geometrically stable state Φ_i having energy E_i^0, one obtains equations for the amplitudes $D_f(t)$:

$$i\hbar\partial D_f^1/\partial t = \langle\Phi_f|H_{\text{int}}^1|\Phi_i\rangle \exp(i[E_f^0 - E_i^0]t/\hbar)\,.$$

Using the expression for the first-order perturbation given earlier

$$H_{\text{int}}^1 = \Sigma_j\{(ie\hbar/m_e c)\mathbf{A}(r_j,t) \bullet \nabla_j\} + \Sigma_a\{(iZ_a e\hbar/m_a c)\mathbf{A}(R_a,t) \bullet \nabla_a\}$$

and writing explicitly the time- and space-dependence of the vector potential

$$2\mathbf{A}_0 \cos(\omega t - \mathbf{k} \bullet \mathbf{r}) = \mathbf{A}_0\{\exp[i(\omega t - \mathbf{k} \bullet \mathbf{r})] + \exp[-i(\omega t - \mathbf{k} \bullet \mathbf{r})]\}\,,$$

gives the following expression for the D_f amplitudes:

$$\begin{aligned}
D_f^1(T) = (i\hbar)^{-1}&\{\langle\Phi_f|\Sigma_j\{(ie\hbar/m_e c)\exp[-i\mathbf{k} \bullet \mathbf{r}_j]\mathbf{A}_0 \bullet \nabla_j \\
&+ \Sigma_a(iZ_a e\hbar/m_a c)\exp[-i\mathbf{k} \bullet \mathbf{R}_a]\mathbf{A}_0 \bullet \nabla_a|\Phi_i\rangle\} \\
&\times [\exp(i(\omega + \omega_{ff,i})T - 1]/i(\omega + \omega_{f,i}) \\
&+ (i\hbar)^{-1}\{\langle\Phi_f|\Sigma_j\{(ie\hbar/m_e c)\exp[i\mathbf{k} \bullet \mathbf{r}_j]\mathbf{A}_0 \bullet \nabla_j \\
&+ \Sigma_a(iZ_a e\hbar/m_a c)\exp[i\mathbf{k} \bullet \mathbf{R}_a]\mathbf{A}_0 \bullet \nabla_a|\Phi_i\rangle\} \\
&\times [\exp(i(-\omega + \omega_{ff,i})T - 1]/i(-\omega + \omega_{f,i})\,,
\end{aligned}$$

where $\omega_{f,i} = [E_f^0 - E_i^0]/\hbar$.

Introducing the short-hand for the non-time-dependent part

$$\begin{aligned}
\alpha_{f,i} = \langle\Phi_f|\Sigma_j&\{(e/m_e c)\exp[-i\mathbf{k} \bullet \mathbf{r}_j]\mathbf{A}_0 \bullet \nabla_j \\
&+ \Sigma_a(Z_a e/m_a c)\exp[-i\mathbf{k} \bullet \mathbf{R}_a]\mathbf{A}_0 \bullet \nabla_a|\Phi_i\rangle
\end{aligned}$$

allows D_f^1 to be written as

$$D_f^1(T) = \{\alpha_{f,i}[\exp(i(\omega + \omega_{ff,i})T - 1]/i(\omega + \omega_{f,i})$$
$$+ \alpha_{f,i}^*[\exp(i(-\omega + \omega_{ff,i})T - 1]/i(-\omega + \omega_{f,i})\}\,.$$

The second term in this expression is the one that has the potential of displaying "resonance character" because its denominator can approach zero as ω approaches $\omega_{f,i}$, which is a positive quantity.

Retaining only the resonant component, the modulus squared $|D_f|^2$ gives the *probability* of finding the anion in the state Φ_f at time $T > 0$:

$$|D_f^1(T)|^2 = 2|\alpha_{f,i}|^2[1 - \cos((\omega - \omega_{f,i})T)]/(\omega - \omega_{f,i})^2$$
$$= 4|\alpha_{f,i}|^2[\sin^2(1/2(\omega - \omega_{f,i})T)]/(\omega - \omega_{f,i})^2\,.$$

The quantity $[\sin^2(1/2(\omega-\omega_{f,i})T)]/(\omega-\omega_{f,i})^2$ has the ω-dependence shown in Fig. 11; the width decreases and the peak height increases as the time of interaction T with the electromagnetic field grows.

The above expression for the probability of transitions from Φ_i to Φ_f has to be modified in two ways before it is explicitly pertinent to the photodetachment case. First, it must be realized that any experimental light source does not subject the anion to a perfectly monochromatic perturbation. Even lasers produce a distribution of frequencies which is usually characterized by a so-called line-shape function $g(\omega)$. As a result, the correct

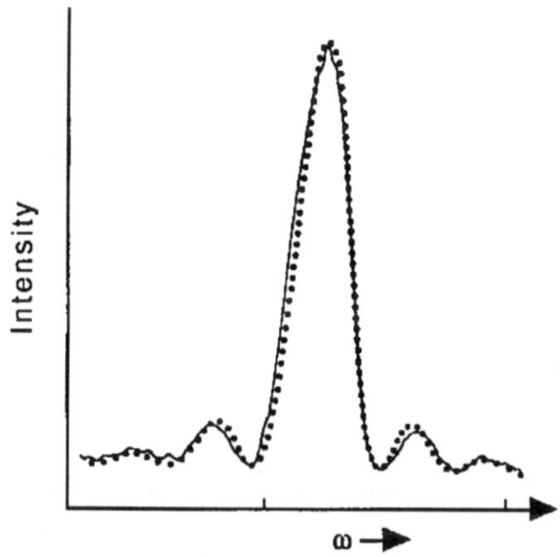

Fig. 11. Plot of $[\sin^2(1/2(\omega - \omega_{f,i})T)]/(\omega - \omega_{f,i})^2$ vs ω for one particular value of T (taken from Simons and Nichols book listed in Ref. 18).

equation for the probability of transitions averaged over the distribution of frequencies presented by the light source is

$$|D_f^1(T)|^2_{\text{ave}} = 4|\alpha_{f,i}|^2 \int g(\omega)[\sin^2(1/2(\omega - \omega_{f,i})T)]/(\omega - \omega_{f,i})^2 d\omega$$

$$= 2|\alpha_{f,i}|^2 T \int g(\omega)$$

$$\times [\sin^2(1/2(\omega - \omega_{f,i})T)]/[1/4(\omega - \omega_{f,i})^2 T^2]\, d(T\omega/2)\,.$$

The latter form of the equation is used for the following reason. If the light source's $g(\omega)$ profile is relatively constant over the narrow range of frequencies where the $[\sin^2(1/2(\omega - \omega_{f,i})T)]/(\omega - \omega_{f,i})^2$ function plotted in Fig. 11 is "peaked" (which will be the case if T is long enough), $g(\omega)$ can be approximated by $g(\omega_{f,i})$ and factored out of the integral and the identity $\int \sin^2(x)/x^2 dx = \pi$ can be used:

$$|D_f^1(T)|^2_{\text{ave}} = 2|\alpha_{f,i}|^2 T g(\omega_{f,i})$$

$$\times \int [\sin^2(1/2(\omega - \omega_{f,i})T]/[1/4(\omega - \omega_{f,i})^2 T^2]\, d(T\omega/2)$$

$$= 2g(\omega_{f,i})|\alpha_{f,i}|^2 T \int \sin^2(x)/x^2 dx = 2\pi g(\omega_{f,i})|\alpha_{f,i}|^2 T\,.$$

2.2. *The Rate and Cross-Section*

The *rate* at which transitions occur is obtained by taking the time derivative of this probability

$$R = 2\pi g(\omega_{f,i})|\alpha_{f,i}|^2$$

this gives the number of transitions per second. If one wants the *cross-section*, one divides this rate by the intensity I (number of photons per cm^2 per sec) of the light source

$$\sigma = 2\pi g(\omega_{f,i})|\alpha_{f,i}|^2/I$$

which has units of cm^2.

The second necessary modification of the rate (and cross-section) expression is to introduce the fact that the final state Φ_f is not a discrete state but, rather, is member of a continuum of states. In particular, the neutral plus ejected electron final state wavefunction is characterized by the momentum $\mathbf{p}$ (having magnitude p and orientation angles α and β relative to a molecule-fixed coordinate system) and the kinetic energy $E = p^2/2m_e$ of the ejected electron. The density of states (number of states per unit energy) for the ejected electron is

$$\rho(E) = 4\pi m_e L^3 (2m_e E)^{1/2}/\hbar^2 .$$

Introducing this density into the above expressions for the rate and cross-section gives the following expression appropriate to the photodetachment process:

$$R = 2\pi \int g(\omega_{E,i})|\alpha_{E,i}|^2 \rho(E)\, dE$$

$$\sigma = 2\pi/I \int g(\omega_{E,i})|\alpha_{E,i}|^2 \rho(E)\, dE$$

where the index f has been replaced by the label E because the final state Φ_f is more appropriately labeled Φ_E according to the kinetic energy of the ejected electron.

The next step in deriving more explicit expressions for photodetachment from molecular anions is to introduce the so-called long wavelength approximation into the expression for the $\alpha_{E,i}$ matrix element

$$\alpha_{E,i} = \langle \Phi_E | \Sigma_j (e/m_e c) \exp[-i\mathbf{k} \bullet \mathbf{r}_j] \mathbf{A}_0 \bullet \nabla_j + \Sigma_a (Z_a e/m_a c) \exp[-i\mathbf{k} \bullet \mathbf{R}_a] \mathbf{A}_0 \bullet \nabla_a | \Phi_i \rangle$$

and to keep only the electric dipole terms. The factors $\exp[-i\mathbf{k} \bullet \mathbf{r}_j]$ and $\exp[-i\mathbf{k} \bullet \mathbf{R}_a]$ are expanded as:

$$\exp[-i\mathbf{k} \bullet \mathbf{r}_j] = 1 + (-i\mathbf{k} \bullet \mathbf{r}_j) + 1/2(-i\mathbf{k} \bullet \mathbf{r}_j)^2 + \cdots$$

$$\exp[-i\mathbf{k} \bullet \mathbf{R}_a] = 1 + (-i\mathbf{k} \bullet \mathbf{R}_a) + 1/2(-i\mathbf{k} \bullet \mathbf{R}_a)^2 + \cdots$$

where $|\mathbf{k}| = 2\pi/\lambda$. Because the photon's wavelength λ is usually much longer than the dimensions of the molecular anion, $\mathbf{k} \bullet \mathbf{r}_j$ and $\mathbf{k} \bullet \mathbf{R}_a$ are less than unity in magnitude. If only the terms linear in $|\mathbf{k}|$ are retained, one obtains what is called the electric-dipole approximation to the transition matrix elements:

$$\alpha_{E,i} = \langle \Phi_E | \Sigma_j (e/m_e c) \mathbf{A}_0 \bullet \nabla_j + \Sigma_a (Z_a e/m_a c) \mathbf{A}_0 \bullet \nabla_a | \Phi_i \rangle$$

which can also be expressed in the so-called "length" form (rather than the "velocity" operator form shown above)

$$\alpha_{E,i} = \omega_{E,i}\mathbf{A}_0 \bullet \langle\Phi_f|\Sigma_j(e/\hbar c)\mathbf{r}_j + \Sigma_a(Z_a e/\hbar c)\mathbf{R}_a|\Phi_i\rangle$$
$$= (\omega_{E,i}/\hbar c)\mathbf{A}_0 \bullet \langle\Phi_f|\mu|\Phi_i\rangle$$

where μ is the electric dipole moment operator for the electrons and nuclei:

$$\mu = \Sigma_j e\mathbf{r}_j + \Sigma_a Z_a e\mathbf{R}_a\,.$$

Within this approximation, the overall rate of transitions is given by:

$$R = 2\pi \int g(\omega_{E,i})|\alpha_{E,i}|^2 \rho(E)\,dE$$
$$= 2\pi \int g(\omega_{E,i})(\omega_{E,i}/\hbar c)^2 |\mathbf{A}_0 \bullet \langle\Phi_E|\mu|\Phi_i\rangle|^2 \rho(E)\,dE\,.$$

Recalling that $\mathbf{E}(\mathbf{r},t) = -1/c\,\partial\mathbf{A}/\partial t = \omega/c\mathbf{A}_0 \sin(\omega t - \mathbf{k}\bullet\mathbf{r})$, the magnitude of $\mathbf{A}_0$ can be replaced by that of the electric field strength $\mathbf{E}$, and this rate expression becomes

$$R = 2\pi \int g(\omega_{E,i})(1/\hbar)^2 |\mathbf{E} \bullet \langle\Phi_E|\mu|\Phi_i\rangle|^2 \rho(E)\,dE\,.$$

The intensity I, in photons per cm^2 per sec, is related to the electric field strength by

$$I = c|\mathbf{E}|^2/(8\pi\hbar\omega)$$

so, the rate can be written in terms of the intensity as:

$$R = 16\pi^2 \int g(\omega_{E,i}) I(\omega/c\hbar)|\mathbf{e} \bullet \langle\Phi_E|\mu|\Phi_i\rangle|^2 \rho(E)\,dE$$

where $\mathbf{e}$ is a unit vector pointing in the direction of the electric field. The cross-section can also be written in this form as:

$$\sigma = 16\pi^2 \int g(\omega_{E,i})(\omega/c\hbar)|\mathbf{e} \bullet \langle\Phi_E|\mu|\Phi_i\rangle|^2 \rho(E)\,dE\,.$$

2.3. *Electronic, Vibrational and Rotational Components*

Finally, the most common way to describe the anion and neutral plus free electron wavefunctions is to write them as products of electronic ψ, vibrational χ, and rotational $D^J_{M,K}$ functions:

$$\Phi_i = \psi_i \chi_v D^J_{M,K}$$

$$\Phi_E = \psi_E \chi_{v'} D^{J'}_{M',K'}$$

and to assume that the electronic functions consist of single Slater determinant functions which differ by Φ_i having the spin-orbital ϕ_i where Φ_E has the continuum spin-orbital ϕ_E.

In the event that the electronic wavefunctions ψ_i and ψ_E possess nonzero angular momentum, it is necessary to properly combine this angular momentum with the rotational angular momentum. In the initial anion state, where the extra electron's motion is "locked" to the framework of anion, the electronic angular momentum is added, component by component, to the rotational angular momentum's components in a molecule-fixed coordinate system (usually taken to be those of the three eigenvalues of the moment of inertia tensor). For example, in the $X^2\Pi$ state of NH^- whose HOMO orbital occupancy π^3 is depicted below in Fig. 12, the electronic wavefunction has one unit of angular momentum along the N–H bond axis and zero components along the two axes perpendicular to this axis.

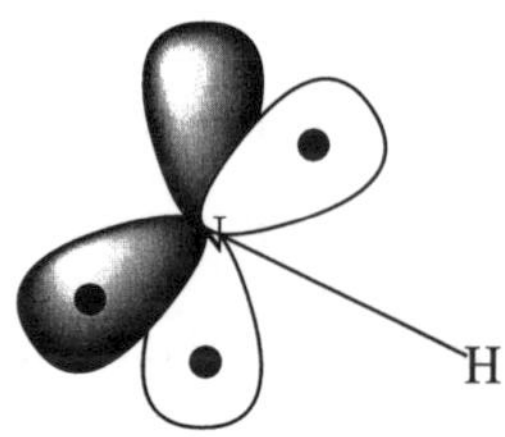

Fig. 12. HOMO orbital occupancy in the $X^2\Pi$ ground state to NH^-.

In this case, the rotational wavefunction's K quantum number (i.e. in $D^J_{M,K}$ labels the electronic function's component along the N–H axis ($K = \pm 1$).

For the neutral plus free electron state Φ_E in which the ejected electron carries away nonzero angular momentum (i.e. is ejected in a *p*-, *d*-, *f*-, or higher wave), it is more proper to vector couple this electron's angular momentum to the rotational angular momentum of the neutral molecule. Vector coupling is appropriate because the ejected electron's motion is no longer "locked" to the molecular framework. Thus, in such cases, the wavefunction Φ_E is replaced by a function of the form:

$$\Phi_E(J, M, K; N, L) = \Sigma_{m,M'} \langle J, M | N, L, M', m \rangle \psi_E(L, m) D^N_{M',K} \,.$$

In this function, the molecule's rotational angular momentum **N** (having component M^{prime} along the laboratory-fixed z-axis and component K along its highest symmetry axis) is vector coupled, to the electronic wavefunction's angular momentum **L** (having component m along the z-axis) to give a total angular momentum **J** (with component M along z and K along the molecular axis).

Once the electronic angular momentum has been coupled to the rotational angular momentum as described above, both the anion and neutral plus free-electron states are expressed as simple products (or sums of such products) of an electronic function, a rotational function, and a vibrational function. In the following decomposition of the rate and cross-section expressions into electronic, vibrational, and rotational components, I will take Φ_i and Φ_E to be of the simple product form for simplicity, realizing that the final results should subsequently be multiplied by vector coupling coefficients (for Φ_E) and summed over m and M' quantum numbers.

With these assumptions, the electric dipole matrix element reduces to

$$\langle \Phi_E | \mu | \Phi_i \rangle = \int D^{*J'}_{M',K'} D^J_{M,K} \int \chi^*_{v'} \chi_v \langle \psi_E | \mu | \psi_i \rangle \, d\mathbf{Q} \sin\alpha \, d\alpha \, d\beta \, d\gamma \,.$$

Here, $\mathbf{Q}$ denotes all $3N - 6$ of the internal (vibrational) coordinates, and α, β, and γ are the Euler angles upon which the rotational wavefunctions depend.

The electronic integral $\langle \psi_E | \mu | \psi_i \rangle$ is assumed to be relatively insensitive to geometry (actually, to vibrational distortions only; this matrix element is, of course, a vector attached to the anion's framework and thus has components directly related to α, β and γ). Therefore, $\langle \psi_E | \mu | \psi_i \rangle$ can be factored out of the integral over $d\mathbf{Q}$, thus yielding the vibrational integrals $\int \chi^*_{v'} \chi_v \, d\mathbf{Q}$ whose squares give the well known Franck–Condon factors between the anion and neutral molecule.

As noted above, $\langle \psi_E | \mu | \psi_i \rangle$ is a vector (i.e. a first-rank tensor) with components attached to the molecule's geometrical frame, while the rotational functions $D^{*J'}_{M',K'} D^J_{M,K}$ are expressed in terms of laboratory fixed coordinates (i.e. α, β, and γ describe the orientation of the molecular frame in a space-fixed frame), angular integral $\int D^{*J'}_{M',K'} D^{J}_{M,K} \langle \psi_E|\mu|\psi_i\rangle$ $\sin\alpha\, d\alpha\, d\beta\, d\gamma$ can be carried out by first reexpressing the components of $\langle \psi_E|\mu|\psi_i\rangle$ in the laboratory-fixed frame using standard rotation matrix algebra:

$$\langle \psi_E|\mu|\psi_i\rangle_m = \Sigma_k \langle \psi_E|\mu|\psi_i\rangle_k (D^1_{m,k})^*\,,$$

where m labels the components of $\langle \psi_E|\mu|\psi_i\rangle$ in the laboratory-fixed coordinate system, k labels components of $\langle \psi_E|\mu|\psi_i\rangle$ in the molecule-fixed coordinate system, and $D^1_{m,k}$ is the transformation matrix between these two coordinate systems. The resultant integral over angles is of the form

$$\int D^{*J'}_{M',K'} D^1_{m,k} D^J_{M,K} \sin\alpha\, d\alpha\, d\beta\, d\gamma$$

(with m and k ranging over -1, 0, and 1) which vanishes unless certain selection rules hold:

$$\begin{aligned} J' &= J, J+1\,,\ \text{or } J-1 \text{ (except if } J = J' = 0 \text{ for which} \\ &\quad\ \text{the integral vanishes)} \\ M' &= M + m \\ K' &= K + k\,. \end{aligned}$$

Thus, the vibrational (Franck–Condon) and rotational (selection rules on the angular momenta of the orbital from which the electron is ejected and the continuum orbital into which the electron is ejected) aspects of the photodetachment rate or cross-section expressions are the same as in conventional photoionization or electronic absorption (or emission) spectra.

2.4. *The Electronic Integral*

There are significant differences among photodetachment, photoionization, and electronic absorption spectra when the electronic transition integral $\langle \psi_E|\mu|\psi_i\rangle$ is considered in further detail. In the photodetachment case, the orbital ψ_E is not a bound-state function (but ψ_i is) but a function that describes an electron moving in the "field" of the underlying neutral molecule. As such, $\psi_E(r,\theta,\phi)$ is orthogonal to all of the occupied bound orbitals of the anion and describes an electron with kinetic energy $E = p^2/2m_e$ for large values of r. The ψ_E orbital appropriate to photodetachment is *not* of the same form as that arising in the theory of photoionization because, in the latter case, the outgoing electron experiences the long-range Coulomb potential of the daughter cation. For anion photodetachment, the ejected electron experiences no long-range potential of the form $-e^2/r$. For anions having significant dipole moments (μ), the longest range potential varies as $-\mu\cos(\theta)/2r^2$, and for non-polar molecules, the longest range potential varies as an even higher power of r. In addition, the ejected electron's orbital angular momentum L affects the form of ψ_E because it contributes the centrifugal potential $L(L+1)/2m_e r^2$ to the effective potential $V_{\text{eff}}(r)$ that shapes the radial form of ψ_E.

There are, however, many similarities among the photoionization (of neutrals), photodetachment, and electronic excitation cases. For example, symmetry can be used to determine (or constrain) the symmetry of the ejected electron's wavefunction. For the electronic matrix element $\langle\psi_E|\mu|\psi_i\rangle$ to be non-vanishing, the direct product of the symmetry of the HOMO ψ_i and the components of the electric dipole operator μ (which transform as x, y, and z) determine the symmetry components of the continuum orbital ψ_E.

Let us consider two examples. The HOMO of NH^- as shown in Fig. 12 is a p-type orbital localized nearly entirely on the nitrogen atom. The direct product (actually, for such essentially atomic cases, we couple the angular momenta of the HOMO and of μ, which has $L = 1$ to determine the angular momentum of ψ_E) of the HOMO's symmetry ($L = 1$) and that of $\mu(L' = 1)$ constrains the symmetry of ψ_E to be $L = 2$, 1, or 0. This means that d-, p- and s-wave electrons will be ejected when NH^- undergoes direct photodetachment.

Figure 7 shows the HOMO of N_2^- (which actually does not exist as a stable species, but we nevertheless use it here as another example). This orbital has predominantly d-character, so the photodetached electron would be expected to possess $L = 2+1, 2$ or $2-1$ character (i.e. to come off in f-, d- and p-waves). For anions whose HOMOs are delocalized over many atomic centers, one must decompose the HOMO into various symmetry components and corresponding amplitudes and subsequently take the direct product of μ's symmetries with each of the HOMO's components to predict the symmetry components and amplitudes of the ejected electron's orbital. An especially nice treatment of such polyatomic anion symmetry analyses was made by Reed, Zimmerman, Andersen, and Brauman.[19]

In addition to being able to use symmetry or angular momentum to predict the directional characteristics (i.e. symmetries) of the ejected electron's orbital ψ_E, the kinetic energy of this electron, KE has a significant effect on the electronic dipole matrix element $\langle\psi_E|\mu|\psi_i\rangle$ and thus on the detachment rate or cross-section. The kinetic energy is determined by the energy $\hbar\omega$ of the photons and the neutral-anion detachment energy DE:

$$\mathrm{KE} = \hbar\omega - \mathrm{DE}\,.$$

Except near thresholds (which are treated more specifically in the next section), the ejected electron departs the molecular framework "promptly" (on the time scales of vibrations and rotations). As the photon frequency ω is increased, the ejected electron carries away more and more kinetic energy. Because the radial form of ψ_E varies as $\exp(ikr)$ for large r, with $\hbar^2k^2/2m_e = \mathrm{KE}$, the orbital ψ_E will be a highly oscillatory function when KE is large (specifically, large enough to make $1/k$ small compared to the dimension of the anion's HOMO). In such cases, the integral $\langle\psi_E|\mu|\psi_i\rangle$ will have a very small magnitude because ψ_E oscillates rapidly while ψ_i and μ are smoothly varying functions of r. Thus, for photon energies far above the detachment threshold, the electron is ejected with large kinetic energy, but the rate and cross-section are small because the rapid oscillation of ψ_E causes the transition matrix element $\langle\psi_E|\mu|\psi_i\rangle$ to be small.

2.5. *Threshold Forms of Cross-Sections*

When the photon energy $\hbar\omega$ is only slightly higher than the minimum energy needed to detach an electron from a particular state (electronic/vibration/rotation) of the anion and to produce a particular (electronic/vibration/rotation) state of the daughter neutral, we say that the detachment is occurring just above threshold. In both the photodetachment and photoionization cases, angular momentum considerations allow one to make important conclusions about the nature of the cross-section (or rate) as a function of the kinetic energy of the ejected electron.

As discussed earlier in Sec. 2.4, the symmetry of the anion's HOMO and that of the electric dipole operator μ limit the range of angular momentum values L that the ejected electron can assume. Among these symmetry-allowed L values, amplitude belonging to the *smallest L will dominate* the ejected electron's flux *near thresholds.* The dominance of the smallest L is a result of the pattern in angular momentum barriers (see Fig. 9) through which the electron must tunnel to detach; the smallest L value has the lowest barrier and thus the highest rate of tunneling.

The cross-section for photodetachment at energies just above a threshold and for ejection of an electron having angular momentum L and kinetic energy KE was shown by Wigner in 1948[20] to depend on these two quantities as:

$$\sigma \propto \mathrm{KE}^{L+1/2}\,.$$

This so-called Wigner law, which applies to photodetachment of anions but not to photoionization of neutrals (because its derivation requires that the ejected electron experience a potential varying no stronger than r^{-2} at large r), is a result of analyzing the KE-dependence of the square of the electronic transition matrix element multiplied by the translational density of states of the ejected electron $|\langle\psi_E|\mu|\psi_i\rangle|^2\rho(\mathrm{KE})$. The density of states vary as $\mathrm{KE}^{1/2}$, and, for small values of KE, the electronic integral squared has been shown to vary as $(\mathrm{KE})^L$ for ejection of L-wave electrons.

For the NH^- and N_2^- examples used in the preceeding section for purposes of illustration, the photoejected electrons are allowed to be ejected in s-, p-, and d-waves (for NH^-) or in p-, d-, and f-waves (for N_2^-). The smallest L values for these cases are $L = 0$ (s-waves for NH^-) and $L = 1$ (p waves for N_2^-). The cross-sections should thus depend on the photon energy (or equivalently on the kinetic energy of the ejected electron) as:

$$\sigma = \mathrm{KE}^{1/2} \text{ for } \mathrm{NH}^-, \text{ and}$$

$$\sigma = \mathrm{KE}^{3/2} \text{ for } \mathrm{N}_2^-\,.$$

The plots of cross-section vs KE have qualitatively different shapes near the thresholds as illustrated below in Fig. 13 which is taken from the work of Engelking and Herrick.[21]

In particular the slopes $d\sigma/d(\mathrm{KE})$ as KE approaches thereshold differ drastically and can be used as diagnostics for determining the minimum L value from experimentally measured cross-sections. The slope generally has the form

$$d\sigma/d(\mathrm{KE}) = (L + 1/2)(\mathrm{KE})^{L-1/2}\,.$$

For s-wave detachment, the slope varies as $(\mathrm{KE})^{-/12}$ near threshold; thus one sees a sharp decline (with essentially infinite slope) in σ as KE approaches zero from above. In contrast, the p-wave cross-section approaches

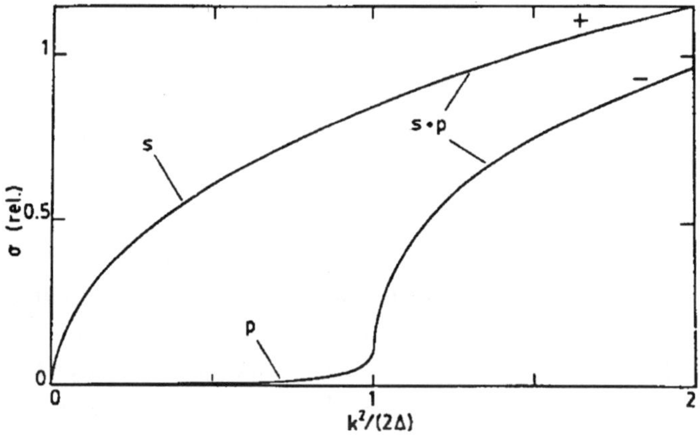

Fig. 13. Cross-sections plotted vs the kinetic energy of the ejected electron for *s*-wave and *p*-wave detachment (taken from Ref. 21). The abscissa is proportional to KE and the ordinate is proportional to the cross-section.

its threshold value (from above) with a small (zero, in principle) slope $d\sigma/d(\text{KE}) = 3/2(\text{KE})^{1/2}$.

Although the Wigner law analysis is very useful and remains a cornerstone for determining the symmetry of the ejected electron's orbital (and, by inference, the symmetry of the anion's HOMO), the shape of the cross-section follows this law only very close to threshold. At higher KE (or photon energy) values, corrections to this simple equation must be made.[22] The explicit form of these corrections depends on the nature of the potential experienced by the ejected electron due to the presence of the underlying neutral molecule core. In the photoionization case, the ejected electron experiences a long-range Coulomb potential which necessitates a different treatment than that used to obtain the Wigner formula. For anion photodetachment, the ejected electron can experience long-range potentials of the form (dispaying only the radial dependence) $-\mu e/r^2$ (for a molecule with dipole moment μ), $-\alpha/2r^4$ (for a molecule with polarizability α), and $-Qe/r^3$ (for a molecule with quadrupole moment Q). In practice, the magnitudes of the molecule's moments (μ, Q, and α) as well as how far above threshold one is (i.e. the KE value), combine to determine the corrections to the Wigner threshold law. A more specific consideration of these corrections is beyond the scope and level of this chapter, so it is sufficient to say that, after making such corrections, the cross-section's KE dependence can be expressed as:

$$\sigma = \sigma_0(\mathrm{KE})^{L+1/2}\{1 + a\,\mathrm{KE} + b\,\mathrm{KE}^2 + c\,\mathrm{KE}^3 + \cdots\}$$

with the correction coefficients $a, b, c, \ldots$ depending on μ, Q, α and other permanent and induced moments of the molecule.

Before closing this section it should be clarified that the Wigner threshold law presented above must be modified, even in its lowest-order (in powers of KE) term for the specific case of detaching an electron to produce a molecule having a large dipole moment.[19,22,23] In such cases, the longest-range potential experienced by the ejected electron having the smallest allowed L value varies as

$$V_{\text{eff}} = L(L+1)/2m_e r^2 - \mu e \cos\theta/r^2$$

where θ is the angle between the molecule's dipole moment vector μ and the ejected electron's position vector $\mathbf{r}$. It is because both the centrifugal potential $L(L+1)/2m_e r^2$ and the longest-range electron molecule potential vary as the same power of r that this is a special case. For photoionization of neutrals (which have long-range $-e^2/r$ potentials), as mentioned earlier, a qualitatively different treatment of the threshold cross-section is needed. However, for detachment of non-polar molecules (which have zero or negligible $-\mu e \cos\theta/r^2$), the conventional Wigner law remains valid.

For the special case of detachment producing a highly polar molecule,[23] it can be shown that the cross-section near threshold varies with KE as

$$\sigma = \sigma_0(\mathrm{KE})^{\lambda+1/2}$$

where $\lambda(\lambda+1) = L(L+1) - 2m_e\mu e$ can be viewed as giving the (non-integer) effective angular momentum value λ in terms of the angular momentum L of the ejected electron and the molecule's dipole moment.

2.6. *What does One Learn from a Photodetachment Experiment?*

2.6.1. *Vibrational (and Maybe Rotational) Energy Levels of the Daughter Neutral*

For an electronically stable and geometrically stable anion, the photodetachment spectrum can be obtained in either of two ways. A fixed-energy

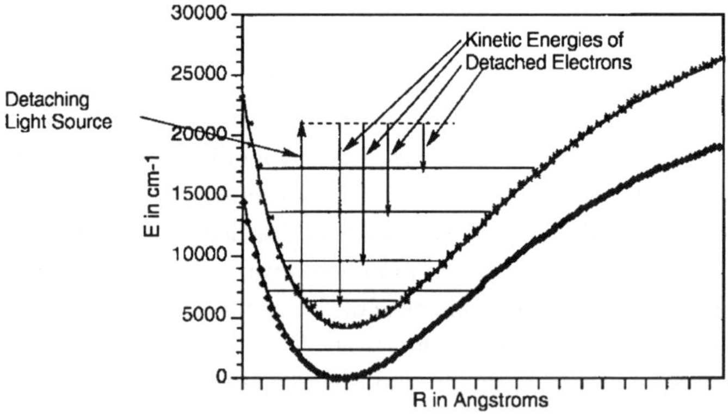

Fig. 14. Fixed frequency photodetachment of an electronically and geometrically stable anion in which the kinetic energies of the detached electrons are monitored.

light source of frequency ω can be used and the kinetic energies of the ejected electrons can be monitored. The kind of transitions probed in such an experiment are described in Fig. 14 for detachment from $v = 0$ of the anion to $v = 0$, 1, 2, and 3 of the neutral.

An example of the kind of data that arises in this kind of experiment is shown below in Fig. 15 taken from work of Leopold, Ho, and Lineberger.[24] A series of peaks corresponding to electrons ejected with differing kinetic energies are the primary data in these experiments.

In such an experiment, the electrons in the peak corresponding to the highest kinetic energies KE_0 determine the adiabatic detachment energy $\mathrm{DE}_{\mathrm{adiabatic}}$ of the anion via the energy balance equation:

$$\hbar\omega = \mathrm{KE}_0 + E_{\mathrm{anion}}(v=0) - E_{\mathrm{neutral}}(v=0) = \mathrm{KE}_0 + \mathrm{DE}_{\mathrm{adiabatic}}\,.$$

The energy balance equation also pertains to detachment to excited vibrational levels of the neutral:

$$\hbar\omega = \mathrm{KE}_v + E_{\mathrm{anion}}(v=0) - E_{\mathrm{neutral}}(v)\,.$$

The spacings $\mathrm{KE}_0 - \mathrm{KE}_v$ thus provide the spacings between the neutral molecule's vibrational levels:

$$\mathrm{KE}_0 - \mathrm{KE}_v = E_{\mathrm{neutral}}(v) - E_{\mathrm{neutral}}(v=0)\,.$$

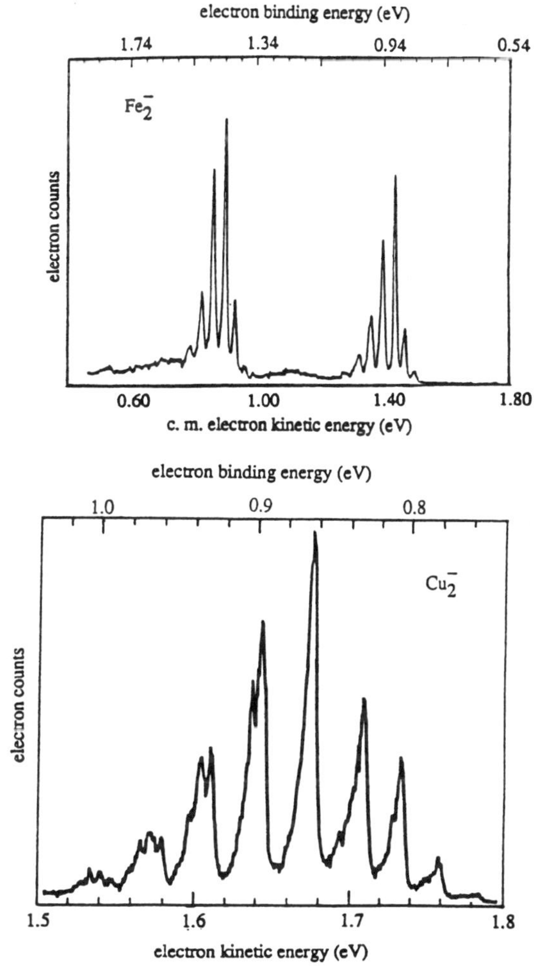

Fig. 15. Photoelectron spectrum of Fe_2^- and of Cu_2^- (taken from Ref. 24).

For the two examples cited above, these spacings give the following information:

1. The two groupings of peaks in Fe_2^- are interpreted as corresponding to transitions from $(4s\sigma)^2(4s\sigma^*)^2(3d)^{13}Fe_2^-$ to $(4s\sigma)^2(4s\sigma^*)^1(3d)^{13}Fe_2$ with the two states giving rise to the two sets of peaks corresponding to different couplings of the $(3d)^{13}$ and $(4s\sigma^*)^1$ orbitals' spins. For both sets of

peaks, removal of an electron from the $4s\sigma^*$ orbital, which is weakly antibonding, gives rise to a slight change in Fe–Fe bond length (i.e. the anion has a slightly longer bond than the neutral) and thus to a Franck–Condon vibrational progression (i.e. the individual peaks within each group) from which the vibrational spacing of 300 cm^{-1} in neutral Fe_2 can be inferred.

2. For Cu_2^-, one vibrational progression is observed with spacings of ca. 265 cm^{-1}, which is the vibrational frequency of neutral Cu_2. The shape of the Franck–Condon envelope can be explained in terms of Cu_2^- having a slightly longer bond length than neutral Cu_2 (2.345 Å compared to 2.220 Å), all of which is consistent with Cu_2 having a $(4s\sigma)^2(4s\sigma^*)^0(3d)^{20}$ orbital occupancy and Cu_2^- having $(4s\sigma)^2(4s\sigma^*)^1(3d)^{20}$ as its configuration.

The intensities of each peak are determined, as discussed earlier in Sec. 2.3, by Franck–Condon factors connecting the anion and neutral vibrational levels as well as by the squared modulus of the electronic integral $|\langle\psi_E|\mu|\psi_i\rangle|^2$ and the density of states of the ejected electrons ρ. For those peaks belonging to electrons ejected with high kinetic energy, the $|\langle\psi_E|\mu|\psi_i\rangle|^2$ factor attenuates the peak intensity because ψ_E is a highly oscillatory function of r, while peaks belonging to electrons with lower kinetic energies have intensities less attenuated. For these reasons, it is difficult to extract from the pattern of peak intensities quantitative information about the geometry differences between the anion and neutral (i.e. by using the Franck–Condon profile alone), although, as illustrated in the preceeding discussion of Fe_2^- and Cu_2^-, it is possible to glean vibrational spacing information and to estimate anion–neutral bond length changes.

Finally, if the resolution of the electron kinetic energy measurement is high enough to detect peaks with rotational structure, the spacings among such peaks can be used to determine rotational energy levels of the neutral molecule (and thus moment of inertia data). Moreover, if the anion sample is warm enough to allow other than its lowest vibrational level to be populated, hot bands may occur, for which the energy balance equation reads:

$$\hbar\omega = \mathrm{KE}_{v'v} + E_{\mathrm{anion}}(v) - E_{\mathrm{neutral}}(v')$$

where v is the quantum number of the anion's vibrational level and v' is the quantum number of the neutral's level.

An alternative experiment involves the use of a light source whose frequency is scanned and the rate of appearance of ejected electrons is monitored as a function of ω (but their kinetic energies are not measured).

In such experiments, one also observes a series of peaks than can be used to determine the anion's adiabatic detachment energy and whose spacings can be used to extract the neutral's vibrational (and perhaps rotational) energy level patterns. Again, the intensities of the peaks seen in this kind of experiment depend on Franck–Condon type factors as well as the squares of electronic matrix elements and densities of states. One strength of the scanning frequency method is that one can more easily examine the near-threshold behavior of the cross-section (i.e. how the cross-section depends on the kinetic energy of the ejected electron). As discussed in Sec. 2.5, the dependence of this cross-section on KE can offer information about the angular momentum carried away by the electron and thus on the symmetry of the anion's HOMO. To study such near-threshold behavior using the scanning (photodetachment) technique is more readily achieved than within the photoelectron spectroscopy method simply because the energy resolution with which one can fix the (laser) light source energy is higher than that with which one can measure the kinetic energy of the ejected electrons.

Three examples of using threshold cross-section shapes to determine HOMO orbital symmetries are shown below in Fig. 16.

The OH^- data[25] clearly show the sharp onset charactistic of $KE^{1/2}$ behavior suggesting that the ejected electron is produced in an $L = 0$ s-wave. Therefore, one can infer that the HOMO must be of p character; indeed the OH^- HOMO is localized largely on the O atom and is of $2p_\pi$ character. In contrast, the O_2^- cross-section[26] rises as $KE^{3/2}$ suggesting ejection of a p-wave electron. This means that the HOMO of O_2^- could be of s, p, or d character; it is now known that this HOMO is the π_g^* orbital consisting of the anti-bonding combination of $2p_\pi$ orbital on the two O atoms. Finally, the NH_2^- cross-section again displays s-wave shape suggesting that the NH_2^- HOMO has p character; in fact, this HOMO is the non-bonding $2p$ orbital localized nearly fully on the N atom.

2.6.2. *Transition-Region Dynamical Information*

In recent years it has been shown that transition regions on the ground (or low-lying) electronic state surfaces of neutral species undergoing chemical reaction can be probed by photodetaching from a locally geometrically stable and electronically stable structure of the corresponding anion species. For example, Dan Neumark's group at Berkeley used photodetachment of

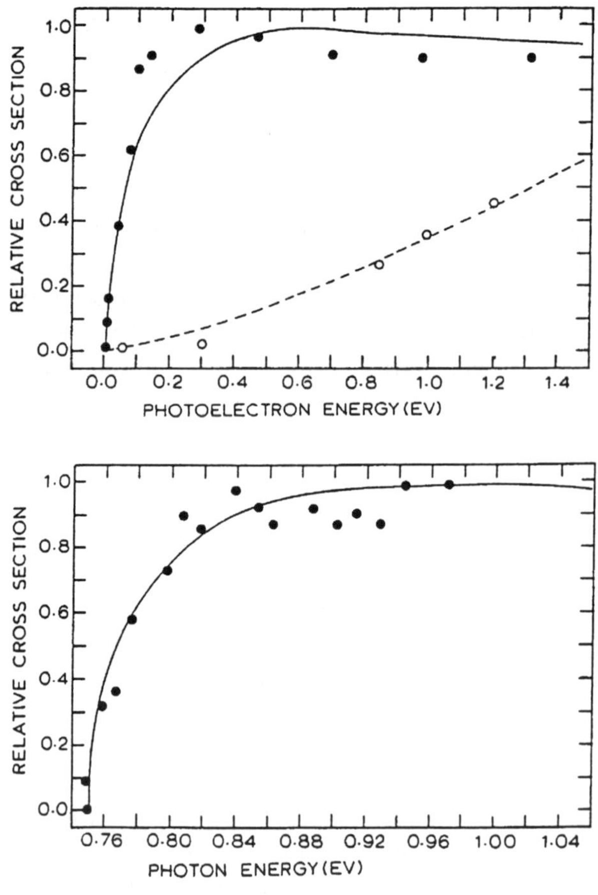

Fig. 16. Photoelectron spectra of OH^- and O_2^- (upper) and of NH_2^- (lower) very close to their respective detachment thresholds (taken from Refs. 25 and 26).

the F^-H–H van der Waals complex,[27] whose equilibrium structure is linear and rather close to the structure of the transition state on the neutral $F + HH \rightarrow FH + H$ reactive surface, to probe the reactive dynamics of the neutral near its transition state.

The basic idea underlying this use of photodetachment spectroscopy is illustrated below in Fig. 17 for the case of the $OH^-(H_2)$ van der Waals complex.[28] In this case, detachment of the $OH^-(H_2)$ complex places the neutral system on the reactive potential energy surface near the transition

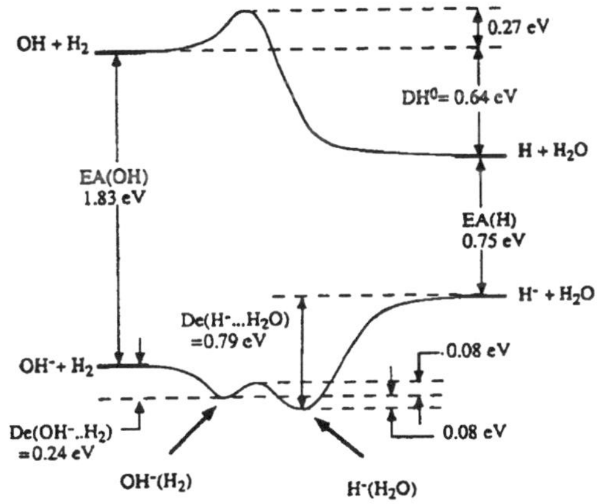

Fig. 17. Energy profiles for $OH^-(H_2)$ in relation to the $OH + H_2 \rightarrow H_2O + H$ reaction's transition state (taken from Ref. 28).

state of the $OH + H_2 \rightarrow H_2O + H$ reaction. Alternatively, detaching the $H^-(H_2O)$ van der Waals complex accesses the neutral surface far on the product side (i.e. away from the transition state near the $H_2O + H$ products).

Another example is provided in the work of Carl Lineberger's group on the photoelectron spectrum[29] of $H_2C{=}C^-$. The anion and neutral energy surfaces as functions of the CH_2 rocking normal mode are depicted in Fig. 18.

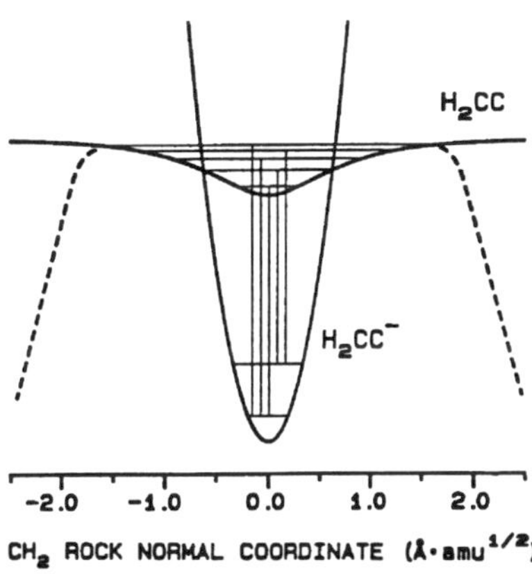

Fig. 18. Anion and neutral energy surfaces along the H_2C rocking mode (taken from Ref. 29). The barrier on the neutral surface connects $H_2C{=}C$ to HCCH.

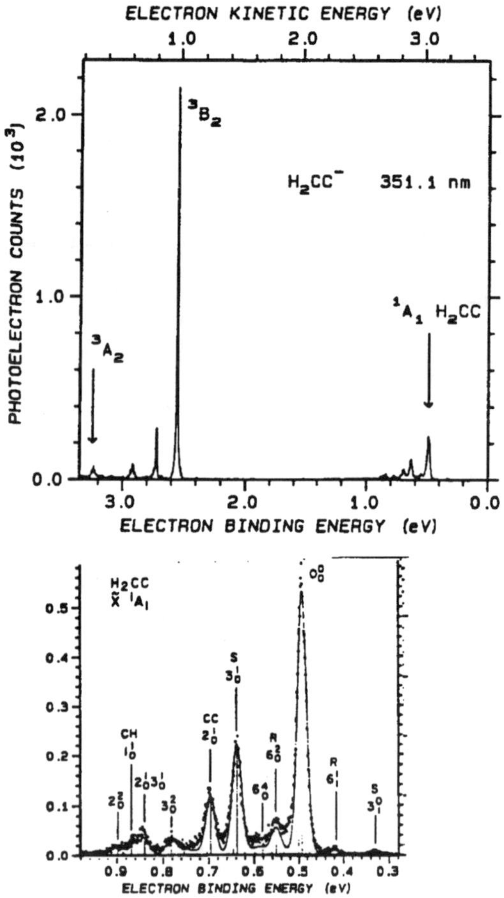

Fig. 19. Photoelectron spectrum of $H_2C{=}C^-$ (taken from Ref. 29) at low resolution (top) and, for the group of peaks belonging to the $^1A_1$ state of the neutral, at higher resolution (bottom) (taken from Ref. 29).

Because detachment of an electron from $H_2C{=}C^-$ (to produce X^1A_1 $H_2C{=}C$), especially when accompanied by excitation of the H_2C rocking mode, places the system in the region of the barrier along the X^1A_1 $H_2C{=}C$ $\rightarrow$ HCCH reaction path, the photoelectron spectrum shown in Fig. 19 has been used to extract valuable information about this transition state and to determine the energy splittings between the ground ($^1A_1$) and excited ($^3B_2$ and $^3A_2$) electronic states of $H_2C{=}C$.

It should be noted that photodetaching an anion van der Waals complex such as $F^-(H_2)$ or $OH^-(H_2)$ does not guarantee that the reactive neutral system is created at a geometry identical to the transition state geometry. However, because the anion species in the cases that have been examined is a rather tightly bound species, it has been argued that the equilibrium geometry (actually, the range of geometries accessed by vibrational motions about this equilibrium) is close enough to the geometry of the neutral's transition state that electron detachment will indeed have a reasonable Franck–Condon probability of sampling the transition state.

2.6.3. *Identification of New Molecular Anions*

One of the advantages of carrying out spectroscopies on anionic (or cationic) species is that mass selection techniques can be used to sort out complex mixtures so one knows precisely the chemical composition of the species one is examining. An example of such a study is provided by the work from Kit Bowen's laboratory at Johns Hopkins University on $H^-(NH_3)_n$ cluster anions[30] from which the photoelectron spectra shown in Fig. 20 are taken. The main peaks (labeled A and A') in $H^-(NH_3)$, and $H^-(NH_3)_2$, respectively, clearly are shifted to higher electron binding energies when compared to the unsolvated H^-, which has an electron binding energy of ca. 0.75 eV. These findings are consistent with the differential solvation of H^- compared to neutral H and with a larger solvation when two NH_3 molecules are present than when one NH_3 is used. The peaks labeled Band B' have been interpreted as due to excitation of an N–H stretching mode when one moves from the anion to the neutral; the weak intensities of these peaks have been used to conclude that the extent of N–H bond elongation is slight (thus the Franck–Condon factors are small).

What does any of this have to do with identification of new species? Although the peaks labeled A and B can straightforwardly be interpreted as belonging to an H^- ion that is "solvated" by a single intact NH_3 molecule, peak C presented a problem. This peak's intensity relative to those of peaks A and B varied from experiment to experiment, depending on specifics of the source conditions used to generate the anions. Moreover, peak C corresponds to a species having NH_4^- stoichiometry but with an electron binding energy of ca. 0.4 eV, which is considerably less even than H^-. Finally, when looked at with as high a resolution as possible, peak C seemed to consist of a single or, at most, a few vibrational sub-peaks, suggesting

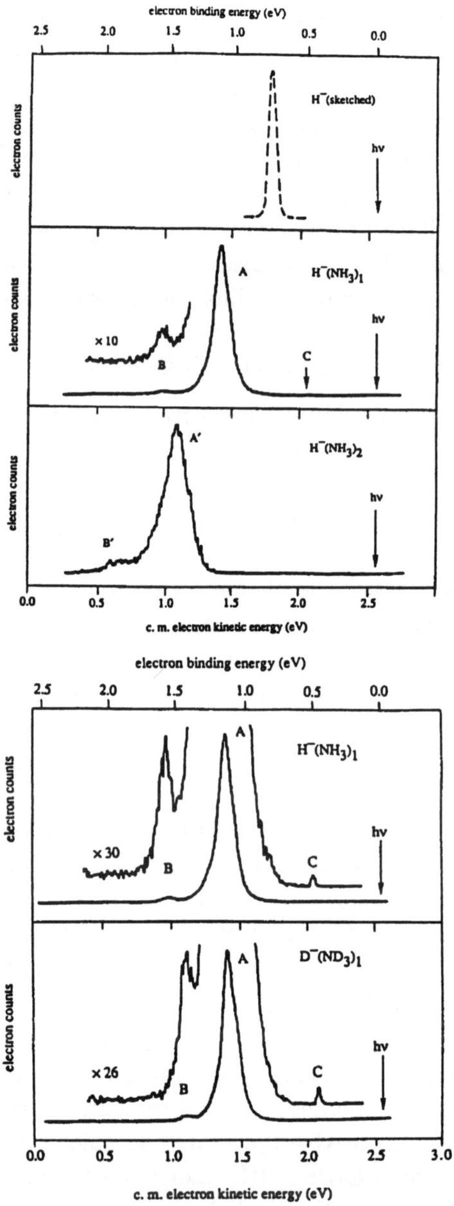

Fig. 20. Photoelectron spectra of H^-, $H^-(NH_3)$, and $H^-(NH_3)_2$ at the top and, at higher resolution for $H^-(NH_3)$ and $D^-(ND_3)$ at the bottom (taken from Ref. 30).

that whatever anion is detached in peak C, the neutral species produced has nearly the same geometry as the anion. Subsequent to the experiments in which peak C was identified, Vince Ortiz[31] carried out a series of *ab initio* theoretical calculations on an unusual isomer of NH_4^- in which the nuclear framework assumes a tetrahedral form, nearly identical to that found in the NH_4^+ cation. Ortiz interpreted the electronic nature of this isomer of the NH_4^- anion in terms of an NH_4^+ core about which a pair of electrons orbits in diffuse Rydberg-like orbitals, and his computed electron binding energy was quite close to the 0.4 eV value found in Bowen's experiments. Subsequently, other so-called double Rydberg anions[32] have been studied by other workers, but Bowen's experiments and Ortiz's calculations allowed the new species of NH_4^- to be discovered.

3. Autodetachment after Photon or Collisional Excitation

Having considered the situation in which a photon induces an *electronic transition* from the anion's HOMO to a continnum orbital relating to the neutral plus free-electron daughter, let us now consider what happens when the anion has its vibration/rotation degrees of freedom excited but does not undergo any immediate change in its electronic structure (i.e. its orbital occupancy). Figure 6 offers a pictorial representation of the kind of excitations that pertain to these cases.

It is possible for an electronically and geometrically stable anion to be promoted to an excited vibrational and/or rotational level by photon absorption or by collision with another species and for this excited anion to live as long or longer than a vibrational period. Subsequent to this excitation process, the anion can eject an electron by one of two processes. If its vibrational/rotational motions cause it to sample geometries at which the anion is no longer electronically stable (e.g. as in the $v = 2$ anion level shown in Fig. 3), an electron can detach whenever the molecular framework moves into such regions. The net rate of detachment will depend on (a) the frequency with which the anion's vibrations enter these regions of electronic instability (this, clearly, will be of the order of magnitude of the vibrational frequencies of the anion) and (b) the rate of electron detachment once the region is accessed. The latter rate will depend on whether the anion's instability is of the shape or Feshbach nature within the regions of instability. Shape resonance decay rates tend to be faster than those of Feshbach resonances with the former usually being comparable to or faster than vibrational periods. As a result, the rate limiting step (i.e. the rate that governs the ultimate rate of appearance of ejected electrons) is often the vibrational frequency for shape resonance cases and the Feshbach decay rate (attenuated by the fraction of time the anion spends in geometrical regions of instability) in these cases.

It is also possible for electron detachment to occur even if the anion and neutral energy surfaces do not cross (i.e. when the anion is not electronically unstable). If the anion's excess vibrational/rotational energy can be converted into electronic energy, an electron can be ejected with a concomitant loss of vibration/rotation energy. This kind of radiationless transition can be induced via so-called non-Born–Oppenheimer couplings as discussed in further detail below. An example of such a case is offered in Fig. 3. This particular enolate, if excited to its $v = 7, 8, 9, \ldots$ torsional vibrational level, posesses greater *total* (i.e. electronic plus vibrational) energy than the $v = 0$ level of the neutral daughter radical. However, at no geometry does the *electronic* energy of an anion cross that of the neutral, so purely electronic mechanisms for ejecting an electron are inoperative. On the other hand, if the anion were able to convert some of its excess torsion vibrational energy into electronic energy, electron ejection might occur. As discussed below and as expected, this intramolecular energy conversion will be most effective as the anion samples geometries where the anion–neutral electronic energy difference is small.

3.1. *Detachment through Shape or Feshbach Resonances*

A good example of detachment through a shape resonance state is offered by the CO_2^- species. The neutral CO_2 is linear in its $X^1\Sigma_g^+$ ground state, but the $X^2\Pi_u$ ground state of the anion, formed by adding an electron to the anti-bonding π_u orbital shown below in Fig. 21, has a bent equilibrium geometry and slightly elongated C–O bond lengths. The energies of CO_2 and CO_2^- and the C–O bond lengths as functions of the bending angle as obtained in recent work of Rod Bartlett's group[33] are shown in Fig. 22.

The formation of CO_2^- in a low vibrational level of the bending mode creates an anion with an equilibrium angle near 130°, where the C–O bond length is ca. 0.04 Å longer than in the neutral at this same angle. As this anion undergoes vibrational motion along its bending coordinate, it samples angles (e.g. greater than ca. 150°; see Fig. 22) where the anion energy

Fig. 21. The anti-bonding π_u orbital of CO_2 into which an electron is attached to form the ground state of CO_2^-.

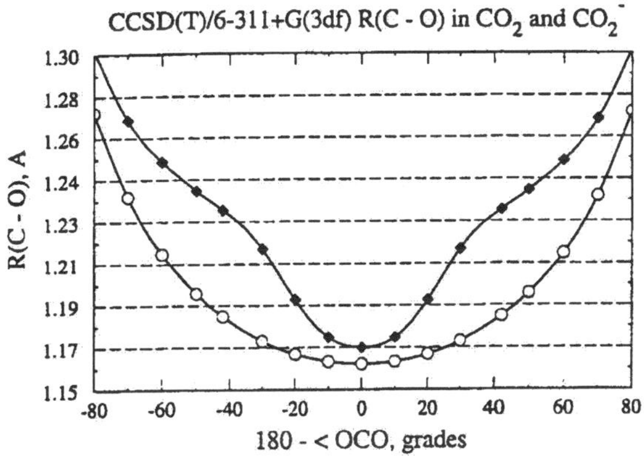

Fig. 22. Energies of ground state CO_2 (open circles) and CO_2^- (filled circles) and C–O bond lengths in the neutral (open circles) and anion (closed circles) as functions of the O–C–O bending angle (taken from Ref. 33).

lies above that of the neutral. Within these regions of anion instability, autodetachment can occur from the π_u orbital. Because the anion and neutral electronic configurations differ only in the occupancy of this single π_u orbital, which has predominantly d and p character, this decay occurs via a shape resonance and produces both d and p wave ejected electrons (n.b. we do not use the optical selection rules $L_{\text{ejected}} = L_{\text{HOMO}} \pm 1$, 0 in this case but $L_{\text{ejected}} = L_{\text{HOMO}}$ because no photon appears in the process).

Even when CO_2^- is formed by electron scattering from neutral CO_2, in which case the anion is created directly in a region (near linear geometry) where the anion curve lies above the neutral, the anion lives long enough to observe vibrational structure in a plot of the transmitted current of electrons vs electron kinetic energy such as is shown[34] in Fig. 23. In this electron transmission data, the spacings between the peaks (ca. 1113 cm^{-1})

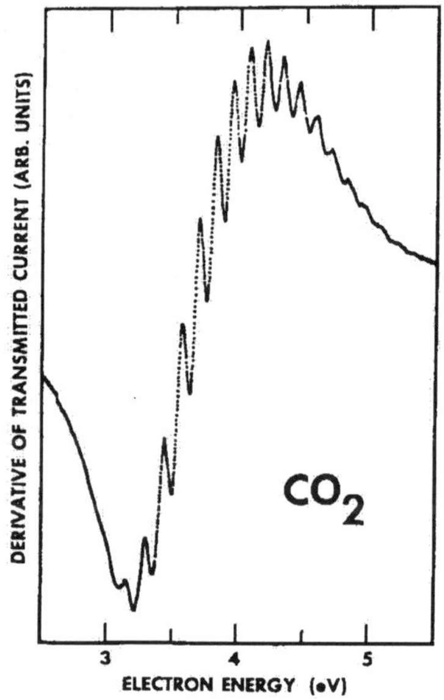

Fig. 23. Transmitted electron current vs kinetic energy of transmitted electron for scattering of electrons from CO_2 (taken from Ref. 34).

corresponds to the symmetric stretch vibrational mode. Structure belonging to the lower frequency bending mode is not seen because the lifetime of this shape resonance state (ca. 2×10^{-14} s) is shorter than the period for this slow bending motion to occur.

3.2. *Detachment through Non-Born–Oppenheimer Coupling*

3.2.1. *What are Non-Born–Oppenheimer Couplings?*

Within the Born–Oppenheimer approximation, the electronic Schrödinger equation

$$h_e(r|Q)\psi_k(r|Q) = E_k(Q)\psi_k(r|Q)$$

is solved to obtain electronic wavefunctions $\psi_k(r|Q)$, which are functions of the molecule's electronic coordinates (collectively denoted r) and atomic coordinates (denoted Q), and the corresponding electronic energies $E_k(Q)$, which are functions of the Q coordinates. The electronic Hamiltonian

$$h_e(r|Q) = \Sigma_i\{-\hbar^2/2m_e\nabla_i^2 + 1/2\Sigma_{j\neq i}e^2/r_{i,j} - \Sigma_a Z_a e^2/r_{i,a}\}$$
$$+ 1/2\Sigma_{a\neq b}Z_a Z_b e^2/R_{a,b}$$

contains, respectively, the sum of the kinetic energies of the electrons, the electron–electron repulsion, the electron–nuclear Coulomb attraction, and the nuclear–nuclear repulsion energy.

Given the solutions to the electronic Schrödinger equation, the solutions of the full Schrödinger equation (i.e. the equation in which all nuclei and electrons are moving)

$$H(r|Q)\Psi(r,Q) = E\Psi(r,Q)$$

are expressed as sums over the (complete set of functions of the electronic coordinates r) electronic functions $\{\psi_k(r|Q)\}$

$$\Psi(r,Q) = \Sigma_k \psi_k(r|Q)\chi_k(Q)$$

with the $\chi_k(Q)$ carrying the remaining Q-dependence. When substituted into the full Schrödinger equation

$$(H - E)\Sigma_k \psi_k(r|Q)\chi_k(Q) = 0$$

then using the fact that the full Hamiltonian H is h_e plus the kinetic energy operator for nuclear motion T

$$H = h_e + T = h_e + \Sigma_a(-\hbar^2/2m_a \nabla_a^2)\,,$$

premultiplying the above Schrödinger equation by ψ_n and integrating over the electronic coordinates gives the set of coupled equations that need to be solved for the $\{\chi_k\}$:

$$\begin{aligned}
\Sigma_k & \int \psi_n^*(r|Q)\{h_e + T - E\}\psi_k(r|Q)\chi_k(Q)\,dr \\
& = \{E_n(Q) - E\}\chi_n(Q) + T\chi_n(Q) \\
& \quad + \Sigma_a\Sigma_k\left\{ \int \psi_n^*(r|Q)(-i\hbar\partial\psi_k/\partial R_a)(-i\hbar\partial\chi_k/\partial R_a)/m_a\,dr \right. \\
& \quad + \left. \int \psi_n^*(r|Q)(-\hbar^2\partial^2\psi_k/\partial R_a^2)/2m_a\,dr\,\chi_k \right\} = 0\,.
\end{aligned}$$

The expression

$$\{E_n(Q) - E\}\chi_n(Q) + T\chi_n(Q) = 0$$

is the equation governing the nuclear motion functions $\{\chi_{k,L}(Q)\}$ in the absence of the *non-Born–Oppenheimer* (*non-BO*) *coupling terms* (i.e. the two terms involving $\partial\psi_k/\partial R_a$ and $\partial^2\psi_k/\partial R_a^2$). Within this model, the vibration-rotation functions $\{\chi_k(Q)\}$ of each specific electronic state labeled k are found by solving the vibration-rotation Schrödinger equation

$$\{T + E_k(Q)\}\chi_{k,L}(Q) = \varepsilon_{k,L}\chi_{k,L}(Q)\,.$$

There are a complete set of functions of Q (i.e. the $\{\chi_{k,L}\}$ for *each* electronic state k.

The two non-BO terms are treated as perturbations (not externally applied, but arising as imperfections within this model of molecular structure) that can induce transitions between unperturbed states each of which is taken to be a specific Born–Oppenheimer product state:

$$\Psi_{k,L}(r,Q) \cong \psi_k(r|Q)\chi_{k,L}(Q)\,.$$

It is reasonably well established that the non-BO coupling term involving second derivatives of the electronic wavefunction contributes less to the coupling than do the terms $(-i\hbar\partial\psi_k/\partial R_a)(-i\hbar\partial\chi_k/\partial R_a)/m_a$ having first derivatives of the electronic and vibration-rotation functions. Hence, it is only the latter terms that will be discussed further.

With this background, the rate R (sec^{-1}) of transition from a Born–Oppenheimer initial state $\Psi_i = \psi_i\chi_i$ to a final state $\Psi_f = \psi_f\chi_f$ is given, via first-order perturbation theory, as[16]:

$$R = (2\pi/\hbar)\int |\langle\chi_i|\langle\psi_i|P|\psi_f\rangle(P/\mu)\chi_f\rangle|^2\delta(\varepsilon_f + E - \varepsilon_i)\rho(E)\,dE\,.$$

Here, $\varepsilon_{i,f}$ are the vibration-rotation energies of the initial (anion) and final (neutral) states, and E denotes the kinetic energy carried away by the ejected electron (e.g. the initial state corresponds to an anion and the final state to a neutral molecule plus an ejected electron). The density ρ of translational energy states of the ejected electron is related to the kinetic energy by $\rho(E) = 4\pi m_e L^3(2m_eE)^{1/2}/\hbar^2$. We use the shorthand notation involving $P\psi P\chi/\mu$ to symbolize the action of the multidimensional derivative operators arising in the non-BO couplings:

$$(P\psi_f)(P/\mu\chi_f) = \Sigma_a(-i\hbar\partial\psi_f/\partial R_a)(-i\hbar\partial\chi_f/\partial R_a)/m_a\,,$$

where R_a runs over the cartesian coordinates (X_a, Y_a, Z_a) of the ath atom whose mass is m_a.

3.2.2. *The Nature of the Electronic Matrix Elements*

The integrals over the anion and neutral plus free-electron electronic states

$$m_{i,f} = \langle \psi_f | P | \psi_i \rangle$$

are known[16] to be large in magnitude only under special circumstances:

1. The orbital of the anion from which an electron is ejected to form the state ψ_f of the neutral (usually the anion's highest occupied molecular orbital (HOMO)) must *be strongly modulated* or affected by movement of the molecule in one or more directions (Q). That is $\partial\psi_i/\partial Q$, which appears in $P\psi_i$, must be significant or the above integral will be small.
2. The state-to-state energy gap $\varepsilon_i - \varepsilon_f$, which is equal to the energy E of the ejected electron, must not be too large; otherwise, the oscillations in the ejected electron's wavefunction ψ_f will be so rapid as to render overlap with $\partial\psi_i/\partial Q$ negligible, again making the above integral small.

Moreover, symmetry can cause $m_{i,f} = \langle \psi_f | P | \psi_i \rangle$ to vanish. In particular, if the direct product of the symmetry of ψ_i and of $\partial/\partial Q$ do not match that of ψ_f, them $m_{i,f}$ will vanish (i.e. if direct product of the HOMO's symmetry and the symmetry of the vibration or rotation motion from which energy is transferred determines the symmetry of the ejected electron's continuum orbital).

The derivatives (i.e. the dynamic responses) of the anion's orbitals to nuclear motions $\partial\psi_i/\partial Q$ arise from two sources:

1. The orbital's LCAO-MO coefficients depend on the positions of the atoms (or, equivalently, on the anion's bond lengths and internal angles). For example, the π^* orbital of an olefin anion that contains the "extra" electron is affected by stretching or twisting the C–C bond involving this orbital because the LCAO-MO coefficients depend on the bond length and twist angle. As the bond stretches or twists, the π^* orbital's LCAO-MO coefficients vary, as a result of which the orbital's energy, radial extent, and other properties also vary.
2. The atomic orbitals (AO) themselves dynamically respond to the motions of the atomic centers. For example, vibration of the $X^2\Pi$ NH^- anion's N–H bond induces d_π character into the $2p_\pi$ orbital containing the extra

$d\Psi/dR = (\Psi(R+\delta) - \Psi(R))/\delta$ causes $2p_x$ to acquire d character

$d\Psi/d\theta = (\Psi(\theta+\delta) - \Psi(\theta))/\delta$ causes $2p_x$ to acquire $2p_z$ character

Fig. 24. Responses of the NH^- $2p_\pi$ orbital to vibrational motion (left) and to rotational motion (right).

electron as shown in Fig. 24. Alternatively, rotation of this anion's N–H bond axis causes the $2p_\pi$ HOMO to acquire some $2p_\sigma$ character (see Fig. 24 again).

In summary, for non-BO coupling to be significant[16] the *anion's HOMO must be strongly modulated* by a motion (vibration or rotation) of the molecule's nuclear framework and *the state-to-state energy gap must not be too large* as to render the HOMO-to-continuum-orbital overlap insignificant. For the HOMO to be strongly modulated, it is helpful if *the anion and neutral energy surfaces approach closely* (n.b. this is not the same as requiring that the state-to-state energy gap $\varepsilon_i - \varepsilon_f$ be small) at some accessible geometries.

Before considering a few specific examples of how non-BO couplings can cause electron ejection, it is important to note that there are qualitative differences between the first-order perturbation theory rate expression shown above for the non-BO case and the analogous expression for the rate of photon absorption:

$$R = (2\pi/\hbar)|\langle \chi_i \langle \psi_i | V | \psi_f \rangle \chi_f \rangle|^2 \delta(\varepsilon_f - \varepsilon_i - \hbar\omega)\,.$$

In the electronic transition spectroscopy case, the perturbation (i.e. the photon's electromagnetic potential V) appears explicitly only in the $\mu_{i,f} = \langle \psi_i | V | \psi_f \rangle$ electronic matrix element because this external field is purely an electronic operator. In contrast, in the non-BO case, the perturbation involves a *product of momentum operators*, one acting on the electronic wavefunction ($P\psi_i$) and the second ($P\chi_i/\mu$) acting on the vibration/rotation wavefunction because the non-BO perturbation involves an explicit exchange of energy and momentum between the electrons and the nuclei rather than an absorption of energy and momentum from a photon. As a result, one has matrix elements of the form $\langle\chi_i|\langle\psi_i|P|\psi_f\rangle(P/\mu)\chi_f\rangle$ in the non-BO case where one finds $\langle\chi_i|\langle\psi_i|V|\psi_f\rangle|\chi_f\rangle$ in the spectroscopy case. A primary difference is that derivatives of the vibration/rotation functions appear in the former case (in $(P/\mu)\chi$) where only χ appears in the latter.

3.2.3. *A Few Examples*

3.2.3.1. *Rotational Detachment in NH⁻*

Depicted in Fig. 1 are the anion and neutral potential curves that are qualitatively illustrative of the $X^2\Pi$ NH^- case mentioned earlier. In this anion, the HOMO is a non-bonding $2p_\pi$ orbital localized almost entirely on the N atom. As such, its LCAO-MO coefficients are not strongly affected by vibration of the N–H bond (because it is a non-bonding orbital). Moreover, the anion and neutral surfaces have nearly identical R_e and ω_e values, and similar D_e values, as a result of which these two surfaces are nearly parallel to one another over a wide range of internuclear distances and are separated by ca. 0.4 eV or more than 3000 cm^{-1} at their minima.

It has been seen experimentally[10] that excitation of NH^- to the low rotational states of the $v = 1$ vibrational level (which lies above $v = 0$ NH of the neutral and thus has enough energy to eject the electron) results in very slow (e.g. ca. 10^8 sec^{-1}) electron ejection, corresponding to ca. one million vibrational periods before detachment occurs. However, excitation to high rotational levels (e.g. $N' = 40$) of $v = 1$ produces much more rapid

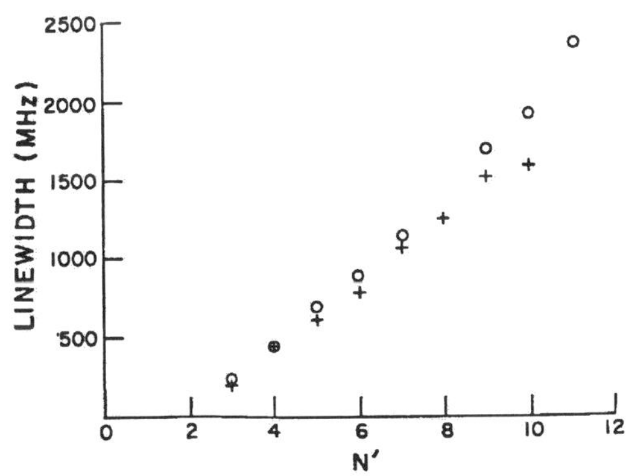

Fig. 25. Electron detachment rates (in MHz) as functions of the anion rotational level N' for transitions between NH^- $(v = 1, N')$ to NH $(v = 0$, all accessible $N)$ + e^- (taken from Ref. 10).

electron ejection (ca. 10^9–10^{10} sec^{-1}). These data, some of which are shown in Fig. 25, have been interpreted as saying that vibrational coupling is weak (i.e. $\partial\psi_i/\partial R$ is small) because of the non-bonding nature of the $2p_\pi$ MO, while rotational coupling becomes significant (i.e. $\partial\psi_i/\partial\theta$ is large) for high N'. A theoretical interpretation of these experimental findings based on the non-BO coupling theory detailed above has also been put forth.[10]

3.2.3.2. *Vibrational Detachment in Enolates*

Figure 2 shows anion and neutral potential curves, as functions of the "twist" angle of the H_2C–C bond in a typical enolate anion such as acetaldehyde enolate H_2CCHO^-. Angles near $\theta = 0$ correspond to geometries where the p_π orbital of the H_2C moiety is delocalized over the two p_π orbitals of the neighboring C and O atoms, thus forming a delocalized π HOMO. At angles near $\theta = 90°$, the p_π orbital of the H_2C group is no longer stabilizated by delocalization; so the HOMO's energy is much higher (as a result of which the anion-neutral surfaces *approach* closely). In this case, excitation of, for example, $v = 7$ in the H_2C–C torsional mode of the anion might be expected to produce electron ejection because $v = 7$ of the anion lies above $v = 0$ of the neutral. However, over the range of θ values accessible to both the $v = 7$ vibrational function of the anion and the $v = 0$ function of the neutral, the anion-neutral energy surface gap is quite large (i.e. $E_{ff}(Q) - E_i(Q)$ is large even though $\varepsilon_i - \varepsilon_f$ is small). In contrast, excitation of $v = 9$ of the anion could produce more rapid electron ejection (to $v = 2$ of the neutral, but not to $v = 0$ of the neutral) because, for the $v = 9 \to v = 2$ transition, there are angles accessed by both $v = 9$ anion and $v = 2$ neutral vibrational functions for which $E_{ff}(Q) - E_i(Q)$ is small and changing; moreover, the state-to-state gap $\varepsilon_i - \varepsilon_f$ is also small in this case.

For the particular case of the acetaldehyde enolate anion, the rates of electron ejection induced by non-BO coupling has been computed[11] for the $v = 9, 10$ and 11 levels decaying to the $v' = 0, 1, 2, 3,$ and 4 levels of the neutral and found to lie between 3×10^9 s^{-1} and 7×10^{10} s^{-1}. Direct experimental probes of these state-to-state rates have not been made, but infrared multiphoton excitation of the parent enolate has been found[35,36] to produce electron ejection. These experimental findings are consistent[11] with the computed rates if one assumes that the IR excitation energy is quickly distributed among all internal vibrational modes of the anion so the probability of observing the anion in a particular torsional level v is given by the statistical density of states probability of this state.

3.2.3.3. *Collisional Detachment in OH⁻*

A third example is offered by the O^- (^3P) + H(^2S) $\rightarrow$ OH($X^2\Pi$) + e^- associative electron detachment process. The pertinent anion and neutral potential energy curves[37] are shown below in Fig. 26. In this process, collisions between gas-phase O^- ions and H atoms take place on the $^1\Sigma$ and $^1\Pi$ curves (as well as on the $^3\Sigma$ and $^3\Pi$ curves which are not shown in the figure). Those collisions that begin on the $^1\Sigma$ curve have some probability of (a) coupling the H 1s electron with the O $2p_\sigma$ electron to form an intact σ bond after which (b) non-BO coupling can eject a $2p_\pi$ electron to produce $X^2\Pi$ OH + e^-. In this situation, the non-BO coupling occurs between an initial $^1\Sigma$ electronic state and a continuum collisional state (rather than an initial bound vibrational state as in the earlier examples) and a final $^2\Pi$+e^- electronic state and a bound vibrational state of OH. Because the electronic energy gap between OH^- and OH is very large (ca. 1.8 eV), the rates of these non-BO induced electron ejections are found to be extremely slow for this case (e.g. rates of the order 10^4 s^{-1} for thermal collision energies).

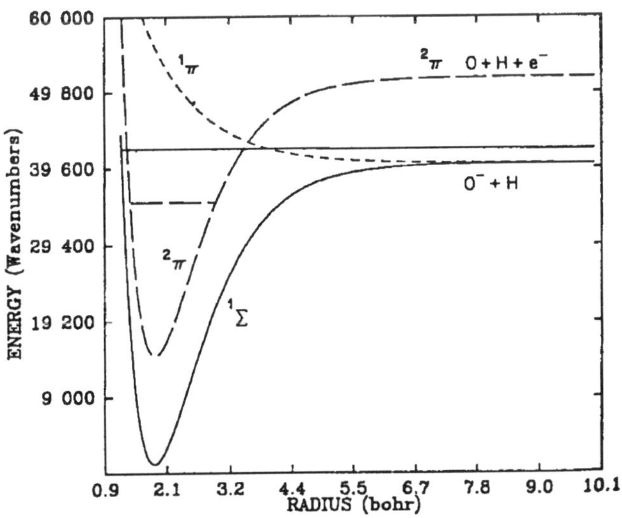

Fig. 26. Anion $^1\Sigma$ and $^1\Pi$ and neutral $^2\Pi$ energy curves for OH^- and OH, respectively (taken from Ref. 37).

3.2.3.4. *Rotational Detachment in Dipole–Bound Anions*

The neutral radicals H_2C–CN and H_2C–CHO both have large dipole moments (ca. 3.71 Debyes for the former and 3.05 Debye for the latter) and they each have an unfilled valence orbital. Therefore, both of these species can attach an electron to form:

1. a valence-bound anion state (H_2CCN^- and H_2CCHO^-) having large electron binding energy (12 500 cm^{-1} for the former and $> 14\,700$ cm^{-1} in the latter) as well as
2. an (excited) dipole-bound state in which the electron is bound very weakly to the underlying neutral radical core (< 66 cm^{-1} in the former and ca. 5 cm^{-1} in the latter).

In dipole-bound anions,[38] the "extra" electron is attracted to the underlying neutral molecule core by a variety of forces, but the potential arising from the charge-permanent dipole interaction, which has the form $-\mu e \cos\theta/r^2$, has a dominant influence and is the longest-range potential. For molecules with dipole moments in the range of 2–10 Debyes, the extra

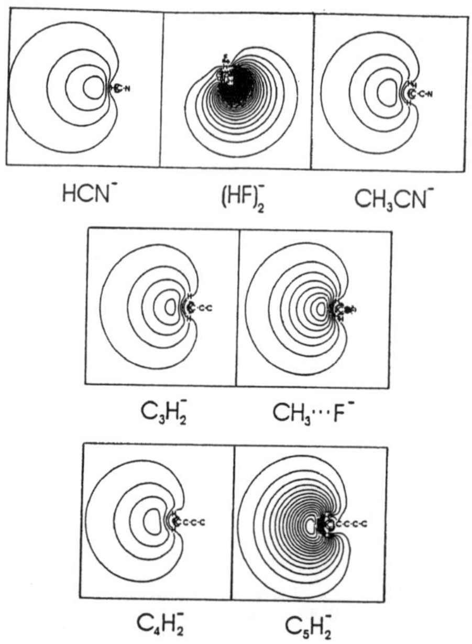

Fig. 27. Molecular structures and contour graphs of dipole-bound orbitals for several dipole-bound anions (taken from the Gutowski *et al.* article mentioned in Ref. 38).

electron is bound by a few to a few hundred cm^{-1} in such states. As a result, the orbital in which the dipole-bound electron resides is very diffuse. Several examples of these orbitals are shown in Fig. 27 in the form of contour plots overlaying the molecular structure shown to scale.

Most experiments[39,40] that probe the non-BO induced electron detachment within the dipole-bound states of these anions begin with the anions in their valence-bound electronic states and use a laser to excite a particular rotational transition from the J, M and K level of the valence-bound state to a well specified level (determined by selection rules) J', M' and K' of the dipole-bound state. The quantum numbers J, M and K refer, as usual, to the total angular momentum, its projection along a laboratory-fixed axis, and its projection along the (near) symmetry axis of the molecule.

As suggested in Fig. 28 for the H_2CCN^- case, subsequent to formation of the dipole-bound anion in a well specified rotational state, autodetachment is observed to occur producing $H_2CCN + e^-$.

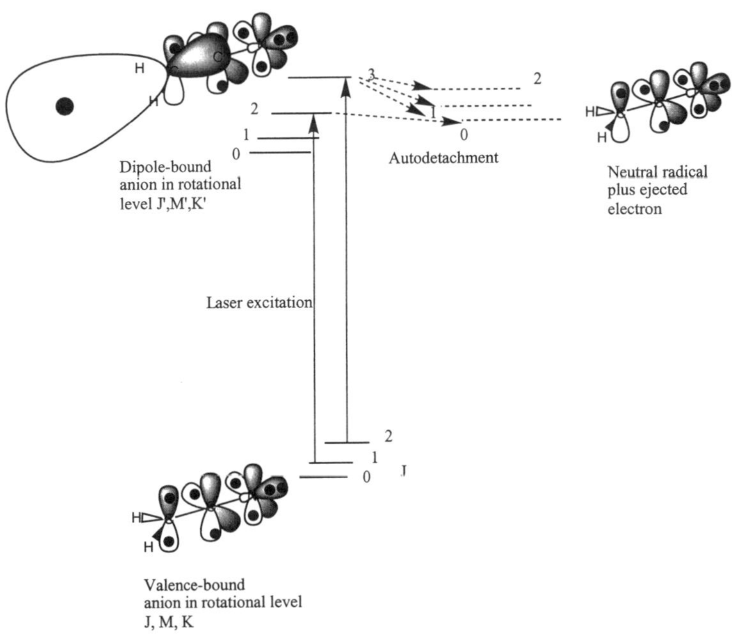

Fig. 28. Depiction of how laser excitation of the valence-bound anion H_2CCN^- to a particular rotational level of the dipole-bound state prepares the latter for subsequent autodetachment to produce $H_2CCN + e^-$.

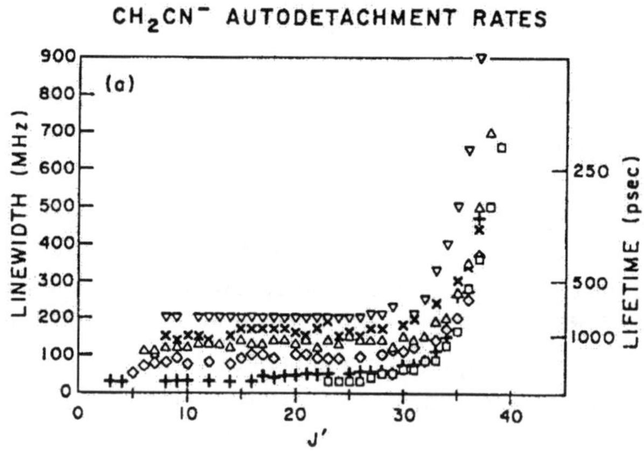

Fig. 29. Decay rates (MHz) of rotational levels of dipole-bound H_2CCN^- vs rotational quantum number J'. The various sets of data correspond to different K' values with larger K' giving only slightly higher decay rates (taken from Ref. 40).

The rate of electron ejection, as inferred by the linewidths of the photodetachment spectrum, is found to depend very strongly on the rotational energy content of the dipole-bound anion as shown below in Fig. 29. More specifically in the above example of H_2CCN^-, the rate is found to grow rapidly with J' beyond ca. $J' = 30$, but in a manner that is relatively insensitive to the component K' of rotational angular momentum along the (near) symmetry axis. This suggests that the rotational energy that does not move the dipole moment vector (i.e. spinning of the molecule about its (near) symmetry axis, which relates to the K' quantum number) is ineffective at electron detachment. Only overall tumbling motion of the molecule, and thus of the dipole moment vector, seems to cause coupling between rotational and electronic energies.

A theoretical model describing the dynamics by which rotational energy and angular momentum of the dipole-bound anion is coverted into electronic energy eventually leading to electron ejection has been presented by Clary[41] and applied to both H_2CCN^- and H_2CCHO^- with good success. In this model, the rate of electron ejection is related to the rate at which the electron moves through the classically forbidden region of an effective radial potential $W(R)$ from small R (where the electron is localized in the dipole-bound orbital and the molecule has rotational quantum number J) to larger and larger R (where the molecule loses angular momentum as the electron

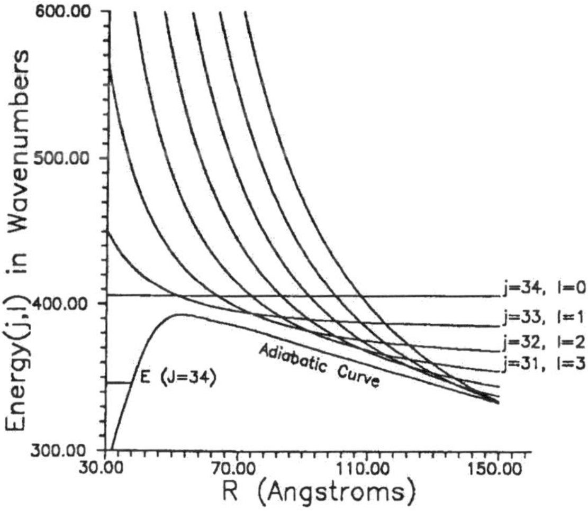

Fig. 30. Effective radial potential (smooth lower curve) arising for $J = 34$ in the case of dipole-bound H_2CCN^-. Also shown are the various j, L components that form the smooth adiabiatic radial potential (taken from the Simons article in Ref. 41).

gains energy and angular momentum), eventually producing the ejected electron with angular momentum L and the neutral molecule with $j < J$. An example of such a radial potential and the J, j, and L values that pertain to $J = 34$ is shown in Fig. 30.

4. Summary

In this chapter, I have tried to both give an overview of the theoretical underpinnings that allow experimental data (e.g. photodetachment or photoelectron rates or cross-sections or electron detachment rates for excited vibration/rotation/translation states of electronically stable anions) to be intepreted in terms of the orbitals and nuclear motion states of the parent anion and daughter neutral species. By defining the various kinds of stability that an anion may possess, I tried to make it clear what kind of species were to be included in (and what kind are excluded from) my discussion. By separating direct photon-induced detachment from other processes, I tried to make distinct the discussions of inherently electronic transitions (i.e. those in which the photon changes the anion's electronic orbital occupancy) and electron detachment induced by nuclear motions (either causing the anion to sample geometries where the anion undergoes spontaneous electron loss or inducing detachment via non-BO internal energy transfer). Finally, I tried to integrate into the discussion examples, taken from the experiences of several of the leading experimental groups who study anion detachment, of the various electron detachment processes that I have treated in this chapter. By so doing, I hoped to make clear what one learns from such experimental data and why carrying out experimental and theoretical research on molecular anions and their diverse electron detachment mechanisms is challenging, exciting to puruse, and filled with important new insights yet to be gained.

Acknowledgments

This work has been supported by NSF Grant CHE-9618904 and by proceeds of the Henry Eyring Endowed Chair.

References

1. G. J. Schultz, *Rev. Mod. Phys.* **45**, 423 (1973).
2. R. S. Berry, *Chem. Rev.* **69**, 533 (1969).
3. K. D. Jordan and P. D. Burrow, *Acc. Chem. Res.* **11**, 341 (1978).
4. J. Kalcher, Annual Reports, *The Royal Society of Chemistry*, Section **C93**, 147 (1995).
5. J. Simons and K. D. Jordan, *Chem. Rev.* **87**, 535 (1987).
6. M. K. Scheller, R. N. Compton and L. S. Cederbaum, *Science* **270**, 1160 (1995).
7. B. K. Janousek and J. I. Brauman, In *Gas Phase Ion Chemistry* **2**, Ed. M. T. Bowers (Academic Press, New York, 1979).
8. H. Hotop and W. C. Lineberger, *J. Phys. Chem. Ref. Data* **14**, 731 (1985).
9. R. N. Compton, In *Negative Ions*, Ed. V. A. Esaulov (Cambridge Univ. Press, 1995).
10. The experimental findings on this anion are detailed in D. M. Neumark, K. R. Lykke, T. Andersen and W. C. Lineberger, *J. Chem. Phys.* **83**, 4364 (1985). Theoretical analysis of this same species appears in G. Chalasinski, R. A. Kendall, H. Taylor and J. Simons, *J. Phys. Chem.* **92**, 3086 (1988).
11. An experimental study on enolates is give in A. H. Zimmerman, K. J. Reed and J. I. Brauman, *J. Am. Chem. Soc.* **99**, 7203 (1977). A theoretical study pertinent to the acetaldehyde enolate system appears in D. O'Neal and J. Simons, *J. Phys. Chem.* **93**, 58 (1989).
12. This particular dianion was examined in H. G. Weikert, L. S. Cederbaum, F. Tarantelli and A. I. Boldyrev, *Z. Phys.* **D18**, 299 (1991). This energy profile was produced in A. I. Boldyrev, M. Gutowski and J Simons, *Acc. Chem. Res.* **29**, 497 (1996).

13. See the chapter in this book by Nora Berrah.
14. J. I. Steinfeld, *Molecules and Radiation* (MIT Press, 1981).
15. P. F. Bernath, *Spectra of Atoms and Molecules* (Oxford Univ. Press, New York, 1995).
16. The theory of such radiationless transitions has been treated (for autoionization of Rydberg states of neutral molecules) in R. S. Berry, *J. Chem. Phys.* **45**, 1228 (1966). This theory has been extended to non-BO induced autodetachment in molecular anions in J. Simons, *J. Am. Chem. Soc.* **103**, 3971 (1981).
17. Good discussions of the differences between shape and Feshbach resonances are contained in Ref. 1 and in H. S. W. Massey, *Negative Ions, 3rd ed.* (Cambridge Univ. Press, New York, 1976).
18. This development follows that found in several text books on molecular spectroscopy including: *Quantum Mechanics in Chemistry*, J. Simons and J. Nichols (Oxford University Press, 1997).
19. K. J. Reed, A. H. Zimmerman, H. C. Andersen and J. I. Brauman, *J. Chem. Phys.* **64**, 1368 (1976).
20. E. P. Wigner, *Phys. Rev.* **73**, 1002 (1948).
21. P. C. Engelking and D. R. Herrick, *Phys. Rev.* **A29**, 2425 (1984).
22. A good treatment of such corrections and of the origins of the Wigner law itself can be found in T. F. O'Malley, *Phys. Rev.* **137**, A1668 (1965).
23. Other good sources of further information about corrections to the Wigner law, to the inclusion of long-range electron-dipole interactions, and to the use of symmetry in polyatomic anions include: R. D. Mead, K. R. Lykke and W. C. Lineberger, *Electronic and Atomic Collisions*, Eds. J. Eichler, I. V. Hertel and N. Stolterfoht (Elsevier, Science, Publishers) see also Ref. 22; D. R. Herrick and P. C. Engleking, *Phys. Rev.* **A4**, 2421 (1984) p. 721.
24. The photoelectron spectrum of Fe_2^- is examined in D. G. Leopold and W. C. Lineberger, *J. Chem. Phys.* **85**, 51 (1986). That of Cu_2^- is covered in D. G. J. Ho and W. C. Lineberger, *J. Chem. Phys.* **86**, 1715 (1987).
25. L. M. Branscomb, *Phys. Rev.* **148**, 11 (1966).
26. S. Burch, S. J. Smith and L. M. Branscomb, *Phys. Rev.* **112**, 171 (1958).
27. S. E. Bradforth, D. W. Arnold, D. M. Neumark and D. E. Manolopoulos, *J. Chem. Phys.* **99**, 6345 (1993).
28. E. de Beer, E. H. Kim and D. M. Neumark, *J. Phys. Chem.* **99**, 13627 (1995).
29. K. M. Ervin, J. Ho and W. C. Lineberger, *J. Chem. Phys.* **91**, 5974 (1989).
30. K. H. Bowen and J. G. Eaton, in *Proceedings of the International Workshop on the Structure of Small Molecules and Ions*, Jerusalem, Israel; S. T. Arnold, J. G. Eaton, D. Patel-Misra, H. W. Sarkas and K. H. Bowen, in *Ion and Cluster Ion Spectroscopy and Structure*, Ed. J. P. Maier (Elsevier, New York, 1989).
31. J. V. Ortiz, *J. Chem. Phys.* **87**, 3557 (1987).
32. J. Simons and M. Gutowski, *Chem. Rev.* **91**, 669 (1991).
33. G. L. Gutsev, R. J. Bartlett and R. N Compton, *J. Chem. Phys.* **108**, 6756 (1998).

34. L. Sanche and G. J. Schulz, *J. Chem. Phys.* **58**, 481 (1973).
35. C. A. Wight and J. L. Beauchamp, *J. Am. Chem. Soc.* **103**, 6501 (1981).
36. R. N. Rosenfeld, J. M. Jasinski and J. I. Brauman, *J. Chem. Phys.* **71**, 1030 (1979).
37. P. K. Acharya, R. A. Kendall and J. Simons, *J. Chem. Phys.* **83**, 3888 (1985).
38. A good overview of recent theoretical work on the electronic structures of dipole-bound anions is given in M. Gutowski, K. D. Jordan and P. Skurski, *J. Phys. Chem.* **A102**, 2624 (1998). An earlier overview is given in K. D. Jordan, *Acc. Chem. Res.* **12**, 36 (1979).
39. R. D. Mead, K. R. Lykke, W. C. Lineberger, J. I. Marks and J. I. Brauman, *J. Chem. Phys.* **81**, 4883 (1984).
40. K. R. Lykke, D. M. Neumark, T. Andersen, V. J. Trapa and W. C. Lineberger, *J. Chem. Phys.* **87**, 6842 (1987).
41. D. C. Clary, *J. Phys. Chem.* **92**, 3173 (1988). Also, see J. Simons, *J. Chem. Phys.* **91**, 6858 (1989).

CHAPTER 18

COMPETITION BETWEEN AUTOIONIZATION AND PREDISSOCIATION IN MOLECULAR RYDBERG STATES

S. T. Pratt

Argonne National Laboratory, Argonne, IL 60439 USA

Contents

1. Introduction

The study of the decay processes of highly excited, gas-phase molecules can provide considerable insight into the mechanisms for intramolecular energy redistribution, and, when the excited molecules are viewed as activated collision complexes, into the mechanisms for bimolecular chemical reactions.[1–4] The focus of the present review is on the decay processes of molecular Rydberg states excited above the first ionization threshold of the molecule. In most instances, this threshold is also above at least one, and often several, of the dissociation thresholds of the molecule. As a result, both ionization and dissociation can be important. The types of resonances discussed here can be thought of as state-selected activated complexes for a large number of different types of reactions. These resonances are particularly relevant to dissociative recombination, i.e. $AB^+ + e^- \rightarrow A + B$, where the neutral products can be in their ground or excited states.[5] The study of the decay of such resonances may ultimately provide insight into the internal state dependence of these reactions.

In general, ionization can result in the population of a number of different rotational, vibrational, and electronic states of the ion, and dissociation can result in the population of a number of different states in the neutral products. Thus, in addition to the overall branching ratio between ionization and dissociation, the product-state distributions for ionization and dissociation are of interest because they provide deeper insight into the molecular decay mechanisms. These decay mechanisms can be defined still more precisely by the angular distributions, velocity distributions, and alignment of the ionization and dissociation fragments.

With all these potential decay channels, it might naively be expected that the photoabsorption, photoionization, and photodissociation spectra of molecules would show nothing but diffuse spectra above the first ionization threshold. However, this is not usually the case, and the photoabsorption spectra of a large fraction of molecules show considerable resonance structure in this region.[1–4,6] This structure results from the excitation of Rydberg series converging to excited rotational, vibrational, and electronic states of the molecular ion. As shown schematically in Fig. 1, although the total energy of the Rydberg states may be well above the lowest ionization and dissociation thresholds, the internal energy is initially distributed among the rotational, vibrational, and electronic degrees of freedom of the molecule such that the amount in any one degree of freedom is insufficient to result in the dissociation or ionization of the molecule. Thus, for the molecule to dissociate, energy must be transferred from the electronic degrees of freedom into the vibrational degrees of freedom and ultimately into translation along the reaction coordinate. For the molecule to ionize, energy must be converted from the rotational and vibrational degrees of

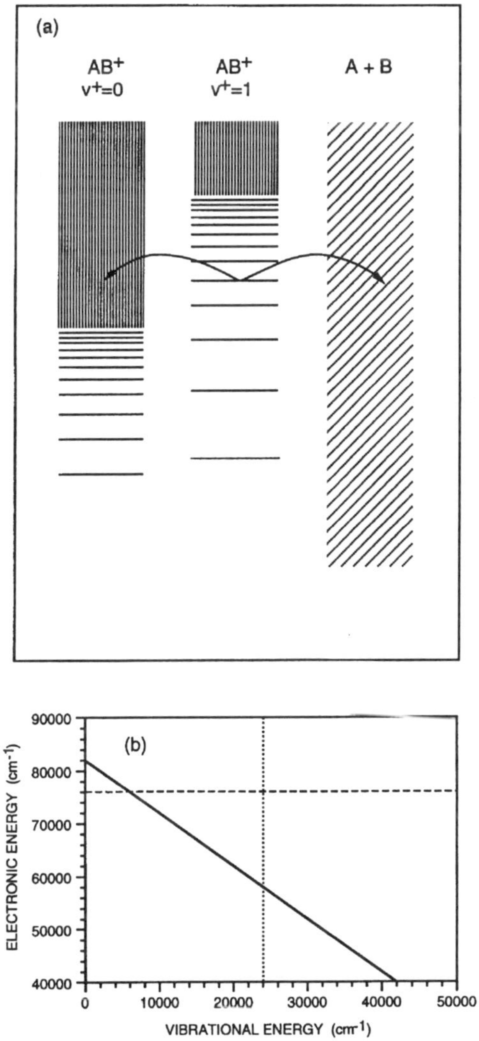

Fig. 1. (a) A schematic energy level diagram for autoionization and predissociation of a Rydberg series converging to a vibrationally excited state of the ion. (b) A schematic diagram of the energy distribution in the electronic and vibrational degrees of freedom for a molecule with an ionization energy of 76 000 cm^{-1}, a dissociation energy of 24 000 cm^{-1}, and 82 000 cm^{-1} total energy. The solid line has constant total energy. By transferring energy between the electronic and vibrational degrees of freedom, the molecule can move along this line. Points on the line above the dashed line are in the ionization continuum, and points to the right of the dotted line are in the dissociation continuum.

freedom into the electronic energy of the excited electron. For doubly excited electronic states, it is also possible that while the total electronic energy is above the ionization threshold, it is shared between two (or more) electrons; ionization in this case requires the transfer of energy between the electronic modes of the molecule. The analogous situation for highly vibrationally excited molecules, in which the total energy in any one mode must exceed the dissociation threshold, is well-known in the study of unimolecular reaction dynamics.[7]

Because the decay of the highly excited Rydberg states requires the conversion/transfer of energy among internal degrees of freedom of the molecule, the study of the decay processes can provide insight into the mechanisms of these conversion processes. This can be particularly interesting when two or more decay channels are present, because it is then possible to determine the effect of different factors on the relative rates of the two decay processes. Such situations are the emphasis of the present review. In particular, the dependence of the decay processes on the principal quantum number, n, the orbital angular momentum of the Rydberg electron, ℓ, its projection of the molecular frame, λ, the vibrational quantum numbers, v_i, and the rotational quantum numbers, J and K_i, can be quite dramatic. Although this is not too surprising in diatomic molecules or in the "small molecule" limit, the general expectation is that such effects are less important in larger molecules where statistical behavior usually dominates. Even so, there is considerable evidence for mode-specific behavior in the decay of the Rydberg states of larger polyatomic molecules, and examples of this evidence are discussed in the present review.

An isolated molecule in a Rydberg state above the ionization and dissociation thresholds can decay in one of three manners, corresponding to fluorescence, autoionization, and predissociation. The dependence of these processes on the principal quantum number is of particular interest. One of the most useful aspects of Rydberg series is that, in the absence of perturbations, the wavefunctions at small r are, aside from a normalization factor, essentially independent of n.[6,8] For a hydrogenic Rydberg state the normalization factor is $1/n^{*3/2}$, and this n dependence is transferred to the matrix elements for the three decay processes. As a result, in the absence of perturbations, the fluorescence, autoionization, and predissociation rates are expected to decrease as $1/n^{*3}$. From the time-dependent perspective, this scaling can be understood in terms of the n^{*3} dependence of the classical orbital period of the Rydberg electron.[8] In classical terms, the Rydberg electron spends most of its orbit moving at low velocities at large distances from the ion core. It is only when it is accelerated by the potential of the core that it gains sufficient momentum to influence the motion of the much heavier nuclei or the much more strongly bound electrons. Because the orbital frequency, and thus the Rydberg electron–ion core collision frequency scales as $1/n^{*3}$, so too do the decay rates. In most instances, the rate for fluorescence is considerably smaller than the rates for autoionization and predissociation and, as a result, fluorescence is generally not an important decay channel; for this reason, it will not be emphasized in the present review. However, as is well known in H_2, situations can arise in which autoionization and predissociation are in some sense forbidden, and in these instances fluorescence can become an important decay channel.[9–11]

In smaller molecules, the autoionization and predissociation pathways and mechanisms are relatively straightforward, but in larger molecules, the greater number of internal degrees of freedom can introduce additional complexity. In particular, the possibility exists for nonionizing radiationless transitions (i.e. internal conversion and intersystem crossing) followed by intramolecular vibrational energy redistribution, resulting in highly vibrationally excited molecules in the ground or low-lying electronic manifolds. Because the total vibrational energy in this situation will ultimately be considerably greater than the dissociation energy, such molecules are expected to undergo unimolecular dissociation to neutral fragments. However, there may be a significant time delay between the radiationless transition and the actual fragmentation of the molecule. In such a case, the branching ratio between ionization and dissociation will be determined by the relative rates of autoionization and the initial nonionizing radiationless transition. As a result, the study of this branching ratio and its dependence on the internal quantum numbers of the Rydberg state is expected to provide insight into the radiationless transition itself. For example, it is worth investigating if it is possible to correlate the excited vibrational modes of Rydberg states with small ionization yields with promoting modes for the radiationless transition. Such a connection would provide a powerful tool for the study of radiationless transitions.[12,13]

Another advantage of studying dissociation dynamics in high Rydberg states is that, because the orbital period scales as n^{*3}, the timescale for dissociation can be considerably longer than in the low-lying Rydberg states. For example, the dissociation rate for an $n = 12$ Rydberg state is expected to be a factor of 64 longer than that for an $n = 3$ Rydberg state. Often this will result in considerable sharpening of the linewidths of the pump transitions, allowing significantly better selectivity of the quantum numbers of the excited state, and thus providing more detailed information on the decay dynamics. Nevertheless, because the small r region of the Rydberg state wavefunctions depend on n only through the $1/n^{*3/2}$ normalization factor, in the absence of perturbations much of the information obtained at higher n is directly relevant to the low-energy region of the spectrum.

Resonances above the lowest dissociation and ionization thresholds are sometimes referred to as super-excited states. Nakamura and coworkers[14] have separated these states into two distinct classes. In their scheme, states with multiple electronic excitation or states with inner shell excitation correspond to super-excited states of the first kind, whereas states with a single excited electron and vibrational or rotational (but no electronic) excitation of the ion core are called super-excited states of the second kind.[14] In this scheme, the distinction between the first and second types of states is that the former can decay by electronic autoionization (transfer of electronic energy of the ion core to the Rydberg electron, which then escapes), while the latter decay by vibrational or rotational autoionization (transfer of rovibrational energy of the ion core to the Rydberg electron, which then escapes). For the most part, the present review is focused on the latter states, that is, Rydberg states converging to excited rovibrational levels of ions in their ground electronic state. In most instances, however, the possible presence of series converging to electronically excited states of the ion must be considered. In addition, many of the electronic states responsible for predissociation of the super-excited states can be considered to be low-lying members of Rydberg series converging to excited states of the ion. Configuration interaction with such states results in states with mixed character, in which both electronic and rovibrational autoionization mechanisms are possible.[15]

Although there have been extremely detailed studies of the photoionization and photodissociation dynamics of the high Rydberg states of diatomic molecules like H_2 and NO, such detailed studies do not exist for larger polyatomic molecules. Even so, the work that has been done clearly points to the presence of strong mode-specific effects in the branching between ionization and dissociation. This review presents a selected survey of this work in an attempt to call attention to this evidence and to suggest areas that are suitable for future research. The review is clearly not meant to be exhaustive; some experiments are only discussed briefly and many interesting studies are not discussed at all. However, it is hoped that the review provides a feel for the types of experiments that can and have been done, and provides an appreciation for mode-specific decay processes in molecular Rydberg states.

The structure of the remainder of this review is as follows. First, although this review is not intended to provide a detailed description of the relevant experimental techniques, a description of the basic aspects of each of the different types of measurements is provided in Sec. 2. The mechanism for vibrational autoionization, in particular the vibrational propensity rule, is discussed in Sec. 3. This provides a framework for the discussion of specific molecules in later sections. In Sec. 4, the results from studies of the photoionization and photodissociation of the high Rydberg states of NO are discussed. Mode-specific effects are not surprising in diatomic molecules, and NO provides a good example of the competition between ionization and dissociation. Work on the triatomic molecules HCO and NO_2 is discussed in Sec. 5. Grant and coworkers[16–26] have performed extensive studies on these systems, and their work should be consulted for a comprehensive discussion of their findings. Only a small portion of their results are discussed, but this provides insight into the competition between ionization and dissociation in triatomic molecules. In Sec. 5, larger polyatomic molecules are discussed, with a particular emphasis on double-resonance ionization spectra on ammonia[27] and diazobicyclo-octane[28] (Dabco). A number of larger molecules show evidence for mode-specific effects in the branching ratio between ionization and dissociation, but the double-resonance ionization and fluorescence-dip experiments on Dabco performed by Fujii *et al.*[29–31] are particularly noteworthy. Finally, a brief summary is presented in Sec. 7.

2. Experimental Approach

2.1. *Preparation of the Excited States and Absorption Spectra*

The high-resolution of photon-based techniques makes them the preferred method for preparing the highly excited Rydberg states discussed here. Originally, most studies of excited state decay dynamics relied on single photon excitation from the ground electronic state of the molecule of interest. However, the development of double- and multiple-resonance techniques provided a set of extremely powerful tools to prepare excited states selectively and to monitor their decay dynamics. The latter techniques are emphasized in this review.

While most of the studies of high Rydberg states have focused on the detection of ionization products and, to a lesser extent, dissociation products, one of the most valuable measurements is that of the absorption spectrum for the transition to the high Rydberg states. Even if it is not recorded on an absolute intensity scale, such a spectrum allows the assessment of the relative initial populations of the high Rydberg states. These relative populations can then be compared with the intensities in the ionization and dissociation experiments to provide insight into the branching ratios for the different decay pathways. Unfortunately, direct absorption spectra are often difficult to measure because the sample densities are small and the effective pathlength through the sample is short. This is particularly true in double-resonance experiments with pulsed lasers, where only a few thousand molecules may be excited to the intermediate level on any given laser shot, and the shot-to-shot fluctuations in the pulse energy of the laser can be substantial.

When the intermediate level fluoresces with a reasonable efficiency, one solution to these difficulties is to use fluorescence-dip spectroscopy.[32–34] In this approach, the pump laser is used to drive a transition to a selected level of the intermediate state and the fluorescence from that state is monitored as the probe laser is scanned. When the probe laser is resonant with a transition from the intermediate level to a higher lying Rydberg state, the intermediate state population is transferred to the Rydberg state and the fluorescence intensity from the intermediate level is reduced. In the linear regime, the magnitude of this "fluorescence-dip" is proportional to the absorption probability for the probe laser, resulting in an absorption spectrum from the intermediate level. This technique has been applied to a number of different molecules and is perhaps the most popular method for recording excited state absorption spectra.[32–34]

While these results are quite useful and lead to some important results that will be discussed in Secs. 3–5, the technique does have some drawbacks. First, it is very easy to saturate the probe transition and thus make it difficult to extract meaningful intensities. Second, because there is substantial shot-to-shot variation in the laser intensity, the fluorescence signal upon which the dips appear can be quite noisy. Finally, stimulated probe transitions from the intermediate state to lower-lying states of the molecule also result in fluorescence-dips that must be distinguished from the transitions to higher lying states. In spite of these disadvantages, fluorescence-dip spectroscopy remains a powerful tool for recording excited state absorption spectra.

Ion-dip spectroscopy is a variation on fluorescence-dip spectroscopy that is particularly sensitive to transitions to dissociative states.[35] In this technique, the signal resulting from ionization of the intermediate level by the pump laser is monitored instead of the signal from the fluorescence of the intermediate state. The probe laser is then scanned, and when it is resonant with a transition to a dissociative state of the molecule, there is a decrease in the ionization signal produced by the pump laser. Because the detection efficiency of ions is generally higher than with photons, the signal-to-noise ratio can be significantly better than in fluorescence-dip spectroscopy, and the technique can be applied in systems where the intermediate state does not fluoresce. Unfortunately, the technique is not necessarily sensitive to transitions to autoionizing states, because these transitions do not lead to a dip in the ion signal. Depending on the relative intensities of the pump and probe laser, transitions to autoionizing levels may result in an increase in the ion signal, but in any event it is difficult to assess the relative importance of the ionization and dissociation channels. Thus, ion-dip spectroscopy is not very useful for the present purposes.

An alternative technique for recording excited state absorption spectra is two-color laser-induced grating spectroscopy (TC-LIGS).[36] This technique is one of a large class of techniques based on resonant four-wave mixing, which is described in a number of excellent recent reviews.[37–40] Only one simplified variation of the technique will be described here. In particular, when a time delay is introduced between the pump and probe lasers, TC-LIGS can be visualized in terms of the formation and scattering off of a diffraction grating in the sample. In this scheme, the pump beam is split into two parts that are crossed at a small angle in the sample, resulting in an interference pattern and alternating regions of high and low laser intensity. When this laser beam is tuned to a transition in the sample, this spatial variation of the laser intensity results in the spatial variation of the ground and excited state populations in the sample, and thus in the index of refraction of the medium; i.e. a population grating is formed. The beam from the second laser is then introduced to probe the grating. When this beam is resonant with a transition from either the ground or excited state of the pump transition, the spatial variation of the index of refraction results in the scattering of the beam off the grating. The resulting signal beam is essentially a 'laser' beam with low divergence and a well defined scattering angle, and can be detected by using a photomultiplier or photodiode. Because the scattering is a resonant process, there is essentially no signal when the probe laser is tuned off resonance (of course, there is no population grating when the pump laser is tuned off resonance). In other words, unlike fluorescence-dip spectroscopy, TC-LIGS is a background-free technique, and high-quality two-color spectra can be obtained. In addition, because the scattering off the grating is independent of the fate of the upper level of the probe transition, the TC-LIGS spectrum contains the information of the absorption spectrum of the intermediate state. However, because TC-LIGS is a nonlinear technique, the extraction of this information is somewhat more difficult than from fluorescence-dip spectra, but in some cases the improved signal-to-noise ratio merits this additional effort. Only a limited number of LIGS studies of high Rydberg states have been reported to date.[41,42] Nevertheless, a large number of variations on this approach have been developed, and the full utility of these techniques is still under development.

2.2. *Detection of Ionization Products*

The ions produced by direct photoionization or autoionization are typically detected by using a quadrupole or time-of-flight mass spectrometer. In most of the experiments discussed here, the parent ion is the dominant species produced by photoionization, but in some instances dissociative ionization as well as photodissociation of parent ions can occur. The use of a mass spectrometer provides the means to begin to unravel these processes. More important, however, is that when the parent ion is detected, the mass spectrometer serves to identify the photoabsorbing species. As a result, mass resolved multiphoton ionization techniques have provided considerable new spectroscopic data on a wide variety of molecules.

While the identification of photoions can provide insight into the photoionization event, photoelectron spectroscopy can provide considerably more detail with respect to the product state distributions in the photoions. Both electrostatic and time-of-flight electron spectrometers have been employed with considerable success for the study of photoionization dynamics.

In general, time-of-flight instruments are better matched to pulsed excitation sources and provide better resolution for slow photoelectrons, but electrostatic analyzers have the advantages of being easier to energy calibrate and maintaining constant resolution as a function of electron kinetic energy. Both types of electron spectrometers can be used to record photoelectron angular distributions, which provide additional information on the partial-wave composition of the outgoing photoelectron.

The magnetic bottle electron spectrometer,[43] which combines a magnetic lens with time-of-flight energy analysis, is particularly useful because it can be operated with 50% collection efficiency (as opposed to $\sim 0.1\%$ efficiency for electrostatic and time-of-flight spectrometers) while maintaining the high-resolution of time-of-flight analyzers for slow photoelectrons. The high collection efficiency of the magnetic bottle spectrometer makes it ideal for constant-ionic-state (CIS) spectroscopy, as well as for the study of dilute samples or weak processes. The principal disadvantage of the magnetic bottle design is that the photoabsorption and decay processes take place in a 1 Tesla magnetic field, which in some instances can influence the decay dynamics. In addition, although it is in principle possible to extract information on the photoelectron angular distribution by using a position-sensitive detector,[43] in practice the magnetic bottle electron spectrometer has not been used to record these distributions.

The resolution of the electron spectrometers currently in use is such that the electronic and vibrational state of the product ion can usually be determined quite readily. In a few systems, for example the hydrides NH,[44] OH,[45,46] HCl,[47] and H_2O,[48,49] the rotational state distributions of the photoion can also be determined; however, this generally requires either a molecule with a large rotational constant or the selective photoionization of high J states, which are spaced sufficiently far apart to resolve. The additional information contained in the rotational distributions is considerable, and such studies provide some of the most stringent tests in the current understanding of photoionization dynamics. As an alternative to photoelectron spectroscopy, it should be mentioned that laser-induced fluorescence has also been used to determine the ionic rotational state distributions following photoionization of a number of small molecules.[50–54]

There has been considerable recent interest in zero kinetic energy-photoelectron spectroscopy[55] (ZEKE-PES), mass-analyzed threshold ionization[56]

(MATI), and related threshold photoionization techniques. These techniques can be implemented relatively easily and can provide rotationally state-resolved information on the photoionization dynamics at threshold. Because they can be applied to essentially any molecule, these techniques have become enormously popular. Unfortunately, because these are threshold techniques, they are not generally applicable for the study of autoionization and ionic branching ratios for autoionizing states. As a result, these techniques will not be emphasized in the present review.

2.3. *Detection of Dissociation Products*

The neutral products of photodissociation are usually detected by using either laser-induced fluorescence or resonantly enhanced multiphoton ionization. In both cases, the fragments are excited following a suitable delay from the initial excitation of the parent molecule. Direct sample fluorescence can also serve as a valuable probe if the fragment of interest is left in an electronically excited state. Laser-induced fluorescence is particularly useful for fragments with low-lying electronic states, but it has also been employed with vacuum ultraviolet light to detect fragments with no low-lying states. Two-photon laser-induced fluorescence is another alternative in such situations. Resonantly enhanced multiphoton ionization is also quite effective for fragments with no readily accessible low-lying electronic states and for systems in which the fluorescence yield is small. In addition, mass-selective detection following multiphoton ionization usually allows the unambiguous assignment of the fragment identity. In general, both laser-induced fluorescence and resonantly enhanced multiphoton ionization provide the means to characterize the electronic, vibrational, and rotational state distributions of the photofragment.

Although substantial information on the dissociation dynamics is contained in the photofragment state distributions, considerable additional information is obtained by determining the velocity distributions, angular distributions, and alignment parameters of the photofragments.[57] Only a limited number of such experiments employing these techniques have been performed on the dissociation dynamics of high Rydberg states, and will not be discussed further in this review. Nevertheless, ion imaging techniques[58] have recently been developed that can provide high-resolution translational energy distributions for state-selected photofragments. The determination of the translational energy distribution of one fragment can allow the determination of the predissociation branching ratios between two or more different decay channels, and provide significant insight into the decay dynamics. The application of these techniques to the decay of high Rydberg states is expected to be extremely informative.

3. The Vibrational Propensity Rule

Simplified models for autoionization and predissociation can be extremely helpful for understanding the rates and product branching ratios of these decay products. Perhaps the best known example of such a model is the propensity rule first described by Berry.[59–63] This rule provides a framework for the discussion of the relative rates and branching ratios for vibrational autoionization, that is, autoionization of a Rydberg state converging to an excited vibrational level of a given electronic state into the continuum of a lower vibrational level of the same electronic state of the ion. The propensity rule[59–63] says that for a given autoionizing state, vibrational autoionization is fastest for processes that require the smallest change in vibrational quantum number and that the autoionization rate decreases rapidly with $|\Delta v|$. Experimental support for the propensity rule in diatomic molecules was first obtained in studies of the vibrational autoionization of molecular hydrogen.[9,64,65] For example, Dehmer and Chupka[65] found that, when it is energetically allowed, vibrational autoionization with $\Delta v = -1$ accounts for 95% of the observed product. When $\Delta v = -1$ autoionization is forbidden, Dehmer and Chupka[65] found that the process with the smallest value of $|\Delta v|$ dominates. More recent laser-based studies on large polyatomic molecules such as aniline,[66,67] Dabco,[29–31] and pyrazine[68,69] have provided additional evidence in support of this propensity rule. Although violations of the propensity rule have been observed for a number of different molecules,[70–73] these exceptions can generally be understood as arising from other decay processes that do not necessarily obey the same propensity rule.[15] For example, in many instances electronic autoionization and autoionization via interactions with dissociative states can occur along with vibrational autoionization, and these processes do not necessarily obey the $|\Delta v| =$ minimum propensity rule.[15] Unless these coexisting processes are recognized, these alternative mechanisms can result in apparent violations of the propensity rule. To date, all of the experimental examples that show a clear deviation from the propensity rule can be understood in terms of 'contamination' of the pure vibrational process by one of these other processes.

Because it involves an interaction between the electronic and vibrational degrees of freedom of the molecule, vibrational autoionization is an inherently non-Born–Oppenheimer process. Early work on the propensity rule in diatomic molecules was based on evaluating the terms of the molecular Hamiltonian that are ignored within the Born–Oppenheimer approximation, and the propensity rule was justified based on a consideration of the relative magnitudes of these terms.[59–63] Within the framework of multichannel quantum defect theory (MQDT), Herzberg and Jungen[74] later made use of the fact that these non-Born–Oppenheimer terms are implicitly contained in the dependence of the quantum defect, μ_k, on the internuclear distance, R, of the diatomic molecule. The quantum defect $\mu_k(R)$ is defined by

$$U_k(R) = U^+(R) - \mathfrak{R}/[n - \mu_k(R)]^2\,, \tag{1}$$

where $U_k(R)$ and $U^+(R)$ are the potential curves of the Rydberg state and ionic state, respectively, $\mathfrak{R}$ is the Rydberg constant, n is the principal quantum number, and k is the electronic symmetry. Herzberg and Jungen[74] then showed that by expanding $\mu_k(R)$ in a Taylor's series, the vibrational autoionization width, Γ, takes the form

$$\Gamma_{n^*,k,v,v'} = 2\pi \frac{2\mathfrak{R}}{n_k^{*3}} \left[\frac{d\mu_k(R)}{dR}\right]^2 \langle \Psi_v | R - R_e^+ | \Psi_{v'} \rangle^2\,, \tag{2}$$

where Ψ_v is vibrational wavefunction for the vth vibrational level of the ion and R_e^+ is the equilibrium internuclear distance. This expression clearly reveals the source of the vibrational autoionization propensity rule. If the vibrational wavefunctions are approximated by harmonic oscillator wavefunctions, the matrix elements are zero unless $\Delta v = \pm 1$. For anharmonic oscillator wavefunctions this is not strictly true, but the magnitude of matrix elements with $|\Delta v| > 1$ is generally considerably smaller than those with $|\Delta v| = 1$. This form of the propensity rule is especially appealing because the necessary parameters, specifically $\mu_k(R)$ and its derivative, can be obtained from experimentally or theoretically derived potential curves for the Rydberg state and the ion. In addition, this form allows an intuitive understanding of the factors that influence the vibrational autoionization rate. For example, if the neutral potential curve is parallel to the ionic curve, it implies that the quantum defect is constant, the derivative of $\mu_k(R)$ is zero, and the autoionization rate in Eq. (2) vanishes. This is readily understood because, if the curves are parallel, the Rydberg electron must only contribute to the electronic energy and does not influence the nuclear motion. On the other hand, if the Rydberg electron modifies the potential curve of the ion (that is, if the Rydberg electron influences the motion of the nuclei), the quantum defect derivative will be nonzero, and the autoionization rate will be finite. Such arguments are particularly valuable in rationalizing the relative autoionization rates for Rydberg states with different electronic symmetries because the R dependence of different electronic states can often be determined qualitatively from correlation diagrams or more quantitatively from accurate calculations of the potential curves.

The extension of these ideas to polyatomic molecules is not completely straightforward.[75–81] With suitable substitutions, most notably the replacement of R and $\mu_k(R)$ by the normal coordinate Q_i and $\mu_k(Q_i)$, Eqs. (1) and (2) can be applied to vibrational autoionization via totally symmetric vibrations in polyatomic molecules.[75,81] Nontotally symmetric vibrations are more complicated. In particular, because the total symmetry of the electron–ion complex must be preserved upon autoionization, a change in the vibrational quantum number of a nontotally symmetric vibration must be accompanied by a corresponding change in the symmetry of the electronic wavefunction.[81] Because, by definition, vibrational autoionization preserves the electronic symmetry of the ion core, a change in the vibrational symmetry of the core requires a change in the symmetry of the Rydberg/ejected electron. This requirement can be incorporated into the MQDT framework by generalizing the quantum defect in Eqs. (1) and (2) into a quantum defect matrix, $\mu_{kk'}^{\Gamma_e}(Q_i)$, where k specifies the particular vibronic channel and Γ_e denotes the overall symmetry. While the diagonal elements of such a matrix are essentially the same as the quantum defects for totally symmetric vibrations used in Eqs. (1) and (2), the off diagonal elements of such a matrix allow for mixing between different electronic channels, for example, between series with different values of ℓ or even between series converging to different electronic states of the ion. Such quantum defect matrices have been used to include interactions between singly and doubly excited states in calculations of the excited states of H_2,[82–85] and to incorporate ℓ mixing in studies of the Rydberg states of NO.[86] The present discussion is limited to the former matrix elements because these provide the means to preserve the total symmetry during vibrational autoionization via nontotally symmetric vibrations. Specifically, the change in symmetry resulting from the change in the quantum number of the nontotally symmetric vibration can be accommodated by a corresponding change in the symmetry of the Rydberg electron through a suitable change in ℓ, with the result that the total symmetry of the electron–ion complex is preserved. In some sense, autoionization via nontotally symmetric vibrations is not 'pure' vibrational autoionization, because it requires electronic ℓ-mixing as well. However, the resulting expressions for the vibrational autoionization rates and, more specifically, the dependence of these rates on the variation of the quantum defect matrix elements as a function of displacement along the appropriate normal coordinate, are quite analogous to those for autoionization via totally symmetric vibrations, and thus merit their simultaneous treatment.[81] Consideration of the off-diagonal quantum defect matrix elements and the autoionization rates leads to the expectation that the vibrational propensity rule will be obeyed for vibrational autoionization via both totally symmetric and nontotally symmetric vibrations, and the factors that determine the relative magnitudes for different vibrational modes have been identified.[81]

4. Autoionization and Predissociation in NO

Studies of the competition between ionization and dissociation in the Rydberg states of NO have been more extensive than those of any other molecule, with the possible exception of H_2. One reason for this situation is the relatively low ionization potential of NO, which greatly simplifies a number of experimental difficulties. More important, however, is the availability of a wealth of spectroscopic information, which is a result of the lifetime of effort by Miescher and his coworkers.[87–91] This knowledge base has provided the foundation for countless laser-based studies of the spectroscopy and dynamics of NO. In NO, the competition between ionization and dissociation reflects the presence of strong Rydberg-valence interactions. These interactions are the same ones that give rise to the spectroscopic perturbations characterized by Miescher and coworkers in the bound portion of the spectrum. This point was made clear by the MQDT analysis of Guisti-Suzor and Jungen,[15] which brought together a number of observations from the single photon absorption and ionization spectra of ground state nitric oxide.

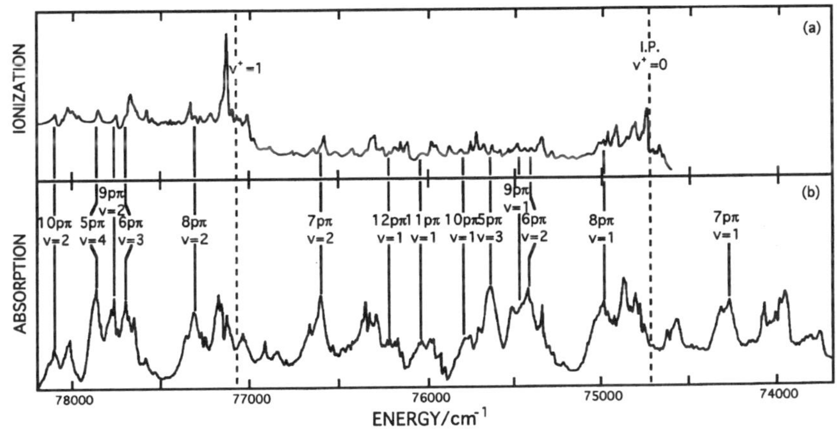

Fig. 2. The photoionization (a) and photoabsorption (b) spectra of NO near the first ionization threshold. The data are from Refs. 70(a) and (b), respectively. The spectra were recorded with comparable spectral resolution but different temperatures, and the baseline in Part (b) has not been properly determined in Ref. 70(b). The assignments of the $np\pi$ Rydberg bands are those of Miescher, which are based on the high-resolution absorption spectrum. (Reprinted with permission from Ref. 15. Copyright 1984 American Institute of Physics.)

Figure 2 shows a portion of the single photon absorption and ionization spectra of NO in the region of the first ionization threshold.[15] Two features of these spectra deserve attention. First, the ionization spectrum shows a prominent step at the $X^1\Sigma^+$, $v^+ = 1$ threshold (and at the $v^+ > 1$ thresholds not shown in Fig. 2), while the absorption spectrum varies smoothly through these thresholds. (The small shifts in the step energies from the true ionization thresholds result from the depression of the ionization thresholds by the electric field used in the mass spectrometer.) Although the step at the $v^+ = 0$ threshold is simply a result of energetics, i.e. ions cannot be produced below this threshold, the steps at the $v^+ \geq 1$ thresholds indicate that a significant fraction of the Rydberg states converging to these thresholds decay via predissociation rather than ionization.[92] More specifically, the ionization efficiency for most of the resonances observed in the ionization spectrum of Fig. 2 is small. Second, Miescher *et al.*[70] pointed out that a significant fraction of the autoionizing peaks that are observed between the $v^+ = 0$ and 1 thresholds can be assigned to Rydberg states converging to vibrational states of NO^+ $X^1\Sigma^+$ with $v^+ > 1$.

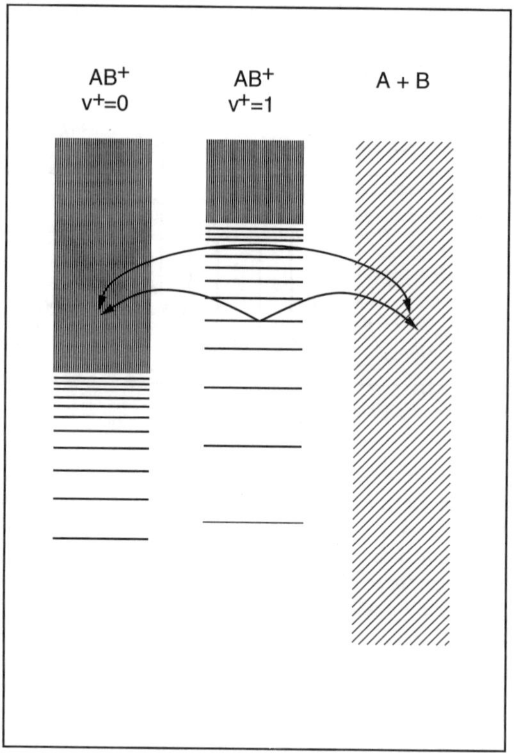

Fig. 3. A schematic diagram of indirect vibrational autoionization via a dissociative valence state. The arrow between the dissociation and ionization continua indicates that interactions between these channels exist, and can serve to indirectly couple the quasi-discrete level to one continuum via an interaction with the other.

Because such states can only autoionize by $|\Delta v| > 1$ processes, this observation reflects a breakdown of the vibrational propensity rule. However, in their MQDT study of the $^2\Pi$ states of NO, Guisti-Suzor and Jungen[15] showed that the autoionization mechanism is not a purely vibrational process, but rather an indirect process involving electronic interactions among the Rydberg states, dissociative valence states, and the ionization continua. This indirect interaction is shown schematically in Fig. 3. For most of the resonances considered by Giusti-Suzor and Jungen,[15] it was found that the ionization efficiency was between 0.3% and 16%, and that the ionization efficiency was essentially independent of the value of Δv required for autoionization. For electronic interactions, the vibrational wavefunctions contribute what is essentially a Franck–Condon factor between the autoionizing state and the continuum, and, in general, this factor will not follow the $|\Delta v|$ = minimum requirement of pure vibrational autoionization.

The work by Guisti-Suzor and Jungen[15] was significant not only because it elucidated this important decay mechanism, but also because it provided a striking example of how information from the study of Rydberg-valence interactions in the bound portion of the spectrum can be used to explain effects in the ionization and dissociation continua. Subsequently, Raoult[93] extended these studies, with a particular emphasis on states below the ionization threshold. In addition, Nakashima *et al.*[94] performed a similar MQDT analysis for NO $(X^1\Sigma^+)n\ell\lambda$ Rydberg series of other symmetries in which the quantum defect parameters were calculated by using *ab initio* techniques. Comparison with experimental data obtained by using double-resonance multiphoton ionization and photoelectron spectroscopy showed that similar indirect autoionization processes, rather than true vibrational autoionization, were the dominant ionization mechanism for the $nd\delta$ and $ns\sigma$ series.

The studies of Giusti-Suzor and Jungen[15] clarified the need for improved measurements of the highly excited Rydberg state spectroscopy and dynamics of NO. The state selectivity of double-resonance techniques makes them ideal for such measurements, and a large number of studies were soon performed. While many of these studies focused on the bound states or predissociating states below the ionization threshold, a significant number were concerned with the super-excited states above the ionization threshold. Perhaps the earliest double-resonance experiments that address some of the issues raised by Giusti-Suzor and Jungen[15] were those of Ito and coworkers.[95–97] In their experiments, double-resonance spectra were recorded by pumping selected rotational levels of the $A^2\Sigma^+$, $v = 1$ state and probing in the region between the NO^+ $^1\Sigma^+$, $v^+ = 0$ and 1 thresholds. The $A^2\Sigma^+$ of NO is nominally a $3s\sigma$ Rydberg state, and thus the strongest transitions from it are expected to be those with $\Delta\ell = 1$ to the $np\sigma$ and $np\pi$ Rydberg states. Although this is indeed the case, a small amount of "*d*" character in the $A^2\Sigma^+$ state results in significant oscillator strength to the nf Rydberg states as well. In addition, the heteronuclear nature of NO also results in some oscillator strength to the even ℓ Rydberg series, and the $ns\sigma$ states are readily observed in double-resonance spectra via the $A^2\Sigma^+$ state. Although transitions from the $A^2\Sigma^+$ state to low-lying members of the $(X^1\Sigma^+)nd$ Rydberg series have been observed,[98] transitions from the $A^2\Sigma^+$ state to the members of these series above the $NO^+ X^1\Sigma^+$ threshold appear to be weak. For the double-resonance experiments discussed below, a double prime refers to the ground state, a single prime to the intermediate state, no prime to the high Rydberg state, and a plus to the ion.

Ito and coworkers[95–97] recorded both ionization and fluorescence-dip spectra and found that although essentially the same peaks were observed in the two spectra, the relative intensities were often quite different. Strong peaks in the fluorescence-dip spectrum, which reflect a large transfer of population to the high Rydberg state, are sometimes correlated with weak peaks in the ionization spectrum, indicating a small ionization efficiency and the presence of competing decay processes. Similarly, weak peaks in the fluorescence-dip spectrum are often correlated with intense peaks in the ionization spectrum, indicating a significantly higher ionization efficiency. These large variations in the relative intensities suggest that the ionization efficiency depends strongly on the quantum numbers of the high Rydberg states. The energetics of these Rydberg states are such that they can predissociate to $N^4S + O^3P$ and to $N^2D + O^3P$, as well as to $N^4S + O^1D$. However, the latter dissociation products give rise only to quartet states of NO, and predissociation of the doublet Rydberg states by such states is spin-forbidden. Thus, dissociation to $N^4S + O^1D$ is expected to be relatively unimportant.

These early double-resonance experiments via the $A^2\Sigma^+$ state were later significantly extended in a series of studies by Fujii and Morita,[99–101] who not only recorded double-resonance ionization and fluorescence-dip spectra, but also used resonantly enhanced multiphoton ionization to determine the relative N^2D, N^4S, and O^3P yields as a function of the probe laser frequency. The N^2D spectra discussed below were obtained by monitoring transitions from the $J'' = 5/2$ level, and the O^3P spectra were obtained by monitoring transitions from the $J'' = 2$ level; the N^4S state has a single level with $J'' = 3/2$. Figure 4 shows a portion of their data, which was recorded by pumping the $A^2\Sigma^+$, $v' = 1$, $N' = 0$ level. Although the intensities in the five spectra are not calibrated with respect to each other, the relative intensities within the four spectra clearly show significantly different behavior. Clearly the product state distributions depend strongly on n and ℓ of the super-excited state.

Although it is difficult to make a quantitative determination of the product branching ratios from relative spectra alone, especially since fluorescence-dip spectra are prone to saturation effects, some qualitative observations are possible.[99–101] In particular, the $(X^1\Sigma^+)11s$, $v = 1$ resonance does not appear in either dissociation channel, and can be used to fix the relative intensities of the ionization and fluorescence-dip spectra. Next, the $(X^1\Sigma^+)10f$, $v = 1$ resonance, which does not appear in the N^4S spectrum, can be used to fix the intensity of the N^2D spectrum relative to the fluorescence-dip and ionization spectra. Finally, the $(X^1\Sigma^+)8f$, $v = 1$ resonance can be used to extract the intensity of the N^4S spectrum relative

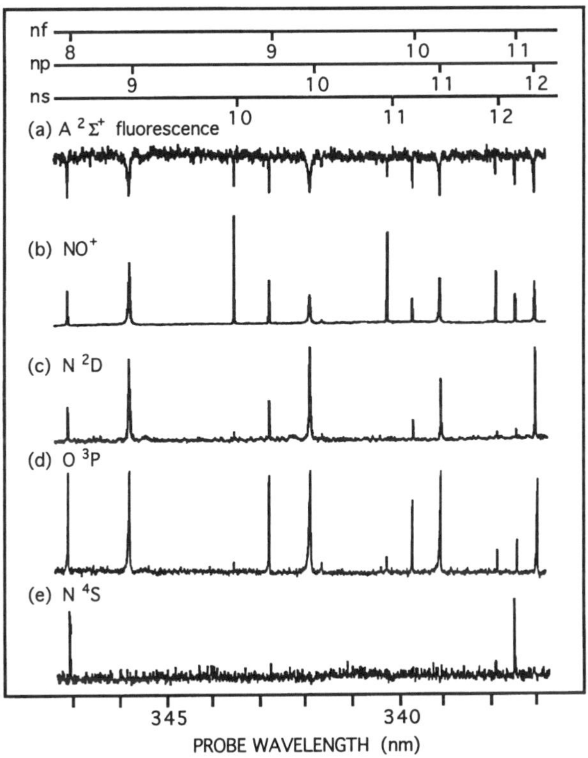

Fig. 4. Excitation spectra obtained by pumping the $A^2\Sigma^+$, $v' = 1$, $N' = 0$ level of NO and probing between the NO^+ $X^1\Sigma^+$, $v^+ = 0$ and 1 ionization thresholds: (a) fluorescence-dip spectrum, (b) $N^2D_{5/2}$ yield spectrum, (c) N^4S yield spectrum, (d) O^3P_2 yield spectrum, and (e) NO^+ yield spectrum. (Reprinted with permission from Ref. 101. Copyright 1994 Gordon and Breach Publishers.)

to the others. This process results in crude scale factors for the spectra of Fig. 4, whereby the intensities in the NO^+, N^2D, and N^4S spectra should be multiplied by factors of 0.25, 1.0, and 0.1, respectively. The resulting intensities are not completely consistent when checked by comparing the resulting intensities of the other resonances (or with the O^3P spectrum), and would be improved by making a least-squares fit of the intensities of all of the resonances; however, a more important source of this inconsistency probably arises from saturation or other effects that affect the relative intensities within one or more of the individual spectra. Nevertheless, the qualitative relative intensities resulting from this procedure are sufficient to provide some insight into the relative importance of the different decay channels. In particular, it is clear that dissociation to $N^2D + O^3P$ is the dominant decay process for most of these resonances. The ionization efficiency for most of the resonances is small, in qualitative agreement with the conclusions of Giusti-Suzor and Jungen.[15] Ionization does appear to be the second most important decay process, however, whereas predissociation to $N^4S + O^3P$ is usually only a minor decay channel.

Fujii and Morita[100] have made a particularly detailed study of the competition between ionization and dissociation in the $(X^1\Sigma^+)7f$, $v = 1$ Rydberg state, which had been studied previously by Anezaki *et al.*[97] and Biernacki *et al.*[102] Figure 5 shows the data of Fujii and Morita[100] that was obtained by pumping the $A^2\Sigma^+$, $v' = 1$, $N' = 1$ level and monitoring four of the channels from Fig. 4. (Note, however, that these four spectra are arranged in a different order than in Fig. 4.) The transitions are labeled by $^{R-N'}\Delta N_{\mathfrak{L}}$ for the $(X^1\Sigma^+)7f$, $v = 1 \leftarrow A^2\Sigma^+$, $v' = 1$, where R is the ion-core rotation quantum number, N is the total angular momentum minus spin, and $\mathfrak{L}$ is the projection of the orbital angular momentum on the axis of core rotation. The important point for the present discussion is that $(X^1\Sigma^+)7f$, $v = 1$ with $\mathfrak{L}$ = even have Σ^- reflection symmetry, while those with $\mathfrak{L}$ = odd have Σ^+ symmetry.

The N^2D predissociation spectrum in Fig. 5 is fairly similar to the fluorescence-dip spectrum, whereas the N^4S and NO^+ spectra show marked differences. These differences provide insight into the predissociation mechanism and are best understood in terms of the potential curves of the doublet electronic states of NO shown in Fig. 6. Figure 6 shows that only the $A'\,^2\Sigma^+$ state dissociates to $N^4S + O^3P$, whereas the $I^2\Sigma^+$, $B^2\Pi$, $L^2\Pi$, and $B'\,^2\Delta$ states all dissociate to $N^2D + O^3P$. As a result, only levels

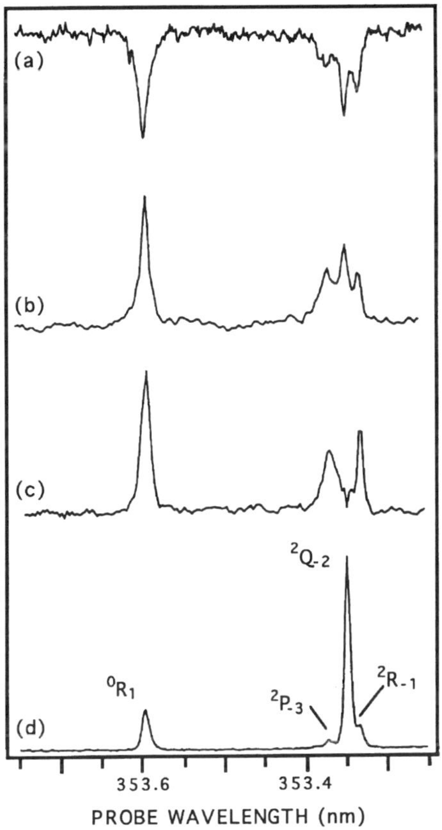

Fig. 5. Excitation spectra obtained by pumping the $A^2\Sigma^+$, $v' = 1$, $N' = 1$ level of NO and probing transitions to the $(X^1\Sigma^+)7f$, $v = 1$ autoionizing level: (a) fluorescence-dip spectrum, (b) $N^2D_{5/2}$ yield spectrum, (c) N^4S yield spectrum, and (d) NO^+ yield spectrum. The transitions are labeled by ${}^{R-N'}\Delta N_{\mathfrak{L}}$. (Reprinted with permission from Ref. 100. Copyright 1992 American Institute of Physics.)

with $\mathfrak{L}$ = odd are expected in the N^4S spectrum, which is exactly what is observed in Fig. 5. The small intensities of the $\mathfrak{L}$ = odd levels in the NO^+ spectrum then indicate that predissociation of these levels via the $A'\,^2\Sigma^+$ state is significantly faster than autoionization. Predissociation via the $B^2\Pi$, $L^2\Pi$, and $B'\,^2\Delta$ states is allowed for levels with either Σ^+ or Σ^- reflection symmetry, and the appearance of both $\mathfrak{L}$ = even and odd transitions in the N^2D spectrum indicates these $\Lambda \geq 1$ states do play a role

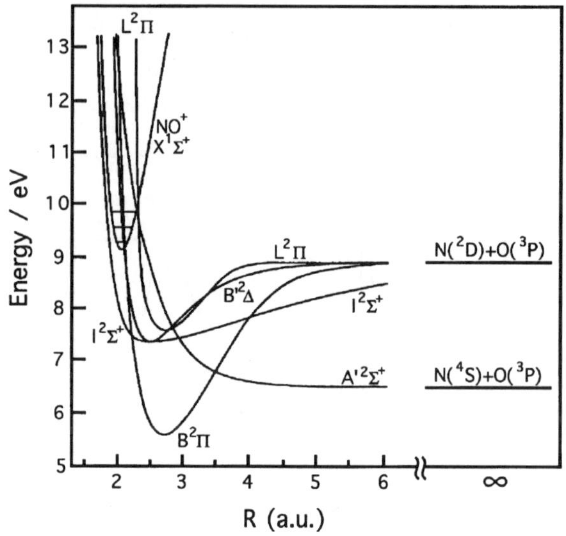

Fig. 6. Potential curves of NO related to the present discussion. (Reprinted with permission from Ref. 100. Copyright 1992 American Institute of Physics.)

in the predissociation of the $(X^1\Sigma^+)7f$, $v = 1$ levels. From an analysis of the relative intensities in the four spectra of Fig. 5, Fujii and Morita[100] made a rough estimate of the relative decay rates for the different processes and found

$$k_A^o \cong 2k_A^e\,, \tag{3}$$

$$k_S^o \cong 4k_A^o\ (\cong 8k_A^e)\,, \tag{4}$$

$$k_D^o \cong 10k_D^e\,, \tag{5}$$

where A, D, and S refer to autoionization and predissociation to N^2D and N^4S, respectively, and e and o refer to $\mathfrak{L}$ = even and odd, respectively.[100] The large difference between k_D^o and k_D^e indicates that the $I^2\Sigma^+$ state plays a substantial role in determining the overall rate of predissociation to $N^2D + O^3P$. This is particularly interesting because, in spite of the large difference between k_D^o and k_D^e, the relative intensities in the N^2D and fluorescence-dip spectra are quite similar, indicating that the branching fraction for dissociation to $N^2D + O^3P$ is essentially the same for $\mathfrak{L}$ = even and odd. Unfortunately, Fujii and Morita[100] were unable to reference the

k_D values to the k_A and k_S values, and thus could not provide a unique interpretation of this behavior.

The strong dependence of the decay rates on the parity of $\mathfrak{L}$ was also observed in earlier measurements by Biernacki *et al.*[102] on the rotational linewidths within the $(X^1\Sigma^+)7f$, $v = 1 \leftarrow A^2\Sigma^+$, $v' = 1$ band. In addition, for the $\mathfrak{L}$ = odd levels, Biernacki *et al.*[102] observed a complex dependence of the linewidth on the rotational level of the $(X^1\Sigma^+)7f$, $v = 1$ state. Fujii and Morita[100] showed that this dependence can be understood in terms of the percentage of Σ^+ character in each rotational level, and concluded that the $(X^1\Sigma^+)7f$, $v = 1$, $\mathfrak{L}$ = odd levels are directly coupled with both the dissociative $A'\,^2\Sigma^+$ and $I^2\Sigma^+$ states.

Although NO is a heteronuclear molecule, the asymmetry of the NO^+ ion core is small, and the resulting nearly good u/g quantum number is often useful in understanding the spectroscopy and dynamics of NO. For example, the even and odd ℓ Rydberg series converging to $NO^+\,^1\Sigma^+$ have approximate g and u symmetry, respectively. As a result, perturbations between even ℓ or odd ℓ series are generally much stronger than those between even ℓ and odd ℓ series. Quantum chemical calculations[103] indicate that the dominant configuration necessary to describe the $A'\,^2\Sigma^+$ state is $(1\sigma)^2(1\bar{\sigma})^2(2\sigma)^2(2\bar{\sigma})^2(3\sigma)^2(1\pi)^4(3\bar{\sigma})^1$, which has nearly pure u symmetry, while that for the $I^2\Sigma^+$ state is $(1\sigma)^2(1\bar{\sigma})^2(2\sigma)^2(2\bar{\sigma})^2(3\sigma)^1(1\pi)^4(1\bar{\pi})^2$, which has nearly pure g symmetry. Thus, while predissociation of the "u" $(X^1\Sigma^+)7f$, $v = 1$ levels by the $A'\,^2\Sigma^+$ state is not surprising, predissociation by the "g" $I^2\Sigma^+$ state indicates that some $u - g$ mixing does occur in these states. As will be discussed below, this observation is important to the understanding of the decay processes of some of the higher lying states as well.

Although the experiments of Fujii and Morita[99–101] provide considerable information of the ionization and dissociation yields for super-excited states of NO, they do not provide information of the product-state distributions produced by autoionization. In the region between the $NO^+\ X^1\Sigma^+$, $v^+ =$ 0 and 1 thresholds, only the $v^+ = 0$ level of the ion can be populated; however, in general there are several different rotational levels that can be populated. The population distribution among these levels provides extremely detailed information on the character of the outgoing partial waves and thus on the autoionization mechanism. Recently, Park, Leahy, and Zare[104,105] have reported rotationally resolved photoelectron spectra for autoionizing levels in this region that were accessed in double-resonance transitions via the $A^2\Sigma^+$, $v' = 1$ level. Because the rotational spacings in the ground state of the ion vary as $2BN^+$, where B is the rotational constant, the photoelectron energy resolution required for such measurements can be reduced by studying levels with large values of N^+. Although such measurements are still quite difficult, Park *et al.*[104,105] have succeeded in recording rotationally resolved photoelectron spectra for a large number of s, p, and f autoionizing levels.

Figure 7 shows four of the photoelectron spectra of Park *et al.*[104,105] which were obtained by pumping the $A^2\Sigma^+$, $v' = 1$, $N' = 19$, $J' = 18.5$ level and tuning the probe laser to different autoionizing resonances. The labels used for these resonances are given in Hund's case d notation.[106] Figure 7(a) corresponds to the $(X^1\Sigma^+)14p$, $v = 1$, $N = 19$, $R = 19$, $\mathfrak{L} = 0$ autoionizing level, which in the absence of ℓ mixing has pure Π^- symmetry. Figure 7(b) corresponds to two overlapping $(X^1\Sigma^+)14p$, $v = 1$, $R = 19$ resonances with $N = 18$ and 20 and $\mathfrak{L} = -1$ and $+1$, respectively;

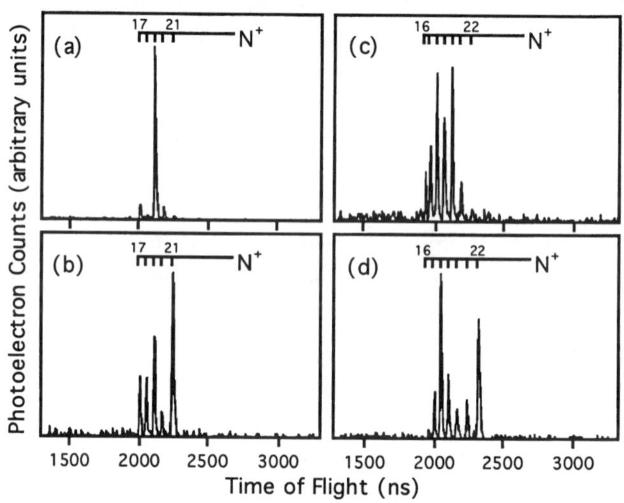

Fig. 7. Time-of-flight photoelectron spectra from vibrational autoionization of (a) the $(X^1\Sigma^+)14p$, $v = 1$, $N = 19$, $R = 19$, $\mathfrak{L} = 0$ level, (b) the $(X^1\Sigma^+)14p$, $v = 1$, N, $R = 19$, $\mathfrak{L} = \pm 1$ level, (c) the $(X^1\Sigma^+)14f$, $v = 1$, N, $R = 19$, $\mathfrak{L}$ level, and (d) the $(X^1\Sigma^+)12f$, $v = 1$, N, $R = 21$, $\mathfrak{L}$ level. The value of N^+ corresponds to the value of the rotational quantum number of the NO^+ $X^1\Sigma^+$, $v^+ = 0$ level produced following autoionization. (Reprinted with permission from Ref. 104. Copyright 1996 American Physical Society.)

these levels were not resolved in the experiments of Park *et al.*[104,105] and both correspond to linear combinations of $p\Sigma^+$ and $p\Pi^+$ Hund's case b states. Figures 7(c) and (d) correspond to the $(X^1\Sigma^+)14f$, $v = 1$, $R = 19$ and $(X^1\Sigma^+)12f$, $v = 1$, $R = 21$ resonances; the different N (and thus $\mathfrak{L}$) components of these resonances were not resolved in the excitation process.

The four photoelectron spectra in Fig. 7 all show markedly different rotational distributions. The significance of these distributions can best be understood in light of the selection rules for excitation of high Rydberg states and photoionization via the $A^2\Sigma^+$ state to the $X^1\Sigma^+$ state of NO^+. For excitation of the high Rydberg states, these rules can be summarized as follows:[107,108]

$$N - N' = 0,\ \pm 1\,, \tag{6}$$

$$J - J' = 0,\ \pm 1\,, \tag{7}$$

$$\begin{aligned} &+ \leftrightarrow -\,,\\ &+ \not\leftrightarrow +\,,\\ &- \not\leftrightarrow -\,, \end{aligned} \tag{8}$$

$$R - N' + \ell = \text{odd}\,, \tag{9}$$

where $\pm$ refers to the total parity. The selection rules for photoionization are the same as those for excitation of the Rydberg states, but with R replaced by N^+ and where it must be remembered that the $\pm$ and N values refer to the electron–ion complex. For the $A^2\Sigma^+$, $v = 1$, $N' = 19$, $J' = 18.5$ intermediate level used in all of the spectra of Fig. 7, Eq. (12) implies that even (odd) rotational levels of NO^+ $X^1\Sigma^+$ can only be produced with even (odd) ℓ partial waves. The allowed range of N^+ is also restricted by the magnitude of ℓ. In addition, for odd ℓ, transitions with $N - N' = 0$, ± 1 access Rydberg states with Λ^- and Λ^+ symmetry, respectively, while for even ℓ, transitions with $N - N' = 0$, ± 1 access states with Λ^+ and Λ^- symmetry, respectively. All of these selection rules are displayed graphically in Fig. 8. By requiring that autoionization preserves N, J, and parity, Fig. 8 can also be used to determine the final states allowed for autoionization of a given resonance.

The $(X^1\Sigma^+)14p$, $v = 1$, $N = 19$, $R = 19$, $\mathfrak{L} = 0$ resonance corresponds to a pure $p\Pi^-$ level, and Fig. 8 shows that autoionization of this resonance

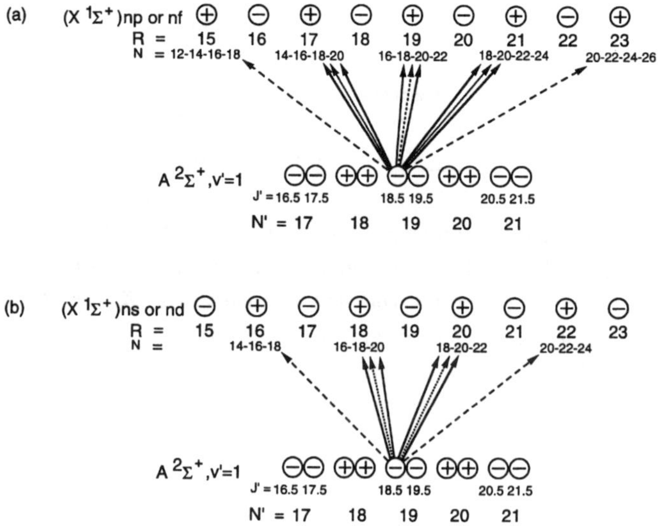

Fig. 8. Schematic diagrams for the selection rules for excitation from the $A^2\Sigma^+$, $v' = 1$, $N' = 19$, $J' = 18.5$ level to the high Rydberg states and ionization continua of NO. (a) odd partial waves; (b) even partial waves. In (a) solid lines (——) indicate transitions to both *np* and *nf* states with Λ^+ symmetry, whereas long dashed lines (– –) indicate transitions limited to *nf* states with Λ^+ symmetry. Similarly, small dashed lines (– – – – –) indicate transitions to both *np* and *nf* states with Λ^- symmetry, whereas dotted lines (······) indicate transitions limited to *nf* states with Λ^- symmetry. In (b), solid lines (——) indicate transitions to both *ns* and *nd* states with Λ^+ symmetry, whereas long dashed lines (– –) indicate transitions limited to *nd* states with Λ^+ symmetry. Small dashed lines (– – – – – –) indicate transitions to *nd* states with Λ^- symmetry. (Reprinted with permission from Ref. 109. Copyright 1998 American Institute of Physics.)

with the ejection of a *p* electron can only result in the population of a single rotational level of NO^+, corresponding to $X^1\Sigma^+$, $v^+ = 0$, $N^+ = 19$. The spectrum in Fig. 7(a) indeed shows a single intense photoelectron peak for $N^+ = 19$, indicating that the autoionization mechanism is relatively simple. The presence of much weaker features for $N^+ = 17$, 18, 20, and 21 must result from a small amount of other partial waves, and may simply reflect the imperfect description of the $A^2\Sigma^+$ and $(X^1\Sigma^+)14p$ states in terms of single electron configurations.

The other photoelectron spectra in Fig. 7 are not so simple. Figure 8 shows that if the overlapping $(X^1\Sigma^+)14p$, $v = 1$, $R = 19$ resonances with $N = 18$ and 20 decay by the ejection of a *p* electron, the former can populate the $N^+ = 17$ and 19 levels of $NO^+\,^1\Sigma^+$, $v^+ = 0$ and the latter can populate the $N^+ = 19$ and 21 levels. Taken together, these decay processes could account for the strong $N^+ = 17$, 19, and 21 peaks in Fig. 7(b), but cannot account for the significant intensity in the $N^+ = 18$ and 20 peaks, which can only be produced through the ejection of even ℓ photoelectrons. This indicates that significant even-odd ℓ mixing occurs in the autoionizing resonances or in the decay process.[104,105] In this respect, the photoelectron spectra for the $(X^1\Sigma^+)14f$, $v = 1$, $R = 19$ and $12f$, $v = 1$, $R = 21$ resonances are even more impressive. Figure 8 indicates that the decay of these resonances into the NO^+ $X^1\Sigma^+$, $v^+ = 0$ continuum with the ejection of an f electron will result in the production of $N^+ =$ even levels, with N^+ restricted to 15, 17, 19, 21, and 23. (Additional restrictions on N^+ would be made if individual N levels could be resolved.) In both cases, however, the integrated intensity of these odd N^+ photoelectron peaks appears to be smaller than the integrated intensity of the even N^+ photoelectron peaks, which can only be produced by the ejection of photoelectrons with even ℓ. Although it must be remembered that the photoelectron spectra are angle-resolved and must be corrected by using the photoelectron angular distributions to determine the true rotational population distributions in the NO^+ $X^1\Sigma^+$, $v^+ = 0$ state, the large intensity of the even N^+ peak does appear to imply a significant fraction of even ℓ photoelectrons.

The great extent of this ℓ mixing is even more surprising in light of the observation that the spectroscopy of the $(X^1\Sigma^+)np$ and $(X^1\Sigma^+)nf$ Rydberg states of NO indicates that mixing with even ℓ levels is relatively unimportant in the description of the resonance positions and structures. Park *et al.*[104,105] pointed out that such ℓ mixing could arise in a simple vibrational autoionization mechanism through an internuclear distance dependence of the quantum defect matrix elements that are off-diagonal in ℓ, i.e. elements $\mu_{\ell\ell'}(\ell \neq \ell')$ for which the values of $d\mu_{\ell\ell'}/dR$ are large. Alternatively, Park *et al.*[104,105] suggested that this ℓ mixing could result from indirect autoionization via dissociative valence states that couple the odd ℓ Rydberg states with even ℓ continua. Park *et al.*[104,105] concluded that additional information was necessary to distinguish between these mechanisms, and suggested that both mechanisms may be necessary to explain the observed data.

The substantial ℓ mixing observed by Park *et al.*[104,105] in the rotationally resolved photoelectron spectra of autoionizing levels is interesting in relation to the ℓ mixing deduced by Fujii and Morita[99–101] from the importance of the "g" $I^2\Sigma^+$ state in the predissociation of the "u" $(X^1\Sigma^+)7f$, $v = 1$ state. Fujii and Morita[99,101] have shown that the Rydberg states in the region studied by Park *et al.*[104,105] are strongly predissociated, as evidenced in the discussion of Figs. 4 and 5. Unfortunately, Fujii and Morita[99,101] confined their studies to very low rotational levels, and direct evidence on the resonances studied by Park *et al.*[104,105] was not available. Recently, however, Pratt[109] performed experiments similar to those of Fujii and Morita for the same autoionizing resonances studied by Park *et al.* Figure 9 shows the fluorescence-dip, NO^+, and N^2D spectra obtained by pumping the $A^2\Sigma^+$, $v' = 1$, $N' = 19$, $J' = 18.5$ level and probing in the region of interest. (Note that the fluorescence-dip spectrum has been inverted to simplify comparison with the other spectra and that the

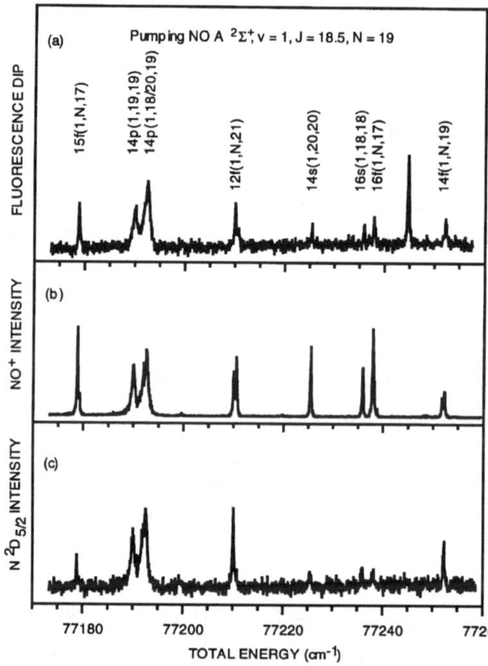

Fig. 9. Two-color ionization spectra of NO obtained by pumping the $A^2\Sigma^+$, $v' = 1$, $N' = 19$, $J' = 18.5$ level and probing transitions between the NO^+ $X^1\Sigma^+$, $v^+ = 0$ and 1 thresholds. The upper states of the transition are labeled by $n\ell(v, N, R)$. (a) $A^2\Sigma^+$ fluorescence (fluorescence-dip spectrum); (b) NO^+ ion intensity; (c) N^+ ion intensity obtained by using two-photon resonant, three-photon ionization of $N^2D_{5/2}$. (Reprinted with permission from Ref. 109. Copyright 1998 American Institute of Physics.)

extra peak at 77 245 cm^{-1} in the fluorescence-dip spectrum results from a transition that is driven by residual light from the fundamental of the frequency-doubled light used for the probe transition.) As in the spectra of Fujii and Morita,[99,101] the relative intensities in the three spectra are quite different, but the N^2D spectrum shows significant similarities to the fluorescence-dip spectrum.

Figure 10 shows a detailed image of the $(X^1\Sigma^+)12f$, $v = 1$, $R = 21$ resonance from Fig. 9. The selection rules outlined above restrict transitions from the $A^2\Sigma^+$, $v' = 1$, $N' = 19$, $J' = 18.5$ level to $(X^1\Sigma^+)12f$, $v = 1$, $R = 21$ levels with $N = 18$, 19, and 20 ($\mathfrak{L} = -3, -2$, and -1, respectively). The structure of an unperturbed nf complex can be approximated by $-3/2k\mathfrak{L}^2$, and the transitions to the $\mathfrak{L} = -1$, -2, and -3 levels have been assigned to features at 77 210.8, 77 210.5, and 77 210.0 cm^{-1}, respectively, with $k = 0.07$ cm^{-1}. The relative intensity of the $\mathfrak{L} = -3$ transition is greatest

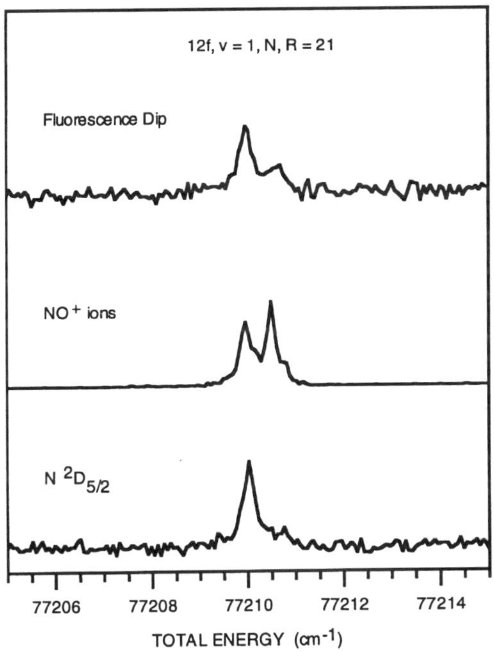

Fig. 10. An expanded portion of Fig. 9 showing the $(X^1\Sigma^+)12f$, $v = 1$, N, $R = 21 \leftarrow A^2\Sigma^+$, $v' = 1$, $N' = 19$, $J' = 18.5$ transition. The upper states of the transition are labeled by $nf(v, N, R)$. (Reprinted with permission from Ref. 109. Copyright 1998 American Institute of Physics.)

in the N^2D spectrum, almost as great in the fluorescence-dip spectrum, and by far the smallest in the NO^+ spectrum. In contrast, the relative intensities of the $\mathfrak{L} = -2$ transitions show the opposite behavior. As in the spectra of Fujii and Morita[100] for the $(X^1\Sigma^+)7f$, $v = 1$ state, the $\mathfrak{L} =$ odd levels in Fig. 10 have Λ^+ symmetry, and those with $\mathfrak{L} =$ even have Λ^- symmetry. Thus, just as in the $(X^1\Sigma^+)7f$, $v = 1$ state, Fig. 10 clearly indicates that predissociation by $^2\Sigma^+$ states is also quite important for the higher rotational levels of higher lying $(X^1\Sigma^+)nf$, $v = 1$ states.

Enhanced predissociation of the $\mathfrak{L} =$ odd levels to N^2D implies a significant role for the $I^2\Sigma^+$ state in the decay dynamics of the $(X^1\Sigma^+)12f$, $v = 1$, $R = 21$ state. Because the $I^2\Sigma^+$ state has "g" symmetry, indirect autoionization of the "u" $(X^1\Sigma^+)12f$, $v = 1$, $R = 21$ state through the $I^2\Sigma^+$ state can reasonably be expected to result in the production of even ℓ photoelectrons. If this is the case, the photoelectron spectra of Park *et al.*[104,105] for the $(X^1\Sigma^+)12$ and $14f$, $v = 1$ resonances imply that this indirect mechanism accounts for a large fraction of the total ionization signal. Similar arguments would suggest that this mechanism also plays a role in the population of even NO^+ $X^1\Sigma^+$, $v = 0$ rotational levels in the decay of the $(X^1\Sigma^+)14p$, $v = 1$, $N = 18$ and 20, $R = 19$ resonances.

As in the spectra of Fujii and Morita,[99,101] Fig. 9 can be used to provide an upper limit to the ionization efficiencies of the observed resonances. If it is assumed that the ionization efficiency of the resonance with the largest ratio of ionization intensity to fluorescence-dip intensity, which corresponds to the $(X^1\Sigma^+)14s$, $v = 1$, $N = 20$, $R = 20$ state, is 100%, the ionization efficiencies of the other resonances range from 30% for the two $(X^1\Sigma^+)14p$ resonances to 80–85% for the $(X^1\Sigma^+)16s$, $v = 1$, $N = 18$, $R = 18$ and $(X^1\Sigma^+)16f$, $v = 1$, $R = 17$ resonances. However, even the $(X^1\Sigma^+)14s$, $v = 1$, $N = 20$, $R = 20$ resonance predissociates to some extent, as can be seen in Fig. 9(c). A better estimate of the ionization efficiency can be obtained by using the ionization efficiencies deduced by Giusti-Suzor and Jungen[15] for low-lying $(X^1\Sigma^+)np\pi$, $v = 1$ resonances to scale the observed data. Giusti-Suzor and Jungen gave ionization efficiencies of ~ 2–3% for the $8p\pi$, $9p\pi$, and $10p\pi$ levels; because, in the first approximation, predissociation and autoionization both scale as $1/n^{*3}$, the ionization efficiency of the pure $\pi(X^1\Sigma^+)14p$, $v = 1$, $N = 19$, $R = 19$ resonance should be similar. If this is the case, the ionization efficiency of even the $(X^1\Sigma^+)14s$, $v = 1$, $N = 20$, $R = 20$ resonance is only 7–8%. This is still, however, a factor of ~ 3 greater that of the $(X^1\Sigma^+)np$ resonances.

The significant variation in the ionization efficiency for different autoionizing resonances, as well as the significant intensity variations in the N^2D and N^4S spectra, show that autoionization and predissociation in the Rydberg states of NO can show significant mode-specific effects. While this is perhaps not too surprising for a relatively simple diatomic molecule, the following sections show how similar effects can arise in the high Rydberg states of triatomic and larger polyatomic molecules.

5. Triatomic Molecules: HCO and NO_2

One of the most intriguing aspects of vibrational autoionization in polyatomic molecules is how the decay rates depend on the normal mode that is excited. For example, in a triatomic molecule, how do the vibrational autoionization rates via the asymmetric stretching vibration, the symmetric stretching vibration, and the bending vibration compare? While the formalism described above suggests that the important factors are the Q_i dependence of the quantum defect matrix elements and the magnitude of the displacement along the normal coordinate,[75,81] quantitative estimates of these factors are not always available. Over the last few years, Grant and coworkers have performed a series of experiments to address this question for HCO[24–26] and NO_2,[16–23] by using multiple-resonance experiments with ionization detection to study vibrational autoionization of Rydberg series converging to selected vibrational levels of HCO^+ and NO_2^+. By comparing the widths of autoionizing resonances with the same value of n but converging to different vibrational states of the ion, Grant and coworkers have begun to reveal some interesting trends in the mode dependence of vibrational autoionization. These trends are discussed in a number of recent publications,[16–28] to which the interested reader is referred. The discussion here focuses on their studies of vibrational autoionization via the bending vibrations, and how they relate to questions about the competition between autoionization and predissociation.

Figure 11 shows the spectrum of HCO obtained by Mayer *et al.*[26] by monitoring the HCO^+ ion signal following double-resonance excitation via the $N' = 2$ level of the upper π component of the $3p\pi\,^2\Pi$ (020) Rydberg state. Excitation via this state is expected to result in strong transitions to Rydberg series converging to the $X^1\Sigma^+$ (020) state of HCO^+, that is, the Rydberg series are expected to have two quanta in the bending vibration. The x-axis in Fig. 11 corresponds to the energy of the second photon in the

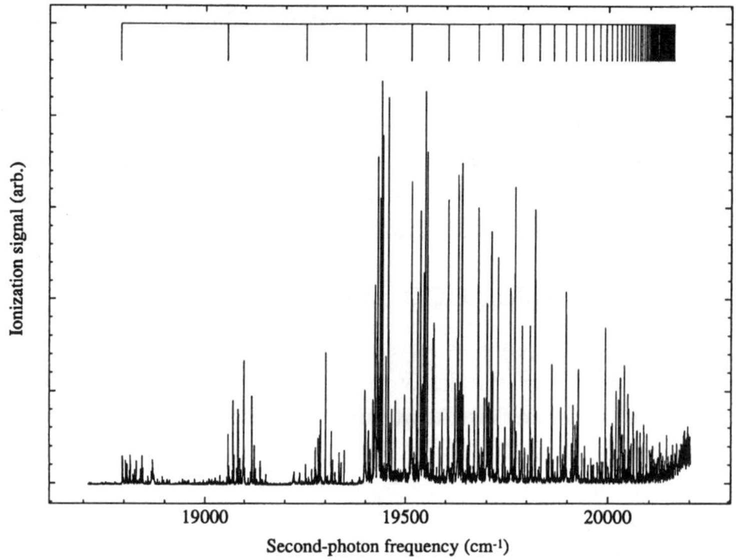

Fig. 11. Spectrum of HCO obtained by pumping the $N' = 2$ level of the upper π component of the $3p\pi\,^2\Pi(020)$ system. On this energy axis, the HCO^+ $X^1\Sigma^+(000)$, (010), and (020) ionization thresholds are at approximately 18 570 cm^{-1}, 19 400 cm^{-1}, and 20 170 cm^{-1}, respectively. (Reprinted with permission from Ref. 26. Copyright 1998 American Institute of Physics.)

double-resonance process. On this scale, the HCO^+ $X^1\Sigma^+$ (020) threshold is at approximately 20 170 cm^{-1}, and the $X^1\Sigma^+$ (000) and (010) thresholds are approximately 18 570 cm^{-1} and 19 400 cm^{-1}, respectively.[26]

Although quite complex, considerable progress has been made in assigning the transitions observed in Fig. 11. However, several interesting observations can be made even without this analysis. In particular, there is a dramatic decrease in the intensity of the Rydberg series below 19 400 cm^{-1}, which is accompanied by a significant narrowing of the linewidths below this energy. In simulations of the spectrum of Fig. 11, Mayer *et al.*[26] found good agreement by assuming the intensities decreased by a factor of 20 and the linewidths by a factor of 5 below the HCO^+ $X^1\Sigma^+$ (010) threshold. The decrease in linewidth is consistent with the vibrational propensity rule, in that above this threshold vibrational autoionization can occur with $\Delta v = -1$, whereas below this threshold autoionization with $\Delta v = -2$ is required. Thus, the comparison of the linewidths indicates that $\Delta v = -1$ autoionization is at least a factor of 5 faster that $\Delta v = -2$ autoionization.[26]

Although the decrease in linewidth below the $X^1\Sigma^+$ (010) threshold can simply be assumed to result from the propensity rule for vibrational autoionization, in the absence of competing decay processes the intensity of a transition should not be decreased as a result of a decrease in the linewidth of a transition. Indeed, given the instrumental linewidth is smaller than the natural linewidth both above and below the $X^1\Sigma^+$ (010) threshold, the intensity of the peaks below the $X^1\Sigma^+$ (010) threshold are actually expected to increase in peak intensity. Such behavior would certainly be expected in the absorption spectrum from the HCO $3p\pi\,^2\Pi$ (020) intermediate state. Unfortunately, no such absorption spectrum is currently available. Mayer *et al.*[26] observed that the narrower linewidths and lower intensities in the (020) spectrum below the (010) threshold suggest an increase in branching to neutral fragments via predissociation. More specifically, the discontinuity and decrease in the intensity of the observed series at the $X^1\Sigma^+$ (010) threshold in the ionization spectrum imply that a second decay process competes effectively with $\Delta v = -2$ vibrational autoionization, but not so effectively with $\Delta v = -1$ autoionization. Fluorescence from the Rydberg state can be ruled out as a decay mechanism. Based on the simulation parameters reported by Mayer *et al.*[26] a crude estimate of the linewidths for the $n = 11$ manifold of the (020) series, which is below the $X^1\Sigma^+$ (010) threshold, gives a value of 0.14 cm^{-1}, corresponding to a lifetime of 0.2 ns, considerably shorter than the expected fluorescence lifetime. Thus, the competing process will be referred to as predissociation. Because HCO is such a small molecule, this predissociation is most likely a direct process resulting from a crossing between the Rydberg state potential surface and that of a dissociative state. However, it is possible that the Rydberg state undergoes rapid internal conversion or intersystem crossing followed by dissociation on a longer timescale. In either case, the predissociation rate is expected to scale as $1/n^{*3}$, and should not be sensitive to the position of the HCO^+ $X^1\Sigma^+$ (010) threshold.

The factor of five increase in the linewidth above the $X^1\Sigma^+$ (010) threshold indicates that $\Delta v = -1$ autoionization is at least five times faster than predissociation, and that ionization is the dominant decay channel. This conclusion also indicates that the ionization mechanism above the $X^1\Sigma^+$ (010) threshold is relatively pure vibrational autoionization. However, the factor of 20 decrease in the ionization intensity below the $X^1\Sigma^+$ (010)

threshold indicates that there, predissociation is the dominant decay mechanism. This observation also raises the possibility that indirect autoionization via the dissociative state(s) may contribute to the mechanism for the observed $\Delta v = -2$ autoionization. Unfortunately, the evidence for these dissociative processes is indirect, and more detailed experiments are clearly necessary. In particular, excited state absorption measurements from the intermediate state are expected to be informative. In addition, monitoring neutral H or CO fragments produced by the same double-resonance excitation shown in Fig. 11 is expected to provide considerable insight into the character of the predissociating state(s).

The Rydberg states of NO_2 display rather different behavior from HCO.[16–26] As with HCO^+, the ground electronic state of NO_2^+ is linear,[19] as are the Rydberg series converging to it. The Rydberg states of interest here were studied by Matsui and Grant[21] by using triple-resonance excitation via selected vibrational levels of the $3p\sigma^2\Sigma_g^+$ electronic state at the two-photon energy. Excitation via the $3p\sigma$ state provides access to the (100), (010), (02^00), and (02^20) vibronic levels that are used to access the corresponding Rydberg series of interest here. The superscript 0 and 2 for the bending quantum number indicates the vibrational angular momentum, l_v. From the $3p\sigma^2\Sigma_g^+$ electronic state in NO_2, Rydberg series converging to the (02^00), and (02^20) levels of the ground state of NO_2^+ can be selectively excited.[21] (The intermediate state used by Mayer *et al.*[26] for the (020) Rydberg series of HCO had mixed (02^00), and (02^20) character, resulting in the simultaneous excitation of series converging to both levels of the ion.)

Figure 12 shows the spectra obtained by Matsui and Grant[21] via the (02^00), (02^20), and (010) levels of the $3p\sigma$ intermediate state. In this figure, the (000), (010), (02^00), and (02^20) thresholds occur at approximately 77 316 cm^{-1}, 77 937 cm^{-1}, 78 554 cm^{-1}, and 78 567 cm^{-1}, respectively.[21] Several aspects of Fig. 12 are quite striking, particularly in comparison with the HCO (020) spectrum in Fig. 11. For example, neither the (02^00) spectrum nor the (02^20) spectrum of NO_2 show signs of a change in intensity at the (010) threshold. In addition, although the linewidths of the (02^00) spectrum are on average considerably greater than those of the (02^20) spectrum, the linewidths in both spectra vary smoothly through the (010) threshold. [This is readily apparent from the figure for the (02^00) spectrum, but is also true upon closer examination of the (02^20) spectrum.] The smooth variation of the linewidths through the (010) threshold

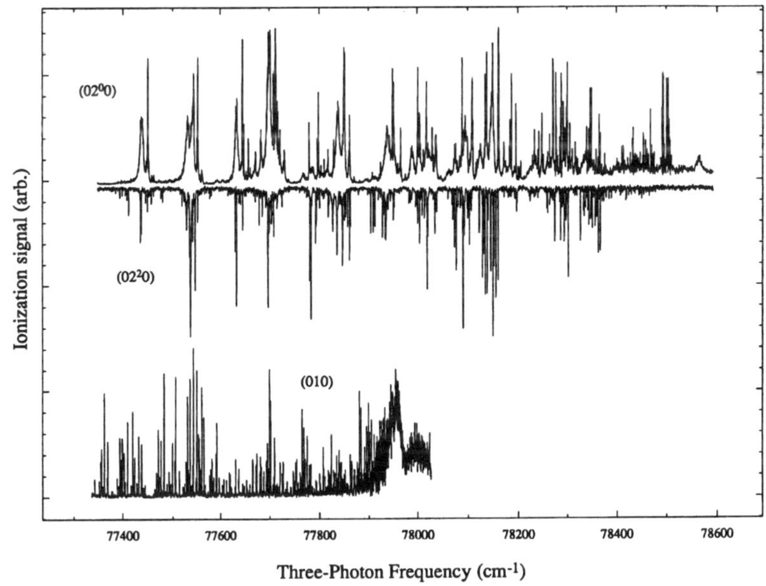

Fig. 12. Triple-resonance ionization spectra of NO_2 via two-photon resonances with the (02^00), (02^20), and (010) levels vibrational levels of the $3p\sigma\,^2\Sigma_g^+$ electronic state. The (000), (010), (02^00), and (02^20) thresholds occur at approximately 77 316 cm^{-1}, 77 937 cm^{-1}, 78 554 cm^{-1}, and 78 567 cm^{-1}, respectively, but in this figure, the energy axis for the (02^20) spectrum has been shifted to align its ionization limit with that of (02^00). (Reprinted with permission from Ref. 21. Copyright 1996 American Institute of Physics.)

implies that the vibrational autoionization rate for the $\Delta v = -2$ processes below the (010) threshold is not significantly different from that for the $\Delta v = -1$ process that becomes allowed above this threshold.[21] From the appearance of the ionization spectrum alone it is difficult to assess the role of predissociation in these states, and it may in fact be negligible or small. Direct probing of the potential fragments will be required to assess the predissociation of these states. If predissociation does compete with autoionization for these states, the competition also appears to vary smoothly with energy over this entire region of the spectrum.

The lack of observable changes in the (02^00) and (02^20) spectra at the (010) threshold is certainly at odds with the vibrational propensity rule, which would predict a large increase in the vibrational autoionization rate for $\Delta v = -1$ processes. Matsui and Grant[21] have suggested that this discrepancy can be explained in terms of the Fermi resonance between the (02^00) and (100) levels in the ground state of NO_2^+ ion as well as in the corresponding Rydberg states. By mixing (100) character into the (02^00) states, the nominal (02^00) Rydberg states can decay by $\Delta v = -1$ throughout the spectrum in Fig. 12. Indeed, Matsui and Grant[21] have shown that the widths in the (02^00) spectrum are comparable to the widths of the analogous peaks in the (100) spectrum. While no definitive spectroscopic assignments of the (100) and (02^00) Rydberg spectra have been made, the similarity of the widths of states with the same value of principal quantum number supports this interpretation. Unfortunately, this explanation does not explain the appearance of the (02^20) Rydberg states, which have the wrong symmetry to interact with the (100) states. Matsui and Grant have suggested,[21] however, that the Coriolis interaction between the (02^00) and (02^20) Rydberg states would mix these two symmetries and, as a result, indirectly mix in character from the (100) state. This admixture would be small, which is consistent with the considerably reduced widths in the (02^20) Rydberg series, providing an attractive interpretation of the observed results.

If the mixing with the (100) Rydberg series is responsible for vibrational autoionization of the (02^00) and (02^20) series, the photoelectron spectra both below and above the (010) threshold should show a strong photoelectron peak corresponding to the production of the (000) level of NO_2^+. Without this admixture, $\Delta v = -1$ vibrational autoionization of the (02^00) and (02^20) series is expected to produce NO_2^+ (000) below the (010) threshold and NO_2^+ (010) above the (010) threshold. Thus, a determination of the photoelectron branching ratios of the autoionizing states should provide a valuable test of the proposed model.

It is clear from these examples that HCO and NO_2 display distinctly different dynamics in their Rydberg series converging to bending vibrational levels of the ground electronic state of the corresponding ion. In HCO, the competition between predissociation and vibrational autoionization is quite obvious from the appearance of the ionization spectrum, while in NO_2 there is no clear evidence for predissociation. Even in HCO, the extent of information that can be obtained from ionization alone is limited. More specifically, analysis of steps in ionization spectra only provides information on the average ionization efficiency of all of the series converging to a given limit. In practice, as was illustrated by the results on H_2 and

NO, the ionization efficiency can depend strongly on the electronic symmetry, principal quantum number, vibrational quantum numbers, and even the rotational quantum numbers of the Rydberg state. Thus, for example, an overall ionization efficiency of 50% determined from the analysis of a threshold step size could result from each of the optically allowed series having an ionization efficiency of 50%, or from some of the series having very high ionization efficiencies and others having very low efficiencies, with an average ionization efficiency of 50%. This situation has been discussed in some detail by Eland.[110] In the simplest case of a single autoionizing series converging to a single threshold, Eland has shown that, in the absence of competing decay processes, the expected intensity, h_r, of an autoionizing resonance (above the direct ionization continuum signal) can be related to the height, h_c, of the step at the threshold to which it converges by the expression

$$h_r = 2\mathfrak{R}h_c/g\,, \tag{10}$$

where g is a reduced width given by

$$\Gamma = gn^{*-3}\,, \tag{11}$$

and Γ is the width of a single resonance in the series. Incomplete autoionization results in an observed h_r that is smaller than that given by Eq. (3). In the general case of several series, Eland has written[110]

$$\sum_i g_i \alpha_i h_i = 2\mathfrak{R}h_c\,, \tag{12}$$

where α_i is essentially the inverse of the autoionization fraction of the ith series. Unfortunately, in the case of two or more series, the α_i cannot be uniquely determined by the ionization data alone. Determining the α_i for individual series thus requires an alternative form of measurement such as fluorescence-dip spectroscopy or the direct detection of predissociation fragments. Such experiments on HCO and NO_2 can be expected to provide an improved understanding of the competition between ionization and dissociation in these molecules.

6. Larger Polyatomic Molecules

6.1. NH_3

In recent years, there has been considerable progress in understanding the spectroscopy and dynamics of the Rydberg states of ammonia converging to the ground electronic state of NH_3^+. In particular, the multiphoton ionization studies of Nieman,[111] Colson,[112,113] and Glownia *et al.*[114,115] provided significant new insight into the Rydberg states of NH_3, and the more recent studies of Ashfold and coworkers[116–119] have greatly extended this work and addressed many detailed aspects of the predissociation dynamics. Relatively little work has been done above this threshold, and the analysis of the spectroscopy and dynamics in this region has proceeded much more slowly. The single photon ionization spectrum[110] shows a series of distinct steps above the first ionization threshold, corresponding to a vibrational progression in the ν_2 umbrella motion of NH_3^+. This progression is a result of the Franck–Condon factors between the pyramidal ground state of the neutral and the planar ground state of the ion. The observation of steps implies substantial predissociation of the Rydberg states converging to these limits.[92] Thus, as in NO, it might be expected that indirect autoionization via the predissociative electronic states may occur along with pure vibrational autoionization. Unfortunately, evidence for such a process is minimal. Eland[110] reported a high-resolution single photon ionization spectrum of NH_3, in which he found evidence for two Rydberg series converging to each threshold of the ν_2 progression. Because the members of the series converging to $NH_3^+\ X^2A_i''(0, v_2, 0, 0)$ were only observed above the $NH_3^+\ X^2A_i''(0, v_2, -1, 0, 0)$ threshold, Eland concluded that the propensity rule for vibrational autoionization was obeyed.[110] This suggests that the indirect autoionization mechanism may not be so important in NH_3. However, using an analysis based on Eq. (3), Eland[110] also concluded that the ionization efficiency of the more intense series was only about 17%. This implies that, although indirect autoionization via dissociative states is not important, both dissociation and ionization are significant decay mechanisms.

Cramb and Wallace[27] have recently reported a two-color ionization spectrum of NH_3 via the $\tilde{C}'\,{}^1A_1'(0100)$ state in the region between the $NH_3^+\ X^2A_1''(0000)$ and (0100) thresholds. This spectrum, which was obtained by pumping two overlapped rotational levels in the intermediate state, is shown in Fig. 13. The strong continuum signal throughout the spectrum is produced by one-color ionization by the pump laser alone. In addition, the extremely broad feature centered at $\sim$ 5668 Å results from stimulated emission from the $\tilde{C}'\,{}^1A_1'(0100)$ level to the strongly predissociated $\tilde{A}^1A_1''(0100)$ level.[27] The Rydberg orbital of the $\tilde{C}'\,{}^1A_1'$ state is the

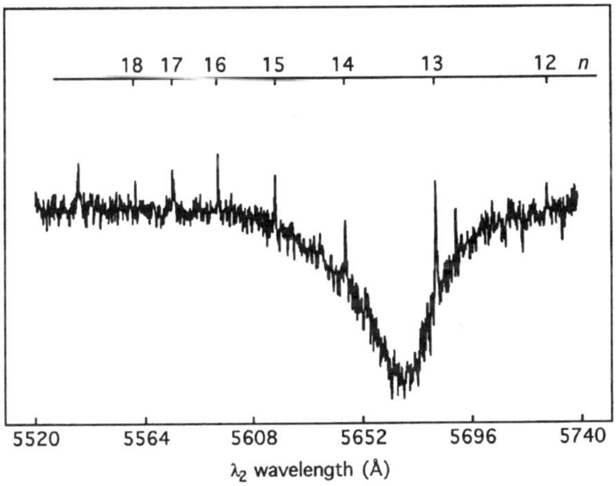

Fig. 13. Two-color ionization spectrum of NH_3 via the $\tilde{C}'\,{}^1A_1'(0100)$ state in the region between the NH_3^+ $X^2A_1''(0000)$ and (0100) thresholds. The spectrum was obtained by pumping two overlapped rotational levels in the intermediate state. The large continuum signal results from two-photon resonant, three-photon ionization by the pump beam alone, and the large dip centered at $\sim$ 5668 Å results from stimulated emission from the $\tilde{C}'\,{}^1A_1'(0100)$ level to the strongly predissociated $\tilde{A}^1A_1''(0100)$ level. (Reprinted with permission from Ref. 27. Copyright 1994 American Institute of Physics.)

same ($3pa_2''$) as that of the HOMO of the X^1A_1 state ($2pa_2''$). As a result, transitions from the $\tilde{C}'\,{}^1A_1'$ state are expected to access Rydberg series of the same symmetry as from the ground state. However, because the $\tilde{C}'\,{}^1A_1'$ state is planar, the Franck–Condon factors are expected to be strongly peaked for transitions with $\Delta v = 0$. Cramb and Wallace[27] identified a single Rydberg series in Fig. 13 and assigned it as the $(\tilde{X}^2A_1'')nsa_1'$ series. At least two additional peaks are observed in Fig. 13 and whether they belong to other series converging to $\tilde{X}^2A_1''(0100)$ or to series converging to higher thresholds is uncertain. Thus, insight into the decay dynamics is limited at this time. Double-resonance experiments are currently underway to extend these results to higher n and to additional intermediate levels.[120] Preliminary spectra show that the series of Cramb and Wallace can be extended to significantly higher $n(\sim 45)$ and that additional series can be observed. Fluorescence-dip spectroscopy using the $\tilde{C}'\,{}^1A_1' \rightarrow \tilde{A}^1A_1''$ fluorescence reported previously is planned to assess the ionization efficiency, and photoelectron spectroscopy is planned to elucidate the ionization mechanism. It is hoped that such experiments will bring together the single photon results of Eland[110] with the double-resonance results of Cramb and Wallace,[27] and bring the understanding of the super-excited states up to that of the Rydberg states below the ionization threshold.

6.2. *Dabco*

As in the case of NO_2, HCO, and NH_3, the principal source of data on the super-excited states of larger molecules is two-color photoionization spectroscopy, and in only a few cases has fluorescence-dip spectroscopy been performed as well. One particularly interesting example in which both ionization and fluorescence-dip experiments have been performed is Dabco.[29–31] Dabco is a 20-atom molecule with the chemical formula $C_6N_2H_{12}$ and 54 vibrational degrees of freedom. Fisanick *et al.*[28] performed the initial double-resonance experiments on Dabco in a static gas cell and observed Rydberg states with n up to 41 based on the ground electronic state of the ion. Later, Fujii *et al.*[29–31] performed much more extensive experiments using a seeded molecular beam to minimize collisional effects and reduce the sample temperature. The cold sample made it possible for Fujii *et al.* to excite individual vibronic levels of the S_1 state of Dabco selectively, and to perform both fluorescence-dip and ionization experiments via these selected states.

Although the comparison of the fluorescence-dip and ionization spectra of Dabco is most interesting, the ionization spectra have a number of interesting features of their own. For example, essentially all of the ionization spectra reported by Fujii *et al.*[29–31] show a sharp rise at the vertical ionization thresholds, regardless of the vibronic level excited within the S_1 state. That is, for a given set of v_i in S_1, a sharp step is observed at the $v_i^+ = v_i$ threshold of the ion. This implies that the geometry of the S_1 state is similar to that of the ion, and that dissociation (or a nonionizing radiationless transition) is very much ($> 10\times$) faster than autoionization. As in the case of NH_3, however, there is also evidence that the autoionization mechanism obeys the vibrational propensity rule. Consider, for example, the spectrum shown in Fig. 14, which Fujii *et al.*[30] obtained by pumping the $2^1 27^1$ vibronic level of the S_1 state (where the notation implies one vibrational quantum in ν_2 and one in ν_{27}) and probing between the vibrationless ionization threshold and the $2^1 27^1$ threshold. The 27^1, 2^1, and $2^1 27^1$ ionization thresholds lie approximately 453 cm^{-1}, 879 cm^{-1}, and 1332 cm^{-1}, respectively above the vibrationless threshold. In the spectrum obtained via

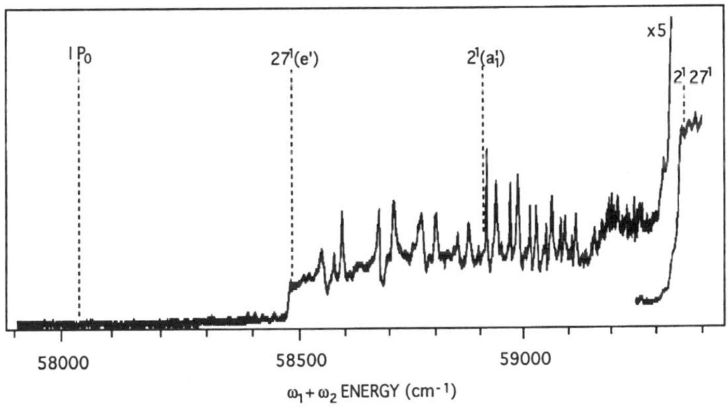

Fig. 14. Two-color ionization spectrum of Dabco obtained by pumping the $2^1 27^1$ vibronic level of the S_1 state and probing in the region just above the first ionization threshold. The ionization thresholds for the vibrationless ground state and the 27^1, 2^1, and $2^1 27^1$ vibronic levels of the ion are indicated by broken lines. (Reprinted with permission from Ref. 30. Copyright 1984 American Chemical Society.)

the $S_1 2^1 27^1$ level, essentially no autoionizing structure was observed below the 27^1 ionization threshold, where autoionization requires $\Delta v = -2(\Delta v_2 = -1$ and $\Delta v_{27} = -1)$. However, above the 27^1 threshold, autoionization can occur by $\Delta v = -1(\Delta v_2 = -1$ and $\Delta v_{27} = 0)$, and several autoionizing Rydberg series are observed. Above the $v_2 = 1$ threshold, autoionization by $\Delta v_2 = 0$ and $\Delta v_{27} = -1$ becomes allowed, and the intensities in the same autoionizing series observed above the 27^1 threshold become still more intense. These observations clearly show that the $\Delta v = -1$ processes are more efficient that the $\Delta v = -2$ processes. This conclusion is supported by additional results from spectra via other S_1 vibronic levels, and led Fujii *et al.*[30] to conclude that vibrational autoionization is the dominant ionization mechanism. Although it has not been quantified, a comparison of the relative intensities of the autoionizing series above and below the 2^1 threshold should allow a determination of the relative rates of vibrational autoionization via $\Delta v_2 = 0 \Delta v_{27} = -1$ and $\Delta v_2 = -1 \Delta v_{27} = 0$. Such a comparison would provide insight into the mode-dependence of the inelastic electron–ion scattering process and on the variation of the relevant quantum defects along the ν_2 and ν_{27} normal coordinates.

The combination of large steps at vertical ionization thresholds and autoionization that obeys the vibrational propensity rule suggests that, although both ionization and dissociation occur, there is little indirect autoionization via dissociative states. One possible explanation for this is that the dissociation itself occurs through an indirect mechanism. For example, the dissociation mechanism could involve internal conversion or intersystem crossing to a lower lying electronic state with significant vibrational excitation, followed by statistical or quasistatistical unimolecular decomposition. As a result of their lower electronic energy, the states populated following the initial radiationless transition would be unlikely to ionize, because autoionization would require an extremely large value of $|\Delta v|$. This scenario would imply, however, that the interactions responsible for internal conversion or intersystem crossing do not contribute significantly to any indirect autoionization mechanism. Although additional experiments are necessary to address these issues, the existing results suggest that their study may ultimately lead to a better understanding of the general mechanisms of radiationless transitions.

The fluorescence-dip spectra reveal a number of additional interesting aspects of the dynamics of the Rydberg states of Dabco.[30,31] Figure 15 shows the fluorescence-dip and ionization spectra obtained by Fujii *et al.*[30] by pumping the $S_1 20^1$ vibronic level and probing between the vibrationless ionization threshold and the 20^1 threshold. Although both spectra appear to show three series converging to the 20^1 threshold, only the first two series (corresponding to the two lowest energy peaks in each triplet) are the same in both spectra. The third series (corresponding to the highest energy peak in each triplet) in the fluorescence-dip spectrum actually has a somewhat larger quantum defect than the third series in the ionization spectrum, and Fujii *et al.*[30] have assigned them as different series. Because the third series in the ionization spectrum is not clearly visible in the fluorescence-dip spectrum, it must have a relatively high ionization efficiency, and the ionization efficiency of the other series must be small. The latter conclusion is consistent with the large step at the 20^1 threshold.

A more detailed comparison of the relative intensities of the first two series in the fluorescence-dip and ionization spectra provides more information on the relative ionization efficiencies. For example, in the triplet between 58 200 cm^{-1} and 58 300 cm^{-1}, the second peak is approximately a factor of two smaller than the first in the fluorescence-dip spectrum, but somewhat more intense than the first peak in the ionization spectrum, indicating at least a factor of two higher ionization efficiency for the second

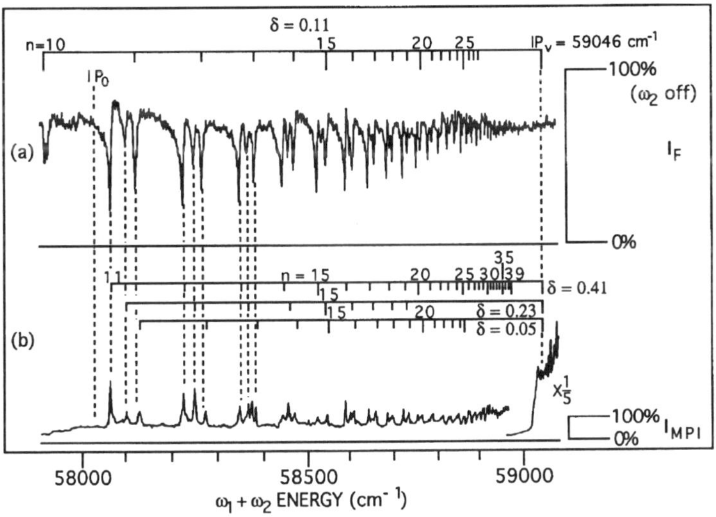

Fig. 15. Two-color fluorescence-dip (a) and ionization (b) spectra of Dabco obtained by pumping the 20^1 vibronic level of the S_1 state and probing in the energy region just above the first ionization threshold. The scale to the right of the fluorescence-dip spectrum indicates the percentage of depletion of the fluorescence signal and shows that there is a strong continuous absorption throughout this energy region. (Reprinted with permission from Ref. 30. Copyright 1984 American Chemical Society.)

series. Similar patterns occur throughout the spectrum. Although these differences have not been fully characterized, they suggest a strong ℓ dependence of the decay dynamics.

Perhaps the most striking difference between the fluorescence-dip and ionization spectra of Dabco in Fig. 15 is that the lineshapes are radically different,[30] even though the homogeneous linewidths must be the same. More specifically, the profiles in the fluorescence-dip spectrum tend to be quite asymmetric with relatively small negative Fano q parameters.[121] By varying the probe laser intensity, Fujii *et al.*[30] confirmed that these lineshapes were not a result of a saturation effect. In contrast, the profiles in the ionization spectrum are far more symmetric, and with what appear to be large positive Fano q parameters. A related observation is that, although there is essentially no continuum signal in the ionization spectrum, the fluorescence-dip spectrum shows a relatively strong continuous absorption throughout the energy range of Fig. 15, suggesting a rather strong direct photodissociation continuum. These observations are consistent with a picture based on Fano's analysis of a discrete state interacting with a continuum.[121] The lineshape for such a state is parameterized in terms of

$$q = \frac{\langle \Psi_i | r | \Phi \rangle}{\pi V_{E'} \langle \Psi_i | r | \Psi_{E'} \rangle}, \tag{13}$$

where $\langle \Psi_i | r | \Phi \rangle$ is the transition moment from the initial state to the discrete state mixed into the continuum (in this case, the dissociation or ionization continuum), $\langle \Psi_i | r | \Psi_{E'} \rangle$ is the transition moment to the unperturbed continuum alone, and $V_{E'}$ is the interaction matrix element between the discrete and continuum state. In the ionization spectrum of Dabco, the continuum matrix element is small, resulting in a large value of q, whereas in the fluorescence-dip spectrum, the dissociation continuum matrix element is significantly larger, resulting in a smaller value of q. The different signs of the q parameter in the two spectra implies that either the signs of the discrete-continuum interactions for ionization and dissociation are different, or the signs of the transition moments to the direct dissociation and ionization continua are different.

The results on Dabco discussed here show clear evidence for mode-specific behavior, including a dependence of the autoionization efficiency on the quantum numbers of the Rydberg states and (although not quantified) a dependence on the vibrational mode excited. Although these observations might at first appear to be surprising for a molecule the size of Dabco, this type of mode dependence has also been observed in other larger molecules, including pyrazine[68,69] and aniline.[66,67] For example, Hager *et al.*[66,67] recorded double-resonance ionization spectra of aniline ($C_6H_5NH_2$) via a number of different vibronic levels of the S_1 intermediate state. For many vibronic levels, a large step is observed at the vertical ionization threshold, but no resonant autoionizing structure is observed below this threshold. This observation implies that nonionizing radiationless transitions are much faster than autoionization for these vibrational modes. For some vibronic levels, however, resonant structure is observed. For example, excitation of the inversion mode (the umbrella motion of the N atom) with $v_I = 1$, 2, or 3 in the S_1 state results in significant resonant structure that obeys the $\Delta v = -1$ propensity rule. That is, the structure only appears above the $v_I = -1$ threshold. Similar behavior was observed for a small number of other vibrational modes.[67] The observation of autoionization for some normal modes and not for others implies either autoionization is faster for the former modes, competing decay processes are slower, or a combination of the two. In any case, the further study of this mode-specific behavior should provide insight into the coupling of the vibrational and electronic motion in polyatomic molecules, and, more generally, into intramolecular energy flow within isolated molecules.

7. Summary

In this review, selected examples of the competition between autoionization and predissociation in molecular Rydberg states have been presented. These examples show that this competition can show strong mode-specific behavior in both small and large molecules. In diatomic molecules such as NO, detailed studies of both neutral and charged products have been performed, and this information is providing a more unified and complete picture of how the decay processes depend on the quantum numbers of the Rydberg states. Although the comparison of spectra in different detection channels provides valuable insight, it is quite difficult to put the relative intensities in different channels on an absolute scale. In this regard, ion imaging techniques[58] may prove valuable because they allow the simultaneous recording of the spectra of several neutral dissociation channels.

Although the information on larger molecules is certainly less comprehensive, convincing evidence for mode-specific effects has been observed in a number of systems. The comparison between fluorescence-dip spectra and ionization spectra is particularly valuable, but considerable insight has been gained from double-resonance ionization spectra alone. Perhaps the largest deficiency has been that there have been few studies of neutral dissociation products from the decay of super-excited states of larger molecules. Fortunately, a huge body of work exists on the photodissociation dynamics of low-lying states of both small and large molecules, and the techniques that have been developed are readily applicable above the ionization threshold. With the application of these techniques to the super-excited states of larger molecules, it is hoped we can go beyond the observation and cataloging of mode-specific effects in the branching between ionization and dissociation, and develop a fuller and more predictive understanding of these effects. The connection of these decay processes with many fundamental chemical processes ensures that such an understanding will have significant utility both inside and outside the field of molecular photoionization and photodissociation dynamics.

Acknowledgments

This work was supported by the U. S. Department of Energy, Office of Science, Office of Basic Energy Sciences, Division of Chemical Sciences, under Contract W-31-109-Eng-38. I would like to thank Costa Raptis for preparing a number of figures and for helpful comments on the manuscript. I would like to thank Dr. Ch. Jungen, Dr. T. Ebata, Dr. N. Morita, Professor S. C. Wallace, and Professor E. R. Grant for permission to use figures from their publications; I also thank the American Institute of Physics, the American Physical Society, the American Chemical Society, and Harwood Academic Publishers for their permission to use figures from their publications.

References

1. J. L. Dehmer, D. Dill and A. C. Parr, in *Fundamental Processes of Atomic Dynamics*, Eds. J. S. Briggs, H. Kleinpoppen and H. O. Lutz (Penum, New York, 1988), p. 541.
2. J. L. Dehmer, A. C. Parr and S. H. Southworth, in *Handbook on Synchrotron Radiation*, Ed. G. V. Marr (North-Holland, Amsterdam, 1987), p. 241.
3. S. T. Pratt, *Rep. Progr. Phys.* **58**, 821 (1995).
4. I. Powis, T. Baer and C. Y. Ng, editors, *High-Resolution Laser Photoionization and Photoelectron Studies* (Wiley, New York, 1995).
5. M. Larsson, *Ann. Rev. Phys. Chem.* **48**, 151 (1997).
6. J. Berkowitz, *Photoabsorption, Photoionization, and Photoelectron Spectroscopy* (Academic, New York, 1979).
7. T. Baer and W. L. Hase, *Unimolecular Reaction Dynamics* (Oxford University Press, New York, 1996).
8. T. F. Gallagher, *Rydberg Atoms* (Cambridge University Press, Cambridge, 1994).
9. P. M. Dehmer and W. A. Chupka, *J. Chem. Phys.* **65**, 2243 (1976).
10. J. Breton, P. M. Guyon and M. Glass-Maujean, *Phys. Rev.* **A21**, 1909 (1980).
11. M. Glass-Maujean, J. Breton and P. M. Guyon, *Z. Phys.* **D5**, 189 (1987).
12. P. Avouris, W. M. Gelbart and M. A. El-Sayed, *Chem. Rev.* **77**, 793 (1977).
13. M Gruebele and R. Bigwood, *Int. Rev. Phys. Chem.* **17**, 91 (1998).
14. M. Hiyama, N. Kosugi and H. Nakamura, *J. Chem. Phys.* **107**, 9370 (1997).
15. A. Giusti-Suzor and Ch. Jungen, *J. Chem. Phys.* **80**, 986 (1984).
16. F. X. Campos, Y. Jiang and E. R. Grant, *J. Chem. Phys.* **93**, 2308 (1990).
17. F. X. Campos, Y. Jiang and E. R. Grant, *J. Chem. Phys.* **94**, 5897 (1992).
18. G. P. Bryant, Y. Jiang and E. R. Grant, *J. Chem. Phys.* **96**, 4827 (1992).
19. G. Bryant, Y. Jiang and E. R. Grant, *Chem. Phys. Lett.* **200**, 495 (1992).
20. G. P. Bryant, Y. Jiang, M. Martin and E. R. Grant, *J. Chem. Phys.* **101**, 7199 (1994).

21. H. Matsui and E. R. Grant, *J. Chem. Phys.* **104**, 42 (1996).
22. H. Matsui, J. M. Behm and E. R. Grant, *Int. J. Mass Spectrom. Ion. Proc.* **159**, 37 (1996).
23. H. Matsui and E. R. Grant, in *Laser Techniques for State-Selected and State-to-State Chemistry III*, Ed. J. W. Hepburn, *Proc. SPIE* **2548**, 24 (1995).
24. E. E. Mayer and E. R. Grant, *J. Chem. Phys.* **103**, 10513 (1995).
25. E. E. Mayer, H. G. Hedderich and E. R. Grant, *J. Chem. Phys.* **108**, 1886 (1998).
26. E. E. Mayer, H. G. Hedderich and E. R. Grant, *J. Chem. Phys.* **108**, 8429 (1998).
27. D. T. Cramb and S. C. Wallace, *J. Chem. Phys.* **101**, 6523 (1994).
28. G. J. Fisanick, T. S. Eichelberger IV, M. B. Robin and N. A. Kuebler, *J. Phys. Chem.* **87**, 2240 (1983).
29. M. Fujii, T. Ebata, N. Mikami and M. Ito, *Chem. Phys. Lett.* **101**, 578 (1983).
30. M. Fujii, T. Ebata, N. Mikami and M. Ito, *J. Phys. Chem.* **88**, 4265 (1984).
31. M. Fujii, K. Sato and K. Kimura, *J. Phys. Chem.* **91**, 6507 (1987).
32. R. W. Field, *Discussions Faraday Soc.* **71**, 111 (1981).
33. R. W. Field, *J. Physique. (Paris)* **C7**, 17 (1987).
34. C. R. Vidal, *Adv. At. Mol. Phys.* **23**, 1 (1988).
35. D. E. Cooper, C. M. Klimcak and J. E. Wessel, *Phys. Rev. Lett.* **46**, 324 (1981).
36. M. A. Buntine, D. W. Chandler and C. C. Hayden, *J. Chem. Phys.* **102**, 2718 (1995).
37. J. T. Fourkas and M. D. Fayer, *Acc. Chem. Res.* **25**, 227 (1992).
38. R. L. Farrow and D. J. Rakestraw, *Science* **257**, 1894 (1992).
39. G. Hall and J. Whitaker, *J. Chem. Soc. Faraday Trans.* **90**, 1 (1994).
40. P. H. Vaccaro, in *Molecular Dynamics and Spectroscopy by Stimulated Emission Pumping*, Eds. H. L. Dai and R. W. Field; *Advances in Physical Chemistry Series*, Ed. C. Y. Ng (World Scientific, Singapore, 1995), Vol. 4, p. 1.
41. J. Geng, T. Kobayashi and M. Takami, *Chem. Phys. Lett.* **266**, 290 (1997).
42. E. F. McCormack, F. Di Teodoro, J. M. Grochocinski and S. T. Pratt, *J. Chem. Phys.* **109**, 63 (1998).
43. P. Kruit and F. H. Read, *J. Phys.* **E16**, 313 (1983).
44. K. Wang, J. A. Stephens, V. McKoy, E. deBeer, C. A. deLange and N. P. C. Westwood, *J. Chem. Phys.* **97**, 211 (1992).
45. E. deBeer, C. A. deLange, J. A. Stephens, K. Wang and V. McKoy, *J. Chem. Phys.* **95**, 714 (1991).
46. E. deBeer, M. P. Koopmans, C. A. deLange, Y. Wang and W. A. Chupka, *J. Chem. Phys.* **94**, 7634 (1991).
47. E. deBeer, W. J. Buma and C. A. deLange, *J. Chem. Phys.* **99**, 3252 (1993).
48. W. L. Glab, P. T. Glynn, P. M. Dehmer, J. L. Dehmer, K. Wang and B. V. McKoy, *J. Chem. Phys.* **106**, 5779 (1997).

49. W. L. Glab, M. S. Child and S. T. Pratt, *J. Chem. Phys.*, (in press).
50. L. F. DiMauro and T. A. Miller, *Chem. Phys. Lett.* **138**, 175 (1987).
51. A. Fujii, T. Ebata and M. Ito, *Chem. Phys. Lett.* **161**, 93 (1989).
52. T. Ebata, A. Fujii and M. Ito, *J. Phys. Chem.* **91**, 3125 (1987).
53. A. Fujii, T. Ebata and M. Ito, *J. Chem. Phys.* **88**, 5307 (1988).
54. J. Xie and R. N. Zare, *Chem. Phys. Lett.* **159**, 399 (1989).
55. K. Muller-Dethlefs and E. W. Schlag, *Ann. Rev. Phys. Chem.* **42**, 109 (1991).
56. L. Zhu and P. M. Johnson, *J. Chem. Phys.* **94**, 5769 (1991).
57. R. Schinke, *Photodissociation Dynamics* (Cambridge University, Cambridge, 1993).
58. A. J. R. Heck and D. W. Chandler, *Ann. Rev. Phys. Chem.* **46**, 335 (1995).
59. R. S. Berry, *J. Chem. Phys.* **45**, 1228 (1966).
60. J. N. Bardsley, *Chem. Phys. Lett.* **1**, 229 (1967).
61. R. S. Berry and S. E. Nielsen, *J. Chem. Phys.* **49**, 116 (1968).
62. S. E. Nielsen and R. S. Berry, *Chem. Phys. Lett.* **2**, 503 (1968).
63. R. S. Berry and S. E. Nielsen, *Phys. Rev.* **A1**, 383 (1970).
64. J. Berkowitz anf W. A. Chupka, *J. Chem. Phys.* **51**, 2341 (1969).
65. P. M. Dehmer and W. A. Chupka, *J. Chem. Phys.* **66**, 1972 (1977).
66. J. Hager, M. A. Smith and S. C. Wallace, *J. Chem. Phys.* **83**, 4820 (1985).
67. J. Hager, M. A. Smith and S. C. Wallace, *J. Chem. Phys.* **84**, 6771 (1986).
68. A. Goto, M. Fujii and M. Ito, *J. Phys. Chem.* **91**, 2268 (1987).
69. A. Goto, M. Fujii and M. Ito, *Chem. Phys. Lett.* **135**, 407 (1987).
70. (a) Y. Ono, S. H. Linn, H. F. Prest, C. Y. Ng and E. Miescher, *J. Chem. Phys.* **73**, 4855 (1980); (b) E. Miescher, Y. T. Lee and P. Gürtler, *J. Chem. Phys.* **68**, 2753 (1978).
71. J. Berkowitz and J. P Greene, *J. Chem. Phys.* **81**, 4328 (1984).
72. J. Kimman, J. W. J. Verschuur, M. Lavollee, H. B. van Linden van den Heuvell and M. J. van der Wiel, *J. Phys.* **B19**, 3909 (1986).
73. Y. Achiba and K. Kimura, *Chem. Phys.* **129**, 11 (1989).
74. G. Herzberg and Ch. Jungen, *J. Mol. Spectrosc.* **41**, 425 (1972).
75. F. Chateauneuf and H. Lefebvre-Brion, *J. Chem. Phys.* **98**, 7657 (1993).
76. A. Staib, W. Domcke and A. L. Sobolewski, *Z. Phys.* **D16**, 49 (1990).
77. A. Staib and W. Domcke, *Z. Phys.* **D16**, 275 (1990).
78. A. Staib and W. Domcke, *Chem. Phys. Lett.* **204**, 505 (1993).
79. J. A. Stephens and C. H. Greene, *Phys. Rev. Lett.* **72**, 1624 (1994).
80. J. A. Stephens and C. H. Greene, *J. Chem. Phys.* **102**, 5808 (1995).
81. Ch. Jungen and S. T. Pratt, *J. Chem. Phys.* **106**, 9529 (1997).
82. S. C. Ross and Ch. Jungen, *Phys. Rev. Lett.* **59**, 1297 (1987).
83. S. C. Ross and Ch. Jungen, *Phys. Rev.* **A49**, 4353 (1994).
84. S. C. Ross and Ch. Jungen, *Phys. Rev.* **A49**, 4364 (1994).
85. Ch. Jungen and S. C. Ross, *Phys. Rev.* **A55**, R2503 (1997).
86. S. Fredin, D. Gauyacq, M. Horani, Ch. Jungen, G. Lefevre and F. Masnou Seeuws, *Mol. Phys.* **60**, 825 (1987).
87. Ch. Jungen and E. Miescher, *Can. J. Phys.* **47**, 1769 (1969).

88. E. Miescher, *Can. J. Phys.* **49**, 2350 (1970).
89. E. Miescher, *Can. J. Phys.* **54**, 2074 (1976).
90. See, for example, E. Miescher and K. P. Huber, in *International Reviews of Science, Physical Chemistry*, Series 2, Vol. 3, Spectroscopy, Ed. D. A. Ramsay (Butterworths, London, 1976), p. 37, and references therein.
91. E. Miescher and F. Alberti, *J. Phys. Chem. Ref. Data* **5**, 309 (1976).
92. W. A. Chupka, in *Ion-Molecule Reactions*, Vol. 1, Ed. J. L. Franklin (Plenum, New York, 1972), p. 33.
93. M. Raoult, *J. Chem. Phys.* **87**, 4736 (1987).
94. K. Nakashima, H. Nakamura, Y. Achiba and K. Kimura, *J. Chem. Phys.* **91**, 1603 (1989).
95. T. Ebata, Y. Anezaki, M. Fujii, N. Mikami and M. Ito, *J. Phys. Chem.* **87**, 4773 (1983).
96. Y. Anezaki, T. Ebata, N. Mikami and M. Ito, *Chem. Phys.* **89**, 103 (1984).
97. Y. Anezaki, T. Ebata, N. Mikami and M. Ito, *Chem. Phys.* **97**, 153 (1985).
98. W. Y. Cheung, W. A. Chupka, S. D. Colson, D. Gauyacq, P. Avouris and J. J. Wynne, *J. Phys. Chem.* **90**, 1086 (1986).
99. A. Fujii and N. Morita, *Chem. Phys. Lett.* **182**, 304 (1991).
100. A. Fujii and N. Morita, *J. Chem. Phys.* **97**, 327 (1992).
101. A. Fujii and N. Morita, *Laser Chem.* **13**, 259 (1994).
102. D. T. Biernacki, S. D. Colson and E. E. Eyler, *J. Chem. Phys.* **89**, 2599 (1988).
103. H. H. Michaels, in *Advances in Chemical Physics*, Vol. 45, Part 2, Ed. J. W. McGowan (Wiley, New York, 1986), p. 225.
104. H. Park, D. J. Leahy and R. N. Zare, *Phys. Rev. Lett.* **76**, 1591 (1996).
105. H. Park and R. N. Zare, *J. Chem. Phys.* **106**, 2239 (1997).
106. G. Herzberg, *Spectra of Diatomic Molecules* (Van Nostrand Rheinhold, Princeton, 1950).
107. S. N. Dixit and V. McKoy, *Chem. Phys. Lett.* **128**, 49 (1986).
108. J. Xie and R. N. Zare, *J. Chem. Phys.* **93**, 3033 (1990).
109. S. T. Pratt, *J. Chem. Phys.* **108**, 7131 (1998).
110. J. H. D. Eland, *J. Chim. Physique* **77**, 613 (1980).
111. G. C. Nieman, *J. Chem. Phys.* **75**, 584 (1981).
112. G. C. Nieman and S. D. Colson, *J. Chem. Phys.* **68**, 5656 (1978).
113. G. C. Nieman and S. D. Colson, *J. Chem. Phys.* **71**, 571 (1979).
114. J. H. Glownia, S. J. Riley, S. D. Colson and G. C. Nieman, *J. Chem. Phys.* **72**, 5998 (1980).
115. J. H. Glownia, S. J. Riley, S. D. Colson and G. C. Nieman, *J. Chem. Phys.* **73**, 4296 (1980).
116. M. N. R. Ashfold, R. N. Dixon and R. J. Strickland, *Chem. Phys.* **88**, 463 (1984).
117. M. N. R. Ashfold, C. L. Bennett and R. N. Dixon, *Chem. Phys.* **93**, 293 (1985).
118. M. N. R. Ashfold, R. N. Dixon, K. N. Rosser, R. J. Stickland and C. M. Western, *Chem. Phys.* **101**, 467 (1986).

119. M. N. R. Ashfold, C. L. Bennett, R. N. Dixon, P. Fielden, H. Rieley and R. J. Stickland, *J. Mol. Spectrosc.* **117**, 216 (1986).
120. C. A. Raptis and S. T. Pratt, *Chem. Phys. Lett.* **303**, 281 (1999).
121. U. Fano, *Phys. Rev.* **A124**, 1866 (1961).

CHAPTER 19

ELECTRON CAPTURE PROCESSES BY FREE AND BOUND MOLECULES

Eugen Illenberger

Institut für Physikalische und Theoretische Chemie der Freien Universität Berlin, Takustrasse 3, D-14195 Berlin, Germany

Contents

1. Introduction

Capture of low-energy electrons by molecules is an ubiquitous phenomenon in many fields of natural science, in the gas-phase as well as in the condensed phase. We mention here is any kind of plasma where electrons in exceeding numbers are present which are involved in ionization, excitation and attachment processes.[1,2] A further aspect concerns the slowing down of high energy radiation in matter[3,4] which is, e.g. related radiation damage in biological material. From the molecular point of view it is clear that such reactions are essentially the result of the interaction of low-energy electrons with the corresponding material.[5,6] These electrons are the result of the many primary and secondary ionization events along the radiation track. Although initially produced with more or less kinetic energy, they are quickly slowed down in the course of inelastic scattering events.

This volume is dedicated to photoionization. A photoinduced process is usually initiated by the excitation of one or more electron(s) within the corresponding molecule or by the removal from it. The excited or ionized compound then evolutes according to the new electronic configuration. In that sense a *photon* induced reaction is always initiated by the displacement of electrons in the corresponding system, including electron removal or addition. In a photon induced electron transfer reaction, e.g. an excited electron is transferred from the electron donating to the electron accepting (attaching) component of the transfer system. Electron transfer is considered as an elementary and very important step in many chemical and biochemical reactions.[7,8] A related field concerns the behaviour of excess electrons in liquids or solids where they can react with dissolved molecules with considerable rate constants.[4,9] As mentioned above, excess electrons are, e.g. always created in the degradation of high energy radiation.

A particularly interesting phenomenon is the transfer of an electron from a substrate to an adsorbed molecule. It has been recognized quite recently that the photochemical behaviour of adsorbed molecules can largely be controlled by the photoinduced electron transfer from the substrate.[10,11] Molecular anions thus formed are considered as the driving force for the respective photochemical reaction.

In the present contribution, we review processes which are induced in molecules by the capture of *free* electrons in an energy range typically below the ionization threshold. As we shall demonstrate, the electron-target interaction as well as the induced chemical reaction can dramatically change with the degree of molecular aggregation. We shall present and discuss prototypical results from gas-phase molecules, clusters and molecules adsorbed and condensed on a solid surface. In the case of isolated molecules we will show that the cross-section for dissociative electron attachment (DA) varies on a large scale which is even true within structural similar molecules like CF_3X (X = Cl, I). The basis for such a behaviour is the competition between dissociation and autodetachment (re-emission of the captured electron) which is inherently present in any decay of a *transient negative ion* (TNI). This phenomenon in turn can lead to dramatic temperature effects in the DA cross-section in both the absolute value and its electron energy dependence.

Clusters provide an ideal medium to generate relaxed molecular anions (via intermolecular collisions) which cannot be obtained in electron attachment to gas-phase molecules. The increasing possibilities of energy dissipation generally leads to a decrease of the DA cross-section in favour of the production of undissociated anions. In this contribution, however, we shall primarily focus on electron attachment reactions in binary van der Waals clusters, namely the study of intracluster electron transfer processes and ion–molecule reactions, i.e. bimolecular nucleophilic displacement ($S_N 2$) reactions. We shall finally study capture processes by adsorbed and condensed molecules associated with the *desorption* of fragment ions. Although the cross-section for desorption via DA at the surface is generally much below the gas-phase DA cross-section, there is evidence for a particular medium catalyzed mechanism which can dramatically enhance the DA cross-section of a molecule at the surface.[12]

1.1. *Electron Capture by Molecules*

Low-energy electrons can be very reactive due to the ability of many molecules to effectively capture free electrons resonantly at a particular energy ε_o

$$e^-(\varepsilon_o) + \mathrm{M} \rightarrow \mathrm{M}^{-(*)} . \tag{1}$$

If the temporary negative ion (TNI) dissociates into thermodynamically stable fragments

$$\mathrm{M}^{-(*)} \rightarrow \mathrm{R} + \mathrm{X}^- , \tag{1a}$$

the overall process (1) and (1a) is known as *dissociation attachment* (DA). For many halogenated molecules, DA is an extremely effective process (at least in the gas-phase) with cross-sections sometimes considerably above 100 Å^2.[9,13–18,33] Since a molecular anion formed by free electrons is thermodynamically unstable, it can also decay by re-emission of the extra electron

$$\mathrm{M}^{-(*)} \rightarrow \mathrm{M}^{(*)} + e^-(\varepsilon_1) , \tag{1b}$$

leaving the neutral molecule in its ground state or in an excited state. This process is called *autodetachment* (AD).

If the target molecule possesses a positive electron affinity, the molecular anion formed by free electron attachment can principally be stabilized into its ground state by radiative cooling

$$\mathrm{M}^{-(*)} \rightarrow \mathrm{M}^- + h\nu . \tag{1c}$$

Since this process is generally much slower than the competitive reactions (1a) and (1b) it is usually not operative in the gas-phase under single collision conditions. In the case when collision partners are available during the lifetime of the negative ion, however, energy can be transferred to a collision partner leaving the ion in a thermodynamically stable state

$$M^{-(*)} \to M^- + \text{Energy}\,. \tag{1d}$$

Such stabilization processes are hence operative in free molecular clusters and certainly in the condensed phase. They can also occur between single molecules in the gas-phase when the time between collisions is comparable to the lifetime of the parent anion $M^{-(*)}$. A chemical fragmentation via dissociative attachment can then be quenched in favour of processes leading to undissociated anions M^-. We assign these latter reactions as *associative attachment* (AA).

We will review here recent results concerning electron attachment to gas-phase molecules to free molecular clusters (generated in supersonic beams) and to molecules adsorbed and condensed on a solid surface. The experimental methods include crossed electron/molecular beam experiments with mass spectrometric detection, and UHV experiments where the interaction of free electrons with adsorbed and condensed molecules is studied.

1.2. *Stable Negative Ions and Temporary Negative Ions* (TNIs)

Before considering the above mentioned processes in detail, we will first recall some basic facts concerning the properties and stability of negative ions as opposed to neutral molecules or positive ions. In particular, we have to distinguish between a thermodynamically stable anion (M^-) and the unstable species ($M^{-(*)}$) formed by *free* electron capture which we have called a *temporary negative ion* (TNI). The discussion, however, will only be extended to a degree relevant to the results presented here.

The basic quantity relating a neutral molecule with its anion is expressed by the electron affinity (EA). The EA of a molecule is defined as the energy difference between the neutral molecule (M) and the corresponding anion (M^-) in their respective ground states. By convention, the electron affinity of M is considered *positive* if the ground state of M^- lies *below* that of M, and *negative*, if the ground state energy of M^- lies *above* M. A positive value for the electron affinity indicates the existence of a stable anion in which the extra electron exists in a bound state. The electron affinity of a neutral particle thus corresponds to the ionization energy (or detachment energy) of the anion.

In molecules which are characterized by a considerable geometry change between the neutral and the anion, one has to distinguish between the (adiabatic) electron affinity (EA) of M and the vertical detachment energy of M^-. The latter is the number which is experimentally obtained in a Franck–Condon transition in photodetachment from anions.[19,20]

A lot of molecules, among them such ubiquitous compounds like N_2, CO_2, CH_4, ethylene, benzene, etc. are not able to form *thermodynamically stable anions*. By capture of free electrons, however, they can form unstable anions according to process (1).[21–26] With the definition from above, they are characterized by a negative electron affinity.

Consider first an isolated molecule in the gas-phase interacting with electrons of defined but variable energy. This interaction is commonly divided into two major classes, *direct* and *resonance* scattering.[15,27–30] In the first case, the electron collides with the target molecule and will be deviated from its original direction. If the scattering process is inelastic, the electron will lose some of its initial energy thereby leaving the target molecule in an excited state. Resonant scattering is considered to occur when the incoming electron is temporarily trapped in a potential created by its interaction with the neutral molecule thus forming a *temporary negative* ion (TNI) according to Eq. (1). Synonymously, such an ion is called a *resonance* since its formation represents an electronic transition from a continuum state ($M+e^-$) to a discrete state ($M^{-(*)}$). If this transition generates the electronic and vibrational ground state of the anion, the molecule is characterized by a negative EA value.

Of course, low-energy electron attachment can also lead to (electronically or vibrationally) excited states of the anion. SF_6, for example, captures electrons at 0 eV to form the anion in its electronic ground state, but vibrationally excited.[31–33] In this case, the vibrational energy is comprised of the electron affinity.

At this point we will not discuss the classification and genealogy of resonances to any great extend, but rather refer the reader to some relevant review articles.[21,27–30] For the purpose of this contribution it is sufficient to distinguish between *single particle* ($1p$) and *two particle-one hole* ($2p-1h$) resonances. In the picture of molecular orbitals (MOs) a single particle resonance is formed when the incoming electron temporarily occupies one of the normally empty (or virtual) MOs without affecting the configuration of the other electrons.

The trapping mechanism is described by the *shape* of the interaction potential between the electron and the neutral molecule: the combination of the attractive polarization interaction

$$V_a(r) = -\alpha e^2/2r^4 , \tag{2a}$$

[α: polarizability of the target molecule, r: distance between the incoming electron and the molecule, e: elementary charge] and the repulsive centrifugal term

$$V_\ell = \frac{\hbar^2 \ell(\ell+1)}{2r^2\mu} , \tag{2b}$$

[ℓ: angular momentum quantum number of the extra electron, μ: reduced mass of the electron/molecule system ($\mu \approx m_e$)] produces a centrifugal barrier in the resulting potential in which the additional electron can temporarily be trapped (Fig. 1). Since it is the particular shape of the potential which is responsible for the trapping, these states are also called *shape resonances.*

A two particle-one hole ($2p-1h$) resonance is formed when the incoming electron concomitantly excites one of the electrons of the target molecule resulting in one hole and two electrons in normally unfilled MOs. This is analogous to the formation of $1p-2h$ (non-Koopmans') states in positive ions.[34,35] If the energy of the core excited resonance lies *below* that of the associated excited neutral state, the extra electron is simply trapped in the field of the excited molecule (Feshbach resonance); if it lies *above*, again the *shape* of the interaction potential is responsible for the trapping and we speak of a *core-excited shape resonance.*

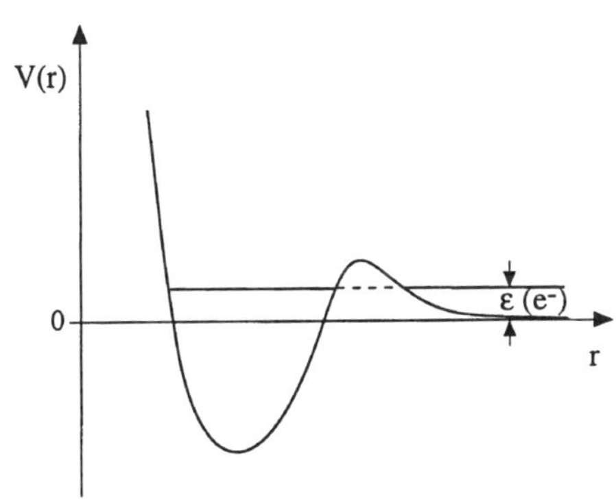

Fig. 1. Schematic potential energy diagram for the interaction of an electron of angular momentum quantum number ℓ and a neutral molecule (from Ref. 18).

The short range attractive interaction Eq. (2a) between and electron and a neutral molecule is also responsible for the low electron binding energy as well as the very restricted number of bound electronically excited states in anions as opposed to neutral or positively charged atoms or molecules.[15,36]

1.3. *Decomposition Pathways of Temporary Negative Ions*

The competitive pathways of a TNI are (i) dissociation into stable fragments (DA), (ii) autodetachment (AD) and (iii) stabilization into a thermodynamically stable parent anion (AA). The latter process is usually only operative when collision partners are available.

Under single collision conditions the autodetachment lifetime can vary over a wide range, depending on the energy of the resonance and the size of the molecule. It extends from less than a vibrational period (10^{-14} s) in small molecules like N_2^-[37] or CF_3Cl^-[38,39] to the micro- or millisecond scale for larger molecules such as SF_6^- or polycyclic hydrocarbons.[40–43] These metastable anions are usually generated at energies close to 0 eV and preferably when DA channels are not yet accessible. In this context, the term *metastable* means that the anion is at least detectable on the time scale of a mass spectrometric experiment, i.e. on the μs scale.

A remarkable exception is made by the fullerenes. Electron impact to C_{60}, e.g. generates undissociated, metastable anions C_{60}^{-*} up to about 14 eV,[44–48] (see Sec. 3.1.2).

The time for the competitive DA reaction can also vary on a large scale, depending on the mechanism of the unimolecular decomposition. It can range from 10^{-14} s for a direct electronic dissociation along a repulsive surface up to the μs regime for metastable decays. These slow decompositions usually involve energy redistribution and sometimes rearrangement processes in the precursor ion.

As already mentioned, DA reactions are very effective in many halogenated compounds. One reason is that the DA channel

$$M^{-(*)} \rightarrow R + X^- , \qquad (3)$$

($X =$ halogen atom) is energetically very favourable due to the large electron affinity of the halogen atoms.

Figure 2 illustrates a typical situation for target molecules of the class $M = RX$ with $R = CF_3$ or C_6F_5 and $X =$ Cl, Br, I in terms of a simplified one-dimensional Born–Oppenheimer potential energy diagram. For these

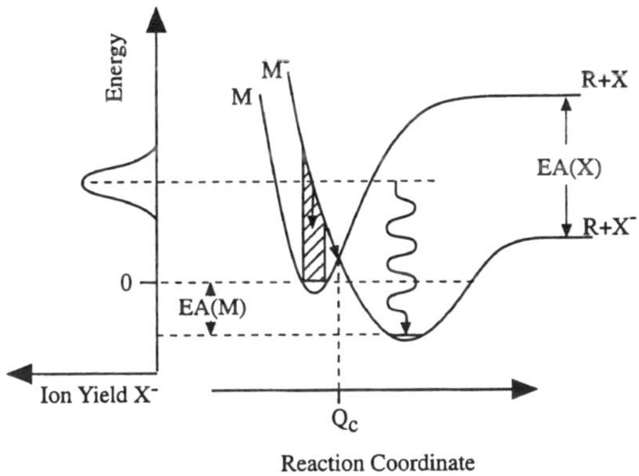

Fig. 2. Hypothetical potential energy diagram for molecules of the type RX with R = CF_3, C_6F_5 and X = Cl, Br, I illustrating the reflection principle. The curved arrow symbolizes transfer from the particular coordinate into other (vibrational) degrees of freedom (from Ref. 18).

particular molecules, the reaction coordinate Q is, to a good approximation, represented by the distance R–X. Although this class of molecules have appreciable positive *adiabatic* EAs, the relaxed (thermodynamically stable) anion RX^- cannot directly be generated by *free* electrons. Free electron attachment generates the TNI via a Franck–Condon transition with the transition probability peaking at the energy ε_o. DA into R + X^- can occur, if the TNI is generated at energies above the asymptotic limit R + X^-. If the entire resonance generated by the Franck–Condon transition is above the limit R + X^- (as in Fig. 2) and dissociation into R + X^- is the only possible channel, then the resonance profile of fragment X^- also peaks at ε_o and is a mirror of the initial Franck–Condon transition. Since the vibrational wave function of the neutral state is reflected at the anionic potential energy surface (Fig. 2), this behaviour is known as *reflection principle.*[14,15,36,49]

During dissociation, however, the TNI is unstable towards autodetachment until it passes the intersection, Q_c, between the two potential surfaces.

The autodetachment rate can be very different. In the case of CF_3Cl^- (Sec. 3.1.1) it is in the range $> 10^{16}$ s^{-1} so that only a ratio of less than 10^{-2} of the initially formed TNIs *survive* dissociation. In this case the DA resonance is shifted to energies below ε_o. The reason is the strong energy dependence in the autodetachment rate which makes DA processes at lower energies more probable.

For the $C_6F_5X^-$ anions, the autodetachment rate is much lower ($\leq 10^6$ s^{-1}, Sec. 3.1.2) and by far most of the initially formed TNIs will survive the separation until Q_c. In perfluorobenzene we have a particular situation in that the lowest dissociation channel ($C_6F_5 + F^-$) is energetically *above* the entire resonance so that the system may vibrate back along Q into the unstable region $Q \leq Q_c$. However, by transferring energy from the relevant coordinate Q into other vibrational degrees of freedom (marked by the curved arrow in Fig. 2) autodetachment can be delayed to an extent that $M^{-(*)}$ becomes observable with mass spectrometric techniques. This is the prototype situation for formation of long-lived (metastable) parent anions.

In term of molecular orbitals (MOs), the vertical Franck–Condon transition can be described as capture of the additional electron into one of the lowest unoccupied molecular orbitals (LUMOs) thus generating a single particle shape resonance. For the present prototype molecules CF_3X and C_6F_5X the associated MO has some σ^* (R–X) character which also results in the larger equilibrium distance and a lower R–X bond dissociation energy in the anion as opposed to the neutral molecule.

In the following chapter we will discuss how the situation changes if a free electron of defined energy interacts with a molecular cluster or molecules in the condensed phase.

1.4. *Formation and Evolution of Temporary Negative Ions* (TNIs) *under Aggregation*

The interaction of electrons with aggregated molecules is principally a multiple scattering problem: since in the energy range relevant here, the de Broglie wavelength of the incoming electron is larger or comparable to the intermolecular distances (the de Broglie wavelength of a 1 eV electron is 12.4 Å), the electron wave is simultaneously scattered at different targets.

Furthermore, an additional electron in a cluster or in condensed matter may no longer be localized on an individual anion M^- which itself is coupled to the environmental molecules.

Water clusters, e.g. can capture low-energy electrons forming $(H_2O)_n^-$, $n \geq 11$.[50] At least in larger H_2O clusters the electronically and geometrically relaxed cluster ion may then be viewed as an excess electron trapped in the field of the oriented H_2O dipoles.[51–53] Solvated electrons have, in fact, been well known to exist in polar liquids for more than 100 years.[54] A similar phenomenon exists when sodium is dissolved in ammonia: the Na atoms will be ionized and the blue colour of the solution is due to optical transitions of the excess electrons trapped by the NH_3 dipoles.[55]

Excess electrons do not only exist as solvated electrons trapped through the interaction with polar molecules, they can also be localized in condensed nonpolar dielectric media. The trapping mechanism then results from electronic polarization of the environmental molecules.[56]

Although principally a multiscattering problem, the DA cross-section in clusters and in the condensed phase often exhibits resonances which can be described on a molecular site, i.e. formation and decay of an individual TNI, coupled to the environmental molecules of the cluster or the molecular film.

While in isolated molecules resonant capture of an electron directly generates either a one particle ($1p$) shape resonance at low energies ($0- \approx 3$ eV) or a core excited resonance at higher energies (in the vicinity of the electronically excited states of the neutral), the situation becomes considerably more complex in clusters. This concerns (1) the *formation mechanism*, (2) the *energy* and (3) the *evolution* of the TNI.

(1) At low energies we expect direct formation of an individual shape resonance in clusters. At higher impact energies, e.g. in the vicinity of electronically excited states, negative ion formation can principally proceed via two mechanisms, namely direct electron capture through formation of a core excited resonance which relaxes into its ground state by evaporation of the target cluster or inelastic scattering of the energetic electron by one molecule and transfer of the slowed down electron to another molecule within the *same* cluster. This latter mechanism has been assigned as *self-scavenging* if it occurs in a homogeneous cluster and *autoscavenging* if it occurs between different molecules within a heterogeneous cluster.[58] Such scavenging mechanisms play a particular role in heterogeneous clusters composed of molecules having a different electron scattering/attachment behaviour.[18]

(2) With respect to the gas-phase, the energy of a TNI is lowered by the polarization interaction with the environmental medium. The polarization energy consists of two components, the *fast electronic polarization* which is already operative at the instant of electron attachment and *polarization*

by reorientation of the molecules in the vicinity of the excess charge occurring on the much slower time scale of molecular relaxation processes ($\approx 10^{-12}$–10^{-10} s). In negative ions, one additionally has to consider the short range *repulsive exchange interaction* (Pauli repulsion) which has a purely quantum mechanical origin. In particular cases, i.e. in media with very low polarizabilities the exchange interaction can prevail.[59–63]

(3) If we have a situation like in Fig. 2, the TNI can either decompose via DA, i.e. emission of a free ionic fragment X^- or formation of a solvated ion $X^- \cdot M_n$. It can also relax into its stable configuration M^- thereby evaporating the target cluster (*evaporative attachment*[57]). In Fig. 3, some possibilities of negative ion formation following electron capture by clusters are illustrated. Not included are intracluster ion–molecule reaction, e.g. the reaction of the DA fragment X^- with a second molecule of the target cluster. In Sec. 3.2.3 we will present some results concerning particular ion–molecule reactions, namely bimolecular nucleophilic substitution (S_N2) reactions in clusters.

An interesting point arises when autodetachment strongly competes with dissociation: since the anionic potential energy surface is usually more

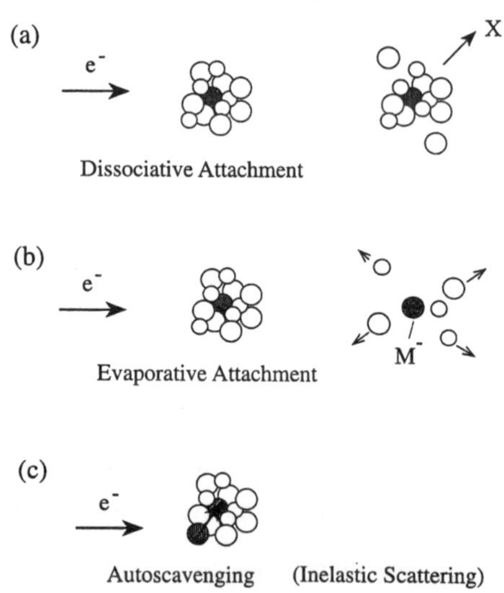

Fig. 3. Some possible processes in clusters induced by low-energy electrons, (a) DA associated with the ejection of a free ionic fragment, (b) evaporative attachment, (c) autoscavenging (from Ref. 18).

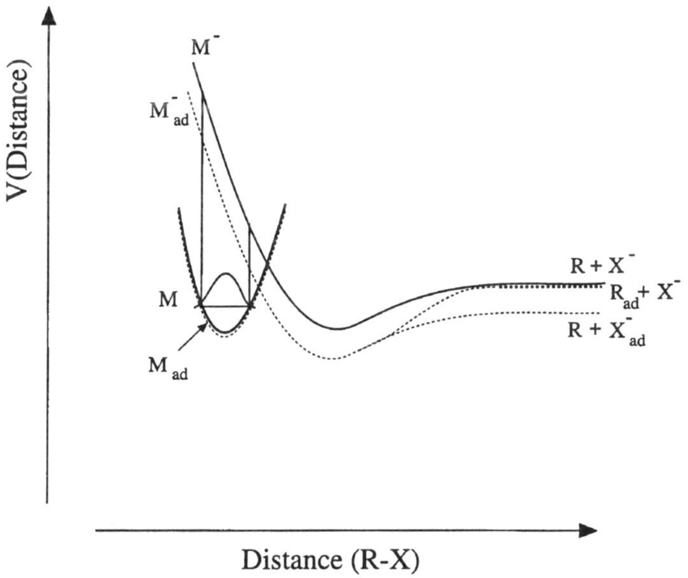

Fig. 4. Schematic potential energy diagram illustrating DA in the gas-phase and in the condensed phase. The subscript (ad) assigns a particle bound to the surface (from Ref. 18).

lowered in the cluster with respect to the neutral, the time the system needs to escape from the autodetachment region is shortened and, consequently, the cross-section for negative ion formation is enhanced (see also Fig. 4).

While clusters consist of a countable number of molecules (finite system), the condensed phase is usually described as consisting of an infinite number of microscopic particles. Nevertheless, low-energy electron interaction with condensed molecules can also lead to resonances, which can be described on a molecular site. If they are of sufficiently repulsive nature, they can also be observed via DA and subsequent desorption of fragment ions as has been shown by Sanche.[64] In the condensed phase, however, many quenching processes like transfer of the excess electron into the metal and relaxation of the TNI strongly compete with desorption.

If DA takes place at or near the surface, desorption of a fragment ion is subjected particular energetic and structural constraints. From the structural point of view it is clear that those orientations of the molecule are preferred allowing an unhindered desorption of the fragment ion. The energetic requirement is then that the fragment must gain sufficient translational energy to overcome the solvation energy to leave the surface. For a DA process yielding two fragments we have the alternate pathways

$$M_{ad}^{-(*)} \rightarrow R_{ad} + X^- , \tag{4a}$$

$$\rightarrow R + X_{ad}^- , \tag{4b}$$

where the subscript *ad* refers to a fragment remaining at the surface. Of course, relaxation of the undissociated ion (M_{ad}^-) and fragmentation at the surface (R_{ad} + X_{ad}^-) is also possible. Figure 4 compares DA from a molecule in the gas-phase with DA from a condensed phase molecule. Due to polarization interaction, the ionic curve is generally more shifted than the neutral one. Stabilized molecular anions (M_{ad}^-) as well as DA fragments remaining at the surface (X_{ad}^-) cannot be observed directly. They contribute to charging of the film under exposure to the electron beam. In fact, charge trapping preferentially occurs at energies where resonances are present.[64–68]

It should be emphasized that in the condensed phase only the emission of a *free* fragment can be directly recorded. In a cluster experiment, the analogue of DA leading to X_{ad}^- [reaction (4b)] is the formation of solvated ions $X^- \cdot M_n$. Accordingly, the analogue to stabilization at the surface (M_{ad}^-) is formation of undissociated ions M_n^-, $n \geq 1$ which both can certainly be observed in a cluster experiment (Sec. 3.2).

At that point it should be noted again that photoinduced electron transfer from the substrate can play a major role in the photochemical behaviour of adsorbed molecules. This is particularly the case for molecules having large cross-sections for low-energy electron attachment like most halogenated compound. Decomposition of transitory ions formed by electron transfer can lead to the desorption of the halogenated fragment X^- [charge transfer photodissociation (CT-PDI)] mechanism as shown by Polanyi and coworkers.[69–71] In other cases the TNI can act as a driving force for the desorption of undissociated molecules.[72,73] In the spirit of the Menzel–Gomer–Redhead (MGR) model[60] and its modification by Antoniewicz[75] the TNI is first accelerated towards the surface due to the image force induced in the metallic solid. After transfer of the electron into the solid, the neutral molecule-substrate system may find itself energetically above the desorption limit.

Thus, molecules having large capture cross-sections for slow electrons can be photochemically active when adsorbed on surfaces. In the case of halogenated methanes this is already the case at photon energies near or even below the threshold for photoemission,[10] i.e. in a wavelength region where the gas-phase molecule is photochemically inactive.

1.5. *Overview on Previous Work*

Electron attachment processes to gas-phase molecules have been studied as long as about four decades. The experimental techniques include beam experiments combined with mass spectrometry, electron transmission, electron cyclotron resonance (ECR), electron drift and flowing afterglow Langmuir probe (FALP) experiments.

Here we only refer to a few monographs concerning the general field of negative ions[9,13,36] and a few reviews dealing with electron beams,[14,15,16(a),76] electron transmission[21,24] and the FALP technique.[77]

To date experiments on attachment of *free* electrons to clusters have been reported by five groups: Oak Ridge,[78–80] Innsbruck,[58,81–98] Konstanz.[99–102] Nowosibirsk,[103–105] and Berlin[106–126] and also some collaborative studies between the laboratories.[127–129] In all these experiments free electrons of (more or less) defined energy interact with pre-existing clusters in a supersonic beam produced by adiabatic gas expansion.

It has been shown that in clusters composed of molecules with positive EA the availability of collision partners in the cluster leads to the observation of relaxed parent anions which cannot be formed under collision free conditions.

Echt, Recknagel and coworkers[99] observed $(H_2O)_n^-$, $n \geq 11$ by attachment of near-zero energy electrons to pre-existing water clusters. It has been shown before by Haberland *et al.*[130–133] that solvated electrons of the form $(H_2O)_n^-$ and $(NH_3)_n^-$ can readily be formed in supersonic beams.

In the latter experiment, negative cluster ions were formed by electron impact in the expansion zone of a supersonic beam. This method was extensively applied (and developed) by Haberland *et al.* to study the minimal number of H_2O and NH_3 molecules or Xe atoms necessary to support a bound state for an excess electron. These systems can probably not form stable negative ions as isolated particles. The minimal numbers were found to be $n = 2$ for $(H_2O)_n^-$,[130–132] $n = 35$ for $(NH_3)_n^-$[133,134] and $n = 1$ in the case of Xe_n^-.[135] However, for Xe^- it is not yet clear whether the detected species is really associated with ground state Xe (which would indicate a positive electron affinity) or whether it represents an electronically excited (metastable) Xe^{*-} ion.

We also mention here a related method which extends the electron energy scale to values below zero, namely electron transfer from highly excited atoms to molecules or clusters. In this technique, Rydberg states are prepared either by electron impact (Kondow *et al.*[136–138]) or state selectively by laser excitation from the metastable intermediate state (Hotop *et al.*[33,139–142] Dunning *et al.*[143,144] Schermann *et al.* and Desfrançois *et al.*[145–155]). As shown by Desfrançois *et al.* electron transfer to dipole bound states exhibits a significant dependence on the principal quantum number n. This dependence itself is sensitive on the magnitude of the dipole moment of the electron acceptor[150,151] and can be used to obtain structural information about weakly bound complexes (dimers) of polar molecules. In the case of nucleic acid base pairs having different geometrical isomers of nearly the same energy, it is possible to identify and separate them by use of this *non destructive ionization technique.*[152] In the meantime new experimental and theoretical work from that group gave evidence for the existence of quadrupolare bound anions, i.e. an extra electron weakly bound by the quadrupole field of the neutral molecule.[154,155]

Energy resolved DA reactions in the condensed phase were first reported in 1984 by Sanche (Sherbrooke)[156] from an oxygen film via desorption of O^- from the strongly repulsive $O_2^{-*}(^2\Pi_u)$ state near 6 eV. The Sherbrooke group has investigated DA reactions in many condensed molecules[64,156–170] including electron transmission through ordered and disordered molecular films,[171–179] electron trapping in condensed molecules[65–68,180–182] and electron-energy-loss spectroscopy of condensed molecules.[183–188]

In the meantime, DA from adsorbed and condensed molecules is also performed at Orsay (Azria, Tronc *et al.*[189–193]) and by our laboratory,[194–198] also within joint projects between these laboratories.[12,199–204]

In the following we will show in the case of some prototype molecules how the electron attachment behaviour changes on proceeding from gas-phase molecules under single collision conditions to free molecular clusters and, finally to adsorbed and condensed molecules.

We will primarily present results emerged from our own laboratory, although other techniques and results will occasionally be mentioned and discussed.

2. Experimental Considerations

2.1. *Gas-Phase Experiments: Molecules and Clusters*

2.1.1. *Principle and Limitation of the Method*

Electron attachment reactions in single molecules and clusters is studied by crossing an electron beam with a molecular beam. The negative ions formed are recorded mass spectrometrically as a function of the electron energy. Figure 5 shows a schematic representation of the gas-phase experiment.

The electron beam is crossed either with a molecular beam *effusing* from a capillary which is directly fitted to the reaction chamber or with a supersonic beam containing clusters. By recording a particular (fragment) ion, e.g. X^-, as a function of the electron energy, one obtains the relative cross-section for X^- formation or, synonymously, the X^- *ion yield curve.*

It must be emphasized that the study of a DA reaction in a *single molecule* yielding two fragments, viz.

$$e^- + \mathrm{M} \rightarrow \mathrm{M}^{-(*)} \rightarrow \mathrm{R} + \mathrm{X}^- \,, \tag{5}$$

is a direct way to study the energetics of the respective unimolecular decomposition. Since the internal energy of the precursor $M^{-(*)}$ is uniquely

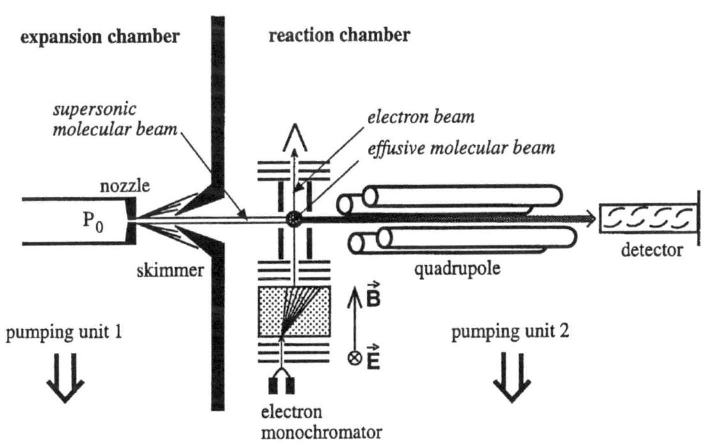

Fig. 5. Schematic of the experimental setup for the study of electron attachment to gas-phase single molecules and free molecular clusters. The capillary for effusive gas inlet is directly fitted to the reaction chamber perpendicular to the cluster beam (from Ref. 122).

controlled by the energy of the primary electron, determination of the excess translational energy of fragment X^- (and, hence, through momentum conservation of $R + X^-$) yields information on the distribution of the total excess energy (translational/internal) of the unimolecular process. In photoionization, e.g. analogous information requires the use of coincidence techniques.

In van der Waals aggregates the situation is less straightforward in the way that *a priori* neither the precursor nor the number of final particles is exactly known. While in single molecules the precursor is known and fragmentation into more than two products can often be ruled out from thermodynamic reasons, in a van der Waals cluster, M_n, electron attachment yielding X^- can occur between the following limiting pathways

$$e^- + M_n \rightarrow R \cdot M_{n-1} + X^- \,, \tag{6a}$$

$$\rightarrow R + (n-1) \cdot M + X^- \,, \tag{6b}$$

(6a) is emission of fragment X^- leaving the rest of the target cluster undissociated while (6b) is complete evaporation of the remaining neutral clusters. Due to the weak van der Waals binding energies it is usually not possible to conclude from the energy dependence of X^- formation on the exact size (n) of the target cluster or the degree of evaporation.

However, since the average size of the cluster distribution can be varied to some extent by changing the expansion conditions one can witness the evolution of a particular quantity (attachment energy, product distribution, kinetic energy release, etc.) by changing the degree of aggregation.

2.1.2. *Supersonic Beam and Electron Beam*

The supersonic beam is formed by adiabatic expansion of the gas through an 80 μm nozzle which can be cooled down to liquid nitrogen temperature.

For conditions when the mean free path of the molecules is much smaller than the nozzle diameter, $\chi \ll d$, we have a *free jet expansion*. The expansion converts the enthalpy associated with *random* motion into *directed* mass flow.[15,205–207] For a monoatomic gas the terminal velocity of the particles in the jet is given by

$$v_t = \left(\frac{5kT_o}{M}\right)^{1/2} \,, \tag{7}$$

with T_o the *temperature in the source.* The (translational) temperature in the jet is determined by the width of the velocity distribution around v_t. Since the degree of cooling is a function of the heat capacity ratio $\kappa = C_p/C_v$, cooling is less effective when a gas of polyatomic molecules is expanded. To establish cluster formation in a polyatomic gas, one applies the *seeded beam* technique, where the polyatomic gas is diluted in an inert carrier gas (Ar, He) which acts as a refrigerant to polymerize the molecules. The molecules are usually diluted in small amounts in the carrier gas, the velocity of the particles in the beam can still be approximated by Eq. (7) with *M now the mass of the carrier gas.* This can lead to considerable values for the translational energy of a cluster in the laboratory frame of reference. For example, if C_6F_5I is seeded in He, the dimer will move with a translational energy of nearly 10 eV in the beam! To avoid discrimination in the draw-out probability of such product ions we have chosen to place the axis of the quadrupole *in line* with the molecular beam (Fig. 5).

In addition, this configuration allows to separate products arising from reactions in the jet from those arising from scattered background molecules by applying a time-of-flight (TOF) technique (see Secs. 2.1.3 and 3.2.3).

After expansion from the nozzle, the beam passes a skimmer (0.8 mm diameter) which separates the expansion chamber from the main chamber. Approximately 7 cm downstream, the molecular beam is crossed at right angles with an electron beam generated by a trochoidal electron monochromator (TEM). This monochromator uses a combination of a magnetic and electric field.[208,209] Its operation principle is briefly as follows: the electrons emitted from a heated filament are aligned by the homogeneous magnetic field ($\approx$ 100 G) which is generated by a pair of Helmholtz coils mounted outside the vacuum system. They enter a deflection region defined by the crossing of the magnetic field with an electric field ($\approx$ 1 V cm^{-1}). The latter is produced by a potential difference between two parallel plates. It is directed into the plane of paper (shaded area in Fig. 5). Under the influence of both fields the electrons describe a trochoidal or cycloidal motion depending on whether or not they possess velocity components perpendicular to the beam axis when entering the deflection region. The trochoidal or cycloidal path implies motion of its guiding center with constant velocity perpendicular to both the magnetic and the electric field. This results in a dispersion of the electrons according to their axial velocity and only those electrons which reach the exit aperture (displaced by some distance with respect to the entrance aperture) are further transmitted. These energy selected electrons are then accelerated or decelerated by subsequent electrodes before entering the reaction volume.

The combination of a TEM with a quadrupole mass filter has proven to be particularly suited for studying electron attachment reactions since it combines some important and necessary properties. (1) The axial magnetic field prevents spreading of the electron beam at low energies so that reasonable intensities (10–100 nA) can be achieved down to very low (thermal) energies. (2) The TEM can be operated in the continuous mode: due to the presence of the magnetic field, the energy of the electrons entering the reaction volume is not significantly disturbed by the (continuous) ion draw-out field. This field only causes a slight deflection of the electron beam along a plane of constant electric potential. (3) The alignment of the electrons by the magnetic field also prevents scattered electrons from reaching the ion detector (secondary electron multiplier) thus producing some background signal.

If not otherwise stated the TEM was operated in the gas-phase experiments at an energy resolution of 0.1–0.2 eV (FWHM) and a current of 50–100 nA. Negative ions created in the interaction region are extracted from the reaction volume by a small electric field, analyzed by the quadrupole mass filter and detected by single event pulse counting techniques.

2.1.3. *Time-of-Flight (TOF) Analysis of Product Ions*

Recording flight times of product ions on their ways between the point of generation and the detector can be used to obtain information on the kinetic excess energy release of the associated reaction and (in favourable cases) to separate reactions occurring in the jet from those occurring in scattered background monomers.

The principle to obtain information on the translational energy release is easiest described in the case of a static target or for conditions where the kinetic energy release of the observed fragment (X^-) is *large* compared to the kinetic energy of the precursor (in the effusive beam or in the jet). The principle of TOF analysis in combination with a quadrupole has been described earlier.[210,211] In brief, the electron beam is now pulsed (pulse width < 1 μs) and the flight time of the ions (generated within the short time of the electron pulse) from the reaction volume through the quadrupole to the detector is measured (Fig. 5). The TOF arrangement, in principle, corresponds to a two-field TOF spectrometer, consisting of a draw-out region ($E_1 = 4$ Vcm^{-1}), an acceleration region ($E_2 = 30$ Vcm^{-1}), and the quadrupole mass filter which acts as the drift tube. This allows space focusing by adjusting E_1/E_2 to a certain value.[15,211] Since the quadrupole guarantees the mass separation, one can use a low draw-out field (4 Vcm^{-1} or less) in order to visualize excess translational energies on the time scale.

Ions formed with low kinetic energy (e.g. thermal ions) exhibit one peak in the TOF spectrum while ions generated with considerable kinetic energy produce two peaks due to 'direct' and 'turn around' ions. In the latter case discrimination against velocity components perpendicular to the flight tube axis results in a separation (in time) of ions ejected parallel (direct) or antiparallel (turn around) to the axis. The turn around ions are decelerated, reversed, and then accelerated and reach the detector by some time delay (ΔT) with respect to the direct ions. From the experimentally determined flight time difference (ΔT), the initial kinetic energy release of the product ion can be calculated as

$$E_T^i = \frac{(\Delta T q E_1)^2}{8m_i}, \tag{8a}$$

where E_1 is the draw-out field introduced above, q the elementary charge, and m_i the mass of the ion. For a decomposition into *two* particles one can calculate the total translational energy E_T, by taking into account the principle of linear momentum conservation as

$$E_T = E_T^i \cdot \frac{M}{m}, \tag{8b}$$

[M: mass of the parent molecule, m: mass of the neutral fragment].

Conversely, for ions with low kinetic energy (< 0.1 eV for the present configuration) there is no discrimination against perpendicular velocity components and all ions are transmitted to the quadrupole. For a Maxwell–Boltzmann velocity distribution, the corresponding TOF peak has a Gaussian shape. Its width is directly related to the average kinetic energy.[15,211]

In the case of percursors moving in the supersonic beam, the TOF spectrum of a product ion will be shifted to lower flight times due to its initial velocity towards the detector. Since all particles in the jet move with approximately the same velocity given by the mass of the seed gas atoms (Eq. 7), one expects appreciable shifts in the case of large product ions.

As already mentioned, one inherent shortcoming of the present cluster experiment is its insensitivity to identify the size of the precursor. By recording flight times, however, it is at least possible to distinguish reactions occurring in the beam from those occurring in scattered background monomers (see Sec. 3.2.3).

TOF spectra are recorded by use of a time-to-pulse height converter (TPHC) followed by analog/digital (ADC) conversion.

2.2. *Condensed Phase Experiment*

Electron stimulated desorption of anions from adsorbed and condensed molecules is studied in a UHV experiment schematically shown in Fig. 6. UHV conditions are necessary to avoid fast contamination of the surface due to background gas. The arrangement consists of an electron monochromator (TEM, similar to the gas-phase experiment), a quadrupole mass spectrometer, a manipulator which carries the metallic substrate (polycrystalline gold or platinum) and a closed cycle He-refrigerator.

The metallic substrate (polycrystalline gold) is mounted on the sample holder of the manipulator. The substrate is *thermally connected* to the head of the refrigerator by a flexible copper cord, but *electrically insulated* by a sapphire plate.

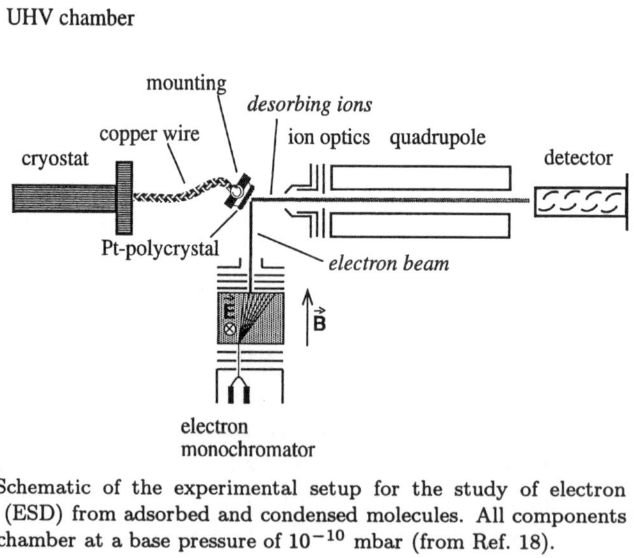

Fig. 6. Schematic of the experimental setup for the study of electron stimulated desorption (ESD) from adsorbed and condensed molecules. All components are housed in a UHV chamber at a base pressure of 10^{-10} mbar (from Ref. 18).

The substrate can thus be heated by resistive heating and cooled down to approximately 25 K. For some of the investigated molecules (CF_4, CF_3Cl etc.) such low temperatures are necessary in order to condense them under UHV conditions. The base pressure is in the range of 10^{-10} mbar.

The molecules are condensed onto the metallic substrate by exposing the cold surface to a volumetrically calibrated gas quantity. The dosing capillary is located less than 1 cm from the surface. Thus, by assuming a sticking coefficient near unity, one obtains a reasonable estimate for the thickness of the molecular film. Some of the experiments were also performed on molecules adsorbed in submonolayers on a multilayer noble gas film which acts as a *spacer* between the molecules and the metallic substrate.

2.3. *Calibration of the Electron Energy Scale in Gas-Phase and Condensed Phase Experiments*

In the gas-phase, calibration of the energy scale is fairly straightforward, since many molecules capture electrons very effectively near 0 eV. A usual standard is SF_6 yielding long-lived SF_6^{-*}. Electron attachment to SF_6 can be described as *s-wave capture* as established in an ultrahigh resolution experiment where free electrons are created via laser photoionization of noble gas atoms.[33] In that case the energy dependence of the attachment cross-section near threshold ($\varepsilon \to 0$) behaves as $\sigma \sim \varepsilon^{-1/2}$ (see also Sec. 3.1.2). Similarly, CCl_4 can be used which also captures electrons near 0 eV, in this case, however, dissociatively yielding Cl^-.

Thus, the measured SF_6^- or Cl^-/CCl_4 resonance yields the zero for the energy scale and its width is an estimate of the energy resolution of the electron beam. Strictly speaking, the experimental resonance profile is folded with the onset curve of the electron beam which may complicate the situation to some degree. However, taking the peak maximum of the SF_6^- or Cl^-/CCl_4 resonance as zero yields a reasonable calibration of the energy scale within an estimated uncertainty of better than 0.1 eV for the spectra presented in this article.

In the case of condensed molecules determination of the vacuum level (zero electron energy) is not as straightforward as in the gas-phase. Resonant associative or dissociative capture at 0 eV usually does not lead to desorbed ions. In this case, the onset of the electron injection curve into the metallic substrate serves to identify the vacuum level. This method is only applicable if the bottom of the conduction band of the dielectric film (V_o) is below the vacuum level, i.e. by definition $V_o < 0$. In fact, most dielectrics have V_o values between 0 and -1 eV.[212,213]

For $V_o < 0$ we expect the onset of the electron injection curve at zero. At electron energies below electronic excitation and in ranges where resonances are not present, no effective energy loss processes are operative and one expects quasi elastic transmission of electrons through the molecular film to the metallic substrate without appreciable attenuation.

We note that the trapping of excess charges leads to a charging of the film which creates a retarding potential for the incoming electrons. Extended exposure of the molecular film to the electron beam thus can cause a shift of the injection curve to higher energies with respect to the monochromator potential. As the onset of the shifted injection curve still gives the vacuum level *for electrons arriving at the surface*, the problem itself becomes prohibitively complicated in the way that the originally sharp onset smoothens and new structures appear near the onset.[195] These effects are probably related to inhomogeneous charging across the area where the electron beam impinges the surface.

Thus, to avoid extended charging of the molecular film, all desorption spectra presented here were always obtained by one single energy scan on a newly deposited substrate.

3. Results and Discussion

This section is arranged as follows: in Sec. 3.1 we shall present and discuss some prototype reactions in single gas-phase molecules — the compounds CF_3I and CF_3Cl (3.1.1) which both undergo prompt DA, however, with a cross-section value differing by more than 4 orders (!) of magnitude. Section 3.1.2 deals with larger molecules of the type C_6F_5X (X = F, Cl, Br, I) and SF_6 where the formation of long-lived metastable parent anions M^{-*} competes with DA in the mass spectrometric observation window. In the fullerenes C_{60} and C_{70}, finally, DA is completely suppressed and associative attachment (AA) takes place within an unusually broad energy range extending up to about 13 eV. Section 3.1.3 describes the temperature dependence of electron attachment reactions for the previously studied prototype molecules. Changing the temperature of the neutral target gas molecules can lead to dramatic effects in both the absolute value and the energy profile of the DA cross-section. Sections 3.2.1 and 3.2.2 are dedicated to molecular clusters of those molecules studied before as isolated species. In particular, the molecules C_6F_5X show very strong solvation effects in the way that any clustering quantitatively quenches dissociation (DA). In Sec. 3.3.3 we use DA to study nucleophilic displacement (S_N2) reactions in mixed clusters for some prototype systems.

In Sec. 3.3, finally, electron stimulated desorption of anions from adsorbed and condensed molecules is described. The systems studied include CF_4, CF_3Cl and CF_3I, where appreciable repulsive precursor states are involved. CF_3I shows a particular surface enhancement in the way that the cross-section for CF_3^- desorption exceeds the corresponding gas-phase DA cross-section by orders of magnitude.

3.1. *Gas-Phase Molecules*

3.1.1. *Dissociative Attachment to $CF_3X(X = Cl, I)$: Structural Similar Molecules with Very Different Cross-Sections*

Figures 7 and 8 display the energy profiles of fragment ions from CF_3I and CF_3Cl obtained under single collision conditions. Although both compounds have a positive electron affinity[108,214] only fragments are formed under these conditions. Both molecules exhibit a low-energy resonance which exclusively decomposes by the breaking of the C–X bond:

$$e^- + CF_3X \rightarrow CF_3 + X^- . \tag{9a}$$

The energy thresholds $\Delta H_o = 0.12 \pm 0.13$ eV (X = Cl) and $\Delta H_o = -0.67 \pm 0.13$ eV (X = I) are obtained from the fairly established bond dissociation energies [D(CF_3–Cl) = 3.74 ± 0.13 eV, D(CF_3–I) = 2.39 ± 0.13 eV[215]] and the very accurately known electron affinities [EA(Cl) = 3.615 ± 0.004 eV, EA(I) = 3.059 ± 0.001 eV[216]].

In both compounds DA also occurs at higher energies. While in CF_3I all further fragments (F^-, FI^-, CF_3^-) appear at the same energy (peak maximum at 3.8 eV, Fig. 7), the peak maxima in CF_3Cl differ substantially for the different fragments, from 4.8 eV (Cl^-) to 3.4 eV (CF_3^-).[217] It thus appears that CF_3I has one additional, isolated resonance at higher energy, while CF_3Cl possesses different, energetically overlapping excited resonances.

Note that in CF_3Cl the high energy Cl^- yield (at 4.8 eV) is more intense than that at low-energy (1.4 eV).

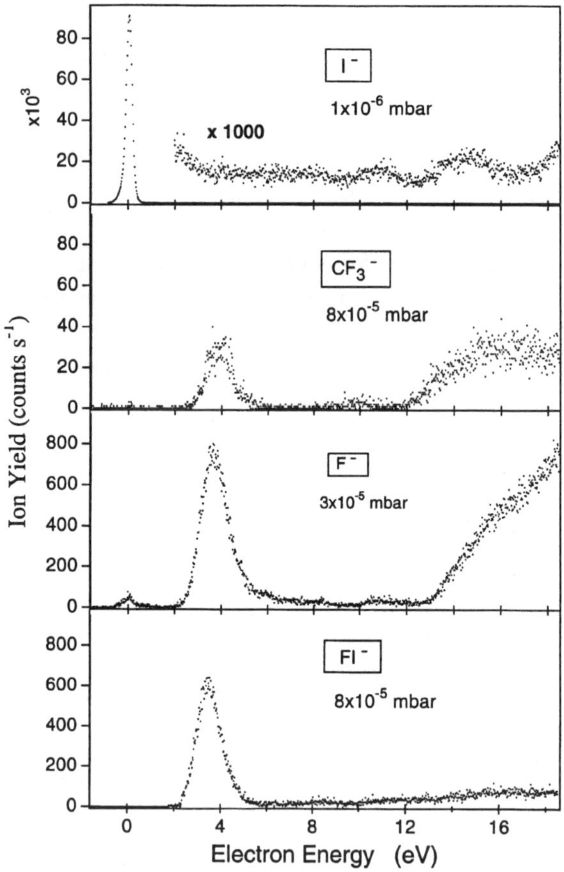

Fig. 7. Negative ions observed in electron attachment to CF_3I under single collision conditions. Note that the I^- spectrum was recorded at a reduced gas pressure (from Ref. 194).

The thermodynamic limits for the complementary reaction to (9a)

$$e^- + CF_3X \to CF_3^- + X\,, \tag{9b}$$

can be calculated by taking $EA(CF_3) = 1.7 \pm 0.2$ eV[14] as $\Delta H_o = 2.0 \pm 0.3$ eV (X = Cl) and $\Delta H_o = 0.7 \pm 0.3$ eV (X = I).

As can be seen from Figs. 7 and 8 by just comparing the count rates, the DA reaction (9a) is by orders of magnitude more intense in CF_3I as compared to CF_3Cl. Note that the I^- spectrum was recorded with considerably reduced pressure. It is further surprising that the low-energy Cl^-

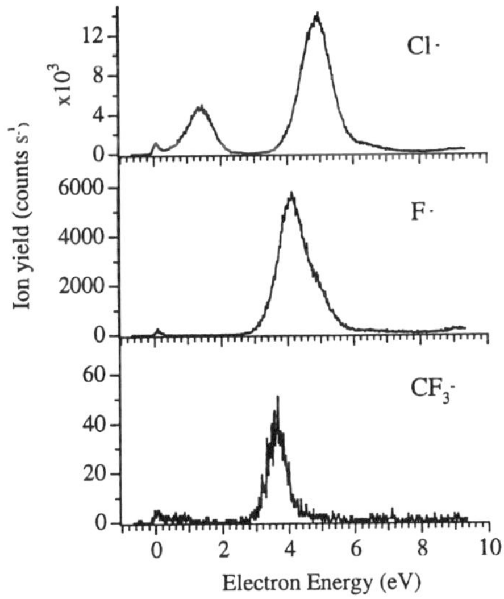

Fig. 8. Negative ions observed in electron attachment to CF_3Cl under single collision conditions (from Ref. 18).

DA resonance is less intense than that at 4.8 eV. From very basic principles one expects an *increase* in the attachment cross-section with *decreasing* electron energy.[13,15,36,218–220]

Absolute cross-section data are available from electron swarm and beam methods. In electron swarm techniques one generally measures the decrease of the electron intensity in the presence of the electron attaching gas diluted into an inert buffer without knowing the fate of the anions formed. In the case of CF_3I the pulse radiolysis swarm method yields an attachment rate of $k = 2.7 \times 10^{-7}$ cm^3 s^{-1} at room temperature.[221] Rate constant k and cross-section are related by

$$k = \int_o^{oo} \sigma(v) f(v) v dv \,,$$

with $f(v)$ the normalized velocity distribution of the electrons. For the *averaged* quantities we can write

$$k = \bar{\sigma} \cdot \bar{v} \,.$$

Thus, the rate constant of 2.7×10^{-7} cm^3 s^{-1} at 300 K ($\bar{v} \approx 1 \times 10^7$ cm s^{-1}) corresponds to an average attachment cross-section of 2.7×10^{-14} cm^2. A recent electron beam experiment yielded 1.6×10^{-14} cm^2 at $\varepsilon = 0$ eV.[222]

The thermal electron attachment cross-section in CF_3I is thus close to that in SF_6 which has one of the highest cross-sections known (2.5×10^{-14} cm^2 at 26 meV[33]).

In the case of CF_3Cl, unfolded electron swarm data give an attachment cross-section of 1.7×10^{-18} cm^2 at the peak (1.5 eV) of the low-energy resonance[223] and the beam experiment[222] of 3.9×10^{-18} cm^2, both taken at room temperature.

By comparing Figs. 7 and 8 it is striking that in CF_3I the high energy resonance at 4 eV does not decompose into I^- while in CF_3Cl the fragment Cl^- appears with even more intensity from the resonance at higher energy.

It is very likely that the low-energy feature in both compounds can be pictured as a single particle shape resonance with appreciable σ^* (C–X) nature. This is also manifested by TOF experiments indicating that most of the available excess energy appears as translational energy of the fragments $CF_3 + X^-$.[14,194] Figure 9 illustrates this in the case of CF_3I.

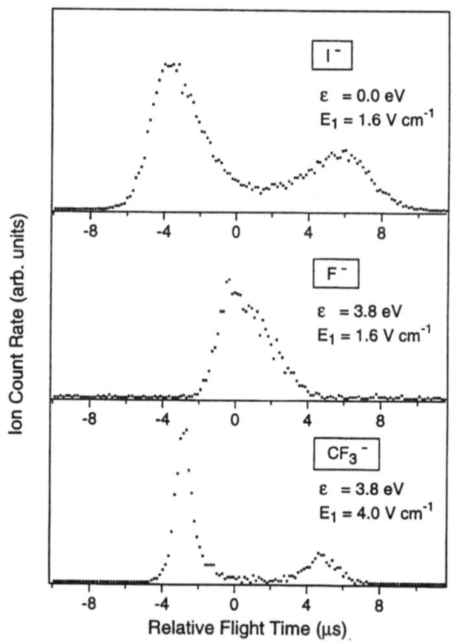

Fig. 9. TOF spectra of the fragments, I^-, F^- and CF_3^-. The CF_3^- spectrum was recorded at a higher ion draw-out field (4 V cm^{-1}) to prevent the turn around ions striking the back electrode of the reaction chamber (from Ref. 194).

I^- ions recorded near 0 eV electron energy produce a TOF doublet when subjected to an ion draw-out field of $E_1 = 1.6$ Vcm^{-1}. This corresponds to a total translational energy $E_T = 0.58$ eV for both fragments [Eqs. (8a) and (8b)]. The total available excess energy (at $\varepsilon = 0$ eV) is 0.67 eV, i.e. 87% of this energy appears as translational energy of the two products. In the case of the low-energy resonance the situation in CF_3Cl is very similar.[14]

The high energy resonance in CF_3I is likely to be of the core excited type, i.e. the incoming electron concomitantly excites one of the electrons in the target molecule and we have two electrons in normally empty MOs. The TOF spectrum (Fig. 9) indicates that the CF_3^- ion is formed with high translational energy while the competing ions F^- [Fig. 9(b)] and also FI^- (not shown here) appear with much less translational energy. Note that the CF_3^- TOF spectrum was recorded with a higher draw-out field (4.0 V cm^{-1} instead of 1.6 V cm^{-1}) in order to prevent the energetic CF_3^- turn around ions from striking the back electrode of the reaction chamber!

The evaluation of the TOF spectra leads to $E_T = 2.6$ eV for the kinetic energy of $CF_3^- + I$ which is also approximately 85% of the total available excess energy at $\varepsilon = 3.8$ eV.

To characterize the electronic structure of the excited CF_3I^{*-} resonance one has to consider the electronically excited neutral states. From photodissociation studies, it is well known that excitation of CF_3I in the A band (350–200 nm $\widehat{=}$ 3.5–6.2 eV) results in prompt dissociation along the C–I bond due to the repulsive nature of the excited state.[224–227] The A band of molecules of the type CX_3I (X = H, F) arises from an $n \rightarrow \sigma^*$ transition *localized in the* C–I *bond*,[228] and the dissociation dynamics is then discussed in the simple model of 'collinear pseudotriatomic dissociation':[229] the molecule is considered as a triatomic A–B–C, with the A–B and B–C bonds being distinct normal coordinates. The electronic chromophore is localized at the A–B bond and the B–C bond is 'spectator' during excitation. It eventually receives vibrational energy as a result of repulsion during dissociation. Due to the C_{3v} symmetry, vibrational excitation in CF_3 is confined to the ν_2 umbrella vibration of CF_3.[224]

The A band consists of three overlapping transitions from the ground N state to the $^3Q_1$, $^3Q_0$, and $^1Q_1$ states with absorption maxima located around 4.1, 4.7, and 5.2 eV, respectively.[230] Recent photodissociation studies in the 304 nm region have shown[225] the dominant formation of spin-orbit excited $I^*(^2P_{1/2})$ due to the $N \rightarrow {}^3Q_0$ transition which carries most of the

A band oscillator strength[230] and which directly correlates with spin-orbit excited I*. In this direct dissociation process it was found that only 5% of the available excess energy (0.9 eV) appears as vibrational energy of CF_3. In photodissociation we have thus a similar situation as in low-energy electron capture. In the first process a *nonbonding* electron is promoted to the lowest unoccupied MO (LUMO) with localized σ^* (C–I) character, while in the latter an electron is *added* to the same MO concomitantly exciting an electron. However, from the TOF analysis we definitely can conclude that the neutral iodine atom must exclusively be formed in its ground state, $I(^2P_{3/2})$ in contrast to photodissociation.

The very strong repulsive σ^* ($I–CF_3^-$) nature of the excited resonance, as documented by the unusually high translational energy release in the CF_3 product, suggests that two electrons reside in the σ^* (C–I) MO, one promoted from the I(5*p*) state and the other one added.

The observation that no I^- contribution appears from this second resonance can then be rationalized by the fact that the positive charge on the iodine atom makes a decomposition into $I^- + CF_3$ most unlikely.

The observation that CF_3^- is less abundant than the competing fragments I^- and FI^- (which appear with low kinetic energy release) remains in question. A possible explanation is that the autodetachment rate along the C–I coordinate is particularly high as indicated in Fig. 10 by the width of the

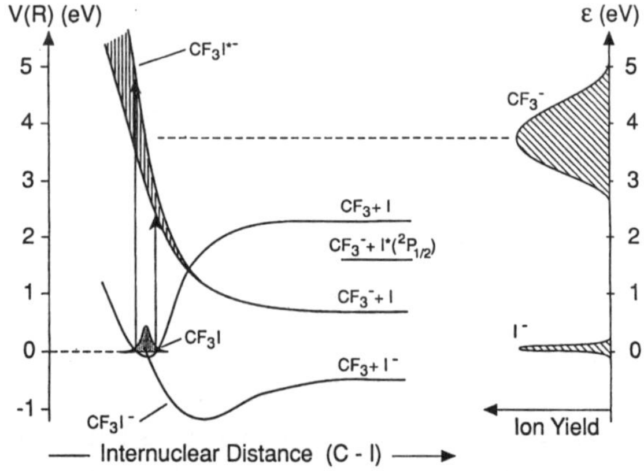

Fig. 10. Schematic potential energy curves for CF_3I and CF_3I^- relevant in low-energy electron attachment. The CF_3I^- curve is shown as blurred indicating its limited autodetachment lifetime at internuclear distances smaller than the crossing point with the curve of neutral CF_3I (from Ref. 194).

associated potential energy curve. It can, in fact, be shown below that at electron energies around 4 eV CF_3I molecules scatter electrons effectively down to low-energy, possibly through resonant scattering (Sec. 3.2.1 and Fig. 26). In the condensed phase, on the other hand, the cross-section for desorption is surprisingly high which is most likely due to a medium enhanced lifetime of the precursor ion (see Sec. 3.3.2). Although the molecule possesses a positive (adiabatic) electron affinity, (Fig. 10) a relaxed molecular anion cannot be formed in electron attachment under single collision conditions.

In CF_3Cl the situation is less straightforward as obvious from the resonance profiles of fragment ions at higher energies (Fig. 8). TOF measurements have shown[231] that from the second resonance all fragments are formed with low kinetic energy indicating no particular valence character of the electronic state. The molecule does not absorb photons in measurable quantities below 5.7 eV and the maximum of the first absorption band is located above 8 eV,[232–234] and thus far above the resonance position (around 4.5 eV). Excitation by electron impact, on the other hand, exhibited a weak band at 7.5 eV which was assigned as a (single or triplet) $n \rightarrow \sigma^*$ transition.[235]

In any case, in spite of the much less pronounced repulsive nature of the excited states in CF_3Cl^{*-} leading to Cl^-, the associated DA cross-section is larger compared to the low-energy feature. In view of the general rule that the attachment cross-section should increase with decreasing electron velocity, this behaviour is even more surprising.

To elucidate this phenomenon we consider further data from electron scattering which will also reveal the physical basis for the huge difference in the low-energy DA cross-section of about 4 order of magnitude between these two compounds.

Figure 11 compares (a) the total scattering cross-section[236] and (b) the partial inelastic scattering cross-section[38] with (c) the DA result,[39] the latter normalized to the above mentioned absolute number.[222] The partial inelastic cross-section is due to excitation of the ν_3 (C–Cl symmetric stretch) vibration in neutral CF_3Cl, i.e. the ν_3 'excitation function'. It is the dominant component in the total inelastic cross-section which was estimated as $\approx 5 \times 10^{-16}$ cm^2. The difference to the *resonantly enhanced* component in the total scattering cross-section [$\approx 12 \times 10^{-16}$ cm^2, Fig. 11(a)] can then be assigned to the resonant elastic scattering cross-section ($\approx 7 \times 10^{-16}$ cm^2).

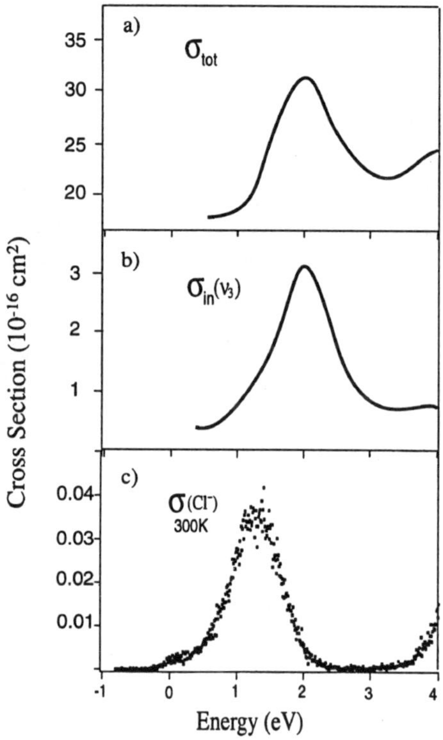

Fig. 11. Comparison of the DA cross-section for the reaction $e^- + CF_3Cl \rightarrow CF_3 + Cl^-$ with data from electron scattering in the region of the shape resonance; (a) total electron scattering cross-section;[236] (b) partial inelastic cross-section, i.e. excitation function of the ν_3 mode[38]; (c) DA cross-section of the process $e^- + CF_3Cl \rightarrow CF_3 + Cl^-$[39] (from Ref. 18).

From that it is clear that DA cross-section is by more than two orders of magnitude smaller than the attachment cross-section. Furthermore, while both scattering cross-sections peak at 2.0 eV, the Cl^- ion yield curve peaks at 1.4 eV.

From these results we must conclude that the low-energy resonance decays predominantly by autodetachment. From the numbers in Fig. 11 (resonant component of the total electron scattering cross-section $\approx 12 \times 10^{-16}$ cm^2, DA cross-section $\approx 3.9 \times 10^{-18}$ cm^2) we can estimate the branching ratio between autodetachment and dissociative attachment as $\sigma_{AD}/\sigma_{DA} \approx 300$. This also explains the shift between the 'primary' and the DA resonance: the potential energy curve for the negative ion is shown as blurred by a width $\Gamma(R)$ (see, e.g. Fig. 21) which expresses the energy uncertainty due to the limited lifetime with respect to autodetachment (uncertainty principle). Since in a shape resonance the electron is bound within a potential barrier, Γ should strongly increase (most likely exponentially) with electron energy (or decreasing internuclear distance), which makes DA from transitions at a low energies (large distances) more probable. This causes a shift of the DA resonance to lower energy with respect to the primary resonance. The ν_3 excitation function, on the other hand, directly reflects the shape of the primary resonance since most of the TNIs formed by a Franck–Condon transition undergo autodetachment.

3.1.2. *Electron Attachment to Larger Molecules: C_6F_5X ($X = F, Cl, Br, I$), SF_6, C_{60} and C_{70}. Competition between Dissociative and Associative Processes*

In electron attachment to larger molecules the electronic energy of the electron attaching system (comprised of the energy of the incoming electron and the electron affinity of the molecule) can be dispersed over the larger number of vibrational degrees of freedom in the molecule. If then DA channels are energetically inaccessible at low electron energies we can have the situation that effective transfer of electronic into vibrational energy delays autodetachment to an extent that the parent anion can be observed on the mass spectrometric time scale. A well-known molecule in that respect is SF_6 which possesses one of the highest attachment cross-sections known forming a long-lived parent anion at low electron energies (see below).

The molecules C_6F_5X are of particular interest since they also have very high cross-sections for the attachment of thermal electrons and possess appreciable positive (adiabatic) electron affinities (Table 1), i.e. the parent anions $C_6F_5X^-$ are thermodynamically stable. On the other hand, for X = Cl, Br, I the thermodynamic threshold (ΔH_o) of the complementary dissociation reactions

$$C_6F_5X^- \rightarrow X^- + C_6F_5\,, \qquad (10a)$$

$$\rightarrow X + C_6F_5^-\,, \qquad (10b)$$

is in the vicinity of 0 eV while in the perfluorocompound it is endothermic by about 1.5 eV (Table 1). As will be shown, this opens the interesting

Table 1. Thermochemical data relevant in DA to C_6F_5X. EA: electron affinity, *D*: bond dissociation energy, ΔH_o: thermoydnamic threshold for the DA channel $C_6F_5X^-$.

C_6F_5X	$D(C_6F_5\text{–}X)$(eV)	ΔH_o (eV)	$EA(C_6F_5X)$(eV)[a]
C_6F_6	4.9[b]	1.5	0.52 ± 0.1
C_6F_5Cl	4.0 ± 0.1[c]	0.4 ± 0.1	0.82 ± 0.1
C_6F_5Br	≈ 3.4[d]	≈ 0	1.15 ± 0.1
C_6F_5I	2.87 ± 0.05[e]	-0.19 ± 0.05	1.41 ± 0.1
EA(eV)[f]			
F 3.399 ± 0.003			
Cl 3.615 ± 0.004			
Br 3.364 ± 0.004			
I 3.059 ± 0.001			

[a]References 237 and 238.
[b]Reference 239.
[c]Reference 215.
[d]Reference 238 (estimated from Br^- formation at low-energy).
[e]Reference 240.
[f]Reference 216.

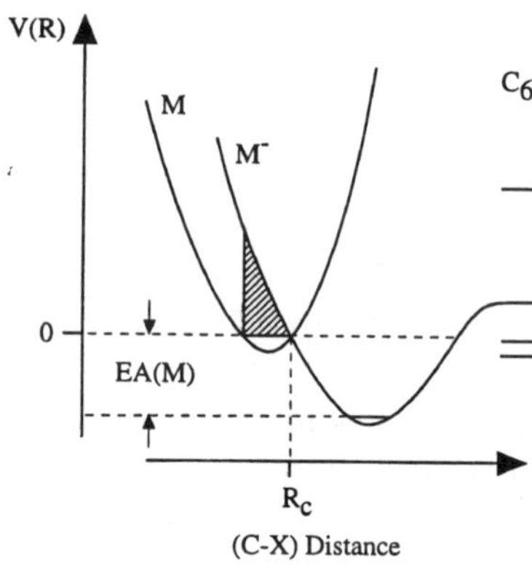

Fig. 12. Schematic potential energy diagram for the DA reaction in molecules of the type C_6F_5X (X = Cl, Br and I). The energy of the asymptotic limit $C_6F_5 + X^-$ is close to the energy of the initial resonance (from Ref. 18).

possibility of studying the competition between metastable $C_6F_5X^{-*}$ formation and dissociation.

This situation is schematically illustrated in Fig. 12 in terms of a one dimensional potential energy diagram. For X = Cl, Br, I the MO involved possesses some σ^* (C–X) nature (see below) which somehow justifies the one dimensional representation.

As mentioned, all molecules C_6F_5X under consideration possess a low-energy resonance, i.e the anionic potential energy surface is accessible via a low-energy Franck–Condon transition as indicated by the shaded area in Fig. 12. In C_6F_6 the lowest DA channel (F^- + C_6F_5) is *energetically inaccessible* from the low-energy resonance and hence autodetachment is the only decay channel. However, removal of energy from the relevant coordinate can transfer the anion temporarily into configurations where the system cannot autodetach ($R > R_c$ and $V(R) < V(R_c)$ in Fig. 2). This is the prototype situation for the formation of long-lived parent anions.

The dissociation channel (10a) becomes energetically more favourable in the order X = Cl, Br, I (Table 1) and is eventually accessible from the low-energy resonances, i.e. dissociative attachment (DA) then competes with associative attachment (AA). By AA we mean, in this context, formation of a long-lived (mass spectrometrically observable) parent anion $C_6F_5X^{-*}$.

Absolute numbers for the rate of thermal electron attachment are available from FALP[77,241] and pulse radiolysis experiments.[16(b)] The rates correspond to cross-sections in the 10^{-14} cm^2 region for all 4 compounds.

What is the fate of the $C_6F_5X^{-*}$ formed by electron capture? Figures 13–15 display the ion yield curves for the different ionic products obtained under single collision conditions. (The results for the system C_6F_5Br are not shown here as this compound directly follows the trend along the line Cl, Br, I described below.[121]) The ion count rates are plotted in absolute numbers to give an idea of the relative intensities of the various products and the changes upon substituting X. Although these spectra are recorded under approximately the same conditions (beam intensity, ion draw-out conditions, etc.) they only roughly mirror the respective relative cross-section since any mass spectrometric detection is subjected to different discrimination effects.

The following features are apparent:

(a) All four molecules show a dominant resonance near 0 eV and further comparatively weak resonant features above 3 eV.

(b) In C_6F_6 thermal electron capture exclusively yields the metastable parent anion. Substitution of one F atom by X = Cl, Br, I opens DA channels accessible from the low-energy resonance increasing in intensity in that order. In C_6F_5I, finally, no metastable parent ion is formed.

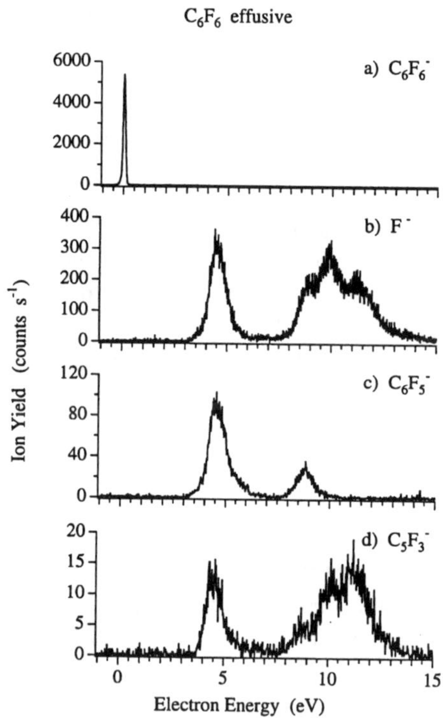

Fig. 13. Negative ion formation from C_6F_6 under single collision conditions (from Ref. 121).

(c) The branching ratio between the complementary channels (10a) and (10b) also changes dramatically by 4 orders of magnitude (!) along the line X = Cl, Br, I in favour of (10b).

(d) Fragmentation into F^- is surprisingly low in all four compounds.

We assign the low-energy TNI as a single particle shape resonance formed by accommodation of the extra electron into the LUMO. In benzene itself the lowest resonance is at 1.35 eV and the LUMO is the doubly degenerate $1e_{2u}(\pi^*)$ orbital (D_{6h}) (see, e.g. Refs. 21 and 23). For perfluorobenzene it was suggested[242] that $C_6F_6^-$ in its electronic ground state exists as a planar carbon skeleton containing out-of-plane C–F bonds. In terms of MOs the extra electron then resides in a $\pi^* - \sigma^*$ combination MO with π^* the dominant component.[243]

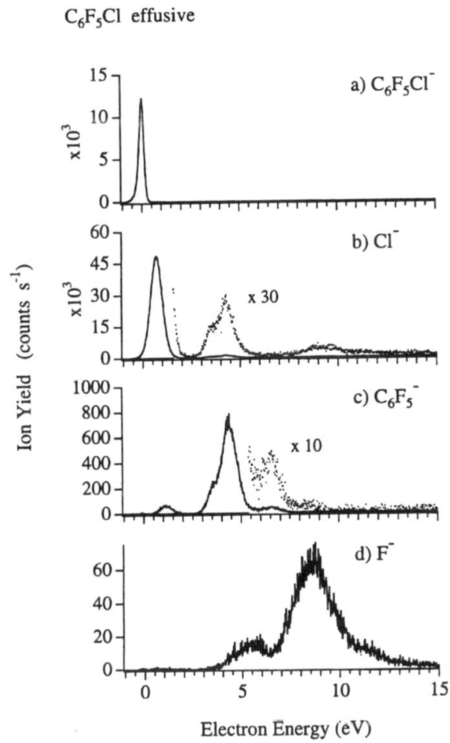

Fig. 14. Negative ion formation from C_6F_5Cl under single collision conditions (from Ref. 121).

On substituting one F atom by X = Cl, Br, I the symmetry is lowered from D_{6h} to C_{2v} which also removes the degeneracy of the lowest two MOs. As already mentioned above and discussed earlier,[238,244] the LUMO in molecules of the type C_6F_5X (X = Cl, Br, I) has π^* character with considerable σ^* (C–X) admixture.

On inspecting the low-energy X^- and $C_6F_5^-$ profiles in Figs. 14 and 15, it is apparent that the maximum of the first resonance is continuously shifted to lower energy on going from Cl over Br (not shown here) to I. In that order the dissociation limit

$$C_6F_5X^- \to C_6F_5 + X^-\,, \tag{10c}$$

also becomes energetically more favourable.

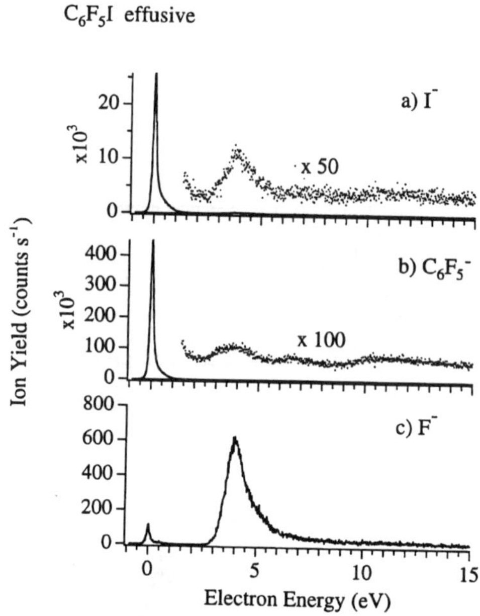

Fig. 15. Negative ion formation from C_6F_5I, No long-lived parent anion $C_6F_5I^-$ is observed under single collision conditions (from Ref. 121).

Interestingly, the low-energy resonance does not result in an observable $C_6F_5I^-$ signal. This means that the decay of the $C_6F_5I^-$ resonance (dissociation and autodetachment) is faster than the observation time in the present experiment (65 μs for $C_6F_5I^-$). Strictly speaking, under the present experimental conditions the $C_6F_5I^-$ ion resides 23 μs in the draw-out system and 42 μs in the quadrupole. If fragmentation occurs within the first 23 μs one will detect the fragment ion. If it decays in the quadrupole, no signal will be detected.

Figures 14 and 15 indicate further resonances in all four compounds visible through the DA products F^-, X^- and $C_6F_5^-$ with a pronounced contribution around 4–4.5 eV. Near that energy benzene itself (at 4.8 eV [21,23]) and the fluorobenzenes (at 4.5 eV [245]) also exhibit a negative ion state which was assigned as a single particle shape resonance involving the π^* system.

In conclusion, for the molecules C_6F_5X it can be seen that they all have very large cross-sections for thermal electron attachment. In perfluorobenzene this results exclusively in the generation of long-lived parent anions. Although the maximum in the transition probability is at 0.42 eV as shown in an electron transmission experiment,[245] long-lived anions are only formed within a very narrow resonance close to 0 eV. We interpret this by the rapidly increasing autodetachment rate with energy in connection with the detection time window of a mass spectrometric experiment.[120] We shall come back to that problem in clusters (Sec. 3.2.2) when effective additional pathways become operative. In the compounds C_6F_5X (X = Cl, Br, I) the DA channels $X^- + C_6F_5$ and $X + C_6F_5^-$ become accessible from the low-energy resonance and thus compete with formation of long-lived $C_6F_5X^{-*}$.

Owing to its very high cross-section for the attachment of thermal electrons SF_6 is frequently used as effective electron scavenger in high power, high voltage technical devices.[9,246–248] The molecule SF_6 and its anion SF_6^- hence represent probably the most extensively studied electron–molecule system.[248–251] Figure 16 shows that a long-lived parent anion is observed

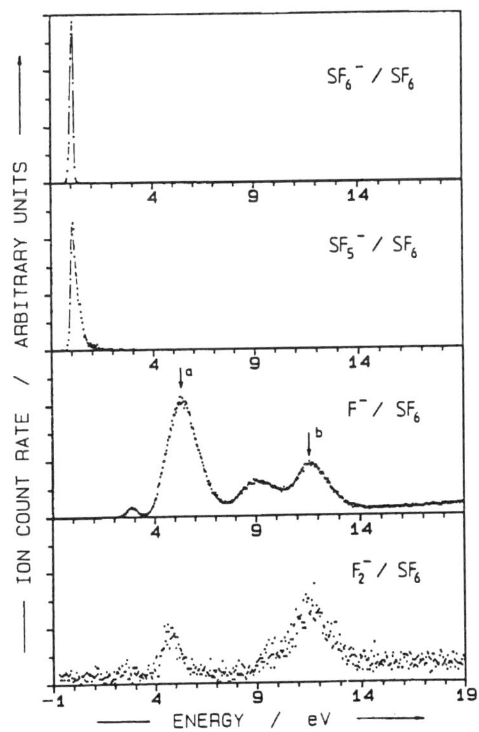

Fig. 16. Negative ion formation in SF_6 (from Ref. 31).

in a narrow energy range near zero eV while various DA processes proceed via different resonances at higher energies. Only the DA channel leading to SF_5^- (which is endothermic by 0.2 eV) is also operative at low energies. Due to its endothermicity, this reaction shows a strong temperature effect[251] in that the SF_5^- signal increases by 2 orders of magnitude on going from room temperature to 880 K.

The attachment cross-section towards very low energies has been the subject of many investigations using beam and swarm methods.[9,13,14] The most accurate results were obtained by the *rare gas photoionization method* where the sharp $^2P_{1/2}$ photoionization threshold of rare-gas atoms is used to generate continuum low-energy electrons with high energy resolution.[33,252] In the case of laser photoionization from the metastable $Ar(4s^3P_2)$ state low-energy electrons down to 0.1 meV at a resolution of 0.1 meV can be generated.[33]

From basic arguments it follows that the cross-section for a reaction near its threshold is controlled by the long-range potential between the reacting particles, no matter what the mechanism is.[218,219] In a reaction between a *charged* and a *neutral* particle (ion–molecule reaction, electron attachment, photodetachment from anions), the long-range interaction is dominated by the repulsive centrifugal potential between the particles.

In the case of electron attachment, the energy dependence of the cross-section near threshold is given by

$$\sigma_o(\varepsilon, \ell) \sim \varepsilon^{2\ell-1/2}\,, \tag{12a}$$

which is a special case of the Wigner threshold laws.[219] Since at low-energy only partial waves $\ell = 0$ contribute, the electron attachment cross-section near threshold should vary as

$$\sigma_o(\varepsilon) \sim \varepsilon^{-1/2}\,. \tag{12b}$$

Such an energy dependence was already predicted for ion–molecule reactions in 1905 by Langevin[253] from classical arguments. In ion–molecule and electron–molecule collisions the long-range interaction is dominated by the ion-induced dipole potential (see also Sec. 2.1)

$$V(r) = -\alpha e^2/2r^4\,,$$

[α: polarizability of the molecule], it can easily been shown that for a given collision energy there is a critical impact parameter b_c such that for $b < b_c$ the two particles are drawn together (i.e. they surmount the potential barrier created by the attractive polarization potential and the repulsive centrifugal potential) to form an unstable collision complex.[254] Thus, the *classical* capture cross-section can be expressed as

$$\sigma_{cap} = \pi b_c^2 , \tag{13a}$$

which has a collision energy dependence of the form

$$\sigma_L = \pi(2\alpha e^2/\varepsilon)^{1/2} . \tag{13b}$$

This expression is known as the *Langevin cross-section* which shows the same collision energy dependence as predicted by the Wigner threshold law. A fully quantum mechanical calculation on the basis of the polarization potential lead to the expression which is twice as large as the Langevin cross-section (Vogt and Wannier[220])

$$\sigma_{VW} = 4\pi(\alpha e^2/2\varepsilon)^{1/2} . \tag{14a}$$

The above equations refer to the electrostatic system in which $e^2 = 27.2$ eV a_0 with a_0 the Bohr radius (0.53Å) and α expressed in units of r^3. For practical use it is convenient to express (14a) as

$$\sigma_{VW} = 4\pi a_0^2(\alpha/2\varepsilon)^{1/2} \tag{14b}$$

which directly yields the cross-section in a_0^2 when α is expressed in units of a_0^3 and ε in atomic units (27.2 eV).

On this basis Klots[255] proposed the following analytical form

$$\sigma_K = (\pi a_0/2\varepsilon)\{1 - \exp[-4(2\alpha\varepsilon)^{1/2}]\} , \tag{14c}$$

which agrees with the capture cross-section derived by Vogt and Wannier in the limit $\varepsilon \to 0$ and yields the 'geometrical' cross-section $\pi/k^2 = \pi\lambda\!\!\!^{-2}$ (*s-wave* capture cross-section) at larger electron energies ($\varepsilon = \hbar^2k^2/2\mu$).

The high-resolution laser experiment[33] indicated that in the energy range below 100 meV the attachment cross-section in SF_6 is well described by the Klots expression (14b). It should be noted that at thermal energy (26 meV) the attachment cross-section is in the range of some 10^{-14} cm^2 but exceeds 10^{-13} cm^2 at energies below 3 meV. It thus appears that in a conventional beam experiment the measured cross-section depends largely on the quality of the electron beam, i.e. the energy resolution and how far the electron energy can be pushed down towards '0 eV'.

It has first been shown by Lezius, Scheier and Märk[43] that electron impact to C_{60} and C_{70} generates long-lived C_{60}^- and C_{70}^- anions within an unusually broad energy range extending from 0 eV to about 13 eV. This is a remarkable situation since *conventional* molecules (like SF_6 and the compounds C_6F_5X discussed above) generate long-lived parent anions only very close to 0 eV. Only in some polycyclic aromatic hydrocarbons[256] and in benzoquinone[13] a metastable anion has been observed at energies appreciably above 0 eV.

Figure 17 displays the C_{60} and C_{70} attachment spectra together with the SF_6^- calibration peak. These spectra were recorded with a modified effusive beam arrangement[45] shown in Fig. 18. The sample containing a

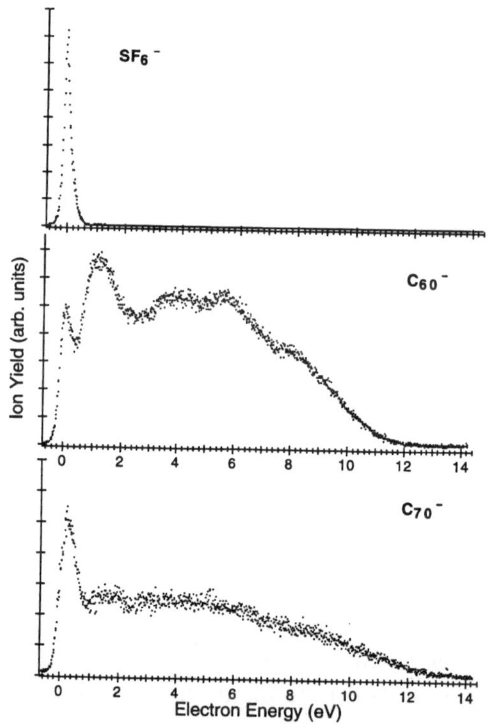

Fig. 17. Comparison of the electron attachment profiles yielding metastable parent anions between SF_6 and the two fullerenes C_{60} and C_{70} (from Ref. 45).

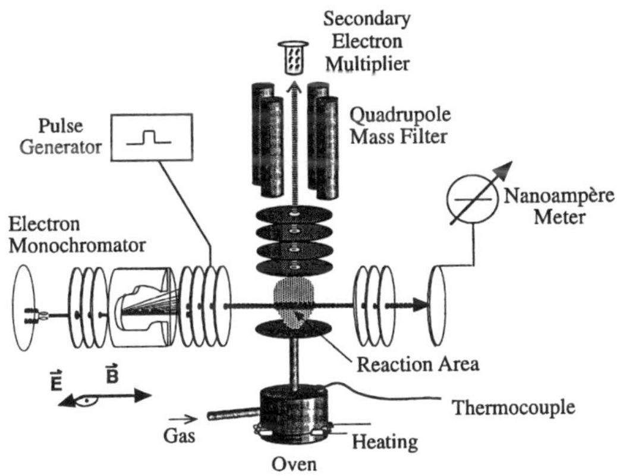

Fig. 18. Experimental setup for the study of electron attachment to gaseous C_{60} and C_{70} (from Ref. 45).

mixture of C_{60} and C_{70} is heated in an oven fitted to the reaction chamber. Heating the sample to 850–900 K results in sufficient vapour pressure to form an effusive beam.

From Fig. 17 it can be seen that long-lived C_{60}^- and C_{70}^- are formed up to impact energies of $\approx$ 13 eV. No anionic fragments were observed up to 50 eV impact energy, although the energy threshold for the process

$$C_{60}^- \to C_2^- + C_{58}\,, \tag{15}$$

is near 6 eV, as predicted from the dissociation energy of the cationic system $C_{60}^+ \to C_{58}^+ + C_2$ (9.5 eV[257]) and the electron affinity of C_2 (3.39 eV[216]). There has been some debate on the correct value of the dissociation energy in C_{60}. While theory predicted values above 10 eV,[258] from early experiments much lower values around 7 eV[259] were derived. These 'experimental' values, however, depend critically on the model of evaluation and the complete consideration of all possible effects [inclusion of radiative cooling etc. (see e.g. Ref. 257)].

From the present beam experiment the position of the first peak was determined at 140 meV[45] at an energy resolution near 200 meV. Another beam experiment by Huang *et al.*[260] also located the first peak slightly above zero while recent beam experiments by Elhamidi *et al.*[47] and Vasil'ev *et al.*[48] showed that the first peak essentially coincides with the SF_6 calibration peak (Fig. 19). It must be noted that in the very low-energy domain

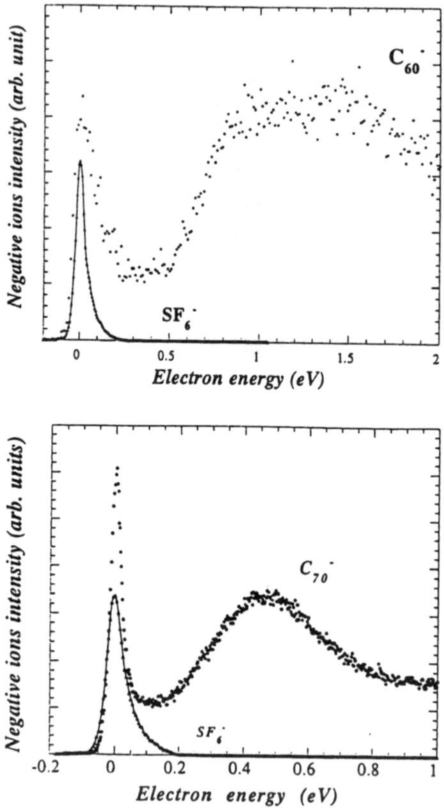

Fig. 19. Detailed attachment profile in the low-energy region for C_{60} and C_{70} in comparison to SF_6 recorded at high energy resolution ($\approx$ 0.04 eV FWHM) (Adapted from Elhamidi *et al.* Ref. 47).

both the observed intensity and also the peak position is extremely sensitive on the ion draw-out conditions.

The structures superimposed on the broad background in the C_{60}^- spectrum toward higher electron energies indicate that electronic excitation is also involved in the attachment process forming long-lived C_{60}^-. Following the discussions in relation to electron energy loss experiments[261–263] it appears that the assignment to particular electronic transitions is not yet definite. The broad and unstructured background is likely to arise from *nonresonant electron capture*, i.e. direct coupling of the incoming electron to the many vibrational degrees of freedom in C_{60}.

In a simple approach the C_{60} electron *scattering cross-section* may be calculated by approximating the C_{60} potential by a rectangular spherical potential with a well depth of $V_o = 4.68$ eV (corresponding to a bound state of 2.65 eV in C_{60}^-[264]) and the C_{60} hard sphere radius $R_o = 5.26$ Å. From elementary quantum mechanics the scattering cross-section can then be expressed as

$$\sigma = \frac{4\pi}{k^2}\sum_{\ell=0}^{\infty}(2\ell+1)\sin^2\delta_\ell\,, \tag{16}$$

with $k^2 = 2\mu\varepsilon/\hbar^2$ and δ_ℓ the phase shift of the electron wave due to the potential. For a rectangular spherical potential the phase shifts δ_ℓ can be calculated analytically.[265] In the energy range between 0.5 and about 15 eV it turned out that it is sufficient to consider the partial waves $1 \leqslant \ell \leqslant 3$. It thus appears that the broad unstructured background is due to p, d and f wave attachment. This would mean that virtually every scattered electron efficiently loses energy and is captured to C_{60} supported by the large number of vibrational degrees of freedom. The modulations on the broad background in the experimental attachment spectrum may then be due to additional electronic excitations related to those seen in the electron energy loss spectrum.

The question arises on the origin of the low-energy peak close to 0 eV. FALP studies by Smith *et al.*[266] on the attachment rate within the *electron temperature* between 500 and 4500 K yielded an activation barrier of 260 meV for electron attachment. According to a theoretical analysis by Tosatti and Manini[267] *s-wave* capture to C_{60} is forbidden. For symmetry reasons, attachment of an electron with $\ell = 0$ cannot form C_{60}^- in its electronic ground state and the activation energy was hence attributed to the p wave centrifugal barrier.

On the other hand in studies of the electron transfer in Rydberg atom C_{60} collisions[260,268,269] low-energy attachment was observed with the n-dependence in the transfer rate indicating *s-wave capture*. In addition, the relative intensity of the zero energy peak strongly increases with the temperature of neutral C_{60}.[270] This would indicate an activation energy with respect to the *internal energy* of C_{60} instead of the electron energy proposed from the FALP studies (see also the following section).

At energies above ≈ 8 eV the experimental spectrum falls distinctly below the calculated scattering cross-section which was attributed to thermal electron detachment from C_{60}^-. This can be substantiated by using a TOF technique where the intensity of the fullerenes was recorded for different flight times through the quadrupole at different primary electron energies.[45]

An analysis of these data indicates that in the time window covered by the present experiment, the decrease in ion intensity with the electron energy ε for both C_{60}^- and C_{70}^- can be described by an exponential expression of the form

$$N(t,\varepsilon) = N_o(\varepsilon)\exp[-t/\tau(\varepsilon)]\,, \tag{17}$$

where $\tau(\varepsilon)$ is the mean lifetime of the anion (formed at the energy ε). The evaluation of a series of data sets showed that in both ions the lifetime (in the electron energy range between about 8 and 12 eV) can be approximated by the simple exponential form $\tau(\varepsilon) = \tau_o \exp(-\varepsilon/\varepsilon_o)$ with ε_o and τ_o constants, the latter expressing the lifetime at $\varepsilon = 0$ eV. The following constants were derived for C_{60}: $\tau_o = 0.1$ s, $\varepsilon_o = 1.43$ eV (see Fig. 20); for C_{70}: $\tau_o = 0.25$ s, $\varepsilon_o = 1.49$ eV. At $\varepsilon = 10$ eV, for example, one obtains $\tau(C_{60}^-) = 92$ μs and $\tau(C_{70}^-) = 304$ μs. These values agree well with those obtained by Elhamidi *et al.*[47]

By increasing the electron energy we thus have a situation in which the anions C_{60}^- and C_{70}^- become increasingly unstable toward detachment. In the light of the huge kinetic shift of more than 34 eV observed[259] for the process $C_{60}^+ \to C_{58}^+ + C_2$ the present result is not surprising. A statistical treatment of neutral C_{60} resulted in a rate for electron emission which strongly dominates that for unimolecular decomposition.[259]

Similarly, in electron attachment a long-lived C_{60}^{-*} is formed up to impact energies near 13 eV. At any electron energy, the anion is unstable with respect to autodetachment

$$C_{60}^{-*} \to C_{60} + e^-\,, \tag{18a}$$

while the threshold for the lowest dissociation channel

$$C_{60}^{-*} \to C_{58} + C_2^-\,, \tag{18b}$$

is near 6 eV. Reaction (18a) is much faster thus suppressing reaction (18b) below the detection limit.

In conclusion, it can be seen that C_{60} is a remarkably efficient electron scavenger over the wide energy range up to 12 eV. Above that energy, effective detachment from the hot C_{60}^- is seen on a mass spectrometric

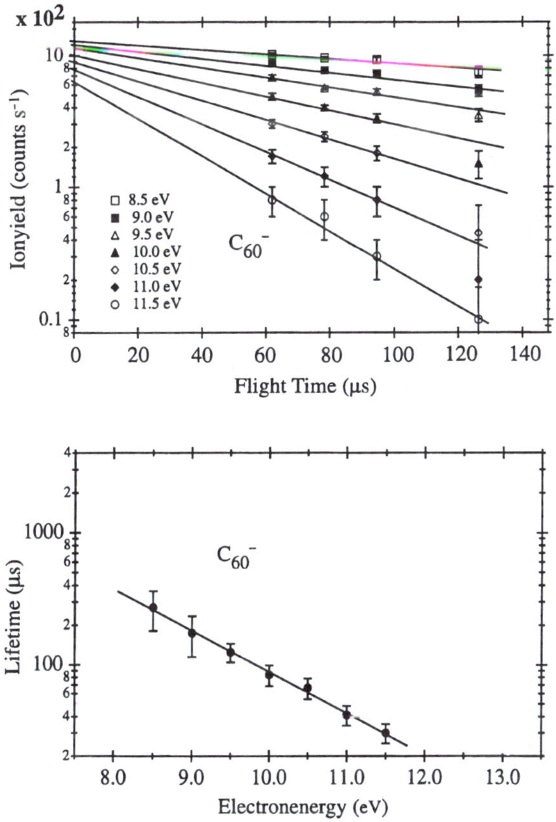

Fig. 20. Evaluation of a set of data recorded at different transmission energies for C_{60}^-. The horizontal scale is the flight time corresponding to a particular transmission energy. One line represents the dependence of C_{60}^- intensity as a function of flight time for a fixed electron energy. Below: plot of the average lifetime of C_{60}^- as function of the electron energy (from Ref. 45).

time scale. The broad background of the electron attachment spectrum between 0.5 and 13 eV can reasonably well be described by *p*, *d* and *f* wave attachment via a square well potential with a radius corresponding to the hard sphere of C_{60} and a well depth according to the electron affinity. Above 7 eV electron detachment competes on the observation time scale of the present experiment. Attachment towards very low energies is most likely due to *s*-wave attachment, subjected to some activation energy with respect to the internal energy of C_{60}.

3.1.3. *Temperature Dependence of Dissociative Attachment Reactions*

It has been known since more than 3 decades that the temperature may have a strong effect on DA. In the case of the oxygen molecule it was shown that the DA cross-section increases by a factor of 3 when the gas temperature is increased from 300 K to 3000 K.[271] At the same time the maximum of the DA cross-section curve is downshifted from 6.4 eV to 4.6 eV. This is much more than the thermal energy of the hot molecule [kT (3000 K) = 0.26 eV]. In a semiclassical model this kind of behavior was explained by the energy dependent competition between DA and AA in the decay of the O_2^{*-} TNI:[272] heating of the molecule and thus population of vibrationally excited states extends the Franck–Condon region thus allowing transitions to the strongly repulsive ionic potential energy curve at lower electron energies and hence higher dissociation probabilities.

It will be shown below that strong autodetachment is the essential key to explain the observed temperature dependence.

We mention here two further experimental techniques which have been applied to measure the temperature dependence of electron attachment processes, namely the electron drift technique[273] and the flowing afterglow Langmuir probe (FALP) method.[75] Both techniques operate at higher pressure, usually above 1 mbar and thus beyond single collision conditions. From both techniques absolute values for the electron attachment rate can be obtained.

Returning to Sec. 3.1.1 it was shown that the value for the DA cross-section to the structural similar molecules CF_3Cl and CF_3I differs by more than 4 orders of magnitude.

The physical basis for this large difference in the DA cross-section is primarily due to an exceptionally strong competition of autodetachment (AD) in the case of DA into $Cl^- + CF_3$ which is also the basis for the strong temperature dependence in this DA channel. The situation for the CF_3Cl system is illustrated in Fig. 21: the potential energy curve for the negative ion is blurred by the width $\Gamma(R)$ which expresses the energy uncertainty due to the autodetachment lifetime according to the Heisenberg uncertainty principle. $\Gamma(R)$ is directly connected to the autodetachment rate $W(R)$ by

$$W(R) = \Gamma(R)/h\,. \tag{19}$$

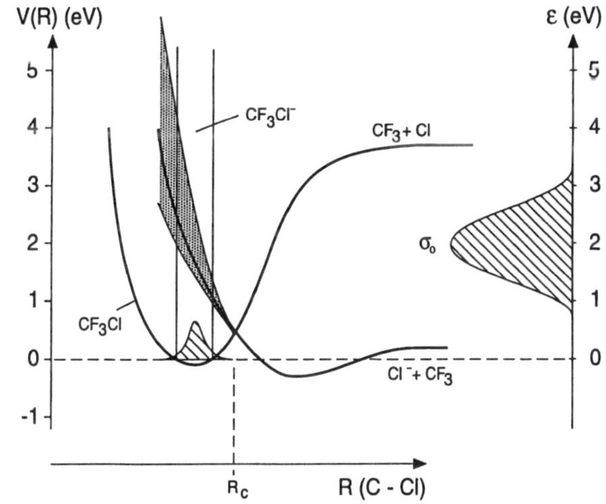

Fig. 21. Schematic potential energy diagram illustrating DA to CF_3Cl (from Ref. 18).

The cross-section for DA can be expressed as

$$\sigma_{DA} = \sigma_o \cdot P\,, \tag{20}$$

with σ_o the attachment cross-section and P the probability of dissociation into Cl^- + CF_3. P is called the *survival probability* and can be written as

$$P = \exp(-\tau_D/\tau_{AD})\,, \tag{21}$$

with τ_D the time it takes the products to dissociate until autodetachment becomes impossible and τ_{AD} the autodetachment lifetime τ_D is given by

$$\tau_D = \int_R^{R_c} \frac{dR'}{v(R')}\,. \tag{22}$$

with R the distance where the electronic transition takes place, R_c the crossing point between the two potential curves (for $R > R_c$, $P = 1$) and $v(R')$ the radial velocity between Cl^- and CF_3. The autodetachment lifetime is given by

$$\tau_{AD} = \frac{h}{\Gamma(R)}\,. \tag{23}$$

Since in a shape resonance the electron is bound within a potential barrier, Γ should strongly increase (most likely exponentially[274,275]) with electron energy (or decreasing internuclear distance) which makes DA from transitions at low energies (large distances) more probable. As already mentioned in Sec. 3.1.1 this also explains the shift of the DA resonance (1.4 eV) with respect to the primary resonance (2.0 eV) as seen in electron scattering: transitions at lower energy (larger intermolecular distance) have a higher dissociation probability which shifts the DA resonance to lower energy (*survival probability shift*).

A closer inspection of the experimental DA cross-section (Fig. 8) shows that at 0 eV a weak "threshold" signal in the Cl^- curve is visible. This threshold signal mirrors transitions from vibrationally excited neutral molecules near the crossing point between the neutral and ionic potential energy surface. Such transition have dissociation probabilities near unity. In addition, the cross-section σ_o for *electron attachment* increases with decreasing electron energy. At low energies, *s-wave* attachment dominates with the cross-section increasing with decreasing electron energy. Thus survival probability near unity and diverging attachment cross-section contribute to the intensity of the threshold signal. It should be noted that the overall discussion here is still done within the Born–Oppenheimer approximation which may no longer rigorously be applicable at very low energies.

Figure 22 shows the temperature dependence of the Cl^- yield indicating an increase in intensity and concomitant shift to lower energy of the first resonance and a strong increase of the threshold peak with temperature. The highest temperature (700 K) corresponds to an average energy of 60 meV and thus only few of the relevant ν_3 (C–Cl) normal modes $[\hbar\omega(\nu_3) = 59 \text{ meV}]$ are excited.

The energy of the crossing between the two potential energy curves can be viewed as the "activation energy" for DA. The variation of the threshold signal with gas temperature should then reflect the activation barrier via an Arrhenius-like behavior. A plot of the logarithm of the Cl^- threshold intensity versus $1/T$, in fact, shows a linear dependence leading to $E_a = 0.26 \pm 0.05$ eV.[276]

The observed temperature effects can reasonably be described within a classical model using exponentially varying model functions for $\Gamma(R)$, and Morse potentials for the neutral and anionic system having a crossing point at 330 meV.[274,275]

While the low-energy Cl^- signal shows a strong temperature effect, the other channels including Cl^- from the second resonance are virtually unaffected by increasing the gas temperature.[277] The same is true for electron

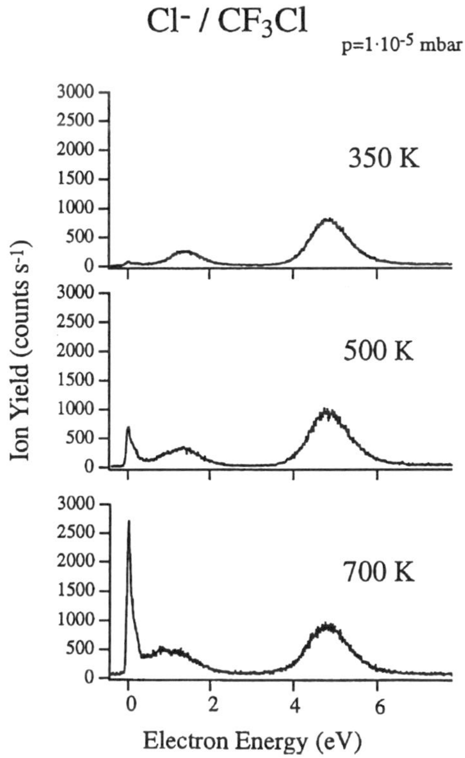

Fig. 22. Temperature dependence of the Cl^- yield from CF_3Cl (from Ref. 276).

attachment to CF_3I where the intensity of most fragment ions as well as the peak positions are virtually insensitive on temperature (with the exception of the F^- channel[277]).

The insensitivity of the I^- signal is simply a result of the disposition between the anionic and neutral potential energy curve. In the system CF_3I the anionic curve crosses that of the neutral molecule near its equilibrium configuration at 0 eV (Fig. 10). Transitions near 0 eV have dissociation probabilities near unity. Increasing the temperature does not depopulate the $v = 0$ level to an appreciable extent resulting in a DA cross-section which is insensitive to temperature.

Figures 23 and 24, finally, illustrate the temperature dependence of electron attachment to C_6F_5Cl. In Sec. 3.1.2 it was shown that for this compound, AA and DA competes on the mass spectrometric observation

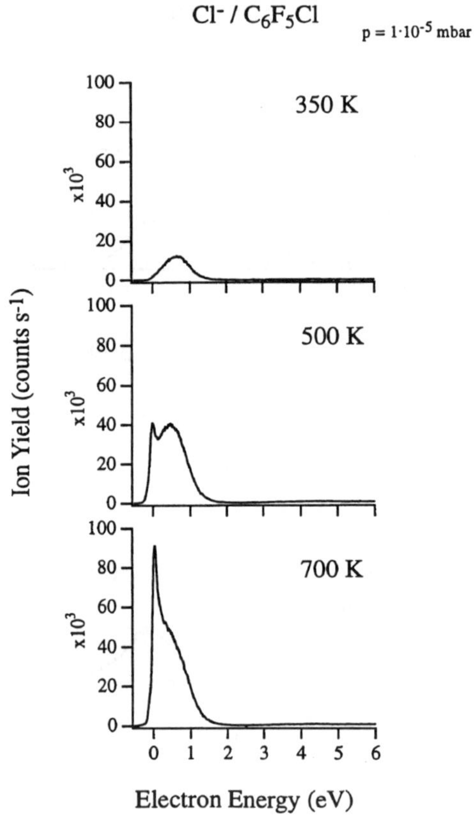

Fig. 23. Temperature dependence of the Cl^- DA channel in C_6F_5Cl (from Ref. 276).

window. In addition to the low-energy complementary DA channels

$$C_6F_5Cl^- \to Cl^- + C_6F_5\,, \tag{24a}$$

$$24b \to Cl + C_6F_5^-\,, \tag{24b}$$

we have thus additionally to consider the decay of the long-lived parent anion

$$C_6F_5Cl^- \to C_6F_5Cl + e^-\,, \tag{24c}$$

i.e. the temperature dependence of the $C_6F_5Cl^-$ signal may also be affected by autodetachment. In fact, from Figs. 24 and 25 the following features are apparent:

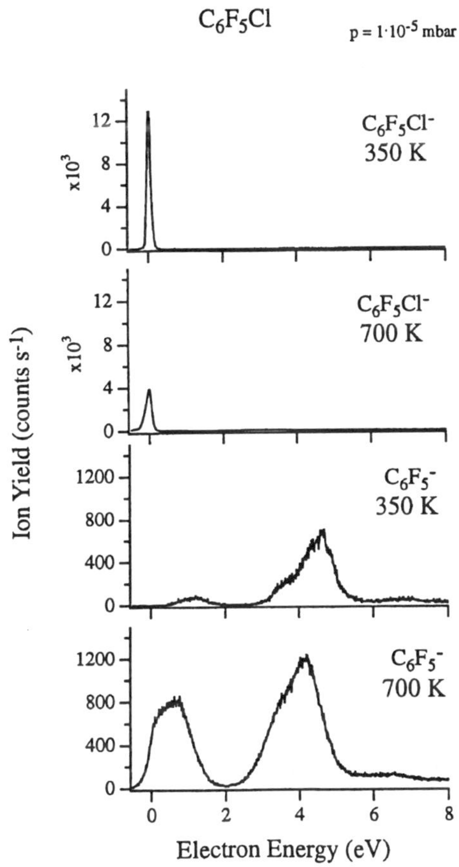

Fig. 24. Temperature dependence of associative attachment and the DA channel yielding $C_6F_5^-$ in C_6F_5Cl (from Ref. 276).

(a) The cross-section for the endothermic DA processes (24a) and (24b) increases with temperature leading to the two features discussed above, the shift of the resonance maximum and the appearance of a threshold peak.

(b) The cross-section for associative attachment decreases with temperature.

The total ion yield also increases by more than a factor of 2 which is in particular due to the strong increase of the DA channel leading to Cl^-.

From the Arrhenius plots for the threshold peaks[276,277] the activation energy for the DA channels are derived as E_a (Cl^-) $= 0.42 \pm 0.1$ eV and E_a ($C_6F_5^-$) $= 0.8 \pm 0.2$ eV. In the present case, the activation energy corresponds to the energy of the dissociation limit while the primary step of electron attachment proceeds without activation energy (represented by the crossing point of the potential energy surfaces). This is obvious from the fact that the parent anion $C_6F_5Cl^-$ is formed near 0 eV. The activation energy for Cl^- from the Arrhenius plot is very close to the energy of the dissociation limit C_6C_5 + Cl^- derived from established thermodynamical data (0.4 eV, see Table 1).

The decreasing intensity of the parent ion signal with temperature must be due to its shorter lifetime. Since this ion is exclusively formed close to 0 eV, it is primarily the *autodetachment lifetime*. The threshold peaks for the dissociation channels, on the other hand, arise from transitions of vibrationally excited neutrals at the level of the dissociation exit. Their appreciable intensity with respect to the resonant contribution is due to the reciprocal energy dependence of the attachment cross-section. We assume that any threshold DA reaction is fast compared to the metastable time scale where $C_6F_5Cl^-$ is detected. In that sense we interpret the decreasing $C_6F_5Cl^-$ intensity primarily by a shortening of its autodetachment lifetime and not by the competition of DA channels with temperature.

In summary, it can be seen that the cross-section for DA can strongly vary with the temperature of the target molecule. For processes having an activation barrier (CF_3Cl) an increase of the temperature results in a shift and concomitantly an increase of the resonance and the appearance of a threshold feature. The threshold peak arises from transitions close to the crossing point having survival probabilities near unity. Its temperature dependence simply mirrors the population of vibrational states according to the Boltzmann distribution at the level of the crossing point and hence obeys an Arrhenius-like behavior. The activation barrier for dissociative attachment into CF_3 + I^- is zero and hence the process *is insensitive* to temperature. In C_6F_5Cl the intensity of the parent anion decreases with temperature, due to a decrease of the autodetachment lifetime.

3.2. *Gas-Phase Clusters*

In the previous sections we have demonstrated that in small molecules like halogenated methanes, formation of a TNI usually leads to a prompt dissociation of the molecular anion into stable, negatively charged and neutral fragments (DA). Although stable ground state parent anions exist, they are not formed in the gas-phase since stabilization mechanisms are not operative. In clusters, on the other hand, we have a *built-in* many-body stabilization mechanism which may open an effective way to prepare gas-phase molecular anions which cannot be formed in electron attachment under single collision conditions. Although van der Waals clusters are coupled by weak forces, the intermolecular collision frequency is high enough ($10^{12}-10^{13}$ s^{-1}) to provide effective relaxation on a much shorter time scale. This is particularly true in ionic clusters (as formed in electron attachment) where — compared to neutral clusters — stronger intermolecular forces are present which in turn enhances the intermolecular collision frequency. The halogenated methanes belong to molecules where the TNIs often decompose directly along a repulsive surface with a shallow minimum. On going to clusters, we expect a competition between relaxation forming undissociated molecular anions (associative attachment, AA) and emission of free ionic fragments (dissociative attachment, DA).

In molecules of the type C_6F_5X (X = Cl, Br, I), on the other hand, we have shown that DA from the low-energy resonance (Sec. 3.1.2) occurs on the metastable time scale. In this case, we expect a much more effective quenching of DA reactions in clusters.

As we shall demonstrate, the cluster environment does not only affect the evolution of a TNI but also the primary step of electron localization.

3.2.1. *CF_3Cl and CF_3I Clusters: Formation of Relaxed Parent Anions and Inelastic Scattering prior to Electron Localization*

In Sec. 3.1.1 we have shown that the DA cross-section for the channel $CF_3 + X^-$ is by more than 4 orders of magnitude lower in X = Cl compared X = I. This is mainly due to the unusually low survival probability of the transient negative ion CF_3Cl^- formed at low-energy (around 1.4 eV).

We will now investigate how aggregation affects ion formation in the case of the weak electron scavenger CF_3Cl as opposed to the very strong scavenger CF_3I. Under collision free conditions, both compounds capture electrons purely dissociatively.

In electron attachment to clusters of CF_3Cl one obtains the DA products already known from single molecules (Cl^-, F^-, FCl^-, CF_2Cl^- and CF_3^-) but also undissociated ions M_n^- (M = CF_3Cl) and solvated ions of the form $Cl^- \cdot M_n$, $F^- \cdot M_n$ and $CF_3^- \cdot M_n$. In the ion yield curves, these latter products more or less carry the signature of the particular fragment ion. These phenomena are described and discussed in some detail in Ref. 109 and we will focus here only on Cl^- formation in relation to the dramatic change of the relative intensity between the first Cl^- resonance (around 1.4 eV) and the second Cl^- resonance (around 4.8 eV) on going to clusters.

The observation is that the peak ratio between the first and second Cl^- resonance increases on going to clusters. In single molecules at room temperature it is 1:2 (Fig. 8), in clusters the ratio increases with stagnation pressure up to 3:1 at 2 bar. Above that stagnation pressure the ratio continuously decreases again until it becomes 1:1 at 5 bar.

In Fig. 25 we display the ion yield curves of Cl^-, M^- and Cl^- at a stagnation pressure of 2 bar. The observed behaviour indicates some enhancement in the low-energy DA cross-section, possibly related to an enhancement of the autodetachment lifetime of the TNI in a cluster.

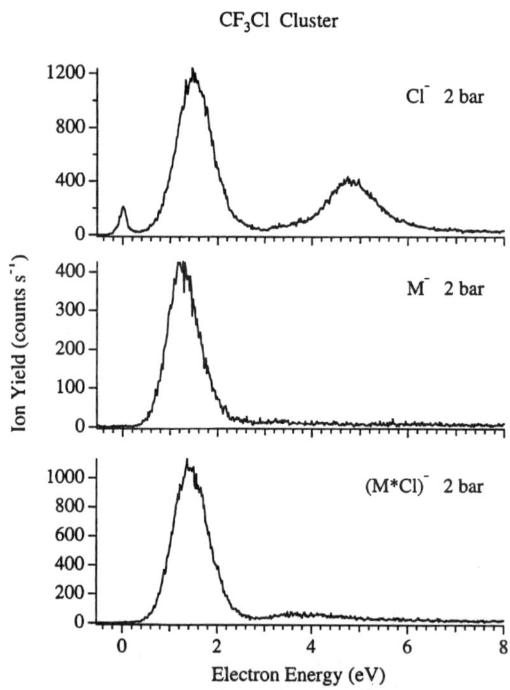

Fig. 25. Ion yield curves of Cl^-, M^- and $Cl^- \cdot M$ ($M = CF_3Cl$) obtained in electron attachment to CF_3Cl clusters at a stagnation pressure of 2 bar (from Ref. 18).

The detailed problem, however, is complicated due to the interplay of different, partly counterbalancing and hardly separable effects which is discussed to some extent in Ref. 18. It was concluded that the Cl^- DA cross-section from the low-energy resonance is increased in clusters by at least one to two orders of magnitude at a stagnation pressure of 2 bar and this increase is virtually attributed to an increase of the autodetachment lifetime when the TNI is coupled to environmental molecules.

Compared to CF_3Cl the compound CF_3I is a much stronger electron scavenger which leads to a very different attachment behaviour in clusters. As above, at low energies we observe the stabilized monomeric anion (evaporative attachment) and also DA reactions yielding *free* fragment ions I^- and *solvated* fragment ions $I^- \cdot M_n$ (Fig. 26).

Returning to Fig. 7 it can be seen, that I^- formation from single CF_3I is exclusively restricted to the energy range near 0 eV. In clusters, on the other hand, I^- also appears with appreciable intensity from resonant-like

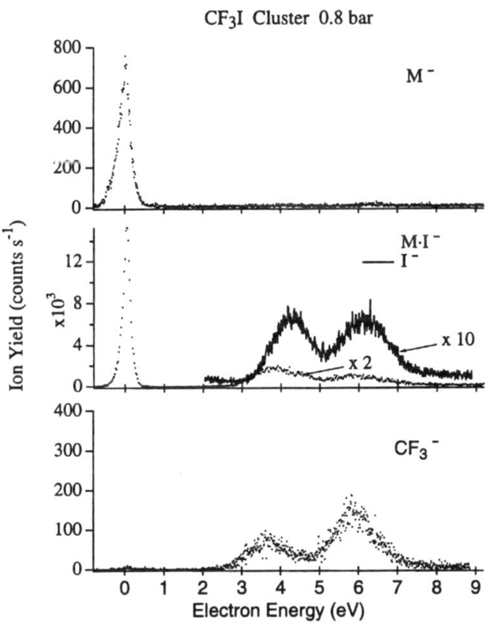

Fig. 26. Ion yield curves of some selected products following electron attachment to CF_3I clusters. Stagnation pressure 0.8 bar; the signal of I^- and $I^- \cdot M$ is normalized to the same height at 0 eV and multiplied by 10 and 2, respectively, above 2 eV. The absolute count rate refers to the product $M \cdot I^-$ (from Ref. 18).

structures around 4 and 6 eV (Fig. 26). These new features are also visible on the solvated ions and on the DA product CF_3^-. Figure 26 also shows that the CF_3^- fragment in the cluster experiment is more intense than from single molecules. This is in contrast to the other DA products which are less intense in the cluster experiment (at a stagnation pressure up to 2 bar) compared to effusive gas inlet (at 5×10^{-5} mbar).

These 'new' features near 4 and 6 eV are due to self-scavenging, i.e. inelastic scattering from one molecule and capture of the slow electron by another molecule of the same cluster. Self-scavenging is a general feature in clusters composed of molecules having large capture cross-sections.[18,122] In the present case, capture of the inelastically scattered electron is predominantly dissociative, as evident from the repulsive nature of the '0 eV' resonance.

The question is whether the inelastic scattering mechanism is *direct* or *resonant*. Since these new features carry the signature of the electronic excitation of the neutral molecule[18] it is likely that most of the self-scavenging intensity arises from *direct* scattering.

3.2.2. C_6F_5X Clusters ($X = F, Cl, Br, I$): Quantitative Quenching of Chemical Fragmentation

These compounds possess very large cross-sections ($\sigma > 10^{-14}$ cm^2) for the attachment of low-energy electrons (< 1 eV) (Sec. 3.1.2). While in perfluorobenzene only a metastable anion is observed at these energies, the other compounds additionally decompose into the complementary DA channels $X^- + C_6F_5$ (10a) and $X + C_6F_5^-$ (10b). The rate for these dissociation processes, however, is very low so that the metastable parent anions $C_6F_5Cl^{-*}$ and $C_6F_5Br^{-*}$ can be observed *in competition* with the DA products on the mass spectrometric time scale. The physical basis for this slow dissociation reaction is the above mentioned delicate balance between *resonance energy* and *energy of the dissociation limits* (10a) and (10b) together with the comparatively large electron affinity (Fig. 12). This provides the mechanism of effective energy redistribution in the TNI on the way from its origin (Franck–Condon transition) to the DA products.

Metastable anions $C_6F_5X^{-*}$ are detected for X = F, Cl, Br. They only appear within a very narrow energy range near 0 eV.

Figures 27 and 28 show a selection of ion yield curves obtained from C_6F_5X clusters (X = Cl and I): by just looking at the absolute count rates

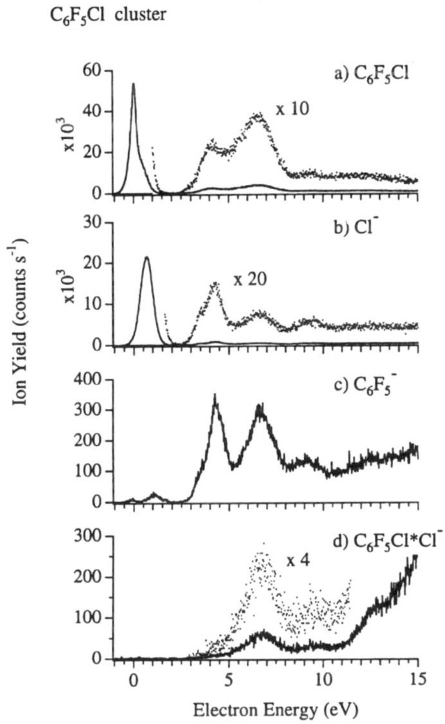

Fig. 27. Ion yield curves of some selected products obtained from electron attachment to C_6F_5Cl clusters (p_o = 1.0 bar). Formation of the solvated ion is quantitatively suppressed near 0 eV (from Ref. 121).

it is obvious that the DA channels associated with the low-energy resonance are strongly quenched in clusters. This effect is most obvious in C_6F_5I. While in isolated C_6F_5I the parent anion could not be observed at all, it is by far the dominant product from clusters.

It can be shown that in clusters quenching of the DA channels from the low-energy resonance is *quantitative* and that the remaining X^- and $C_6F_5^-$ intensity near 0 eV in Figs. 28 and 29 is solely due to DA from scattered background molecules. This was done by (i) recording ion yield curves of the solvated fragments $X^- \cdot M$ and $C_6F_5^- \cdot M$ and (ii) performing a time-of flight (TOF) analysis of the products.

(i) Solvated fragments are necessarily products of an attachment process to a *cluster*. They indicate chemical fragmentation and are usually very

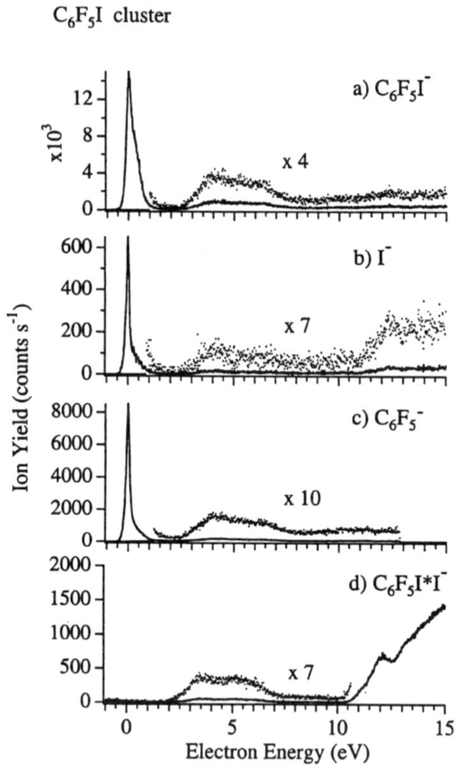

Fig. 28. Ion yield curves of some selected products obtained from electron attachment to C_6F_5I clusters ($p_o = 2.2$ bar). Formation of the solvated ion $M \cdot I^-$ is quantitatively suppressed near 0 eV (from Ref. 121).

prominent and ubiquitous in electron attachment to van der Waals clusters. In the present molecules, solvated ions $X^- \cdot M_n$ and $C_6F_5^- \cdot M_n$ can, in fact, be observed at higher energies. From the low-energy resonance, however, these ions are virtually absent as can be seen at the bottom of Figs. 27 and 28. It is thus evident that DA yielding free X^- from the low-energy resonance is quantitatively quenched in clusters.

(ii) A more direct proof of the quantitative quenching of chemical fragmentation is possible through TOF spectrometry since a product arising form a precursor traveling in the seeded beam can be distinguished from a product arising from a scattered background molecule. As discussed in Ref. 121 it can be unambiguously shown by means of the applied TOF experiments that the complementary DA channels near 0 eV are *quantitatively quenched in clusters.*

It must be emphasized that we have a situation where *the chemistry induced by a slow electron is completely changed when the molecule* C_6F_5X *is coupled to a medium,* in this case molecules of the same species. While the reaction in single C_6F_5X molecules is completely dissociative, at least on a time scale $\geq 10^{-4}$ s (chemical fragmentation by cleavage of the C–X bond), it becomes completely associative in clusters in the sense that free ionic fragments are not formed. A small signal due to solvated ions indicate that some 'intracluster' DA processes are still operative on an intensity scale several orders of magnitude lower than DA in the gas-phase. The reason for the strong quenching is the energy balance between position of the resonance with respect to the dissociation limits in combination with an appreciable molecular electron affinity. In isolated C_6F_5I, e.g. only transitions close to 0 eV are possible and the TNI decomposes exclusively into fragments according to

$$e^-(\approx 0\ \text{eV}) + C_6F_5I \rightarrow C_6F_5I^{-*} \rightarrow C_6F_5 + I^-\,, \tag{25a}$$

$$\rightarrow C_6F_5^- + I\,, \tag{25b}$$

on a time scale $\geq 1\ \mu$s. $C_6F_5I^{-*}$ is the anion in its electronic ground state, but vibrationally excited. The excitation energy equals the EA of the molecule (1.41 eV). Channel (25a) is by only 0.19 eV exothermic while the exact value of (25b) is not known.

In clusters, on the other hand, intermolecular collisions can remove energy from the TNI within its lifetime, leaving a thermodynamically stable species. In the case of a neutral target dimer M_2 ($M = C_6F_5X$) we can write

$$e^-(\approx 0\ \text{eV}) + M_2 \rightarrow M^{-*} \rightarrow M^{-(*)} + M^{(*)}\,. \tag{26}$$

Both final particles have to share the excess energy in the form of internal and translational energy. The excess energy is composed of the electron affinity of M (1.41 eV in the case of X = I) *minus* the bond dissociation energy of the neutral dimer (in the order of ≈ 0.1–0.2 eV). As mentioned above, M^{-*} is stable with respect to autodetachment and dissociation, if its energy is below the energy of the intersection, Q_c, between the anionic and neutral potential energy surface and below the lowest dissociation limit, respectively (Fig. 2).

M^- (and M_n^-) formation also occurs at energies above 3 eV via self-scavenging. In the case when the monomer is detected, the excess energy is used to either liberate M^- from the target cluster or to evaporate the target cluster more or less completely.

3.2.3. *Nucleophilic Displacement (S_N2) Reactions in Binary van der Waals Clusters Induced by Resonant Electron Capture*

In this section we use heterogeneous van der Waals clusters to study bimolecular nucleophilic displacement (S_N2) reactions of the form

$$X^- + RY \rightarrow [XRY]^- \rightarrow XR + Y^- . \tag{27}$$

The idea is to generate a molecular beam containing mixed neutral clusters composed to the components R_1X and R_2Y. The nucleophile can then be released by dissociative electron attachment (DA) from the first component

$$e^-(\varepsilon_1) + R_1X \rightarrow R_1X^{-(*)} \rightarrow R_1 + X^- , \tag{28}$$

which then undergoes the exchange reaction (27) with the 'substrate' molecule, visible by the observation of the leaving group Y^-. The first step [reaction (27)] is generally *resonant* in nature and is operative at a particular energy (ε_1) which is specific for the particular target molecule R_1X.

The necessary requirement for an unambiguous observation of a substitution reaction is then that the substrate molecule R_2Y does either *not* undergo DA to form Y^- at all or at least in an energy range different from ε_1. If then at ε_1 we observe Y^- anions from a mixed cluster, we have strong evidence that they are the result of an S_N2 reaction.

The primary step of electron capture inherently implies that (on the basis of the underlying thermodynamics) the energy of the reaction is completely controlled by the energy of the electron. The excess energy can hence be varied within the width of the resonance. Moreover, in the case of repulsive DA precursor ions R_1X^-, the translational energy of the nucleophile can eventually be varied on a particular scale.

Since in most halogenated hydrocarbons the F^- DA resonances are usually *above* those yielding Cl^-, Br^- or I^- we have to consider also possible inelastic electron scattering events in the target cluster as possible sources for the generation of the 'leaving group' Y^-. As frequently discussed above a DA reaction is principally subjected to the competition of autodetachment (AD), i.e. re-emission of the extra electron prior to dissociation of the transient negative ion (TNI). If the energy spectrum of the resonantly scattered electrons include ε_2, Y^- may also be a product of DA to R_2Y via inelastically scattered electrons.

Inelastic scattering can, in fact, considerably contribute to negative ion formation in homogeneous and heterogeneous clusters (*supra vide*), in particular, in the presence of an effective electron scavenger. However, a negative ion formed by attachment of secondary electrons usually carries the signature of the electronically excited state(s) of the neutral scatterer and not that of its negative ion resonance, i.e. the relevant *inelastic* scattering events proceed rather through *direct* scattering than *resonant* scattering.

S_N2 reactions have played a significant role in the development of concepts in physical organic chemistry.[278–280] While in solution educts and products are separated by an activation barrier (with its height largely controlled by the nature of the solvent), a gas-phase S_N2 reaction proceeds through a double minimum potential (Fig. 29) as first suggested by Olmstead and Brauman.[281] The minima represent the ion–molecule complex before ($X^- \cdot RY$) and after electron transfer and rearrangement of the (R–Y) and the (R–X) bonds ($XR \cdot Y^-$). They are still separated by a central activation barrier, which, however, is considerably lowered with respect to solution.

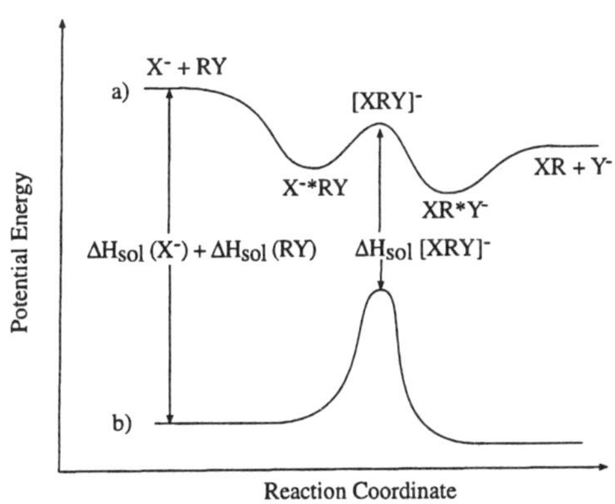

Fig. 29. Comparison of the potential energy surface in the gas-phase and in the condensed phase along the reaction coordinate of the collinear S_N2 reaction $F^- + CH_3Y \rightarrow CH_3F + Y^-$ (from Ref. 125).

Gas-phase S_N2 reactions reflect the *intrinsic* properties of the involved ions and molecules. In solution, on the other hand, these intrinsic properties interfere with solvent effects.

According to the model introduced by Shaik and Pross,[278–280] the central barrier arises from an avoided crossing between the two electronic curves containing reactant-like (X^-RY) and product-like (XRY^-) Heitler–London valence bond (VB) forms. These *diabatic* potential energy curves describe the transfer of the radical R without simultaneous electron transfer. The potential energy curve relevant for the S_N2 reaction is then obtained by introducing an electron transfer matrix element which splits the diabatic curves. The two low lying electronic configurations are then asymptotically identified as $X^- + RY$ and $XR + Y^-$.

It has been known for many years that in the absence of solvation the rate constant can dramatically increase. For the prototype system $R{=}CH_3$, $X{=}Cl$ and $Y{=}Br$, the gas-phase rate constant (2.1×10^{-11} cm^{3}s^{-1}) is by about 15 orders of magnitude (!) larger than that in water.[282,283]

Gas-phase S_N2 reactions have been studied since more than 2 decades. The experimental techniques include ion cyclotron resonance (ICR) mass spectrometry,[281,284,285] high pressure mass spectrometry,[286] and flow or drift techniques.[287–289] In a drift tube, the mean collision energy can be varied by changing the drift voltage.[289]

We mention here related studies, namely, reaction dynamics upon optical excitation of the ion–molecule complex $X^- \cdot RY$ which is considered the intermediate for the S_N2 reaction (27),[290,291] reactions of solvated nucleophiles with CH_3Y molecules in the flowing afterglow,[292] and S_N2 reactions in mixed cationic cluster systems like $(NH_3/C_6H_5Cl)^+$ generated by resonantly enhanced multiphoton ionization (REMPI).[293]

In the gas-phase the displacement reactions of the systems F^-/CH_3Y (Y = Cl, Br, I) are characterized by an appreciable exothermicity. On the basis of the well established thermodynamical data (bond dissociation energies and electron affinities[215,216,294,295]), the reaction enthalpies are −1.3 eV (Cl), −1.6 eV (Br) and −1.9 eV (I). A particular interesting problem is the interaction of low-energy electrons with mixed clusters composed of C_2F_6 and CH_3Cl. While isolated CH_3Cl has an unmeasurably low DA cross-section into Cl^-,[296,297] it generates this fragment ion with appreciable intensity when CH_3Cl is coupled to C_2F_6 molecules via the corresponding S_N2 reaction.

The Displacement Reaction $F^- + CH_3Y \rightarrow CH_3F + Y^-(Y = Cl, Br, I)$

Before identifying possible displacement reactions in binary clusters it is necessary to study the behaviour of the substrate molecules with respect to DA. This is shown in Fig. 30 for the three compounds. The spectra represent the Y^- ion yield obtained from a pure expansion of CH_3Y seeded in Ar at a stagnation pressure of 2.0 bar in each case with the count rates as before given in absolute numbers. The behavior of CH_3Cl with respect to DA at low-energy is still under discussion. This is related to the unusually low cross-section which requires an extraordinarily high purity of the sample and the gas inlet system. According to recent experimental and theoretical studies,[296,297] the cross-section for the DA reaction

$$e^- + CH_3Cl \rightarrow CH_3Cl^{-*} \rightarrow CH_3 + Cl^- , \tag{29}$$

is most likely below 10^{-23} cm^2. This is many orders of magnitude lower than that of other chlorinated compounds and e.g. about 10 orders of magnitude (!) lower than the DA cross-section in CCl_4.[298,299] It thus appears that in earlier experiments (including those of our laboratory[300]) any detected low-energy Cl^- signal was obviously due to impurities.

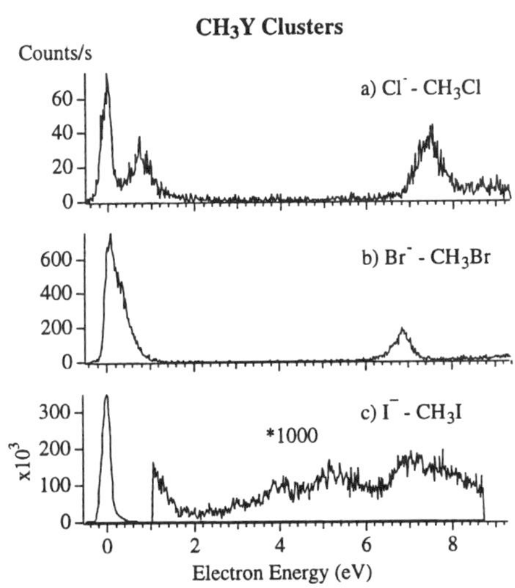

Fig. 30. Y^- ion yields (Y = Cl, Br and I) obtained from the pure expansion of the respective substrate molecules CH_3Y seeded in Ar at a stagnation pressure of 2.0 bar. The following mixing ratios were used: (a) CH_3Cl:Ar = 1:10; (b) CH_3Br:Ar = 1:10; (c) CH_3I:Ar = 1:20 (from Ref. 125).

Accordingly, we assign the low-energy (< 2 eV) Cl^- signal in Fig. 30(a) due to impurities in the sample and/or the gas inlet system. This, however, does not affect any of the further results or conclusions. Under the present experimental conditions and for a cross-section in the range of 10^{-23} cm^2 we would expect a count rate much below 1 s^{-1}.

There is a further resonant contribution yield near 7.4 eV (CH_3Cl) and 6.9 eV (CH_3Br). In the recent DA studies to CH_3Cl[296] the cross-section for this second Cl^- contribution was determined as 5×10^{-20} cm^2. This high energy signal (Cl^-, Br^-) is most likely the result of a multiple fragmentation reaction involving a core excited resonance. In the system CH_3I some I^- contribution is already observed above 3.5 eV which is not seen from the isolated molecule. Like in the cluster systems studied above, this signal is due to *self-scavenging*, i.e. inelastic scattering of the fast electron creating an electronically excited molecule and dissociative attachment by a second molecule of the same cluster.

C_2F_6 is chosen as the F^- donating compound for the following reason: from Fig. 31(a) it can be seen that F^- is formed from a DA resonance peaking at 3.8 eV and hence at an energy where the substrate molecule does not

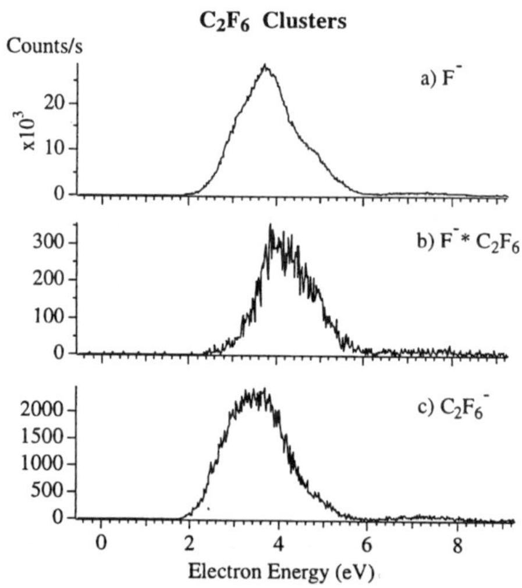

Fig. 31. Some ion yields obtained from a pure expansion of the nucleophile delivering component C_2F_6 seeded in Ar(1:10) at a stagnation pressure of 2.5 bar. (a) isolated DA product F^-, (b) solvated fragment ion $F^- \cdot C_2F_6$ and (c) stabilized parent ion $C_2F_6^-$ (from Ref. 125).

undergo DA at all. Figures 31(b) and (c) show further products resulting from electron attachment to a pure C_2F_6 expansion seeded in Ar(1:10 at a stagnation pressure of 2.5 bar). While isolated C_2F_6 is exclusively subjected to DA with F^- by far the dominant ion,[301] from C_2F_6 clusters ionic complexes of the form $(C_2F_6)_n^-$ including the monomer and also solvated fragment ions $F^- \cdot (C_2F_6)_n$ are observed.

The F^- signal obtained from a pure C_2F_6 expansion, on the other hand, is the result of DA reactions from (i) C_2F_6 clusters traveling in the molecular beam, (ii) C_2F_6 monomers traveling in the molecular beam and, finally, n(iii) C_2F_6 monomers representing the scattered background gas. As already mentioned above it is in favourable cases possible to separate products resulting from reactions in the molecular-beam from those in the scattered background gas by means of a time-of-flight (TOF) technique (see below).

Figure 32 presents an overview for the 3 substrate molecules concerning potential S_N2 reactions from the mixed C_2F_6/CH_3Y expansion. For comparison, Fig. 32(a) again shows the F^- contribution from a pure C_2F_6 expansion. From the mixed clusters the F^- profile is virtually identical to that from the pure expansion with roughly 50% of the original intensity.

However, from each of the mixed cluster systems we observe in the Y^- yield now a relatively strong contribution in the vicinity around 3.8 eV which we ascribe to the S_N2 reaction

$$F^- + CH_3Y \rightarrow CH_3F + Y^- . \tag{30}$$

As mentioned, apart from the reaction (30) we have to consider two further possible processes which could be a source of some Y^- signal, namely (a) inelastic electron scattering from C_2F_6 in the mixed cluster and capture of the slow electron by the second component which then undergoes DA forming Y^- and (b) reactions occurring in the scattered background gas, i.e. either an S_N2 reaction between F^- (generated from an isolated C_2F_6) and gas-phase CH_3Cl or the inelastic scattering sequence described in (a) occurring between uncorrelated monomers.

The latter possibility (b) can easily be checked by effusing the two compounds at a pressure corresponding to the background pressure of the supersonic beam experiment (10^{-5} mbar). This did not result in any detectable Y^- signal at energies near 4 eV, not even by increasing the pressure into the 10^{-4} mbar region. We will show below by means of TOF measurements, that in fact any Y^- signal detected in the vicinity of 4 eV must be attributed to reactions occurring in the jet.

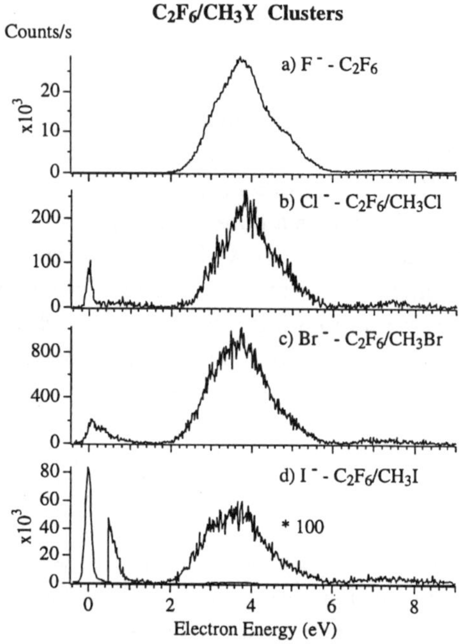

Fig. 32. Energy profile of the nucleophile F^- observed from a pure C_2F_6 expansion compared to the energy profiles of the respective naked product ions Y^- (Y = Cl, Br and I) observed from expansion of the mixture C_2F_6/CH_3Y seeded in Ar at a stagnation pressure of 2.5 bar. The following mixing ratios were used: (a) C_2F_6:Ar = 1:10; (b) C_2F_6:CH_3Cl:Ar = 3:1:40; (c) C_2F_6:CH_3Br:Ar = 10:1:100; (d) C_2F_6:CH_3CI:Ar = 10:1:100 (from Ref. 125).

Since the Y^- ion in every case carries the profile of the C_2F_6 DA resonance, only inelastic scattering via the $C_2F_6^-$ resonance has to be considered.

There are particular molecules where resonances in the DA cross-section have their counterpart in the inelastic scattering cross-section. This is generally the case when autodetachment strongly competes with dissociation in the decay of the transient negative ion as in the case of CF_3Cl discussed in Sec 3.1.1. However, the intensity of the inelastically scattered electrons decreases rapidly with the amount of energy loss which is a general phenomenon in electron scattering. For the present problem the primary electron has to be scattered nearly completely down from about 4 eV (to be dissociatively captured by the second component) which is a very unlikely event. In addition, only CF_3I is a reasonable electron scavenger and the inelastic scattering cross-section of C_2F_6 does not show any pronounced enhancement near 4 eV.[302]

We therefore conclude that the observed Y^- signal around 4 eV in fact is the result of the exchange reaction. For the CH_3Cl molecule we have thus the remarkable phenomenon that the interaction of single molecules with low-energy electrons does not result in a detectable Cl^- signal while from mixed C_2F_6/CH_3Cl clusters effective generation of Cl^- via the ion–molecule reaction (30) takes place.

Interestingly, the count rate of the leaving group is in the same order of magnitude for all the three systems with the intensity ratio Y^-/F^- in the order of a few percent. Superficially one would expect that the effectivity of a reaction in a mixed cluster is low. This number, however, is not a direct mirror of the reaction efficiency in the target cluster as Y^- represents the 'naked' product leaving the target cluster and accordingly F^- represents the 'naked' DA product from C_2F_6.

We have performed time-of-flight experiments in order to get some idea to which degree observed ions arise from events in the background gas or in the jet. Since the quadrupole axis is mounted *in line* to the supersonic beam, ions arising from reactions in the beam carry the additional velocity component and can be distinguished from the randomly moving background gas. These TOF experiments showed that under the present experimental conditions most of the F^- signal arise from scattered background C_2F_6 molecules. This is always the observation for an expansion of a gas of low condensation temperature (195 K for C_2F_6 under standard conditions) when no cooling of the nozzle is applied as in the present experiments, it also indicates the weak tendency of C_2F_6 to form aggregates (although in the expansion mixture used in considerable excess). Apart from mixed clusters, the molecular beam may also contain homogeneous clusters and probably monomers. We can hence conclude that an F^- DA event in a mixed C_2F_6/CH_3Y cluster must have a much higher probability to undergo an S_N2 reaction than suggested by the weak intensity ratio Br^-/F^-. From the TOF spectra of the naked products Br^-, on the other hand it is clear that they virtually must arise from precursors in the supersonic beam.

To illustrate the TOF principle, Fig. 33 shows TOF spectra for the Br^- product from a C_2F_6/CH_3Br mixture recorded at different inlet conditions and electron energies. Figures 33(a) and (b) show the TOF spectra

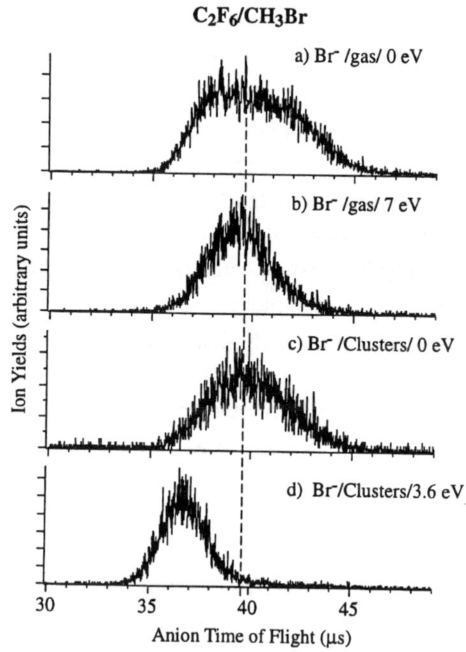

Fig. 33. TOF distributions of the product ion Br^- obtained from effusive gas inlet (a and b) and from an expansion of the C_2F_6:CH_3Br:Ar mixture (3:1:40) at a stagnation pressure of 2.5 bar (c and d) recorded at the indicated electron energies. Ion draw-out field 2.8 V cm^{-1} (see the text, from Ref. 125).

obtained at an incident energy of 0 eV and 7 eV, respectively, when the gas is effusing from the capillary directly fitted to the reaction volume. Near 0 eV one observes an uncompletely resolved TOF doublet due to ions emitted along the quadrupole axis either directly towards or directly away from the detector. This indicates that near 0 eV the precursor ion CH_3Br^- obviously dissociates impulsively via direct electronic decomposition along a repulsive potential energy surface. Near 7 eV, on the other hand, only one peak appears which indicates the formation of Br^- ions with only thermal or near thermal energy. The associated decomposition reaction is obviously less direct and most likely associated with the appearance of more than one neutral fragment. Figures 33(c) and (d) show the TOF spectra obtained from the mixed expansion through the nozzle at a stagnation pressure of 2.5 bar near 0 eV and 3.6 eV, respectively. The spectrum near 0 eV resembles that from the effusive inlet although the unresolved doublet is less pronounced which may be related to the fact that the reaction volume (defined by the crossing of the electron beam with the gas) is now less localized with respect to the effusive gas inlet by means of the capillary.

Near 3.6 eV, however, the Br^- TOF spectrum is now distinctly shifted to lower flight times. This is due to the additional velocity component of the particles in the jet (see Experimental). This is a further proof that Br^- ions formed around 3.6 eV arise from *reactions in the molecular-beam and not from scattered background molecules.* Br^- formation from the mixed expansion close to 0 eV, on the other hand, is essentially due to DA from scattered background CH_3Br molecules.

In the mixed cluster systems studied here, a variety of ion–molecule complexes arising from electron attachment to the respective target cluster can be observed. Figure 34 shows the energy profiles of ions with the stoichiometric composition $(FCH_3Br)^-$ and $(BrCH_3Br)^-$ together with that of the

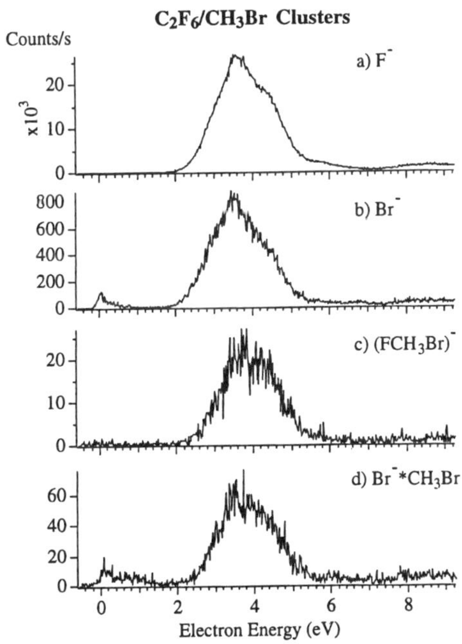

Fig. 34. Energy profiles of two ion–molecule complexes obtained from the expansion of C_2F_6:CH_3Br seeded in Ar(10:1:100) at a stagnation pressure of 2.5 bar. For comparison the energy profiles of the nucleophile (F^-) and the naked product ion (Br^-) is also shown (from Ref. 125).

nucleophile F^- and the product ion Br^-. In both complexes the localization of the excess charge is *a priori* not known. In the case of $(BrCH_3Br)^-$ it is rather $Br^- \cdot CH_3Br$ than $Br \cdot (CH_3Br)^-$ due to the much larger electron affinity of Br as compared to CH_3Br. In any case, since the energy profile of the complex $Br^- \cdot CH_3Br$ resembles that of the nucleophile F^-, it must be formed in the course of a completed S_N2 reaction in the target cluster. The ion $(FCH_3Br)^-$, on the other hand, is either the ion–molecule complex before $(F^- \cdot CH_3Br)$ or after the S_N2 reaction $(FCH_3 \cdot Br^-)$.

Energy Dependence of the Reaction Efficiency

Reactions of F^- with CH_3Y (Y = Cl, Br, I) as prototype S_N2 reactions have been the subject of numerous investigations in the gas-phase.[279,304] The rate of these exothermic reactions decreases with collision energy and is only weakly sensitive on the temperature of the substrate molecule. We mention here, e.g. a recent study on the system F^- with CH_3Cl[303] indicating that the *rate constant* for the S_N2 reaction decreases from 1×10^{-9} cm^3 s^{-1} to $\approx 2 \times 10^{-10}$ cm^3 s^{-1} on increasing the collision energy from 0.3 eV to 1 eV.

It appears that in the gas-phase the internal barrier for the S_N2 reactions studied here are in each case below the energy of the entrance channel which generally explains why the reactivity decreases with collision energy.

The question arises whether from the present results information on the energy dependence of the reaction efficiency can be extracted. We note that the amount of energy deposited in the system and hence the excess energy of the S_N2 reaction is (on the basis of the underlying thermodynamics for the reaction) entirely controlled by the energy of the attaching electron. This energy can be varied over the range of the respective DA resonance, in the present case between approximately 2 eV and 6 eV. From gas-phase measurements the distribution of the excess energy in the DA reaction of the nucleophile donating compound in often known.[14,15,18] In C_2F_6 most of the available excess energy appears as internal energy of the neutral radical (C_2F_5) while the translational energy of the nucleophile only slightly increases over the resonance up to 250 meV.[305] There are, however, also systems (e.g. CF_4[14]) which decompose into F^- via strongly repulsive precursor states and hence with the release of appreciable kinetic energy.

The question is then whether from the explicit comparison between the F^- and the Y^- energy profile the energy dependence of the respective S_N2 reaction can be extracted. To reveal this problem we return to the system

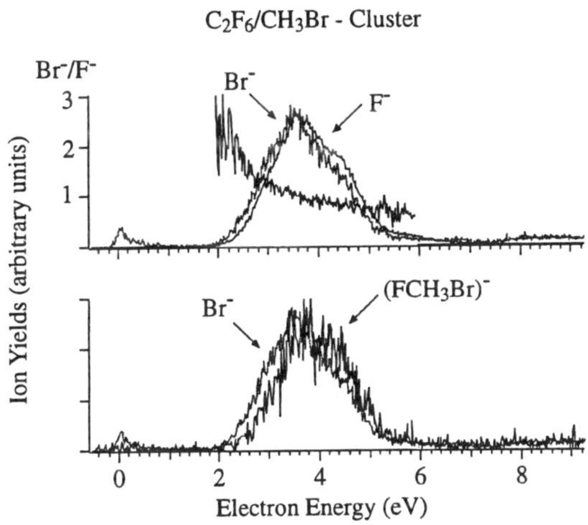

Fig. 35. (a) Comparison of the energy profile of the nucleophile (F^-) with the naked product ion (Br^-) and energy dependence of the ratio Br^-/F^-; (b) comparison of the energy profile between the naked product ion Br^- and the ion–molecule complex $(FCH_3Br)^-$ (see the text, from Ref. 125).

C_2F_6/CH_3Br. Figure 35(a) shows that by increasing the energy across the resonance the relative intensity the reaction product (Br^-) decreases with respect to that of the nucleophile (F^-). This suggests a *negative energy dependence* of the S_N2 reaction. In Fig. 35(b) it can be seen that the ratio Br^-/F^- drops to below 25% when the electron energy is increased from 2 eV to 6 eV.

To which extend does this number reflect the energy dependence of the underlying S_N2 reaction occurring in the respective *target cluster*? It must be remembered that F^- represents the free DA product from C_2F_6 and Br^- the 'naked' product emitted from the target cluster. It has already been discussed above that in the case of F^- only a minor amount of the detected signal arises from DA processes in the cluster beam. In any case, the observed F^- ion represents a *free* DA fragment, either formed from scattered background molecules or ejected from clusters moving in the jet. Electron capture by an aggregated molecule, on the other hand, is subjected to polarization interaction with the neighbouring molecules which will modify the DA resonance profile to some extend. We can thus conclude that the F^- resonance profile obtained in the present experiment is not an *exact* mirror of the formation probability of the nucleophile within the cluster.

Similar arguments apply for the energy dependence of Br^- which represents a particular product ion of a completed S_N2 reaction. Other products are the ion–molecule complexes $Br^- \cdot CH_3F$, isolated or solvated by methylbromide molecules. These latter products cannot be distinguished by mass spectrometry from their isomers $F^- \cdot CH_3Br$ *before* the reaction. Interestingly the formation probability of this unspecified complex $(FCH_3 \cdot Br)^-$ shows a slight shift to *higher* energies with respect to the isolated product ion Br^-. From simple arguments (energy dependence of Langevin capture cross-section to form the initial complex $F^- \cdot CH_3Br$ and energy dependence for the stabilization of such a complex) one would rather expect the opposite behaviour, namely a shift of the resonance of the ion–molecule complex to lower energy with respect to the nucleophile. One has to be aware, however, that formation of the *isolated* ion–molecule complex $(FCH_3Br)^-$ as in Fig. 35(b) requires evaporation of a target cluster of eventually appreciable size. This probability depends on the cluster size and is expected to increase with the excess energy deposited in the system. In contrast to that, owing to the appreciable excess energy in the S_N2 reaction we do not expect that ejection of the isolated product ion Br^- is generally accompanied with a *complete* evaporation of the target cluster.

From this short discussion we may conclude that the energy dependence of the ratio Br^-/F^- *alone* is no reliable measure of the energy dependence of the corresponding S_N2 reaction. A more detailed information would eventually be available via extended experiments of the energy profiles of all product ions related to the reaction in dependence of the stagnation pressure.

3.3. *Adsorbed and Condensed Molecules*

In this last paragraph we shall briefly discuss processes which are induced by low-energy electron collisions with molecules in the condensed phase. These are multilayer molecular films condensed on a metallic substrate or submonolayers of molecules adsorbed either directly on the metal or on a multilayer rare gas film which acts as a spacer between the metal and the substrate. The present work focuses on the observation of desorbed fragment ions as a function of the primary electron energy.

Consider a TNI which in the gas-phase decomposes into two fragments $R + X^-$. If it is generated at or near the surface, it is subjected to the following alternate pathways (see also the Introduction and Fig. 4)

$$M_{ad}^{-(*)} \rightarrow R_{ad} + X^- , \tag{31a}$$

$$\rightarrow R + X_{ad}^- , \tag{31b}$$

$$\rightarrow R_{ad} + X_{ad}^- , \tag{31c}$$

$$\rightarrow M_{ad}^- , \tag{31d}$$

which become energetically more favourable in the order (31a)–(31d). If desorbed fragment anions are detected, one is restricted to the energetically most unfavourable channel (31a).

It should be emphasized that in clusters the analogues to the channels (31b)–(31d) can directly be observed and also identified. There, channels (31b) and (31c), e.g. correspond to the formation of solvated ions $X^- \cdot M_n$ and (31d) corresponds to (evaporative) AA generating M_n^-.

With the present experimental arrangement, we have no *direct* access to the reactions (31b)–(31d) in the condensed phase. Since they lead to accumulation of negative charge in the adsorbate, they can *indirectly* be observed through the shift of the electron injection curve due to the accumulated charges.

Observation of a desorbed ionic fragment following DA to an adsorbed or condensed molecule is subjected to two major constraints related to the *structure* and the *energy* of an ion at a surface. It is obvious that those orientations are preferred which allow a direct desorption of the fragment ion avoiding post dissociation interactions (PDI).

From the energetic point of view it is clear that the fragment ion must gain sufficient kinetic energy to overcome the polarization energy necessary for desorption. In addition, there will always be some transfer of energy either into internal degrees of freedom in the decomposition of the resonance itself or via post dissociation interactions (PDI) of the dissociation products prior to desorption which may further shift the desorption feature to higher energies. Thus, in spite of the downshift of the primary resonance due to solvation, the desorption resonance can be shifted to higher energies with respect to the gas-phase DA resonance.

In the following we will study anion desorption from adsorbed and condensed CF_3X molecules (X = F, Cl, Sec. 3.3.1). In Sec. 3.3.2, finally, we present and discuss two remarkable enhancement effects caused by the surrounding molecules and the substrate, respectively.

Of course, the geometry of the precursor ion at the surface is also essential for the desorption behaviour of the fragment ions. With the present experimental configuration, we cannot *directly* obtain information on the adsorbate structure. However, from the particular desorption behaviour, some indirect conclusions can be drawn.

3.3.1. $CF_3X(X = F, Cl)$: *Desorption of Ionic Fragments via Precursor Resonances*

Figure 36 presents a comparison of gas-phase and condensed phase ion yields for the system CF_4. It can be seen that in the gas-phase both F^- and CF_3^- ions are formed with comparable intensity while only F^- is detected as desorption product. However, the desorption product is observed within approximately the same energy region as the gas-phase DA ions.

A more detailed analysis shows,[14,210,306] that CF_4 possesses two overlapping resonances with maxima near 6.8 and 8 eV. The 6.8 eV state dissociates into the complementary channels

$$CF_4^- \to F^- + CF_3\,, \tag{32a}$$

$$\to F + CF_3^-\,, \tag{32b}$$

with the branching ratio F^-: $CF_3^- \approx$ 10:6 while the 8 eV excited state only couples with channel (32b).

CF_4 represents a molecule with a pronounced *impulsive* DA mechanism as mirrored in the remarkably high translational excess energy of the products [Fig. 36(d)]. While the thermodynamic limit (ΔH_o) for F^- formation is 2.25 ± 0.15 eV,[14] the experimental appearance energy is higher by more than 2 eV (Fig. 36). This and the large width of the resonance suggests a Franck–Condon transition to a strongly repulsive area of the anionic potential energy surface. Kinetic energy analysis[14,210,306] showed that from the low-energy resonance both F^- and CF_3^- appear with considerable kinetic energy.

What is the physical origin that desorption of CF_3^- is completely suppressed?

There are essentially two effects which control the desorption energetics of a charged fragment (1) the ionic state at the surface is lowered by the

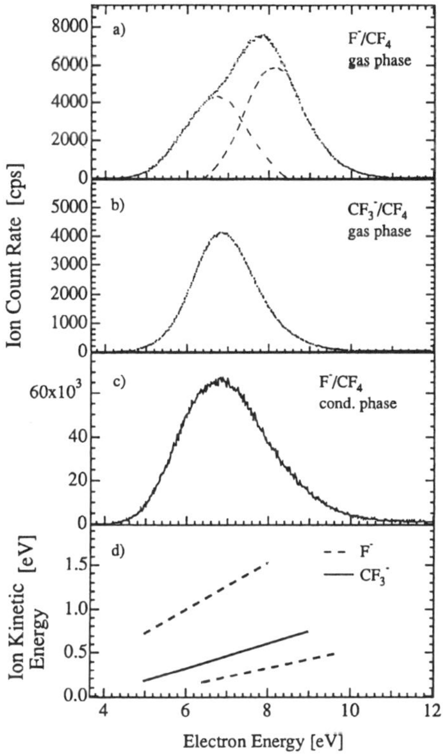

Fig. 36. Dissociative attachment of CF_4 in the gas-phase (a), (b); F^- desorption from condensed CF_4 **(c)**. The desorption spectrum is recorded from a 0.3 monolayer CF_4 adsorbed on a 7 monolayer Xe film as spacer between the metallic substrate; (d) mean translational energy released to F^- and CF_3^- in the gas-phase (from Ref. 125).

polarization energy, and (2) the negative fragment ion arising from the decomposition at the surface must acquire sufficient kinetic energy in order to escape into vacuum.

The maximum value for the translational energy of an ionic fragment, $E_T^i(\max)$ in the *gas-phase* is given by

$$E_T^i(\max) = \frac{m}{M}(\varepsilon - \Delta H_o) = (1 - m_i/M)(\varepsilon - \Delta H_o)\,, \tag{33}$$

with m, m_i and M the mass of the neutral fragment, the ionic fragment and the parent ion, respectively, and ΔH_o the thermodynamic threshold for the corresponding reaction.

Equation (33) is simply the result of the principle of linear momentum conservation and indicates that in a unimolecular decomposition leading to 2 fragments the light particle gains more kinetic energy.

The expected decrease of the resonance energy due to polarization is not visible in the F^- desorption yield. On the other hand, in a recent charge trapping experiment, a shift of the 6.8 eV gas-phase resonance to 5.8 eV could be seen in the 'trapping spectrum'[68] which was attributed to the solvation energy. This means that the solvation shift in the condensed phase is overcompensated by the desorption dynamics of the TNI at or near the surface. The charge trapping experiment also indicated that the DA cross-section of adsorbed CF_4 is enhanced by 4–7 times with respect to the gas-phase value ($\sigma = 1.58 \times 10^{-18}$ cm^{-2}). This was interpreted by the reduced dissociation time (and hence enlarged average autodetachment lifetime) of the solvated resonance.

Some idea for the value of the polarization energy can be obtained from the Born equation[308] in which the environmental medium is treated as a continuum

$$V_p = \frac{q^2}{8\pi\varepsilon_o R}\left(1 - \frac{1}{\varepsilon_d}\right) . \qquad (34)$$

Here, R denotes the ionic radius, ε_o the vacuum permitivity and ε_d the dielectric constant of the medium. The dielectric constant of CF_4 can be estimated from the molecular polarizability[295] within the Clausius–Mosotti model[309] as $\varepsilon_d \approx 1.7$. With $R(F^-) = 1.33$ Å[295] we calculate $V_p = 2.2$ eV which refers to the *bulk* value of the polarization energy. For a particle at the surface this reduces to approximately the half (1.1 eV) which also corresponds to the number derived from the trapping experiments (1 eV, *supra vide*). Thus, only ions with translational energy above ≈ 1 eV can escape into vacuum.

To get a quantitative idea how this energy filter affects ions of different mass, we can, to a first approximation, describe the dissociation of the TNI as *unimolecular*, i.e. no energy or momentum exchange with the medium during dissociation.

The approaching electron is subjected to polarization and its kinetic energy is increased by V_p. For a given electron energy, ε, the maximum translational energy of the ionic fragment in the gas-phase is given by Eq. (33). Accordingly, for a TNI decomposing in the condensed phase, the maximum translational energy of the ionic fragment *prior to desorption or interaction with the environmental molecules* is given by

$$E_T^i(\max) = \frac{m}{M}(\varepsilon - \Delta H_o + V_p)\,, \tag{35}$$

with ε the electron energy with respect to the vacuum level.

Compared to Eq. (33), this expression is modified by the polarization energy V_p of the negative charge. Note that V_p is considered positive if the energy of the solvated ion is lowered with respect to the gas-phase. It it also assumed that V_p is the same for the TNI and the fragment ion. For a given electron energy, ε, the kinetic energy of the fragment ion is higher compared to the gas-phase.

The *energetic* requirement for desorption is that the translational energy of the ionic fragment must exceed the polarization energy $E_T^i > V_p$. With Eq. (35) this leads to the energy threshold, ε_{th}, for desorption

$$\varepsilon_{th} = \frac{m_i}{m} V_p + \Delta H_o\,. \tag{36}$$

This expression directly shows that desorption of a heavy ion is energetically less favourable. If we take $V_p = 1$ eV the threshold for F^- desorption becomes $\varepsilon_{th}(F^-) = 2.52 \pm 0.15$ eV and that for CF_3^- desorption $\varepsilon_{th}(CF_3^-) = 7.5 \pm 0.3$ eV.

For a desorption process at the energetic threshold all the available energy must be transferred into kinetic energy of the dissociating fragments with no energy transfer into internal degrees of freedom of the fragments and no transfer to the solid.

In conclusion for the system CF_4 it can be seen that the F^- DA profile from gaseous CF_4 is very similar to the F^- desorption profile from condensed CF_4. The significant difference between the phases is the strong suppression of CF_3^- desorption from condensed CF_4 and also the enhancement of the DA cross-section in the condensed phase as obtained from the trapping experiments.

There exist many molecules having large gas-phase DA cross-sections which, however, do not show such intense ionic desorption yields as observed in the case CF_4. Halogenated molecules with π electrons usually decompose in DA reactions with low kinetic energy release[310–312] i.e. the unimolecular process can be described as 'statistical' in the way that randomization of the excess energy in the precursor ion takes place prior to fragmentation[313,314] and that the reaction coordinate receives energy according to the statistical treatment of all degrees of freedom.

The reason is that in molecules with π electrons, the π^* system is usually involved in low-energy electron capture. Some admixture of σ^* (C–X) components can then result in comparatively slow unimolecular DA processes as extensively discussed in Sec. 3.1.2. In clusters, these slow DA processes are strongly quenched in favour of AA (Sec. 3.2.2). On going to the condensed phase, these π bonded systems are virtually desorption *inactive* at energies where the π^* system is involved via single particle resonances.

Returning to Sec. 3.1.1 and Fig. 8 it was shown that the molecule CF_3Cl possesses a surprisingly low DA cross-section into Cl^- via the low-energy resonance at 1.4 eV (one particle shape resonance) even lower than that through the second resonance near 4.8 eV (core excited resonance). The reason is the very short lifetime of the low-energy precursor towards autodetachment. The Cl^- desorption yield (not shown here) is also weak with a shape roughly resembling the gas-phase Cl^- yield. On the other hand, the *charge trapping cross-section* (Fig. 37) indicates that trapping preferentially occurs in the vicinity of the low-energy resonance with a cross-section exceeding the gas-phase DA cross-section by more one order

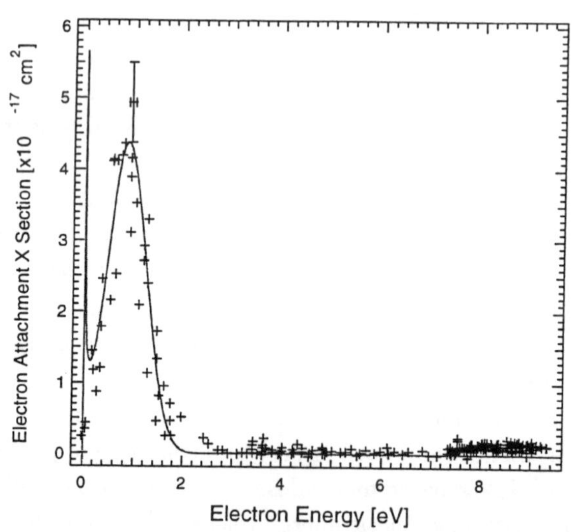

Fig. 37. Charge trapping cross-section of a 0.1 monolayer CF_3Cl on a 15 monolayer Kr film (from Ref. 204).

of magnitude.[204] The charge trapping cross-section was obtained at the Sherbrooke laboratory with a spectrometer particular dedicated to obtain differential trapping cross-sections. This is done by recording the shift of the electron injection curve after exposing the substrate to a defined number of electrons at a particular energy. The shift of the onset curve in the reading of the monochromator voltage is then a measure of the accumulated charge creating a potential which the impinging electrons have to overcome to be injected into the metallic sample holder. From Fig. (37) it follows that although the DA cross-section is lower than the gas-phase DA cross-section, the capture cross-section is enhanced. The latter does not *a priori* contain information on whether the excess electron is localized on the DA fragment (Cl^-) at the surface or on a relaxed parent anion CF_3Cl^-.

The enhancement in the charge trapping cross-section was quantitatively modeled by treating the system as a diatomic-like molecule and using *ab initio* potential energy curves and model functions for the autodetachment width.[204] The measured enhancement can simply be explained by the shift of the anionic curve (1 eV), leading to a shorter dissociation time of the TNI in the autodetachment region (i.e. until the crossing of the two potential energy curves). The peak position of the low-energy trapping resonance (0.8 eV) is then the result of the interplay between the *solvation shift* and the *survival probability shift.*

Recent charge trapping experiments of CF_3Cl *within* the solid Kr lattice revealed that the cross-section with respect to CF_3Cl at the surface is further enhanced by a factor of 3 to 4.[315] This enhancement was attributed to the effect of increased induced polarization on the transient CF_3Cl^- ion in the lattice.

By inspecting the gas-phase DA results,[14] it is obvious that Cl^- is released with appreciable kinetic energy and desorption of Cl^- should energetically be possible for electron energies > 1 eV. The Cl^-/CF_3Cl desorption cross-section from the low-energy resonance[204] (not shown here), however, is more than two orders of magnitude lower than in the system F^-/CF_4 described above and in the order of less than 10^{-20} cm^2. This suggests a preferential orientation of the molecule at the surface which does not allow a direct emission of Cl^- into vacuum, i.e. orientations with the C–Cl axis pointing away from the surface are unlikely. Thus, formation of the shape resonance at the surface preferentially leads to charge accumulation.

3.3.2. *Medium Enhanced Dissociative Attachment at the Surface*

In the previous paragraph we have already seen that DA via ground state shape resonances having repulsive character can lead to an enhancement of the attachment cross-section at the surface as measured by the charge trapping experiments. In this last section we shall demonstrate that in particular cases the coupling of *excited* resonances to a neighbouring medium can lead to a much more dramatic enhancements in the DA cross-section.

Figure 38 compares desorption of I^- and CF_3^- from condensed (multilayer) CF_3I with the corresponding gas-phase DA channels (see Fig. 7).

Remarkably, the four-atomic fragment CF_3^- is now the most abundant product, by even one order of magnitude more intense than desorption of the light atomic F^- fragment (not shown here). furthermore, the intensity ratio I^-: CF_3^- changes from 2×10^5: 1 in the gas-phase (from Fig. 7, normalized to the same pressure) to $1{:}2 \times 10^2$, i.e. by more than 7 orders of magnitude (!).

The maxima of the desorption resonances are distinctly *above* those from the gas-phase. For the resonant state itself (i.e. the vertical attachment energy) one expects a *lowering* in energy on going to the condensed phase due to polarization interaction of the excess charge with the surrounding molecules.

The reaction of interest here is the decomposition of the second resonance into $CF_3^- + I$. In the gas-phase (Sec. 3.1.1) we have shown that 85% of the available excess energy appears as kinetic energy of the two DA products, and that the intensity of the CF_3^- fragment is surprisingly low (more than 5 orders of magnitude lower than I^- and more than one order of magnitude lower than the other two fragments (F^-, FI^-) also appearing at that energy. This is a fairly puzzling situation as the high translational energy of CF_3^- indicates a *direct impulsive dissociation mechanism* while the other two fragments appear with low kinetic energy reflecting a less direct mechanism.

This high desorption energy and the observation that the signal ratio CF_3^-/I^- is enhanced by more than 7 orders of magnitude in the condensed phase suggests a preferential orientation of the molecule with the CF_3 group pointing away from the surface. Such a preferential orientation would explain the high intensity and high kinetic energy of desorbed CF_3^- (as measured in a collaborative study with the Orsay laboratory[12]) and the low intensity of the I^- desorption yield.

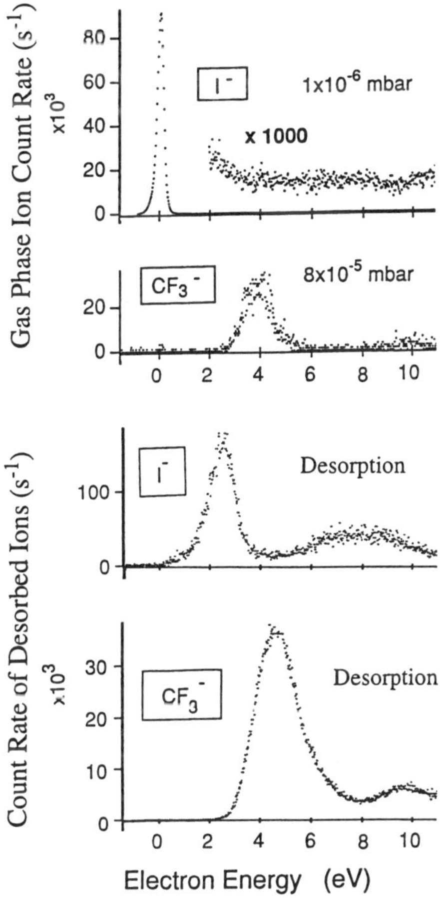

Fig. 38. Top: relative DA cross-section for I^- and CF_3^- formation from gas-phase CF_3I; bottom: relative desorption cross-section I^- and CF_3^- from an 8 monolayer CF_3I film (from Ref. 12).

The picture of a preferential orientation is corroborated by previous results on the thermal chemistry of CF_3I on Ag(111)[316] by means of temperature programmed desorption (TPD) and reflection absorption infrared spectroscopy (RAIRS). It was found that at low coverages (≤ 0.15 ML) the C–I bond direction is parallel to the surface while at higher coverages the molecular axis is tilted from the surface plane. At high coverages, the CF_3 radical desorbs at ~ 110 K leaving iodine atoms at the surface.

A conservative estimation[12] indicates that the CF_3^- *desorption cross-section* is enhanced by more than two orders of magnitude with respect to the *gas-phase* DA *cross-section.* Note that the enhancements discussed in the previous section refers to the charge trapping cross-section. The desorption cross-section for those systems in any case remains orders of magnitude below the corresponding gas-phase DA cross-section which is the usual situation in desorption, in particular for charged particles. For the present system we proposed[12] the following medium enhanced decomposition mechanism for CF_3I^{*-}.

Since the DA reaction yielding CF_3^- is characterized by a strong competition of autodetachment, the process is subjected to a *survival probability shift.* In the gas-phase this effect shifts the observed DA resonance below that of the *primary resonance* formed by the Franck–Condon transition. For the present system a two particle (core excited) resonance is involved which decays via a single electron process into associated electronically excited states of neutral CF_3I^*.

In the condensed phase the TNI is subjected to polarization interaction. This downshift of the precursor ion does not necessarily result in a lowering of the *desorption resonance* as the desorption probability itself depends on the initial energy. The solvation shift in fact reduces the dissociation time in the autodetachment region (Fig. 3) which can increase the DA cross-section to some extent. It is, however, very unlikely that this effect alone is responsible for the dramatic increase of the CF_3^- desorption cross-section.

The essential point then is that under solvation the resonance is shifted to an energy *below* the excited neutral state resulting in a dramatic increase of its lifetime since autodetachment is no longer possible through a one electron process. In other words, *solvation converts a short lived open channel resonance into a long-lived Feshbach resonance* and we propose that this mechanism is responsible for the dramatic enhancement of the desorption cross-section (Fig. 39).

Figure 40 finally shows desorption of F^- from SF_6 adsorbed in submonolayers (0.05 ML) on a Kr and Xe substrate, respectively.[198] In addition to the desorption features observed from a multilayer SF_6 (which have some relation to the gas-phase DA resonances, Fig. 16), from the noble gas substrate sharp enhancements at 7.95 eV (Xe) and 9.82 eV (Kr) are seen. These features are interpreted to arise from 'electron/exciton complexes', i.e. an extra electron bound to an exciton in the noble gas solid. The

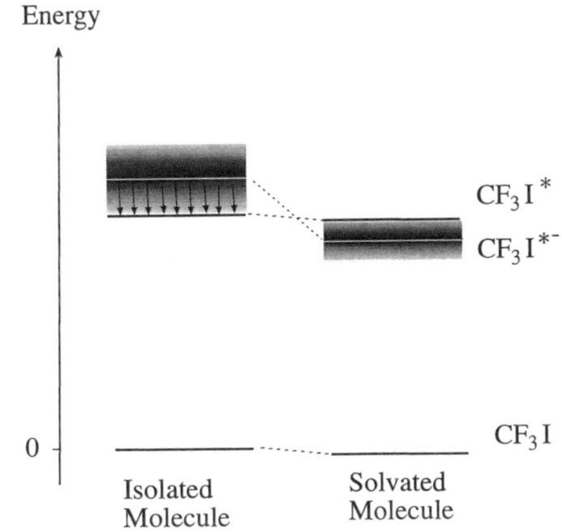

Fig. 39. Schematic representation of the conversion of a short lived open channel resonance into a Feshbach resonance under solvation. The width of the open channel resonance is due to Franck–Condon broadening *and* lifetime broadening. The arrows indicate effective autodetachment via one electron transition into the associated electronically excited neutral state (from Ref. 12).

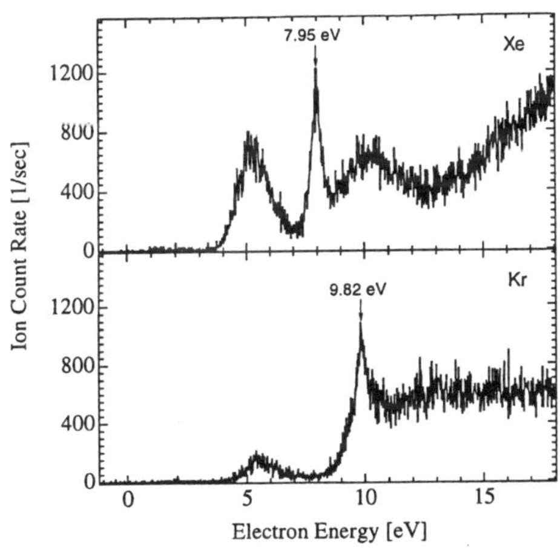

Fig. 40. Desorption yield of F^- from 0.05 monolayer SF_6 adsorbed on a 22 monolayer Kr and Xe substrate, respectively (from Ref. 198).

electron/exciton complex transfers its electron *and* energy to a dissociative state of SF_6^{*-} which decomposes into F^-. The corresponding negative ion states are well known for isolated rare-gas atoms (e.g. Kr^- $(4p^55s^2)$ at 9.52 eV). They are observed ≈ 0.5 eV below the lowest excited state of the neutral atom. Similar observations have been made for D^-/D_2O and D^-/C_2D_6.[317,318] Since SF_6 does not undergo DA to form F^- at low-energy, inelastic scattering by the noble gas substrate and dissociative capture by the slow electron can be ruled out as a possible mechanism for the sharp enhancements in the DA cross-section.

Acknowledgments

This work has been supported by the Deutsche Forschungsgemeinschaft, the Alexander von Humboldt-Stiftung, the Volkswagen-Stiftung, the Freie Universität Berlin, the NATO Research Council and the Fonds der Chemischen Industrie. I am grateful to all my current and previous students, colleagues and coworkers for valuable contributions and discussions. Their names appear in the original publications cited here. Particularly, I appreciate fruitful collaboration with colleagues from Orsay (Roger Azria, Michel Tronc, Yvonick Le Coat, Robert Abouaf), Innsbruck (Tilmann D. Märk, Alexander Stamatovic), Bratislava (Stefan Matejcik, Jan D. Skalny), Villetaneuse (Jean–Pierre Schermann), Sherbrooke, Québec (Léon Sanche, Michael Huels), Keele (David Smith, Patrick Spanel), Washington (Loucas G. Christophorou) and Moscow (Alexei G. Khrapak, Boris M. Smirnov).

References

1. J. W. Coburn and H. F. Winters, *J. Vac. Sci. Technol.* **16**, 391 (1979).
2. D. L. Flamm in *Plasma Etching*, Eds. D. M. Manos and D. L. Flamm (Academic Press, Boston, 1989).
3. M. Kimura, N. Inokuti and M. A. Dillon, in *Advances in Chemical Physics*, Vol. LXXXIV, Eds. I. Prigogine and S. A. Rice (John Wiley and Sons, 1993).
4. *Excess Electron in Dielectric Media*, Eds. C. Ferradini and J.-P. Jay–Gerin (CRC Press, Boca Raton 1991).
5. *Radiation Damage in DNA: Structure/Function Relationship in Early Times*, Eds. A. F. Fuciarelli and J. D. Zimbrick (Battele Press, Columbus-Richland, USA, 1995).
6. L. Sanche, *J. Chimie Physique* **94**, 216 (1997).
7. R. A. Marcus, *Angew. Chem.* **105**, 1161 (1993).
8. G. J. Kavarnos, *Fundamentals of Photoinduced Electron Transfer* (VCH Publishers, New York, 1993).
9. L. G. Christorphorou (Editor), *Electron-Molecule Interactions and Their Applications*, Vols. I and II (Academic, Orlando, FL, 1984).
10. St. J. Dixon-Warren, E. T. Jensen and J. C. Polanyi, *Phys. Rev. Lett.* **67**, 2395 (1991).

11. S. A. Buntin, L. J. Richter, D. S. King and R. R. Cavanagh, *J. Chem. Phys.* **91**, 6429 (1989).
12. Y. Le Coat, N. M. Hedhili, R. Azria, M. Tronc, O. Ingólfsson and E. Illenberger, *Chem. Phys. Letters* **296**, 208 (1998).
13. L. G. Christophorou, *Atomic and Molecular Radiation Physics* (Wiley Interscience, London, 1971).
14. T. Oster, A. Kühn and E. Illenberger, *Int. J. Mass Spectrom. Ion Proc.* **89**, 1 (1989).
15. E. Illenberger and J. Momigny, *Gaseous Molecular Ions. An Introduction to Elementary Processes Induced by Ionization.* Steinkopff Verlag, Darmstadt/Springer-Verlag, New York 1992.
16. (a) A. Chutjian, A. Garscadden and J. M. Wadehra, *Physics Reports* **264**, 393 (1996). (b) H. Shimamori, Y. Tatsumi and T. Sunagawa, *J. Chem. Phys.* **99**, 7787 (1993).
17. *Linking the Gaseous and Condensed Phases of Matter: The Behaviour of Slow Electrons*, Eds. L. G. Christophorou, E. Illenberger and W.-F. Schmidt (Plenum, New York, 1994).
18. O. Ingólfsson, F. Weik and E. Illenberger, *Int. J. Mass Spectrom. Ion Proc.* **155**, 1 (1996).
19. D. M. Wetzel and J. I. Braunman, *Chem. Rev.* **87**, 607 (1987).
20. G. Markovich, S. Pollack, R. Giniger and O. Chesnowsky, *J. Chem. Phys.* **101**, 9344 (1994).
21. M. Allan, *J. Electron Spect.* **48**, 219 (1989).
22. M. Berman, H. Estrada, L. S. Cederbaum and W. Domcke, *Phys. Rev.* **A28**, 1363 (1983).
23. L. Sanche and G. J. Schulz, *J. Chem. Phys.* **58**, 479 (1973).
24. K. D. Jordan and P. D. Burrow, *Acc. Chem. Res.* **11**, 341 (1978).
25. K. D. Jordan and P. D. Burrow, *Chem. Rev.* **87**, 557 (1987).
26. J. B. Hasted and D. Mathur, *Electron-Molecule Resonances*, in Ref. 9.
27. G. J. Schulz, *Rev. Mod. Phys.* **45**, 378 (1973).
28. G. J. Schulz, *Rev. Mod. Phys.* **45**, 423 (1973).
29. L. Sanche and G. J. Schulz, *Phys. Rev.* **A5**, 1672 (1972).
30. L. Sanche and G. J. Schulz, *Phys. Rev.* **A6**, 69 (1972).
31. M. Fenzlaff, R. Gerhard and E. Illenberger, *J. Chem. Phys.* **88**, 149 (1988).
32. D. Smith, P. Spanel, S. Matejcik, A. Stamatovic, T. D. Märk, T. Jaffke and E. Illenberger, *Chem. Phys. Lett.* **240**, 481 (1995).
33. (a) D. Klar, M.-W. Ruf and H. Hotop, *Aust. J. Phys.* **45**, 263.
(b) A. Schramm, J. M. Weber, J. Kreil, D. Klar, M.-W. Ruf and H. Hotop, *Phys. Rev. Lett.* **81**, 778 (1998).
34. J. H. D. Eland, *Photoelectron Spectroscopy* (Butterworth, London, 1984).
35. J. W. Rabalais, *Principles of Ultraviolet Photoelectron Spectroscopy* (Wiley, New York, 1977).
36. H. S. W. Massey, *Negative Ions* (Cambridge University Press, Cambridge, 1976).

37. D. T. Birtwistle and A. Herzenberg, *J. Phys.* **B4**, 53 (1971).
38. A. Mann and F. Linder, *J. Phys.* **B25**, 1621 (1992).
39. I. Hahndorf, E. Illenberger, L. Lehr and J. Manz, *Chem. Phys. Lett.* **231**, 460 (1994).
40. W. T. Naff, R. N. Compton and C. D. Cooper, *J. Chem. Phys.* **54**, 212 (1971).
41. C. Lifshitz, A. M. Peers, R. Grajower and M. Weiss, *J. Chem. Phys.* **53**, 4605 (1970).
42. S. Süzer, E. Illenberger and H. Baumgärtel, *Org. Mass Spectrom.* **19**, 292 (1984).
43. S. Tobita, M. Meinke, E. Illenberger, L. G. Christophorou, H. Baumgärtel and S. Leach, *Chem. Phys.* **161**, 501 (1992).
44. M. Lezius, P. Scheier and T. D. Märk, *Chem. Phys. Lett.* **203**, 232 (1993).
45. T. Jaffke, E. Illenberger, M. Lezius, S. Matejcik, D. Smith and T. D. Märk, *Chem. Phys. Lett.* **226**, 213 (1994).
46. S. Matejcik, T. D. Märk, P. Spanel, D. Smith, T. Jaffke and E. Illenberger, *J. Chem. Phys.* **102**, 2516 (1995).
47. O. Elhamidi, J. Pommier and R. Abouaf, *J. Phys. B: At. Mol. Opt. Phys.* **30**, 4633 (1997).
48. Y. V. Vasil'ev, R. F. Tuktorov and V. A. Mazunov, *Rap. Comm. Mass Spectrom.* **11**, 757 (1997).
49. T. F. O'Malley, *Phys. Rev.* **150**, 14 (1966).
50. M. Knapp, O. Echt, D. Kreisle and E. Recknagel, *J. Phys. Chem.* **91**, 2601 (1987).
51. U. Landman, R. N. Barnett, C. L. Cleveland, D. Scharf and J. Jortner, *J. Phys. Chem.* **91**, 4890 (1987).
52. J. V. Coe, G. H. Lee, J. G. Eaton, S. T. Arnold, H. W. Sarkas, K. H. Bowen, C. Ludewigt, H. Haberland and D. R. Worsnop, *J. Chem. Phys.* **92**, 3980 (1990).
53. H. Haberland and K. H. Bowen, in *Clusters of Atoms and Molecules*, Vol. II, Ed. H. Haberland, *Springer Series in Chemical Physics* **56**, (Springer-Verlag, Berlin, 1994).
54. *Electrons in Fluids*, Eds. J. Jortner and N. R. Kestner (Springer, Berlin, 1973).
55. D. Neiholls, *Inorganic Chemistry in Liquid Ammonia: Topics in Inorganic and General Chemistry* (Elsevier, Amsterdam, 1979).
56. L. Sanche, *IEEE Trans. Electr. Insul.* **28**, 789 (1993).
57. C. E. Klots, *Radiat. Phys. Chem.* **20**, 51 (1982).
58. M. Foltin, V. Grill and T. D. Märk, *Chem. Phys. Lett.* **188**, 427 (1992).
59. I. T. Iakubov, *Electron Scattering in Dense Gases and Liquids and Related Phenomena*, in Ref. 9.
60. I. T. Iakubov and A. G. Khrapak, *Rep. Progr. Phys.* **45**, 697 (1982).
61. M. Rosenblit and J. Jortner, *Phys. Rev. Lett.* **75**, 4079 (1995).
62. A. G. Khrapak and I. G. Iakubov, *Electrons in Denses GAses and Plasmas* (Nauka, Moscow, 1981).

63. A. G. Khrapak, W. F. Schmidt and K. F. Volykhin, *Phys. Rev.* **E51**, 4804 (1995).
64. L. Sanche, *J. Phys. B: At. Mol. Opt. Phys.* **23**, 1597 (1990).
65. L. Sanche and M. Dechênes, *Phys. Rev. Lett.* **61**, 2096 (1988).
66. A. D. Bass and L. Sanche, *J. Chem. Phys.* **95**, 2910 (1994).
67. R. M. Marsolais, M. Dechênes and L. Sanche, *Rev. Sci. Instrum.* **60**, 2724 (1989).
68. A. D. Bass, J. Gamache, L. Parenteau and L. Sanche, *J. Phys. Chem.* **99**, 6123 (1995).
69. St. J. Dixon-Warren, E. T. Jensen and J. C. Polanyi, *J. Chem. Phys.* **98**, 5938 (1993).
70. St. J. Dixon-Warrent, E. T. Jensen, J. C. Polanyi, G. Q. Xu, S. H. Yang and H. C. Zeng, *Faraday Discuss. Chem. Soc.* **91**, 1 (1991).
71. J. C. Polanyi and H. Rieley, in *Dynamics of Gas-Surface Interactions*, Eds. C. T. Rettner and M. N. R. Ashfold (Royal Society of Chemistry, Cambridge, 1991).
72. J. W. Gadzuk, L. J. Richter, S. A. Buntin, D. S. King and R. R. Cavanagh, *Surf. Science* **235**, 317 (1990).
73. T. Hertel, M. Wolf and G. Erti, *J. Chem. Phys.* **102**, 3414 (1995).
74. D. Menzel and R. Gomer, *J. Chem. Phys.* **41**, 3311 (1964).
75. P. Antoniewicz, *Phys. Rev.* **B21**, 3811 (1980).
76. J. L. Franklin, in *Gas-Phase Ion Chemistry*, Vol. II, Ed. M. T. Bowers (Academic Press, New York, 1979).
77. D. Smith and P. Spanel, *Adv. At. Mol. Opt. Phys.* **32**, 307 (1994).
78. (a) C. E. Klots and R. N. Compton, *J. Chem. Phys.* **67**, 1779 (1977); (b) C. E. Klots and R. N. Compton, *J. Chem. Phys.* **69**, 1636 (1978).
79. (a) C. E. Klots and R. N. Compton, *J. Chem. Phys.* **69**, 1644 (1978); (b) C. E. Klots and R. N. Compton, *Chem. Phys. Lett.* **73**, 589 (1980).
80. C. E. Klots, *J. Chem. Phys.* **71**, 4172 (1979).
81. T. D. Märk, *Int. J. Mass Spectrom. Ion Proc.* **107**, 143 (1991).
82. T. D. Märk, K. Leiter, W. Ritter and A Stamatovic, *Phys. Rev. Lett.* **55**, 2559 (1985).
83. T. D. Märk, K. Leiter, W. Ritter and A. Stamatovic, *Int. J. Mass Spectrom. Proc.* **74**, 265 (1986).
84. G. Walder, D. Margreiter, C. Winkler, A. Stamatovic, Z. Herman and T. D. Märk, *J. Chem. Soc. Faraday Trans.* **86**, 2395 (1990).
85. G. Walder, D. Margreiter, C. Winkler, V. Grill, T. Rauth, P. Scheier, A. Stamamtovic, Z. Herman, M. Foltin and T. D. Märk, *Z. Phys.* **D20**, 201 (1991).
86. T. D. Märk, P. Scheier and A. Stamatovic, *Chem. Phys. Lett.* **136**, 177 (1987).
87. A. Stamatovic, P. Scheier and T. D. Märk, *Z. Phys.* **D6**, 351 (1987).
88. T. D. Märk, A. Stamatovic and P. Scheier, *Z. Phys.* **D12**, 303 (1989).
89. A. Stamatovic, P. Scheier and T. D. Märk, *J. Chem. Phys.* **88**, 5894 (1988).

90. A. Stamatovic, K. Leiter, W. Ritter, K. Stephan and T. D. Märk, *J. Chem. Phys.* **83**, 2942.
91. A. Stamtovic, K. Stephan and T. D. Märk, *Int. J. Mass Spectrom. Ion Proc.* **63**, 37 (1985).
92. T. Rauth, M. Foltin and T. D. Märk, *J. Phys. Chem.* **96**, 1528 (1992).
93. M. Lezius, T. Rauth, V. Grill, M. Foltina and T. D. Märk, *Z. Phys.* **D24**, 289 (1992).
94. T. Rauth, O. Echt and T. D. Märk, *Chem. Phys. Lett.* **201**, 345 (1993).
95. S. Matejcik, A. Kiendler, A Stamatovic, P. Stampfli and T. D. Märk, *Phys. Rev. Lett.* **77**, 3771 (1966).
96. P. Scheier, Y. Chu, G. Senn, S. Matejcik, P. Stampfli, A. Stamatovic, E. Illenberger and T. D. Märk, in *Physics of Clusters*, Ed. V. D. Lakhno (World Scientific, Singapore, 1998).
97. S. Matejcik, A. Kiendler, P. Stampfli, J. D. Skalny, A. Stamatovic and T. D. Märk, *Z. Phys.* **D40**, 70 (1997).
98. S. Matejcik, A. Kiendler, P. Cicman, P. Stampfli, J. D. Skalny, E. Illenberger, Y. Chu, A. Stamatovic and T. D. Märk, *Plasma Sources and Technology* **6**, 140 (1997).
99. M. Knapp, O. Echt, D. Kreisle and E. Recknagel, *J. Chem. Phys.* **85**, 636 (1986).
100. M. Knapp, D. Kreisle, O. Echt, K. Sattler and E. Recknagel, *Surf. Science* **156**, 313 (1985).
101. M. Knapp, O. Echt, D. Kreisle, T. D. Märk and D. Recknagel, *Chem. Phys. Lett.* **126**, 225 (1986).
102. O. Echt, M. Knapp, C. Schwarz and E. Recknagel, in *Large Finite Systems*, Eds. J. Jortner, A. Pullman and B. Pullman (D. Reidel, Dordrecht, 1987).
103. A. A. Vostrikov, D. Yu Dubov and M. R. Predtechenskii, *Sov. Phys. Tech. Phys.* **32**, 459 (1987) and references therein.
104. A. A. Vostrikov, D. Yu Dubov and M. R. Predtechenskii, *Sov. Phys. Tech. Phys.* **31**, 825 (1986).
105. A. A. Vostrikov, D. Yu Dubov and A. A. Agarkov, *Sov. Tech. Phys. Lett.* **21**, 515 (1995).
106. R. Hashemi, A. Kühn and E. Illenberger, *Int. J. Mass Spectrom. Ion Proc.* **100**, 753 (1990).
107. J. Lotter, A. Kühn and E. Illenberger, *Chem. Phys. Lett.* **157**, 171 (1989).
108. A. Kühn and E. Illenberger, *J. Phys. Chem.* **93**, 7060 (1989).
109. A. Kühn and E. Illenberger, *J. Chem. Phys.* **93**, 357 (1990).
110. J. Lotter and E. Illenberger, *J. Phys. Chem.* **94**, 8951 (1990).
111. R. Hashemi, A. Kühn, J. Lotter and E. Illenberger, *Ber. Bunsenges. Phys. Chem.* **94**, 1334 (1990).
112. T. Oster and E. Illenberger, *Ber. Bunsenges. Phys. Chem.* **94**, 1287 (1990).
113. R. Hashemi and E. Illenberger, *J. Phys. Chem.* **95**, 6402 (1991).
114. R. Hashemi and E. Illenberger, *Chem. Phys. Lett.* **187**, 623 (1991).
115. R. Hashemi, M. Meinke, T. Jaffke and E. Illenberger, *Ber. Bunsenges. Phys. Chem.* **96**, 440 (1992).

116. E. Illenberger, *Ber. Bunsenges. Phys. Chem.* **96**, 1148 (1992).
117. R. Hashemi, T. Jaffke and E. Illenberger, *Ber. Bunsenges. Phys. Chem.* **96**, 1151 (1992).
118. T. Jaffke, R. Hashemi, L. G. Christophorou and E. Illenberger, *Z. Phys. D: Atoms, Molecules and Clusters* **25**, 77 (1992).
119. T. Jaffke, R. Hashemi, L. G. Christophorou and E. Illenberger, *J. Phys. Chem.* **96**, 10605 (1992).
120. O. Ingólfsson and E. Illenberger, *Int. J. Mass Spectrom. Ion Proc.* **149–150**, 79 (1995).
121. P. Tegeder, L. Lehmann, O. Ingólfsson and E. Illenberger, *Zeitschr. Phys. Chem.* **195**, 21 (1996).
122. E. Illenberger, *Chem. Rev.* **92**, 1589 (1992).
123. O. Ingólfsson and E. Illenberger, *Chem. Phys Lett.* **241**, 180 (1995).
124. L. Lehmann, S. Matejcik and E. Illenberger, *Ber. Bunsenges. Phys. Chem.* **101**, 287 (1997).
125. L. Lehmann and E. Illenberger, *Int. J. Mass Spectrom. Ion Proc.* **185–187**, 463 (1998).
126. P. Tegeder and E. Illenberger, *J. Radioanal. Nuclear Chem.* **232**, 53 (1998).
127. S. Matejcik, P. Cicman, A. Kiendler, J. D. Skalny, E. Illenberger and T. D. Märk, *Chem. Phys. Lett.* **261**, 437 (1996).
128. S. Matejcik, P. Cicman, A. Kiendler, G. Senn, E. Illenberger and T. D. Märk, *Chem. Phys. Lett.* **267**, 329 (1997).
129. Y. Chu, G. Senn, P. Scheier, A. Stamatovic, T. D. Märk, F. Brüning, S. Matejcik and E. Illenberger, *Phys. Rev.* **A57**, R697 (1998).
130. H. Haberland, H. Langosch, H.-G. Schindler and D. R. Worsnop, *J. Phys. Chem.* **88**, 3903 (1984).
131. H. Haberland, C. Ludewigt, H.-G. Schindler and D. R. Worsnop, *Surf. Sci.* **156**, 157 (1985).
132. H. Haberland, C. Ludewigt, H.-G. Schindler and D. R. Worsnop, *Phys. Rev.* **A36**, 967 (1987).
133. H. Haberland, H.-G. Schindler and D. R. Worsnop, *Ber. Bunsenges. Phys. Chem.* **88**, 270 (1984).
134. P. Stampfli and K. H. Bennemann, *Phys. Rev. Lett.* **58**, 2635 (1987).
135. H. Haberland, T. Kolar and T. Reiners, *Phys. Rev. Lett.* **63**, 1219 (1989).
136. K. Mitsuke, T. Kondow and K. Kuchitsu, *J. Phys. Chem.* **90**, 1552 (1986).
137. K. Mitsuke, T. Kondow and K. Kuchitsu, *J. Phys. Chem.* **90**, 1505 (1986).
138. K. Tsakuda and T. Kondow, *Chem. Phys. Lett.* **185**, 511 (1991).
139. T. Kraft, M.-W. Ruf and H. Hotop, *Z. Physik* **D14**, 179 (1989).
140. T. Kraft, M.-W. Ruf and H. Hotop, *Z. Physik* **D20**, 13 (1991).
141. T. Kraft, M.-W. Ruf and H. Hotop, *Z. Physik* **D18**, 403 (1991).
142. D. Klar, B. Mirbach, H.-J. Korsch, M.-W. Ruf and H. Hotop, *Z. Phys.* **D31**, 235 (1994).
143. F. B. Dunning, *J. Phys. Chem.* **91**, 2244 (1987).
144. F. B. Dunning, *J. Phys.* **B28**, 1645 (1995).

145. J. P. Schermann, J. P. Astruc, C. Desfrançois and R. Barbé, in *Photoinduced Electron Transfer*, Eds. M. A. Fox and M. Chanon (Elsevier, 1987).
146. C. Desfrançois, N. Khelifa and J. P. Schermann, *J. Chem. Phys.* **91**, 5853 (1989).
147. C. Desfrançois, N. Khelifa, A Lisfi and J. P. Schermann, *J. Chem. Phys.* **96**, 5009 (1992).
148. C. Desfrançois, N. Khelifa, A. Lisfi and J. P. Schermann, *J. Chem. Phys.* **90**, 1267 (1993).
149. C. Desfrançois, B. Baillon, J. P. Schermann, S. T. Arnold, J. H. Hendricks and K. H. Bowen, *Phys. Rev. Lett.* **72**, 48 (1994).
150. C. Desfrançois, H. Abdoul-Carime, N. Khelifa, J. P. Schermann, V. Brenner and P. H. Millié, *J. Chem. Phys.* **102**, 1 (1995).
151. C. Desfrançois, *Phys. Rev.* **A51**, 3667 (1995).
152. C. Desfrançois, H. Abdoul-Carime, C. P. Schulz and J. P. Schermann, *Science* **269**, 1707 (1995).
153. C. Desfrançois, H. Abdoul-Carime and J. P. Schermann, *J. Chem. Phys.* **104**, 7792 (1996).
154. C. Desfrançois, H. Abdoul-Carime and J. P. Schermann, *Int. J. Mod. Phys.* **B10**, 1339 (1996).
155. H. Abdoul-Carime and C. Desfrançois, *European Physics Journal 1998* (in print).
156. L. Sanche, *Phys. Rev. Lett.* **53**, 1638 (1984).
157. L. Sanche, *Primary Interactions of Low Energy Electrons in Condensed Matter*, in Ref. 4.
158. L. Sanche, *Comments At. Mol. Phys.* **26**, 321 (1991).
159. P. Rowntree, L. Parenteau and L. Sanche, *J. Phys. Chem.* **95**, 523 (1991).
160. P. Rowntree, L. Parenteau and L. Sanche, *J. Chem. Phys.* **94**, 8570 (1991).
161. M. A. Huels, L. Parenteau, P. Cloutier and L. Sanche, *J. Chem. Phys.* **103**, 6775 (1995).
162. L. Parenteau, J.-P. Jay-Gerin and L. Sanche, *J. Phys. Chem.* **98**, 10277 (1994).
163. M. A. Huels, L. Parenteau, M. Michaud and L. Sanche, *Phys. Rev.* **A51**, 337 (1995).
164. L. Sanche, A. D. Bass, P. Ayotte and I. J. Fabrikant, *Phys. Rev. Lett.* **75**, 3668 (1995).
165. D. Klyachko, P. Rowntree and L. Sanche, *Surf. Sci.* **346**, L49 (1996).
166. P. Rowntree, C. Dugal, D. Hunting and L. Sanche, *J. Phys. Chem.* **100**, 4546 (1996).
167. M. Akbulut, T. E. Madey, L. Parenteau and L. Sanche, *J. Chem. Phys.* **105**, 6043 (1996).
168. I. J. Fabrikant, K. Nagesha, R. Wilde and L. Sanche, *Phys. Rev.* **B56**, 1 (1997).
169. W. C. Simpson, M. T. Siegel, G. A. Kimmel, L. Parenteau, L. Sanche and T. M. Orlando, *J. Chem. Phys.* **107**, 8668 (1997).

170. W. C. Simpson, M. T. Sieger, G. A Kimmel, K. Nagesha, L. Parenteau, T. M. Orlando and L. Sanche, *J. Chem. Phys.* **108**, 5027 (1998).
171. G. Perluzzo, G. Bader, L. G. Caron and L. Sanche, *Phys. Rev.* **26**, 3976 (1982).
172. L. Sanche, G. Perluzzo and M. Michaud, *J. Chem. Phys.* **83**, 3837 (1983).
173. G. Perluzzo, L. Sanche, C. Gaubert and R. Baudoing, *Phys. Rev.* **B30**, 4292 (1984).
174. M. Michaud and L. Sanche, *Phys. Rev.* **B30**, 6067 (1984).
175. R. M. Marsolais and L. Sanche, *Phys. Rev.* **B38**, 11118 (1988).
176. E. Keszei, J.-P. Jay-Gerin, G. Perluzzo and L. Sanche, *J. Chem. Phys.* **85**, 7396 (1986).
177. I. T. Steinberger, A. D. Bass, R. Shechter and L. Sanche, *Phys. Rev.* **B48**, 8290 (1993).
178. L. Sanche, *Phys. Rev. Lett.* **75**, 2904 (1995).
179. K. Nagesha, J. Gamache and L. Sanche, *Rev. Sci. Instrum.* **68**, 3883 (1997).
180. A. D. Bass, J. Gamache. P. Ayotte and L. Sanche, *J. Chem. Phys.* **104**, 4258 (1996).
181. P. Ayotte, J. Gamache, A. D. Bass, I. I. Fabrikant and L. Sanche, *J. Chem. Phys.* **106**, 749 (1997).
182. M. Lezius, L. Parenteau, A. D. Bass and L. Sanche, *J. Phys.* **B30**, 3527 (1997).
183. P. Swiderek, M. Michaud and L. Sanche, *J. Chem. Phys.* **98**, 8397 (1993).
184. P. Swiderek, M.-J. Fraser, M. Michaud and L. Sanche, *J. Chem. Phys.* **100**, 70 (1994).
185. P. Swiderek, M. Michaud and L. Sanche, *J. Chem. Phys.* **103**, 8424 (1995).
186. P. Swiderek, M. Michaud and L. Sanche, *J. Chem. Phys.* **105**, 6724 (1996).
187. A. T. Wen, M. Michaud and L. Sanche, *Phys. Rev.* **A54**, 4162 (1996).
188. P. Swiderek, M. Michaud and L. Sanche, *J. Chem. Phys.* **106**, 9403 (1997).
189. R. Azria, Y. Le Coat, J.-P. Ziesel, J.-P. Guillotin, B. Mharzi and M. Tronc, *Chem. Phys. Lett.* **220**, 417 (1994).
190. Y. Le Coat, R. Azria, B. Mharzi and M. Tronc, *Surf. Sci.* **331–333**, 360 (1995).
191. S. Lacombe, F. Cemic, K. Jacobi, M. N. Hedhili, Y. Le Coat, R. Azria and M. Tronc, *Phys. Rev. Lett.* **79**, 1146 (1997).
192. K. B. K. Tang, R. Azria, Y. Le Coat, M. Tronc and R. E. Palmer, *Z. Phys.* **D38**, 41 (1996).
193. L. Siller, M. N. Hedhili, Y. Le Coat, M. Tronc and R. Azria, *Chem. Phys. Lett.* **288**, 776 (1998).
194. T. Oster, O. Ingólfsson, M. Meinke and E. Illenberger, *J. Chem. Phys.* **99**, 5141 (1993).
195. M. Meinke and E. Illenberger, *J. Phys. Chem.* **98**, 6601 (1994).
196. F. Weik and E. Illenberger, *J. Chem. Phys.* **103**, 1406 (1995).
197. O. Ingólfsson, F. Weik and E. Illenberger, *Int. Rev. Phys. Chem.* **15**, 133 (1996).

198. F. Weik and E. Illenberger, *J. Chem. Phys.* **109**, 6079–6085 (1998).
199. M. Meinke, L. Parenteau, P. Rowntree, L. Sanche and E. Illenberger, *Chem. Phys. Lett.* **205**, 213 (1993).
200. P. Rowntree, L. Parenteau, L. Sanche, M. Meinke, F. Weik and E. Illenberger, *J. Chem. Phys.* **101**, 4248 (1994).
201. M. N. Hedhili, L. Parenteau, M. A. Huels, R. Azria, M. Tronc and L. Sanche, *J. Chem. Phys.* **107**, 7577 (1997).
202. M. Tronc, R. Azria, Y. Le Coat and E. Illenberger, *J. Phys. Chem.* **100**, 14745 (1996).
203. N. M. Hedhili, Y. Le Coat, R. Azria, M. Tronc, F. Weik and E. Illenberger, *Int. J. Mass Spectrom. Ion Proc.* **164**, 231–239 (1997).
204. F. Weik, E. Illenberger, K. Nagesha and L. Sanche, *J. Phys. Chem.* **B102**, 842 (1998).
205. A. Kantrowitz and J. Grey, *Rev. Sci. Instr.* **22**, 328 (1951).
206. J. B. Anderson, in *Molecular Beam and Low Density Gas Dynamics*, Ed. P. P. Wegener (Marcel Dekker, New York, 1974).
207. D. R. Miller, in *Atomic and Molecular Beam Methods*, Ed. G. Scoles (Oxford University Press, Oxford, 1988).
208. A Stamatovic and G. J. Schulz, *Rev. Sci. Instr.* **39**, 1752 (1968).
209. A Stamatovic and G. J. Schulz, *Rev. Sci. Instr.* **41**, 423 (1970).
210. E. Illenberger, *Chem. Phys. Lett.* **80**, 153 (1981).
211. E. Illenberger, *Ber. Bunsenges. Phys. Chem.* **86**, 247 (1982).
212. L. G. Christophorou, in *The Liquid State and its Electrical Properties*, Eds. E. E. Kuhnhardt and L. G. Christophorou (Plenum, New York, 1984).
213. W.-F. Schmidt, *Liquid State Electronics of Insulating Liquids* (CRC Press, Boca Raton, 1997).
214. R. N. Compton, P. W. Reinhardt and C. D. Cooper, *J. Chem. Phys.* **68**, 4360 (1978).
215. D. F. McMillen and D.M. Golden, *Ann Rev. Phys. Chem.* **33**, 493 (1982).
216. R. D. Mead, A. E. Stevens and W. C. Lineberger, in *Gas-Phase Ion Chemistry*, Vol. III, Ed. M. T. Bowers (Academic, New York, 1984).
217. E. Illenberger, H.-U. Scheunemann and H. Baumgärtel, *Chem. Phys.* **38**, 21 (1979).
218. H. A Bethe, *Phys. Rev.* **47**, 747 (1935).
219. E. P. Wigner, *Phys. Rev.* **73**, 1002 (1948).
220. E. Vogt and G. H. Wannier, *Phys. Rev.* **95**, 1190 (1954).
221. H. Shimamori and Y. Nakatani, *Chem. Phys. Lett.* **150**, 109 (1988).
222. T. Underwood-Lemons, T.-J. Gergel and J. H. Moore, *J. Chem. Phys.* **102**, 119 (1995).
223. S. M. Spyrou and L. G. Christophorou, *J. Chem. Phys.* **82**, 2620 (1985).
224. P. Felder, *Chem. Phys.* **143**, 414 (1990).
225. H. Guo and G. C. Schatz, *J. Chem. Phys.* **93**, 393 (1990).
226. H. J. Hwang and M. A. El-Sayed, *J. Phys. Chem.* **96**, 8728 (1992).
227. G. N. A. van Veen, T. Baller, A. E. de Vries and M. Shapiro, *Chem. Phys.* **93**, 277 (1985).

228. R. S. Mulliken, *J. Chem. Phys.* **3**, 506 (1935).
229. M. Shapiro and R. Bersohn, *J. Chem. Phys.* **73**, 3810 (1980).
230. A. Gedanken, *Chem. Phys. Lett.* **137**, 462 (1987).
231. E. Illenberger, *Ber. Bunsenges. Phys. Chem.* **86**, 252 (1982).
232. M. B. Robin, *Higher Excited States of Polyatomic Molecules*, Vol. III (Academic, Orlando, 1985).
233. C. Hubrich and F. Stuhl, *J. Photochem.* **12**, 93 (1980).
234. C. Sandorfy, *Atmospheric Environment* **10**, 343 (1976).
235. G. J. Verhaart, W. J. van der Hart and H. H. Brongersma, *Chem. Phys.* **34**, 161 (1978).
236. R. K. Jones, *J. Chem. Phys.* **84**, 813 (1986).
237. S. Chowdhury, E. P. Grimsrud, T. Heinis and P. Kebarle, *J. Am. Chem. Soc.* **108**, 3630 (1986).
238. G. W. Dillow and P. Kebarle, *J. Am. Chem. Soc.* **111**, 5592 (1989).
239. K. Y. Choo, G. D. Mendenhall, D. M. Golden and S. W. Benson, *Int. J. Chem. Kinet.* **6**, 813 (1974).
240. M. J. Kreech, S. J. W. Price and W. F. Yared, *Int. J. Chem. Kinet.* **6**, 257 (1974).
241. C. R. Herd, N. G. Adams and D. Smith, *Int. J. Mass Spectrom. Ion Proc.* **87**, 331 (1989).
242. K. Hiraoka, S. Mizuse and S. Yamabe, *J. Phys. Chem.* **94**, 3689 (1990).
243. A. P. Hitchcock, P. Fischer, A. Gedanken and M. B. Robin, *J. Phys. Chem.* **91**, 531 (1987).
244. C. Glidewell, *Chimica Scripta* **25**, 142 (1985).
245. J. R. Frazier, L. G. Christophorou, J. G. Carter and H. C. Schweinler, *J. Chem. Phys.* **69**, 3807 (1978).
246. *Gaseous Dielectrics VI*, Eds. L. G. Christophorou and D. R. James (Plenum, New York, 1995).
247. *Gaseous Dielectrics IV*, Eds. L. G. Christophorou and M. O. Pace (Pergamon, New York, 1984).
248. K. M. Bansal and R. W. Fessenden, *J. Phys. Chem.* **80**, 1743 (1976).
249. J. D. Chen and D. A. Armstrong, *J. Chem. Phys.* **48**, 2310 (1968).
250. D. Hildenbrand, *J. Phys. Chem.* **77**, 897 (1973).
251. C. L. Chen and P. J. Chantry, *J. Chem. Phys.* **71**, 3897 (1979).
252. A. Chutjian and S. H. Allajajian, *Phys. Rev.* **A31**, 2885 (1985).
253. P. Langevin, *Ann. Chim. Phys.* **5**, 245 (1905).
254. G. Gioumousis and D. P. Stevenson, *J. Chem. Phys.* **29**, 294 (1958).
255. C. E. Klots, *Chem. Phys. Lett.* **38**, 61 (1976).
256. S. Tobita, M. Meinke, L. G. Christophorou, E. Illenberger, H. Baumgärtel and S. Leach, *Chem. Phys.* **161**, 501 (1992).
257. J. Laskin, B. Hadas, T. D. Märk and C. Lifshitz, *Int. J. Mass Spectrom.* **177**, L9 (1998).
258. A. D. Boese and G. E. Scuseria, *Chem. Phys. Lett.* **294**, 233 (1998).
259. M. Foltin, M. Lezius, P. Scheier and T. D. Märk, *J. Chem. Phys.* **98**, 9624 (1993).

260. J. Huang, H. S. Carman, Jr. and R. N. Compton, *J. Phys. Chem.* **99**, 1719 (1995).
261. R. Abouaf, J. Pommier and S. Cvejanovic, *Chem. Phys. Lett.* **213**, 503 (1993).
262. C. Bulliard, M. Allan and S. Leach, *Chem. Phys. Lett.* **209**, 434 (1993).
263. A. W. Burose, T. Dresch and A. M. G. Ding, *Z. Phys.* **D26**, 294 (1993).
264. L.-S. Wang, J. Conceicao, C. Jin and R. E. Smalley, *Chem. Phys. Lett.* **182**.
265. N. F. Mott and H. S. W. Massey, *The Theory of Atomic Collisions*, 3rd Edition (Oxford University Press, Oxford, 1965).
266. D. Smith, P. Spanel and T. D. Märk, *Chem. Phys. Lett.* **213**, 202 (1993).
267. E. Tosatti and N. Manini, *Chem. Phys. Lett.* **223**, 61 (1994).
268. C. D. Finch, R. A. Popple, P. Nordlander and F. B. Dunning, *Chem. Phys. Lett.* **244**, 345 (1995).
269. J. M. Weber, M.-W. Ruf and H. Hotop, *Z. Phys.* **D37**, 351 (1996).
270. Y. Vasil'ev (private communication, 1998).
271. W. R. Henderson, W. L. Fite and R. T. Brackmann, *Phys. Rev.* **183**, 157 (1969).
272. T. F. O'Malley, *Phys. Rev.* **155**, 59 (1967).
273. L. G. Christophorou, *Contrib. Plasma Phys.* **27**, 238 (1987).
274. L. Lehr and W. H. Miller, *Chem. Phys. Lett.* **250**, 515 (1996).
275. L. Lehr, PhD Thesis, Fachbereich Chemie, Freie Universität Berlin, 1996.
276. I. Hahndorf and E. Illenberger, *Int. J. Mass Spectrom. Ion Proc.* **176–177**, 78 (1997).
277. I. Hahndorf, PhD Thesis, Fachbereich Chemie, Freie Universität Berlin, 1996.
278. S. S. Shaik and A. Pross, *J. Am. Chem. Soc.* **104**, 2708 (1982).
279. S. S. Shaik, A. B. Schlegel and S. Wolfe, *Theoretical Aspect of Physical Organic Chemistry: The S_N2 Mechanism* (Wiley Interscience, New York, 1992).
280. A. Pross, *Theoretical and Physical Principles of Organic Reactivity* (Wiley, Chichester, 1995).
281. W. N. Olmstead and J. I. Brauman, *J. Am. Chem. Soc.* **99**, 4219 (1977).
282. J. H. Bowie, *Acc. Chem. Res.* **13**, 6 (1980).
283. J. M. Riveros, S. M. José and K. Takashima, in *Advances in Physical Organic Chemistry*, Eds. V. Gold and D. Bethell (Academic Press, London, 1985).
284. M. J. Pellerite and J. I. Brauman, *J. Am. Chem. Soc.* **105**, 2672 (1983).
285. S. L. Craig and J. I. Brauman, *J. Am. Chem. Soc.* **118**, 6786 (1996).
286. G. Caldwell, T. F. Maguera and P. Kebarle, *J. Am. Chem. Soc.* **106**, 959 (1984).
287. D. K. Bohme, G. I. Mackay and S. D. Tanner, *J. Am. Chem. Soc.* **102**, 407 (1980).
288. C. H. De Puy, S. Gronert, A. Mullin and V. M. Bierbaum, *J. Am. Chem. Soc.* **112**, 8650 (1990).

289. A. A. Viggiano, R. A. Morris, J. S. Paschkewitz and J. F. Paulson, *J. Am. Chem. Soc.* **114**, 10477 (1992).
290. D. M. Cyr, L. A. Posey, G. A. Bishea, C. C. Han and M. A. Johnson, *J. Am. Chem. Soc.* **113**, 9697 (1991).
291. D. M. Cyr, C. G. Bailey, D. Serxner and M. A. Johnson, *J. Chem. Phys.* **101**, 10507 (1994).
292. D. K. Bohme and A. K. Raksit, *Can. J. Chem.* **63**, 3007 (1985).
293. B. Brutschy, *Chem. Rev.* **92**, 1567 (1992).
294. A. Radzig and B. M. Smirnov, *Reference Data on Atoms, Molecules and Ions* (Springer-Verlag, Berlin, 1985).
295. *CRC Handbook of Chemistry and Physics*, 76th Edition, Ed. D. R. Lide (CRC Press, Boca Raton, 1995).
296. D. M. Pearl and P. D. Burrow, *J. Chem. Phys.* **104**, 2940 (1994).
297. D. M. Pearl, P. D. Burrow, I. Fabrikant and G. A. Gullup, *J. Chem. Phys.* **102**, 2737 (1995).
298. T. Kraft, PhD Thesis, Fachbereich Physik, Universität Kaiserslautern, 1990.
299. S. Matejcik, A. Kiendler, A. Stanmatovic and T. D. Märk, *Int. J. Mass Spectrom. Ion Proc.* **149–150**, 311 (1995).
300. H.-U. Scheunemann, E. Illenberger and H. Baumgärtel, *Ber. Bunsenges. Phys. Chem.* **84**, 580 (1980).
301. S. R. Hunter and L. G. Christophorou, *J. Chem. Phys.* **80**, 6150 (1984).
302. F. Linder, private communication.
303. J. V. Seeley, R. A. Morris, A. A. Viggiano and J. F. Paulson, *J. Phys. Chem.* **A101**, 4598 (1997).
304. T. Su, R. A. Morris, A. A. Viggiano and J. F. Paulson, *J. Phys. Chem.* **94**, 8426 (1990).
305. P. W. Harland and J. L. Franklin, *J. Chem. Phys.* **61**, 1621 (1973).
306. Y. Le Coat, J. P. Ziesel and J. P. Guillotin, *J. Phys. B: At. Mol. Opt. Phys.* **27**, 965 (1994).
307. R. S. Hunter and L. G. Christophorou, *J. Chem. Phys.* **80**, 6150 (1984).
308. M. Born, *Z. Phys.* **1**, 45 (1920).
309. H. A. Stuart, *Molekülstrucktur*, 3rd Edition (Springer, Berlin, 1967).
310. M. Heni, E. Illenberger, H. Baumgärtel and S. Süzer, *Chem. Phys. Lett.* **87**, 244 (1982).
311. E. Illenberger, H. Baumgärtel and S. Süzer, *J. Electron Spectrosc. Relat. Phenom.* **33**, 12 (1984).
312. M. Heni and E. Illenberger, *J. Electron Spectrosc. Relat. Phenom.* **41**, 453 (1986).
313. P. J. Robinson and K. A. Holbrook, *Unimolecular Reactions* (Wiley-Interscience, London, 1971).
314. T. Baer and W. L. Hase, *Unimolecular Reaction Dynamics* (Oxford University Press, New York, 1996).
315. K. Nagesha and L. Sanche, *Phys. Rev. Lett.* **78**, 4725 (1997).
316. A. Szabó, S. E. Converse, S. R. Whally and J. M. White, *Surf. Sci.* **364**, 345 (1996).

317. P. Rowntree, L. Parenteau and L. Sanche, *Chem. Phys. Lett.* **182**, 479 (1991).
318. P. Rowntree, H. Sambe, L. Parenteau and L. Sanche, *Phys. Rev.* **B47**, 4537 (1993).

CHAPTER 20

VISUALIZATION OF ELECTRON CORRELATIONS IN DOUBLY AND TRIPLY EXCITED STATES OF ATOMS

C. D. Lin and Toru Morishita

Department of Physics, Cardwell Hall
Kansas State University, Manhattan, Kansas 66506

Contents

1. Introduction

In the absorption of electromagnetic radiation by atoms and molecules, the coupling of the photon to the electron is weak. The photoionization or photodetachment processes can thus be treated usually by perturbation theory. Experimental photoabsorption spectra thus serves to reveal the structure of atoms and molecules. When photoionization takes place from the ground state of an atoms, single electron processes dominate where a single photon is absorbed by one-electron which is then ejected to the continuum states while all the other electrons remain in the same original orbitals. This single electron picture is conceptually very simple and has been used to explain a wealth of experimental data. It also fits well with the conventional description of atoms and molecules which is based on the independent particle model. In this model, each electron occupies an atomic or a molecular spin-orbital, and the electrons are arranged to occupy these orbitals in accordance with the Pauli exclusion principle. This simple conceptual description of course has its limitations. The orbitals in a many electron system are determined in some variational procedure such as the Hatree–Fock approximation where the full electron–electron interaction is not treated completely.

Since the early synchrotron radiation works,[1,2] it has been shown that the simple single electron process depicted above is incomplete. The absorption of a single photon can result in the ejection of two or more electrons, and the photoabsorption spectra often exhibit pronounced sharp structures which are attributed to the formation of unstable bound states or resonances. Among the many electron processes, the simplest is the formation of doubly excited states of a two-electron atom such as He. Since the very first spectra of doubly excited states of He obtained using synchrotron radiation more than three decades ago, this simple atom has been investigated in every laboratory whenever a new generation of light source becomes available.[3–6] In particular, doubly excited states of He force theorists to contemplate new description of atoms when two electrons are excited at the same time. The precise observation that can be achieved in atoms thus serves as the test ground for the understanding of multiply excited states of the general few-body systems. In this article, we will review the new understanding achieved for the description of doubly excited states which is fairly complete now,[7–10] and report the recent progress made on the triply excited states.

To appreciate the need of a new description of the doubly excited states, we first consider the He atom specifically. According to the independent electron model, the ground state of He is designated as $1s^2\,{}^1S^e$ where the two electrons are both in the $1s$ orbital and they are coupled to form spin singlet with zero total orbital angular momentum and even parity. After the absorption of a single photon, the doubly excited states are expected to form Rydberg series such as $2snp\,{}^1P^o$, $2pns\,{}^1P^o$, and $2pnd\,{}^1P^o$ where each series converges to the $2s$ or $2p$ limit of the He^+ ion. The formation of doubly excited states in a single photon process is weak, but most surprisingly is the observed fact[1,2] already in 1963 that instead of finding three Rydberg series converging to the degenerate $He^+(2s)$ and $He^+(2p)$ limits, only one prominent Rydberg series was observed, with a second weak series barely visible and a third series not seen at all. This experimental fact caught the attention of Fano and coworkers.[11] They immediately pointed out that the observed spectra is an illustration of the failure of the independent electron model for describing doubly excited states. They proclaimed that the observed pronounced Rydberg series should be designated as the $(2snp + 2pns)\,{}^1P^o$ series and the very weak series be designated as the $(2snp - 2pns)\,{}^1P^o$ series, while the unobserved third series be called the $2pnd\,{}^1P^o$ series. Historically these three series have been abbreviated as "+", "−", and "*pd*" series. They are the most famous doubly excited states.

The designations of the observed doubly excited states by Fano and coworkers are easy to understand from the viewpoint of the traditional atomic and molecular structure theory. For example, the $(2s3p+2p3s)\,{}^1P^o$

state has an equal admixture of $2s3p\,^1P^o$ and $2p3s\,^1P^o$ states, but this is also true for the $(2s3p - 2p3s)\,^1P^o$ state, 'except' that the difference is only in the sign of the mixing coefficient. However, the sign dictates the difference between the "+" and the "−" series, and dictates which series will be prominently excited from the ground state, and which series will not. The initial success of Fano and coworkers for the doubly excited states of He that converge to the $\mathrm{He}^+(N = 2)$ threshold opens up a long list of new questions concerning doubly excited states in general. For example, according to the independent electron model, there will be five $^1P^o$ Rydberg series, $3snp$, $3pns$, $3pnd$, $3dnp$, and $3pnf$, that will converge to the $\mathrm{He}^+(N = 3)$ threshold. How to generalize the "+", "−", and "*pd*" designation for the expected five series in this case? Another looming critical question is why the "+" series is so predominantly populated in the photoabsorption process and the observed widths are so broad, while the other two series are hardly found? Answers to these questions are not transparent from the "+", "−", and "*pd*" designation.

The designation of "+", "−", and "*pd*" series is a declaration of the complete failure of the independent electron model. While a notation such as $2s3p$ is understood to mean that one of the electrons is in the $2s$ orbital while the other is in the $3p$ orbital, the designation of $2s3p + 2p3s$ implies that it is no longer possible to talk about an electron is in which orbital. Thus while it is possible to obtain relatively accurate wavefunctions of doubly excited states, there exists no conceptual framework for the description of these states. If the wavefunctions for members of the "+" and "−" series can be expressed as $(2s3p + 2p3s)\,^1P^o$ and $(2s3p - 2p3s)\,^1P^o$, respectively, what are the major features built into these wavefunctions which in turn determine the gross difference of these states? It is desirable to visualize these two-electron wavefunctions. However, besides the spins which can be factored out for two-electron systems, each wavefunction is a function of six spatial coordinates which cannot be easily displayed on a two-dimensional surface. Partial display of one or two coordinates, unless chosen correctly, may not be able to reveal the distinct features of the states. Thus the urgent need in the study of doubly excited states, as well as other multiply excited states, is to identify special coordinates which can reveal the major features that distinguish these states. The identification of these distinct features is an essential step in order to establish new approximate quantum numbers that can be used to classify them. This task has been accomplished for the doubly excited states of He owing to the complementary viewpoints developed from different groups.[12–15] For the triply excited states the task is only in the beginning stage.[16–22]

2. Visualization of Electron Correlation in Doubly Excited States of Atoms

2.1. *Configuration Mixing*

The designation of the "+", "−", and "*pd*" series by Fano and coworkers originates from the configuration mixing or configuration interaction (CI) viewpoint. The improved wavefunction which is an approximate eigenstate of the two-electron Hamiltonian is expressed as the linear combination of the configurations constructed from the independent electron model. Consider the two-electron Hamiltonian (atomic units will be used throughout this article)

$$\begin{aligned} H &= -\frac{1}{2}\nabla_1^2 - \frac{1}{2}\nabla_2^2 - \frac{Z}{r_1} - \frac{Z}{r_2} + \frac{1}{r_{12}} \\ &= H_0 + \frac{1}{r_{12}}, \end{aligned} \tag{1}$$

where $r_i (i = 1, 2)$ is the distance of electron i from the nucleus and r_{12} is the distance between the two electrons, with Z being the charge of the nucleus. In the simplest form of the independent electron model, the full electron–electron interaction is neglected and one can construct $nln'l'$ configurations from H_0, which are then coupled to form eigenstates of the total orbital (L) and spin (S) angular momenta and parity (π). Thus configurations such as $2s3p\,^1P^o$, $2p3s\,^1P^o$, and $2p3d\,^1P^o$, for example, which are eigenstates of H_0, can be formed. Within the subspace of these three configurations, the eigenstates are obtained by diagonalizing the Hamiltonian H. The results are

$$\begin{aligned} |\,1\rangle &= 0.68\,|\,2s3p\rangle - 0.70\,|\,2p3s\rangle + 0.21\,|\,2p3d\rangle\,, \\ |\,2\rangle &= 0.63\,|\,2s3p\rangle + 0.71\,|\,2p3s\rangle + 0.32\,|\,2p3d\rangle\,, \\ |\,3\rangle &= 0.37\,|\,2s3p\rangle + 0.09\,|\,2p3s\rangle - 0.92\,|\,2p3d\rangle\,, \end{aligned} \tag{2}$$

where the energies of the three states are $E_1 = -0.584$, $E_2 = -0.504$, and $E_3 = -0.470$ a.u. From the mixing coefficients, one can identify that $|1\rangle$ belongs to a member of the "−" series, $|2\rangle$ belongs to the "+" series and

$|3\rangle$ belongs to the "*pd*" series of Fano and coworkers. The CI method, together with many other approaches based initially on the independent electron approximation, allows practitioners to obtain reasonable accurate wavefunctions to describe doubly excited states. However, accurate doubly excited state wavefunctions often cannot be represented by a dominant single configuration. Since the validity of the independent electron model relies on the existence of a single dominant configuration, the strong mixing renders the designation in terms of independent particle quantum numbers such as $2s3p\,^1P^o$ invalid. While it is possible to label each eigenstate by specifying the CI coefficients in principle, this is clearly awkward and the coefficients themselves depend on how the configurations are defined.

2.2. *Pauli Correlation and Visualization of Electron Correlation*

The approximate eigenfunctions listed above provide nearly adequate description of the doubly excited states. However, the nature of each of these eigenstates is not transparent. Recall that the spatial wavefunction of a single atomic orbital is given by $R_{nl}(r)Y_{lm}(\hat{r})$, where $R_{nl}(r)$ is the radial function and $Y_{lm}(\hat{r})$ is the spherical harmonics. The functional form of each of these functions is well understood and the properties of each function are implicitly represented by the quantum numbers n, l, and m. For the two-electron wavefunctions such as those in Eq. (2), an immediate question is whether it is possible to visualize them such that their distinct properties are revealed. However, this cannot be easily done since each wavefunction is represented by a complex function of six variables which cannot be fully displayed on a two-dimensional surface. Thus it is desirable to average over some nonessential variables, and then look for suitable variables that can better display the properties of the states.

The lack of visualization of a two-electron wavefunction is already evident in the independent electron approximation. For the $1s2s\,^1S^e$ and $1s2s\,^3S^e$ singly excited states, there are no angular dependence and the spatial wavefunctions are given by

$$|\,1s2s^{1,3}S^e\rangle = \frac{1}{\sqrt{2}}\left[R_{10}(r_1)R_{20}(r_2) \pm R_{10}(r_2)R_{20}(r_1)\right], \tag{3}$$

where the $+$ sign is for the $1s2s\,^1S^e$ state and the $-$ sign is for the $1s2s\,^3S^e$ state. For the present purpose, we approximate $R_{nl}(r)$ to be the hydrogenic

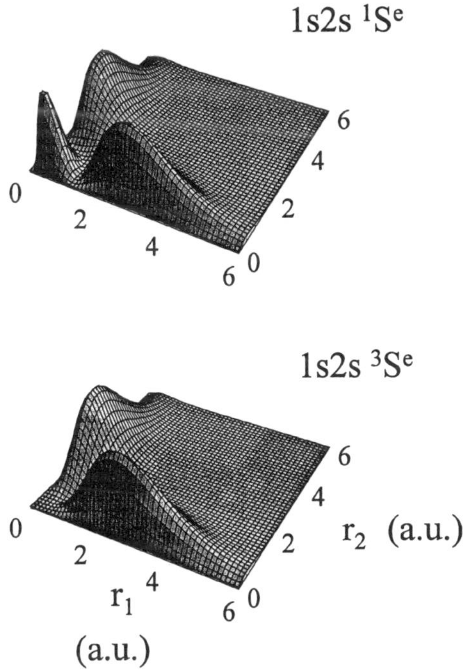

Fig. 1. Densities of the two-electron wavefunctions on the (r_1, r_2) plane for the $1s2s^1S^e$ and $1s2s^3S^e$ states of He under the single configuration approximation.

radial function with charge $Z = 2$ for He. Since each of the functions in Eq. (3) depends only on r_1 and r_2, these two-electron wavefunctions can be displayed directly. In Fig. 1 we show the modulus square, or the density, of these two functions on the (r_1, r_2) plane. Clearly the difference between these two functions is in the region of small r_1 and r_2 (< 1.5 a.u.). The $^1S^e$ has larger density in this region and the $^3S^e$ has negligible density there. In fact the wavefunction for $^1S^e$ is symmetric with respect to the $r_1 = r_2$ line (the diagonal line), and the wavefunction for the $^3S^e$ is antisymmetric and vanishes along $r_1 = r_2$. When these wavefunctions are used to evaluate the expected value of the Hamiltonian H in Eq. (1), the larger density for the $^1S^e$ state in the small r_1 and r_2 region results in larger Coulomb repulsion between the two electrons. Consequently the $1s2s\,^1S^e$ state has higher energy than the $1s2s\,^3S^e$ state.

The need of "+" and "−" combinations in Eq. (3) arises from the Pauli exclusion principle where the spatial wavefunction for the singlet (triplet) state is symmetric (antisymmetric) under the exchange of the two electrons. The quantum statistics is reflected in the boundary conditions on the many-electron wavefunction in the multidimensional space. For the present states, that $r_1 = r_2$ is an antinodal line for $1s2s\,{}^1S^e$ and a nodal line for $1s2s\,{}^3S^e$.

2.3. *Body-Fixed Frame Wavefunctions*

For configurations where the total orbital angular momentum L is not zero, the total wavefunction depends on the magnetic quantum number M which is the projection of L with respect to a space-fixed axis. In the independent electron model, the wavefunctions for $1s2p\,{}^1P^o$ and $1s2p\,{}^3P^o$ can be written down explicitly as

$$|\,1s2p^{1,3}P^o\rangle = \frac{1}{\sqrt{2}}[R_{10}(r_1)R_{21}(r_2)\mathcal{Y}_{01}^{1M}(\hat{\mathbf{r}}_1,\hat{\mathbf{r}}_2) \pm R_{10}(r_2)R_{21}(r_1)\mathcal{Y}_{01}^{1M}(\hat{\mathbf{r}}_2,\hat{\mathbf{r}}_1)]\,, \tag{4}$$

where the "+" sign is for the $1s2p\,{}^1P^o$ state and the "−" sign is for the $1s2p\,{}^3P^o$ state, and $\mathcal{Y}_{ll'}^{LM}(\hat{\mathbf{r}}_1,\hat{\mathbf{r}}_2)$ is the coupled total orbital angular momentum wavefunction constructed from two spherical harmonics $Y_{lm}(\hat{\mathbf{r}}_1)$ and $Y_{l'm'}(\hat{\mathbf{r}}_2)$. The differences in these two functions clearly are not in their r_1 and r_2 distributions. By integrating the modulus square of each function over the four spherical angles, $\hat{\mathbf{r}}_1$ and $\hat{\mathbf{r}}_2$, the two resulting functions in r_1 and r_2 are identical. Thus the difference between these two states lies in the angular coordinates.

The wavefunctions presented in Eq. (4) are given in the laboratory-fixed frame. One can rewrite the laboratory-fixed wavefunction in terms of the body-fixed frame wavefunction,

$$\psi_{LM}(\mathbf{r}_1,\mathbf{r}_2) = \sum_{Q=-L}^{L} f_{LQ}(r_1,r_2,\theta_{12})D_{QM}^{(L)}(\omega)\,, \tag{5}$$

where $D_{QM}^{(L)}(\omega)$ is the rotation matrix with ω being the Euler angles relating the body-fixed frame to the laboratory-fixed frame. At the moment we have not specified the quantization axis of the body-fixed frame where Q is the projection of L along that axis. The rotational component f_{LQ} does not depend on M; it depends on the internal coordinates of the atom. For the present two-electron atom, we have the three internal coordinates r_1,

r_2, and θ_{12}, the latter being the angle between the two electrons with respect to the nucleus. Thus by using the body-fixed frame, we can reduce the independent coordinates for the two-electron system from six to three, i.e. the three degrees of freedom describing the overall rotation of the atom with respect to the space-fixed frame can be discarded since they have nothing to do with the correlation of the two electrons.

The specific form of each of the rotational components f_{LQ} above depends on the choice of the quantization axis of the body-fixed frame. We define the two-electron density to be

$$\begin{aligned}\rho_I(r_1, r_2, \theta_{12}) &= \int |\psi_{LM}(\mathbf{r}_1, \mathbf{r}_2)|^2 d\omega \\ &= \frac{1}{8\pi^2} \sum_{Q=-L}^{L} |f_{LQ}(r_1, r_2, \theta_{12})|^2 ,\end{aligned} \tag{6}$$

which is independent of the choice of the body-fixed frame quantization axis. This is equivalent to a rotation averaged density or shape function for each state since its gives the distribution of the relative positions of the two electrons. This function is used to demonstrate the correlation pattern between the two electrons.

The density function in Eq. (6) still depends on three variables. Instead of displaying the distribution of the second electron while holding the first electron fixed, as done by Berry and collaborators,[23] it is preferable to treat the two identical electrons symmetrically. Note that the interchange of the coordinates of the two electrons leaves $(r_1^2 + r_2^2)$ unchanged. Thus instead of using r_1 and r_2, we can use hyperspherical coordinates where the hyperradius R and the hyperangle α are defined by

$$\begin{aligned}R &= \sqrt{r_1^2 + r_2^2} , \\ \alpha &= \tan^{-1}(r_2/r_1) .\end{aligned} \tag{7}$$

The hyperradius R measures the size of the atomic state and has nothing to do with the correlation of the two electrons which is measured by the relative positions of the particles in the system. In Fig. 2 we plot the charge densities ρ_I for $1s2p\,^1P^o$ and $1s2p\,^3P^o$ states on the (α, θ_{12}) plane at four values of the hyperradius, $R = 0.3$, 1.0, 2.0, and 3.0 a.u. First we note that the R-dependence is very smooth. Second, the major difference in the two states is that the two functions have different θ_{12} dependence for smaller

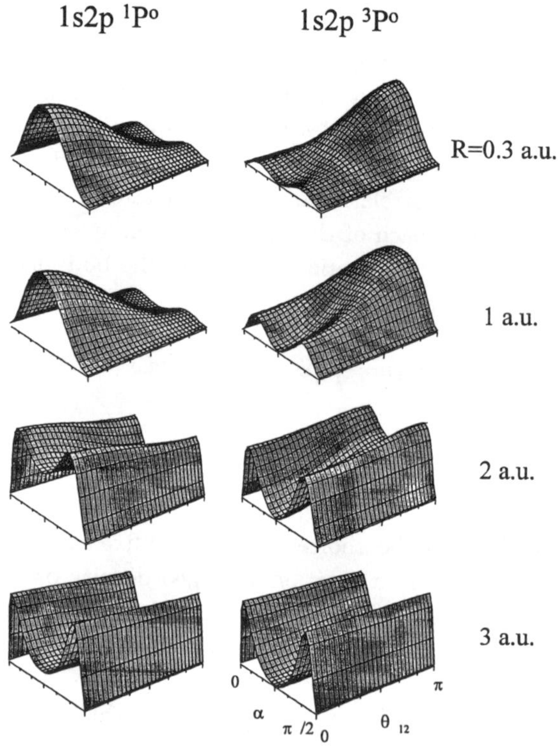

Fig. 2. Surface density plots on the (α, θ_{12}) plane for four values of the hyperradius R for the $1s2p\,^1P^o$ and $1s2p\,^3P^o$ states of He within the independent electron approximation. The plots show that these two states have different angular correlation as the result of the exchange symmetry.

R's. For the $1s2p\,^1P^o$ state, the density for the two electrons being in the small θ_{12} region is larger. When these wavefunctions are used to evaluate the expectation value of H, the larger density for the $^1P^o$ state in the small θ_{12} region results in larger Coulomb repulsion between the two electrons. Thus the energy for the $1s2p\,^1P^o$ state is higher than for the $1s2p\,^3P^o$ state.

The above two examples illustrate that it is possible to understand the relative energies of a two-electron atom in terms of the two-electron wavefunctions when these wavefunctions are displayed in suitable coordinates. For the two-electron atom, by averaging over the rotation of the whole atom, the total density reduces to a function of three variables. Using hyperspherical coordinates, the correlation between the two electrons can then be easily displayed in the two angles (α, θ_{12}) for each hyperradius. By displaying the density on the (α, θ_{12}) plane in the region of the hyperradius where the wavefunction is near the maximum, we look for distinct features of the distributions which characterize the nature of correlation for the two electrons.

2.4. *Electron Correlations in Doubly Excited States of Atoms*

The smooth variation of the shape density with R in Fig. 2 demonstrates that each wavefunction can be expanded in the hyperspherical adiabatic form,

$$\psi(\hat{\mathbf{r}}_1, \hat{\mathbf{r}}_2) = \sum_{\mu} F_{\mu}(R)\Phi_{\mu}(R; \Omega), \tag{8}$$

where $\Phi_{\mu}(R; \Omega)$ is expected to vary slowly with R. We choose the convention that each function $\Phi_{\mu}(R; \Omega)$ is normalized over the space of the five angles $\Omega = (\alpha, \hat{\mathbf{r}}_1, \hat{\mathbf{r}}_2)$ at each R, and thus $F_{\mu}(R)$ can be evaluated. In Fig. 3 we show the surface densities on the (α, θ_{12}) plane at the values of R indicated for the four ${}^1P^o$ doubly excited states which could be called $2s2p$, $(2s3p - 2p3s)$, $(2s3p + 2p3s)$, and $2p3d$ states, following the notation of Fano and coworkers.

From Fig. 3 we can see that the surface densities for the $2s2p$ and the $(2s3p + 2p3s)$ states are very similar — they belong to the same channel, or Rydberg series. Clearly these two states differ mostly in their size but at small R the $(2s3p + 2p3s)$ state has identical correlation pattern as the $2s2p$ state. From Fig. 3, the surface density for the $(2s3p - 2p3s)$ state has an approximate nodal line along $\alpha = \pi/4$ (i.e. $r_1 = r_2$). These surface plots clearly demonstrate that the "+" and "−" classification of Fano and coworkers is related to the approximate symmetry of the wavefunction at $\alpha = \pi/4$. For the "+" states, the two electrons can stay at about the same distance from the nucleus, while for the "−" states the two electrons are not allowed to maintain at the same distance from the nucleus. In terms of classical interpretation, the two electrons in the "+" states can be viewed as oscillating toward or away from the nucleus in phase, while for the "−" states such oscillation is out of phase. For the $2p3d$ state, its surface density shows that the region near $\alpha = \pi/4$ is not occupied, meaning that the two electrons never get close to each other. This is similar to singly excited states where the outer electron rarely penetrates the core region.

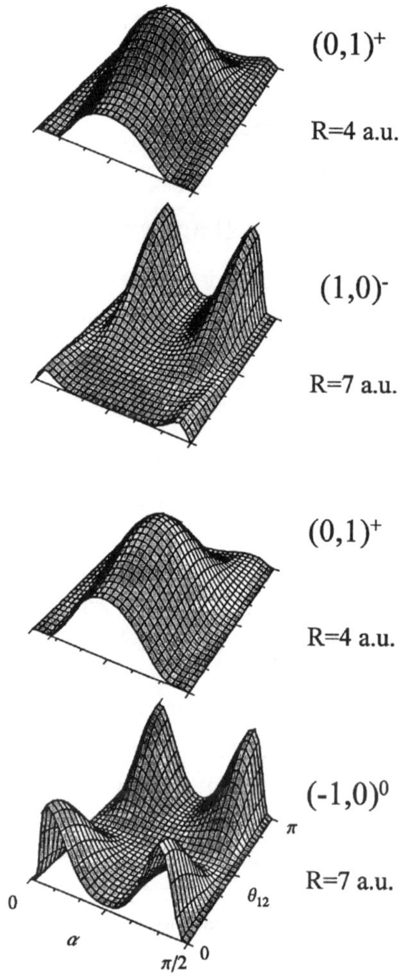

Fig. 3. Surface density plots on the (α, θ_{12}) plane for the four $^1P^o$ states of He which are called $2s2p$, $(2s3p - 2p3s)$, $(2s3p + 2p3s)$, and $2p3d$ according to the designation of Fano and coworkers[11] at the values of R indicated. These states are designated in terms of the K, T, and A quantum numbers as shown.

The surface density plots in Fig. 3 also show that there are significant differences in the θ_{12} distributions. For the $(2s3p - 2p3s)$ state, the maximum density is close to $\theta_{12} \approx \pi$, i.e. the two electrons are on opposite sides of the nucleus. For the $2s2p$ and $(2s3p+2p3s)$ states, on the other hand, the maximum density is at $\theta_{12} \approx \pi/2$. In other words, the two electrons and the nucleus almost form a right triangle as in a water molecule. Such angular dependence is not contained in the designation of Fano and coworkers. In fact, the density plots shown in Fig. 3 were obtained with a much larger CI calculation. The fact that the density plots are relatively localized implies strong mixing of different angular momentum states and thus designation in terms of independent particle model or even the "+" and "−" type designation is inadequate. Using the new quantum numbers to be discussed below, the new designation of the $2s2p$, $(2s3p - 2p3s)$, $(2p3s + 2p3s)$, and $2p3d$ states are ${}_2(0,1)_2^+$, ${}_3(1,0)_2^-$, ${}_3(0,1)_2^+$, and ${}_3(-1,0)_2^0$, as indicated in Fig. 3 (see Sec. 2.6 below).

2.5. *Electron Correlation in Higher Doubly Excited States*

We next consider the five He$({}^{1,3}P^o)$ states formed from $3s4p$, $3p4s$, $3p4d$, $3d4p$, and $3d4f$ configurations. In Fig. 4 we show the surface density plots at the values of R indicated. Extensive CI calculations have been carried out to obtain the resulting plots. They are arranged such that each pair has nearly identical θ_{12} dependence, and the major difference between each pair (except the last pair) is that in one case the density peaks at $\alpha = \pi/4$, while in the other the density nearly vanishes there. In other words, they are like the "+" and "−" states in the examples of Fig. 3. The last pair has essentially identical density plots. This set shows the striking similarities correlation patterns of some doubly excited states.

As another example, consider the following states which are designated using the traditional independent electron model. Actual extensive CI calculations show that each state is a strong mixing of other configurations but only the major components are listed:

$$
\begin{aligned}
|\,3s3s\,{}^1S^e\rangle &= 0.72\,|\,3s3s\rangle + 0.61\,|\,3p3p\rangle + 0.15\,|\,3d3d\rangle + \cdots\,,\\
|\,3s3p\,{}^3P^o\rangle &= 0.87\,|\,3s3p\rangle + 0.39\,|\,3p3d\rangle + \cdots\,,\\
|\,3p3p\,{}^1D^e\rangle &= 0.71\,|\,3p3p\rangle - 0.58\,|\,3s3d\rangle + 0.15\,|\,3d3d\rangle + \cdots\,,\\
|\,3p3d\,{}^3F^o\rangle &= 0.91\,|\,3p3d\rangle + \cdots\,,\\
|\,3d3d\,{}^1G^e\rangle &= 0.80\,|\,3d3d\rangle + \cdots\,.
\end{aligned}
\tag{9}
$$

From the CI coefficients in Eq. (9) no one would have imagined that these states are related. In fact, the correlation patterns between the two electrons for all these states are essentially identical, as demonstrated by the

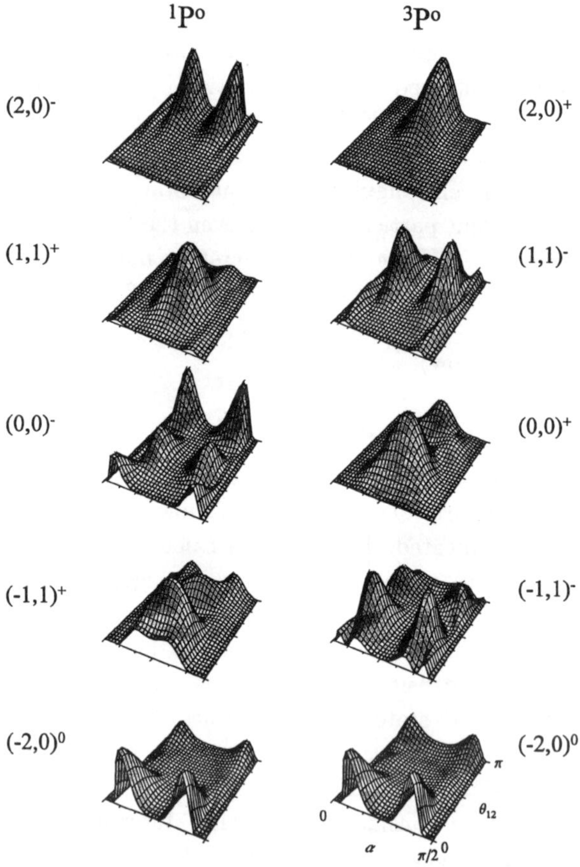

Fig. 4. Comparison of the surface density plots on the (α, θ_{12}) plane for the $3l4l'\,^1P^o$ and $^3P^o$ doubly excited states of He. The states are designated using the $(K,T)^A$ quantum numbers. All the "+" states are plotted at $R = 10$ a.u. and all the "−" and "0" states are plotted at $R = 15$ a.u.

surface density plots depicted in Fig. 5. If we are to use a new set of approximate quantum numbers which can describe the correlation of the two electrons, these quantum numbers obviously should be identical for all these states. Thus even though the CI approach (or many other approaches) can obtain accurate wavefunctions of doubly excited states, the nature of electron correlation is not transparent. On the other hand, according to the $(K,T)^A$ classification scheme below, each of these states has the same $(2,0)^+$ designation.

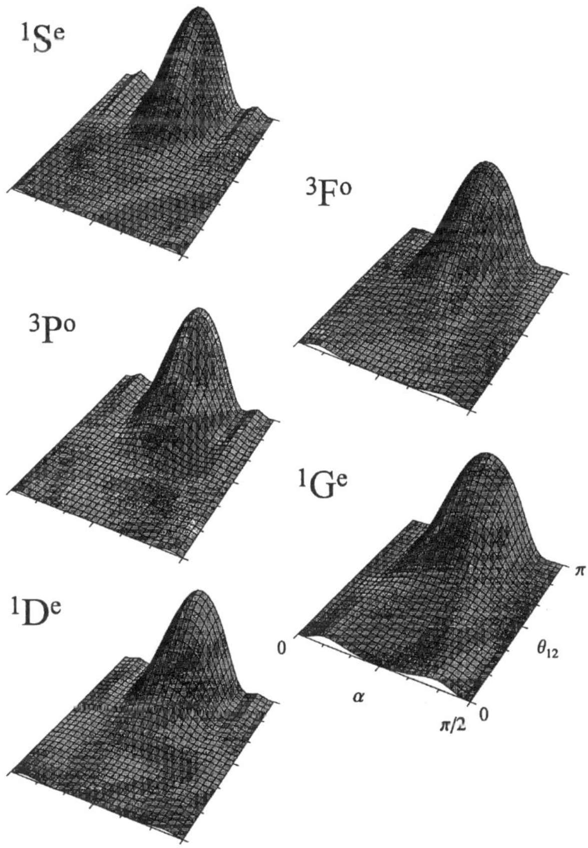

Fig. 5. Surface density plots on the (α, θ_{12}) plane for the five states listed in Eq. (9). These states have nearly identical correlation patterns and all are designated as $(2,0)^+$ in terms of the correlation quantum numbers. The plots are for He at $R = 10$ a.u.

2.6. *Classification of Doubly Excited States*

The visualization of electron correlation for doubly excited states as shown above does not lead itself to a new set approximate quantum numbers. To obtain such quantum numbers one has to start with models which approximately describe doubly excited states. The first success in this respect is based on the algebraic method. This method is very similar to an approximate CI approach. Consider the simplest CI within the configurations such as $|Nlnl'>$ where N and n are fixed but all the l and l' that can couple to the desired symmetry are included. Diagonalization of the two-electron

Hamiltonian in Eq. (1) within this subspace will give approximate doubly excited states. However, even in such a limited CI calculation, it is still possible to obtain the approximate quantum numbers algebraically. Herrick and Sinanoğlu[12] found that the eigenstates that diagonalize the operators

$$\mathbf{B}^2 = (\mathbf{b}_1 - \mathbf{b}_2)^2 = [(N+K)^2 - 1 - T^2 - L(L+1)]^2\,, \tag{10}$$

and

$$(\mathbf{B}\cdot\mathbf{L})^2 = T^2(N+K)^2\,, \tag{11}$$

give wavefunctions which are good approximations to the doubly excited state wavefunctions. These doubly excited state basis functions (DESB) can be written down explicitly

$$|NnPT\rangle = \sum_{ll'} |Nlnl'\rangle\langle Nlnl'|NnPT\rangle\,, \tag{12}$$

where

$$\langle Nlnl'|NnPT\rangle = (-1)^{l'}\sqrt{(P+T+1)(P-T+1)(2l+1)(2l'+1)} \left\{\begin{matrix} (N-1)/2 & (n-1)/2 & (P+T)/2 \\ (N-1)/2 & (n-1)/2 & (P-T)/2 \\ l & l' & L \end{matrix}\right\}, \tag{13}$$

where the curly bracket is the 9-j symbol, and P and K are related by $P = n-1-K$. The N and n are the principal quantum numbers of the inner and the outer electrons, respectively. The transformation sets the range of K and T to be $K = N-1-T, N-3-T, \ldots, -(N-1-T)$ and $T = 0, 1, 2, \ldots, L$. For example, the DESB functions for the first two states listed in Eq. (9) are:

$$\begin{aligned} \psi(3s3s\,{}^1S^e) &= 0.75\,|\,3s3s\rangle + 0.65\,|\,3p3p\rangle + 0.17\,|\,3d3d\rangle\,, \\ \psi(3s3p\,{}^3P^o) &= 0.91\,|\,3s3p\rangle + 0.41\,|\,3p3d\rangle\,, \end{aligned} \tag{14}$$

which are to be compared to those from the limited CI calculations including only the $3l3l'$ configurations for each state:

$$\begin{aligned} \psi(3s3s\,{}^1S^e) &= 0.78\,|\,3s3s\rangle + 0.61\,|\,3p3p\rangle + 0.13\,|\,3d3d\rangle\,, \\ \psi(3s3p\,{}^3P^o) &= 0.93\,|\,3s3p\rangle + 0.36\,|\,3p3d\rangle\,. \end{aligned} \tag{15}$$

The algebraic DESB approach provides a new set of K and T quantum numbers which replace the l and l' orbital angular momentum quantum numbers for individual electrons. Since Eq. (13) is merely a unitary transformation from the $Nlnl'$ subspace to the new $NnKT$ subspace, some aspect of the shell structure of the independent electron model is maintained in terms of the number of states within each manifold of doubly excited states. The DESB approach does not lend itself directly to the interpretation of the correlation of the two electrons. However, the energy levels calculated using the DESB functions reveal structure similar to the ro-vibrational energies of a linear triatomic molecule. This has stimulated the discussion of doubly excited states in terms of the bending vibrations of a linear molecule where the nucleus is at the center and the two electrons are on the opposite sides of the nucleus. This analogy allows one to draw the connection between K and the bending vibrational quantum number v, and T as the projection of L along the interelectronic axis.[13,24]

The DESB and the molecular approach address the qualitative classification of doubly excited states. However, these descriptions in general are good only for doubly excited states that have large θ_{12} or those low-lying states within the manifold. In fact, not all the doubly excited states behave like the ro-vibrations of a triatomic linear molecule. From the classification viewpoint, the molecular quantum numbers do not incorporate the shell model concept such that the number of doubly excited states in a manifold is not restricted. On the other hand the DESB quantum numbers are limited by the shell model and thus can be used to label the doubly excited states. Thus the K and T quantum numbers are the preferred approximate quantum numbers for labeling doubly excited states. To emphasize the nature of the wavefunctions at $r_1 = r_2$ or $\alpha = \pi/4$, a new quantum number $A = +1, -1$, and 0 was introduced.[14] The $+1$ and -1 were introduced in conjunction with the "+" and "−" series of Fano and coworkers. It also implies that the wavefunction for the + state has an antinode at $\alpha = \pi/4$ and that for the − state has an approximate node at $\alpha = \pi/4$. To account for states that do not behave like that of a triatomic molecule, $A = 0$ was introduced to label these states.[14] For each pair of K and T, the auxiliary quantum number A is assigned as

$$\begin{aligned} A &= \pi(-1)^{S+T} \quad \text{if} \quad K > (L-N)\,, \\ &= 0 \qquad\qquad\quad \text{otherwise}\,, \end{aligned} \tag{16}$$

which can be understood if the rotational components $f_{LQ}(r_1, r_2, \theta_{12})$ of the body-fixed frame wavefunctions are analyzed. Watanabe and Lin[25] have shown that if the body-fixed frame quantization axis is chosen to be along the line joining the two electrons, and if these is only one dominant Q-component [see Eq. (5)], then $T = |Q|$ is an approximate good quantum number and then the symmetry of the rotational component $f_{LQ}(r_1, r_2, \theta_{12})$ under the interchange of the two electrons is given by the first line of Eq. (16). When the wavefunction in the body-fixed frame has only one Q-component, T and A are good quantum numbers, and K is then related to the number of nodes m in θ_{12} or to the bending vibrational quantum number v. The relation between these quantum numbers are[25] $v = 2m + T$, $K = N - 2m - T - 1$. In this limiting case, it is possible to classify doubly excited states entirely in terms of the quantum numbers of a linear triatomic molecule. However T is only an approximate good quantum number and the degree of mixing from different Q-components depends on the specific states, see Chen *et al.*.[26] For each pair of K and T, if $A = +1$ or -1 [according to Eq. (16)], then T, as well as A and K, are approximate good quantum numbers and thus K, T, and A can be used to designate doubly excited states. When the different rotational components of the wavefunction in the body-fixed frame are of comparable magnitude, then T is not a good quantum number and thus K and A are not good quantum numbers. In this case we use $A = 0$ to label these states, and the corresponding K and T are just labels without meaning. For singly excited states that can be designated as $1snl$, the wavefunction in the body-fixed frame has no dominant Q-components. For $A = 0$ doubly excited states, this is true too but in general the CI mixing for these states is also quite large.

2.7. *Ro-Vibrational Spectra of Doubly Excited States*

The surface density plots in Fig. 5 show that all the five states have similar distributions, or shapes. Note that the density peaks at $\theta_{12} = \pi$, thus each doubly excited state resembles a linear XYX triatomic molecule with the nucleus at the center. Since each state has a different angular momentum L, the differences in the energies in these states are mostly due to the rotational excitations.

If the doubly excited states can be viewed as the ro-vibrational motion of a linear triatomic molecule, then states that have different K and T can be viewed as having different ro-vibrational energies. It is interesting to compare the relative energy levels of the $3l3l'$ doubly excited states according to the independent electron model as well as to the new classification scheme based on the K, T, and A quantum numbers. Following the independent electron model, the possible intrashell $3l3l'$ states are $3s^2\,{}^1S^e$, $3s3p({}^1P^o, {}^3P^o)$, $3p^2({}^1S^e, {}^3P^e, {}^1D^e)$, $3s3d({}^1D^e, {}^3D^e)$, $3p3d({}^1P^o, {}^3P^o, {}^1D^o, {}^3D^o, {}^1F^o, {}^3F^o)$, $3d^2({}^1S^e, {}^3P^e, {}^1D^e, {}^3F^e, {}^1G^e)$. For each configuration such as $3p3d$, if the independent electron approximation is valid, then the relative energy ordering, according to the Hund's rule, the state with the highest spin has the lowest energy and for each spin, the one with the higher L has the lower energy. The Hund's rule says nothing about the relative energies for states from different configurations. According to the new classification scheme based on the K, T, and A quantum numbers, the states that have the same $(K,T)^A$ form a rotor series with higher L states at higher energies. Since smaller K means higher bending vibrational excitations, the smaller the K is, the higher the energy and for a given K, the larger T has the higher energy.

We can look at the calculated energies of the $3l3l'$ doubly excited states of Li^+ as an example.[27] In the upper frame of Fig. 6 these energies are arranged and ordered according to their configurations using the independent electron model (this may not be unique since mixing is large in general) and in the lower frame in terms of the $(K,T)^A$ classification which are further separated into three groups. From this figure it is seen that Hund's rule is often violated since configuration mixing is large for most of the states. On the other hand, for each $(K,T)^A$ the states are arranged in the fashion of a rotor series, i.e. the energy separation between neighboring states increases with L since the rotation energy of each state for a rotor is proportional to $L(L+1)$. Furthermore, comparing different rotor series, the series with smaller K has higher base energies since smaller K implies higher bending vibrational excitations. Thus the order of the relative energies of doubly excited states is better displayed if they are arranged in terms of the K, T, and A quantum numbers. We remark that the levels are separated into three groups in the lower frame in Fig. 6. The first group (the leftmost one) has no nodes in the θ_{12} dependence, i.e. $m = 0$. For the next two groups, $m = 1$ and $m = 2$, respectively.

The regularity of doubly excited states when ordered according to $(K,T)^A$ can also be seen in the autoionization widths.[28] Similar order has also been shown for the $A = -1$ states. For the $A = 0$ states, on the other hand, their order is similar to that for singly excited states. These issues were addressed elsewhere.[7]

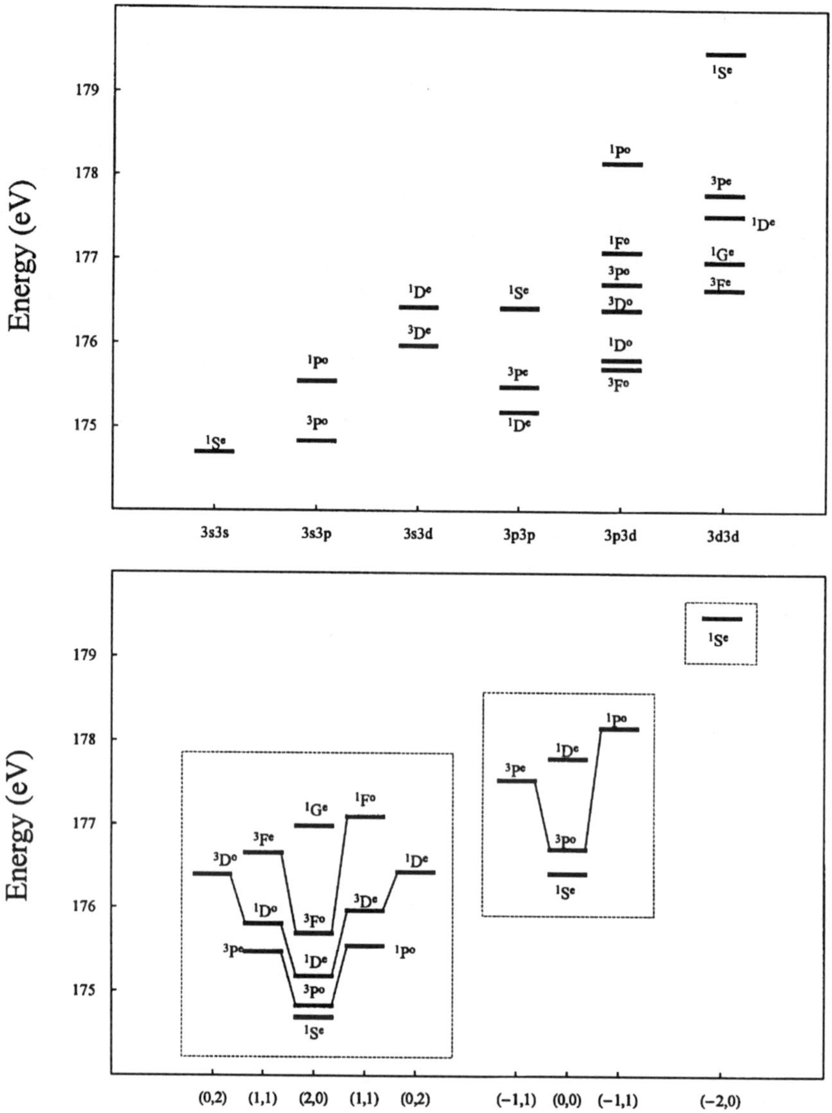

Fig. 6. Energies of $3l3l'$ doubly excited states of Li^+ ordered according to the independent electron model for each individual dominant configuration (top frame) and in terms of the $(K,T)^A$ classification scheme (bottom frame). In the lower frame the wavefunction in the body-fixed frame for each state in the three subgroups have 0, 1, and 2 nodes in the θ_{12} coordinate, respectively.

2.8. *The Hyperspherical Approach*

The discussion of doubly excited states above is based on the more conventional configuration mixing approach which is an extension from the independent electron model. We have shown that by casting the wavefunction in hyperspherical coordinates, one can display the correlation between the two electrons on the (α, θ_{12}) plane at a given value of R. Such a presentation is meaningful only if the surface density does not change significantly with R. Actual calculations such as those shown in Fig. 2 confirm indeed that the surface density does not change with R significantly in the region where the wavefunction is not small. This slow variation of the channel function underlies the validity of the treatment of solving the two-electron atomic problem using the adiabatic approximation in hyperspherical coordinates. This approach has been adopted since the pioneering work of Macek.[29] If the hyperradius is treated as an adiabatic parameter, then we can calculate the hyperspherical potential curves which are analogous to the electronic energy curves of a diatomic molecule in the Born–Oppenheimer approximation. The potential curves that have similar shapes are shown to have nearly identical correlation patterns and thus are assigned to have identical K, T, and A quantum numbers.

The hyperspherical approach has now been extended much beyond the adiabatic approximation and the coupling among the adiabatic channels can be accounted for to obtain accurate numerical results.[30–34] Photoionization cross-sections, as well as electron–hydrogen atom scattering cross-sections[35] and electron impact ionization of atomic hydrogen,[34] have been accurately calculated using the hyperspherical approach. The hyperspherical approach has also been used extensively in other few-body systems[36] such as positron-atom collisions[37–40] as well as helium trimer[41] and other exotic triatomic molecules.[42]

3. Doubly and Triply Excited States of Three-Electron Atoms

3.1. *Introduction*

The success of describing doubly excited states of a two-electron atom poses two immediate new questions. First, how the doubly excited states of a multielectron atom is to be described? We recall that the major modification to the singly excited state of a many electron atom from the excited states of a hydrogen atom is the introduction of a quantum defect which accounts for the non-Coulombic effective potential as the excited electron penetrates the ionic core. In what way do the properties of doubly excited states of a multielectron atom be modified? Second, can one make the next major step to elucidate the correlation of the three electrons for triply excited states of atoms? Both questions can be dealt with directly by studying the doubly and triply excited states of a three-electron system, such as the Li atom.

Similar to doubly excited states in a two-electron atom, the nature of correlation of doubly excited states and triply excited states in a three-electron atom is implicitly contained in the wavefunctions of these states. Indeed, it is possible to construct accurate wavefunctions for three-electron atoms starting from the CI approach or similar methods.[43–46] However, a three-electron wavefunction cannot be easily visualized. Recall that the density distribution of a two-electron wavefunction can be reduced to three variables if the overall rotation of the atom is averaged out. If we choose to display the surface density at a fixed hyperradius, then the density distribution is reduced to a function of two angles α and θ_{12}. For a three-electron atom, there are many new complications. First the spin wavefunction and the spatial wavefunction do not separate in general and then each spatial wavefunction is a function of nine variables. By averaging over the overall rotation and summing over the spin components, the density function at each fixed hyperradius is still a function of five variables. Clearly such a function cannot be easily displayed such that the correlation among the electrons can be visualized. We comment that the standard procedure of calculating the two-body or even three-body correlation functions is not very useful unless one happens to have the 'right' variables for displaying their essential features. Thus the search for understanding electron correlation amounts to finding the ultimate methods of visualizing the multidimensional wavefunctions. Different variables and different procedures have to be used to display the correlation of doubly and triply excited states. This is clear since for doubly excited states the correlation is between the two excited electrons, while for triply excited states the correlation among the three electrons has to be treated altogether.

We will analyze the correlation in a three-electron wavefunction using the hyperspherical coordinates within the adiabatic approximation instead of the pedagogical approach adopted for the two-electron atoms. The spectral density for a three-electron atom is very high and very few reliable methods are available for generating both the doubly and triply excited states. Thus we will first discuss the computational procedure using the hyperspherical approach[20] and then focus on the analysis of the wavefunctions to obtain information on electron correlation.[21,22]

3.2. *The Hyperspherical Approach*

The Schrödinger equation for a three-electron atom in the independent particle coordinates is given by

$$\left[\sum_{i=1}^{3}\left(-\frac{1}{2}\nabla_i^2-\frac{Z}{r_i}\right)+\sum_{i<j}\frac{1}{|\hat{\mathbf{r}}_i-\hat{\mathbf{r}}_j|}-E\right]\Psi(\hat{\mathbf{r}}_1,\hat{\mathbf{r}}_2,\hat{\mathbf{r}}_3)=0\,, \tag{17}$$

where Z is the nuclear charge and E is the total energy measured from the triple ionization threshold. The hyperspherical method replaces the radial distances r_1, r_2, and r_3 of the three electrons from the nucleus by the hyperradius R and two angles α_1, α_2 defined by,

$$\begin{aligned} r_1 &= R\sin\alpha_2\cos\alpha_1\,,\\ r_2 &= R\sin\alpha_2\sin\alpha_1\,,\\ r_3 &= R\cos\alpha_2\,, \end{aligned} \tag{18}$$

where R stands for the size of the system, and α_1 and α_2 measure the relative distances of the electrons from the nucleus — they are the variables for measuring radial correlations. In the calculation of the wavefunctions, we use the spherical angles $\hat{\mathbf{r}}_i = (\theta_i, \phi_i)$, $(i = 1, 2, 3)$ of each electron in the space-fixed frame. Thus the nine dimensional spatial coordinates of the three electrons in the space-fixed frame are represented by the hyperradius R and eight hyperangles $\{\alpha_1, \alpha_2, \hat{\mathbf{r}}_1, \hat{\mathbf{r}}_2, \hat{\mathbf{r}}_3,\}$, where we shall use Ω to denote all the eight hyperangles collectively. In hyperspherical coordinates the Schrödinger equation for the rescaled wavefunction $\psi = Rr_1r_2r_3\Psi$ is

$$\left[-\frac{1}{2}\frac{\partial^2}{\partial R^2}+H_{ad}(\Omega;R)-E\right]\psi(R,\Omega)=0\,, \tag{19}$$

where $H_{ad}(\Omega; R)$ is the adiabatic Hamiltonian defined at fixed hyperradius.[20]

Within the adiabatic approximation[29] the total wavefunction for the n-th state in channel μ can be written as

$$\psi(R,\Omega)=F_\mu^n(R)\left(\sum_{S_{12}}\Phi_{\mu LM}^{S,S_{12}}(\Omega,R)\chi_{S_{12}}^S\right)\,, \tag{20}$$

where $F_\mu^n(R)$ is the hyperradial function, the parenthesis containing Φ_μ is the hyperspherical adiabatic channel function, which is the product of the spin function and the spatial function, and the latter provides information about electron correlation for states within channel μ. The spin function $\chi_{S_{12}}^S = [\{\chi(1)\chi(2)\}^{S_{12}}\chi(3)]^S$ which has intermediate spin S_{12}. The spatial function Φ_μ and its associated adiabatic potential $U_\mu(R)$ are obtained by solving the adiabatic eigenvalue problem at each R,

$$[H_{ad}(\Omega; R) - U_\mu(R)]\Phi_\mu(\Omega; R) = 0\,. \tag{21}$$

Equation (21) is a partial differential equation of the eight angles Ω. To solve this equation, we use the eigenfunctions of the adiabatic Hamiltonian which contains only the nucleus–electron attractive potential as the basis set. These eigenfunctions can be written in the separable form,

$$\hat{\Phi}_\mu^\gamma(\Omega; R) = g_\mu^\gamma(\alpha_1, \alpha_2; R)\mathcal{Y}_\gamma^{LM}(\hat{\mathbf{r}}_1, \hat{\mathbf{r}}_2, \hat{\mathbf{r}}_3)\,, \tag{22}$$

where $\gamma = \{l_1 l_2 l_3 l_{12}\}$, and $\mathcal{Y}_\gamma^{LM}(\hat{\mathbf{r}}_1, \hat{\mathbf{r}}_2, \hat{\mathbf{r}}_3) = [\{Y_{l_1}(\hat{\mathbf{r}}_1)Y_{l_2}(\hat{\mathbf{r}}_2)\}^{l_{12}}Y_{l_3}(\hat{\mathbf{r}}_3)]^{LM}$ is the coupled angular momentum of the three electrons. The function $g_\mu^\gamma(\alpha_1, \alpha_2; R)$ satisfies the two-dimensional eigenvalue equation with respect to α_1 and α_2 which is then solved using basis functions consisting of direct products of discrete variable representation (DVR) basis functions.[47] In Eq. (20), the spatial part of the channel function $\Phi_{\mu LM}^{S,S_{12}}$ is coupled to the spin function. In constructing the total wavefunction ψ the Pauli exclusion principle has to be accounted for. To obtain the correct symmetry under exchange of any pair of electrons, we apply the antisymmetrization operator to the basis functions in Eq. (22), and then diagonalize the adiabatic Hamiltonian with these antisymmetrized basis functions.

3.3. *Hyperspherical Adiabatic Potential Curves*

To see the global features of the eigenstates of a three-electron atom, we first examine the adiabatic potentials. In Fig. 7 we show the potential curves $U_\mu(R)$ as functions of R calculated for Li($^2P^o$) as an example. The top figure gives the overall view of the potential curves. At large R, each curve approaches the two-electron (Li^+) limits. The lowest curve goes to the ground state of Li^+ in the asymptotic limit. This curve supports the singly excited states of Li. All the singly excited states which are normally designated as $1s^2np\,^2P^o$ are obtained by solving the hyperradial equation from this curve.

The next group of curves approach the singly excited states of Li^+ asymptotically. In the lower-left frame we show the first few of them, where the first six curves converge to the $1s2s\,^3S^e$, $1s2s\,^1S^e$, $1s2p\,^3P^o$, and $1s2p\,^1P^o$ limits of Li^+ at large R. These curves support doubly excited states which are designated as $1s2lnl'$. The higher adiabatic potentials

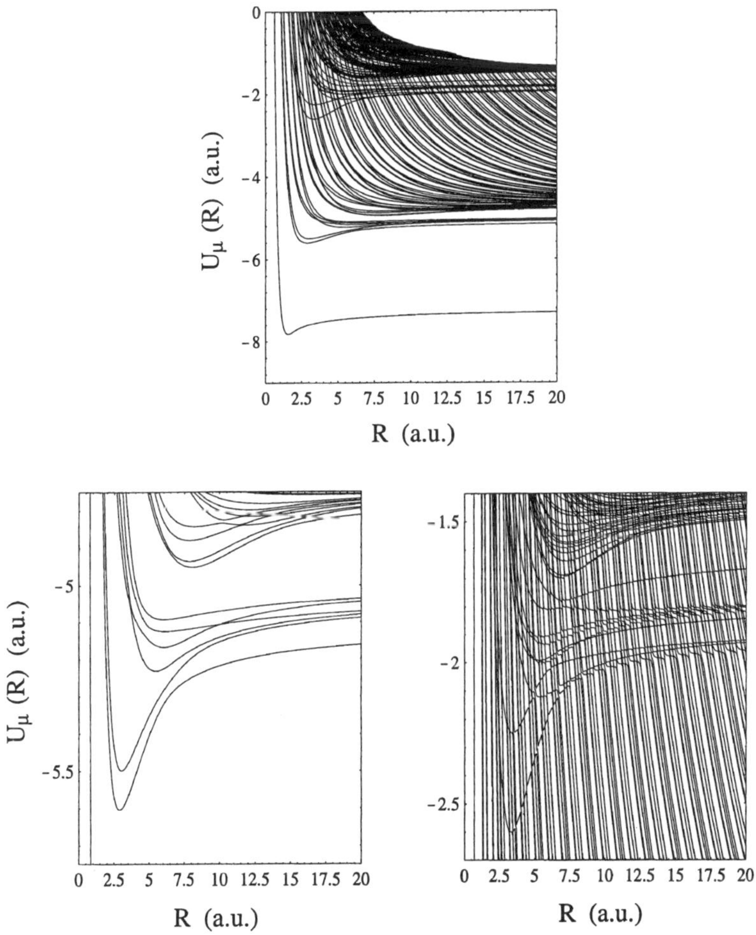

Fig. 7. Adiabatic hyperspherical potentials for the $^2P^o$ symmetry of Li. Top frame: the complete set; lower-left frame: expanded view of the adiabatic potentials that support $1s2lnl'$ doubly excited states and some curves that support $1s3lnl'$ doubly excited states; lower-right frame: expanded view of the adiabatic potentials that support $2l2l'2l''$ triply excited states and some curves that support $2l3l'nl''$ triply excited states.

of this group support doubly excited states of the type $1s3lnl'$, $1s4lnl'$, ..., etc.

On the lower-right corner of Fig. 7 we show the third group of adiabatic potentials. These potentials exhibit numerous sharp avoided crossings with the curves that support doubly excited states. The doubly excited curves drop rapidly with increasing values of R, while the curves in the third group show clear attractive wells in the small R region. These curves approach the doubly excited states of Li^+ asymptotically, and they support triply excited states of Li. In the third group, the lowest nine curves approach the $2l2l'$ doubly excited states of Li^+ such that they support $2l2l'nl''$ triply excited states of Li.

From the discussion above it is clear that the hyperspherical adiabatic potentials can be used to separate singly, doubly, and triply excited states. In fact, for the low-lying members of doubly excited states and triply excited states, the adiabatic potentials allow us to separate the eigenstates into $(1,1,n)$, $(1,2,n)$, $(1,3,n), \ldots$, and $(2,2,n)$, $(2,3,n), \ldots, (3,3,n)$ manifolds. Within each manifold, there are many channels or Rydberg series. The adiabatic potentials clearly illustrate that there are hierarchical ordering in the energy levels — that the states are separated into manifolds and then into different Rydberg series. Our major goal is to identify features that characterize the different manifolds as well as features that distinguish the different channels within the same manifold.

3.4. *Radial Correlations and Separations into Manifolds*

We first define the radial density function,

$$\rho^{\mu}_{rad}(\alpha_1, \alpha_2; R) = \sum_{S_{12}} \int |\Phi^{S,S_{12}}_{\mu LM}(\Omega; R)|^2 d\hat{\mathbf{r}}_1 d\hat{\mathbf{r}}_2 d\hat{\mathbf{r}}_3 \,. \tag{23}$$

This function gives the radial distribution of the three electrons at each hyperradius R. We first show that the manifolds are distinguished by distributions of the radial density function $\rho^{\mu}_{rad}(\alpha_1, \alpha_2; R)$. In Fig. 8(a) we divide the (α_1, α_2) plane into six domains, or sextants, which are labeled by (I)–(VI) and separated by the dotted lines where the radial distances of two of the electrons are identical [indicated explicitly in Figs. 8(b)–(d)]. The point where the three dotted lines intersect is for $r_1 = r_2 = r_3$. Since the radial density function is symmetric under the exchange of any two electrons, the density function distribution within each domain is equivalent to that for the other five.

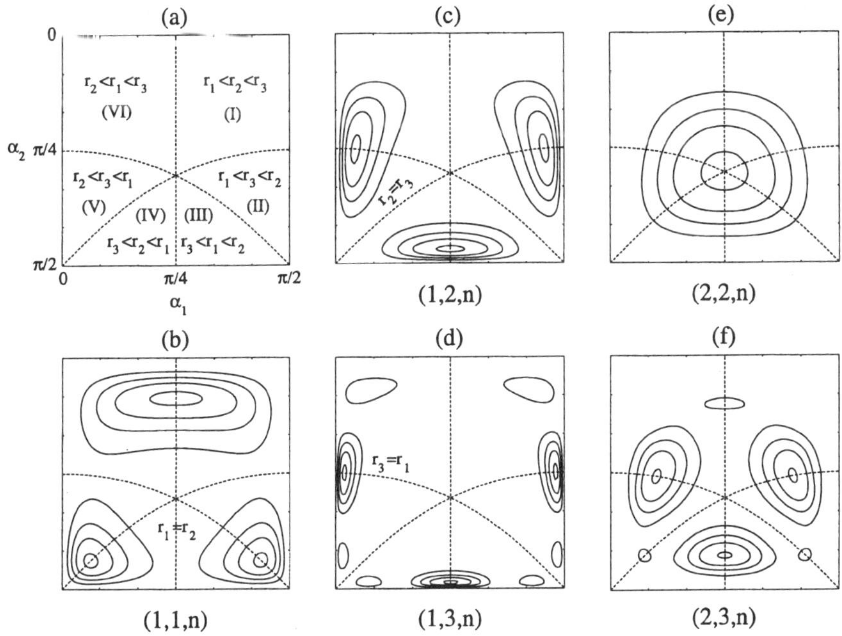

Fig. 8. Radial density functions $\rho^{\mu}_{rad}(\alpha_1, \alpha_2; R)$ on the (α_1, α_2) plane at fixed R. (a) the six equivalent domains where the relative magnitudes of the radial distances of the electrons are indicated. The dashed lines indicate where the radial distances of two of the electrons are equal, see (b)–(d). The point where the three dashed lines cross is for $r_1 = r_2 = r_3$. The radial densities $\rho_{rad}(\alpha_1, \alpha_2; R)$ are shown; (b) for the $1s^2nl$ singly excited states at $R = 2$ a.u. (c) for the $1s2lnl'$ doubly excited states at $R = 3.5$ a.u. (d) for the $1s3lnl'$ doubly excited states at $R = 8$ a.u. (e) for the $2l2l'nl''$ triply excited states at $R = 3.5$ a.u. and (f) for the $2l3l'nl''$ triply excited states at $R = 7$ a.u.

In Fig. 8(b) we first show the density plot for the $1s^2nl$ singly excited states at $R = 2$ a.u. where the potential (the lowest curve of Fig. 7) is near the minimum. We will focus on sextant (I) or (VI), where r_3 is larger than both r_1 and r_2. The peak of this distribution is at $(\alpha_1, \alpha_2) \approx (\pi/4, \pi/12)$. Since $R = 2$ a.u., this gives $r_1 = r_2 \approx 0.4$ a.u. and $r_3 \approx 1.9$ a.u., which gives roughly the radial distances of the three electrons in the $1s^22p$ state.

We can proceed to understand the other figures. In Fig. 8(c), it is more convenient to examine sextant (III) or (IV), where r_3 is the smallest. The peak occurs at $(\alpha_1, \alpha_2) \approx (\pi/4, 17\pi/36)$. Since $R = 3.5$ a.u., this gives $r_3 \approx 0.3$ a.u. and $r_1 = r_2 \approx 2.5$ a.u., thus representing a $1s2l2l'$ doubly excited state. In Fig. 8(d), the peaks in sextant (III) or (IV) indicate that the maximum of the density is at $r_3 \approx 0.3$ a.u., $r_1 = r_2 \approx 5.7$ a.u. Therefore this represents the radial distribution of a $1s3l3l'$ doubly excited state. The same procedure can be used to understand that Fig. 8(e) represents $2l2l'2l''$ triply excited states where the density peaks at $r_1 = r_2 = r_3 \approx 1.9$ a.u. and that Fig. 8(f) represents a $2l3l'3l''$ triply excited state, see sextant (I) or (VI) where the peak occurs at $r_1 = r_2 \approx 4.8$ a.u. and $r_3 \approx 1.8$ a.u.

The results in Fig. 8 clearly establish that the different manifolds of singly, doubly, and triply excited states can be distinguished by the radial distributions of the wavefunctions. Thus to characterize the radial correlations of the three electrons, it is possible to use the manifold designations.

3.5. *Correlations in Doubly Excited States of Li*

We next address what distinguish the different channels in a given manifold. Although there are some differences in the radial density function $\rho^{\mu}_{rad}(\alpha_1, \alpha_2; R)$ among the channels within the same manifold on the (α_1, α_2) plane, the major differences are in the relative angles among the electrons. In this section, we focus on the $1s2lnl'$ doubly excited states of Li.

In Fig. 9(a) we show the six ${}^2P^o$ adiabatic potential curves that support the $1s2lnl'$ doubly excited states in more details. The six Rydberg series can be designated as $1s2s({}^3S^e)np$, $1s2s({}^1S^e)np$, $1s2p({}^3P^o)ns$, $1s2p({}^3P^o)nd$, $1s2p({}^1P^o)ns$, and $1s2p({}^1P^o)nd$ series in the independent electron approximation. The designation above is inadequate since the mixing of the series is expected to be very strong. In fact, we can compare the six adiabatic potentials above with the three ${}^1P^o$ and three ${}^3P^o$ adiabatic potentials of He, shown in Fig. 9(b), where each curve can be labeled by the $(K,T)^A$ quantum numbers.[12,14] Clearly the adiabatic potentials in the two systems are very similar in the small R region where the two outer electrons are strongly correlated, and approximate quantum numbers used to describe the doubly excited states of He can be used to label the doubly excited states of Li. However, for the higher states of each Rydberg series, the outermost electron decouples from the two inner ones which form the $1s2s\,{}^3S^e$, $1s2s\,{}^1S^e$, $1s2p\,{}^3P^o$ or $1s2p\,{}^1P^o$ series limits of Li^+. For the He atom, the asymptotic limit is either the $2s$ or the $2p$ states of He^+ which are degenerate. Thus the adiabatic potentials in the large R region in Figs. 9(a) and (b) are different.

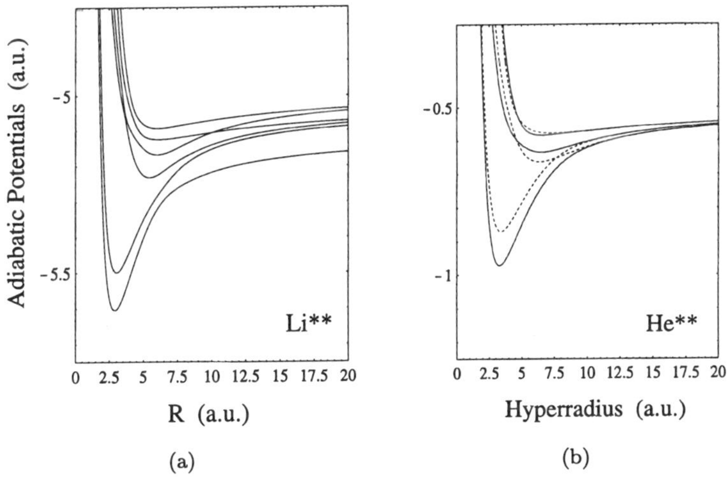

Fig. 9. Comparison of (a) the $1s2lnl'\,{}^2P^o$ adiabatic potentials of Li with (b) the $2lnl'\,{}^3P^o$ (dashed lines) and $2lnl'\,{}^1P^o$ (solid lines) doubly excited potential curves of He. Note the similarities of the curves in the small R region. The three ${}^3P^o$ curves for He are labeled as $(1,0)^+$, $(0,1)^-$, $(-1,0)^0$ starting from below, while the three ${}^1P^o$ curves are labeled as $(0,1)^+$, $(1,0)^-$, and $(-1,0)^0$ starting from below. The similarity of the adiabatic potentials suggests that the $1s2lnl'$ doubly excited states of Li can be labeled in terms of the quantum numbers of the $2lnl'$ doubly excited states of He.

To show the correlation of doubly excited states in a three-electron atom, it is more important to focus on the two excited electrons. This is done by fixing $r_3(r_3 \ll r_1, r_2)$ [or fixing $\alpha_2 \sim 0$, see Eq. (18)] and integrating over $\hat{r}_3$. We define the two-electron density function for the two excited electrons on the (α_1, θ_{12}) plane by

$$\rho_{2e}(\alpha_1, \theta_{12}; \alpha_2, R) = \sum_{S_{12}} \int |\Phi_\mu^{S,S_{12}}(\Omega; R)|^2 d\hat{\mathbf{r}}_3 d\omega\,, \tag{24}$$

where θ_{12} is the angle between $\mathbf{r}_1$ and $\mathbf{r}_2$, and $d\omega$ is the volume element for the three Euler angles used to describe the overall rotation of the whole atom. The resulting two-electron density plots for the six $1s2lnl'$ doubly excited series of Li are shown in Fig. 10. They are essentially the same as the correlation patterns of doubly excited states of a two-electron atom like He. For example, the three frames that have been labeled as analogous to the ${}^1P^o$ channels can be compared directly to the three frames in Fig. 2. The correlation properties of doubly excited states of Li are shown to be

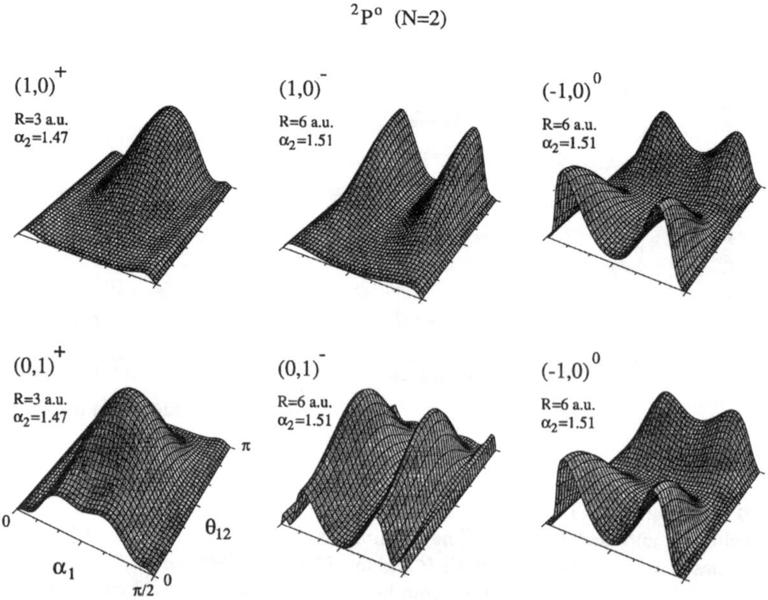

Fig. 10. The rotation-averaged density of the two excited electrons $\rho_{2e}(\alpha_1, \theta_{12}; \alpha_2, R)$ on the (α_1, θ_{12}) plane for the six $1s2lnl'\,^2P^o$ channels of Li. Each density is evaluated at the values of R and α_2 shown and labeled by the $(K,T)^A$ quantum numbers used previously for the doubly excited states of two-electron atoms.

similar to those in the doubly excited states in He and can be labeled by the same set of K, T, and A quantum numbers.

4. Correlation in Triply Excited States of Li

4.1. *Body-Fixed Frame Analysis of the Channel Wavefunctions*

Since correlation is a property of the relative motion among the electrons, the overall rotation of the system does not play a role. Thus, we analyze the wavefunctions in the body-fixed frame of the atom.

The channel function $\Phi^{S,S_{12}}_{\mu LM}(\Omega; R)$ in the space-fixed frame can be expressed in terms of channel functions $\varphi^{S,S_{12}}_{\mu LQ}(\Omega_I; R)$ in the body-fixed frame by a general rotation

$$\Phi^{S,S_{12}}_{\mu LM}(\Omega; R) = \sum_{Q=-L}^{L} \varphi^{S,S_{12}}_{\mu LQ}(\Omega_I; R) D^{(L)}_{QM}(\omega) , \tag{25}$$

where $D^{(L)}_{QM}(\omega)$ is the rotation matrix for the frame transformation, and Q and M are the azimuthal components of L in the body-fixed frame and in the space-fixed frame, respectively. Here we use ω for the three Euler angles which represent the orientation of the atom, and Ω_I for the five internal angles which represent the 'shape' of the system. There are many different ways to choose the internal coordinates. Mathematical details on the internal coordinates for the general four-body system has been studied by Littlejohn and Reinsch.[48] In the present analysis of the triply excited states of a three-electron atom, we use α_1, α_2, and three relative angles among the electrons. Let us define the three relative angles θ, η, and ϕ.[17,21,22] For $r_1 = r_2 = r_3$, the three electrons lie on the surface of a sphere. We define a σ plane formed by the three electrons. This plane makes an angle θ with respect to the nucleus, see Fig. 11(a). On the σ plane, the three electrons lie on a circle. The angle between electrons 1 and 2 is defined to be 2η, choosing along the arc containing electron 3. The angle between electron 3 and the line bisecting electrons 1 and 2 is defined to be ϕ, see Fig. 11(a). The range of the angles are $(0 \le \theta \le \pi, 0 \le \eta \le \pi, -\eta \le \phi \le \eta)$. These three angles specify a definite shape of the triangle and the position of the triangle with respect to the nucleus.

To visualize the collective motion of the three electrons, we introduce the three-electron density function $\rho^{\mu}_{3e}(\Omega_I; R)$ which is defined as the rotation-averaged density distribution for each channel function,

$$\rho^{\mu}_{3e}(\Omega_I; R) = \sum_{S_{12}} \int |\Phi^{S,S_{12}}_{\mu LM}(\Omega; R)|^2 d\omega \,. \tag{26}$$

This density represents the probability for the three electrons to take specific shapes. From Fig. 8(e) it has been shown that each intrashell state wavefunction peaks at $r_1 = r_2 = r_3$ at the value of R where the potential is near the minimum. Thus the correlation of the three electrons at $r_1 = r_2 = r_3$ is expected to represent the dominant correlation properties of intrashell triply excited states. In Fig. 11(c) we show the density for an intrashell state of Li at $r_1 = r_2 = r_3$ as an example.

To display the global feature of a function in three dimensions we use the 'equidensity surfaces' which is a surface of constant electronic density. The plots in Figs. 11(b) and (c) represent the contour surfaces where the density is 60% of the maximum. A contour surface of higher density would fit inside the surface. Such contour surfaces would provide information

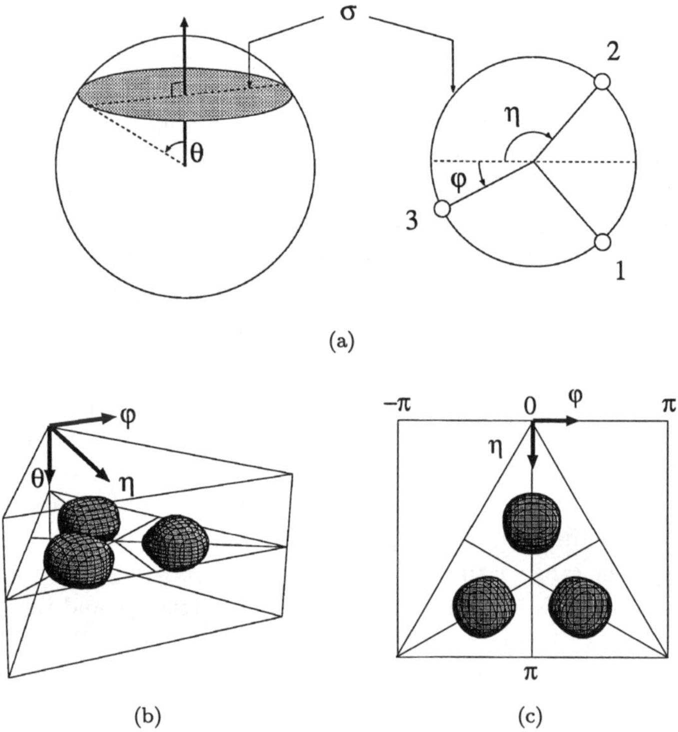

Fig. 11. (a) Definition of the three angles used to describe the three electrons on a sphere. The three electrons form a σ plane. On the plane (the right figure) the three electrons are confined to a circle. (b) Side view and (c) top view of the equidensity surface plots of the three-electron density functions $\rho^{\mu}_{3e}(\Omega_I; R)$ for a typical intrashell triply excited states of Li at $r_1 = r_2 = r_3$. The surface represents 60% of the maximum density.

on the most probable shape of the three electrons in each intrashell triply excited state.

4.2. *Rotational Excitations and Energies of $2l2l'2l''$ States of Li*

There are eight intrashell $2l2l'2l''$ triply excited states for Li. In Fig. 12 we show the equidensity surfaces of each of these states. These surfaces can be grouped into three types. In type I, we found that there are four states which have similar equidensity surfaces. They are the $^2P^o$, $^4P^e$, $^2D^e$, and $^2D^o$ states. Each surface is a contour where the density is 60% of

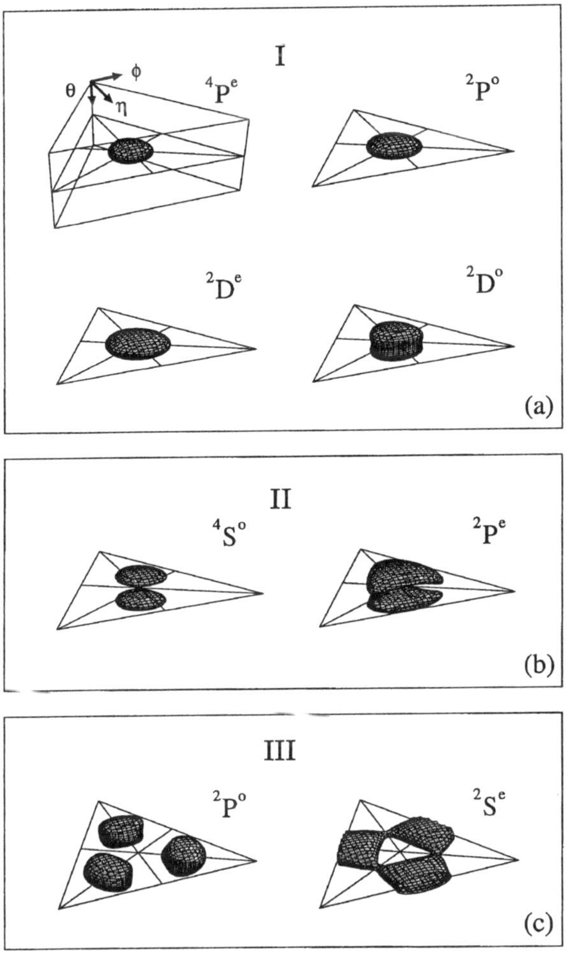

Fig. 12. The equidensity surface plots of the three-electron wavefunctions for the eight intrashell states $r_1 = r_2 = r_3$. The surface represents 60% of the maximum density. Each 'slice' represents the whole range of the three angles $(0 \leq \theta \leq \pi, 0 \leq \eta \leq \pi, -\eta \leq \phi \leq \eta)$. See text for detailed discussion of these plots.

the maximum. The similarity of the contour surfaces for these four states shows that they have nearly identical internal shape. In fact, the maximum of the density for each figure occurs at $\theta = \pi/2$, $\eta = 2\pi/3$, and $\phi = 0$; that is, the most favorable geometry is a coplanar equilateral triangle with the nucleus at the center of the triangle. This geometry is similar to that of the planar BF_3 molecule. Since these states basically have the same shape, their relative energies are expected to be similar to the rotational excitation of an oblate symmetric top.[16,17] For an oblate symmetric top, the rotation energy is

$$E(L,T) = \frac{1}{2I}\left[2L(L+1) - T^2\right], \tag{27}$$

where I is the moment of inertia and T is the projection of L along the direction perpendicular to the plane.

To show that this is indeed the case, in Fig. 13(a) we display the energy levels of these eight states, arranged according to their configurations. In

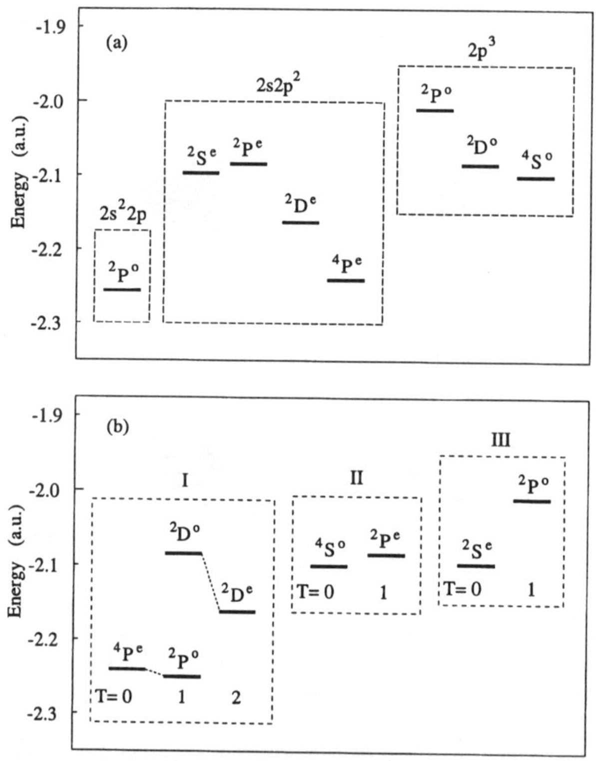

Fig. 13. The energy levels of the eight $2l2l'2l''$ intrashell triply excited states of Li. (a) the states are arranged in terms of the electronic configurations according to the independent electron model. (b) The states are rearranged according to their internal wavefunctions. The states within each group (I, II, III) have similar shapes and their energies are similar to the rotational excitation of a symmetric top.

the relative energy ordering, the Hund's rule is satisfied except between $^2S^e$ and $^2P^e$ states. In Fig. 13(b) we regroup these eight states according to their internal density distributions. The $^2P^o$, $^4P^e$, $^2D^e$, and $^2D^o$ states have the most probably shape of a coplanar equilateral triangle and their relative energies indeed appear to be similar to that of the rotational excitation of an oblate symmetric top.

In Fig. 12(b) we show the equidensity surfaces for the $^4S^o$ and $^2P^e$ states. The $\theta = \pi/2$ plane is clearly forbidden to the three electrons, indicating that the coplanar geometry is not allowed for these two states. Above or below this plane, the three electrons still prefer to form an equilateral triangle. This geometry is similar to the antisymmetric state of the pyramidal NH_3 molecule where the wavefunction vanishes when the plane of the three hydrogen atoms coincides with the nitrogen atom. The existence of a nodal surface when the σ plane intersects the nucleus implies higher excitation energy since minimum electrostatic repulsion occurs when the three electrons and the nucleus are coplanar. Note that in Fig. 12(b) the energy levels of group II are higher than the levels in group I except for the highest $^2D^o$ state which has higher rotational energy.

The remaining two states are shown in Fig. 12(c). Both states have the maximum density at the $\theta = \pi/2$ plane, i.e. the three electrons are coplanar. However, the density vanishes in the middle, indicating that the three electrons cannot form an equilateral triangle. We are not aware of any examples of low-lying vibrational states of any polyatomic molecules exhibiting such modes.

4.3. *Rotational Decomposition of the Body-Fixed Frame Wavefunctions*

Next we consider the rotational decomposition of the wavefunctions of triply excited states. In Eq. (25), the channel function of each Q-component $\varphi^{S,S_{12}}_{\mu LQ}$ depends on the choice of the body-fixed frame axes. Following Watanabe and Lin[16] we define our body-fixed frame by

$$\begin{aligned} \mathbf{S}_z &= \mathbf{r}_1 \times \mathbf{r}_2 + \mathbf{r}_2 \times \mathbf{r}_3 + \mathbf{r}_3 \times \mathbf{r}_1 \,, \\ \mathbf{S}_y &= \frac{\sqrt{3}}{2}(\mathbf{r}_1 - \mathbf{r}_2) \,, \\ \mathbf{S}_x &= \mathbf{S}_y \times \mathbf{S}_z \,. \end{aligned} \tag{28}$$

This defines the z-axis on the body-fixed frame to be perpendicular to the plane spanned by the three electrons. In the body-fixed frame, the channel functions satisfy the following relation,[16]

$$\varphi_{\mu L-Q}^{S,S_{12}}(\Omega_I;R)=\pi(-1)^{L+Q}[\varphi_{\mu LQ}^{S,S_{12}}(\Omega_I;R)]^*\,, \tag{29}$$

where $\pi=\pm1$ is the parity of the system. Then we use $T=|Q|(0\leq T\leq L)$ for analysis in what follows. Quantum symmetries often impose boundary conditions on each rotational component wavefunction. More specifically, each quantum state with well-defined rotational symmetry, well-defined parity, as well as Pauli exchange symmetry between each pair of electrons has nodal surfaces on the internal wavefunctions. For the body-fixed frame defined by Eq. (29), Watanabe and Lin,[16] and Bao *et al.*[17] have shown the following conditions that the internal wavefunctions vanish because of their quantum symmetries:

(1) For

$$\pi(-1)^T=-1\,, \tag{30}$$

the wavefunction vanishes when the three electrons and the nucleus are coplanar ($\theta=\pi/2$).

(2) For

$$\begin{aligned} T&=1,2\,, \quad mod\ 3 \quad (S=3/2)\,,\\ T&=0\,, \quad mod\ 3 \quad (S=1/2)\,, \end{aligned} \tag{31}$$

the wavefunction has a nodal surface when the three electrons form an equilateral triangle ($\eta=2\pi/3$, $\phi=0$) with $r_1=r_2=r_3$.

(3) For

$$\pi(-1)^L=1 \quad (T=0,\ S=3/2)\,, \tag{32}$$

the wavefunction has a nodal surface when the three electrons form an isosceles triangle ($\phi=0$) with $r_1=r_2=r_3$.

These nodal surfaces will be seen in the density function if there is only one rotational component represented by

$$\begin{aligned} \rho_{3e}^{\mu,T}(\Omega_I;R)&=\sum_{S_{12}}|\varphi_{\mu LT}^{S,S_{12}}(\Omega_I;R)|^{12} \quad (T=0)\,,\\ \rho_{3e}^{\mu,T}(\Omega_I;R)&=\sum_{S_{12}}(|\varphi_{\mu L-T}^{S,S_{12}}(\Omega_I;R)|^{12}+|\varphi_{\mu LT}^{S,S_{12}}(\Omega_I;R)|^2) \quad (T\neq0)\,. \end{aligned} \tag{33}$$

We note that there are no examples for condition 3 above for the $2l2l'2l''$ states.

For a rigid body T is a good quantum number. One of the possible measures of how 'rigid' the three electrons are moving together can be obtained by determining whether T is nearly a good quantum number for the given state. We define the partial normalization constant of each rotational component by integrating over all the five internal coordinates,

$$N_T(R) = \frac{1}{8\pi^2} \int \rho_{3e}^{\mu T}(\Omega_I; R)\, d\Omega_I \,, \tag{34}$$

where

$$\sum_{T=0}^{L} N_T(R) = 1 \,, \tag{35}$$

at each R. For the calculation of the above integral, we take the five coordinates to be $(\alpha_1, \alpha_2, \cos\theta_{12} = \hat{\mathbf{r}}_1 \cdot \hat{\mathbf{r}}_2, \cos\theta_{31} = \hat{\mathbf{r}}_3 \cdot \hat{\mathbf{r}}_1, \phi_{23} = \phi_3 - \phi_2)$, which are different from the variables for the density plots. In this coordinate system, $d\Omega_I = d\cos\alpha_1 d\alpha_2 d\cos\theta_{12} d\cos\theta_{31} d\phi_{23}$ and the $5D$ integration is performed numerically. There is no *a priori* reason to expect that the three electrons to move like a rigid body. However, if there is a T-component which is nearly pure, meaning that $N_T(R)$ is nearly equal to one, then there is a possibility that the motion of the three electrons will be close to that of a rigid body.

4.4. *Demonstration of Rotor Structure of Intrashell Triply Excited States*

In Fig. 14 we show the rotational decomposition into the T-components for each state. Each plot in Fig. 14 represents the contour surface where the density is 60% of the maximum at fixed $r_1 = r_2 = r_3 = 2.02$ a.u. ($R = 3.5$ a.u.). The fraction of each T-component with respect to the total density for each state is also indicated in the figure. Let us consider the first four states, $^4P^e$, $^2P^o$, $^2D^o$, and $^2D^e$. The total density $\rho_{3e}^{\mu}(\Omega_I; R)$ for each state, as shown on the rightmost column of Fig. 14, which are the same as in Fig. 12, clearly indicates that they are essentially identical. Each state has a single dominant T-component, and this T-component has the coplanar equilateral triangle geometry for the three electrons. The existence of a single dominant T-component and that this component has nearly identical shape underlies the basis of the classification of these states as similar to

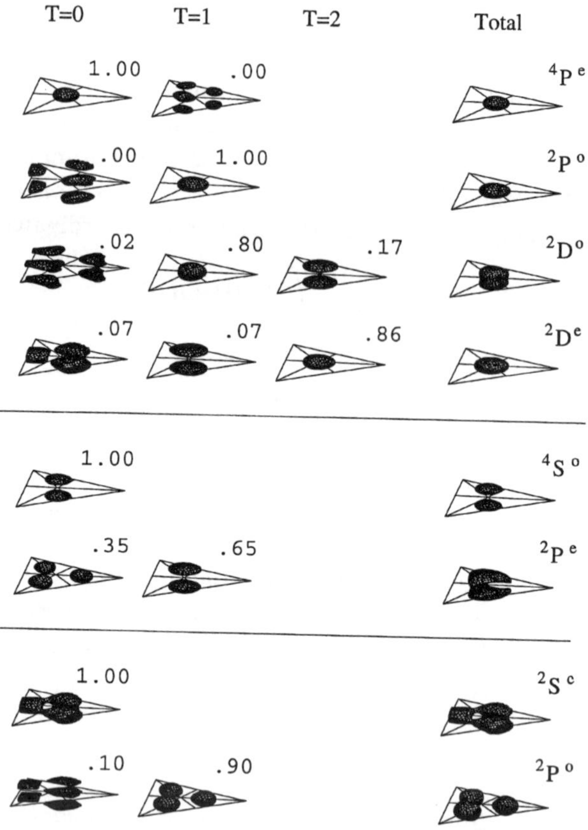

Fig. 14. The equidensity surface plots of the three-electron density functions $\rho^{\mu}_{3e}(\Omega_I; R)$ for the eight intrashell $2l2l'2l''$ triply excited states of Li at $r_1 = r_2 = r_3 = 3.5$ a.u. The total density is shown on the right-most column while the density for each rotational T-component is shown to the left. The fraction of each T-component is indicated to denote the purity of that component for a given state. The surface represents 60% of the maximum density.

the rotational structure of an oblate symmetric top. From Fig. 14, it is clear that the purity is very good for the low-lying $^4P^e$ and $^2P^o$ states. For the highest $^2D^o$ state the fraction of the nondominant T-components is not small, and the total density for the $^2D^o$ state is somewhat 'irregular'. From the molecular physics viewpoint, this large mixture is interpreted as due to the rotation-vibrational coupling which becomes important for states with higher angular momentum.

The next group consists of the $^4S^o$ state and the $^2P^e$ state. The $^4S^o$ state has the $T = 0$ component only and it has a nodal plane at $\theta = \pi/2$, i.e. this state satisfy Eq. (30). For the $^2P^e$ state, the $T = 1$ component has a node at $\theta = \pi/2$ but the $T = 0$ component has a node at the coplanar equilateral triangle geometry [see Eq. (31)]. Since each component has a nodal surface, the two T-components have nearly identical composition and thus the purity of this state is not good, as shown by the fraction of each rotational component and the 'irregular shape' in the total density.

The last group consists of the $^2S^e$ state and the second $^2P^o$ state. The $^2S^e$ state has one nodal surface at the coplanar equilateral triangle geometry. This is also true for the $T = 1$ component of the $^2P^o$ state. We note that the surface of the second $^2P^o$ state is not due to the symmetry of the state but the excitation of the angular mode.[17] The $^2P^o(T = 0)$ component contains two nodal surfaces due to symmetry. Thus in comparison the $T = 0$ component is less easily excited and it contributes only about 10% to the total density.

4.5. *Higher Triply Excited States*

We have analyzed the correlation of $2l2l'2l''$ triply excited states. The three types of correlation discussed are due to constraints imposed by the quantum symmetry on the rotational components of the body-fixed frame wavefunctions. We emphasize that these constraints are due to the symmetry of the system and have nothing to do with the interaction potentials. On top of these basic modes, combinations of these modes as well as excitation of these basic modes can be expected for the higher intrashell states. Considering a model of intrashell states where the three electrons are confined to move on the surface of a sphere, the wavefunction on the body-fixed can be described by the three angles θ, η, and ϕ as defined in Fig. 11(a). These higher excited modes are related to the bending vibrational quantum numbers associated with the wavefunction in each of these angular coordinates. Some features of these higher modes are already evident in the smaller Q-components for the $2l2l'2l''$ triply excited states, as shown in Fig. 14.

For higher intrashell states such as $3l3l'3l''$ states, the lower members of the states in the manifold are expected to fall into the same three groups as in Fig. 13(b) except that the number of states within each group will be larger. The higher members in the manifold will have more nodal surfaces in their rotational component wavefunctions, some of these nodal surfaces are due to the dynamic excitations (in contrast to nodal surfaces imposed by the quantum symmetry). However, the analysis of these higher states turns out to be quite complicated.[17] On the one hand, the wavefunctions of these states often do not have a single dominant Q-component such that classification in terms of a symmetric top is questionable. On the other hand, the number of intrashell $nlnl'nl''$ states grows rapidly with n, and it would be too tedious to analyze the individual Q-components of these wavefunctions.

Similar difficulties are encountered for attempt to visualize intershell triply excited states such as $2l2l'nl''\,(n > 2)$ and $2l3l'nl''\,(n \geq 3)$ states. For them, it appears that correlation has to be analyzed in different regions. The complexities and the large number of states within each manifold seem to indicate that it would be a daunting task to entangle correlation in these states. On the other hand, unless some handle can be made on these states, a complete classification of triply excited states will still be a distant goal.

5. Summary and Conclusions

In this article we show how to visualize the wavefunctions of doubly and triply excited states of atoms. By expressing the wavefunctions on the body-fixed frame and using hyperspherical coordinates, we have shown that the correlation between the two electrons in doubly excited states can be easily displayed. The visualization of the two-electron densities allows us to identify the different modes of the joint motion of the two electrons and concludes that doubly excited states resemble the rotation and vibration excitations of a linear triatomic molecule. A new classification scheme for doubly excited states has emerged and new spectral regularity has been found. For the triply excited states, we have shown that the intrashell states can be understood in terms of the basic modes of the motion of an XY_3 triatomic molecule. By focusing only on the dominant part of the wavefunction, we have been able to visualize the different modes of the joint motion of the three electrons. On the other hand, the higher members of intrashell triply excited states as well as intershell triply excited states remain too complicated and have not been analyzed.

Acknowledgments

This work is supported in part by the US Department of Energy, Office of Basic Energy Research, Division of Chemical Sciences.

References

1. R. P. Madden and K. Codling, *Phys. Rev. Lett.* **10**, 516 (1963).
2. R. P. Madden and K. Codling, *Astrophy. J.* **141**, 364 (1965).
3. D. W. Lindle, T. A. Ferrett, P. A. Heimann and D. A. Shirley, *Phys. Rev.* **A36**, 2112 (1987).
4. M. Domke, C. Xue, A. Puschmann, T. Mandel, E. Hudson and D. A. Shirley, *Phys. Rev.* **A36**, 2112 (1987).
5. M. Domke, G. Remmers and G. Kaindl, *Phys. Rev. Lett.* **69**, 1171 (1992).
6. K. Schulz, G. Kaindl, M. Domke, J. D. Bozek, P. A. Heimann, A. S. Schlachter and J. M. Rost, *Phys. Rev. Lett.* **77**, 3086 (1996).
7. C. D. Lin, in *Review of Fundamental Processes and Applications of Atoms and Ions*, Ed. C. D. Lin (World Scientific, 1993, Singapore), p. 357.
8. C. D. Lin, *Adv. Atom. Mol. Phys.* **22**, 77 (1986).
9. J. M. Rost, K. Schulz, M. Domke and G. Kaindl, *J. Phys.* **B30**, 4663 (1997).
10. D. R. Herrick, *Adv. Chem. Phys.* **52**, 1 (1983).
11. J. W. Cooper, U. Fano and F. Prats, *Phys. Rev. Lett.* **10**, 518 (1963).
12. D. R. Herrick and O. Sinanoğlu, *Phys. Rev.* **A11**, 97 (1975).
13. M. E. Kellman and D. R. Herrick, *Phys. Rev.* **A22**, 1536 (1980).
14. C. D. Lin, *Phys. Rev.* **A29**, 1019 (1984).
15. J. M. Rost and J. Briggs, *J. Phys.* **B24**, 4293 (1991).
16. S. Watanabe and C. D. Lin, *Phys. Rev.* **A36**, 511 (1987).
17. C. G. Bao, X. Z. Yang and C. D. Lin, *Phys. Rev.* **A55**, 4168 (1997).
18. X. Yang, C. G. Bao and C. D. Lin, *Phys. Rev. Lett.* **76**, 3096 (1996).
19. T. Morishita, O. I. Tolstikhin, S. Watanabe and M. Matsuzawa, *Phys. Rev.* **A56**, 3559 (1997).
20. T. Morishita and C. D. Lin, *Phys. Rev.* **A57**, 4268 (1998).
21. T. Morishita, Y. Li and C. D. Lin, *Phys. Rev.* **A58**, 4214 (1998).
22. T. Morishita and C. D. Lin, *Phys. Rev.* **A59**, 1835 (1999).
23. J. E. Hunter and R. S. Berry, *Phys. Rev.* **A36** 3042 (1987).
24. D. R. Herrick, M. E. Kellman and R. D. Poliak, *Phys. Rev.* **A22**, 1517 (1980).
25. S. Watanabe and C. D. Lin, *Phys. Rev.* **A34**, 823 (1986).
26. Z. Chen, C. G. Bao and C. D. Lin, *J. Phys.* **B25**, 61 (1991).
27. K. T. Chung and C. D. Lin, *At. Data Nucl. Data Tables* **69**, 101 (1998).
28. Z. Chen and C. D. Lin, *Phys. Rev.* **A40**, 6712 (1989).
29. J. Macek, *J. Phys.* **B1**, 831 (1968).
30. J. Z. Tang, S. Watanabe and M. Matsuzawa, *Phys. Rev.* **A46**, 2437 (1992).
31. J. Z. Tang, S. Watanabe, M. Matsuzawa and C. D. Lin, *Phys. Rev. Lett.* **69**, 1633 (1992).
32. J. Z. Tang and I. Shimamura, *Phys. Rev.* **A50**, 1321 (1998).
33. O. I. Tolstikhin, S. Watanabe and M. Matsuzawa, *J. Phys.* **B29**, L389 (1996).
34. D. Kato and S. Watanabe, *Phys. Rev.* **A56**, 3687 (1997).
35. M. K. Chen, C. D. Lin and J. Z. Tang, *Phys. Rev.* **A56**, 2435 (1997).
36. O. I. Tolstikhin, S. Watanabe and M. Matsuzawa, *Phys. Rev. Lett.* **74**, 3573 (1995).

37. Y. Zhou and C. D. Lin, *Phys. Rev. Lett.* **75**, 2296 (1995).
38. Y. Zhou and C. D. Lin, *J. Phys.* **B27**, 5065 (1994).
39. C. D. Lin, *Phys. Rept.* **257**, 1 (1995).
40. J. Yuan, B. D. Esry, T. Morishita and C. D. Lin, *Phys. Rev.* **A58**, R4 (1998).
41. B. D. Esry, C. D. Lin and C. H. Greene, *Phys. Rev.* **A54**, 394 (1996).
42. J. Yuan and C. D. Lin, *J. Phys.* **B31**, L637 (1998).
43. N. A. Piangos and C. A. Nicolaides, *Phys. Rev.* **A48**, 4142 (1993).
44. K. T. Chung and B. C. Gou, *Phys. Rev.* **A52**, 3669 (1995).
45. S. Diehl, D. Cubaynes, E. T. Kennedy, F. J. Wuilleumier, J.-M. Bizau, L. Journel, L. Voky, P. Faucher, A. Hibbert, C. Blancard, N. Berrah, T. J. Morgan, J. Bozek and A. S. Schlachte, *J. Phys.* **B30**, L595 (1997).
46. K. Berrington and S. Nakazaki, *J. Phys.* **B31**, 313 (1998).
47. D. O. Harris, G. G. Engerholm and W. D. Gwinn, *J. Chem. Phys.* **43**, 1515 (1965); A. S. Dickinson and P. R. Certain, *J. Chem. Phys.* **49**, 4209 (1968); J. C. Light and R. B. Walker, *J. Chem. Phys.* **65**, 4272 (1976); J. C. Light, I. P. Hamilton and J. V. Lill, *J. Chem. Phys.* **82**, 1400 (1985).
48. R. G. Littlejohn and M. Reinsch, *Phys. Rev.* **A52**, 2035 (1995).

CHAPTER 21

HIGH-RESOLUTION ANGLE-RESOLVED STUDIES OF ATOMS AND MOLECULES USING ADVANCED ELECTRON SPECTROSCOPY AT THE ALS

Nora Berrah

Western Michigan University
Physics Department
Kalamazoo MI 49008

Contents

1. Introduction

The last thirty years have seen significant improvements in the understanding of the electronic structure and electron dynamics of light atoms using photoionization experiments, especially in the case of outer-shell photoionization.[1–4] The photoionization process by which a low-intensity electromagnetic field is absorbed by a light atom has allowed satisfactory agreement between experiments and theory (based on the one-electron approximation) for photoionization parameters such as cross-sections, $\sigma(\omega)$, and for photoelectron angular distribution asymmetry parameters $\beta(\omega)$. The cross-section parameters provide direct insight into the orbital or electronic structure of atoms and molecules and the angular distribution parameters provide information on the partial-wave matrix elements and the phase shifts between the two partial waves of the outgoing electron. However, these parameters are often complicated by the presence of electron correlation leading to phenomena which are outside the single particle model.[5,7] Therefore, the structure and dynamics of most atoms and molecules are still not well understood and in some cases not explored at all. This is especially true for photoionization with excitation of atoms heavier than

Li, of inner-shells atoms and of molecules. Fundamental understanding of complex processes in atoms and molecules through gas-phase VUV/soft X-ray interaction can be achieved since these type of interactions have an immense potential to unravel the many-body problem. These interactions are excellent probes of various weak and strong effects in this photon energy range as well as for tests of various theoretical models.

Recently, there has been renewed interest in photoexcitation and photoionization experiments for the investigation of the structure and dynamics of atoms and molecules due to the rapid technological development of high-resolution spectrometers coupled with third-generation synchrotron radiation (SR) sources. Many accounts of recent advances in electron spectrometry can be found in the literature.[7–14] Technological advances have permitted the improvement of instruments, the development of spectroscopies and new methods for the quantum control of atomic and molecular processes. These continued improvements allow ever more difficult atomic and molecular experiments to be performed with a level of detail that is unprecedented. For example, inner-valence photoelectron spectra in molecules exhibit complex bands due to many-electron processes in addition to the main bands expected from an independent electron picture. A good description of these spectra generally requires consideration of configuration mixing in the final state as well as in the initial state.[15]

The aim of this chapter is to report on recent photoexcitation and photoionization measurements on atoms and molecules. These illustrate how advanced photoelectron spectroscopy techniques, coupled with photons from the high-resolution atomic and molecular (HRAM) undulator beamline of the Advanced Light Source (ALS) at Lawrence Berkeley National Laboratory allow new and detailed structure to be measured that lead to deep insight into, and sometimes new understanding of physical processes. Some of the following sections include theoretical interpretation or speculation since, in general, this area of research benefits from a close collaboration between theorists and experimentalists. Recent photoionization with excitation measurements tested and guided various models and predictions, observed new processes, and stimulated development of atomic and molecular theories. For example, the left part of Fig. 1 shows the decay spectrum of the carbon $1s \rightarrow 2\pi^*$ resonance in the CO molecule recently measured at the HRAM beamline at the ALS for three vibrational levels while the right part of Fig. 1 shows the calculated spectra convoluted to

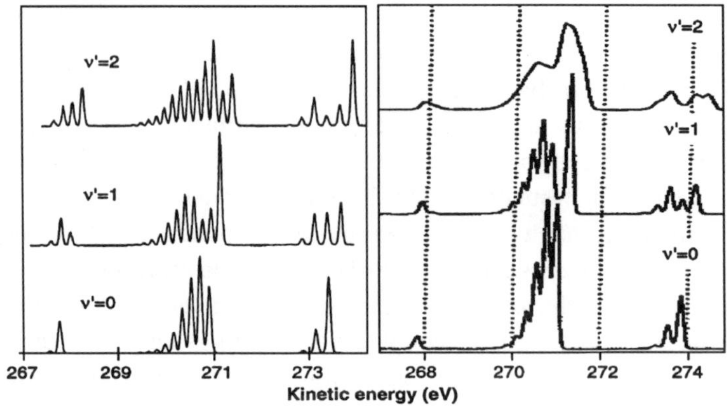

Fig. 1. Electron emission spectra at the $\nu' = 0$, 1 and 2 vibrational levels of the C $1s \to \pi^*$ resonance of the CO molecule. Left panel presents recent experiment conducted at the HRAM beamline of the ALS, right panel represents convoluted calculation[16] (see text for detail).

match the experimental spectra measured at NSLS.[16] This illustrates very clearly the progress achieved in just one year in photoelectron spectroscopy (PES) with advanced instrumentation since this new measurement is definitively fully vibrationally resolved. These measurements are also angle resolved, thus a more complete and deeper understanding of the dynamics of this molecule may be achieved (see Sec. 6.4 for details).

This chapter will illustrate specifically the recent progress in areas such as high-resolution angle-resolved atomic autoionization, the resonant Auger process and its use under Raman conditions, doubly and triply excited states in a heavier elements than He, and finally core excitation measurements of various molecules under Resonant Raman conditions will be presented. Also presented are the photoelectron spectroscopy methods used in all of the measurements presented and the main features of the experimental systems, i.e. an electron imaging apparatus based on angle-resolved time-of-flight spectrometry, an SES-200 hemispherical electron analyzer, and a cylindrical mirror analyzer.

2. Photoelectron Spectroscopy

Over the last twenty years PES has proven to be one of the most powerful techniques for the study of many-electron processes in atoms and molecules. PES has the unique capacity to study separately the different decay channels, in particular, partial photoionization cross-sections and angular distributions of the emitted electrons can be measured. The success of PES is, in great part, due to the development of intense SR sources since the low density of atoms, the small interaction volume in a PES experiment, and the low cross-section for weak processes must be counterbalanced by a high flux of photons. The great advantage of electron spectrometry is that it can characterize the individual processes contributing to the different ionic yields. Thus, experimental advances toward a highly differentiated analysis have been achieved by energy analyzing the electrons and by determining the electron angular distribution providing physical insight into the decay dynamics.

2.1. *Differential Photoionization Cross-Section*

In the dipole approximation, valid for low energy photons $h\nu$, the differential photoionization cross-section $d\sigma_{if}(h\nu)/d\Omega$ resulting from photoionization of randomly oriented atoms in the initial state $|i>$ by linearly polarized photons leaving the ions in the final state $|f>$ is given by:[17]

$$\frac{d\sigma_{if}}{d\Omega} = \frac{\sigma_{if}}{4\pi}[1 + \beta_{if} P_2(\cos\theta)], \tag{1}$$

where σ_{if} is the photoionization cross-section for producing state $|f>$ of the ion, θ is the angle between the photon's polarization vector and the photoelectron's momentum direction, $P_2(x) = (3x^2 - 1)/2$, and β_{if} is the electron angular distribution parameter with $-1 \leq \beta_{if} \leq 2$. For linear polarization degrees P_1 less than 1, and for emission of electrons perpendicular to the photon beam directions, Eq. (1) has to be rewritten as:[18]

$$\frac{d\sigma_{if}}{d\Omega} = \frac{\sigma_{if}}{4\pi}\left[1 + \frac{\beta_{if}}{4}(1 + 3P_1 \cos 2\theta)\right]. \tag{2}$$

2.2. *Two-Dimensional Time-of-Flight Spectroscopy*

One of the experimental method used to produce the reported data consists of electron time-of-flight spectroscopy that utilizes two-electron time-of-flight spectrometers (TOF)[19] (built at Western Michigan University) in conjunction with photons from the ALS synchrotron radiation source. Two TOF spectrometers, as shown in Fig. 2, are housed in a rotating chamber and are mounted in a plane perpendicular to the photon beam direction.

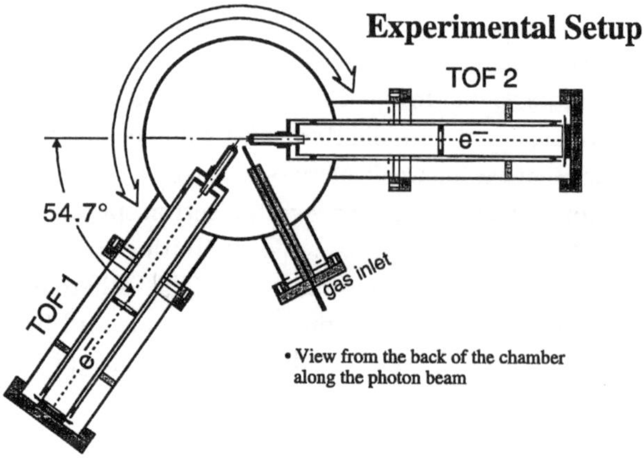

Fig. 2. Schematic diagram of the experimental setup showing two time-of-flight (TOF) electron spectrometers mounted to a rotatable chamber. The incident synhrotron radiation goes through the interaction region, perpendicular with respect to the drawing plane.

The spectrometers are mounted with a relative spacing of 125.3°, each equidistant from the gas inlet valve. The spectrometers' apertures and gas inlet needle are each approximately 20 mm from the interaction region on the axis of the chamber. For the measurements described in this chapter, the TOF analyzers were fixed at O° and 54.7° (magic angle) with respect to the electric field vector of the incident radiation. This configuration, shown in Fig. 2, allows for two spectra to be taken simultaneously at different angles. Electrons of all kinetic energies are detected which allow measurements independent of the fluctuations of the target density or the photon flux. As a result, angular anisotropy can be measured since the β parameters can be extracted. The interior of the chamber is shielded with two layers of magnetic shielding which reduces the local fields, especially the earth's magnetic field. The chamber is pumped with a turbo pump of 1000 $1/s$, which is capable of reducing the base pressure to 10^{-8} torr. The gas is passed into the interaction region through a needle to a pressure of about 3×10^{-5} torr. The whole chamber including analyzers, gas inlet, turbo pumps can be rotated around the synchrotron beam to allow measurement of spectra at different angles.

Figure 3 shows one of the analyzers in detail. Electrons created in the source volume fly straight to the microchannel plates (MCPs) where they

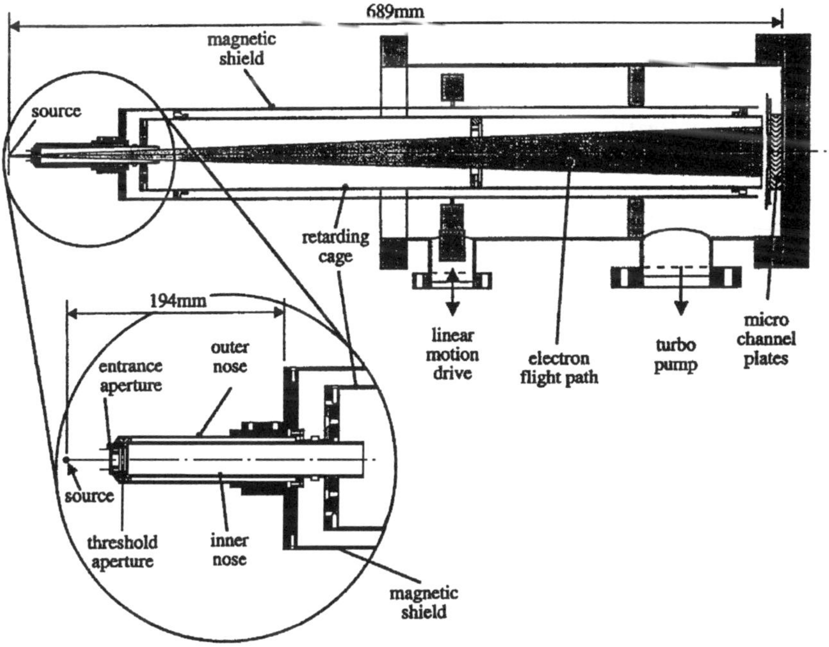

Fig. 3. Detail schematic of one of the time-of-flight electron spectrometer built[19,20] and used in the reported measurements with an enlarged part of its entrance.

are detected. The signal is amplified, connected to a constant fraction discriminator (CFD), and then sent to the start of the time-to-amplitude-converter (TAC). The stop signal of the TAC is provided by the bunch marker signal from the synchrotron radiation ring. The analog pulse generated is then transformed by the analog to digital converter (ADC) which digitizes the pulse and sends it to the multichannel analyzer (MCA). The MCA then histograms the counts by arrival time. The resulting counts versus time spectrum is then stored and displayed on a PC using a labview data acquisition program that we developed. The labview acquisition program, in addition to recording spectra, monitors the synchrotron radiation's current and the chamber pressure. Further details of this experimental system can be found in the literature.[20]

In order to measure the flight times of the electrons, we have to use the light source in few bunch operation. The ALS ring was operated at either 1.9 or 1.5 GeV and delivered 40 mA in double bunch operation at injection. In all the experiments presented here, we used double bunches with a timing difference of 328 ns between the bunches. Each TOF spectrometer consists of a pair of microchannel plates and a detection plate mounted at the end of a 689-mm flight path. By measuring the arrival times of the ejected electrons relative to the photon pulse, complete photoelectron spectra can be measured. The spectra are then converted from time to kinetic energy as described below. In order to improve the resolution of our spectra, a retarding potential was often applied at the retarding cage and the inner nose. We applied different retarding potentials for different experiments. To prevent the slower electrons from being affected by electric and magnetic stray field, we also used a magnetic shield in the drift tube of the analyzer. The resolution of the analyzers can be 1% of the kinetic energy of the incoming electrons. The time-to-energy conversion curve was obtained by using calibration spectra of the Ne $2s$ and $2p$ photoelectron lines of well-known binding energy. In order to obtain the relative detection efficiency of the two analyzers, the Ne calibration spectra are also used for all the photoelectrons and the constant ionic state spectra are corrected for the analyzers' transmission and detector's efficiency. Also, the scans were normalized to the incident photon flux. The degree of linear polarization P_1 of the monochromatic light in the resonance region was determined by us to be $P_1 = 0.99$ by comparing measured angular distributions of the He $1s(\beta = 2)$ and the Ne $2p$.[21] The tilt of the polarization plane with respect to the experimental system was measured to be zero.

In gas-phase experiments, the photoexcitation and decay mechanisms vary as the photon energy is varied. However, measuring the different processes in one picture is not a simple matter. Most current experimental ionization spectroscopic methods are one-dimensional in nature; that is, they either scan the photon energy, giving in actuality the sum of all processes, or they set the photon energy fixed and collect the electrons as a function of kinetic energy. In order to improve our detection system, we added to our angle resolved method an imaging technique to build a two-dimensional (2D) image of the ionization processes. Figure 4 shows an example of a 2D image where the Ar $2p \to ns, nd$ resonances are depicted. This method combines the ability to measure and record kinetic energy resolved TOF spectra (x-axis) with the ability to scan the photon energy by making small-automated photon energy steps (y-axis) and extracted the electron yield. To produce a 2D electron emission data sets, a photon energy range and a kinetic energy window covering the majority of the

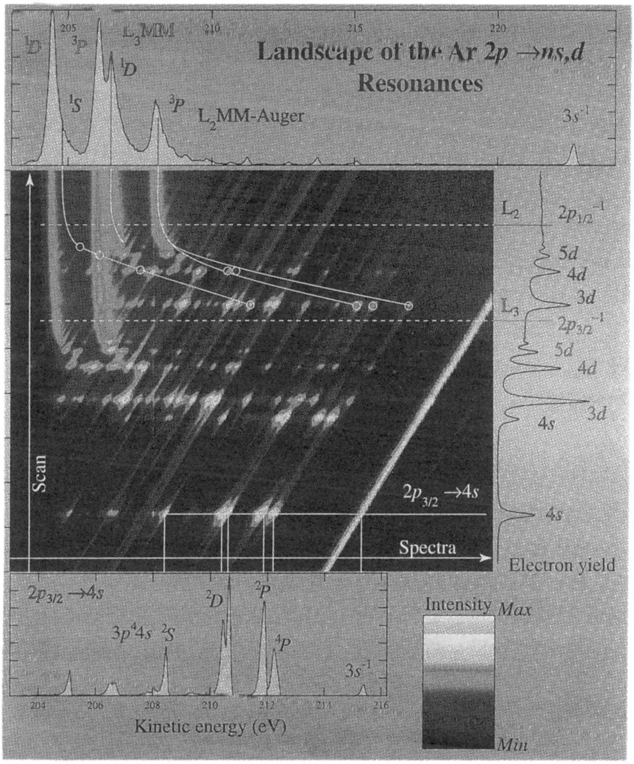

Fig. 4. 2D landscape of the Ar $2p \to ns$, d resonances. The top portion of the figure shows the LMM Auger spectrum at 252 eV photon energy. The bottom portion of the figure shows the spectrum measured at the peak of the $2p \to 4s$ resonance and the right side portion of the figure shows the scan (electron yield) recorded between 244–252 eV. The middle part of the figure shows the 2D map depicting various processes as described in the text of Sec. 4.1.

ejected electrons is chosen to allow the investigation of the inner-shell ionization processes. In this example, the data is collected by setting the photon energy to the beginning of the range of interest (244 eV) and PES are accumulated simultaneously in each analyzer for a particular time (that time varies depending upon the level of statistics needed for the measurements). For example, in the case shown in Fig. 4, $t = 10$ s. The photon energy is then incremented in steps of 1/4 the photon resolution and another pair of PES was collected. By repeating this procedure over the photon energy range of interest (244 eV–252 eV in the example shown), which can take a total of two and a half hours, we are able to produce comprehensive and detailed 2D images. This new method of looking at the data collected allows examination of all of the data in order to spot areas of interest and in order to follow the progression of various processes. As can be seen in Fig. 4, various spin-orbit split resonances are observed leading to their continua. It also allows experimental artifacts, such as peaks due to higher order or background electron, to be identified and removed from the 2D data wherever possible. In addition to allowing swift data collection, this technique also provides substantially improved ways of viewing such processes as post collision interaction (PCI) (seen in Fig. 4 as curves toward higher kinetic energy close to both ionization thresholds) or complicated decay pathways. This 2D imaging method was used in most sections of this chapter.

2.3. *The High-Resolution Hemispherical Analyzer*

The second experimental method used to measure the data reported here uses a new end station built for high-resolution electron spectroscopy to complement the TOF experimental technique. The system is built around a state-of-the art commercial Scienta SES-200 hemispherical electron energy analyzer, as shown in Fig. 5 and is optimized for gas-phase studies of atoms

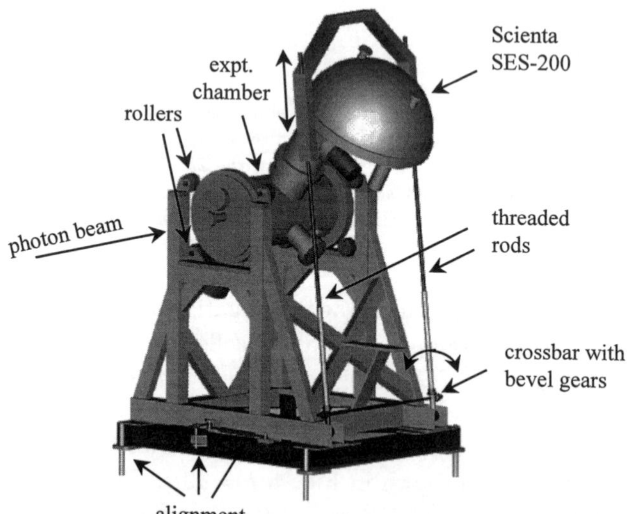

Fig. 5. A schematic view of the new end station for high-resolution gas-phase electron spectroscopy at the ALS, based on a Scienta SES-200 electron energy analyzer.

and molecules. The design maximizes the performance with gaseous samples by using differential pumping stages both in the experimental chamber and in the connection to the beamline. The analyzer, electron lens and experimental chamber are surrounded by μ-metal layers providing magnetic shielding. The number and diameter of access ports has been kept to minimum in order to reduce the penetration of the magnetic field into the interaction region. Details about this apparatus can be found in Ref. 20.

The sample gas is fed into a gas cell, which is mounted directly to the lens of the analyzer. The pressure in the cell can be raised to about 10^{-3} torr. Electrodes are placed inside the gas cell at both ends along the photon path to correct for small gradients in the potential and to improve the energy resolution. These electrodes can also be used to collect positive ions from the gas cell permitting the simultaneous measurement of the total ion yield and energy dispersed electron spectrum.

A key feature of the system is its ability to record angle-resolved data, which requires the analyzer and experimental chamber to be rotatable around the axis of the photon beam. This is a nontrivial mechanical problem because of the high precision required and heavy weight of the analyzer (about 200 kg). Our solution, which allows for maximum portability and compactness, is to support the analyzer from below with two threaded rods, which pivot around their bottom fixture. The analyzer is moved along the rods by rotating them. The experimental chamber is equipped with tapered disks at each end, which are clamped between V-shaped small diameter rollers, allowing rotational motion of the chamber. The linear motion of the attachment points of the analyzer thus translates into the rotation of the whole system within a range of 90 degrees.

The analyzer is operated in a constant pass energy mode, which can be selected according to the conditions of the experiment. Under the most stringent settings (5 eV pass energy), an energy resolution better than 5 meV is achieved.

2.4. *The Electron Cylindrical Mirror Analyzer Spectrometer*

High transmission of electron spectrometers and high detection efficiency are essential in metal vapors, low target-density experiment. The cylindrical mirror analyzer (CMA) spectrometer, used in conditions of intermediate focusing, was shown to be the best choice in the early experiments when the SR intensity was quite low.[22] For PES measurements on hollow lithium

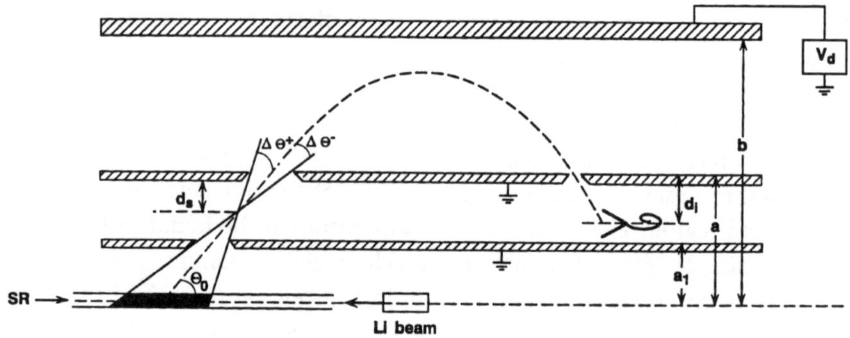

Fig. 6. A schematic of the CMA electron spectrometer.

atoms, a CMA whose schematic is shown in Fig. 6 was used. It is made of three cylinders: in addition to the external and internal cylinders separating the regions with and without field, a third cylinder closer to the axis allows the source volume to be defined. The photon beam defines the electron emitting region, and the two slits define the part of the region seen by the analyzer.

In this geometry, the channeltron used to detect the electrons is placed on the axis of the CMA after the exit slits. The integrated area under a photoelectron line is proportional to the partial cross-section σ_{nl}. Thus, using a photoelectron spectrum, one can compare on a relative scale various subshell partial cross-sections. Once relative values of the different partial cross-sections are known, their absolute values are obtained by normalization to the total photoabsorption cross-section. In the case of lithium, the absolute values of all cross-sections were established by reference to the absorption data of Mehlman *et al.*[23] at 112 eV. In the experiments, reported in Sec. 5.2, the resolution was about 0.039 eV (resolving power of 3800), in order to keep the photon flux close to 10^{13}/sec since the vapor density of Li had to be maintained to about 10^{10}–10^{12} atoms/cm^2. In some cases, photoelectron spectra for Li measurements were recorded with a resolution of 0.019 eV. Details of this experimental technique has been described elsewhere.[25]

2.5. *The High-Resolution Atomic and Molecular Beamline*

All PES experiments described in this chapter were performed on the undulator 9.0.1 (HRAM) beam line of the ALS. In the optical design chosen for this beam line,[24] all optics have a spherical profile. The exit slit translates and the radius of the refocusing mirror is simultaneously changed to keep the exit slit at focus. Three gratings from the spherical grating monochromator cover the photon energy from 20 to 350 eV and deliver approximately 10^{12} photons/sec at an average resolving power of 10 000. However, resolving power of 1000 000 have however been obtained in absorption experiments. Detailed on this beamline can be found in the literature.[24]

3. Atomic Autoionization

The discovery of a series of autoionizing states in He[26] and its explanation[27] were crucial for the understanding of the coupling of bound states with continua. According to Starace,[28] dramatic changes in the parameter β are an indication of strong effects due to the manifestation of certain types of electron–electron correlation. The determination of the anisotropy parameter β in the resonance region is an important source of information on the dynamics of resonance photoionization which cannot be obtained by absorption or ion mass spectrometry.

With the availability of third generation synchrotron light sources, detailed experiments are now possible by combining various spectroscopic methods with high-resolution photons and high photon flux. Highly resolved data can now be obtained to provide the basis for critical tests of different theoretical models. Theoretical work has shown that the electron angular distribution[29] and the shape of the autoionization resonances[30] are crucial to the understanding of certain types of electron–electron correlation. In the case described below, detailed measurements and analysis were needed in order to test calculations by Taylor[31] for the angular distribution of autoionization resonances.

3.1. *Angular Distribution and R-Matrix Calculations of the Ne $2s^{-1} \to np$ Autoionization Resonances*

The absorption spectrum of Ne was studied by Codling *et al.*[32] and previous determinations of the angular distribution anisotropy parameter β have been reported by Schmidt,[33] and by Becker and Shirley[34] with low-resolution and with no detailed analysis. Hall *et al.*[35] obtained a resolved β parameter variation, up to the $n = 4$ resonance, using a 100 meV photon source; however, no analysis was performed. The most recent measurements prior to the present quantitative work was done by Chan *et al.*[36]

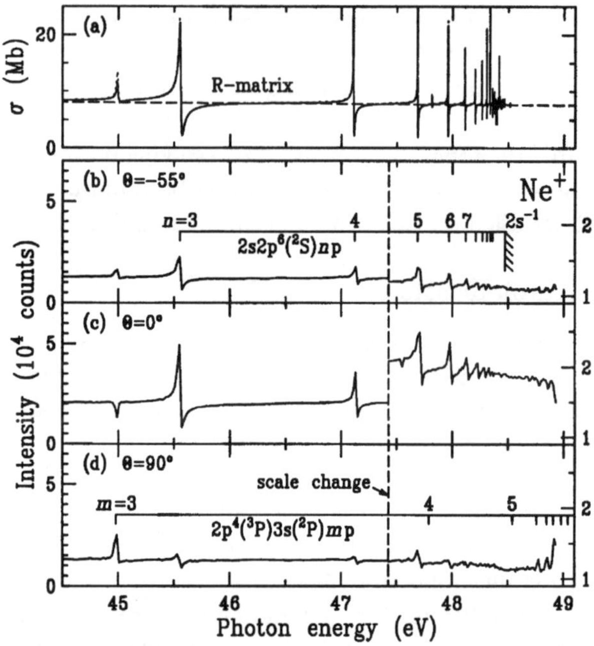

Fig. 7. Photoelectron yield scan of the Ne $2p$ line in the region of the autoionization resonances below the $2s$ threshold recorded at (b) the magic angle (54.7°) (c) 0° and (d) 90° with respect to the polarization of the synchrotron radiation. (a) Shows our theoretical results in length (dotted curve) and velocity (full curve) form; the broken line is a linear interpolation of the photoionization cross-sections from Bizau and Wuilleumier.[44]

but no angular distribution measurement was done. Autoionization resonances result from the decay of the excited discrete state Ne* $2s2p^6np$ into the continuum state $\text{Ne}^+3s^23p^5 + e^-(ks,\ kd)$. Since the continuum can also be reached by direct photoionization, both paths add coherently, giving rise to interferences that produce the characteristic Beutler–Fano line shape.[37] Figure 7(b, c and d) shows a photoelectron yield scan of the Ne $2s^22p^6 \to 2s2p^6np$ ($n = 3$–10) autoionizing resonances and the $2s^22p^6 \to 2p^43s3p$ doubly excited resonance in the photon energy range between 44 eV and 48.5 eV measured at three different angles using three time-of-flight spectrometers. With 20 μm entrance and 30 μm exit slits, the monochromator had a resolving power of $h\nu/\Delta h\nu = 2500$ which corresponds to a photon bandpass between 18.0 and 19.4 meV in the resonance region. Figure 7(a) shows our R-matrix calculation[38] that agrees very well with the measurements.

In order to analyzer in details the measurements, we have used a parameterization of the variation in β over autoionizing resonances introduced by Kabachnik and Sazhina.[39] It is based on the Fano[40] parameterization for the total photoionization cross-section σ_{if}:

$$\sigma_{if} = \sigma_a \frac{(q+\varepsilon)^2}{1+\varepsilon^2} + \sigma_b \,, \tag{3}$$

where q is the line profile index and the reduced energy ε is given by:

$$\varepsilon = \frac{h\nu - E_r}{\frac{1}{2}\Gamma} \,, \tag{4a}$$

and

$$\rho^2 = \frac{\sigma_a}{\sigma_a + \sigma_b} \,, \tag{4b}$$

where E_r is the position of the resonance and Γ its width. The cross-sections σ_a and σ_b are slowly varying background cross-sections and ρ^2 is the correlation coefficient. Also listed[31,39] below are the expressions used to derive β which are given by:

$$A_0 = \frac{\sigma_{if}}{2\pi} \quad A = \frac{\sigma_a + \sigma_b}{4\pi} \quad B = \frac{2q\sigma_a}{4\pi} \quad C = \frac{\sigma_a q^2 + \sigma_b}{4\pi} \,, \tag{5}$$

Equation (5) can also be written as:

$$A_0 = \frac{A\varepsilon^2 + B\varepsilon + C}{1+\varepsilon^3} \,. \tag{6}$$

Additionally, for

$$A_2 = -\sigma_{if}\beta_{if}/8\pi \,, \tag{7}$$

Kabachnik and Sazhina[39] derived a similar resonance formula

$$A_2 = \frac{X\varepsilon^2 + Y\varepsilon + Z}{1+\varepsilon^2} \,, \tag{8}$$

where X, Y and Z are constants. Finally, using Eqs. (6), (7) and (8) the resonance behavior of the anisotropy parameter β_{if} can be expressed in terms of A, B, C, X, Y and Z as follows:

$$\beta = -2\frac{A_2}{A} \frac{\dfrac{X\varepsilon^2 + Y\varepsilon + Z}{1+\varepsilon^2}}{\dfrac{A\varepsilon^2 + B\varepsilon = C}{1+\varepsilon^2}} = \frac{X\varepsilon^2 + Y\varepsilon + Z}{A\varepsilon^2 + B\varepsilon + C} \,, \tag{9}$$

with X, Y and Z being considered as free parameters in the fit to our data. Far away from the resonance position, the ε^2 term dominates and thus the background β is given by the ratio X/A, whereas $\beta = Z/C$ at the resonance position ($\varepsilon = 0$).

The results for the total photoionization cross-section are shown in Fig. 8 for one double excitation and three single excitation autoionization resonances. The data are plotted along with a fit (represented by the solid line going through the data), as well as our R-matrix calculations. The photon bandpass was (a) 18.0 meV, (b) 18.2 meV, (c) 18.8, and (d) 19.1 meV. R-matrix calculations were performed using the latest version of the Belfast R-matrix codes,[41] as written for the Iron Project,[42] with modifications to include certain energy shifts described in the literature.[38] In order to take into account the monochromator bandpass, we convoluted the line shape calculated on the basis of a Fano shape (Eq. 3) with the slit function before comparing it to the normalized data. The slit function was a normalized Gaussian curve with FWHM equal to the measured bandpass. The fitting parameters, the width Γ, the line profile q and the correlation coefficient ρ^2, for the photoionization cross-section in the region of the first 4 resonances are listed on Table 1. Also listed are previous quantitative measurements[32]

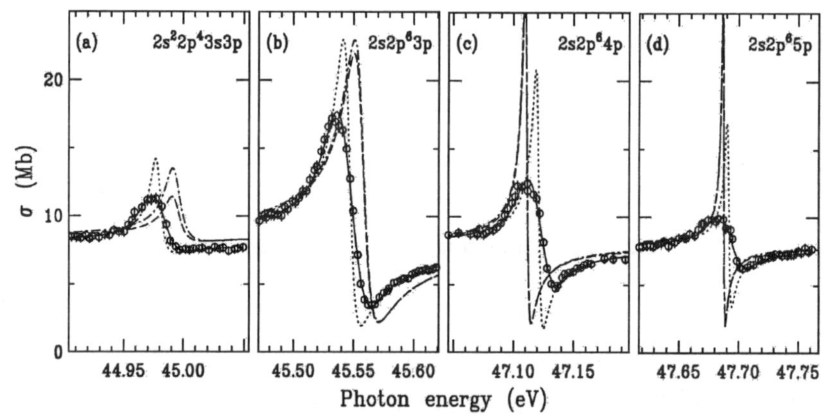

Fig. 8. Cross-section behavior of the Ne $2s^22p^6 \rightarrow$ (a) $2s^22p^4(^3P)3s(^2P)3p\,^1P_1$ and (b)–(d) $2s2p^6np$ ($n = 3$–5) autoionization resonances. The circle are experimental data taken at the magic angle and normalized to the nonresonant cross-section from Bizau and Wuilleumier.[44] The full curves represent a Fano fit to our data assuming a photon resolution power of $h\nu/\Delta h\nu = 2500$; the dotted curve are created using the Fano parameters without bandpass; the chain and the broken curves are our theoretical R-matrix calculation in length and velocity form, respectively.

Table 1. Fano profile parameters for the Ne $2s^22p^6 \rightarrow 2s^22p^4(^3P)3s(^2P)3p\,^1P_1$ and $2s2p^6np(n = 3\text{–}5)$ autoionization resonances compared with previous measurements and calculations.

	Method	$2s^22p^43s3p$ Pos(eV)	Γ(meV)	q	ρ^2	$2s2p^63p$ Pos(eV)	Γ(meV)	q	ρ^2
I	CI(L)						11.7	−1.6	0.76
	CI(V)							−1.3	0.77
II	abs.	44.979(8)	10(3)	−2.0(3)	0.17(5)	45.546(8)	13(2)	−1.6(2)	0.70(7)
present	theo.(L)	44.993	16.3	−3.76	0.04	45.557	18.6	−1.59	0.72
results	theo.(V)		16.3	−3.05	0.04		18.6	−1.53	0.73
	exp.		11(1)	−2.93(1)	0.09(1)		13.2(10)	−1.58(1)	0.75(5)

	Method	$2s2p^64p$ Pos(eV)	Γ(meV)	q	ρ^2	$2s2p^65p$ Pos(eV)	Γ(meV)	q	ρ^2
I	CI(L)		3.8	−1.75	0.76				
	CI(V)			−1.46	0.77				
II	abs.	47.121(5)	4.5(15)	−1.6(3)	0.70(7)	47.692(5)	2(1)	−1.6(5)	0.70(14)
present	theo.(L)	47.111	4.3	−1.88	0.72	47.687	1.8	−1.90	0.74
results	theo.(V)		4.3	−1.82	0.73		1.8	−1.87	0.75
	exp.		5.7(10)	−1.47(1)	0.78(11)		3.6(18)	−1.46(5)	0.6(2)

I. Burke and Taylor (1975a)
II. Codling, Madden, and Ederer (1967)

along with the present R-matrix results and the earlier results of Burke and Taylor.[43] As can be seen from Table 1, there is good agreement among the measurements.

The pronounced behavior of the angular distribution β as a function of the photon energy is shown in Fig. 9 for the $2s2p^6(^2S)np$ resonances, the strong $2s^22p^4(^3P)3s(^2P)3p\,^1P_1$ and the weak $2p^4nln'1'$ resonances. Schmidt[33] reported a β variation of approximately 0.95 to 1.05 while Becker and Shirley[34] reported a variation of 0.92 to 1.1 compared to our variation of about 0.0 to 1.5 over the energy range of 44 to 49 eV due to a better photon source. The parameter β is obtained by applying Eq. 2 to three simultaneously recorded electron-yield signals measured at $\Theta = 0°$, the magic angle $\Theta = 54.7°$, and 90°. The angular distribution parameter β is then deduced by the following expressions:

$$\beta_{ab} = 4\frac{R_{ab} - 1}{K_a - R_{ab}K_b}, \tag{10}$$

where $R_{ab} = I_a/I_b$ is the ratio of a pair of our normalized intensities I under 0°, 54.7° and 90° and

$$K_0 = 1 + 3P_1 \; K_{54.7} = 1 - P_1 \; K_{90} = 1 - 3P_1\,,$$

where P_1 is the degree of linear polarization. Finally, we took the average of the β parameters obtained for the different angle combinations. In Fig. 9(b), angular distribution data are shown for the $2p^2 \to 3s3p$ resonance as well as the $2s \to np$ ($n = 3$–5) resonances and are compared with the fit. As can be seen, the parameterization of Kabachnik and Sazhina[39] gives a very good fit to the experimental angular distribution data. Figure 9(a) is our R-matrix calculation[38] which agrees very well with the measurements. For our data and calculations we used the cross-section values from Bizau and Wuilleumier[44] when calculating the A, B and C parameters. The angular distribution is also compared in Fig. 10 with the untested R-matrix results of Taylor[31] and our R-matrix results.[38] From the figures we note that the entire Rydberg series $2s2p^6np$ generated by our calculations lines up well with the experimental values, indicating that our computed quantum defect μ is quite reasonable. Also, the widths of these resonances, the degree of interference with the background, and the shape of the background itself all show good agreement with experiment. We point out that the positions of the low-lying $2s^22p^4(^3P)3s(^2P)3p\,^1P_1$ and $2s2p^63p$ resonances

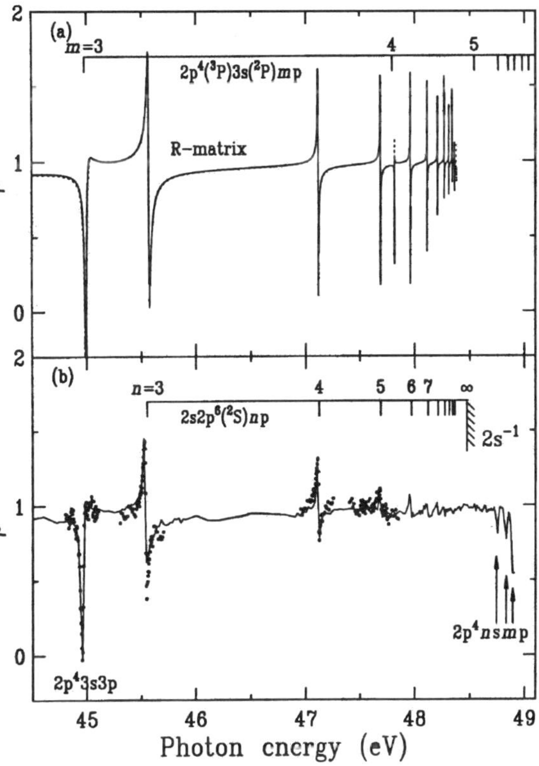

Fig. 9. (a) Theoretical calculation and (b) experimental results of the angular distribution anisotropy parameter β for the Ne $2p$ line in the region of the autoionization resonances below the $2s$ threshold. In (b), the dotted curve and the full curve represents the β parameter as obtained from the scans shown in Fig. 7; the circles show a second set of data taken with higher data point density.

were aligned with experiment by performing the shift in the bound part of the Hamiltonian. We also note good agreement with Taylor's[31] calculations which we shifted so it matches the data since he did not give the energy position of the lines. In all cases, the dotted lines are deconvoluted fits. As can be seen from the figure, agreement between the data and both theories is excellent. The parameters X, Y and Z produced by our fit and our calculations are listed in Table 2 along with the results from Taylor[31] available only for $n = 3$–4. It is worth pointing out that although both the length and the velocity form of Taylor's[31] calculation coincide (as shown in Fig. 10), the parameters X, Y and Z are slightly different. This difference

Table 2. Angular distributions profile parameters for the Ne $2s^22p^6 \to 2s^22p^4(^3P)3s$ $(^2P)3p\,^1P_1$ and $2s2p^6np(n = 3\text{–}5)$ autoionization resonances compared with theoretical calculations.

	Method	$2s^22p^43s3p$ X	Y	Z	$2s2p^63p$ X	Y	Z
Taylor	CI(L)				−0.35	0.23	−1.27
(1977)	CI(V)				−0.26	0.96	−0.84
present	theo.(L)	−0.28	−0.14	0.35	−0.28	1.06	−1.02
results	theo.(V)	−0.28	−0.14	0.21	−0.28	1.12	−1.17
	exp.	−0.291(1)	−0.22(1)	0.23(3)	−0.285(1)	1.08(6)	−1.06(4)
	Method	**$2s2p^64p$ X**	**Y**	**Z**	**$2s2p^65p$ X**	**Y**	**Z**
Taylor	CI(L)	−0.33	1.33	−1.37			
(1977)	CI(V)	−0.29	1.04	−0.93			
present	theo.(L)	−0.27	1.14	−1.23	−0.27	1.16	−1.32
results	theo.(V)	−0.27	1.12	−0.97	−0.27	1.14	−1.28
	exp.	−0.312(1)	1.02(15)	−0.92(10)	0.301(1)	0.7(3)	−0.8(3)

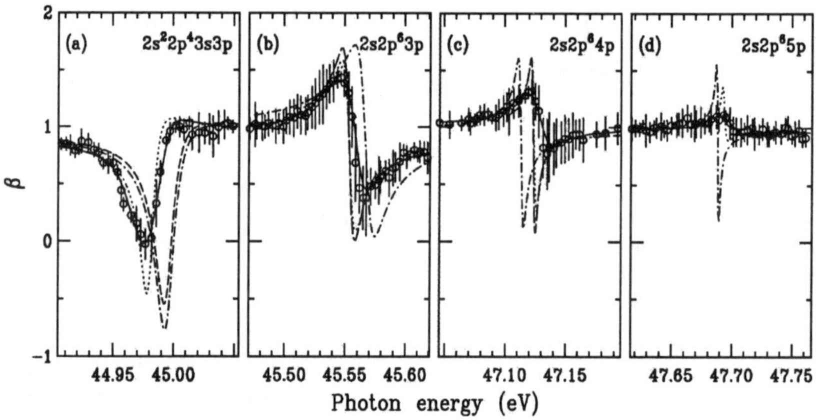

Fig. 10. β parameter as a function of photon energy across the resonances of the Ne $2s^22p^6 \to$ (a) $2s^22p^4(^3P)3s(^2P)3p\,^1P_1$ and (b)–(d) $2s2p^6np$ ($n = 3\text{–}5$) autoionization resonances. The circles are our experimental data; the full curves represent a fit to our data using Kabachnik and Sazhina's[39] formula and assuming a photon resolution power of $h\nu/\Delta h\nu = 2500$; the dotted curves are created using the results of the fits without bandpass; the chain [in (a)–(d)] and the short-broken [only in (a)] curves are our theoretical R-matrix calculations in length and velocity forms, respectively; the long-broken curves in (b) and (c) which sit almost exactly on top of our deconvoluted results (dotted curves) are taken from Taylor[31] and shifted to the resonance position.

stems from the contribution of the cross-section parameters, discussed in the literature.[38]

3.2. *Angular Distribution and R-Matrix Calculations of the Ar $3s^{-1} \rightarrow np$ Autoionization Resonances*

Like in the case of Ne, autoionization resonances in Ar result from the decay of the excited discrete states Ar* $3s3p^6\,np$ into the continuum state Ar^+ $3s^23p^5 + e^-$ (ks, kd). As in the case of neon, because the continuum can also be reached by direct photoionization, the two paths give rise to interference effect that produce the characteristic Beutler–Fano line shape.[37] Detailed measurements of the shape of this autoionization series were conducted previously.[45–48] Also, angular distributions for the first three autoionization resonances have been measured.[49–51] These autoionization resonances have been calculated using many-body perturbation theory,[52] multichannel quantum-defect theory,[53] the eigenchannel R-matrix method,[54] the K-matrix procedure,[55] and the random-phase approximation with exchange (RPAE).[29]

Measurements[56] at the ALS allowed a quantitative analysis of highly resolved photoelectron measurements and stimulated R-matrix calculations of the angular-distribution parameters from $n = 4$ to $n = 9$ of the Ar

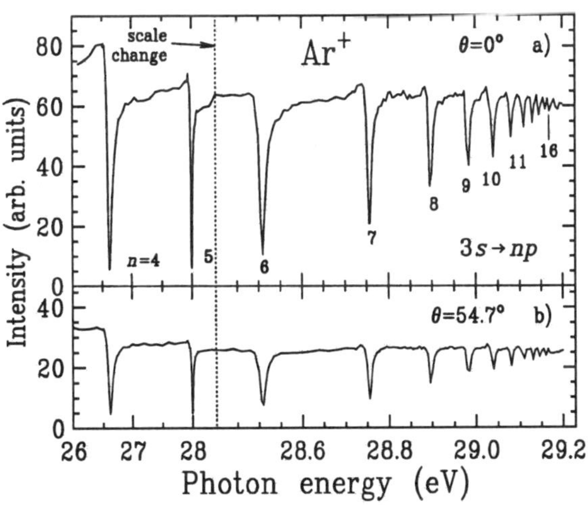

Fig. 11. Photoelectron yield scan of the Ar $3s^23p^6 \rightarrow 3s3p^6np$ autoionization resonances at (a) 0° and (b) at the magic angle with respect to the polarization of the synchrotron light.

$3s^2 3p^6 \rightarrow 3s3p^6\, np$ Rydberg series. These measurements[56] allowed a comparison with untested calculations[31] and with previous measurements[45,47] for the lower members of the series.

Figure 11 shows a scan of the electron yield measured at an angle of 0° and 54.7° with respect to the electric field vector of the linearly polarized synchrotron beam. This figure shows the Ar $3s^2 3p^6 \rightarrow 3s3p^6\, np$ Rydberg series of *window* type resonances from $n = 4$ to $n = 16$. The monochromator bandpass was 6 meV at 27 eV with 40 μm entrance and 60 μm exit slit widths and the scan was normalized to the incident photon flux.

The theoretical work done in conjunction with these measurements was obtained using the R-matrix method,[57,58] in close analogy to the calculations of Burke and Taylor[43] and Taylor,[31] with one important addition to the wavefunction, as detailed in the literature.[56]

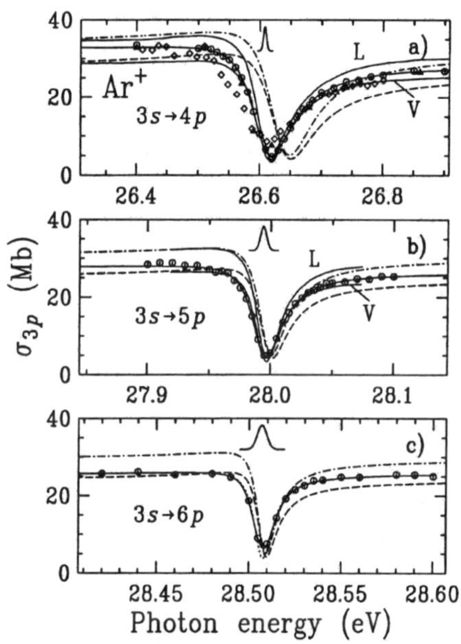

Fig. 12. Ar $3s^2 3p^6 \rightarrow 3s3p^6 np$ autoionization resonances for $n = 4$, 5 and 6. The open circles are present data; the open diamonds are from Codling *et al.*;[32] the full triangles are the $2p_{3/2}$ data from Svensson *et al.*[50] scaled by a factor of 7.3. The full curve through the data is our fit. The two full curves are the length (L) and velocity (V) calculations of Burke and Taylor,[43] and the broken and chain curves are the velocity and length form, respectively, of our calculation. The dotted curves are our deconvoluted fits in all figures. The Gaussian curve at the resonance position represents a monochromator bandpass of 6 meV.

The total photoionization cross-section data and the angular distribution parameter were fit in the same manner as for the case of Ne. The results for the total photoionization cross section are plotted in Figs. 12 and 13 for five resonances, along with their fit (represented by the solid line going through the data) as well as previous measurements for the lower resonances[49,50] and calculations.[43] The fitting parameters, the width Γ, the

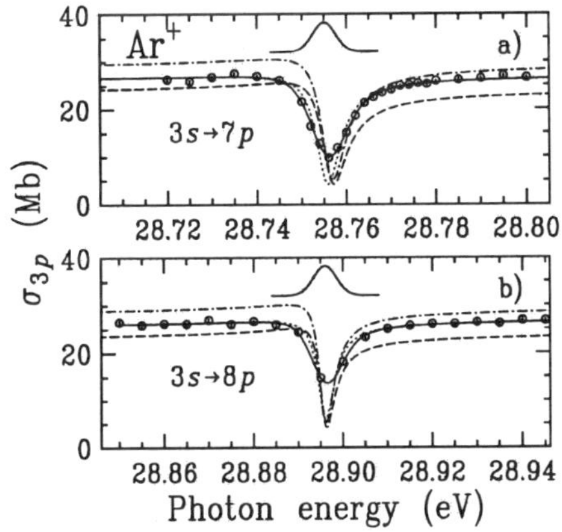

Fig. 13. Ar $3s^2 3p^6 \rightarrow 3s3p^6 np$ autoionization resonances for $n = 7$ and $n = 8$. The open circles are present data. The full curve through the data is our fit. The broken and chain curves are the velocity and length form, respectively, of our calculation. The dotted curves are our deconvoluted fits in all figures. The Gaussian curve at the resonance position respresents a monochromator bandpass of 6 meV.

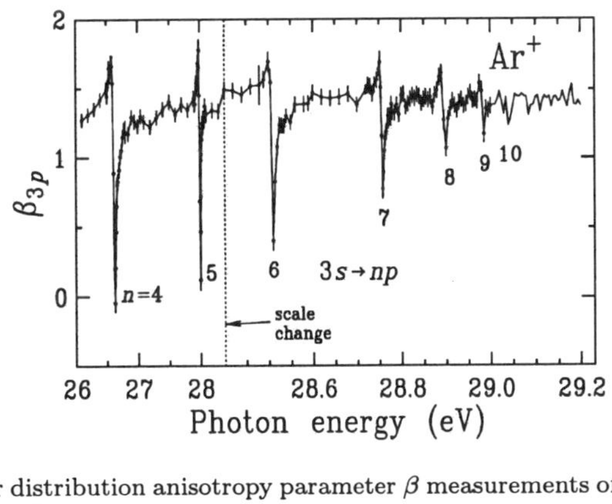

Fig. 14. Angular distribution anisotropy parameter β measurements of the Ar $3s^2 3p^6 \rightarrow 3s3p^6 np$ autoionization resonances.

Table 3. Profile parameters for the first six members of the $3s3p^6np$ series resonances compared with previous measurements and calculations.

	Method	$3s \to 4p$ Pos (eV)	$3s \to 4p$ Γ (meV)	$3s \to 4p$ q	$3s \to 4p$ ρ^2	$3s \to 5p$ Pos (eV)	$3s \to 5p$ Γ (meV)	$3s \to 5p$ q	$3s \to 5p$ ρ^2	$3s \to 6p$ Pos (eV)	$3s \to 6p$ Γ (meV)	$3s \to 6p$ q	$3s \to 6p$ ρ^2
I	CI(L)		68	−0.33	0.855		24.3	−0.26	0.85				
	CI(V)			−0.29	0.861			−0.22	0.86				
II	RPAE		85	−0.27	0.89								
	MCDF		12	−0.70	0.24								
III	Ion yield	26.606	76(5)	−0.249(3)	0.890(5)	27.993	25(7)	−0.168(5)	0.873(7)	28.506	16(7)	−0.200	0.526(7)
IV	Abs	26.614	80(5)	−0.22(5)	0.86(4)	27.996	28.2(13)	−0.21(2)	0.83(2)	28.509	12.6(12)	−0.17(3)	0.85(3)
V	Electron loss	26.605	76.1(50)	−0.31(3)	0.87(5)	27.994	28.0(21)	−0.21(3)	0.86(5)	28.509	13.0(33)	−0.18(3)	0.87(14)
Present	Theory (L)	26.633	83.8	−0.383	0.843	27.997	27.4	−0.292	0.825	28.508	12.4	−0.262	0.824
results	Theory (V)			−0.433	0.846			−0.342	0.829			−0.312	0.827
	Experiment		80.3(7)	−0.286(4)	0.840(3)		28.5(3)	−0.177(3)	0.848(3)		12.2(3)	−0.135(9)	0.852(9)

	Method	$3s \to 7p$ Pos (eV)	$3s \to 7p$ Γ (meV)	$3s \to 7p$ q	$3s \to 7p$ ρ^2	$3s \to 8p$ Pos (eV)	$3s \to 8p$ Γ (meV)	$3s \to 8p$ q	$3s \to 8p$ ρ^2	$3s \to 9p$ Pos (eV)	$3s \to 9p$ Γ (meV)	$3s \to 9p$ q	$3s \to 9p$ ρ^2
III	Ion yield	28.757	9(7)	−0.129(6)	0.453(8)	28.898	10(10)	−0.104(3)	0.38(1)				
Present	Theory (L)	28.756	6.65	−0.249	0.823	28.896	3.97	−0.240	0.824	28.928	2.57	−0.235	0.823
results	Theory (V)		6.66	−0.299	0.827			−0.291	0.827			−0.286	0.826
	Experiment		6.6(1)	−0.125(4)	0.846(9)		4.5(2)	−0.132(4)	0.77(2)		4.1(2)	−0.115(8)	0.63(3)

I Burke and Taylor (1977)
II Amusia and Kheifets (1982)
III Sorensen *et al.* (1994)
IV Madden *et al.* (1969)
V Wu *et al.* (1995)

line profile q and the correlation coefficient ρ^2, for the photoionization cross-section in the region of the first six resonances are listed in Table 3. Also listed are previous quantitative measurements[45,47,48] and calculations.[43,50] As can be seen from the table, there is good agreement among the measurements. It is also clear that there is excellent agreement with Burke and Taylor's[43] CI theory. In particular, in the case of $n = 4$, one observes that the CI and RPAE theories agree very well the measurements, while the multiconfiguration Dirac–Fock (MCDF) calculation of Sorensen[47] available only for $n = 4$ does not agree with the data. Also, from this table one can note that the q value given by the RPAE method agrees better with the measurements than do the R-matrix calculations. We interpret this as being due to the fact that the RPAE method[17] gives a more accurate nonresonant background than the R-matrix with CI gives in general. Note disagreement between length and velocity R-matrix cross-sections whereas the RPAE method gives identical results.[59] However, we also note that the present R-matrix calculations with the appropriate CI yield widths, Γ, in better agreement with measurements than the RPAE method. The variation of β as a function of photon energy is shown in Fig. 14 where again, the photon source was instrumental in obtaining such detail. The angular-distribution data for five resonances of the series $3s3p^6np\,^1P^\circ n = 4, 8$ are shown in Figs. 15 and 16 and are compared with the fit.

The angular-distribution data is also compared to the untested theory of Taylor[31] based on the R-matrix method and our R-matrix calculations.[56] Taylor[31] used in the first approximation a single-configuration (SC) wavefunction to represent the $^2P^\circ$ and $^2S^e$ ionic states of the direct and indirect photoionization paths. He also made a more sophisticated approximation (CI)[43] where multiconfigurational wavefunctions are used in the representation of these states. Both calculations were done in the length and velocity form. The CI calculations improved the agreement between the length and velocity results for the only two calculated resonances $n = 4$ and $n = 5$ by Taylor;[31] the length and velocity results were coincident. Comparison of the $n = 4, 5$ resonances with the CI calculation of Taylor[31] is shown in Figs. 15(a) and (b). Figures 15 and 16 show the comparison of the $n = 6, 7$ and 8 resonances with our R-matrix calculation.[56] As can be seen, the calculation appears shifted to higher energies with respect to the measurements for the $n = 4$ resonance but this shift almost vanishes for the $n = 8$ resonance. We would like to point out that in the calculations the $3s^23p^6$

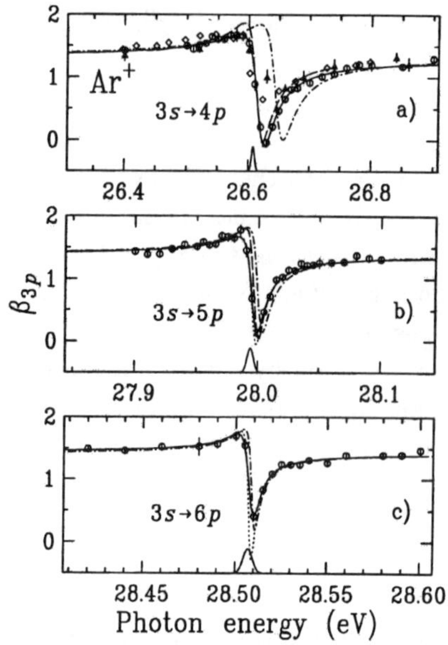

Fig. 15. Angular distribution of the Ar $3s^2 3p^6 \rightarrow 3s3p^6 np$ autoionization resonances for $n = 4, 5$ and 6 fitted to the Kabachnik and Sazhina[39] model. They are compared with the calculation of Taylor[43] for $n = 4$ and $n = 5$ (broken curves) and with our calculations (chain curves) for $n = 6$. The dotted curves are deconvoluted fits. The triangles are $2p_{3/2}$ data from Svensson *et al.*[50] and the full diamonds are taken from Codling *et al.*[32] The Gaussian curve at the resonance represents a monochromator bandpass of 6 meV.

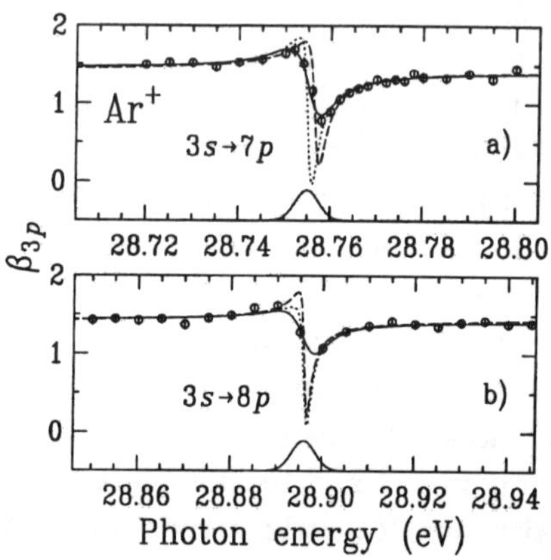

Fig. 16. Angular distribution of the Ar $3s^2 3p^6 \rightarrow 3s3p^6 np$ autoionization resonances compared with calculations (broken curves) from $n = 7$ and $n = 8$. The dotted curves are deconvoluted fits.

Table 4. Angular distributions profile for the first six members of the $3s3p^6np$ series resonances compared with theoretical calculations.

	Method	$3s \to 4p$			$3s \to 5p$			$3s \to 6p$		
		X	Y	Z	X	Y	Z	X	Y	Z
Taylor	SC(L)	−1.53	−2.88	−1.21						
(1978)	SC(V)	−0.81	−0.81	−0.09						
	CI(L)	−1.79	1.85	−0.44	−1.77	1.33	−0.14			
	CI(V)	−1.47	1.44	−0.31	−1.46	1.01	−0.06			
Present	Theory (L)	−1.73	1.99	−0.56	−1.67	1.55	−0.41	−1.66	1.40	−0.34
results	Theory (V)	−1.41	1.73	−0.54	−1.36	1.35	−0.39	1.35	1.22	−0.32
	Experiment	−1.567(12)	1.49(2)	−0.322(12)	−1.483(11)	1.00(2)	−0.158(13)	−1.46(2)	0.95(4)	−0.10(2)

	Method	$3s \to 7p$			$3s \to 8p$			$3s \to 9p$		
		X	Y	Z	X	Y	Z	X	Y	Z
Present	Theory (L)	−1.65	1.36	−0.35	−1.64	1.22	−0.26	−1.63	1.29	−0.33
results	Theory (V)	−1.34	1.19	−0.33	−1.33	1.08	−0.25	−1.32	1.13	−0.31
	Experiment	−1.507(10)	0.97(4)	−0.15(3)	−1.489(14)	0.79(7)	−0.13(6)	−1.49(2)	0.56(9)	−0.46(8)

energy was shifted but we did not shift the calculated resonance energies to match the data. However, we did shift the calculations of Taylor[31] so it matches the data since he did not give the energy position of the lines. In all cases, the dotted lines are deconvoluted fits (6 meV). As can be seen from the figure, agreement between the data and both theories is excellent. The parameters X, Y and Z produced by the fit are listed in Table 4 for $n = 4$ to $n = 9$ along with the results from Taylor.[31] From Table 4, one can note the excellent agreement between the data and the CI calculations. It is clear that the CI calculation is indeed a better model for the data because, in the case of $n = 4$, the SC calculation did not even predict the correct sign of the Y parameter. We also would like to point out that the R-matrix calculations as described in the literature[56] tried, unlike others, to obtain the energy positions. The result is fair for the low resonances but agrees better with the measurements for the higher resonances. Finally, this work has tried to address the issue that Amusia and Kheifets[59] made, namely that there has been insufficient experimental resolution to distinguish between the RPAE and the R-matrix approach. The present measurements detailed indicate that the RPAE results give a more reliable prediction for the q value, but less reliable prediction for Γ. The RPAE treats the nonresonant photoionization process more accurately that the present R-matrix calculations, seen by the large discrepancy between length and velocity nonresonant cross-sections. Thus the q value, which is a measure of the interference between resonances and this nonresonant background, is more accurate in the RPAE method. The large amount of CI included in the present R-matrix calculation leads, however, to a better resultant width.

4. The Resonant Auger Process

The Auger decay of core excited states in rare-gases has been subject to rather intense investigation over a number of years due to its complex nature. For example, the $4d^{-1}np$ transitions in Xe have been studied[60–64] with particular attention focused on the role of shake probabilities.[60] Similar work has been performed on the Ar $2p^{-1}nl$ transitions[63,65–73] (described in more detail below). Angular distributions have been measured for the Xe $4d^{-1}6p$ and $4d^{-1}7p$[64,67,74–76] resonances and low-resolution angular resolved measurements of Ar, Kr and Xe were reported by Carlson *et al.*[67] Theoretical studies have also been made on Ar,[66,70,77–79] Kr[78–80] and Xe.[78–80]

4.1. *Angle-Resolved Ar* $2p^{-1}_{1/2,3/2}ns$*, d Resonant Auger Decay*

After the excitation of an inner-shell electron, an atom can relax nonradiatively through one of the following mechanisms, summarized here for the Ar $2p^{-1}4s$ resonance:

$$2p^5 4s \to 2p^6 3s^2 3p^4 4s + e^- \quad \text{Spectator Auger decay}, \tag{a}$$

$$2p^5 4s \to 2p^6 3s^2 3p^5 + e^- \quad \text{Participator Auger decay — shake down}, \tag{b}$$

$$2p^5 4s \to 2p^6 3s^2 3p^4 4p + e^- \quad \text{Participator Auger decay — shake up}, \tag{c}$$

$$2p^5 4s \to 2p^6 3s^2 3p^4 + 2e^- \quad \text{Participator Auger decay — shake off}. \tag{d}$$

In previous nonangular-resolved studies of the intensity distributions, it has been reported that the decay of the Ar $2p^{-1}4s$ states can be described well by the spectator model, while the decay of the Ar $2p^{-1}3d$ and $2p^{-1}4d$ states is dominated by shake up processes.[63,65,66,68–70,72] In its purest form, the strict spectator model assumes no interaction between the spectator electron, which was resonantly excited from an inner-shell, and the core. Previous angular-resolved measurements of Auger decays, observed at the Ar $2p^{-1}4s$ resonance, have suffered from poor resolution in the photoelectron spectra;[67] and such measurements of the $2p^{-1}3d$ and $2p^{-1}4d$ resonances have been unobtainable due to insufficient photon resolution.

The Auger decays of the $2p^{-1}4s$ resonances were first described theoretically by Cooper[77] who applied angular momentum transfer theory and treated the resonant decay as a single-step process. Hergenhahen *et al.*[78] using the spectator model, treated the resonant Auger transition as a two-step process involving excitation and subsequent Auger decay. Chen[80] has shown the importance of including the effect of intermediate coupling in the calculation of angular distributions of the $2p^{-1}$ hole of normal Auger decay. Later, using the multiconfiguration Dirac–Fock (MCDF) method, in a two-step model, Chen[80] reported calculations of the angular anisotropy of the Ar $2p^{-1}4s$ resonant Auger decays.

Recent measurements of the intensity distributions and angular distributions for the Auger decay spectra following the $2p \rightarrow 4s$, $3d$, $4d$ resonant excitations in argon have been carried out using the TOF system described in Sec. 2.2 at the ALS. A detailed analysis of the angular distribution of the resonant Auger electrons resulting from the decay of the $2p_{1/2}^{-1}4s$, $2p_{3/2}^{-1}4s$, $2p_{1/2}^{-1}3d$, $2p_{3/2}^{-1}3d$, $2p_{1/2}^{-1}4d$ and $2p_{3/2}^{-1}4d$ states and on the angular distribution anisotropy parameters (β parameters) of nearly all possible $3p^4nl$ states has been accomplished.[81] In the case of the $2p^{-1}3d$ and $2p^{-1}4d$ resonances, the angular resolved measurements have been made for the first time. In addition, these results[81] are compared with theoretical work and demonstrate best agreement with the two-step process calculations. They point out that a separation in excitation and decay mechanisms, which are characterized by alignment and intrinsic anisotropy, is justified. In order to determine the photon energies necessary to excite the resonant transitions, a 2D map was recorded using the TOF system. It is shown in Fig. 4 where the y-axis shows the total electron-yield as a function of photon energies and the x-axis shows the kinetic energy spectrum as described in Sec. 2.2.

4.1.1. *The $2p^{-1}4s$ Resonant Auger Spectra*

A comparison of the resonant Auger spectra with normal nonresonant Auger spectra (e.g. Meyer *et al.*[66]) shows that, as predicted by the spectator model, the resonant Auger spectrum looks much like the normal Auger spectrum with only a slight energy shift (due to the screening effect of the spectator electron). This results supports the validity of the spectator model for predicting relative intensities.

Figures 17 and 18 depict portions of the $2p_{1/2}^{-1}4s$ and $2p_{3/2}^{-1}4s$ spectra, respectively. Each of the prominent features is labeled with a number, and each individual LS term is labeled in all figures and tables by a numbered letter.[81] Here, the same binding energy region is viewed from two different angles. The β values are plotted in Figs. 17(c) and 18(c) for the peaks in this region. It can be observed from Figs. 17(b) and 18(b) (note the different scales) that the relative populations of the final ionic states (viewed at different angles) are significantly different, although the β parameter for each state is approximately the same. This illustrates rather well the requirement of the resolving power of these type of measurements.

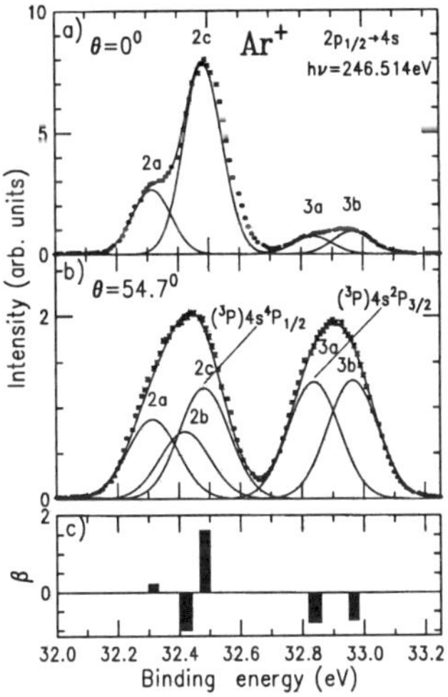

Fig. 17. Auger decay spectra for Ar $2p_{1/2}^{-1}4s$ resonance shown at (a) $\theta = 0°$ and (b) $\theta = 54.7°$. The peak labels correspond to those states listed in Table 5. The angular distribution anisotropy parameters (β) for the $3p^4(^3P)4s^4P$ and $3p^4(^3P)4s^2P$ lines are shown in (c). These spectra were recorded with a 201 V retarding potential applied to the flight tube.

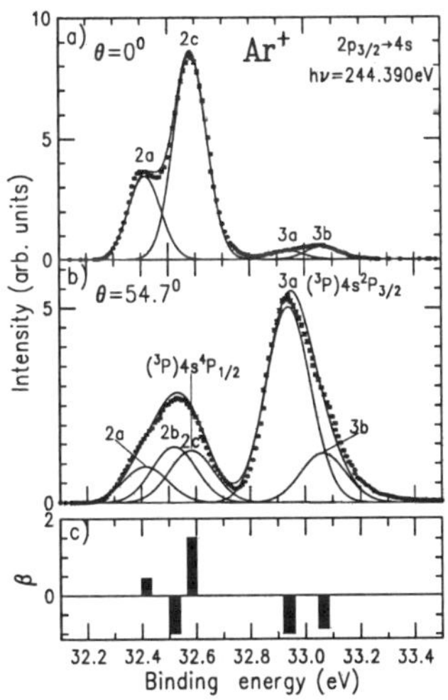

Fig. 18. Auger decay spectra for Ar $2p_{3/2}^{-1}4s$ resonance shown at (a) $\theta = 0°$ and (b) $\theta = 54.7°$. The peak labels correspond to those states listed in Table 6. The β parameters for the $3p^4(^3P)4s^4P$ and $3p^4(^3P)4s^2P$ lines are shown in (c). These spectra were recorded with a 201 V retarding potential applied to the flight tube.

Table 5. Experimental and theoretical angular distribution anisotropy parameters (β) for the Ar $2p_{1/2}^{-1}4s$ resonance. The uncertainty in the last two digits is given in parentheses.

Final State $3p^4nl$		β Parameters: $2p_{1/2} \rightarrow 4s$					
		Experiment				Theory	
		β	β	β		β	β
No	LS-term	This work	Carlson[67]	Menzel[83]		Hergenhahn[78]	Chen[79]
2a	$3p^4(^3P)4s\,^4P_{5/2}$	0.21(4)					0.19
2b	$3p^4(^3P)4s\,^4P_{3/2}$	−1.00(5)					0.81
2c	$3p^4(^3P)4s\,^4P_{1/2}$	1.61(15)					1.8
2a,b,c	$\mathbf{3p^4(^3P)4s\,^4P_{5/2,3/2,1/2}}$	**0.54(7)**		**0.56**	**0.69**	–	**0.62**
3a	$3p^4(^3P)4s\,^2P_{3/2}$	−0.79(4)					−0.79
3b	$3p^4(^3P)4s\,^2P_{1/2}$	−0.73(4)					−0.72
3a,b	$\mathbf{3p^4(^3P)4s\,^2P_{3/2,1/2}}$	**−0.75(4)**	**−0.37**	**−0.57**	**−0.66**	–	**−0.79**
2a,b,c	$\mathbf{3p^4(^3P)4s\,^4P_{5/2,3/2,1/2}}$	**−0.09(2)**	**−0.17**	**0.01**	**0.01**	**0.00**	**−0.007**
3a,b	$\mathbf{3p^4(^3P)4s\,^2P_{3/2,1/2}}$	**−0.09(2)**	**−0.17**	**0.01**	**0.01**	**0.00**	**−0.007**
7a	$3p^4(^1D)4s^2D_{3/2}$	−1.00(5)					
7b	$3p^4(^1D)4s^2D_{5/2}$	0.05(3)					
8a	$3p^4(^3P)3d^2F_{7/2}$	α					
8b	$3p^4(^3P)3d^2F_{52}$	−0.52(3)					
9a	$3p^4(^3P)3d^2D_{3/2}$	−0.15(3)					
9b	$3p^4(^3P)3d^2D_{5/2}$	−0.04(2)					
7a,b	$\mathbf{3p^4(^1D)4s\,^2D_{5/2,3/2}}$						
8a,b	$\mathbf{3p^4(^3P)3d\,^2F_{7/2,5/2}}$	**−0.03(2)**	**−0.22**	**0.01**	**0.00**	**0.00**	**0.006**
9a,b	$\mathbf{3p^4(^3P)3d\,^2D_{5/2,3/2}}$						
12a	$3p^4(^1S)4s^2S_{1/2}$	0.07(3)	−0.20	0.00	0.00	0.00	0.001

β Menzel *et al.*[83] using two different energies, 246.0 eV and 246.5 eV.
α. Strongly positive β.

A complete listing of β parameters for nearly all $3p^4nl$ final ionic states is given in Table 5 for the $2p_{1/2}^{-1}4s$ decays and Table 6 for the $2p_{3/2}^{-1}4s$ decays. In order to compare our results with previously unresolved work, we have listed in the table both the individual β components and their average β value (shown in boldface) which includes the sum of all overlapping peaks in a given LS term. It can be seen that most β parameters are small positive or negative, the latter indicating a preferred emission in the direction perpendicular to the electric field vector of the radiation. This is due to the enhanced population of final ionic states which can be reached via parity unfavored transitions only through the resonant Auger processes.[82] These results are compared with previous lower resolution measurements from

Carlson *et al.*[67] and Menzel,[83] and are compared with theoretical calculations of Cooper,[77] Hergenhahn *et al.*[78] and Chen.[79] In this work, Cooper has treated the Auger decay as a single-step process, Hergenhahn *et al.*[78] have considered excitation and decay of the resonant state as a two-step process and predicts the resonant Auger decay based on the spectator model. Chen used a multiconfiguration Dirac–Fock method in intermediate coupling with configuration interaction, also a two-step model. In general, our results differ considerably from those of Carlson *et al.*[67] and we attribute this to the lack of resolution in the previous experiments. However, our results agree quite well with those of Menzel[83] who did not resolve the individual components. When compared with calculations, we find fair agreement with

Table 6. Experimental and theoretical angular distribution anisotropy parameters (β) for the Ar $2p_{3/2}^{-1}4s$ resonance. The uncertainty in the last two digits is given in parentheses.

Final State $3p^4nl$		β Parameters: $2p_{3/2} \to 4s$					
		Experiment			Theory		
		β	β	β	β	β	β
No	LS-term	This work	Carlson[67]	Menzel[83]	Hergenhahn[78]	Chen[79]	Cooper[77]
2a	$3p^4(^3P)4s\,^4P_{5/2}$	0.45(.06)				0.25	
2b	$3p^4(^3P)4s\,^4P_{3/2}$	−1.00(.05)				0.88	
2c	$3p^4(^3P)4s\,^4P_{1/2}$	1.51(.13)				1.65	
2a,b,c	$\mathbf{3p^4(^3P)4s\,^4P_{5/2,3/2,1/2}}$	**0.32(.05)**	**0.37**	**0.30**	**0.53**	**0.34**	**0.88**
3a	$3p^4(^3P)4s\,^2P_{3/2}$	−1.00(.05)				−0.98	
3b	$3p^4(^3P)4s\,^2P_{1/2}$	−0.85(.05)				−0.87	
3a,b	$\mathbf{3p^4(^3P)4s\,^2P_{3/2,1/2}}$	**−0.94(.06)**	**−0.86**	**−0.93**	**−0.94**	**−0.95**	**−0.83**
2a,b,c **3a,b**	$\mathbf{3p^4(^3P)4s\,^4P_{5/2,3/2,1/2}}$ $\mathbf{3p^4(^3P)4s\,^2P_{3/2,1/2}}$	**−0.50(.03)**	**−0.65 −0.53**	**−0.47**	**−0.50**	**−0.47**	**−0.18**
7a	$3p^4(^1D)4s\,^2D_{3/2}$	0.66(.09)					
7b	$3p^4(^1D)4s\,^2D_{5/2}$	0.50(.06)					
8a	$3p^4(^3P)3d\,^2F_{7/2}$	*					
8b	$3p^4(^3P)3d\,^2F_{52}$	−0.35(.04)					
9a	$3p^4(^3P)3d\,^2D_{3/2}$	−0.51(.06)					
9b	$3p^4(^3P)3d\,^2D_{5/2}$	−0.55(.06)					
7a,b **8a,b** **9a,b**	$\mathbf{3p^4(^1D)4s\,^2D_{5/2,3/2}}$ $\mathbf{3p^4(^3P)3d\,^2F_{7/2,5/2}}$ $\mathbf{3p^4(^3P)3d\,^2D_{5/2,3/2}}$	**0.58(.07)**	**0.45 0.36**	**0.62**	**0.42**	**0.46**	**0.73**
12a	$3p^4(^1S)4s\,^2S_{1/2}$	0.93(.09)	0.66 0.39	1.09	1.00	0.98	1.00

Large positive β value.

Cooper's work[77] (Table 6), good agreement with Hergenhahn[78] and very good agreement with Chen.[79] It should therefore be noted that the β parameters agree best with the two-step process calculations.[78,79]

One can also note that the spectator model[78] agrees relatively well with the measured β parameters. It should be noticed that from Tables 5 and 6 that, as predicted by the spectator model, the combined anisotropy of the ^2P and ^4P transitions is very close to zero in the case of the $2p_{1/2}^{-1}4s$ resonance and to -0.5 in the case of the $2p_{3/2}^{-1}4s$ resonance. The spectator model, however, demonstrates rather poor agreement with the measured relative intensities when compared to Chen's model[79] which seems to perform better on both fronts. One possible explanation may be that Chen's model[79] calculates the full state vector and takes into account the mixing of all the states, which is important in the determination of the intensities. The spectator model instead uses simple assumptions[78] by relating the resonant and nonresonant spectra. It uses good Coulomb matrix elements but perhaps insufficient state vectors. This deficiency, however, does not affect the calculated β values significantly because β is a relative quantity describing the character of a certain transition independent of its strength. The spectator model describes the β values in terms of the most simple geometric coupling situation yielding good values for the sum of the multiplet components, whereas the refined calculations of Chen[79] give also the correct relative intensities of the different multiplets with respect to each other, and even good values for the individual multiplet components with respect to both β and the relative intensities. The spectator model allows a good first approximation of the β parameters for the $2p^{-1}4s$ resonance and gives a simple physical explanation for the differences in the angular distribution of resonant and nonresonant Auger electrons.

4.1.2. *The $2p^{-1}3d$ and 4d Resonant Auger Spectra*

The complete $2p_{3/2}^{-1}3d$ and $2p_{1/2}^{-1}3d$ spectra as a test case for the validity of the spectator model have also been measured and contrary to the $2p^{-1}4s$ case described above, this resonant transition does not obey the predicted behavior of the spectator model, i.e. the observed spectra are much different from the normal nonresonant Auger spectra.[66] Instead, the spectra are dominated by shake-up processes. Reasons for this have been discussed by Aksela *et al.*[65] and Mursu *et al.*[69] who have demonstrated the importance of the shake up process arising from the different degree of collapse of the

$3d$ wavefunction in the initially excited $2p^{-1}3d$ state, as compared to the final ionic state. Another explanation has been provided by Meyer *et al.*[66] who has described in some detail the special properties of the wave function which leads to a high shake-up probability. A closer look at the corresponding resonant Auger spectra, however, revealed that the occurrence of strong shake transitions did not completely alter the character of the spectrum, in particular concerning the angular distribution of the resonant Auger lines. Hence, a shake-modified spectator model was introduced. Concerning the angular distribution, the relationship between resonant and nonresonant Auger electrons could be further specified, and as a result, the strict[64,78,84] spectator models were introduced. In the strict spectator model, the LS final states are completely determined by the coupling of the spectator electron to the core in the kj coupling scheme, whereas in the gross spectator model, only the coupling of the core remains fixed. The strict spectator model could successfully be applied to predict the angular distribution anisotropy parameter β for many prominent resonant Auger lines, e.g. of the Ar $2p^{-1}4s$ resonance, but in other cases it failed again. These examples were considered as the real breakdown of the spectator model. On the other hand, there has to be a smooth transition between the resonant and the nonresonant cases along the principal quantum number n. Tulkki, Aksela and Kabachnik[85] studied this evolution of the Auger anisotropy with n. They came to the conclusion that the recovery of the spectator model occurs only at very high n and fail for low n, even for a show case of the spectator model such as the Ar $2p^{-1}ns$ excitations. However, they studied Auger transition cases in which proving their prediction was difficult because there was no experimental evidence for deviations from the spectator model predictions even for the case[86] of $n = 4$.

Figures 19 and 20 show resonant Auger spectra of the $2p_{1/2}^{-1}3d$ and $2p_{1/2}^{-1}4d$ excitations, respectively, that were used in order to demonstrate a possible recovery of the strict spectator model. A complete list of relative intensities and β parameters can be found in the literature.[81] The β parameter values for the $2p^{-1}3d$ resonances are the only reported results and demonstrate small positive or negative β parameters. This supports Cooper's theory[77] which was only applied to the case of the $2p^{-1}4s$ resonances, but which seems to be also valid for the $2p^{-1}3d$ and $2p^{-1}4d$ resonances. The similarity of β values at the two different spin-orbit resonances is again demonstrated here.

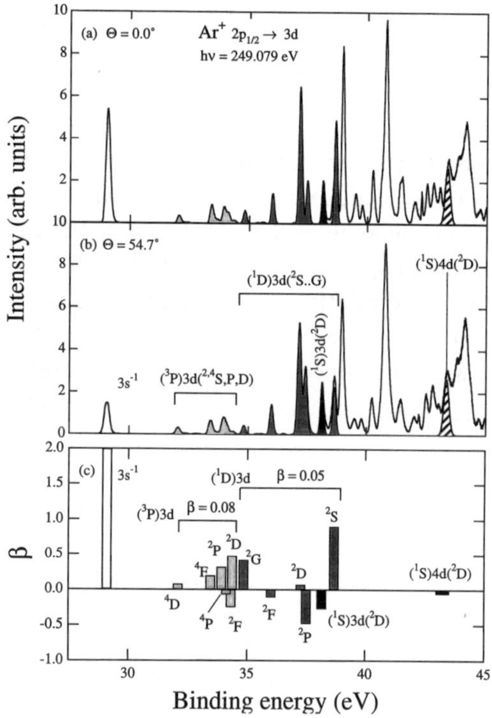

Fig. 19. High-resolution resonant Auger spectra taken on the Ar $2p_{1/2} \to 3d$ resonance. The light-grey shaded curves correspond to the ^3P core, the dark-grey shaded curves to the ^1D core, and the heavy shaded curves to the ^1S core. The β values in (c) are shown by bars shaded correspondingly. The average β of all lines with the same core parent term are shown above or below by their numbers.

No theoretically predicted β values are available, therefore no comparison is possible at present. However, a comparison with the spectator model can be made if we restrict it to the so-called 'gross spectator model' (GSM) and to the $2p_{1/2} \to 3d$ resonance. This seems to be unjustified because of the difference in the structures of the resonant and nonresonant spectrum. But, as mentioned earlier, the validity of the spectator model may be discussed in the context of two different situations: the presence of a strong shake-up probability, and the existence of strong configuration interaction and singlet–triplet mixing. It is the second situation that really causes a breakdown of the spectator model.[87] The shake-modified spectator model would predict the same β values for the diagram and shake-up transitions

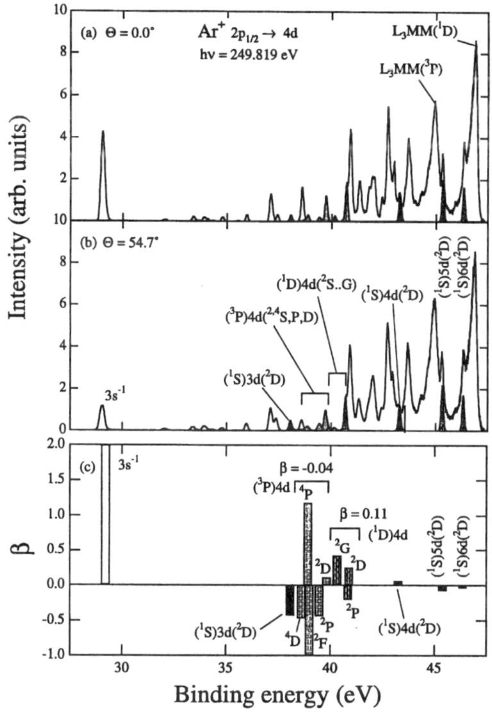

Fig. 20. High-resolution resonant Auger spectra taken on the Ar $2p_{1/2} \rightarrow 4d$ resonance represented in the same way as the $2p_{1/2} \rightarrow 3d$ spectra shown in Fig. 19.

because shake-up probability is basically an overlap factor that does not affect the rest of the Coulomb matrix elements. Thus, the spectrum may look quite different but the angular distribution is similar. In order to prove this without numerical calculations, the test case of the $2p_{1/2} \rightarrow nd$ excitations was chosen because they converge to the same ionization limit as the ns excitations with a spherical $2p_{1/2}$ final ionic state. Therefore, all the arguments for the ns excitation apply also to the nd excitations. Since we have no β values for the different multiplets, we can compare our data only with the sum of the doublet and quartet states. This sum β has to be zero according to the spectator model which in this particular restricted version is known as gross spectator model. A comparison with the ^1S, ^1D and ^3P states shows β values of -0.27, 0.05 and 0.08, respectively, corresponding quite well — except for the ^1S state — with the prediction of the spectator model. The deviation of the ^1S core state is most pronounced for the $3d^2D$

state with its $\beta = -0.27$. The even stronger shake-up state $4d\,^2D$, however, shows already a β value close to zero in accordance with the gross spectator model. The behavior of the $3d$ state could be considered a real breakdown of the spectator models as observed for magnesium by Hergenhahn *et al.*[87] The collapsed $3d$ wavefunction may cause mixing between different parent states but also between the $2p_{1/2}$ and $2p_{3/2}$ core hole states. Both interactions would cause a measurable deviation from the GSM predictions. These deviations, however, seem to disappear already for $n = 4$ suggesting to consider the $3d$ behavior as a collapse induced anomaly rather than a common deviation in resonant Auger decay. In order to prove if this statement holds also for higher n, the spectrum of the $2p \to 4d$ resonance is examined (Fig. 20). Concentrating on the ^1S, ^1D and ^3P parent states with $4d$ spectator electron the sum of the β parameters are the following: 0.06, 0.11 and −0.04. Obviously, the correspondence with the gross spectator model prediction is much more favorable than in the case of the $2p_{1/2} \to 3d$ excitation. All three parent states of the $4d$ spectator electron still agree quite well. In contrast, the $3d$ shake-down state deviates significantly from the predicted vanishing Auger anisotropy. The observed β value is somewhat more negative than the value observed for the $3d$ excitation, but still relatively similar.

If the reasons for these deviations are the same for both resonances the higher shake-up states should recover to the GSM values. The β values for the $5d$ and $6d$ shake-up states indeed nicely confirm this prediction, although their position in the wings of the two LMM Auger lines makes a very reliable determination of the β values difficult. The sum of all (^1S) parent states already starts to approach the GSM β value showing that the β value of the normal Auger lines are approached via a successive inclusion of higher and higher n states in an Auger like line distribution of shake-up states. Further measurements on resonances with even higher n would be helpful to substantiate the predictions on the resurgence of the spectator model made here. Experimentally, for the higher n values, high-resolution measurements in small steps as performed recently by Camilloni *et al.*[72] have to be performed across these resonances following the behavior of each resonant Auger line by angle-resolved electron spectroscopy.

4.2. *Angle-Resolved Kr* $3d^{-1}_{3/2,5/2}np$ *Resonant Auger Decay*

Studies[88–95] in Kr have concentrated on the decay to the $Kr^+4s^24p^{-2}mp$ states. However, the Auger decay spectra of the Kr $3d^{-1}np \to 4s^{-1}4p^{-1}$

$(^1P)mp$ (n = 5–9, m = 5–11) transitions have received much less attention.[96,97] As a results, extensive experimental results on photoexcitation and photoionization processes in Kr, in the region of the $3d$ ionization thresholds[98] using the 2D TOF technique have been carried out. In particular, the $3d^{-1}np \rightarrow 4s^{-1}4p^{-1}(^1P)mp$ ($n = 5$–9, $m = 5$–11) Auger decay transitions have been explored, and for the first time, the angular distributions of the resonant Auger electrons resulting from the decay of $3d^{-1}_{3/2,5/2} \rightarrow np$ states have been measured. In addition, spectator and shake probabilities for the $3d^{-1}np \rightarrow 4s^{-1}4p^{-1}(^1P)mp$ ($n = 5$–9, $m = 5$–11) resonant Auger transition have been obtained for almost all the final $4s^{-1}4p^{-1}(^1P)mp$ ($m = 5$–11) states.

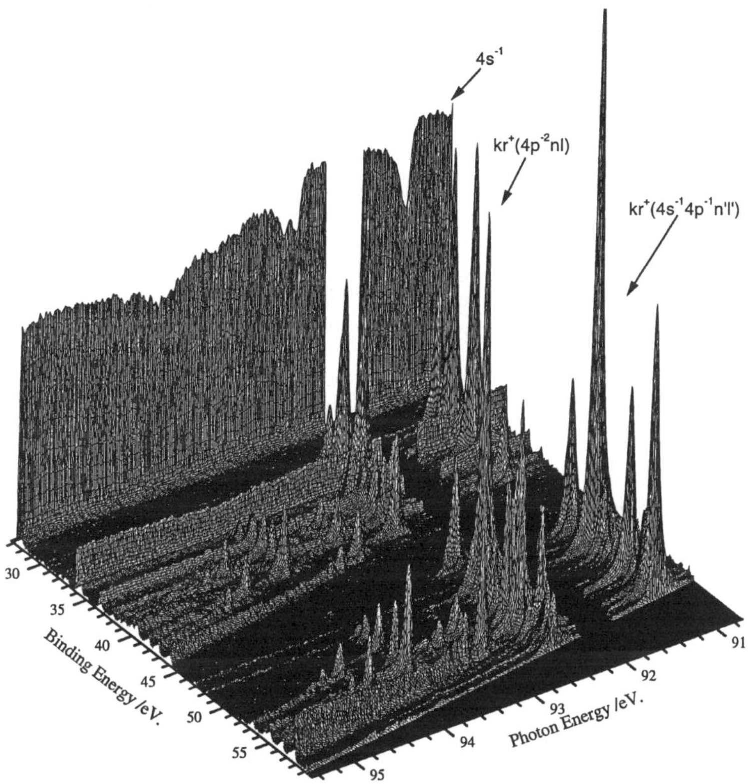

Fig. 21. Three-dimensional mapping of the resonant Auger landscape measured at 0° relative to the polarization plane.

A three-dimensional landscape of the resonant Auger emission is presented in Fig. 21, which was obtained, at an angle of 0° relative to the photon polarization plane. The z-axis represents the intensity, the y-axis the photon energy, and the x-axis the binding energy. This figure gives a sense of the extent of the competition between participator (Kr^+4s^{-1}) and spectator ($Kr^+4s^24p^4nl$ and $4s^14p^5n'l'$) decay processes. The Kr^+4s^{-1} participator channel clearly shows asymmetric line profiles at the lower lying resonances, while the spectator decay channels show up as peaks all over the ionization landscape.

To get a more detailed view of this landscape, the projection of the intensity onto the x-y plane is presented in Fig. 22 as a contour map with gray levels added to indicate higher intensities (increasing from light to dark). The total electron-yield spectrum and the photoelectron spectrum are shown at the side and the top of this 2D map. This 2D contour map

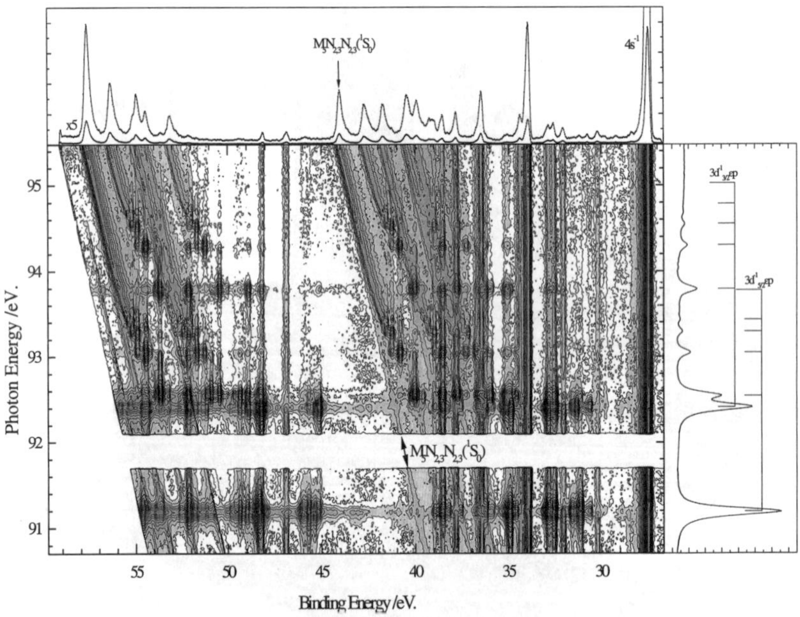

Fig. 22. Two-dimensional (2D) map showing Auger electron emissions of the Rydberg resonances measured at 0° relative to the polarization plane. The intensity of the electron signal is represented as different shades of gray, where darker regions indicate stronger intensities. The total electron-yield spectrum and the photoelectron spectrum are shown on the right and on the top respectively.

allows examination of all of the data in a single figure from which features can be identified for more detailed analysis. In this section, we will concentrate on the evolution of the Auger decay processes of the $3d \rightarrow np$ Rydberg series $n = 5$–11 over a photon energy range of 6 eV. To produce 2D elelctron emission data sets, a photon energy range of 6 eV and a kinetic energy window covering the majority of the ejected elelctrons of interest, corresponding to the Kr^+ $4s^2 4p^4 nl$ and $4s^1 4p^5 n'l'$ ionic states, was chosen. This allows detailed investigation of the Kr $3d$ photoexcitation ionization relaxation processes in the resonance region below the threshold up to a few eV above the ionization limit. The photon energy region covered in the 2D map can be broken down into three parts:

(a) The low-lying $3d^{-1}np$ ($n = 5$–9) resonances.
(b) The merging of the high-lying resonances and the ionization edges to which they converge.
(c) The $3d$ ionization which is followed by Auger decay, which results in spectral lines that exhibit shifting, and broadening due to post collision interaction (PCI) between the two outgoing electrons.

In the case of (a), the low-lying resonances decay preferentially by Auger emission where the angular momentum of the Rydberg electron is unchanged but may undergo changes in its principal quantum number. In the case of (b), the resonant states lying just below the $3d$-ionization threshold can be excited at the same photon energy at which ionization is also possible, due to the essentially constant resonance linewidth (83 meV) observed for these states which is set by the core hole lifetime and not the Rydberg electron. When the photon energy is in the region of the ionization threshold, the shakedown lines becomes more important and merge smoothly into the high kinetic energy tail of the PCI broadened normal Auger lines.[93] Details of how this evolution takes place is not clear from a theoretical point of view.[99]

In the case of (c), the so-called normal Auger decay appear in the 2D map along diagonal lines, which represent the direction of constant kinetic energy. This decay involves the ejection of an inner-shell electron into the continuum followed by the decay of another electron to the inner-shell vacancy and the simultaneous ejection of a second electron, leaving a doubly ionized ion core.[1] The Auger lines however, exhibit shifting, and broadening due to PCI since the slow photoelectron and the much faster Auger electrons can interact while still in the field of the nucleus since the core hole lifetime is so short. The effect is most clearly seen in Fig. 22 for the $M_5N_{2,3}N_{2,3}(^1S_0)$ Auger line where there is also a visible trace of the same line (produced by higher order light) to the higher binding energy side (i.e. lower kinetic energy) throughout the photon energy range covered. Since here the photon energy is much higher, the photoelectron is consequently much further away from the atom before the Auger electron is ejected and so PCI effects are no longer important.

In order to analyze the details and obtain quantitative information, we extracted photoelectron spectra from the 2D map. Figure 23(a) shows a photoelectron spectrum measured at an angle of 54.7° with respect to the electric field vector of the photon beam. The photon energy of 91.2 eV, corresponds to the Kr $3d_{5/2}5p$ resonance. This spectrum displays on resonance the population of the Kr $4s^{-1}$, $4p^{-2}nl$, $4s^{-1}4p^{-1}n'l'$ and $4s^{-2}n''l''$

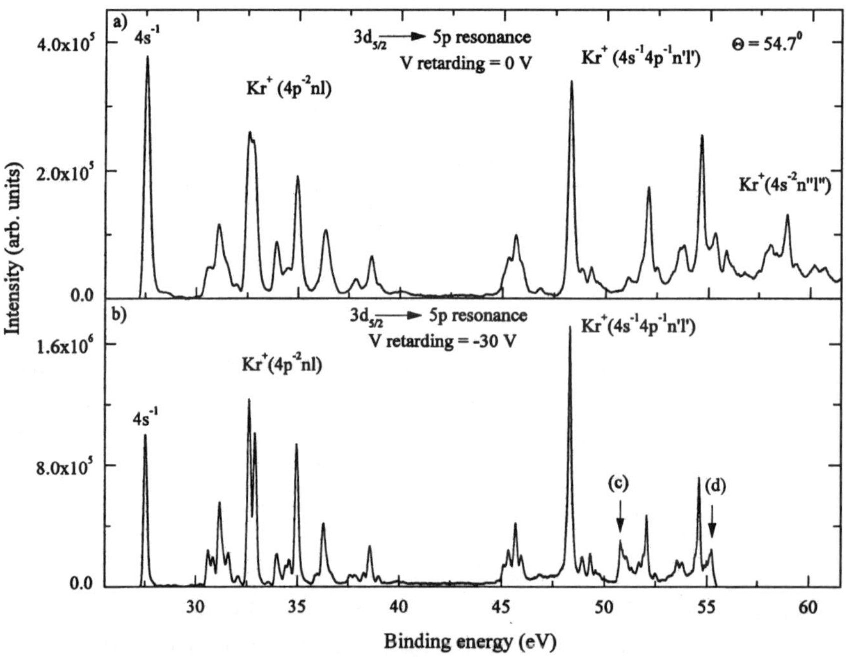

Fig. 23. Photoelectron spectrum of the $3d_{5/2} \to 5p$ resonance taken at the 54.7° with respect to the electric field vector of the incident radiation. (a) Collected with no retarding potential. (b) Collected with −30 V retarding potential. (The two data set were calibrated with two different transmission function due to the two retarding voltage).

ionic states. TOF analyzers have an energy resolution, which is a function of the kinetic energy of the electrons, with the best resolution at lowest kinetic energies. Under the present conditions, the best resolution was 110 meV decreasing to 147 meV at the highest kinetic energies of interest. The photon energy resolution was set to 40 meV at 90 eV. Clearly, one can see that the overall intensity of the lines resulting from transitions to the Kr^+ $(4s^{-1}4p^{-1}n'l')$ states is similar to that of the transitions to the $Kr^+(4p^{-2}nl)$ states (the ratio of the intensities of the two group of lines was 1.2). Similar results have been seen in the case of the Xe $5s^{-1}5p^{-1}n'l'$ and Xe $5p^{-2}nl$ following the Xe $4d_{3/2} \to 6p$ decay.[100] In order to study the $4s^{-1}4p^{-1}n'l'$ ionic states with optimum resolution, we used a retarding voltage of −30 V on the TOF analyzers as shown in Fig. 23(b). It should be noted that, the peak *c* arises from an experimental artifact and peak *d* is at the cut off point of the spectrum caused by our choice of retarding voltage.

Assignments and experimental shake-up probabilities for each $4s^{-1}4p^{-1}$ $(^1P)np$ final state are given in Table 7. The experimental shake-up probabilities were obtained as a percentage of the total intensity of all the $4s^{-1}4p^{-1}(^1P)np$ final states at each resonance. It is very clear from Table 7 that the spectator core coupling is very strong in the decay of $3d^{-1}5p \to 4s^{-1}4p^{-1}(^1P)5p$ and $3d^{-1}6p \to 4s^{-1}4p^{-1}(^1P)6p$ excitations. In the region where the resonance lines are well separated, (n = 7–9), the probabilities of the spectator transition become very small (less than 9%) and shake-up processes ($Kr^+\, np$ $Kr^+(n+1)p$) become very important. A comparison of these results with measurements and theoretical calculations made by Mursu *et al.*[97] showed a similar behavior as indicated in Table 7. Similar behavior has been found in the decay of the $3d^{-1}5p \to 4p^{-2}5p$ and the $3d^{-1}6p \to 4p^{-2}6p$ excitations in Kr by Jauhiainen *et al.*.[92] The probability for the $3d_{5/2}{}^{-1}5p \to 4s^{-1}4p^{-1}(^1P)6p$ excitation was found to be 13%. Significantly higher shake-up probability (21%) can be seen in the case of $3d_{3/2}{}^{-1}5p \to 4s^{-1}4p^{-1}(^1P)6p$ excitation which is also found for the decay to the $4p^{-2}np$ case.[92] This similarity is extended to comparison with the Xe $4d^{-1}6p$ resonances,[101] where we notice that the shake probabilities behaved exactly the same way.

Table 8 shows the experimental β values extracted from the $3d^{-1}np \to 4s^{-1}4p^{-1}(^1P)mp$ (n = 4–9, m = 5–11) resonance Auger transitions. For each β value, the uncertainty in the last digit is given in parentheses.

Table 7. Experimental spectator and shake probabilities (in percentage of the total intensities of all the $4s^{-1}4p^{-1}(^1P)mp$) resonances for the Kr $3d^{-1}np \to 4s^{-1}4p^{-1}(^1P)mp$ ($n = 5$–9, $m = 5$–11) resonance Auger transitions. The statistical uncertainty of the last digits is given in parentheses.

Final ionic state	Resonant state $3d^{-1}np$																			
	$n=5$				$n=6$				$n=7$				$n=8$				$n=9$			
	$3d^{-1}_{5/2}$		$3d^{-1}_{3/2}$		$3d^{-1}_{5/2}$		$3d^{-1}_{3/2}$		$3d^{-1}_{5/2}$		$3d^{-1}_{3/2}$		$3d^{-1}_{5/2}$		$3d^{-1}_{3/2}$		$3d^{-1}_{5/2}$		$3d^{-1}_{3/2}$	
	This work	Ref. 97 Exp./Calc.	This work	Ref. 97 Exp./Calc.	This work	Ref. 97 Exp./Calc.	This work	Ref. 97 Exp./Calc.	This work	Ref. 97 Exp./Calc.	This work	Ref. 97 Exp./Calc.	This work	Ref. 97 Exp./Calc.	This work	Ref. 97 Exp./Calc.	This work	Ref. 97 Exp.	This work	Ref. 97 Exp.
$4s^{-1}4p^{-1}(^1P)5p$	**64.3(4)**	75/77	56.6(5)	18/8	8.2(3)	18/8	9.2(3)	7/8	–	-/-	–	-/-	–	-/-	–	-/-	–	–	–	–
$4s^{-1}4p^{-1}(^1P)6p$	13.0(3)	20/22	21.2(5)	20/23	42.1(7)	29/35	28.7(6)	27/33	9.8(3)	10/10	9.2(3)	9/10	5.7(3)	18/4	5.6(2)	11/4	1.7(3)	9	1.1(3)	10
$4s^{-1}4p^{-1}(^1P)7p$	5.9(3)	5/0	1.7(3)	5/0	41.1(6)	53/56	46.1(5)	66/57	2.2(4)	5/4	2.5(3)	0/4	0.4(4)	4/4	1.6(3)	-/-	1.3(3)	5	1.8(4)	–
$4s^{-1}4p^{-1}(^1P)8p$	5.3(2)		11.0(3)		3.3(3)		5.1(2)		**37.1(8)**	73/73	**47.2(5)**	73/74	1.5(3)	6/2	4.1(2)	3/0	–	6	2.1(3)	–
$4s^{-1}4p^{-1}(^1P)9p$	4.1(2)		3.3(2)		3.1(3)		5.2(3)		25.1(5)	12/10	38.3(7)	18/10	**40.7(5)**	43/57	**44.4(7)**	54/57	4.3(2)	11	1.5(4)	7
$4s^{-1}4p^{-1}(^1P)10p$	5.3(4)		3.0(5)		1.0(4)		2.0(3)		16.0(6)		–		29.0(4)	30/31	41.7(6)	32/31	**67.8(6)**	38	**60.2(5)**	36
$4s^{-1}4p^{-1}(^1P)11p$	2.1(8)		3.2(6)		1.2(5)		3.7(4)		9.8(4)		2.8(4)		22.7(4)	-/-	2.6(3)	-/-	24.9(5)	31	33.3(5)	40

Table 8. Experimental angular distribution anisotropy parameters (β) for the Kr $3d^{-1}np \rightarrow 4s^{-1}4p^{-1}mp$ ($n = 5$–9, $m = 5$–11) resonance Auger transitions. For β value, the uncertainty is given in parentheses.

Final ionic state	Resonant state $3d^{-1}np$									
	$n = 5$		$n = 6$		$n = 7$		$n = 8$		$n = 9$	
	$3d^{-1}_{5/2}$	$3d^{-1}_{3/2}$	$3d^{-1}_{5/2}$	$3d^{-1}_{3/2}$	$3d^{-1}_{5/2}$	$3d^{-1}_{3/2}$	$3d^{-1}_{5/2}$	$3d^{-1}_{3/2}$	$3d^{-1}_{5/2}$	$3d^{-1}_{3/2}$
$4s^{-1}4p^{-1}(^1P)5p$	**−0.18(6)**	**−0.14(5)**	−0.24(8)	−0.26(9)	–	–	–	–	–	–
$4s^{-1}4p^{-1}(^1P)6p$	0.68(16)	0.06(2)	**−0.20(5)**	**0.14(4)**	0.50(12)	0.12(3)	0.37(9)	0.66(15)	0.17(4)	0.14(3)
$4s^{-1}4p^{-1}(^1P)7p$	1.73(17)	0.83(8)	0.86(9)	0.31(3)	**−0.22(2)**	–	−0.03(3)	–	–	–
$4s^{-1}4p^{-1}(^1P)8p$	0.46(11)	0.46(11)	0.28(7)	−0.26(6)	0.02(5)	−0.01(3)	**0.20(5)**	**0.03(8)**	−0.19(5)	−0.33(8)
$4s^{-1}4p^{-1}(^1P)9p$	0.42(10)	0.54(13)	0.18(5)	−0.18(5)	−0.28(7)	0.33(8)	0.77(19)	0.17(4)	**0.43(11)**	**−0.06(2)**
$4s^{-1}4p^{-1}(^1P)10p$	0.01(2)	0.71(17)	0.21(5)	−0.11(3)	−0.60(15)	0.52(13)	0.9(2)	0.62(15)	0.9(2)	0.48(12)
$4s^{-1}4p^{-1}(^1P)11p$	−0.13(3)	−0.18(5)	−0.10(3)	0.03(2)	−0.54(13)	0.10(3)	1.1(3)	0.72(18)	1.1(3)	0.76(19)

These have been estimated by comparing β values for the $3_{5/2}{}^{-1}5p \rightarrow 4s^{-1}4p^{-1}(^1P)5p$ transition from two independent data sets (0 and 30 V retarding potential). From the table, we can note that close to isotropic angular distributions are observed for most of the $3d^{-1}np \rightarrow 4s^{-1}4p^{-1}(^1P)mp$ $(n = m)$ strict spectator decays, while for decays involving shake-up $(m > n)$ or shake-down $(m < n)$ larger anisotropies are seen. According to the prediction of the general strict spectator model, which was introduced by Hergenhahn *et al.*[78] these β values have to be zero, which agrees qualitatively with our experimental observations. When the $3d$ electrons are excited to the $7p$ or higher Rydberg orbitals, the strength of the spectator core coupling diminishes, the shake-up transition becomes dominant, and a fluctuation of β values is observed. This new angular information now awaits theoretical treatment.

4.3. *Angle-Resolved Xe $4d_{5/2} \rightarrow 6p$ Decay under Auger Resonant Raman Conditions*

Auger resonant Raman[102] spectroscopy is a powerful tool for studying the resonant Auger decay processes with a resolution narrower than the natural lifetime width of the initial inner-shell hole state.[103] This effect has been used to analyze branching ratios of resonantly excited atoms[101,104] and molecules.[105] The resonant Raman conditions have been used to study the angular distributions of the spectator decay line of Xe following $4d_{5/2} \rightarrow 6p$ excitation using the TOF system at the ALS. The energy positions and intensities of the peaks in the Xe $4d_{5/2}6p \rightarrow 5p^46p$ decay spectrum, as shown in Fig. 24, were measured with a resolution better than the natural linewidth (106 meV[106]) of the $4d$ inner-shell hole. The angular distribution parameters β of almost all of the possible final ionic $5p^4(^3P, {}^1D, {}^1S)6p$ states has been determined.[75]

The resonant Auger decay spectrum of the Xe $4d_{5/2} \rightarrow 6p$ resonance was first reported by Eberhardt *et al.* in 1978[107] and has been followed by other experimental and theoretical studies.[60,96,108–111] It took more than a decade after the first observation until measurements on the angular distribution were performed by Carlson *et al.*[67] who found anomalously negative β-values in the decay spectrum. Such behavior was first explained theoretically for the decay of the Ar $2p \rightarrow 4s$ resonance by Cooper,[77] who applied angular momentum transfer theory, treating the resonant decay as a single-step process. Earlier work[64,85] were also done at second-generation sources, but was limited by the low-resolution of the photon source.

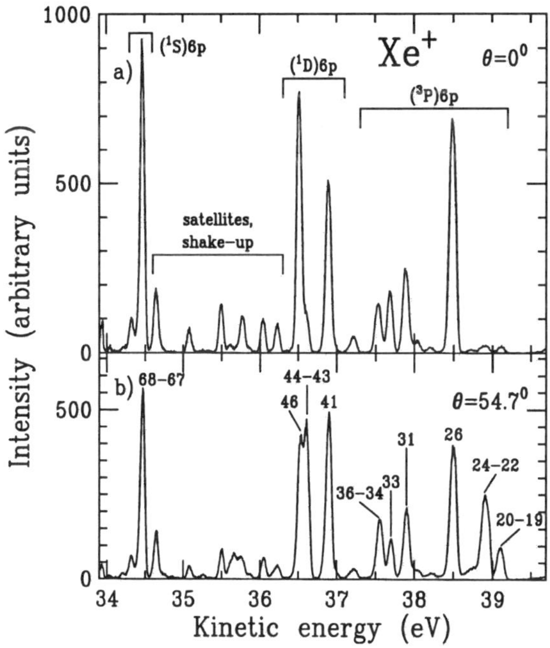

Fig. 24. Xe $5p^4nl$ decay spectra after $4d_{5/2} \to 6p$ resonant excitation at (a) 0° and (b) 54.7° with respect to the polarization of the incident photons. The spectra were recorded with a 30 V retarding potential which corresponds to a spectrometer resolution of between 45 and 70 meV in the display region.

After a Xe $4d \to 6p$ excitation, the decay process can involve (1) an excited electron (participator decay) resulting in an enhancement of the $5p^{-1}$ or $5s^{-1}$ main lines or (2) an excited $6p$ electron that remains in its state during the decay process (spectator decay) leaving the ion in a two-hole, one-electron (satellite) state. The spectator decay is the dominant process (57%), followed by simultaneous emission of two electrons (shake-off), leaving almost no intensity for the participator decay.[60] During the decay, the excited $6p$ electron can also move into the $7p$ orbital (shake-up) enhancing the $5p^4\,7p$ final states. In this section we focus on the strongest spectator decay channels, $5p^4(^3P,\,^1D,\,^1S)6p$.

The results for the relative intensities and the angular distribution parameters β are shown in Table 9 together with theoretical calculations.[78,79,85,112] Chen,[79] Tulkki *et al.*,[85] and Hergenhahn *et al.*[112] used a multiconfiguration Dirac–Fock (MCDF) method in intermediate coupling

Table 9. Intensities and β parameters of the electron spectrum of Xe after $4d_{5/2} \rightarrow 6p^{3/2}$ excitation (65.11 eV). Intensities are normalized relative to the well separated $(^3P)6p(^2P_{3/2})$ line (line 26). The statistical uncertainty of the last digits is given in brackets.

Final ionic state term	Line in Experiment	Kinetic energy (eV)[a]	Relative intensity	β	β theory			
					Ref. 85	Ref. 79	Ref. 112	Ref. 78
$5p^4(^3P)6p\,^4P_{3/2}$	19	39.119	2.0(5)	1.4(6)	1.045	0.984	0.061	1.014
$(^3P)6p\,^4P_{5/2}$	20	39.098	23.2(7)	−0.85(3)	−0.994	−1.000	−0.999	−0.998
$(^3P)6s\,^2D_{3/2}$	21[c]	38.975	0.82(12)	2.0(3)				
$(^3P)6p\,^2D_{5/2}$	22	38.906	37.5(8)	−0.967(1)	−0.994	−1.000	−1.000	−0.998
$(^3P)6p\,^2S_{1/2}$	23	38.886	17.1(6)	−0.69(3)	−0.448	0.215*	0.157*	0.451*
$(^3P)6p\,^4D_{7/2}$	24	38.882	17.1(6)	−0.69(3)	−0.558	−0.974	−0.923	−0.932
$(^1D)5d\,^2GP_{9/2,7/2}$	25[c]	38.738	4.0(2)	−0.16(3)				
$(^3P)6p\,^2P_{3/2}$	26	38.501	100	1.30(2)	1.030	1.018	0.972	1.014
$(^1D)5d\,^2F_{5/2}$	27[c]	38.126	2.9(2)	0.60(11)				
$(^3P)6p\,^2P_{1/2}$	28	37.988	5.0(2)	1.03(7)	0.984	0.962*	0.749*	n.a.[b]*
$(^3P)6p\,^4P_{1/2}$	30	37.955	7.1(3)	−0.13(6)	0.233	0.774*	0.927	1.000
$(^3P)6p\,^2D_{3/2}$	31	37.899	42.8(5)	0.73(3)	0.656	0.653*	0.910*	0.800*
$(^3P)6p\,^4D_{5/2}$	32	37.716	1.3(4)	0.3(6)	−0.188	−0.331	−0.323	0.737
$(^3P)6p\,^4S_{3/2}$	33	37.627	24.0(6)	1.13(5)	0.745	0.955	0.557	−0.861
$(^3P)6p\,^4D_{3/2}$	34	37.570	22.1(5)	−0.14(3)	−0.536	−0.860*	−0.764*	−0.861*
$(^3P)5d\,^4G_{5/2}$	35[c]	37.765	22.1(5)	−0.14(3)				
$(^3P)6p\,^4D_{1/2}$	36	37.535	19.5(4)	0.52(3)	0.593	0.935*	0.817*	1.000*
$(^1D)5d\,^2P_{1/2}$	37[c]	37.232	5.6(2)	1.36(6)				
$(^1D)5d\,^2D_{3/2}$	38[c]	37.169	2.7(2)	0.69(11)				
$(^1D)6p\,^2F_{5/2}$	39	37.001	2.10(14)	−0.85(10)	−0.875	−0.860	−0.914	−0.928
$(^1D)6p\,^2S_{1/2}$	40[c]	36.959	3.0(2)	2.0(3)				
$(^1D)6p\,^2P_{3/2}$	41	36.902	82.7(10)	0.47(2)	0.175	0.073	0.319	−0.399
$(^1D)6p\,^2F_{7/2}$	42	36.853	24.3(5)	−0.11(3)	0.246	0.052	0.116	0.112
$(^1D)6p\,^2D_{3/2}$	43	36.621	39.90(7)	−0.66(2)	−0.553	−0.375	−0.375	−0.399
$(^1D)6p\,^2D_{3/2}$	44	36.587	51.0(8)	−0.65(2)	−0.888	−0.882	−0.930	−0.928
$(^3P)7s\,^2S_{1/2}$	45[c]	36.550	s.n.[d]					
$(^1D)6p\,^2P_{3/2}$	46	36.521	63.1(6)	1.66(2)	1.503	1.307	0.55	0.373
$(^3P)7s\,^2P_{3/2}$	47[c]	36.232	13.9(3)	0.94(4)				
$(^1S)6p\,^2P_{1/2}$	65	34.602	1.6(3)	0.73(4)	0.130	−0.139	−0.035	n.a.[b]
$(^1S)6p\,^2P_{3/2}$	67	34.479	98.9(6)	1.17(4)	0.928	0.847	0.754	0.800

[a] According to Aksela *et al.*[104]
[b] Not allowed
[c] Satellite line
[d] Strongly negative

with configuration interaction, whereas the older calculations of Hergenhahn *et al.*[78] were carried out in jK coupling applying a strict spectator model. Only Tulkki *et al.*[85] include exchange with different continuum channels in their calculations. All the theoretical calculations have in common that both the direct photoionization and the participator decay are neglected, and these approximations have been verified experimentally.[96,108]

Besides the $5p^4 6p$ spectator lines, Table 9 also includes some pure satellite lines. There is fair agreement between our intensities and those reported by Aksela *et al.*[104] (not shown) at least for the most intense lines. For small lines which are close to a strong line, our intensities tend to be larger than those of Aksela *et al.*[104] This may be due to the fact that we used Gaussian profiles which drop more rapidly than the Lorentzian profiles used by Aksela *et al.*[104]

Comparing experimental results to the different calculations, we find that the agreement varies between excellent and poor, depending of the configuration and method used. For some lines (20, 22, 31, 39) there is excellent agreement and for others (24, 43, 44) good agreement between our experimental anisotropy parameters and the results from all four calculations. For other lines (30, 34, 41, 65) the theoretical values are in disagreement with each other and with our experimental values. Finally, there are some configurations where our data agree with one or the other calculation. For instance, Chen[79] comes close to our β value for the $(^3P)6p(^4S_{3/2})$ state (line 33), whereas Tulkki *et al.*[85] and Hergenhahn *et al.*[78] do not even have the correct sign. On the other hand, for the $(^3P)6p(^4D_{1/2})$ peak (line 36), Tulkki *et al.*[85] give almost the same β value as the experiment but the other calculations are off. Interestingly, there is almost perfect agreement between all theories for our reference peak $(^3P)6p(^2P_{3/2})$ (line 26), but the experimental β value is significantly larger. We were able to observe the splitting the $(^1S)6p(^2P_{3/2})$ state (lines 67 and 68), as Aksela *et al.*[104] did, but the fitting procedure was very sensitive to even small changes in the positions and widths of the peaks. Therefore, in Table 9 we give only the average β for those lines. In Table 10 we compare our β results with previous experimental data from Carlson *et al.*,[67] Becker *et al.*[113] and Kämmerling *et al.*[64] There is, in general, good agreement between the latter experiment and these results.

Using Auger resonant Raman conditions to obtain energy resolutions well below the natural linewidth of the $4d$ inner-shell hole, the present

Table 10. β parameters of the electron spectrum of Xe after $4d_{5/2} \rightarrow 6p^{3/2}$ excitation including comparison with previous, lower resolution data.

Line(s)		β			
No.[a]	No.[a]	Present work	I[c]	II[d]	III[b]
19, 20	1a	−0.66(6)	−0.60(3)	−0.67(5)	−0.88
22–24	1b	−0.88(2)	−0.90(2)	−0.93(3)	−0.93
26	1c	1.30(2)	1.31(2)	1.35(6)	0.82
28–31	2a	0.65(4)	0.58(2)	0.89(6)	0.26
32–36	2b	0.52(5)	0.54(3)	0.45(6)	0.16
39–42	3a	0.36(4)	0.23(2)	0.55(5)	−0.02
43–47	3b	0.28(3)	0.33(5)	0.46(5)	−0.09
67, 68	5	1.17(4)	0.83(5)	1.09(6)	0.51

a. According to Aksela *et al.* Ref. 104.
b. Carlson *et al.* Ref. 67.
c. Kämmerling *et al.* Ref. 64.
d. Becker *et al.* Ref. 113.

results appear to remove the existing experimental discrepancy. Comparisons with different theoretical calculations show partly good agreement, but there is room for improvement for some lines.

5. Doubly and Triply Excited States in Atoms

The simultaneous excitation of two and three atomic electrons by a single photon has been the subject of numerous experimental and theoretical investigations. These intermediate resonance states reveal higher-order, interelectron correlation effects, and provide a more stringent testing ground for the accuracy of existing theoretical and computational methods. Helium, due to its relative simplicity, has been studied extensively.[114] However, the case of Li, an open-shell atom, and Ne, the next rare gas after He, are more complicated. These cases are discussed in detail below.

5.1. *Doubly Excited States in Neon: Breakdown of LS Coupling*

The case of doubly excited states in neon, is more complicated than helium due to the occupied $2p$ subshell, which leads to a dense grouping of the Ne^+ $2p^4nl$ states and hence more complex resonances. Nevertheless, many studies have been performed on this system as well, and some have observed spin-orbit effects. Early photoabsorption results demonstrated departure from LS coupling predictions. More recently, the fine-structure splitting of resonance series was observed in greater detail.[115] These results established the existence of certain small spin-orbit effects in the photoionization of neon, but they lacked additional information provided by highly differential measurements with high-resolution photons that can provide an unambiguous signature of these effects.

Recent studies[116] demonstrated that for certain neon resonances, the angular distribution of photoelectrons and the ratio of partial cross-sections to individual fine structure levels both exhibit behavior that deviates markedly from LS coupling predictions. Such observations provide further quantitative determination of the breakdown of LS coupling. Particular attention has been given to the $2p^4 3pns, nd$ doubly excited resonances in photoionization excitation to the Ne^+ $2p^4(^3P)3s(^2P)$ state. These findings were achieved using the TOF system at the ALS.

For photoionization from the $2p^6(^1S)$ ground state to the $2p^4 3s(^2P)$ $\varepsilon p(^1P)$ continuum, the angular momentum transfer is restricted to the single, parity-unfavored value $j_t = 1$ in LS coupling, and the differential cross-section should vanishes at $\theta = 0°$. However, this was not observed as can be seen on Fig. 25. The PES were allowed to accumulate simultaneously at the two angles for 10s; then the photon energy was incremented by 6 meV. This was repeated until the photon energy range of interest was covered. The photon energy resolution was set to 20 meV at 57 eV for the data shown, which is a significant improvement over the previous resolution of 100 meV in earlier differential measurements.[117] First, the vertical lines (b) in Fig. 25 [corresponding to the $2p^4 3s(^4P_{jc})$ satellites at 48.8 eV binding energy] show, on certain resonances, noticeable photoelectron yields at both angles, indicating LS coupling breakdown. Also, by comparing the photoelectron yields at 54.7° and 0° along the vertical lines (c) in Fig. 26 [corresponding to the $2p^4 3s(^2P_{jc})$ satellites at 49.4 eV], it can be seen that several resonances appear at both angles while others disappear in the 0° spectrum. For instance, the $2p^4(^3P)3p(^2P)3d$ resonance at (photon energy) 51.3 eV shows a photoelectron yield that is almost equally as strong at 0° as it is at 54.7°, whereas the $2p^4(^1D)3s(^2D)4p$ resonance at 50.6 eV disappears almost completely in the 0° spectrum. Since these are parity-unfavored transitions in LS coupling, the appearance of any signal at 0° is an immediate indication of the breakdown of LS coupling. The results presented in Fig. 26 indicate

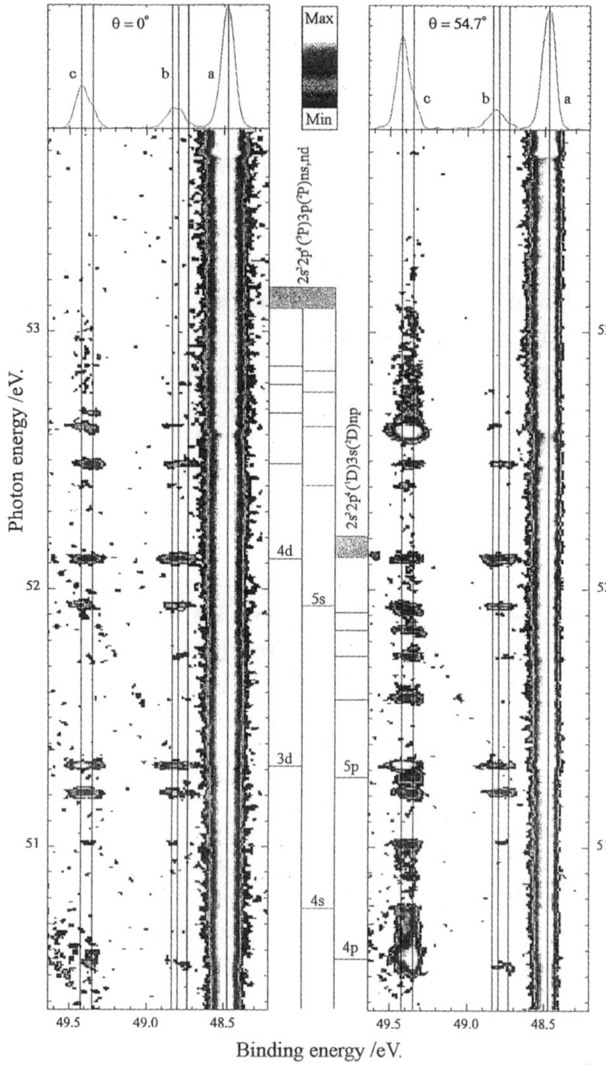

Fig. 25. 2D images of electron emission processes taken at 0° and 54.7°. The images represent the photoelectron yield as functions of the electron binding energy and the photon energy. A color bar is shown to indicate the way in which the yield is represented. The upper graphs show the spectrum at 51.3 eV photon energy, and the vertical bars indicate the positions of the (a) $2s2p^6(^2S)$, (b) $2s^22p^4(^3P)3s(^4P)$, and (c) $2s^22p^4(^3P)3s(^2P)$. The positions of three doubly excited Rydberg series are indicated between the two images.

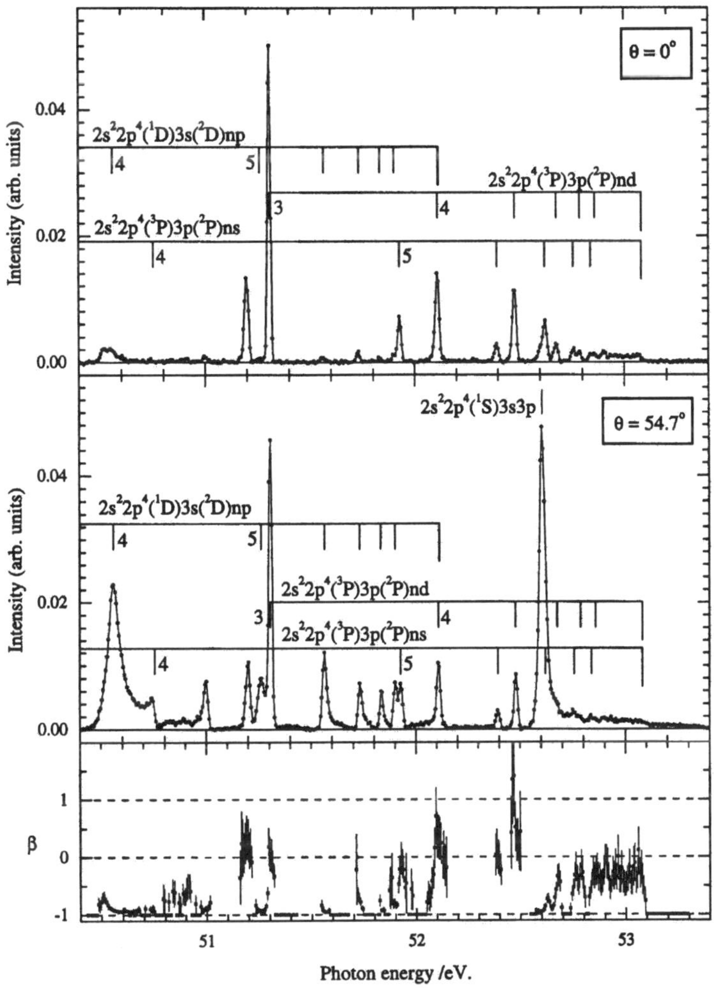

Fig. 26. Photoelectron yield to the $2s^22p^4(^3P)3s(^2P)$ ionic state at 0° and 54.7° along with the derived β values. (β is set to -1 and the error bar is not shown where the intensity at 0° was at the noise level.)

more quantitively that the $2p^4(^3P)3p(^2P)ns, nd$ resonances are essentially the only ones that show a nonzero cross-section at $0°(\beta > -1)$; they further emphasize that spin-orbit effects play a major role in the process we are studying, causing substantial changes in β rather than just small perturbations. The $2p^43pns, nd$ resonances are particularly strong because the

-10% mixing of the $2p^53p$ configuration in the $2p^6$ ground state leads to the one-electron photoexcitation $2p^53p \rightarrow 2p^43pns, nd$.[115] Breakdown of LS coupling for this light system is due to the spin-orbit interaction, which most likely causes mixing between certain LS-allowed $2p^43pns, nd(^1P_1)$ resonances and their $^3S_1$, $^3P_1$, and $^3D_1$ LS-forbidden counterparts.

We have also found that a second LS-coupling prediction does not hold for certain resonances. In the absence of any spin-orbit effects, the $2p^43(^2P)\varepsilon p(^1P_1)$ final continuum decomposes, by statistical weights, into an admixture consisting of 2/3 of the $^2P_{3/2}\varepsilon p_{1/2,3/2}$ and 1/3 of the $^2P_{1/2}$ $\varepsilon p_{1/2,3/2}\, J = 1$ continua. Differential cross-sections to these two levels should therefore assume the exact, energy-independent ratio $r = 2$ in LS coupling. By considering spin-orbit effects, on the other hand, and thus including mixing with the triplet final states, this ratio may fluctuate ($r \neq 2$). For example, photoexcitation of the $^3D_1$ resonances through spin-orbit mixing with the $^1P_1$ LS-allowed resonances enables autoionization to f-wave photoelectrons in the $^2P_{3/2}\varepsilon f_{5/2}(^3D_1)$ states, and these photoelectrons cannot couple to the $^2P_{1/2}$ ionic level. Such differences between photoelectrons coupled to the $^2P_{3/2}$ and $^2P_{1/2}$ ionic levels can then lead to LS-forbidden ratios $r \neq 2$. This can also be seen by looking at the photoelectron intensity along the two vertical lines (c) in Fig. 25 corresponding to the $2p^43s(^2P_{3/2})$ and $2p^43s(^2P)3d$ satellites (at 49.35 eV and 49.42 eV binding energies, respectively). For instance, the $2p^4(^3P)3p(^2P)3d$ resonance shows a ratio $r \ll 2$, indicating a breakdown of LS coupling.[116] High-resolution measurements[116] allowed the separation of the 3/2 and 1/2 levels where the splitting is too small to have been observed in previous measurements.

The detailed measurements of two-electron photoionization processes in neon using an angle-resolved 2D photoelectron imaging technique unambiguously indicated that the photoelectron angular distribution and the fine-structure-resolved partial cross-sections of a parity-unfavored transition exhibited new LS-forbidden behavior. Important implications of these findings are twofold. First, viewing 2D photoionization images at various angles can permit the easy identification of many unexpected (e.g. LS-forbidden) features that might be overlooked using less-detailed probing techniques. Second, detection of prominent spin-orbit effects means that it is not safe to assume the validity of LS coupling, even for low-lying resonances in a system as light as neon, when performing detailed photoionization studies.

5.2. *Photoelectron Spectroscopy Studies of Hollow Lithium States*

A hollow atomic system is an atom or an ion in which at least one inner-shell is empty. Doubly excited states of helium[26] are the first hollow atomic states ever observed. Studies of these states were continuously improved over the years,[118] especially with the use of the ALS in Berkeley.[114,119] In collision experiments of highly charged ions with metallic surfaces,[120–122] the formation of highly excited hollow ions have also been demonstrated.

Lithium, with one additional electron outside the $1s^2$ core, is the simplest open-shell many-electron system. Simultaneous excitation of all three electrons can create hollow atoms of the type $nln'l'n''l''$, with n greater or equal to 2. The third electron makes the theoretical and experimental studies of hollow lithium states an even greater challenge than that of helium. Such states provide the best opportunity for investigating the four-body Coulombic problem, because their formation (single photon absorption leading to three excited electron) relies completely on electron–electron correlations. The great advantage of experiments on isolated atoms as in the case of hollow lithium is that they can provide information on the basic physical processes involved in their decay, without perturbation by the environment.

The photoexcitation and decay dynamics of hollow lithium atoms has recently become the subject of intense experimental and theoretical interest.[123–125] The excited states of Li^+ can be reached either by direct photoionization into the continuum or by autoionization from the triply excited states. Over this energy region, direct and resonant processes are competing, producing destructive interferences in some of the continuum channels and strongly enhanced partial cross-sections in other channels.

PES was first used at Super-ACO in Orsay, France with low spectral resolution to study the decay of the $2s^22p\,^2P^0$ and $2p^2(^3P)3p\,^2P^0$ hollow states into the various continua of the Li^+ ion.[126] However, the high spectral band pass (0.5 eV) limited the critical nature of the comparison with highly-correlated R-matrix calculations. Much better spectrally resolved (19 to 39 meV resolution) results were obtained using the higher photon brightness from ALS in Berkeley.[127–131] These experimental data provided the first definitive test of the *ab initio* R-matrix[126] and Saddle-Point[132] calculations resulting in very good agreement between theory and experiment for the energy dependence of the partial cross-sections, and for the energies and widths of the hollow states, respectively. The first measurements of even-parity hollow lithium states[128] were achieved by triple photoexcitation of lithium atoms which were initially laser-excited into the $1s^2 2p\,^2P_{3/2}$ state.

In the experiments reviewed in this section, three SR beam lines[25] were used to energy-analyze, with the *same* electron spectrometer,[22] the electrons emitted after photoexcitation of neutral lithium atoms over the energy range of the triply excited states. Angle-integrated measurement of the energies and intensities of electrons emitted in the autoionization of the hollow states were achieved with a (CMA) described in Sec. 2.4. For the experiments on laser-excited lithium atoms,[128] the beam of an Ar^+-ion pumped dye laser was focused into the source volume of the CMA.

In order to illustrate the spectacular progress made possible by the use of third generation storage rings, we present in Fig. 27 photoelectron spectra obtained at 152.4 eV photon energy corresponding to the excitation energy of the $2p^2 3p\,^2P_0$ state. From top to bottom, the three panels display the photoelectron spectra obtained at the SA23, SU7, beamlines at Super-ACO (France) and 9.0.1 beam line at ALS, respectively. The use of an undulator at Super-ACO (middle panel) evidently increases considerably the photon flux available, but the resolution is not yet sufficient to resolve the $n = 3$ electron lines. The spectrum measured at ALS (lower panel) shows the dramatic increase in both the resolution and counting rates. These results demonstrate that a high-precision undulator installed on a third-generation storage ring in combination with excellent X-ray optics is able to provide at the same time high-resolution and high intensity monochromatic photon beams, allowing a detailed test of the atomic theories.

To complement the recent experiments on hollow lithium atoms there has been considerably new theoretical developments in order to interpret the results. When correlation effects are dominant, the independent particle approximation is no longer valid; neither is it very satisfactory to consider the strong correlations as a perturbation. For doubly excited states of helium an alternative description has been developed in terms of a six-coordinate configuration space using hyperspherical coordinates.[133] This approach has provided a set of new quantum numbers which replace the quantum numbers used in the independent-particle model. Preliminary application of these ideas to triply excited states of lithium was made some time ago,[134] but it has only been recently[135] that some initial quantitative predictions have been made. R-matrix calculations[126–128] have produced

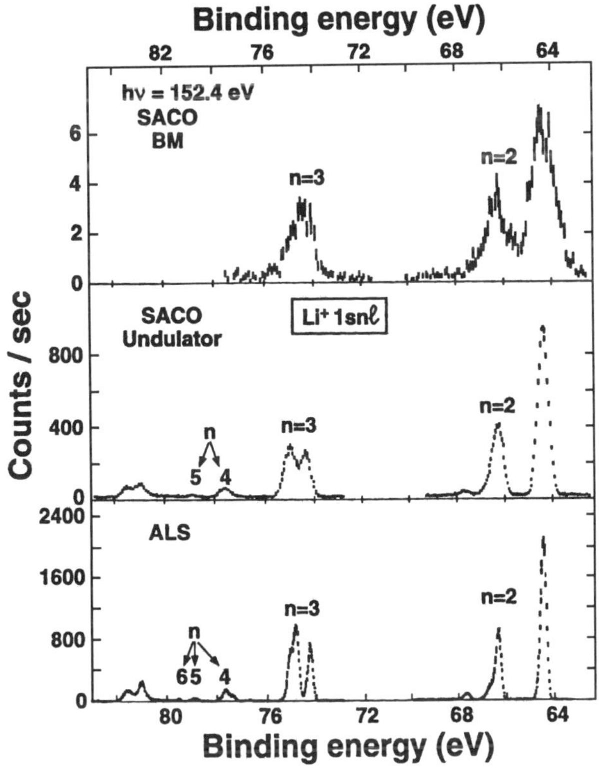

Fig. 27. Electron spectra resulting from photoionization of Li atoms at 152.4 eV photon energy with SR from Super-ACO (binding magnet, upper panel; undulator, middle panel) and from ALS (lower panel).

extensive photoionization cross-sections in the photon energy range corresponding to formation of the hollow lithium states. The Saddle-Point technique[132] has been used to produce the most accurate resonance energies and widths for several resonances. It is anticipated that hollow lithium will continue to prove over the next years to be a useful atomic system for critical evaluation of the various theoretical methods.

The amount of data already measured by PES on photoexcitation and decay of hollow lithium states is quite significant. In this section, we will present only a few selected examples of the results which have been obtained so far.[126–131] Below the energy of the lowest doubly excited $2s^2$ states of Li^+, decay of the triply excited hollow states can occur only to $(1snl)$

final states of Li^+, with the exception of direct decay to Li^{2+} ions with low probability (about 1%). For photon energies in excess of 151.7 eV, new additional decay paths become available [through single electron autoionization decay to *doubly excited hollow* $(nl, n'l')$ Li^+ ionic states] many of which undergo further Auger decay Li^{2+} ground state. Particular advantages of following the hollow–hollow decay routes are that decay via these channels may often be the most important, and the ratio of resonant to direct cross-sections may be very favorable.

5.2.1. *Partial Photoionization Cross-Sections, Energies and Widths*

Partial cross-sections measured at ALS for photoionization of lithium into $1s2s\,^1S + \varepsilon p$ and $ls2p\,^3P + \varepsilon s$ continuum channels (left channels) are shown in Fig. 28. The profiles show a quite different behavior: while in the ^1S channel, the shape of the cross-section has a Fano-type profile, in the ^3P channel one observes a symmetric profile. This is as expected, since the direct photoionization cross-section to the $1s2s\,^{1,3}S$ ionic states has a significant value, allowing interferences to occur over the energy range of the resonance, whereas the cross-section to the $1s2p\,^{1,3}P$ states is weak. We show also on the right panels of Fig. 28 the results of R-matrix calculations,

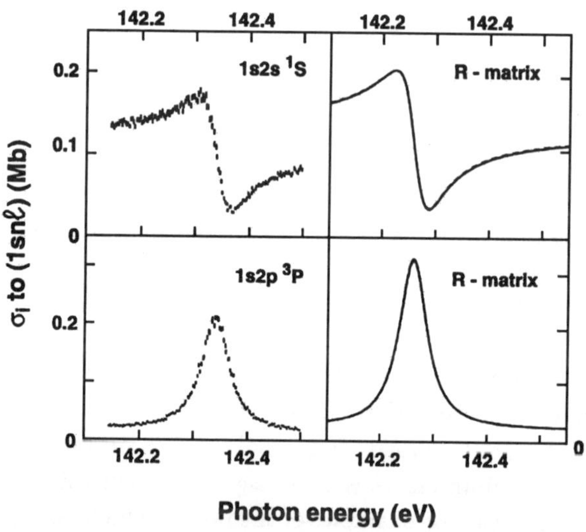

Fig. 28. Measured cross-sections of Li to the $1s2s\,^1S$ and $1s2p\,^3P$ (left panel) Li^+ states and convoluted R-matrix calculations (right panel), at 142.3 eV (from Ref. 127).

after convolution with the instrumental function. These experimental data provided the first definitive test of these calculations, resulting in an extremely good agreement between theory and experiment for the energy dependence of the partial cross-sections. With regard to the absolute values of the cross-sections, the precision of the experimental values is limited by the uncertainty ($\pm$ 25%) in the measured photoabsorption cross-section[23] used for normalization. The calculated R-matrix values differ from the measured energy of the resonance by 0.10 eV; an almost constant shift of 0.10 to 0.20 eV between experimental and R-matrix energy values was observed throughout the whole energy range examined in this work.

Following deconvolution of the spectral profile (from an instrumental Gaussian width of 0.019 eV), a FWHM of 0.118(5) eV was obtained for the $2s^2p\,^2P$ resonance profile. This agrees, within the experimental error-bars, with the high precision calculations using the Saddle Point technique.[132] These results show immediately the advantage of high spectral resolution experiments which succeed in determining the profile width with an uncertainty sufficiently small to test the most recent theoretical calculations.

Figure 29 shows measured partial cross-sections for photoionization of lithium atoms into the $1s3s\,^3S$ $ls4l\,^{1,3}L$ channels,[127] respectively. Apart from an energy shift of 0.1 to 0.3 eV, the detailed agreement (on a relative cross-section basis) in the $1s3s\,^3S$ ionic channel is again a tribute to both the

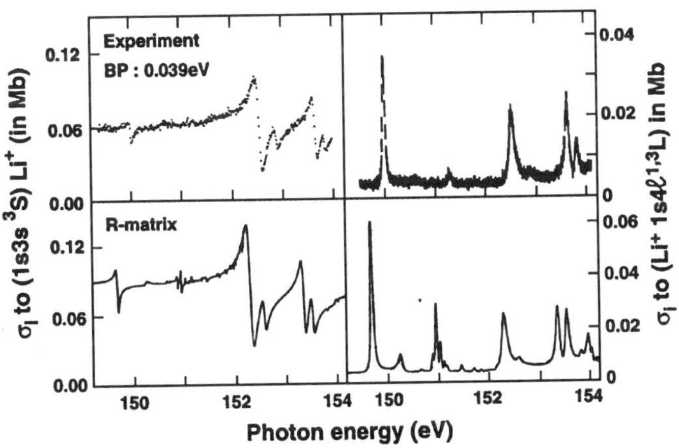

Fig. 29. Experimental (upper panels) and convoluted R-matrix (lower panels) partial cross-sections for photoionization of lithium atoms into the $1s3s\,^3S$ (left) and $1s4l^1\,^3L$ (right) final ionic states (from Ref. 127).

quality of the experimental data and the R-matrix calculations. Measured values for the energies of these states can be found in literature,[127] together with their identification and as can be noted, the profiles of the resonances are different in the different channels.

5.2.2. *Rydberg Series*

In investigating the many hollow lithium states, an important question was whether hollow states which exist can be regarded as helium-like doubly excited Li^+ terms plus a running outer electron or whether the configuration mixing is so great that such Rydberg states are absent. Theoretical R-matrix calculations predicted[126] the existence of eight Rydberg series between 150 eV and 160 eV. The series $1s^2 2s\,^2S \rightarrow [(2s2p\,^3P)ns]\,^2P$ and $[2p^2\,^1S)np]$ were supposed to be the most intense in this energy region. At excitation energies lower than the $2s^2\,^1S$(151.66 eV), the $n = 3$ and $n = 4$ terms of the $[(2s2p\,^3P)ns\,^2P]$ series only decay to the singly excited $1snl$ states, while the $n \geq 5$ terms of this series, lying above the $2s^2\,^1S$ excited state of Li^+, could decay to both singly and doubly excited states of the Li^+ ion. Decay to the $2s^2\,^1S$ Li^+ ionic state produces emission of a very low energy electron. The $2s^2\,^1S$ Li^+ state subsequently decay to the ground state of the doubly charged Li^{2+} ion, with emission of a high-energy ($\cong 70$ eV) easily detected electron.[131]

The intensity of these high-energy electrons as a function of the photon energy has been measured.[25] Since the radiative decay of Li^+ is very weak compared to autoionization, the partial cross-section measured from the second step of the autoionization process is essentially the cross-section for photoionization into the $2s^2\,^1S$ Li^+ final ionic state. Figure 30 shows the measured partial cross-section for photoionization into this state (upper panel). For $n \geq 5$, well-resolved electron lines are observed with intensities decreasing with increasing n. The extremely low associated background ensures that Rydberg states with n extending up to 8 appear very clearly, despite the fact that the partial cross-section at the top of the most intense resonance is less than 15 kilobarns. We show in the lower panel of Fig. 30 the results of the R-matrix calculations for the same partial cross-section where a very good agreement between experiment and theory can be seen. The regular intensity behavior of the observed Rydberg series implies that the relaxation process which leads to ionization of the outer electron does not depends much on the running nl orbital. The partial cross-section for

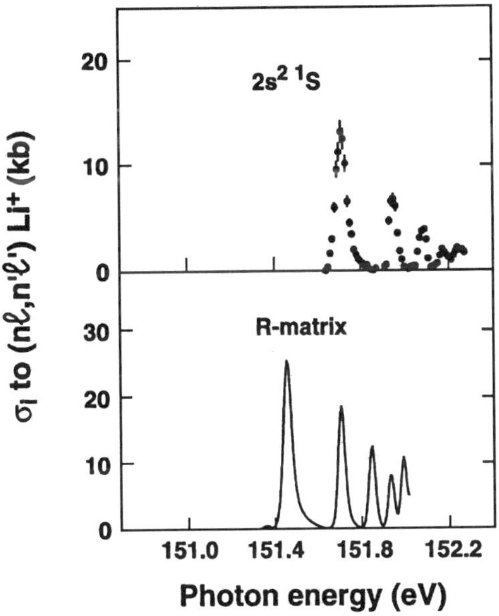

Fig. 30. The $[(2p^2\ ^1S)np]^2P$ Rydberg series as observed in autoionization of its terms into the doubly excited $2s^2\ ^1S$ Li^+ state from Ref. 131.

photoionization into the $1s2s\ ^3S$ is many times greater, nevertheless the Rydbergs are barely noticeable in this channel[131] due to the large continuum arising from the direct nonresonant photoionization process.

5.2.3. *The Lowest Energy Doubly Hollow* $3s^23p\ ^2P$ *State*

At higher photon energies, triply excited states with all three electrons having principal quantum numbers equal to three or above can be created. We named *doubly hollow state* such a state in which both K- and L-shells are empty.[130] Triply excited states with $n \geq 3$ can decay to either singly or doubly excited states of the Li^+ ion. Decay to doubly excited $(2l2l')Li^+$ ionic states again produces emission of a low-energy electron, followed by the decay of this Li^+ state to the ground state of the doubly charged Li^{2+} ion with emission of a high-energy electron. The $3s^23P\ ^2P$ triply excited state[130] was observed by measuring resonant structures in the autoionization profile into three different ionic channels: the $2s^2\ ^1S$, the $2s2p\ ^3P$ and the lowest lying $(2l3l')$ excited states. The experimental results for two of them are presented in Fig. 31. In the $2s2p\ ^3P$ final ionic state,

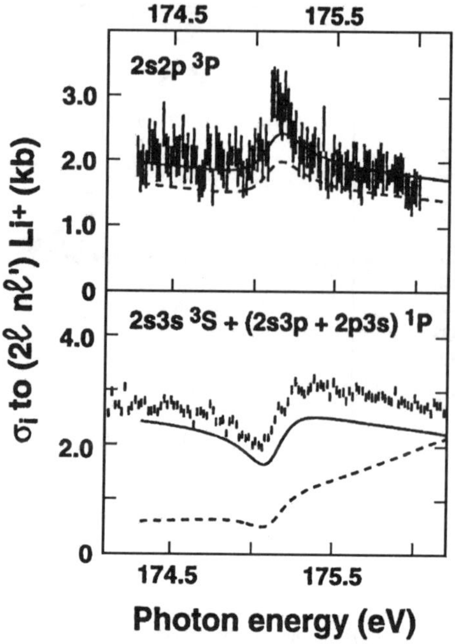

Fig. 31. Measured (upper panels) and convoluted R-matrix (lower panels) partial cross-sections for photoionization of lithium atoms into the hollow $2s2p\,^3P$ and lowest lying $2l3l'$ Li^+ ionic states in the energy range of the $3s^23p\,^2P$ doubly hollow atomic state (from Ref. 130).

the partial cross-section profiles appear to be mostly symmetric, while in the $(2l3l')$ ionic state, the experimental curve presents a quite asymmetric inverted Fano profile. Our measured value for the energy of the $3s^23p\,^2P$ hollow state is 175.165(50) eV, in excellent agreement with the results of the Saddle-Point calculation (175.12 eV). Another experimental results measuring ion yield[136] agrees with our result. The results of the R-matrix calculation for the cross-sections are also shown in Fig. 31. While differences between the length-(dashed line) and velocity-(full line) forms are only about 10% into the $2s2p\,^3P$ channel, fitting rather well with experimental data, both shapes differ strongly in the $(2l, 3l')$ channel, confirming that the wavefunctions for the target states should be improved when all the electrons occupy $n > 2$ orbitals.

While more than thirty years have been needed for high-resolution measurements to be performed on doubly excited states of helium, the amount of such data already obtained on triply excited hollow states in lithium is impressive. This major advance is due to the availability of a high-resolution monochromator on an undulator beamline of the third generation ALS storage ring. New information on higher hollow states and angular distribution of autoionizing electrons is still to be gained. More generally, with the growing number of such storage rings becoming available to users, many more exciting results are expected in the investigation of atomic photoionization.

6. Inner-Shell Excitation in Molecules

Inner-shell processes in small molecules have been studied with growing interest during recent years. Rapid improvements in experimental techniques have revealed new details in Auger and photoelectron spectra, such as anisotropy of electron emission,[86,137–141] molecular field splitting,[142] and postcollision interaction.[143] Below, we illustrate recent measurements, some are preliminary, in core excitation or ionization of molecules that lead to new or improved findings about the structure and dynamics of various molecules.

6.1. *Angle-Resolved Studies of Cl 2p Excitations in HCl Using an Imaging Technique*

Studies on molecules have been performed at only a few selected photon energies. Here, we present a complete angle-resolved two-dimensional experimental picture of electron emission from a diatomic molecule over the preedge structure and into the continuum of a core ionization threshold. This extensive set of data illustrates many phenomena associated with core excitation or ionization.

The absorption spectrum of HCl around the Cl $2p$ ionization threshold[144] can be divided into three regions: (i) excitation to the antibonding σ^* orbital ($2p \rightarrow \sigma^*$), (ii) excitation to the Rydberg orbitals ($2p \rightarrow nl$), and (iii) ionization into the continuum ($2p \rightarrow \varepsilon l$). These three regions are indicated on the total electron yield curve along the right side of Fig. 32. The 2D electron emission map, shown in the main panel of Fig. 32, was obtained at an angle of 90° relative to the polarization plane. In this map, the intensity of the electron signal is represented as different shades of gray. Electrons emitted with the same kinetic energy at different photon energies

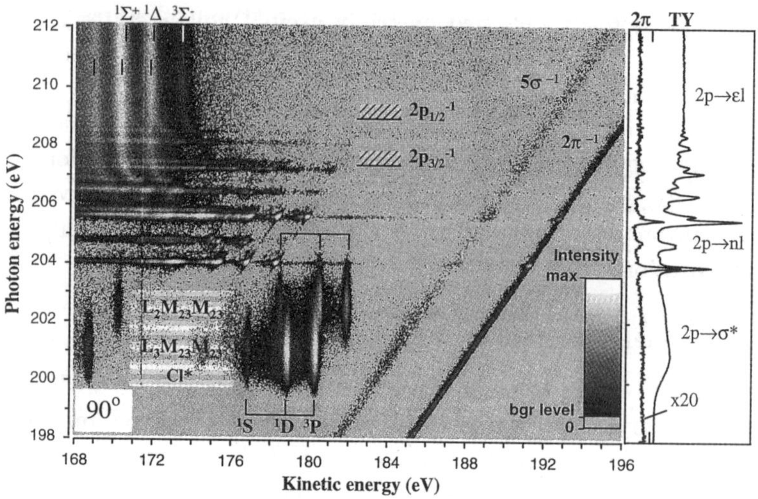

Fig. 32. Two-dimensional (2D) map of electron emission from HCl molecule across the Cl $2p$ ionization threshold, taken at 90° relative to the polarization plane. The total electron yield and partial 2π photoelectron yields are shown on the right.

are aligned vertically in the 2D map. Those emitted with constant binding energy form diagonal lines across the map. The features in the different regions of the 2D map are discussed in detail below.

Broad bands in the total electron yield between 199.0 and 203.5 eV correspond to excitation of the $2p$ electrons to the first unoccupied molecular orbital, σ^*. When occupied, this antibonding orbital causes the core-excited molecule to dissociate. The dissociation, HCl^* $(2p^5\sigma^*) \rightarrow H(1s) + Cl^*(2p^53p^6)$, produces neutral hydrogen and core-excited chlorine atoms and proceeds on a times scale ($\approx 10^{-14}$s) that competes with the Auger decay of the core hole.[145,146] The light hydrogen atom moves quickly away from the Cl atom, so that a large part of the Auger decay occurs in the atomic chlorine as $2p^53p^6 \rightarrow 3p^4$ transitions, seen as intense vertical lines in the 2D map at the photon energies of the $2p \rightarrow \sigma^*$ excitation. These sharp atomic lines have pronounced tails at their low kinetic energy side arising from Auger transitions in the molecular environment prior to the complete dissociation. These tails form a 'molecular background' to the atomic Auger spectrum.[146]

As the photon energy is increased across the breadth of the σ^* band, HCl molecules with decreasing internuclear separation are selected due to the projection of the ground state population onto the dissociative potential energy curve. This shift may influence the time scale of the early stage of the dissociation, which should be reflected in the balance between the molecular background and the sharp atomic peaks in the Auger spectra. In order to investigate this possibility, the intensities of one atomic line and its associated background were monitored as a function of photon energy.[147] The $2p_{1/2}^{-1}3p^6 \rightarrow 3p^4(^3P)$ atomic line at a kinetic energy of 182.0 eV was selected, since it is well separated from neighboring peaks. The molecular tail of this line is assigned to transitions to the $^4\Sigma^-$, $^2\Sigma^-$ and $^4\Pi$ states of HCl^+.[146] The molecular part of the Auger spectrum dominates at photon energies well below the absorption peak maximum. With increasing photon energy, the internuclear distance at which the excitation can take place shifts to smaller values, where the energy of the core excited state is higher. Here, the atomic peak gains intensity over the molecular background, inferring a higher dissociation rate. This effect can be related to the slope of the potential-energy curve of the excited state. At lower photon energies the molecular ground state is projected onto the lower part of the potential-energy curve of the $2p_{1/2}^{-1}\sigma^*$ state, where the curve is less steep. The dissociation rate and Auger line shape depend on the slope of the potential-energy curve at the point of excitation:[146] a smaller slope leads to a smaller initial acceleration of the fragments and consequently to a lower dissociation rate and vice versa. At the photon energies well above the peak maximum, the spectra are strongly affected by the overlap with the $2p_{1/2}^{-1}4s$ resonance and at smaller emission angles also by 4σ, photoelectron satellite lines that start to cross this kinetic energy region. We do not see conclusive evidence that the molecular background is again enhanced at the high-energy side of the resonance due to the shortening of the duration time of the resonant scattering process at larger detunings, as proposed in the literature.[148]

A quite different pattern of Auger decay can be seen in the region of the sharp core-to-Rydberg excitations starting at 204.0 eV photon energy. The spectral structures here arise from molecular Auger decay and are assigned within the simplest spectator model as the $2p^{-1}nl \rightarrow (4\sigma5\sigma2\pi)^{-2}n'l'$ transitions, where nl and $n'l'$ describe the Rydberg electron in the Auger initial and final state, respectively.[149] This region of the 2D map is expanded in Fig. 33 and shown on a binding-energy scale, where transitions to the same final ionic states are aligned vertically.

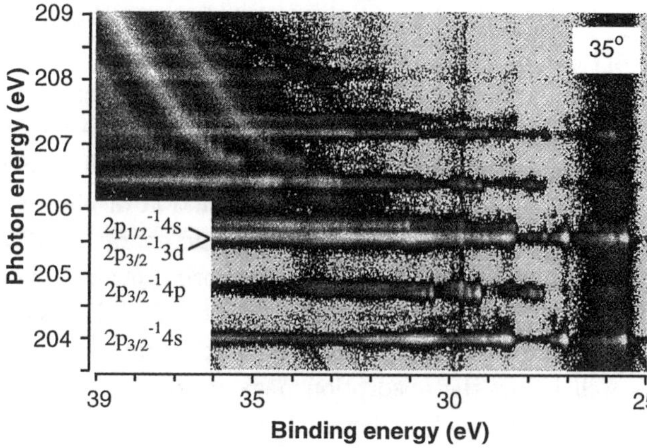

Fig. 33. 2D map of Auger electron emission at the Rydberg resonances measured at 35.3°.

The tail of the $2p_{1/2}{-}1\sigma,^*$ resonance overlaps with the first Rydberg resonance $2p_{3/2}^{-1}4s$ at 204.0 eV[147] (see Fig. 32). The intensity of the Auger lines from the decay of the dissociating $2p_{1/2}^{-1}\sigma^*$ state shows a sharp increase at the $2p_{3/2}^{-1}4s$ resonance. At the magic angle, these Auger lines account for about 6% of the total Auger intensity at this photon energy. About half of their intensity comes from the overlap with the Gaussian shaped tail of the $2p_{1/2}^{-1}\sigma^*$ resonance. The remainder is resonant enhancement indicating the possibility of predissociation on the first Rydberg resonance. The transition from the bound $2p_{3/2}^{-1}4s$ to dissociative $2p_{1/2}^{-1}\sigma^*$ potential energy curve must involve some interaction between the two spin-orbit components of the $2p$ orbital. Detailed calculations of these potential energy curves are needed to investigate this effect further.

The next Rydberg resonance, $2p_{3/2}^{-1}4p$, is broader than the others in the absorption spectrum, apparently consisting of two overlapping peaks of equal intensity, separated by about 100 meV.[144,149] The 2D maps shown in Fig. 33 exhibit a unique pattern for this resonance, suggesting that it is split into two levels that have different Auger electron spectra. On the lower energy side of the resonance, the Auger intensity is concentrated in the binding energy region of the $2\pi^{-2}n'l'$ final states, whereas the higher energy side tends to populate the higher binding energy $5\sigma^{-1}2\pi^{-1}n'l'$ and $5s\sigma^{-2}n'l'$ states. If the two states underlying this resonance correspond to the molecular field splitting of the $2p_{3/2}$ core orbital into $2p_{3/2,1/2}$

and $2p_{3/2,3/2}$ components (using the $nl_{j,mj}$ notation), then the different population of Auger final states can be explained by different orientations of the core hole. In the lower energy state, $2p^{-1}_{3/2,3/2}4p$, the core hole has a purely π character[150] and therefore exhibits a large overlap with the valence π orbital in the Auger matrix elements, whereas the $2p_{3/2,1/2}$ hole has a mostly σ character and overlaps more with the 5σ orbital.

The intensity of the Auger spectra of the higher Rydberg states gradually shifts towards higher binding energies due to the weakening of the influence of the spectator electron and higher shake-up probability. Finally, the $2\pi^{-2}n'l'$ resonant Auger lines merge into the $2\pi^{-2}(^3\Sigma^-,^1\Delta,^1\Sigma^+)$ normal Auger lines. There is no sudden change in the Auger decay pattern at the $2p$ ionization thresholds. Rather there is a smooth evolution from the spectator to normal Auger spectra through a strong post collision interaction (PCI). The energy shift of the normal Auger lines, caused by the PCI between the outgoing slow $2p$ photoelectron and Auger electron, can be seen through the threshold down to the Rydberg states.

Anisotropy in the electron emission is described by a single parameter β, which in the case of direct ionization derives from the shape of the initial- and final state wavefunctions of the photoelectron. The β parameter for the Auger electrons is given as a product of a molecular alignment parameter and an intrinsic anisotropy parameter for the Auger decay.[151] At first, the β values of the 2π and 5σ photolines were determined at resonant photon energies using their branching ratios with some Auger lines that exhibit opposite angular asymmetry from the spectra measured three angles. The 2π photoline intensity at these resonance together with its obtained β value were then used for intensity normalization of the 2D maps so that all other β values could be determined directly by comparing spectra measured at two angles.

The β parameters for the 2π and 5σ photolines are considered first. They have equal values of 1.6(2) below the Rydberg resonances. At the first Rydberg resonance, the β value of the 2π line drops to 0.90(10), whereas the β of the 5σ line shows no change. At the next strong (overlapping) resonances $2p^{-1}_{3/2}3d$ and $2p^{-1}_{1/2}4s$, the value of β for the 2π line drops again to 1.15(10). These changes in the value of β also indicate that a resonant valence ionization channel becomes available.

Auger electron spectra at 35° and 90° relative to the polarization plane, measured at the $2p^{-1}_{3/2}\sigma^*$ resonance and over the first Rydberg series ($2p \rightarrow 4s$) are shown at the bottom of Fig. 34 together with derived β values. The

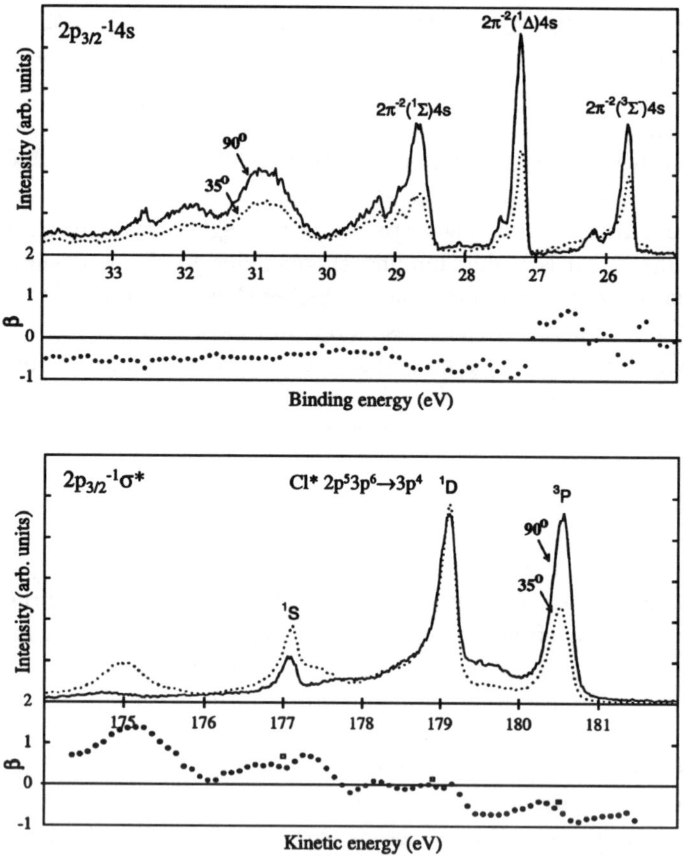

Fig. 34. Auger decay spectra and angular distributions for the Cl $2p \to 4s$ and Cl $2p \to \sigma^*$ excitations in the HCl molecule. The upper panels show the electron spectra taken at 35° and 90° with respect to the electric-field vector of the incident radiation. The lower panels display the corresponding β values. Rectangles are the β values for the $2p_{3/2}^{-1}\sigma^*$ resonance from Ref. 139.

β values of the Auger lines vary between -0.7 and 0.7. Strongly positive β values near 175 eV are assigned to the direct 4σ, ionization. Our results are in good agreement with those of Becker and Menzel,[139] who determined the β parameters for the three atomic lines only. The molecular background also shows large variations of the angular parameter, which, when combined with theoretical calculations, should prove useful for refining the assignment of the spectrum.

The β values for the Auger lines show a markedly different behavior at the Rydberg resonances. At the $2p_{3/2}^{-1}4s$ resonance, all Auger lines have negative β value. The $2\pi^{-2}(^1\Delta)4s$ line at $E_b = 27.3$ eV has the largest anisotropy with $\beta = -0.7(2)$, as can be seen from the top of Fig. 34 which infers that the molecule is strongly aligned after the $2p_{3/2} \rightarrow 4s$ excitation.

Negative β values at the overlapping $2p_{3/2}^{-1}3d$, $2p_{1/2}^{-1}4s$ resonances were observed. All of the molecular Auger lines have similar β values, and for the normal Auger lines β parameters are equal within the error limits, since there is no change in the branching ratios of the Auger lines at different angles.

In the case of the molecules, the 2D electron emission mapping technique has been shown to be a valuable tool for investigating various characteristics over large kinetic and photon energy ranges, since the following effects have been observed: (i) a photon energy dependent dissociation rate of the core-excited molecule as reflected in the Auger electron spectra; (ii) very strong resonant behavior of the 2π photoelectron yield at $2p^{-1}4s$ Rydberg resonances, observed both in the intensity and angular distributions; (iii) a link of the first Rydberg state to the $2p^{-1}\sigma^*$ state, leading to an increased dissociation probability; and (iv) different Auger decay patterns of the split $2p_{3/2}^{-1}4p$ resonance levels, possibly associated with core-hole orientation. In addition, the angular distributions of the Auger electrons at the σ^* and Rydberg excitations were found to be different, with the latter showing uniformly negatives β values.

6.2. *Evidence of a σ_u Shape Resonance in C_2H_2*

Shape resonances in the core excitation of small molecules have been of considerable interest, especially in light of the suggestion that the position of the resonance above threshold is related to the size of the molecule and, thus, could be used as a ruler to measure molecular dimensions.[152] Subsequent discussion[153] has raised questions about the criteria for establishing that a particular structure in a photoabsorption spectrum is a shape resonance and recent experiments by Kempgens *et al.*[154] have called into question the very existence of shape resonances in ethane, ethene, and ethyne — key molecules in establishing the relationship between bond length and position of the resonance.

A shape resonance in C_2H_2 has been predicted for at least fourteen years[152] but recently Kempgens *et al.*[154] have expressed doubt as to the

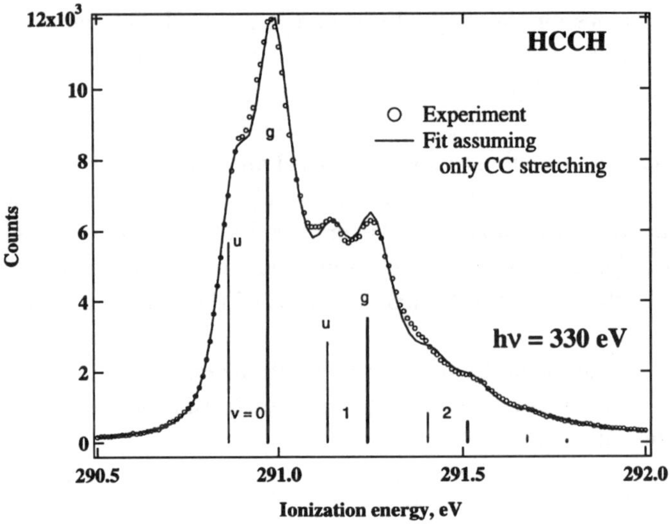

Fig. 35. Carbon 1*s* photoelectron spectrum of HCCH at a photon energy of 330 eV. The points represent the data and the solid curve a fit assuming that carbon–carbon stretching is the only mode contributing to the vibrational structure. The data and the fit extend to both higher and lower energies than shown here. The vertical bars indicate the position and intensity of the different vibrational transitions. The light lines are for the *u* final states and the heavy lines are for the *g* final states.

existence of shape resonances in the photoexcitation and photoionization in C_2H_2 based on their recent measurements.[154]

The unexpected result is that cross-section for the ionization of the $1\sigma_g$ molecular orbital is significantly greater than that for ionization of the $1\sigma_u$ orbital in this energy range as shown in Fig. 35. The most prominent part results from ionization of the $1\sigma_g$ orbital and the shoulder on the left from the $1\sigma_u$ orbital. These measurements were carried out using the SES-200 electron energy analyzer. It is apparent that the cross-section for ionization of the u orbital is less than that for the g orbital. The u–g split shoulder on the high-energy side of the main peak arises from the $\nu = 1$ excitation of the carbon–carbon stretching mode. The solid curve shows a fit to the data as explained in the literature.[155] Furthermore, the ratio of the cross-sections $u|g$ goes through a minimum of about 0.58 at a photon energy of about 317 eV as shown in Fig. 36. This energy corresponds approximately to the maximum in what may be the σ_u shape resonance in ethyne.[155] Furthermore, the energy dependence of the ratio can

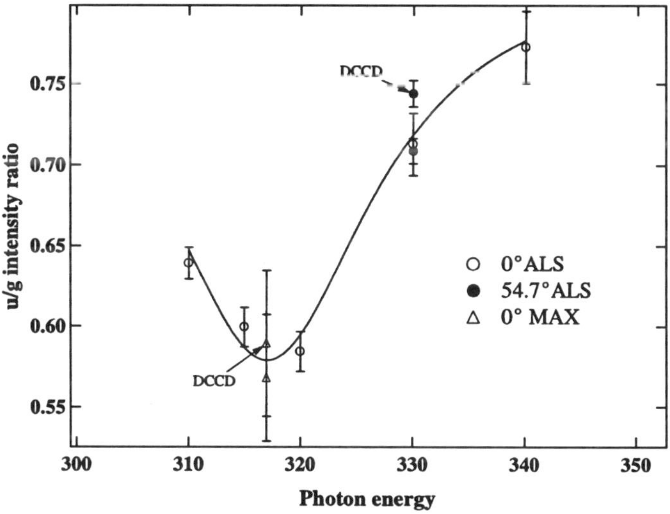

Fig. 36. The ratio of the cross-section $1\sigma_u$ to $1\sigma_g$ for ethyne as a function of photon energy. The open circles represent 0° results for HCCH. The closed circles show 54.7° results for both HCCH and DCCD. The triangles are derived from 0° data for HCCH and DCCD. The solid line represents a Gaussian fit, included to show the general trend of the data.

be qualitatively understood if we assume a smoothly varying cross-section for the direct ionization of the carbon $1s$ electrons, with equal probability of exciting the u and g components, plus a contribution from dipole excitation of a $1\sigma_g$ electron to the $1\sigma_u$ resonance. Since this resonance is above the threshold for core ionization, it couples to the continuum and provides an additional path for $1\sigma_g$ ionization. Thus, these results are significant since they support the existence of a σ_u resonance in the predicted[152] energy region.

6.3. *Angular Distribution of the Sulphur 2p Photoemission in OCS: Variations Revealed by Molecular Field Splitting*

Inner-shell electrons of the atoms constituting molecular systems retain largely their atomic properties. Core orbitals with $l > 0$ are characterized by strong spin-orbit splitting, which is the dominant feature also in molecular inner-shell photoelectron spectra. Also, an additional energy splitting is often present in molecules, which have lower spatial symmetry than free atoms. This so-called molecular field splitting is much smaller than the spin-orbit splitting and is usually superimposed on the vibrational structure of the photoelectron spectrum. The utilization of highly monochromatized synchrotron radiation in combination with high-resolution electron energy analyzers has only recently made it possible to observe the molecular field splitting in core photoelectron spectra for a number of molecules — I 4*d* in HI,[156] Br 3*d* in HBr,[157] Cl 2*p* in HCl,[142], S 2*p* in H_2S,[158,159] SO_2, CS_2 and OCS.[159] The nature of this splitting is rather well understood and quantitative theoretical calculations are available.[160,161]

In the case of molecules, the angular distribution of the electron leaving the atomic site can be altered due to its scattering on the neighboring atoms of the molecule. The pioneering experiments probing such effects in diatomic molecules relied on obtaining information on the orientation of the molecular axis in reference to which the anisotropy could be studied.[162,163] The photoelectrons were detected in these experiments in coincidence with the ionic fragments of the dissociation following the same photoionization event and assuming that their direction still represents the molecular orientation at the moment of the ionization. Using the coincidence method it is, however, very difficult to obtain sufficiently high count rates together with high-energy resolution.

In this section, we show how the molecular field splitting of core orbitals can provide an alternative gateway for studying molecular photoemission, since it breaks down the spherical symmetry of completely filled inner-shells. In linear molecules, for example, *p*-subshells split into components which are oriented preferably either along (σ-character) or perpendicular (π-character) to the molecular axis.[161] It is also known that photoionization of a *p*-electron occurs at preferred orientations of its wavefunction relative to the electric vector of the radiation. These two factors combined can eliminate the isotropy of the molecular orientation, if one can measure resolved individual molecular field split components separately. In such a case, the effects of the noncentral potential of the molecule on the outgoing photoelectrons could be investigated using noncoincidence photoelectron spectroscopy, benefiting from its higher energy resolution and high count rates. Very recently, angle-resolved photoelectron measurements of the sulphur 2*p* electrons in the carbonyl sulphide OCS with completely resolved vibrational and molecular field splitting, have been measured, demonstrating that photoemission from the molecular field split components has different angular distribution, attributable to the effects of the molecular surroundings.

Carbonyl sulphide is a linear molecule with carbon atom surrounded by sulphur and oxygen and with valence configuration of $1\sigma^2 2\sigma^2 3\sigma^{*2} 4\sigma^{*2} 1\pi^4 2\pi^4$.

The photon beam had an estimated 20 meV bandwidth at 200 eV photon energy. The experimental station used was the Scienta SES-200 electron energy analyzer. The electron energy resolution used in this experiment was 26 meV.

A preliminary Sulphur $2p$ photoelectron spectra of the OCS molecule are shown in Fig. 37, measured at three different emission angles relative to the electric vector of the radiation. The ionizing photon energy of 200 eV was about 30 eV above the $2p$ ionization threshold. The lower kinetic energy structure, marked by A in Fig. 37, is assigned to the $2p_{1/2}$ spin-orbit split component of the $2p$ orbital, with the vibrational progression $\nu = 0$, 1 and 2. The higher energy structure of $2p_{3/2}$ is more complex consisting of the

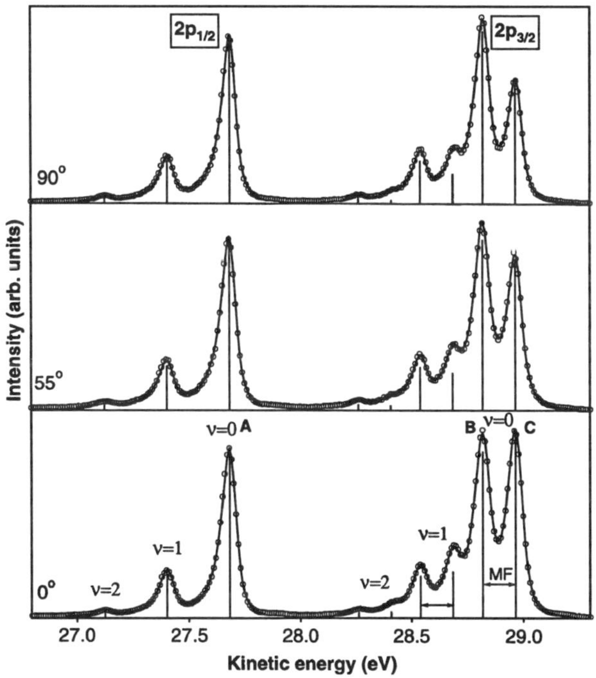

Fig. 37. Sulphur $2p$ photoelectron spectra of the OCS molecule, excited by 200 eV photons. The detection angle relative to the electric vector is shown next to the spectra. Solid lines are curve-fitting results and vertical bars represent the positions and intensities of individual peaks.

vibrational progressions of two molecular-field split components, marked by B and C. The molecular field splitting between peaks B and C was found to be 145(1) meV, which is in excellent agreement with a theoretical value of 144 meV.[160] The spin-orbit splitting is 1.212 eV. The vibrational structure can be assigned to the C–O stretching mode of the core-ionized molecule, with the vibrational constant of 279 meV. The other vibrational modes, C–S stretching and O–C–S bending that have lower frequencies, do not appear in the spectra.

The most remarkable feature of the spectra in Fig. 37 is the change of the branching ratio of the two molecular field split components of the $2p_{3/2}$ peak as a function of the emission angle. The preliminary values of the angular parameter are $\beta(C) = 0.75(20)$, $\beta(B) = 0.40(20)$, $\beta(A) = 0.40(20)$.

In the present case, component $A(2p_{1/2})$ is not affected by the molecular field splitting indicating that it retains its atomic spherical symmetry. This assumption is supported by an anology with the Mulliken population analysis of the Cl $2p$ orbital in the HCl molecule,[161] where the $2p_{1/2}$ orbital was found to have 35% of σ and 65% of π contribution. Variations of the $2p_{1/2}$ photoelectron angular distribution, caused by the molecular surroundings, are therefore averaged over random molecular orientation and cannot be studied.

The two components (B and C) of the $2p_{3/2}$ orbital are separated by the molecular field splitting and have nonisotropic charge distributions with respect to the molecular axis. One of them is represented solely by the $\pi_{3/2}$ molecular orbital, whereas the other has mainly $\sigma_{1/2}$ character with some mixing of $\pi_{1/2}$. Also the continuum waves within a linear molecule can be represented by molecular orbital (λ) notation, so that the molecular selection rule $\Delta\lambda = 0, \pm 1$ applies. It is well known that discrete molecular photoexcitation to bound orbitals results in alignment of the excited molecules; transitions with $\Delta\lambda = 0$ (e.g. $\sigma \to \sigma$) select linear molecules with the axis parallel to the electric vector, while $\Delta\lambda = \pm 1$ transitions occur in the molecules with perpendicular orientation. This can be assumed to be the case also for continuum ionization. The total photoelectron emission is described by a set of one-electron bound-to-continuum transitions $\lambda_b \to \lambda_c$, each favoring either parallel or perpendicular alignment of the molecules. The set $\lambda_b \to \lambda_c$ is different for the two molecular field split components — for example, δ continuum waves are accessible only from π-type bound orbital. Consequently, the two components B and C represent molecules with nonisotropic — and different — molecular alignment. In such a case the influence of the molecular neighborhood is not completely cancelled by averaging over different orientations and can be observed by noncoincidence photoelectron spectroscopy, as the present results indicate.

These experimental results are evidence of changes in the molecular photoelectron angular distribution for different molecular field split components of a p core orbital. For the first time, an intrinsic selector of the molecular orientation — the molecular field itself — has been used instead of detecting the orientation of the molecule by means of an additional coincidence ion detection channel. In this case, the randomness of the molecular orientation, which prevents the detection of any angular effects caused by the molecular geometry, is reduced by separating the molecular field split components of the electron spectrum in energy. The high-resolution, noncoincidence photoelectron spectroscopy method, does not rely on the instantaneous dissociation of the molecule after the photoabsorption, in which case it is assumed that the atomic fragments still carry the information of the original orientation of the molecule. In contrast, the approach presented here is more general and should be easily applicable also to nonlinear molecules, for which the dissociation dynamics is less straightforward. Theoretical studies, to interpret in detail the experimental results in these molecular system unfortunately are lacking.

6.4. *Core Excitation in CO*

It is now possible to observe the evolution of a deexcitation process following the resonant core excitation of a molecule with a resolution less than the natural width of the core excited states. Under Auger resonant Raman (ARR) conditions,[164] one is able to obtain very high-resolution deexcitation spectra with the ultimate limit set only by the lifetime width of the final ionic state, eliminating the contribution from the life time broadening of the core excited state.

The C $1s$ photoabsorption in CO is characterized by the large cross-section of excitations to the π^* molecular orbital. The C $1s \to \pi^*$ transitions are vibrationally resolved in CO and the valence photoelectron spectra of these molecules show large resonant enhancement at the photon energies of these excitations due to the participator Auger decay. Piancastelli *et al.*[16] have recently reported vibrationally resolved $1s \to \pi^*$ decay spectra at both the C and O K-edges. They also report on calculations that allow for a detailed assignment of the electronic final states. We have pushed the resolution even further by measuring this excitation and its decay spectra at $\nu' = 0, 1$ and 2 levels of the core excited state under Raman conditions with a spectrometer resolution of 27 meV, and photon resolution of 30 meV. The highly resolved spectra provide means for studying the unresolved structure of the photoexcitation peaks, the properties of the core excited states, and their excitation and deexcitation dynamics.

High resolution photo- and Auger electron spectra from the C $1s \rightarrow \pi^*$ excitation in CO have been measured using the Scienta SES-200 hemispherical analyzer. These measurements were complemented by two-dimensional mapping of electron emission using the TOF analyzers system that allowed a comprehensive observation of the excitation to the $\nu' = 0, 1$ and

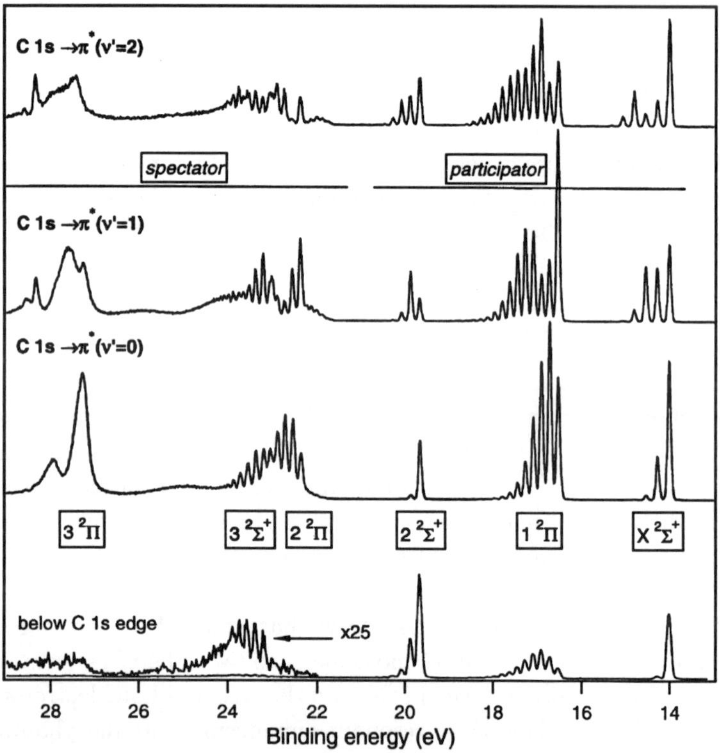

Fig. 38. Electron emission spectra of the CO molecule following the C $1s \rightarrow \pi^*$ resonant core-excitations. Also is shown a photoelectron spectrum well bellow the C $1s$ edge. The spectra have been taken at 54.7° relative to the electric field vector.

2 and investigation of a range of photon-processes that revealed effects that is the subject of another report.[165] Figure 38 shows a preliminary electron spectra were both the spectator and participator decay are observed with unprecedented resolution following photoexcitation to the C $1s^{-1}2\pi^*\, \nu' = 0, 1$ and 2 levels. They are fully vibrationally resolved even for the decay of the $\nu' = 2$ vibrational level of the core excited state, for which the resonant photoabsorbtion cross-section is small. Our measurements are compared to previous calculation[16] as shown in Fig. 1, where it is clear that the new experimental results show more structure than the calculation which has been convoluted with a Gaussion lineshape to approximate the experimental resolution of data reported previously.[16] The analysis of the vibrational progressions will give detailed information on the shape of the potential energy (PEC) curve of the core excited state, especially for the spectra of $\nu' = 2$ level, which start to probe the upper (nonharmonic) part of the PEC. In addition, the 2D maps allow us to trace photon energy dependent effects such as vibrational lifetime interference.

7. Conclusion

This section reported on recent photoexcitation and photoionization investigations of atoms and molecules that used a high-resolution beamline (HRAM) from a third-generation synchrotron radiation source, the ALS, coupled with advanced photoelectron spectroscopy techniques. The measurements, and in some cases the reported calculations, illustrate how advanced technologies have allowed rapid progress in the understanding of the electronic structure and dynamics of simple and complex atomic and molecular systems. The study of the presented phenomena would have been impossible without such instrumentation since most of the weak processes have small cross-sections. It is also worth pointing out that, in many cases, the resolution improved by at least a factor of ten. For example, although the shape of autoionization resonances has been well explored previously, a quantitative analysis of both the shape and the angular distribution of the high-n resonances, based on high-resolution measurements, was lacking. The cases discussed here allowed a detailed comparison with calculations. In the case of the angular distribution, this comparison was done for the first time and excellent agreement was found. The resonant Auger process and in some cases Auger resonant Raman conditions were used to unravel the complex Auger decay spectra of core excited heavy atoms such as Ar, Kr and Xe. A systematic study was done that indicates some similarities among the three targets. Comparison with models and theories demonstrated some successes but also indicates the serious need for refined calculations. In some cases, better resolved as well as more extensive measurements are needed for a better understanding of the evolution of resonance Auger transitions. Doubly and triply excited states in Ne and Li respectively were investigated and strong evidence of breakdown of LS coupling in Ne as well as impressive new information on hollow Li were obtained. These types of measurements have stimulated the extension and development of new theories. Finally, even though investigations of inner-shell excitation of molecules has just begun with the new technologies, some long standing questions have already been answered and new as well as deep insights seem promising. However, theories in this area are in great need.

In general, with the development of advanced storage rings and instrumentation, the field of atomic and molecular science has many opportunities to grow, holds promise and excitement for the future and is in a strong position to uncover fundamental physics.

Acknowledgments

The author would like to thank her team and collaborators for contributing to the above work; B. Langer, A. Farhat, A. Wills, E. Kukk, J. D. Bozek, M. Humphrey, T. W. Gorczyca, Z. Felfi, O. Nayandin, M. Alsherhi, D. Cubaynes, S. Diehl, J.-M. Bizau, F. J. Wuilleumier, D. D. Thomas, T. X. Carrol, J. J. Saethre, R. Wehlitz, O. Hemmers, U. Becker, and J. Cornell. We are indebted to the excellent ALS source, and its technical staff.

This work was supported by the Division of Chemical Sciences, Office of Basic Energy Sciences, Office of Science, U. S. Department of Energy. B. L. and R. W. wish to thank the Alexander von Humboldt Foundation for partial support within the Feodor–Lynen program.

References

1. J. Berkowitz, *Photoabsorption, Photoionization, and Photoelectron Spectroscopy* (Academic Press, 1979).
2. G. V. Marr, *Photoionization Processes in Gases* (Academic Press, 1967).
3. J. A. R. Samson, *Methods of Experimental Physics*, Vol. 7A, Eds. B. Bederson and W. L. Fite (Academic Press, 1968).

4. U. Becker and D. Shirley, *VUV and Soft X-Ray Photoionization* (Plenum Press, 1996).
5. M. Y. Amusia, V. K. Ivanov and L. V. Chernysheva, *Phys. Lett.* **A59**, 191 (1976).
6. M. Y. Amusia and V. K. Ivanov, *Phys. Lett.* **A59**, 194 (1976).
7. T. A. Carlson, *Photoelectron and Auger Spectroscopy* (Plenum Press, 1975).
8. K. D. Sevier, *Low Energy Electron Spectroscopy* (Wiley-Intescience, 1972).
9. M. O. Krause, in *Atomic Inner-Shell Processes*, Vol. II, Ed. B. Caresmann (Academic Press, 1975), p. 33.
10. H. Siegbahn and L. Karlsson, in *Corpuscles and Radiation in Matter*, Ed. W. Mehlhorn, *Handbuch der Physik*, Vol. 31 (Springer-Verlag, 1982), p. 215.
11. M. O. Krause, in *Synchrotron Radiation Research*, Eds. H. Winick and S. Doniach (Plenum Press, 1980).
12. J. A. R. Samson, in *Corpuscles and Radiation in Matter*, Ed. W. Mehlhorn, *Handbuch der Physik*, Vol. 31 (Springer-Verlag, 1982), p. 123.
13. B. Sonntag and P. Zimmermann, *Rep. Prog. Phys.* **55**, 911 (1992).
14. V. Schmidt, *Rep. Prog. Phys.* **55**, 1483 (1992).
15. P. Balzer, M. Lundqvist, B. Wannberg, L. Karlsson, M. Larsson, M. A. Hayes, J. B. West, M. R. F. Siggel, A. C. Parr and J. J. Dehmer, *J. Phys. B: At. Mol. Opt. Phys.* **27**, 4915 (1994).
16. M. N. Piancastelli, M. Neeb, A. Kivimäki, B. Kempgens, H. M. Köppe, K. Maier, A. M. Bradshaw and R. F. Fink, *J. Phys. B: At. Mol. Opt. Phys.* **30**, 5677 (1997).
17. A. F. Starace, *Theory of Atomic Photoionization*, Ed. W. Melhorn *Handbuch der Physik* (Springer-Verlag, Berlin, 1982) pp. 1–121.
18. H. Derenbach and V. Schmidt, *J. Phys. B: At Mol. Opt. Phys.* **17**, 83 (1984).
19. B. Langer, A. Farhat, B. Nessar, N. Berrah, O. Hemmers and J. D. Bozek, in *Proceedings of the 'Atomic Physics with Hard X-Rays from High Brilliance Synchrotron Light Sources' Workshop* (ANL/APS/TM-16, 1996) p. 245 and references therein; N. Berrah, B. Langer, A. A. Wills, E. Kukk and J. Bozek, *Synch. Rad. News*, **11**, 21 (1998); R. Z. Bachrach, F. C. Brown and S. B. Hagström, *J. Vac. Sci. Technol.* **12**, 309 (1982); M. G. White, R. A. Rosenburg, G. Gabor, E. D. Poliakoff, G. Thornton, S. H. Southworth and D. A. Shirley, *Rev. Sci. Instrum.* **50**, 1268 (1979); U. Becker and R. Wehlitz, *J. Electron Spectrosc. Relat. Phenom.* **67** (1993).
20. N. Berrah, B. Langer, A. A. Wills, E. Kukk, J. D. Bozek, A. Farhat and T. W. Gorczyca, *J. Electron Spectrosc. Rel. Phenom.* **101**, 1 (1999).
21. F. Wuilleumier and M. O. Krause, *J. Electron Spectrosc. Rel. Phenom.* **15**, 15 (1979).
22. F. J. Wuilleumier, *Atomic Physics*, **7**, 493 (1980).
23. G. Mehlman, J. W. Cooper and E. B. Saloman, *Phys. Rev.* **A25**, 2113 (1982).
24. P. A. Heimann, T. Warwick, M. R. Howells, W. McKinney, D. Digennaro, B. Gee, D. Yee and B. Kincaid, *Nucl. Instr. Meth.* **A319**, 106 (1992).

25. F. J. Wuilleumier, S. Diehl, D. Cubaynes, J.-M. Bizau and E. T. Kennedy [submitted to *J. Electron Spectrosc. Rel. Phenom.* (1998)].
26. R. P. Madden and K. Codling, *Phys. Rev. Lett.* 516 (1963).
27. J. W. Cooper, U. Fano and F. Prats, *Phys. Rev. Lett.* **10**, 518 (1963).
28. A. F. Starace, *Phys. Rev.* **A16**, 231–242 (1977).
29. M. Y. Amusia, *Atomic Photoeffect* (Plenum Press, 1990).
30. H. Feshbach, *Ann. Phys.* **19**, 287 (1962).
31. K. T. Taylor, *J. Phys. B: At. Mol. Opt.* **10**, L699 (1977).
32. K. Codling, R. P. Madden and D. L. Ederer, *Phys. Rev.* **155**, 26 (1967).
33. V. Schmidt, *Z. Phys.* **D2**, 275–283 (1986).
34. U. Becker and D. A. Shirley, *Phys. Scrip.* **T31**, 56 (1990).
35. R. I. Hall, G. Dawber, K. Ellis, M. Zubek, L. Avaldi and G. C. King, *J. Phys. B: At. Mol. Opt. Phys.* **24**, 4133 (1991).
36. W. F. Chan, G. Cooper, X. Guo and C. E. Brion, *Phys. Rev.* **A45**, 1420 (1992).
37. H. Beutler, *Z. Phys.* **93**, 177 (1935).
38. B. Langer, N. Berrah, R. Wehlitz, T. W. Gorczyca, J. Bozek and A. Farhat, *J. Phys. B: At. Mol. Opt. Phys.* **30**, 593 (1997).
39. N. M. Kabachnik and I. P. Sazhina, *J. Phys. B: At. Mol. Opt. At.* **9**, 1681 (1976).
40. U. Fano, *Phys. Rev.* **124**, 1866 (1961).
41. K. A. Berrington, W. B. Eissner and P. H. Norrington, *Comput. Phys. Commun.* **92**, 290 (1995).
42. D. G. Hummer, K. A. Berrington, W. B. Eissner, A. K. Pradhan, H. E. Saraph and J. A. Tully, *Astron. Astrophys.* **279**, 298 (1993).
43. P. G. Burke and K. T. Taylor, *J. Phys. B: Atom. Mol. Phys.* **8**, 2620 (1975).
44. J. M. Bizau and F. Wuilleumier, *J. Elect. Spectrosc. Relat. Phenom.* **71**, 205–224 (1995).
45. R. P. Madden, D. L. Ederer and K. Codling, *Phys. Rev.* **177**, 136 (1969).
46. M. A. Baig and M. Ohno, *Z. Phys.* **D3**, 369–373 (1986).
47. S. L. Sorensen, T. Åberg, J. Tulkki, E. Rachlew-Källne, G. Sundström and M. Kirm, *Phys. Rev.* **A50**, 1218–1230 (1994).
48. S. L. Wu, Z. P. Zhong, R. F. Feng, S. L. Xing, B. X. Yang and K. Z. Xu, *Phys. Rev.* **A51**, 4494–4500 (1995).
49. K. Codling *et al.*, *J. Phys. B: At. Mol. Phys.* **13**, L693–L697 (1980).
50. A. Svensson, M. O. Krause and T. A. Carlson, *J. Phys. B: At. Mol. Phys.* **20**, L271 (1987).
51. P. van der Meulen, C. D. Caldwell, S. B. Whitfield and C. A. de Lange, *Phys. Rev.* **A46**, 2468–2485 (1992).
52. H. P. Kelly, *X-Ray and Inner-Shell Processes*, AIP Conference Proceedings No. 215, Eds. T. A. Carlson, M. O. Krause and S. T. Manson, American Institute of Physics, New York, 1990, pp. 292–311.
53. M. Seaton, *Rep. Prog. Phys.* **46**, 167 (1983).
54. C. H. Greene and L. Kim, *Phys. Rev.* **A38**, 5953 (1988).

55. J. P. Connerade and A. M. Lane, *Rep. Prog. Phys.* **51**, 1439 (1988).
56. N. Berrah, B. Langer, J. D. Bozek, T. W. Gorczyca, O. Hemmers, D. W. Lindle and O. Toader, *J. Phys. B: At. Mol. Opt. Phys.* **29**, 5351 (1996).
57. P. G. Burke and W. D. Robb, *Adv. At. Mol. Phys.* **11**, 143 (1975).
58. P. G. Burke and K. A. Berrington, *Atomic and Molecular Processes: An R-matrix Approach* (IOP Publishing, Bristol 1993).
59. M. Ya. Amusia and A. S. Kheifets, *Phys. Lett.* **A89**, 437–440 (1982).
60. U. Becker, D. Szostak, M. Kupsch, H, G. Kerkhoff, B. Langer and R. Wehlitz, *J. Phys, B: At. Mol. Opt. Phys.* **22**, 749 (1989).
61. H. Aksela, S. Aksela, G. M. Bancroft, K. H. Tan and H. Pulkkinen, *Phys. Rev.* **A33**, 3867 (1986).
62. U. Becker, T. Prescher, E. Schmidt, B. Sonntag and H. E. Wetzel, *Phys. Rev.* **A33**, 3891 (1986).
63. P. A. Heimann, D. W. Lindle, T. A. Ferrett, S. H. Liu, L. J. Medhurst, M. N. Piancastelli, D. A. Shirley, U. Becker, H. G. Kerkhoff, B. Langer, D. Szostak and R. Wehlitz, *J. Phys. B: At. Mol. Phys.* **20**, 5005 (1987).
64. B. Kämmerling, B. Krässig and V. Schmidt, *J. Phys. B: At. Mol. Opt. Phys.* **23**, 4487 (1990).
65. H. Aksela, S. Aksela, H. Pulkkinen, G. M. Bancroft and K. H. Tan, *Phys. Rev.* **A37**, 1798 (1988).
66. M. Meyer, E. von Raven, M. Richter, B. Sonntang and J. E. Hansen, *J. Electron Spectrosc. Relat. Phenom.* **51**, 407 (1990).
67. T. A. Carlson, D. R. Mullins, C. E. Beall, B. W. Yates, J. W. Taylor, D. W. Lindle and F. Grimm, *Phys. Rev.* **A39**, 1170 (1989).
68. M. Meyer, E. von Raven, B. Sonntag and J. E. Hansen, *Phys. Rev.* **A43**, 177 (1991).
69. J. Mursu, H. Aksela, O.-P. Sairanen, A. Kivimäki, E. Nõmmiste, A. Ausmees, S. Svensson and S. Aksela, *J. Phys. B: At. Mol. Opt. Phys.* **29**, 4387 (1996).
70. M. Meyer, E. von Raven, B. Sonntag and J. E. Hansen, *Phys. Rev.* **A49**, 3685 (1994).
71. H. Aksela and J. Mursu, *Phys. Rev.* **A54**, 2882 (1996).
72. R. Camilloni, M. Zitnik, C. Comicioli, K. C. Prince, M. Zacchigna, C. Crotti, C. Ottaviani, C. Quaresima, P. Perfetti and G. Stefani, *Phys. Rev. Lett.* **77**, 2646 (1996).
73. J. A. de Gouw, J. van Eck, A. C. Peters, J. van der Weg and H. G. M. Heideman, *J. Phys. B: At. Mol. Opt. Phys.* **28**, 2127 (1995).
74. C. D. Caldwell, in *Alignment in Low-energy Photoionization*, 15th International Conference on X-Ray and Inner-Shell Processes, Eds. T. A. Carlson, M. O. Krause and S. T. Manson, American Institute of Physics, Knoxville, 1990, p. 685.
75. B. Langer, N. Berrah, A. Farhat, O. Hemmers and J. D. Bozek, *Phys. Rev.* **A53**, R1946 (1996).
76. C. D. Caldwell and S. Hallman, *Phys. Rev.* **A53**, 3344 (1996).

77. J. W. Cooper, *Phys. Rev.* **A39**, 3714 (1989).
78. U. Hergenhahn, N. M. Kabachnik and B. Lohmann, *J. Phys. B: At. Mol. Opt. Phys.* **24**, 4759 (1991).
79. M. H. Chen, *Phys. Rev.* **A47**, 3733 (1993).
80. M. H. Chen, *Phys. Rev.* **A45**, 1684 (1992).
81. A. Frahat, M. Humphery, B. Langer, N. Berrah, J. D. Bozek and D. Cubaynes, *Phys. Rev.* **A56**, 501 (1997).
82. B. Langer, J. Viefhaus, O. Hemmers, A. Menzel, R. Wehlitz and U. Becker, *Phys. Rev.* **A51**, R882 (1995).
83. A. Menzel, Ph.D. Thesis, Technische Universität Berlin (unpublished, 1990).
84. U. Becker, *Synchrotron Radiation Experiments on Atoms and Molecules* in Proceedings of The Physics of Electronic and Atomic Collisions, Eds. A. Dalgarno, R. S. Freund, P. M. Koch, M. S. Lubell and T. B. Lucatorto, American Institute of Physics, New York, 1990, pp. 162–175.
85. J. Tulkki, H. Aksela and N. M. Kabachnik, *Phys. Rev.* **A50**, 2366–2375 (1994).
86. U. Hergenhahn and U. Becker, *J. Electron Spectrosc. Relat. Phenom.* **72**, 243–247 (1995).
87. U. Hergenhahn, S. B. Whitfield, J. Tulkki, F. Heiser, N. M. Kabachnik, B. Langer and U. Becker, *Phys. Rev.* **A55**, 2050–2066 (1997).
88. S. Svensson, B. Eriksson, N. Mårtensson, G. Wendin and U. Gelius, *J. Electron Spectrosc. Relat. Phenom.* **47**, 327 (1988).
89. B. Kämmerling, B. Krässig, O. Schwarzkopf, J. P. Ribeiro and V. Schmidt, *J. Phys. B: At. Mol. Opt. Phys.* **25**, L5–L10 (1992).
90. H. Aksela, J. Jauhianien, E. Kukk, E. Nmmiste, S. Aksela and J. Tulkki, *Phys. Rev.* **A53**, 290 (1996).
91. A. Kikas, S. J. Osborne, A. Ausmees, S. Svensson, O.-P. Sairanen and S. Aksela, *J. Electron Spectrosc. Relat. Phenom.* **77**, 241 (1996).
92. J. Jauhianien, H. Aksela, O.-P. Sairanen, E. Nmmiste and S. Aksela, *J. Phys. B: At Mol. Opt. Phys.* **29**, 3385 (1996).
93. H. Aksela, M. Kivilompolo, E. Nmmiste and S. Aksela, *Phys. Rev. Lett.* **25**, 4970 (1997).
94. H. Aksela, S. Aksela, A. Mäntykenttä, J. Tulkki, E. Shigmasa, A. Yagishita and Y. Furusawa, *Phys. Scrip.* **T41**, 113 (1992).
95. G. B. Armen, *J. Phys. B: At. Mol. Opt. Phys.* **29**, 677 (1996).
96. H. Aksela, S. Aksela, H. Pulkkinen. G. M. Bancroft and K. H. Tan, *Phys. Rev.* **A33**, 3876 (1986).
97. J. Mursu, J. Jauhiainen, H. Aksela and S. Aksela, *J. Phys. B: At. Mol. Opt. Phys.* **31**, 1973 (1998).
98. A. Farhat, A. A. Wills, B. Langer and N. Berrah [submitted to *Phys. Rev.* (1998)].
99. T. Åberg, *Phys. Scip.* **T41**, 71 (1992).
100. N. Berrah, B. Langer and A. Farhat, *Angle-Resolved Resonant Raman Auger Spectroscopy on Xe*, Proceedings of the Raman Emission by X-Rays

(REX-1) Workshop, Eds. D. L. Ederer and J. H. McGuire (World Scientific, 1996), p. 156.
101. H. Aksela, S. Aksela, O.-P. Sairanen, A. Kivimäki, A. Naves de Brito, E. Nmmiste, J. Tulkki, S. Svensson, A. Ausmees and S. J. Osborne, *Phys. Rev.* **A49**, R4269 (1994).
102. G. S. Brown, M. H. Chen and B. Crasemann, *Phys. Rev. Lett.* **45**, 1937 (1980).
103. B. Crasemann, in *Atomic and Molecular Physics with Synchrotron Radiation*, XVH International Conference on the Physics of Electronic and Atomic Collisions, Eds. L. E. McCarthy, W. R. MacGilliray and M. C. Standish, American Institute of Physics, Brisbane, Australia, 1992, p. 69.
104. H. Aksela, O.-P. Sairanen, S. Aksela, A. Kivimäki, A. Naves de Brito, E. Nmmiste, J. Tulkki, A. Ausmees, S. J. Osborne and S. Svensson, *Phys. Rev.* **A51**, 1291 (1995).
105. Z. F. Liu *et al.*, *Phys. Rev. Lett.* **72**, 621 (1994).
106. S. Masui *et al.*, *J. Phys.* **B28**, 4529 (1995).
107. W. Eberhardt, G. Kalkhoffen and C. Kunz, *Phys. Rev. Lett.* **41**, 156 (1978).
108. V. Schmidt *et al.*, *Phys. Rev.* **A24**, 1803 (1981).
109. U. Becker, T. Prescher, E. Schmidt, B. Sonntang and H. E. Wetzel, *Phys. Rev.* **A33**, 3891 (1986).
110. P. A. Heimann, D. W. Lindle, T. A. Ferrett, S. H. Liu, L. J. Medhurst, M. N. Piancastelli, D. A. Shirley, U. Becker, H. G. Kerkhoff, B. Langer, D. Szostak and R. Wehlitz, *J. Phys. B: At. Mol. Phys.* **20**, 5005 (1987).
111. C. D. Caldwell, in *Alignment in Low-energy Photoionization*, 15th International Conference on X-Ray and Inner-Shell Processes, Eds. T. A. Carlsson, M. O. Krause and S. T. Manson, American Institute of Physics, Knoxville, 1990, p. 685.
112. U. Hergenhahn *et al.*, *J. Phys.* **B26**, L117 (1993); Non-averaged P-values, U. Hergenhahn, private communication (1995).
113. U. Becker *et al.* quoted in Ref. 78; R. Weckwerth, Diplomarbeit (Diploma Thesis), Technical University Berlin (1991), (unpublished).
114. K. Schulz, G. Kaindl, M. Domke, J. Bozek, P. Heimann, A. Schlachter and J. Rost, *Phys. Rev. Lett.* **77**, 3086 (1996).
115. K. Schulz, M. Domke, R. Püttner, A. Gutiérrez, G. Kaindl, G. Miecznik and C. Greene, *Phys. Rev.* **A54**, 3095 (1996).
116. A. A. Wills, N. Berrah, T. Gorczyca, B. Langer, Z. Felfli, M. Alshehri, O. Nayandin and J. Bozek, *Phys. Rev. Lett.* **80**, 5085 (1998).
117. A. A. Wills, A. A. Cafolla, A. Sevensson and J. Comer, *J. Phys.* **B23**, 2013 (1990).
118. M. Domke, G. Remmers, and G. Kaindl, *Phys. Rev. Lett.* **69**, 1171 (1992).
119. A. Menzel, S. P. Frigo, S. B. Whitfield, C. D. Caldwell, M. O. Krause, J. Z. Tang and L. Shimamura, *Phys. Rev. Lett.* **75**, 1479 (1995).
120. J. P. Briand, *Comm. At. Mol. Phys.* **32**, 9 (1996).

121. R. Bruch, G. Paul, J. Andra and L. Lipsky, *Phys. Rev.* **A12** 1808 (1975).
122. M. Rodbro, R. Bruch and P. Bisgaard, *J. Phys.* **B12**, 2413 (1979).
123. L. M. Kiernan, E. T. Kennedy, J.-P. Mosnier, J. T. Costello and B. F. Sonntag, *Phys. Rev. Lett.* **72** 2354 (1994).
124. L. M. Kiernan, M.-K. Lee, B. F. Sonntag, P. Sladeczek, P. Zimmermann, E. T. Kennedy, J.-P. Mosnier and J. T. Costello, *J. Phys.* **B28**, L161 (1995).
125. Y. Azuma, S. Hasegawa, F. Koike, G. Kutluk, T. Nagata, E. Shigemasa, A. Yagishita and I. Sellin, *Phys. Rev. Lett.* **74**, 3768 (1995).
126. L. Jounel, D. Cubaynes, J.-M. Bizau, S. Al Moussalami, B. Rouvellou, F. J. Wuilleumier, L. Voky, P. Faucher and A. Hibbert, *Phys. Rev. Lett.* **76**, 30 (1996).
127. S. Diehl, D. Cubaynes, J.-M. Bizau, L. Joumel, B. Rouvellou, S. Al Moussalami, F. J. Wuilleumier, E. T. Kennedy, N. Berrah, C. Blancard, T. J. Morgan, J. D. Bozek, A. S. Schlachter, L. Voky, P. Faucher and A. Hibbert, *Phys. Rev. Lett.* **76**, 3915 (1996).
128. D. Cubaynes, S. Diehl, J.-M. Bizau, L. Joumel, B. Rouvellou, S. Al Moussalami, F. J. Wuilleumier, E. T. Kennedy, N. Berrah, C. Blancard, T. J. Morgan, J. Bozek, A. S. Schlachter, L. Voky, P. Faucher and A. Hibbert, *Phys. Rev. Lett.* **77**, 2194 (1996).
129. S. Diehl, D. Cubaynes, E. T. Kennedy, F. J. Wuilleumier, J.-M. Bizau, L. Joumel, L. Voky, P. Faucher, A. Hibbert, C. Blancard, N. Berrah. T. J. Morgan, J. Bozek and A. S. Schlachter, *J. Phys.* **B30**, L595 (1997).
130. S. Diehl, D. Cubaynes, K. T. Chung, F. J. Wuilleumier, E. T. Kennedy, J.-M. Bizau, L. Joumel, C. Blancard, L. Voky, P. Faucher, A. Hibbert, N. Berrah, T. J. Morgan, J. Bozek and A. S. Scghlachter *et al.*, *Phys. Rev. A, Rapid Comm.* **56**, R1071 (1997).
131. S. Diehl, D. Cubaynes, F. J. Wuilleumier, J.-M. Bizau, L. Joumel, E. T. Kennedy, C. Blancard, L. Voky, P. Faucher, A. Hibbert, N. Beffah, T. J. Morgan, J. Bozek and A. S. Schlachter, *Phys. Rev. Lett.* **79**, 1241 (1997).
132. K. T. Chung and B. C. Gou, *Phys. Rev.* **A53**, 2189 (1996).
133. C. D. Lin, *Phys. Rev.* **A29**, 1019 (1984).
134. C. H. Greene and C. W. Clark, *Phys. Rev.* **A30**, 2161 (1984).
135. X. Yang, C. G. Bao and C. D. Lin, *Phys. Rev. Lett.* **76**, 3096 (1996).
136. Y. Azuma *et al.*, *Phys. Rev. Lett.* **79**, 2419 (1997).
137. U. Becker, R. Hölzel, H. G. Kerkhoff, B. Langer, D. Szostak and R. Wehlitz, *Phys. Rev. Lett.* **56**, 1455 (1986).
138. O. Hemmers, F. Heiser, J. Eiben, R. Wehlitz and U. Becker, *Phys. Rev. Lett.* **71**, 987 (1993).
139. U. Becker and A. Menzel, *Nucl. Instrum. Methods Phys. Res.* **B99**, 68 (1995).
140. A. Kivimäki, M. Neeb, B. Kempgens, H. M. Köppe and A. M. Bradshaw, *Phys. Rev.* **A54**, 2137 (1996).
141. S. Sundin, S. J. Osbome, A. Ausmees, O. Bjömeholm, S. L. Sorensen, A. Kikas and S. Svensson, *Phys. Rev.* **A56**, 480 (1997).

142. H. Aksela, E. Kukk, S. Aksela, O.-P. Sairanen, A. Yivimiu, E. No-mrniste, A. Ausmees, S. J. Osbome and S. Svensson, *J. Phys.* **B28**, 4259 (1995).
143. M. Neeb, M. Biermann and W. Eberhardt, *J. Electron Spectrosc. Relat. Phenom.* **69**, 239 (1994).
144. D. A. Shaw, D. Cvejanoviö, G. C. King and F. D. Read, *J. Phys.* **B17**, 1984 (1984).
145. P. Morin and I. Nenner, *Phys. Rev. Lett.* **56**, 1913 (1986).
146. E. Kukk, H. Aksela, O.-P. Sairanen, S. Aksela, A. Mvimiu, E. Nommiste, A. Ausmees, A. Kikas, S. J. Osbome and S. Svensson, *J. Chem. Phys.* **104**, 4475 (1996).
147. E. Kukk, A. A. Wills, N. Berrah, J. D. Bozek, O. Nayandin, M. Alsheheri, A. Farhat and D. Cubaynes, *Phys. Rev.* **A57**, R1485 (1998).
148. O. Bjömeholm, S. Sundin, S. Svensson, R. R. T. Marinho, A. Naves de Biito, F. Gel'mukhanov and H. Ägren, *Phys. Rev. Lett.* **79**, 3150 (1997).
149. E. Kukk, H. Aksela, O.-P. Sairanen, E. No-mmiste, S. Aksela, S. J. Osbome, A. Ausmees and S. Svensson, *Phys. Rev.* **A54**, 2121 (1996).
150. K. Ellingsen, T. Saue, H. Aksela and O. Gropen, *Phys. Rev.* **A55**, 2743 (1997).
151. D. Dill, J. R. Swanson, S. Wallace and J. L. Dehmer, *Phys. Rev. Lett.* **45**, 1393 (1980).
152. A. P. Hitchcock, S. Beaulieu, T. Steel, J. Söhr and F. Sette, *J. Chem. Phys.* **80**, 3972 (1984). J. Stöhr, F. Sette and A. L. Johnson, *Phys. Rev. Lett.* **53**, 1684 (1984). F. Sette, J. Stöhr and A. P. Hitchcock, *J. Chem. Phys.* **81**, 4906 (1984).
153. M. N. Piancastelli, D. W. Lindle, T. A. Ferrett and D. A. Shirley, *J. Chem. Phys.* **86**, 2765 (1987); A. P. Hitchcock and J. Stöhr, *J. Chem. Phys.* **87**, 3253 (1987); M. N. Piancastelli, D. W. Lindle, T. A. Ferrett and D. A. Shirley, *J. Chem. Phys.* **87**, 3255 (1987).
154. B. Kempgens, H. M. Köppe, A. Kivimdki, M. Neeb, K. Maier, U. Hergenhahn and A. M. Bradshaw, *Phys. Rev. Lett.* **79**, 35 (1997); B. Kempgens, A. Kivimdki, H. M. Köppe, M. Neeb, A. M. Bradshaw and J. Feldhaus, *J. Chem. Phys.* **107**, 4219 (1997).
155. T. D. Thomas, N. Berrah, J. Bozek, T. X. Caroll, J. Hahne, E. Kukk and L. J. Saether, *Phys. Rev. Lett.* **82**, 1120 (1999).
156. J. N. Cutler, G. M. Bancroft and K. H. Tan, *J. Phys.* **B24**, 4897 (1991).
157. Z. F. Liu, G. M. Bancroft, K. H. Tan and M. Schachter, *J. Electron Spectrosc. Relat. Phenom.* **67**, 299 (1994).
158. S. Svensson, A. Ausmees, S. J. Osbome, F. Gel'mukhanov, H. Ågren, A. Naves de Brito, O.-P. Sairanen, A. Kivimäki, E. Nõmmiste, H. Aksela and S. Aksela, *Phys. Rev. Lett.* **72**, 3021 (1994).
159. M. R. F. Siggel, C. Field, L. J. Saethre, K. J. Borve and T. D. Thomas, *J. Chem. Phys.* **105**, 9035 (1996).
160. K. J. Borve, *Chem. Phys. Lett.* **262**, 801 (1996).
161. K. Ellingsen, T. Saue, H. Aksela and O. Gropen, *Phys. Rev.* **A55**, 2743 (1997).

162. E. Shigemasa, J. Adachi, M. Oura and A. Yagishita, *Phys. Rev. Lett.* **74**, 359 (1995).
163. F. Heiser, O. Gessner, J. Viefhaus, K. Wieliczek, R. Hentges and U. Becker, *Phys. Rev. Lett.* **79**, 2435 (1997).
164. G. S. Brown, M. H. Chen, B. Crasemann and G. E. Ice, *Phys. Rev. Lett.* **45**, 1937 (1980).
165. E. Kukk, J. D. Bozek, A. Wills, W.-T. Cheng, R. F. Fink, A. A. Wills and N. Berrah *J. Chem. Phys.* **111**, (1999).

CHAPTER 22

X-RAY SCATTERING AND FLUORESCENCE FROM ATOMS AND MOLECULES

S. H. Southworth, L. Young and E. P. Kanter

Argonne National Laboratory
Argonne, IL 60439, USA

T. LeBrun

National Institute of Standards and Technology
Gaithersburg, MD 20899, USA

Contents

1. Introduction

Fundamental understanding of X-ray interactions with atoms and molecules provides a basis for applying X-ray methods to complex materials, such as structural determinations by X-ray diffraction and extended X-ray absorption fine structure, Compton scattering measurements of electron-momentum distributions, and determinations of local chemical environments by X-ray absorption near-edge spectroscopy. This report discusses the basic physics of X-ray interactions with atoms and molecules relevant to understanding elastic and inelastic scattering cross-sections far from threshold (Secs. 3–5) and resonance and threshold phenomena in X-ray fluorescence (Secs. 7–9). The ideas are discussed in the context of providing theoretical interpretations of experimental results and are topics of ongoing theoretical developments. We also discuss experimental methods of X-ray spectroscopy using solid-state and wavelength-dispersive spectrometers with excitation by tunable synchrotron radiation. The examples discussed are primarily from the authors' research. We did not attempt to provide a comprehensive review, but several recent review articles and research papers are referenced for further information. The ideas discussed are general and relevant to understanding fundamental studies and applications of X-ray scattering and fluorescence.

2. Photon–Atom Interactions in the X-ray Regime

The primary photon–atom interactions at X-ray energies are photoabsorption and elastic and inelastic scattering,[1] as illustrated in Fig. 1 for a high- and low-Z element. For incident energies below the K-shell ionization threshold, photoabsorption dominates as in the visible and ultraviolet

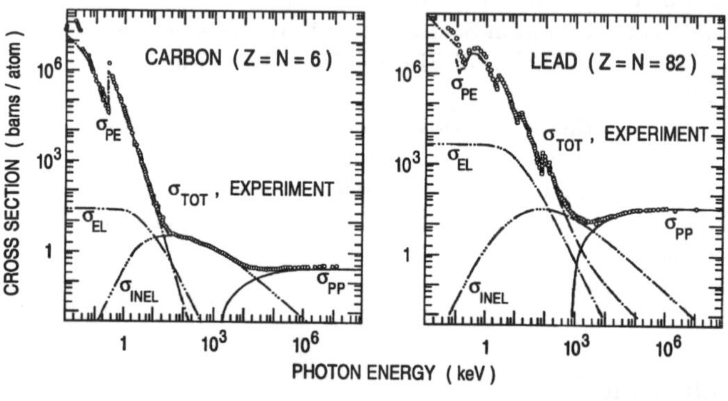

Fig. 1. Cross-sections for the primary photoprocesses: photoabsorption (σ_{PE}), elastic scattering (σ_{EL}), inelastic scattering (σ_{INEL}), and pair production (σ_{pp}) for carbon and lead. (Data from Ref. 4.)

energy ranges. However, with increasing energy, the photoabsorption cross-section falls as $E^{-7/2}$, and, especially for low-Z elements, scattering becomes dominant over a large energy range until the onset of pair production and nuclear resonant processes at MeV energies. Photoabsorption, elastic (Rayleigh) and inelastic (Compton) scattering are the lowest-order interactions between the radiation field and an electron in an atom. The nonrelativistic interaction Hamiltonian is of the form

$$H_{int} = \frac{e^2}{2mc^2}A^2 - \frac{e}{mc}\mathbf{p}\cdot\mathbf{A}\,,$$

where $\mathbf{p}$ is the electron momentum and $\mathbf{A}$ is the vector potential of the electromagnetic field. Photoabsorption, accompanied by photon annihilation, is described by the $\mathbf{p}\cdot\mathbf{A}$ term in the expression. Scattering, however, is a process which conserves photons, and $\mathbf{A}$ therefore must appear an even number of times. Scattering processes (elastic and inelastic) are therefore described in lowest order by the A^2 term. These, as well as higher-order scattering processes, i.e. second-order $\mathbf{p}\cdot\mathbf{A}$ processes, can be visualized by the Feynmann diagrams in Fig. 2. The first-order A^2 scattering is the 'seagull' diagram [Fig. 2(a)] and dominates in regions far from threshold and resonance (Sec. 3). The second and third diagrams result from second-order $\mathbf{p}\cdot\mathbf{A}$ interactions, with emission first [Fig. 2(b)] and absorption first [Fig. 2(c)]. The diagrams account for infrared divergence (Sec. 4) and resonant Raman scattering (Secs. 7–9).[2] The corresponding equation is known as the Kramers–Heisenberg formula.[3]

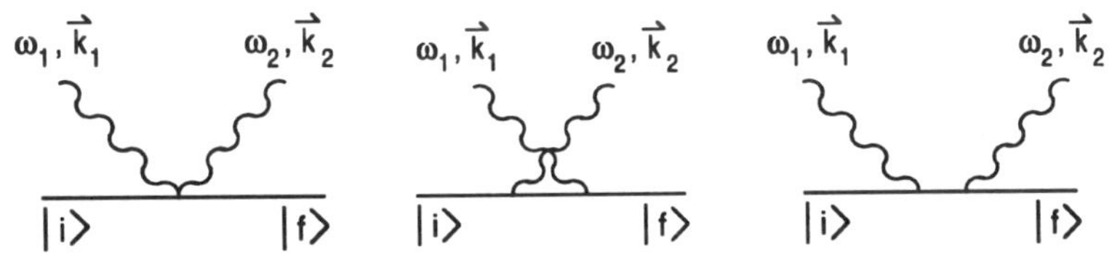

(a) (b) (c)

Fig. 2. Feynmann diagrams for scattering amplitudes in a nonrelativistic framework. For elastic scattering, $\omega_1 = \omega_2$. (a) 'Seagull' diagram representing first-order A^2 term; (b) and (c) represent second-order $\mathbf{p}\cdot\mathbf{A}$ terms.

3. Photon Scattering from Free Atoms

In contrast to the photoabsorption process, relatively little data exists on the scattering of photons from free atoms. The reason for this is clear upon inspection of Fig. 1, where it is seen that the cross-section for scattering is substantially smaller than that for photoabsorption, being on the order of the Thomson cross-section for the scattering from a free electron, $\sigma_T = (8\pi r_e^2)/3 = 0.6652 \times 10^{-24}$ cm^2. The coherent (or Rayleigh) and incoherent (or Compton) scattering processes are commonly described (within the A^2 approximation) by modifying the scattering from a point charge to account for the extended charge distribution through the use of form factors and incoherent scattering factors, $F(k)$ and $S(k)$.[4] For an unpolarized beam

$$\left(\frac{d\sigma}{d\Omega}\right)_{coh} = \left(\frac{d\sigma_T}{d\Omega}\right)|F(k)|^2 = \frac{r_e^2}{2}(1+\cos^2\theta)|F(k)|^2\,,$$

and

$$\left(\frac{d\sigma}{d\Omega}\right)_{inc} = \left(\frac{d\sigma_{KN}}{d\Omega}\right)S(k) = \frac{r_e^2}{2}\left(\frac{\omega_2}{\omega_1}\right)^2\left[\frac{\omega_1}{\omega_2}+\frac{\omega_2}{\omega_1}-\sin^2\theta\right]S(k)\,.$$

Here, $\hbar\mathbf{k} = \hbar(\mathbf{k_2}-\mathbf{k_1})$ and $\hbar\omega = \hbar(\omega_2-\omega_1)$ are the momentum and energy transferred in the scattering, and θ is the angle between incident and outgoing photons. The $F(k)$ and $S(k)$ are given by,[4]

$$F(k) = \sum_{n=1}^{Z}\langle i|\exp(i\mathbf{k}\cdot\mathbf{r}_n)|i\rangle\,,$$

and

$$S(k) = \sum_{m=1}^{Z}\sum_{n=1}^{Z}\langle i|\exp(i\mathbf{k}\cdot(\mathbf{r}_m-\mathbf{r}_n)|i\rangle - |F(k)|^2\,.$$

The form factor approach is important for several reasons. Essentially all tabulations of elastic scattering use this approximation to estimate the Rayleigh scattering contribution.[5] Both $F(k)$ and $S(k)$ can be evaluated in terms of *ground state* wavefunctions and are conveniently tabulated for theoretical calculations of varying complexity.[6] Cross-sections are conveniently defined as a function of only one variable, the momentum transfer k. Finally, physical interpretations can be given to the measured quantities: the elastic scattering form factor $F(k)$ is the Fourier transform of the charge density, and the inelastic scattering factor $S(k)$ is related to the electron pair distribution function (see Sec. 5). Through the impulse approximation, its doubly differential counterpart $S(\mathbf{k},\omega)$ yields the electron momentum distributions in the atom.[7] Both are widely used to study complex materials[8] and charge density distributions.[9] However, the range of validity of the form factor approach is limited at low energies due to the neglect of binding, and, at high energies and large angles, due to the assumption of low momentum transfer $\hbar k < mc$.[10]

Theoretical approaches that go beyond the form factor and incoherent scattering factor have been developed recently. For a review of these second-order S-matrix calculations of Rayleigh scattering see Kissel *et al.*[11] and for Compton scattering see Bergstrom and Pratt.[2] They are generally considered state-of-the-art, and treat both the A^2 and the second-order $\mathbf{p} \cdot \mathbf{A}$ terms simultaneously in a fully relativistic framework. For Rayleigh scattering, the additional dynamical terms correspond to the anomalous scattering factors, which alter the phase and magnitude of the scattered photon field due to absorption in the atom. For Compton scattering these second-order terms correspond to Raman scattering and the infrared divergence. These general methods which treat the interaction to higher order require considerable computational effort, and necessarily treat the atomic wavefunctions using the independent particle approximation (IPA). It is of interest to understand the improvements relative to simpler methods which are able to treat the target more accurately.

Discrepancies between the S-matrix calculations and tabulations based on other theories are fairly substantial in light elements, even at energies high enough where the S-matrix and the tabulations agree in higher-Z elements. Only four early experiments exist for scattering from free atoms,[12–15] the case for which cross-sections are usually calculated. Wollan[12] measured total scattering in various gases relative to H_2. Experimental uncertainties at this early stage of instrumentation made comparison with theory difficult.[14] Chipman and Jennings[13] carefully measured absolute total scattering cross-sections for rare-gases at 17.4 keV and angles $< 32°$ and confirmed form factor calculations using then–current theory to estimate the inelastic component. Ice *et al.*,[14] using synchrotron radiation, measured total cross-sections in H_2 and He, and found evidence for electron correlation effects. Smend and Czerwinski[15] focused on the large angle elastic scattering in Kr and Xe at 59.54 keV, finding better agreement with the S-matrix than with form factor calculations. In light of recent theoretical developments,[2] it is of general interest to understand the regions of validity for the various calculational methods and develop generalized schemes for the prediction of scattering cross-sections, angular distributions, and polarization dependencies. In an initial experiment at Argonne's Advanced Photon Source (APS), Compton and Rayleigh scattering from Ne were measured between 11 and 22 keV, where sizable differences (up to 16%) existed between the S-matrix and form-factor calculations of the elastic scattering channel.[16]

Early measurements did not resolve the Compton and Rayleigh components,[12–14] making detailed comparison with theory difficult. Two modern advances enable higher quality scattering data: (1) high incident photon flux at synchrotron sources coupled with high-resolution crystal monochromators enables gas-phase experiments, and (2) advances in theoretical methods permit scattering cross-sections to be placed with high precision ($\approx 0.5\%$) on an absolute scale by comparison to well-known cross-sections for He. Although crystal spectrometers provide significantly better resolution (e.g. $\Delta E/E \approx 10^{-4}$ for Si (111)), there are advantages to using relatively low-resolution ($\Delta E/E \approx 3\%$) solid state detectors in some situations. First, solid state detectors, such as Si(Li)s, permit simultaneous data collection over the entire usable energy range [Si(Li)s have efficiencies $> 10\%$ from ≈ 1 to 60 keV] as opposed to the small energy window available using crystal spectrometers. Second, because of their high efficiencies, absolute scattering cross-sections can be obtained simply using well-characterized solid state detectors to make measurements relative to He and using theory to put values on an absolute scale. Third, solid state detectors, being compact, can be arranged to subtend large solid angles for efficient coincidence experiments.

For this first experiment, only ratios are measured: (1) total scattering in Ne to He, and (2) Compton to Rayleigh scattering in Ne, thus circumventing several difficult *absolute* measurements, e.g. X-ray flux and detector efficiency. In the energy range 11–22 keV, which is far above the Ne K-edge (870 eV), the Compton and Rayleigh scattering cross-sections are comparable and the IPA may be reasonably well-justified. The total cross-section for X-ray scattering from He in this energy range is dominated by the Compton channel, with the Rayleigh contribution ranging from $\approx$ 5–0.1%. Calculations for X-ray scattering cross-sections from He[4–6] are in good agreement and are assumed to be correct.

In order to compare measured cross-sections with theory, it is important to eliminate uncertainties due to the polarization of the incident X-ray beam and variations therein during the measurement process. This can be done

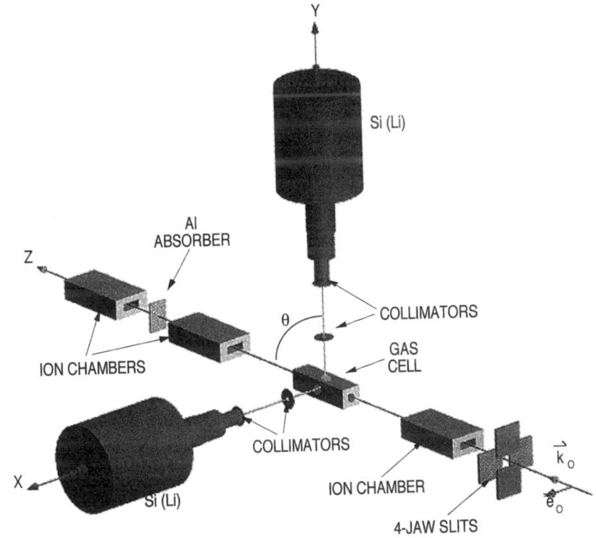

Fig. 3. Experimental setup for measuring polarization-independent scattering cross-sections from gases. The two Si(Li) detectors are placed 90° apart azimuthally around the incident photon beam direction. The scattering angle, $\theta = 90° \pm 3°$, as defined by two collimators. The incident photon beam is defined by slits and its intensity measured by ion chambers. The incident beam is polarized along the X direction.

using the configuration shown in Fig. 3, which measures simultaneously the scattering yield in two detectors placed 90° azimuthally about the photon beam propagation axis. The average yield in the two detectors is the polarization-independent differential scattering cross-section. Monochromatic X-rays incident on the gas sample create a line source viewed by the two well-characterized Si(Li) detectors. Higher harmonics in the incident X-ray beam are not a direct factor here, since they are resolved in the Si(Li) detectors, but can be monitored using ion chambers and suitable attenuators.

An example of the scattering data is shown in Fig. 4, which shows an empty cell spectrum followed by spectra from Ne at 11, 15, 18, and 22 keV. The empty cell spectrum is featureless (no window scattering) and serves as a good measure of background. After background subtraction, the yields were corrected for detector deadtime, window and gas attenuation, X-ray intensity and pressure dependence. The $\approx 1\%$ errors associated with these early measurements of the total scattering cross-sections are composed of:

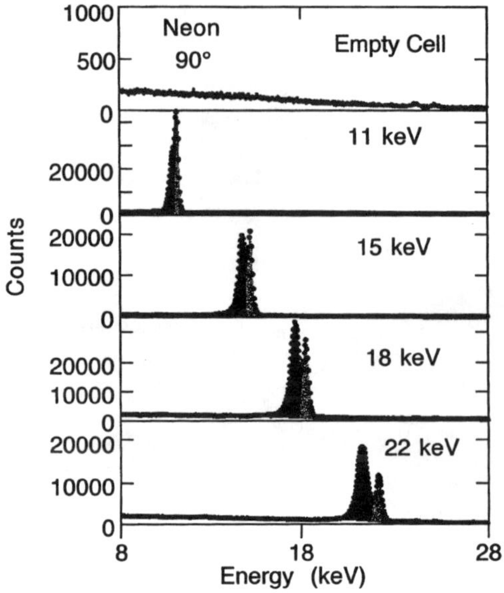

Fig. 4. Typical scattering spectra in the vertical (y) detector. The empty cell spectrum is featureless, indicating no window scattering. The two small features at $\approx$ 25 keV correspond to fluorescence from In and Sn impurities in the detectors. The light shaded areas of each spectrum indicate the Monte–Carlo simulation of the Rayleigh line, and the darker areas correspond to the simulated Compton peak.

statistical error ($\approx$ 0.2%) and systematic errors due to (1) variations in background ($\approx$ 0.2%), (2) extrapolation to zero pressure ($\approx$ 0.4%), and (3) run-to-run variation in He normalization ($\approx$ 0.7%). Reductions in systematic errors are expected in a double-cell arrangement where both the He reference gas and the Ne sample gas can be measured simultaneously.

Separation of the Compton and Rayleigh components in the spectra requires detailed knowledge of both the instrumental response and the lineshape associated with the Compton scattering. For a Si(Li) detector operated in this energy range the response to a monochromatic X-ray input can be modeled as a Gaussian peak with a minor ($\approx$ 0.5%) low-energy exponential tail and a small ($\approx$ 0.1%) escape peak. The detailed response of the detectors can be determined through calibration using radioactive sources.[17] However, the most reliable measurements come from an *in situ* calibration using, e.g. Compton scattering from He or fluorescence from Kr. The *in situ* measurements are important in that they quantify the width of

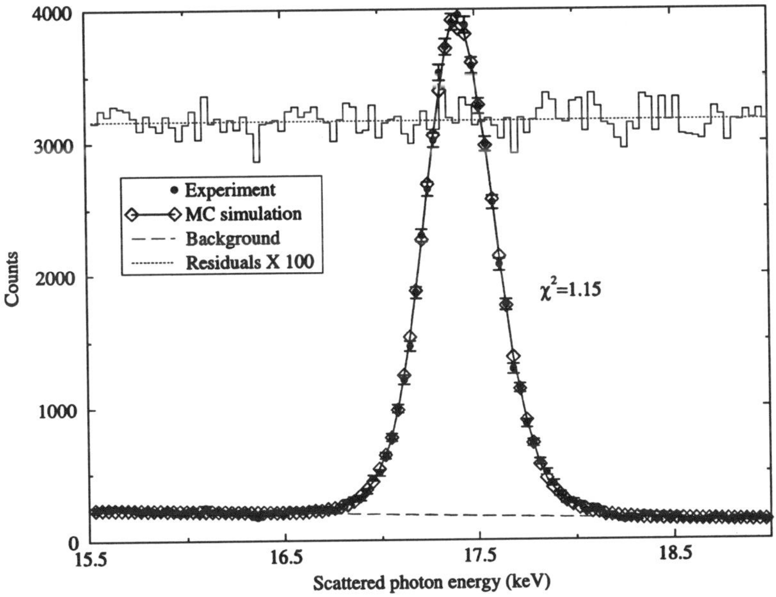

Fig. 5. Spectrum for Compton scattering of 18 keV incident photons on a He target detected at 90° scattering angle. The experimental points are shown as solid circles with error bars. The Monte–Carlo simulation, incorporating the detector response function and the impulse approximation to the Compton profile, and the fit residuals (multiplied by 100) are also shown.

the Gaussian response during the experiment. The Gaussian peak standard deviation is characterized by $\sigma^2 = a + bE$, where E is the incident photon energy and a and b are calibration constants. It was found that the lineshape associated with Compton scattering can be modeled adequately in the peak region using impulse approximation Compton profiles[18] in a Monte–Carlo simulation[17] (see Fig. 5). In using the tabulated Compton profiles to fit the data, it is extremely important to include kinematic cut-offs for each subshell (see Figs. 6 and 7). The amplitudes and positions of the generated Compton and Rayleigh lineshapes were varied in a four-parameter fit to the experimental data (see Fig. 8). While the overlap between the Compton and Rayleigh scattering is substantial at 11 keV, the separation is reliable to $\approx$ 8%. The error in the Compton/Rayleigh ratio decreases to $\approx$ 2% at 22 keV.

With these precise measurements of ratios (total scattering from Ne relative to He and Compton to Rayleigh scattering from Ne), detailed

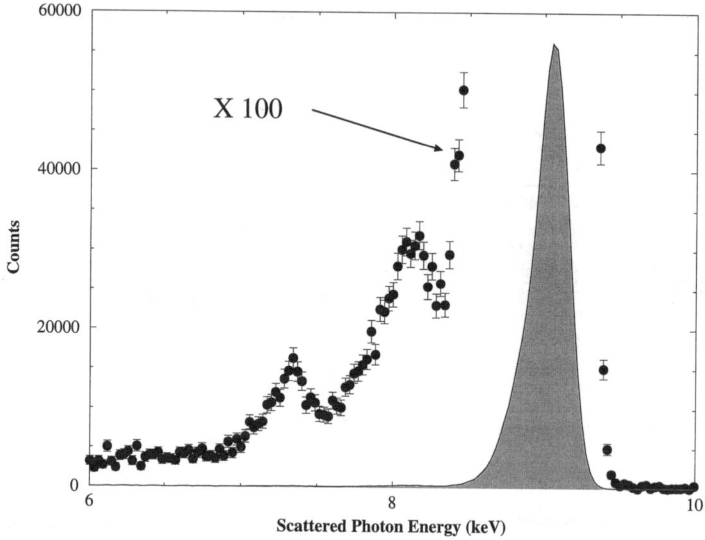

Fig. 6. Spectrum for Compton + Rayleigh scattering at 9 keV on a Ne target (filled solid). When multiplied by 100, the effect of the kinematic cutoff of the K-edge (8.130 keV) on the $1s$ Compton profile is evident in the data. The Si escape peak of the detector is also seen at 7.3 keV. These features are all included in the simulations.

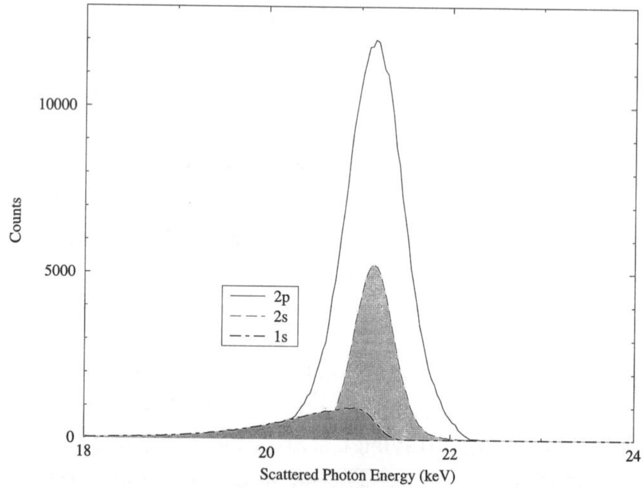

Fig. 7. Simulations of the experimental response to the subshell-specific Compton profiles within the impulse approximation for scattering from the $1s$, $2s$, and $2p$ subshells. The kinematic cutoffs produce asymmetric low-energy tails.

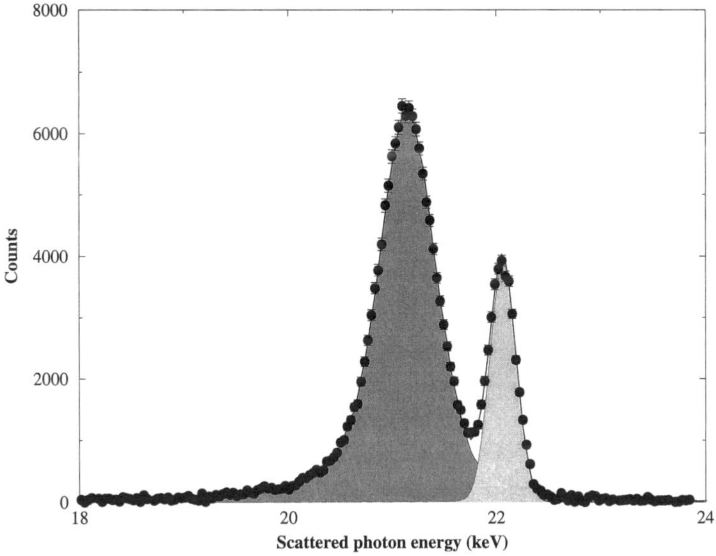

Fig. 8. Spectrum for Compton + Rayleigh scattering at 22 keV on a Ne target for 90° scattering angle (filled circles). The darker filled region is the result of the Monte–Carlo simulation for Compton scattering while the lighter filled region is the same for Rayleigh scattering. The fit to the data (solid line) represents the sum of these simulations. Similar fits are also shown in Fig. 4 for the other energies.

comparisons with theory are feasible. The analysis reveals the importance of including electron exchange and electron correlation in the atomic wavefunctions and the dynamical scattering effects (second-order $\mathbf{p} \cdot \mathbf{A}$ terms) in accurate calculations of scattering cross-sections for low-Z atoms. Since no theory to date includes all of these effects simultaneously, a perturbative method was prescribed to obtain predictions with 1% accuracy.[16]

Outstanding challenges in this area of research are (1) understanding the limitations of this perturbative correction method at lower momentum transfers, where cross-terms between the largest corrections, the exchange (10%) and dynamical effects (5%), will increase; (2) understanding the angular dependence of the anomalous scattering factors, where the common assumption is angle independence, but $\approx$ 40% differences have been predicted[19]; (3) understanding and developing theoretical models for anomalous scattering near absorption edges.[20] Research of this type impacts a number of fields, because form and structure factors are widely used for structure determination and radiation transport calculations. Yet no single theory is presently adequate, due to the need to include both the X-ray interaction and the atomic wavefunction accurately.

4. Subshell-Specific Compton Scattering: Infrared Divergence

The previous section focused on the inelastic scattering spectrum near the Compton peak. In this section, the focus is on the inelastic scattering cross-section $d^2\sigma/d\Omega d\omega$ at low energy, which has been predicted to diverge as $1/\omega$, as $\omega \to 0$, where ω is the scattered photon energy. The implication of an inelastic scattering spectrum which is infrared divergent is that the cross-section is undefined unless a lower energy limit to the detection system is specified. The theoretical situation is clear, with considerations on general grounds,[21] nonrelativistic calculations,[22] and relativistic S-matrix calculations[23] all in agreement about the existence of an infrared divergence (IRD). However, the experimental situation is not as clear. There is not, to date, an unambiguous observation of the onset of the infrared rise. Early reports of an infrared divergent rise for scattering from the L-shell of Zr[24] were questioned by Marchetti and Franck[25] who suggested that the observed rise could be due to interatomic bremsstrahlung following photoeffect. Manninen *et al.*[26] later observed an infrared divergent rise, but concluded that it was due entirely to photoelectron bremsstrahlung in the target. The magnitude of bremsstrahlung is a function of the thickness of the target, and therefore questions have been raised as to the correct procedure for its subtraction.[23] All of the studies above were performed on solid targets, motivating our use of a thin gas target in order to minimize the interatomic bremsstrahlung correction.

The approach taken for this experiment was to isolate the scattering from the K-subshell in gaseous Ar, where the predicted infrared rise of the cross-section was accessible to the energy range of Si(Li) detectors. Compton scattering from the K-shell of Ar was identified by detecting the $K\alpha, \beta$ fluorescence photon in coincidence with the Compton-scattered photon. The layout of the experiment is shown in Fig. 9. In this view, the incident X-ray beam is normal to the page and traverses Ar gas which fills the region between the detectors. Monochromatic 40 keV X-rays from the bending-magnet beamline at the Basic Energy Sciences Synchrotron Radiation Center of the APS were incident on the Ar target. At 40 keV, the ratio of photoabsorption to Compton scattering in Ar is ≈ 7, and the fluorescence quantum yield is $\approx 12\%$, therefore the relative photon yields

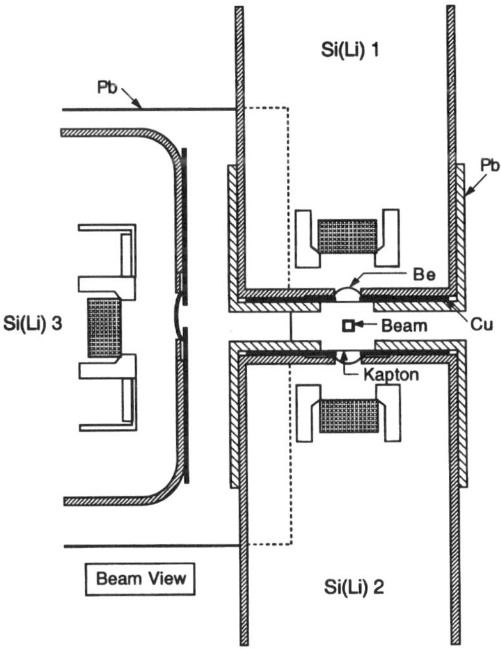

Fig. 9. Detector layout for subshell-specific Compton scattering experiments. The incident X-ray beam points into the page with the electric vector parallel to the axis of Si(Li) 3. The beam bisects the region filled with Ar gas between Si(Li) 1 and Si(Li) 2.

are well-matched for the two processes. Since the incident X-rays are polarized in the horizontal plane, the Compton scattering is maximized in the vertical direction toward Si(Li) 1 and Si(Li) 2 which have the largest solid angles. During the experiment, the energies and relative arrival times of photons hitting Si(Li)s 1, 2, and 3 were recorded in 'event' mode. This permits resorting the data after the experiment to further explore the coincidences and correlations between all three detectors. Secondary scattering into the detectors was minimized by lead shielding, and photoelectron penetration into the detectors was reduced by placing kapton film over the detector entrance apertures.

A spectrum of the difference in the arrival times of photons in Detectors 1 and 2 is shown in Fig. 10. The large spike at zero time corresponds to the 'prompt' coincidences, where both detectors simultaneously detect photons, but contains both 'true' and some 'random' events. The remainder of the spectrum is completely due to random coincidences exhibiting

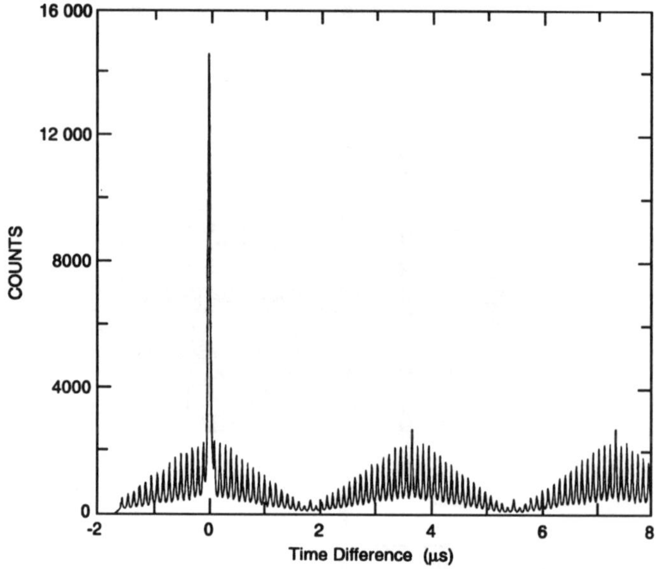

Fig. 10. Time delay between arrival of photons in Detector 2 relative to Detector 1. The 'prompt' spike at $t = 0$ includes the 'true' coincidences. The circulation time for a given electron bunch at the Advanced Photon Source is $\approx 3.6\mu$s, giving rise to the broad envelope structure. The fine structure arises from the $\approx$ 100 ns between filled buckets in the ring.

the complex time structure of the APS storage ring. 'True' coincidences are defined as (prompt — random) coincidences. The periodicity of the broad envelope is due to the revolution time of a single pulse in the storage ring ($\approx 3.6\mu$s), and the fine structure corresponds to the $\approx$ 100 ns between filled buckets.

The correlated energy spectra in Detectors 1 and 2 within the $t = 0$ peak are shown in two dimensions in Fig. 11(a). Making the appropriate cuts on the $K\alpha, \beta$ line ($\approx$ 3 keV) in Detectors 1 and 2 will yield the inelastic scattering spectrum from the K-shell. The $K\alpha, \beta$ lines are only faintly visible in this view, and correspond to horizontal and vertical lines starting from the spot near the lower left corner. It is clear from Fig. 11(a) that substantial contamination of the true coincidences is present. The contamination is due primarily to detector–detector scattering in this particular pair of detectors. For example, the diagonal lines extending from $\approx$ 40 keV from both axes corresponds to the constant sum-energy line.

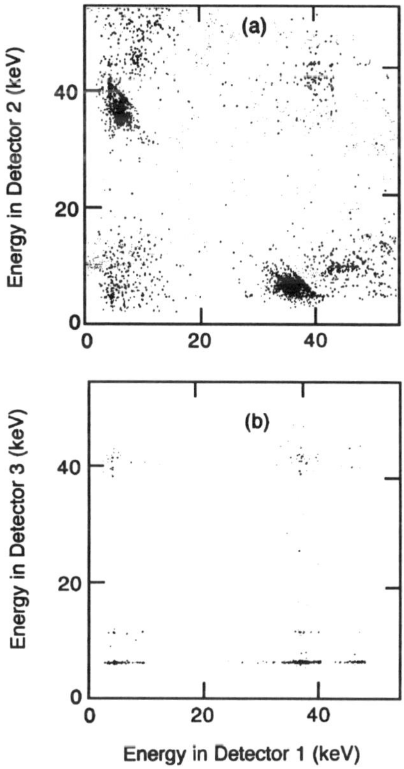

Fig. 11. (a) Energy in Detector 2 vs. energy in Detector 1 after the sorting for events contained in the $t = 0$ peak shown in Fig. 10. (b) Energy in Detector 3 vs. energy in Detector 1 after the sorting for events contained in the $t = 0$ peak shown in Fig. 10. Intensities are shown on log scales in both (a) and (b).

True coincidences along this line can originate from Rayleigh scattering off the target into Detector 1 and subsequent Compton backscattering into Detector 2, leading to the sharp diagonal line. A similar explanation holds for the diffuse spots, which originate from Compton scattering from the target. These crosstalk events vanish in coincidences of Detectors 1 or 2 with the horizontal Detector 3 [see Fig. 11(b)].

Figure 12 shows the resulting true coincidence spectrum after the appropriate time and energy cuts on the data have been made. Overlaid as a solid line is the singles spectrum for comparison. The Rayleigh line is offscale at 40 keV. The full-atom Compton spectrum peaks around 37 keV and is dominated by scattering from outer shells with narrower electron-momentum

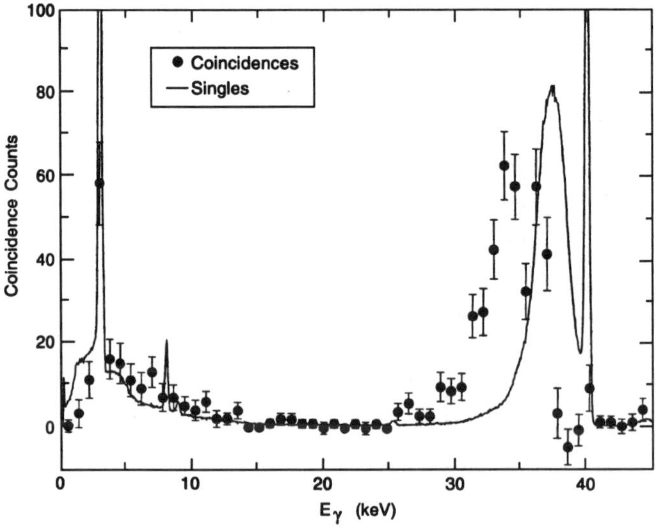

Fig. 12. Coincidence spectrum (trues = prompts − randoms) isolating scattering from the K-shell in Ar. Coincidences in Detector 1 are obtained with a cut on the $K\alpha, \beta$ fluorescence in the other detectors as well as the timing cuts. In the region of the Compton peak, the 'singles' spectrum, shown as a solid line for comparison, is noticeably narrower than the coincidence spectrum since it is dominated by scattering from the outer shells of the atom.

distributions. A comparison of the coincidence spectrum with an impulse approximation[18] K-shell profile is in reasonable agreement, as shown in Fig. 13. The rise at low energy is somewhat larger than predicted for infrared divergence, and is probably mostly photoelectron bremsstrahlung. An improved experiment, with better detector geometry and pressure-dependent measurements, is necessary to isolate the IRD. However, the excess in the coincidence spectrum at the $K\alpha, \beta$ energy (≈ 3 keV) is interesting in that it suggests the creation of a double-K vacancy state. This observation has led to further studies of the cross-section for double-K vacancy production which will be reported elsewhere.

Of general interest are the analyses of pulsed synchrotron sources for inelastic scattering studies using coincidence techniques.[27,28] Although the fluxes from synchrotron sources are extremely high relative to radioactive sources, the measurement time for a given signal-to-noise ratio is not significantly reduced. This results from the fact that the true coincidence rate is proportional to $I\varepsilon_1\varepsilon_2$ whereas the random coincidence rate is proportional

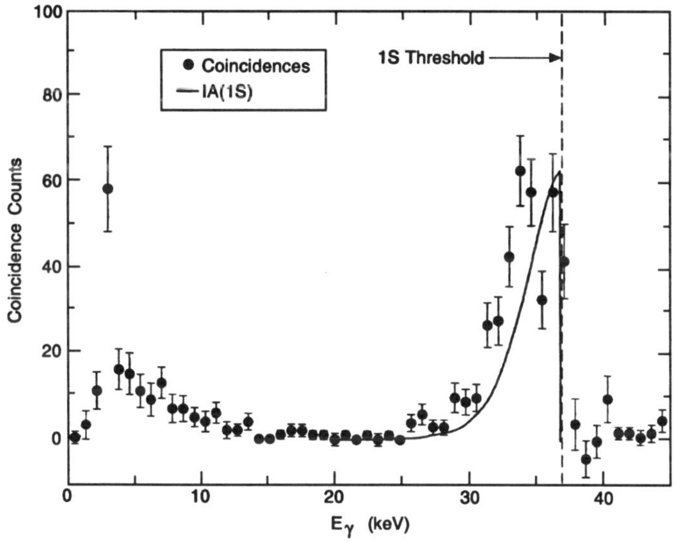

Fig. 13. The coincidence spectrum is compared to the impulse approximation Compton profile from Ref. 18. Note that the onset of scattering is displaced from the incident energy, 40 keV, by the K-shell binding energy (3.2 keV). These data have not been corrected for the detection efficiency cutoff at low energies.

to $I^2\varepsilon_1\varepsilon_2\tau$, where I is the incident X-ray intensity, ε_1 and ε_2 are the detection efficiencies for the scattered and fluorescence photon, and τ is the time resolution of the detector system. Therefore, the ratio of true-to-random events is maximized as $I \to 0$, so the high intensity from third-generation synchrotron sources is not necessarily utilized during the measurement process, albeit being extremely useful during setup of the experiment. The principal advantage of these new sources for such experiments lies in the tunability and narrow bandwidth of the beam.

5. Inelastic Scattering Probes of Electron Pair Correlation and Other Properties

High-energy X-ray scattering is regarded as an effective probe of target properties, because the photon–atom interaction is impulsive, allowing the first Born (A^2) approximation to be made.[29] With this approximation, cross-sections are factored into two terms, dealing with the projectile and the target separately. Target properties are contained in the generalized oscillator strength (GOS) $\frac{df(\mathbf{k},\omega)}{d\omega}$,

$$\frac{df(\mathbf{k},\omega)}{d\omega} = \frac{\omega}{R(ka_o)^2}\sum_f \left|\left\langle f\left|\sum_{n=1}^{Z}\exp(i\mathbf{k}\cdot\mathbf{r}_n)\right|i\right\rangle\right|^2 ,$$

where $\mathbf{k}$ and ω are the momentum and energy transferred in the scattering process. Here, $\langle f|$ and $|i\rangle$ are the final and initial states of the target atom with Z electrons, $\mathbf{r}_n$ the radial vector from the nucleus of the nth electron, a_o the Bohr radius, and R the Rydberg constant. The GOS is closely related to the dynamic structure factor, $S(\mathbf{k},\omega)$, and the doubly-differential inelastic scattering spectrum, $(d^2\sigma/d\Omega d\omega)$. The Bethe surface (a three-dimensional representation of $df(\mathbf{k},\omega)/d\omega$ as a function of $\ln(ka_o)^2$ and ω) gives the best visualization of the GOS. Complete studies of any Bethe surface using inelastic X-ray scattering are difficult due to the small cross-section and the need to reliably remove the intense elastic scattering peak. Some studies have been done using X-ray scattering,[30] but most have used electron scattering on gaseous targets.[31] Many properties can be calculated from the GOS, e.g. polarizability, dielectric response, and the electron pair correlation function.[32]

Here the focus is on the static factor, $S(k)$, and its well-known sensitivity to electron correlation.[5] Originally outlined by Waller and Hartree in 1929, the total (elastic plus inelastic) scattering cross-section can be inverted to yield the electron pair distribution function, $h(\mathbf{r}_{12})$, the radial probability of finding an electron $\mathbf{r}_{12}$ away from another. This can be easily seen by rearrangement of the following expression for $S(k)$,

$$S(k) = Z + 2\int \exp(i\mathbf{k}\cdot\mathbf{r}_{12})h(\mathbf{r}_{12})d\mathbf{r}_{12} - |F(k)|^2 .$$

The pair distribution is strongly affected by Coulomb correlation and therefore measurements of total scattering cross-sections can, in principle, discern IPA from models including correlation. This was demonstrated on H_2 and He in the first study using synchrotron radiation[14] and in others using X-ray tubes.[33] Bonham[34] pointed out that the interpretation of these results was based upon a simplified theory, where the questionable issue is that doubly-differential cross-sections are commonly measured at fixed-angle and not fixed momentum transfer as required by Waller–Hartree theory. He derived formulae for $(d\sigma/d\omega)_{inel} \propto S(k)$, where the quantity

$(d^2\sigma/d\Omega d\omega)_{inel}$ was integrated over energy transfer for the cases of constant scattering angle and constant momentum transfer.

Recently, measurements of $S(k)$ have been obtained by integrating the energy- and angle-resolved X-ray inelastic scattering spectrum on a variety of liquids after correction for background, detector efficiency, and elastic scattering.[35] The absolute scale was then determined using the Bethe sum rule.[29] The variation of the value of the momentum transfer over the spectra was ignored in the analysis. Comparison of $S(k)$ to theoretical calculations using wavefunctions varying in complexity from the simplest independent atom model (IAM) to the configuration interaction singles and doubles (CISD), shows markedly improved agreement when both electron correlation and chemical binding effects are included. However, discrepancies of order 10% remain. Various sources of these discrepancies were suggested: breakdown of the first Born approximation, neglect of molecular vibrations, and relativistic effects. The studies described in Sec. 3[16] show that corrections to cross-sections calculated by the simplest A^2 theory, i.e. first Born approximation, can be substantial. Refined measurements to determine $S(k)$ in gases are currently in progress at Argonne. These are expected to reveal the relative importance of the corrections which should be applied to calculations using the A^2 approximation.

6. Experimental Methods for X-ray Fluorescence Spectroscopy

Ejection of a deep inner-shell electron by photoabsorption or particle impact produces a vacancy state which rapidly relaxes ($\sim 10^{-15}$ s) by a radiative (X-ray fluorescence) or radiationless (Auger electron emission) process. Here we will focus on X-ray fluorescence spectra (XFS) recorded with wavelength-dispersive spectrometers to resolve electronic states. Characteristic XFS can be excited by particle impact or nonresonant X-ray absorption using laboratory sources and has long been studied.[36] However, we focus here on variations of XFS when tunable synchrotron radiation is used to excite inner-shell electrons to Rydberg or valence-like resonant states or to low-energy continuum states near the ionization threshold. The photoexcitation and decay processes are coupled at inner-shell resonances and thresholds, producing several interesting phenomena in XFS of atomic, molecular, and solid-state samples.[37,38] We begin with a discussion of experimental methods.

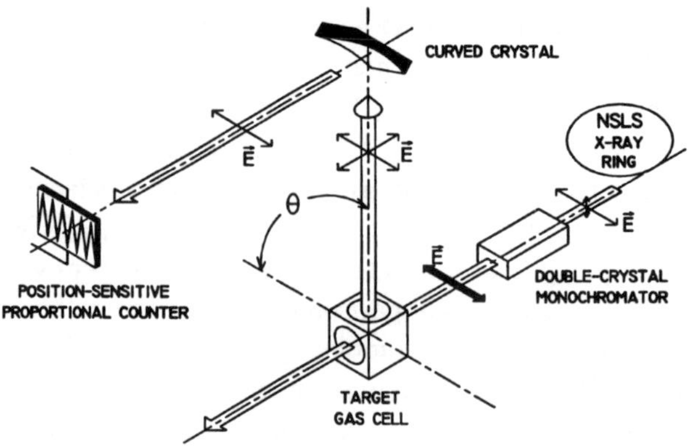

Fig. 14. Diagram of the instrumentation used for the measurement of the angular distribution of polarized X-ray emission from atoms and molecules following excitation by tunable synchrotron radiation.

Deslattes *et al.*[39] performed pioneering experiments on Ar $K\beta$ XFS which demonstrated the potential of tunable synchrotron radiation for studies of inner-shell threshold phenomena. Subsequently, a beamline dedicated to X-ray spectroscopy in the energy range $\approx$ 1–5 keV was developed at Brookhaven's National Synchrotron Light Source.[40] Several XFS experiments have been performed on atoms, molecules, and solids using a curved-crystal spectrometer with position-sensitive detector.[41] A diagram of the instrumentation used for gas-phase studies is shown in Fig. 14. The beamline[40] uses a bending magnet source of synchrotron radiation, followed by a collimating mirror, double-crystal monochromator, and focusing mirror. Using Si(111) crystals in the double-crystal monochromator[42] typically provides a flux $\approx 10^{11}$ s^{-1} at 3.2 keV (the Ar K-edge) at a bandwidth of $\approx$ 0.8 eV.

The focused X-ray beam passes through a target gas cell, and fluorescent X-rays emitted normal to the incident X-ray beam are dispersed by a curved crystal[43] and detected by a position-sensitive proportional counter (PSPC).[44] Thin windows of low-Z material, such as 13 μm polypropylene, are used to contain the sample gas while providing high transmission of X-rays into and out of the gas cell. Ionization chambers or thin-foil photocurrent detectors are used to measure the intensities of the incident and transmitted X-ray beam, providing absorption spectra and normalization for variations in beam intensity. The incident X-rays pass through a path length of ≈ 10 mm of sample gas, and the pressure (≈ 100 torr) is adjusted to absorb 30–50% of the incident X-rays. This produces an extended source of fluorescent X-rays accepted by the curved crystal. By placing the target cell inside the Rowland circle defined by the crystal and detector and using a position-sensitive detector, a range of X-ray energies is recorded simultaneously.[41] Typically, the acceptance window of the PSPC is wide enough to simultaneously record the entire XFS of interest for a given experiment, so there is no need to normalize the recorded spectrum for variations in the intensity of the incident X-ray beam. Parallel recording of the XFS is crucial, since the total counting rates are typically 10–100 s^{-1}. Once the spectrometer angle is set to record a given XFS, the sample gas is replaced with N_2 to measure the dispersion and sensitivity across the PSPC window by elastic X-ray scattering. The target cell is then refilled with sample gas, and variations in the XFS are studied as the incident X-ray energy is scanned through resonances and threshold.

The small collection solid angle of the spectrometer is compensated by use of large sample gas densities to absorb a large fraction of the incident X-rays. Such high sample pressures are incompatible with electron or ion spectroscopy but are suitable for X-ray spectroscopy. Since the energies of the X-ray fluorescence lines are shifted down from the energy of the inner-shell absorption edge, absorption of the fluorescent X-rays by the sample gas is due to shallower subshells (at energies far above their edges) and is much smaller (≈ 10%) than for the incident X-rays. Although the results are not discussed here, measurements of resonant elastic X-ray scattering[45] are sensitive to absorption in the gas cell, so the path length of scattered X-rays through the sample gas is minimized, and the pressure dependence of the scattered intensities is measured. A related issue is the possibility of quenching the vacancy states due to collisions with other sample gas molecules. However, the mean time between collisions for a sample gas at 760 torr and 15°C is $\approx 10^{-10}$ s,[46] much longer than the $\approx 10^{-15}$ s of an X-ray fluorescence process, so collisional effects are negligible.

Synchrotron radiation from a bending magnet source is highly linearly polarized when viewed along the stored electron orbital plane with a narrow vertical acceptance.[47] For the beamline in Ref. 40, 95% linear polarization is calculated for typical use of ±80 μrad vertical acceptance. The degree of linear polarization is enhanced by the double-crystal monochromator when the Bragg angles are near 45°. The linear polarization selectivity of the curved-crystal spectrometer can also be quite high.[41] In particular, several polarization-dependent studies have been made on Cl-containing molecules of Cl K–V X-ray emission, in which a valence electron fills a vacancy in the Cl K-shell.[48–52] For incident X-ray energies near the Cl K-edge ($\approx$ 2820–2880 eV) and for the Cl K–V X-ray fluorescence ($\approx$ 2800–2820 eV), the Bragg angles for Si(111) crystals are $\approx 43°–45°$, giving essentially complete polarization selectivity. Polarization-selective measurements have also been made on S-containing molecules[53,54] and on atomic Ar.[55,56] The curved-crystal spectrometer in Fig. 14 can be rotated about the emitted X-ray direction to measure the fluorescence polarization,[48] or it can be rotated about the incident X-ray beam direction to measure the angular distribution of polarized fluorescence.[51] Such measurements give information on the spatial alignment of resonantly-excited molecules and on the symmetries of the electronic states (Sec. 8).

We have recently performed X-ray spectroscopy experiments using $\approx$ 12–14 keV X-rays on a high-brightness undulator beamline at Argonne's APS. The beamline uses a double-crystal monochromator[57] with a liquid nitrogen cooled, thin Si(111) crystal[58] which maintains resolution and brightness despite the high power density of the undulator beam. The beamline has been used for a variety of experiments over the energy range $\approx$ 3–28 keV, and provides a flux $\approx 10^{13}$ s^{-1} in a bandwidth $\Delta E/E \approx 1.8 \times 10^{-4}$. X-ray fluorescence experiments were performed on the K-shells of Kr and Br_2 using the spectrometer system shown in Fig. 15. Beamline

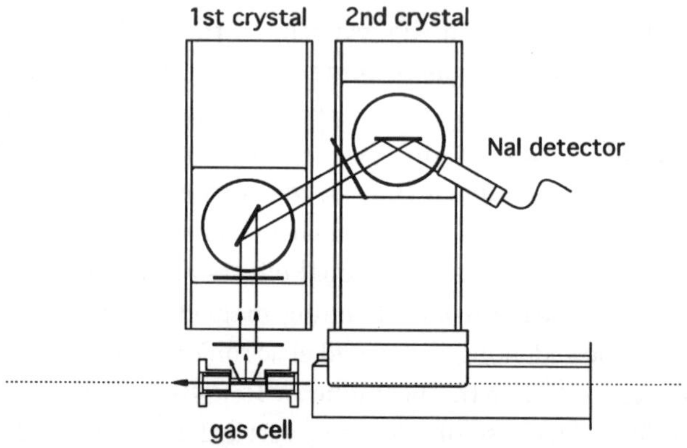

Fig. 15. Diagram of a double-flat-crystal X-ray spectrometer used for high-resolution X-ray spectroscopy of atoms and molecules.

X-rays pass through the target gas cell with ion chambers front and back to measure the intensities of the incident and transmitted beam. An incident X-ray beam passing through a gas target typically produces a line-like source of fluorescent X-rays, and the spectrometer was optimized for this geometry. A 5–10 mm line of X-ray fluorescence produced in the sample cell was collimated with apertures, dispersed by two flat Si(111) crystals oriented in high-resolution dispersive geometry, and detected by a NaI scintillator. The crystals are positioned on precision translational and rotary stages under computer control to allow more complex experimental geometries in the future, including curved-crystal scans along the Rowland circle. In the present geometry, X-ray spectra were recorded by step-scanning the first crystal with the second crystal fixed. This varies the Bragg angle defined by the two-crystal system at one-half the angular speed as the included angle. A double-flat-crystal system was chosen for initial experiments to allow the resolution to approach the theoretical limit for Si(111) ($\Delta E/E = 1.41 \times 10^{-4}$).[59] However, the resolution was limited by the ≈ 4 mm width of the incident X-ray beam, which was used without focusing in early experiments. The initial results obtained on resonance and threshold effects at Kr and Br_2 K-shells complement studies at lower X-ray energies made with the instrumentation in Fig. 14. However, higher resolution is needed to resolve valence molecular orbitals in K–V XFS such as for Br_2 at 13.4 keV. Future versions of the X-ray spectrometer could use a curved-crystal and position-sensitive detector to increase efficiency, but with a high-resolution reflection such as Si(333). Beamline focusing mirrors will also be used to produce an intense, narrow source of fluorescent X-rays in the target gas cell.

7. X-ray Resonant Raman Scattering from Atoms

When an atomic inner-shell vacancy is produced by particle impact or photoionization far above threshold, the resulting XFS has characteristic energies, intensities, and line shapes. A two-step model, in which the vacancy creation process and radiative decay process are treated independently, is adequate to describe the XFS. However, when vacancies are produced by photoabsorption at energies near resonances or the ionization threshold, the XFS is modified and is sensitive to the precise energy and bandwidth of the incident X-ray beam. In that case, the vacancy excitation and decay steps are coupled and described as a one-step, resonant inelastic scattering process.[60] Numerous studies have been reported on this phenomenon, termed X-ray resonant Raman scattering (XRRS), and it continues to be an active field of research.[37,38] Key early experiments include those of Sparks,[61] Eisenberger *et al.*,[62] and Briand *et al.*[63] The variations in X-ray spectral energies, intensities, and line shapes observed in those experiments are well explained by theoretical treatments[62,64,65] based on the resonant term of the Kramers–Heisenberg formulation of photon-scattering processes.[3,45,60] There is much current interest in using XRRS to study electronic structure of solids[66] and the dynamics of molecules excited at inner-shell resonances.[67] Here we simply discuss XRRS from atoms, but the ideas are relevant to understanding experiments on complex targets.

Deslattes *et al.*[39] and Cowan[45] reported variations in Ar X-ray spectra excited below the K-ionization threshold, at the $1s \to 4p$ resonance, and just above threshold which demonstrate XRRS features including linear dispersion, asymmetric line shapes, and line width narrowing. Those observations are consistent with the theoretical study of Tulkki[68] showing the evolution of XRRS into $K\alpha$ and $K\beta$ XFS as the incident X-ray energy is scanned through Ar K edge resonances and threshold. To further characterize atomic XRRS, MacDonald *et al.*[69] measured Xe $L\alpha_{1,2}(L_3M_{4,5})$ and $L\beta_{2,15}(L_3N_{4,5})$ X-ray spectra as the incident X-ray energy was scanned across the L_3 edge. Figure 16 shows a Xe $L\alpha_{1,2}$ spectrum, in which $3d_{5/2}$

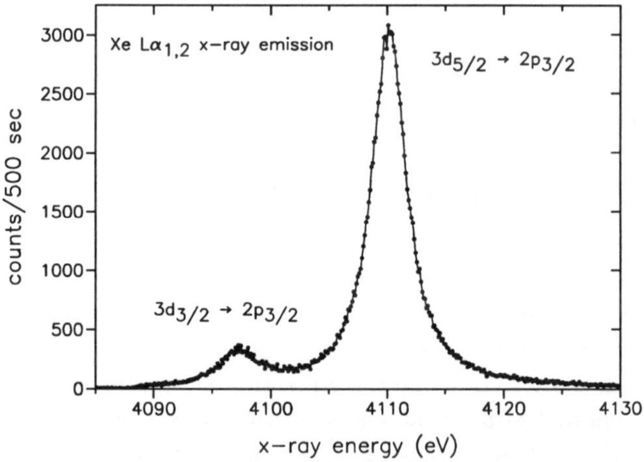

Fig. 16. Xe $L\alpha_{1,2}$ X-ray emission spectrum recorded $\approx$ 27 eV above the L_3 ionization threshold to show the appearance of characteristic fluorescence.

and $3d_{3/2}$ electrons fill the $2p_{3/2}$ vacancy. This spectrum was recorded ≈ 27 eV above the L_3 ionization energy $I(2p_{3/2}) = 4786.3 \pm 0.6$ eV,[70] and the energies, intensities, and line shapes are representative of characteristic fluorescence.[71] The excitation energy is sufficiently above threshold to use the two-step model. First, the photoelectron carries off the excess energy $h\nu - I(2p_{3/2})$, leaving a $2p_{3/2}$ hole state. Second, the radiative transition proceeds with X-ray fluorescence lines appearing at

$$I(2p_{3/2}) - I(3d_{5/2}) = 4110.09 \pm 0.02 \text{ eV},$$

and

$$I(2p_{3/2}) - I(3d_{3/2}) = 4097.38 \pm 0.03 \text{ eV},$$

where $I(nl_j)$ is the subshell ionization energy.[71] Symmetric Voigt line shapes are observed, with Lorentzian widths equal to the sum of the lifetime widths[72] of a $2p_{3/2}$ hole (2.9 eV) and a $3d$ hole (0.5 eV) convoluted with a Gaussian instrumental width (≈ 1.6 eV). The relative intensities of the two lines are determined by the $\langle 3d_j|r|2p_{3/2}\rangle$, $j = 3/2$, $5/2$ dipole matrix elements and multiplicity factors. The XFS in Fig. 16 does not depend on the precise energy or bandwidth of the incident X-rays, as long as the energy is far above threshold.[73]

Figure 17 shows the Xe L_3 absorption edge measured and fit by Breinig *et al.*[70] to components consisting of $2p \to ns$, $n = 6, 7, 8, \ldots$ and $2p \to nd$, $n = 5, 6, 7, \ldots$ Rydberg series leading to the $2p \to \varepsilon s, \varepsilon d$ continuum threshold. Spin-orbit quantum numbers are neglected here for simplicity. Also, as indicated in Fig. 17, the ns Rydberg series is relatively weak and will be neglected. The Rydberg and continuum components have been broadened to reflect the 2.9 eV lifetime width of a $2p$ hole and strongly overlap. In XRRS theory,[60] the Rydberg and continuum excitations indicated in Fig. 17 participate as intermediate states in an inelastic-scattering process:

$$\text{Xe} + \gamma(\text{in}) \to \text{Xe}[2p]5d, 6d, 7d, \ldots, \varepsilon d\} \to \text{Xe}[3d]5d, 6d, 7d, \ldots, \varepsilon d + \gamma(\text{out}),$$

where $[nl]n'l'$ represents a state with a vacancy in the nl and an electron in the $n'l'$ subshell and { } indicates intermediate states. A Rydberg series of bound final states is produced: $[3d]nd$, $n = 5, 6, 7, \ldots$, with a $[3d]\varepsilon d$ continuum state at the series limit, i.e. the same final state as in the X-ray fluorescence process. For the bound final states, the scattered X-ray

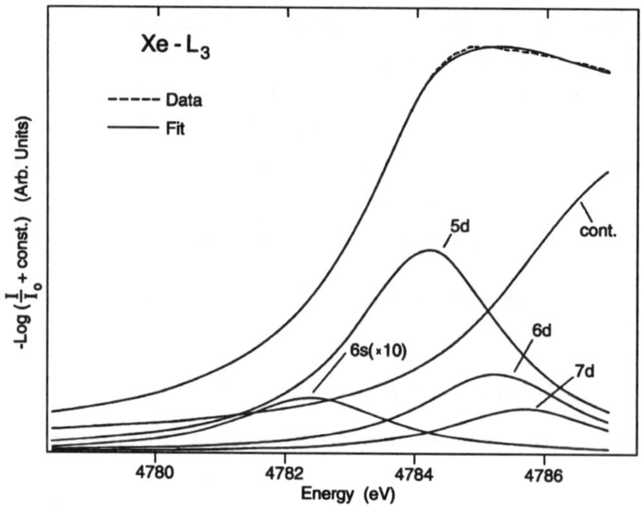

Fig. 17. Xe L_3 absorption edge and fit to Rydberg and continuum components, from Breinig *et al.*[70]

spectrum displays peaks at fixed energy losses:

$$E(\text{out}) = E(\text{in}) - I([3d]nd),\ n = 5, 6, 7, \ldots$$

The bound-state peaks therefore display Raman-like linear dispersion as the incident X-ray energy is varied, and their lifetime widths are that of a $3d$ hole (0.5 eV) rather than a $2p$ hole (2.9 eV). This is the origin of the use of XRRS spectra to study inner-shell electronic structure at higher resolution than the inner-shell lifetime.[66] Of course, the bandwidths of the incident X-ray beam and of the X-ray spectrometer must be narrow to take advantage of this effect. The distribution over intermediate states depends on the precise energy of the incident X-rays and will be modified by convolution over the incident bandwidth.[68] Similarly, structure present in the scattered X-ray spectrum will be broadened by the spectrometer bandwidth.

The distinguishing feature in XRRS spectra associated with the $[3d]\varepsilon d$ continuum final state is that at incident X-ray energies in the range from below the $2p$ ionization threshold to several lifetime widths above threshold, there is insufficient energy to fully develop the X-ray fluorescence peak. This is a simple consequence of energy conservation.[68] How can the continuum final state be observed 'below threshold'? Richtmyer *et al.*[74] showed that the inner-shell lifetime width broadens the continuum edge jump into an arctangent shape extending below the nominal ionization energy, as shown in Fig. 17. LeBrun[75] showed how energy conservation arguments can explain the main features observed in XRRS spectra. In the inelastic-scattering picture, the threshold for producing a $3d_{5/2}$ hole is the ionization energy, 676 eV, which is far below the $2p$ threshold. However, the cross-sections for photon-scattering processes are relatively small in the absence of resonant intermediate states. The cross-sections increase many orders of magnitude when the incident photon energy approaches the $2p$ resonance and threshold region.[3,60]

The Xe $L\alpha_{1,2}$ X-ray spectra recorded as the incident X-ray energy was scanned across the L_3 resonances and threshold are shown in Fig. 18 in a perspective plot. The first eight spectra, recorded at incident energies from 12.8 eV to 2.5 eV below threshold, are plotted overlapping each other in Fig. 19 to show how the lineshape develops with increasing energy. As energy is increased above threshold, the peak positions remain fixed at the characteristic energies and the Lorentzian tails on the high-energy sides increasingly develop. The bandwidths of the incident X-ray beam (≈ 2 eV) and of the X-ray spectrometer (≈ 1.6 eV) blended the contributions of bound and continuum states and limited the observation of detailed

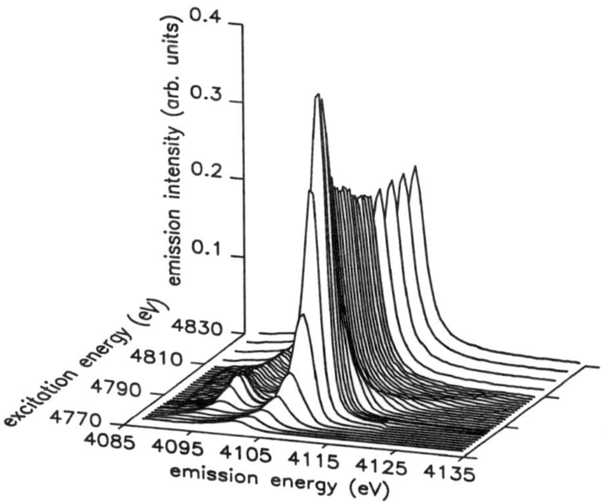

Fig. 18. Variation of Xe $L\alpha_{1,2}$ X-ray emission spectrum across the L_3 edge showing resonant Raman scattering and the onset of characteristic fluorescence.

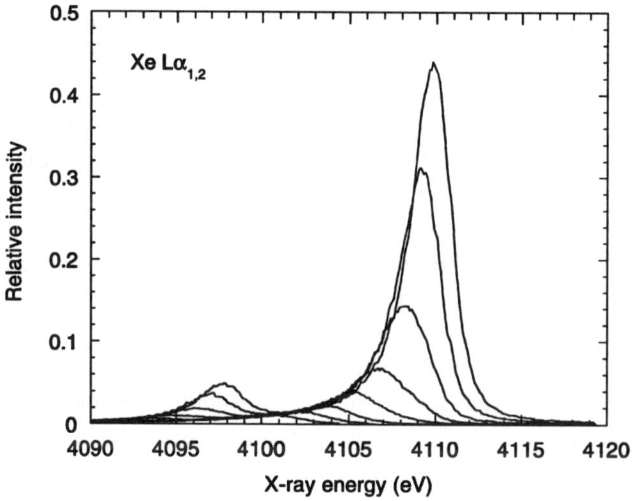

Fig. 19. Xe $L\alpha_{1,2}$ X-ray emission spectra recorded at incident X-ray energies between 12.8 eV and 2.5 eV below the L_3 ionization threshold and plotted overlapping each other to show the evolution of lineshape with increasing energy.

structure which is expected to be present.[68] However, the weak remnant of the bound states can be seen in Fig. 18 dispersing away from the fluorescence peak above threshold. The variations with incident X-ray energy of the Xe $L\alpha_{1,2}$ X-ray spectra in Figs. 18 and 19 and similar variations for Xe $L\beta_{2,15}$ X-ray spectra[69] are consistent within experimental resolution with XRRS features calculated by Tulkki.[68]

In XRRS theory,[60] the cross-sections for various final states are, in principle, sensitive to interference among the intermediate-state excitation amplitudes. In practice, atomic XRRS spectra have been modeled using the independent electron approximation in which interference effects are neglected.[45,68,75] In this approximation, there is a one-to-one correspondence between the final state cross-sections and the lifetime-broadened resonant shapes of the intermediate states indicated in Fig. 17. The implication is that XRRS can be used essentially to measure a high-resolution absorption spectrum.[66] However, in the context of solid samples, Carra *et al.*[76] emphasized that XRRS spectra are sensitive to many-electron effects and must be interpreted with caution. In the case of radiationless decay in atoms, which involves two continuum electrons (photoelectron and Auger electron), post-collision interaction and relaxation effects are relatively strong.[60] Armen *et al.*[77] recently reported an experimental and theoretical study of Auger resonant Raman scattering across the Xe L_3 edge, the radiationless analog to the XRRS study of MacDonald *et al.*[69] Inclusion of relaxation and interference among intermediate-state amplitudes is required to explain the variations of Auger final state intensities with excitation energy.[77] Similar effects have been calculated and measured in Auger resonant Raman scattering at the Ar K-edge.[78,79]

Breinig *et al.*[70] also measured the Kr K absorption edge and fit it to the form of a $1s \rightarrow np$, $n = 5, 6, 7, \ldots$ Rydberg series and ionization threshold. Schaphorst *et al.*[80] remeasured the Kr K edge and found that calculated absorption cross-sections for the Rydberg states and continuum edge underestimate the measured total cross-section. To investigate XRRS at the Kr K edge, we used the double-crystal X-ray spectrometer shown in Fig. 15 to measure $K\alpha_{1,2}$ X-ray emission spectra. A selection of the measured spectra is shown in Fig. 20. Asymmetric line shapes are observed at incident energies below threshold due to truncation of the fluorescent peaks, just as observed in Fig. 19 for Xe $L\alpha_{1,2}$ XRRS spectra. Using Si(111) crystals, the bandwidths of the incident X-ray beam (≈ 4.7 eV) and of the spectrometer

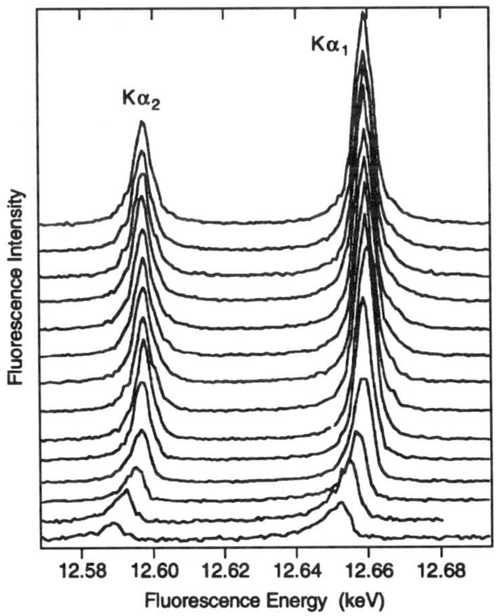

Fig. 20. Variation of Kr $K\alpha_{1,2}$ X-ray emission spectrum at excitation energies across the K edge.

($\approx$ 2.4 eV) were too large to resolve bound final states in the $K\alpha_{1,2}$ X-ray spectra in Fig. 20. However, the method of Hämäläinen *et al.*[66] was used to search for resonance and threshold effects by comparing the high-resolution X-ray fluorescence yield (HRXY) with the total absorption cross-section.

The HRXY is defined here as the relative intensity of emitted X-rays in the spectrometer bandwidth centered on the $K\alpha_1$ characteristic peak energy ($\approx$ 12 649 eV). The structure observed in a HRXY spectrum[66,69] can be understood heuristically by analogy to the Xe $L\alpha_{1,2}$ spectra in Fig. 18. The integral over emission energy of the 2D spectrum in Fig. 18 would give a 1D spectrum analogous to the absorption spectrum in Fig. 17. Note that the large resonant peak in Fig. 18, which corresponds to excitation of the Rydberg series, includes a clearly visible tail, the central energy of which disperses one-to-one with incident X-ray energy. As a function of incident energy, the intensity integrated over emission energy represents a Lorentzian with a natural width corresponding to the core hole lifetime. With better resolution, we expect the integrated peak intensity to mirror the partial contributions indicated by the fit in Fig. 17. This raises two important points. First, the 2D spectra provide an experimental approach to disentangling contributions from different final states in the absorption spectrum and to verifying the assumptions made in model fits to absorption spectra such as in Fig. 17. Second, confining the fluorescence excitation yield to the characteristic fluorescence energy means that we do not observe the tails which display the large Lorentzian lifetime broadening of the core hole. As a result, the contribution of the tails does not appear in the HRXY, and the resonance and threshold structure is manifested at an apparently higher resolution. This is in contrast with the total absorption spectrum, which sums over all final states. Indeed, as shown in Fig. 21, the HRXY has a sharper onset and intensity spike in the resonance and threshold region than does the absorption cross-section. In addition, at $\approx$ 30 eV above threshold, the HRXY appears to be insensitive to multielectron excitation features in the absorption spectrum.[80] Multielectron excitations give rise to satellite peaks in X-ray emission,[39] which presumably are excluded from the fluorescence yield measured here at the $K\alpha_1$ peak energy. We conclude that by selecting a narrow band of final states, the HRXY gives information on resonance and threshold structure which complements total absorption cross-section studies. More detailed information on atomic XRRS will result from future studies using higher resolution of both the incident and emitted X-rays.

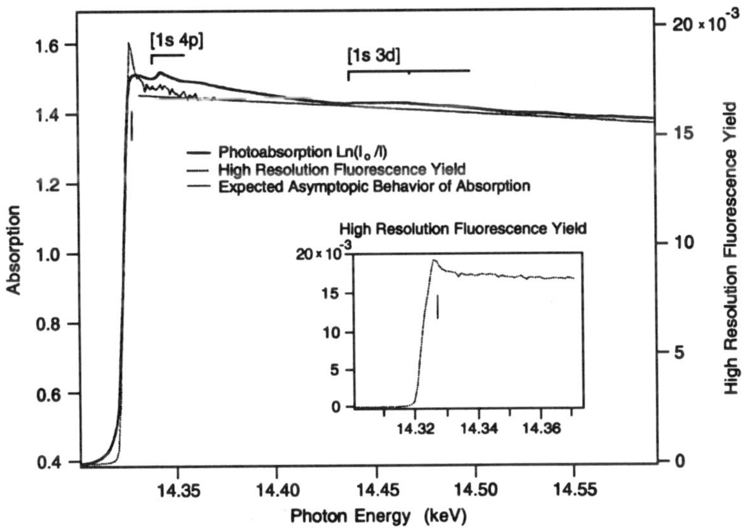

Fig. 21. Comparison of Kr $K\alpha_1$ high-resolution X-ray fluorescence yield (dotted line) with the total photoabsorption cross-section (thick solid line). The thin solid line shows the expected $(h\nu)^{-7/2}$ asymptotic dependence on X-ray energy of the one electron photoionization cross-section,[81] normalized to the high-resolution fluorescence yield. Structures in the absorption cross-section due to [1*s*4*p*] and [1*s*3*d*] multielectron excitations are indicated.[80]

8. X-ray Polarization Anisotropy of Aligned Molecules

The atoms or molecules present in the target gas cell of the X-ray spectrometer in Fig. 14 are randomly oriented in space. However, X-ray absorption will generally produce an anisotropic state due to the transfer of angular momentum from the absorbed photon and due to its directionality and polarization. The anisotropy of the excited state will be reflected in the polarization and angular distribution of emitted X-rays.[82,83] Gel'mukhanov and Mazalov[84] predicted that excitation of molecules near threshold to states of specific symmetry will result in polarized X-ray fluorescence. This effect was observed by Lindle *et al.*[48] in Cl K–V resonant X-ray fluorescence from CH_3Cl and was followed by several experimental studies in which the X-ray spectrometer of Fig. 14 was rotated to measure either the polarization[48–50,52–54] or angular distribution[51,55,56] of emitted X-rays.

As discussed in Sec. 6, the linear polarization of the incident X-ray beam and the polarization selectivity of the curved-crystal spectrometer in Fig. 14 both approach 100% in many cases. Partial polarization of the incident beam and partial selectivity of the curved-crystal can be treated, but for simplicity here we assume 100% for both. We also assume that X-ray absorption and emission are well described as electric–dipole transitions, an approximation that was recently shown to break down in the case of resonant X-ray emission from Cl_2.[52] The Cl_2 case will be discussed in the next section. Here we discuss experiments for which the electric–dipole approximation is expected to be valid.

X-ray absorption produces an anisotropic ensemble of core excited atoms or molecules which has cylindrical symmetry around the polarization vector of the incident X-ray beam in Fig. 14. The photoexcited system is also symmetric with respect to reflection in the plane normal to the incident polarization vector. An excited system with these symmetry properties is said to be 'aligned', and the symmetry properties are retained in the emitted X-ray patterns.[82,83] The angular distributions and polarizations of emitted X-rays can be characterized by measurements of X-rays emitted in the plane normal to the incident X-ray beam, with their polarization vectors oriented either parallel or perpendicular to that plane. Defining θ to be the angle of X-ray emission with respect to the incident polarization vector, the angular distributions for parallel and perpendicular emission polarization are

$$I_{\parallel}(\theta) = I_0\{1 + R[3\sin^2\theta - 1]\}\,,$$

and

$$I_{\perp}(\theta) = I_0\{1 - R\}\,,$$

where I_0 is proportional to the total intensity emitted in all directions and summed over polarizations, and R is the polarization anisotropy. The parameter R characterizes the radiation pattern and relates measurements to theory.[55,85] The symmetry properties of the emitted X-ray pattern are contained in the above expressions. For example, $I_{\perp}$ is isotropic and $I_{\parallel}(0°) = I_{\perp}$, so the emitted X-rays are unpolarized at 0° (along the incident polarization vector). The measurement goal is to determine the parameter R for a given X-ray absorption and emission process. R can be determined by measuring the angular distribution of the parallel component, $I_{\parallel}(\theta)$, or by measuring $I_{\parallel}$ and $I_{\perp}$ at fixed $\theta = 90°$ where

$$R = 2P/(3 - P)\,,$$

and

$$P = (I_{\parallel} - I_{\perp})/(I_{\parallel} + I_{\perp})\,,$$

is the degree of linear polarization. The expression for $I_{\|}(\theta)$ shows that an intensity proportional to the total intensity I_0 is measured by $I_{\|}$ at the 'magic angle'

$$\theta_m = \sin^{-1}(3^{-1/2}) \approx 35.3^\circ .$$

With the curved-crystal spectrometer positioned in parallel orientation at the magic angle, the relative intensities of the peaks in the XFS are proportional to their total emission cross-sections. Measurements of R and I_0 for X-ray emission are analogous to measurements of photoelectron anisotropy parameters β and angle-integrated cross-sections σ for photoionization processes.[86]

A rigorous theory of polarization anisotropies of photon emission from aligned atoms and molecules has been developed by Fano and Macek[82] and by Greene and Zare.[83] A two-step model is used, but angular momentum quantum numbers are treated explicity. The formalism is typically applied to cases involving small numbers of well-defined quantum states, i.e. atomic transitions and rotationally-resolved molecular transitions. This approach was used to understand the small polarization anisotropy $R = 0.07 \pm 0.03$ that was measured for Ar $K\alpha$ X-rays following $1s \to 4p$ resonant excitation.[55,56] Large and distinctly different R values are expected for different final state multiplet components, but the multiplet splittings are small and unresolved in the X-ray spectra, which washes out the anisotropy. This situation appears to generally apply to resonant X-ray spectra of atoms, that is, large R values are washed out by averaging over unresolved final states.

Experimental studies of X-ray polarization anisotropies from small molecules are more feasible than for atoms, because the molecular field splits the valence electronic states by several eV, so they are often resolvable in XFS. Electronic states and transition moments must reflect the symmetry properties of molecular geometries, which provides the basis for producing aligned molecules by resonant X-ray absorption and symmetry-dependent polarization anisotropies in X-ray emission. A simple, classical two-step model[87,88] has been used to qualitatively understand molecular X-ray polarization anisotropies[51] and will be discussed here. However, resonant X-ray emission should be described as a one-step, resonant inelastic scattering process, i.e. as XRRS but with uniquely molecular features. Recently, rigorous theoretical treatments have been developed for polarization anisotropies in molecular XRRS based on the Kramers–Heisenberg formula.[89,90]

In addition to Rydberg-like excitations leading to the ionization threshold as in rare-gas atoms,[70] molecular X-ray absorption spectra display strong subthreshold resonances due to excitation of valence-like states. Perera *et al.*[91] have studied the Cl K absorption spectrum and Cl K–V XFS of CF_3Cl. The electronic states are labeled by the C_{3v} molecular point group symmetries. For example, the Cl K-shell electrons are in the $1a_1$ molecular orbital (MO), and two of the valence MOs observed in the K–V XFS (see Fig. 22) are labeled $7e$ and $10a_1$. A strong, well-resolved subthreshold absorption resonance[91] is assigned as the $1a_1 \rightarrow 11a_1$ transition, where the $11a_1$ MO is the lowest unoccupied MO of the molecular ground state. Transition dipole moments between states of particular symmetries are directed along specific directions with respect to the molecular framework. The $1a_1 \rightarrow 11a_1$ absorption dipole moment points parallel to the molecular figure axis (the C–Cl bond axis). The direction of the emission dipole moment depends on the symmetry of the valence MO which drops into the $1a_1$ vacancy. Valence MOs of a_1 symmetry have emission

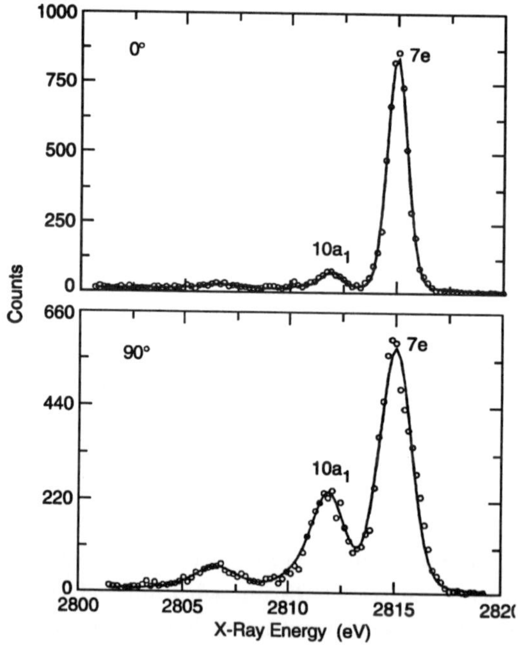

Fig. 22. Parallel-polarized Cl K–V X-ray emission from CF_3Cl excited on resonance at emission angles $\theta = 0°$ (top) and $\theta = 90°$ (bottom).

dipole moments directed parallel to the figure axis, while those for MOs of e symmetry are perpendicular to the figure axis.

Parallel-polarized XFS excited at the $11a_1$ resonance and recorded at 0° and 90° emission angles[51] are plotted in Fig. 22. The relative intensities of the $7e$ and $10a_1$ peaks vary strongly between the two emission angles, so they have very different R values. The variations can be understood using the classical two-step model in which the absorption and emission dipole moments are replaced by classical dipole oscillators treated as rigidly attached to the molecular framework. The expected R values are determined by the relative directions of the absorption and emission oscillators averaged over molecular orientations.[88] Defining γ to be the ensemble average of the angle between the absorption and emission oscillators, the polarization anisotropy is given by

$$R = (3\cos^2\gamma - 1)/5\,.$$

The $1a_1 \to 11a_1$ absorption oscillator is parallel to the molecular figure axis, so molecules in the target gas cell with their figure axes more closely aligned with the polarization vector of the incident X-rays are preferentially excited. A $\cos^2\theta$ distribution of figure axes with respect to the incident polarization vector describes the alignment of the excited system. The $10a_1 \to 1a_1$ emission oscillator is also parallel to the figure axis, so $\gamma = 0°$ and $R = 2/5$. The $7e \to 1a_1$ emission oscillator is perpendicular to the figure axis, so $\gamma = 90°$ and $R = -1/5$. The R values predict that the $10a_1$ MO has its maximum intensity at 90°, while the $7e$ MO has its maximum at 0°. The $10a_1 : 7e$ intensity ratios measured at $\theta = 0°$, 30°, 60°, and 90° for both parallel- and perpendicular-polarization are plotted in Fig. 23 and compared with predictions of the classical two-step model. While the perpendicular-polarized intensity ratio is isotropic, the parallel-polarized ratio varies by a factor of 6. The parallel- and perpendicular-polarized XFS recorded at $\theta = 90°$ in a separate experiment[50] are also consistent with the predicted polarizations

$$P = 3R/(2 + R)\,,$$

which gives 1/2 and −1/3 for the $10a_1$ and $7e$ transitions, respectively.

Large polarization and anisotropy effects are observed in XFS for the large R values which result for resonant absorption to states of well-defined symmetry. The sensitivity of polarized XFS to molecular symmetry is its most useful application, as demonstrated by experimental[53] and

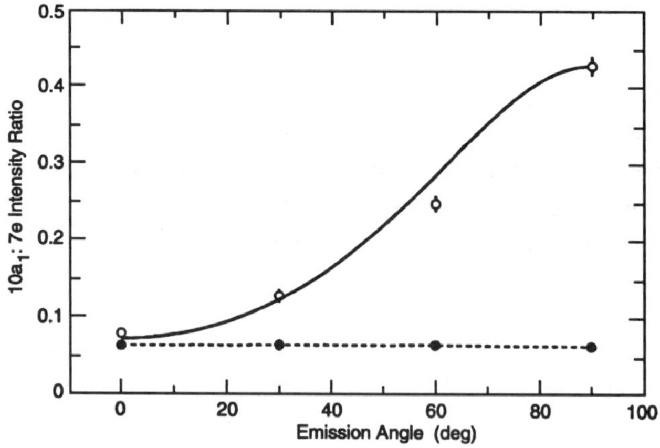

Fig. 23. Angular dependence of the $10a_1$: $7e$ emission intensity ratio measured using parallel polarization (open circles) and perpendicular polarization (closed circles). The solid curve is a theoretical prediction for parallel polarization based on the classical two-step model and has been normalized to the measurement at $\theta = 90°$. The dashed curve simply connects the perpendicular-polarized measurements, for which no variation with angle is expected.

theoretical[89,92] studies of H_2S. Above threshold, degenerate continuum channels of different symmetries usually contribute, so the alignment of the vacancy state and, consequently, the R values and observed polarizations and anisotropies of the XFS, are relatively weak. Nevertheless, in the case of valence photoionization of molecules, the small polarizations produced in VUV fluorescence have been measured and used to study photoionization dynamics.[85,93] Precise measurements of the polarizations and anisotropies of XFS excited above threshold could similarly provide information on photoionization dynamics of inner-shells.

9. Nondipole X-ray Resonant Raman Scattering from Molecules

The electric–dipole approximation to photon–electron interactions in atoms and molecules is usually expected to be valid when the photon wavelength is large compared with the radius of the electron's orbit. This criterion is met for the case considered in the previous section of XRRS near the Cl K-edge by CF_3Cl which contains a single Cl atom. The Cl K ionization threshold ≈ 2833 eV corresponds to an X-ray wavelength ≈ 4.4 Å, which is large compared with the radius $r_{1s} \approx 0.032$ Å of a Cl $1s$ atomic orbital.[52]

Molecular Cl_2, however, contains two equivalent Cl atoms separated by an internuclear distance ≈ 2.0 Å, suggesting that the XRRS process will be sensitive to variations of the phase of the incident and emitted X-rays over the interatomic distance.[52,94] This idea was confirmed experimentally and explained theoretically in an XRRS study of Cl_2 that was recently reported by Mills *et al.*[52] It was necessary to treat the full exponential factor of the photon–electron interaction in the Kramers–Heisenberg formula, which resulted in showing how the XRRS spectrum is sensitive to diffraction-like structure factors.[52] The discussion here will be limited to the experimental observation of nondipole final states in the polarized XRRS spectra of Cl_2. We also discuss XRRS spectra of Br_2, for which the K ionization energy ≈ 13474 eV corresponds to an X-ray wavelength ≈ 0.92 Å, compared with the internuclear distance ≈ 2.3 Å. Predictions for XRRS spectra of Br_2 were also calculated by Mills *et al.*[52]

The previous section discussed how the symmetries of molecular electronic states and transition dipole moments are related to polarization anisotropies of XFS. Among the symmetry properties which characterize states of homonuclear diatomic molecules is inversion through the center of symmetry. The states are either symmetric and labeled g (for *gerade* or even parity) or antisymmetric and labeled u (for *ungerade* or odd parity). A selection rule for electric–dipole transitions is that the parities of atomic or molecular states must change when a photon is absorbed or emitted. For a two-photon process such as XRRS, the dipole selection rule requires that the final state parity be the same as the initial state parity. This parity selection rule has been confirmed[67,95,96] in the cases of XRRS spectra of N_2 and O_2, for which the X-ray wavelengths are ≈ 30 Å and ≈ 23 Å, respectively, i.e. large compared with internuclear distances. Skytt *et al.*[67,97] and Glans *et al.*[96] discuss the case of CO_2, which also has inversion symmetry. However, the parity selection rule applies to the symmetry of the product wavefunction, electronic times vibrational modes, and CO_2 has both even parity (symmetric stretch) and odd-parity (antisymmetric stretch) vibrational modes. Still within the electric–dipole approximation, parity-forbidden electronic levels can be populated via the antisymmetric stretch and result in interesting vibronic coupling effects in XRRS.[97,98] For homonuclear diatomic molecules, however, only even parity symmetric stretch modes are involved, so vibronic coupling cannot account for observations of parity-forbidden electronic states.

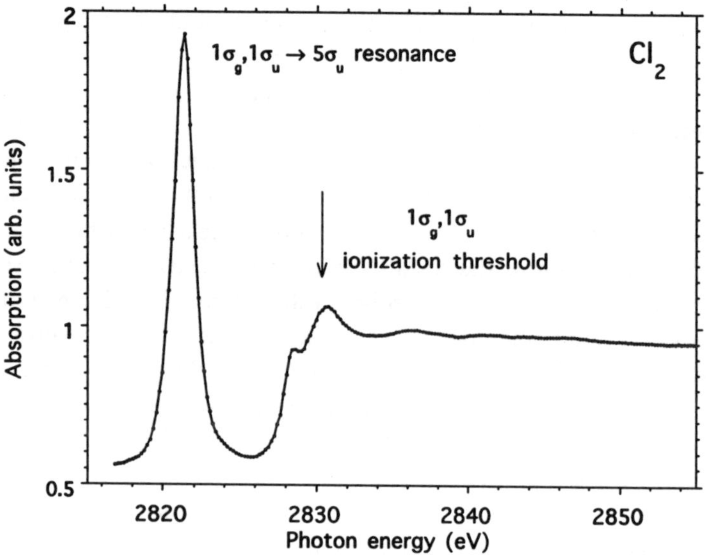

Fig. 24. Photoabsorption spectrum measured at the Cl K edge of Cl_2. The ionization threshold and $5\sigma_u$ subthreshold resonance are indicated.

The Cl K absorption spectrum and polarized K–V X-ray emission spectra were recorded using the instrumentation in Fig. 14. The absorption spectrum is plotted in Fig. 24 and is similar to the spectrum analyzed by Bodeur *et al.*[99] The characteristic XFS measured above the ionization threshold is plotted in the bottom of Fig. 25 and shows partially-resolved peaks due to the $2\pi_g$, $2\pi_u$, and $5\sigma_g$ valence MOs. The above-threshold XFS is not sensitive to the parity selection rule, because the unobserved photoelectron can have the parity required to give the many-electron state allowed total parity in each case. The three occupied valence MOs are built from linear combinations of $3p$ atomic orbitals on the two Cl atoms, and the unfilled $3p$ subshells also results in an unoccupied $5\sigma_u$ MO. A strong resonance occurs in the absorption spectrum (Fig. 24) due to $1\sigma_g \to 5\sigma_u$ and $1\sigma_u \to 5\sigma_u$ transitions, where $1\sigma_g$ and $1\sigma_u$ are the symmetric and antisymmetric linear combinations, respectively, of $1s$ atomic orbitals on the two Cl atoms. The $1\sigma_u \to 5\sigma_u$ transition is dipole forbidden, but becomes significant in the Cl_2 case for which the X-ray wavelength is comparable to internuclear distance.[52] The X-ray emission spectrum recorded on the $5\sigma_u$ resonance with parallel polarization is plotted in the top of Fig. 25.

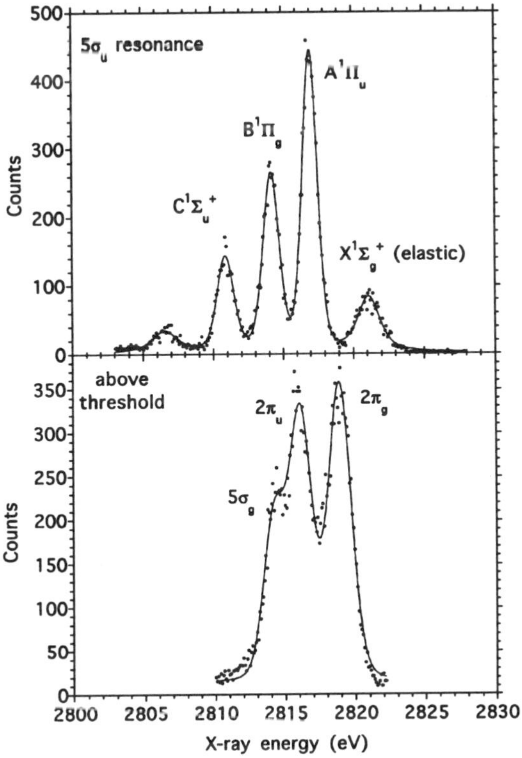

Fig. 25. Cl K–V X-ray emission spectra from Cl_2 recorded with parallel polarization and excited on the $5\sigma_u$ resonance (top) and above the ionization threshold (bottom).

A resonant elastic peak appears in which the molecule has been left in the $X^1\Sigma_g^+$ ground state. The remaining peaks constitute an XRRS spectrum with bound final states described by configurations with holes in the valence MOs and a $5\sigma_u$ spectator electron: $[2\pi_g]5\sigma_u\, A^1\Pi_u$, $[2\pi_u]5\sigma_u\, B^1\Pi_g$, and $[5\sigma_g]5\sigma_u\, C^1\Sigma_u^+$. Comparison with the above-threshold XFS shows that the XRRS peaks are shifted to lower energies, and the $[5\sigma_g]5\sigma_u\, C^1\Sigma_u^+$ configuration splits into three final states.[52] The *gerade* final states, $X^1\Sigma_g^+$ and $B^1\Pi_g$, are dipole allowed, but the parity-forbidden states, $A^1\Pi_u$ and $C^1\Sigma_u^+$, are also strongly populated. The XRRS transitions are also polarized, as shown in Fig. 26. Mills *et al.*[52] calculated a perpendicular-polarized XRRS spectrum for Cl_2 which describes the measured spectrum well. The ideas discussed in the previous section to understand polarized X-ray emission in

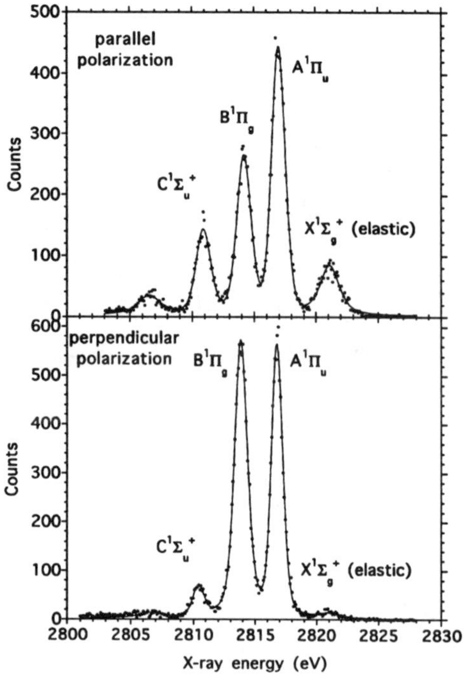

Fig. 26. Cl *K–V* X-ray emission spectra from Cl_2 excited on the $5\sigma_u$ resonance and recorded with parallel polarization (top) and perpendicular polarization (bottom).

the context of the electric–dipole approximation are inadequate to describe polarized XRRS from Cl_2. The full exponential factors in the photon–electron interactions in the Kramers–Heisenberg formula are required and causes the polarization and angular distribution of the XRRS spectrum to depend on diffraction-like structure factors.[52,94]

The halogen molecules F_2, Cl_2, and Br_2 are valence-isoelectronic, i.e. the occupied and lowest-unoccupied MOs are built on $2p, 3p$, and $4p$ atomic orbitals, respectively. Mills *et al.*[52] used this correspondence to show how XRRS spectra for the three molecules vary between the electric–dipole regime (F_2) to the hard X-ray regime (Br_2). Thus, the dipole-forbidden $1\sigma_u \to \sigma_u$ resonant absorption is predicted to be negligible for F_2, significant for Cl_2, and equal to the dipole-allowed $1\sigma_g \to \sigma_u$ resonant absorption for Br_2. However, an important point made by Mills *et al.*[52] is that absorption spectra are not sensitive to nondipole and structure-factor effects while the XRRS spectra are. As a follow-up experiment to the Cl_2 study, *K–V*

X-ray emission from Br_2 was recorded using the double-crystal spectrometer in Fig. 15. The X-ray spectra excited above the K ionization threshold and on the subthreshold resonance[100] are plotted in Fig. 27. At the higher X-ray energies involved, Si(111) crystals were not adequate to resolve the valence final states for detailed analysis as in the Cl_2 case. Two partially resolved peaks were recorded in each spectrum. The larger peaks apparently contain unresolved transitions from valence MOs while the smaller peaks contain inner-valence transitions.[101] The peaks recorded on resonance are narrower and shifted to higher energy, indicating XRRS effects, but the resolution is inadequate to compare with predictions of Mills *et al.*[52] The Br_2 case represents a future direction for development of both experiment and theory. Higher resolution and sensitivity are required to record XRRS spectra with well-resolved transitions from molecules containing higher-Z atoms. Theoretical treatments will need to include spin-orbit interaction as well as nondipole and molecular geometry effects in resonant X-ray scattering.

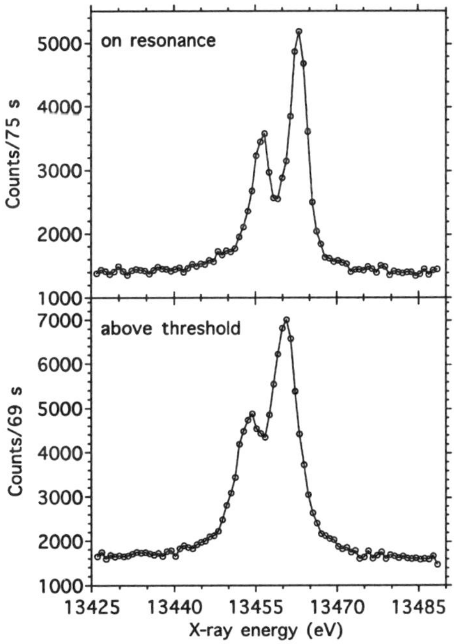

Fig. 27. Br K–V X-ray emission spectra from Br_2 excited on the subthreshold resonance (top) and above the ionization threshold (bottom).

Acknowledgments

We thank our Argonne coworkers R. W. Dunford, D. S. Gemmell, S. Hasegawa, M. Jung, and B. Krässig for their participation in recent experiments conducted at the Advanced Photon Source. We also thank B. Zabransky and the BESSRC staff (M. Beno, C. Kurtz, M. Engbretson, G. Jennings, G. Knapp, P. Montano, and C. Wiley) for support in conducting experiments at the APS. We thank P. M. Bergstrom, Jr., R. Bonham, J. P. J. Carney, M. Inokuti, L. LaJohn, and R. H. Pratt for guidance in understanding X-ray scattering from atoms, and we thank P. W. Langhoff, J. D. Mills, and J. A. Sheehy for explanations of resonant X-ray scattering from molecules. This work was supported by the US. Department of Energy, Office of Basic Energy Sciences under Contract No. W-31-109-ENG-38.

References

1. For a review, see for example B. Crasemann, in *Atomic Molecular and Optical Physics Handbook*, Ed. G. F. Drake, American Institute of Physics, Woodbury, NY, 1996, Chapter 60, p. 701.
2. P. M. Bergstrom, Jr. and R. H. Pratt, *Rad. Phys. Chem.* **50**, 1 (1997).
3. J. J. Sakurai, *Advanced Quantum Mechanics* (Addison-Wesley, New York, 1967), Chapter 2.
4. J. H. Hubbell, W. Veigele, E. A. Briggs, R. T. Brown, D. T. Cromer and R. J. Howerton, *J. Phys. Chem. Ref. Data* **4**, 471 (1975); J. H. Hubbell, H. A. Gimm and I. Overbo, *J. Phys. Chem. Ref. Data* **9**, 1023 (1980).
5. C. T. Chantler, *J. Phys. Chem. Ref. Data* **24**, 71 (1995); J. Wang, R. P. Sagar, H. Schmider and V. H. Smith, *At. Data Nucl. Data Tab.* **53**, 233 (1993); B. L. Henke, E. M. Gullikson and J. C. Davis, *At. Data Nucl. Data Tab.* **54**, 181 (1993).
6. J. Wang, R. O. Esquivel, V. H. Smith, Jr. and C. F. Bunge, *Phys. Rev.* **A51**, 3812 (1995); H. Meyer, T. Müller, A. Schweig, *Chem. Phys.* **191**, 213 (1995) and *Acta Cryst.* **A51**, 171 (1995).
7. P. Eisenberger and P. M. Platzman, *Phys. Rev.* **A2**, 415 (1970).
8. For a recent review, see M. J. Cooper, *Radiat. Phys. Chem.* **50**, 63 (1997).
9. P. Coppens, *X-ray Charge Densities and Chemical Bonding*, International Union of Crystallography, Chester, England, 1997.
10. L. Kissel, R. H. Pratt and S. C. Roy, *Phys. Rev.* **A22**, 1970 (1980).
11. L. Kissel, B. Zhou, S. C. Roy, S. K. Sen Gupta and R. H. Pratt, *Acta Cryst.* **A51**, 271 (1995).
12. E. O. Wollan, *Phys. Rev.* **37**, 862 (1931).
13. D. R. Chipman and L. D. Jennings, *Phys. Rev.* **132**, 728 (1963).
14. G. E. Ice, M. H. Chen and B. Crasemann, *Phys. Rev.* **A17**, 650 (1978).
15. F. Smend and H. Czerwinski, *Z. Phys.* **D1**, 139 (1986).

16. M. Jung, R. W. Dunford, D. S. Gemmell, E. P. Kanter, B. Krässig, T. W. LeBrun, S. H. Southworth, L. Young, J. P. J. Carney, L. LaJohn, R. H. Pratt and P. M. Bergstrom, Jr., *Phys. Rev. Lett.* **81**, 1596 (1998).
17. R. Ali, I. Ahmad, R. W. Dunford, D. S. Gemmell, M. Jung, E. P. Kanter, P. H. Mokler, H. G. Berry, A. E. Livingston, S. Cheng and L. J. Curtis, *Phys. Rev.* **A55**, 994 (1997).
18. F. Biggs, L. B. Mendelsohn and J. B. Mann, *At. Data Nucl. Data Tab.* **16**, 201 (1975).
19. P. M. Bergstrom, Jr., L. Kissel, R. H. Pratt and A. Costescu, *Acta Cryst.* **A53**, 7 (1997).
20. A. N. Hopersky, V. A. Yavna and V. A. Popov, *J. Phys.* **B29**, 461 (1996).
21. F. E. Low, *Phys. Rev.* **110**, 974 (1958).
22. M. Gavrila, *Phys. Rev.* **A6**, 1348 (1972); ibid **6**, 1360 (1972).
23. P. M. Bergstrom, Jr., T. Suric, K. Pisk and R. H. Pratt, *Phys. Rev.* **A48**, 1134 (1993).
24. J. P. Briand, A. Simionovici, P. Chevalier and P. Indelicato, *Phys. Rev. Lett.* **62**, 2092 (1989).
25. V. Marchetti and C. Franck, *Phys. Rev.* **A39**, 647 (1989); V. Marchetti and C. Franck, *Phys. Rev. Lett.* **65**, 268 (1990).
26. S. Manninen, K. Hämäläinen and J. Graeffe, *Phys. Rev.* **B41**, 1224 (1990).
27. V. Marchetti and C. Franck, *Rev. Sci. Instrum.* **59**, 407 (1988).
28. J. Laukkanen, K. Hämäläinen, S. Manninen and V. Honkimäki, *Nucl. Instrum. Methods* **A416**, 475 (1998).
29. A comprehensive review is given in M. Inokuti, *Rev. Mod. Phys.* **43**, 297 (1971); M. Inokuti, Y. Itikawa and J. E. Turner, *Rev. Mod. Phys.* **50**, 23 (1978).
30. N. Watanabe, H. Hayashi and Y. Udagawa, *Bull. Chem. Soc. Jpn.* **70**, 719 (1997).
31. Some examples include: E. N. Lassettre and E. R. White, *J. Chem. Phys.* **60**, 2460 (1974); A. Lahman-Bennani, A. Duguest, H. F. Wellenstein and M. Rouault, *J. Chem. Phys.* **72**, 6398 (1980); R. S. Barbieri and R. A. Bonham, *Phys. Rev.* **A44**, 7361 (1991).
32. H. Hayashi, N. Watanabe, Y. Udgawa and C.-C. Kao, *J. Chem. Phys.* **108**, 823 (1998).
33. K. Nishikawa and T. Iijima, *J. Chem. Phys.* **87**, 3752 (1987); T. Mitsuhashi and T. Iijima, *Chem. Phys. Lett.* **109**, 195 (1984).
34. R. A. Bonham, *Phys. Rev.* **A23**, 2950 (1981).
35. N. Watanabe, H. Hayaski, Y. Udagawa, S. Ten-no and S. Iwata, *J. Chem. Phys.* **108**, 4545 (1998).
36. *X-ray Spectra and Chemical Binding*, Eds. A. Meisel, G. Leonhardt and R. Szargan (Springer-Verlag, New York, 1989).
37. *Resonant Anomalous X-ray Scattering: Theory and Applications*, Eds. G. Materlik, C. J. Sparks and K. Fischer (North-Holland, Armsterdam, 1994).

38. *Raman Emission by X-ray Scattering*, Eds D. L. Ederer and J. H. McGuire (World Scientific, River Edge, NJ, 1996).
39. R. D. Deslattes, R. E. LaVilla, P. L. Cowan and A. Henins, *Phys. Rev.* **A27**, 923 (1983).
40. P. L. Cowan, S. Brennan, R. D. Deslattes, A. Henins, T. Jach and E. G. Kessler, *Nucl. Instrum. Methods* **A246**, 154 (1986).
41. S. Brennan, P. L. Cowan, R. D. Deslattes, A. Henins, D. W. Lindle and B. A. Karlin, *Rev. Sci. Instrum.* **60**, 2243 (1989). Note that Fig. 3 of this reference is mislabeled and should read 'ratio of π to σ polarized light'.
42. P. L. Cowan, J. B. Hastings, T. Jach and J. P. Kirkland, *Nucl. Instrum. Methods* **208**, 349 (1983).
43. A. Henins, *Rev. Sci. Instrum.* **58**, 1173 (1987).
44. B. P. Duval, J. Barth, R. D. Deslattes, A. Henins and G. G. Luther, *Nucl. Instrum. Methods* **222**, 274 (1984).
45. P. L. Cowan, in *Resonant Anomalous X-ray Scattering: Theory and Applications* (Ref. 37), p. 449.
46. E. W. McDaniel, *Atomic Collisions: Electron and Photon Projectiles* (Wiley, New York, 1989), p. 46.
47. G. Margaritondo, *Introduction to Synchrotron Radiation* (Oxford University Press, New York, 1988), p. 38.
48. D. W. Lindle, P. L. Cowan, R. E. LaVilla, T. Jach, R. D. Deslattes, B. Karlin, J. A. Sheehy, T. J. Gil and P. W. Langhoff, *Phys. Rev. Lett.* **60**, 1010 (1988).
49. D. W. Lindle, P. L. Cowan, R. E. LaVilla, T. Jach, R. D. Deslattes, R. C. C. Perera and B. Karlin, *J. Phys.* **C9**, 761 (1987).
50. D. W. Lindle, P. L. Cowan, T. Jach, R. E. LaVilla, R. D. Deslattes and R. C. C. Perera, *Phys. Rev.* **A43**, 2353 (1991).
51. S. H. Southworth, D. W. Lindle, R. Mayer and P. L. Cowan, *Phys. Rev. Lett.* **67**, 1098 (1991).
52. J. D. Mills, J. A. Sheehy, T. A. Ferrett, S. H. Southworth, R. Mayer, D. W. Lindle and P. W. Langhoff, *Phys. Rev. Lett.* **79**, 383 (1997).
53. R. Mayer, D. W. Lindle, S. H. Southworth and P. L. Cowan, *Phys. Rev.* **A43**, 235 (1991).
54. K. E. Miyano, U. Arp, S. H. Southworth, T. E. Meehan, T. R. Walsh and F. P. Larkins, *Phys. Rev.* **A57**, 2430 (1998).
55. S. H. Southworth, D. W. Lindle, R. Mayer and P. L. Cowan, *Nucl. Instrum. Methods* **B56/57**, 304 (1991).
56. S. H. Southworth, *Nucl. Instrum. Methods* **B87**, 247 (1994).
57. M. Ramanathan, M. A. Beno, G. S. Knapp, G. Jennings, P. L. Cowan and P. A. Montano, *Rev. Sci. Instrum.* **66**, 2191 (1995).
58. G. S. Knapp, M. A. Beno, C. S. Rogers, C. L. Wiley and P. L. Cowan, *Rev. Sci. Instrum.* **65**, 2792 (1994).
59. T. Matsushita and H. Hashizume, in *Handbook on Synchrotron Radiation*, Vol. 1, Ed. E.-E. Koch (North-Holland, New York, 1983), p. 261.

60. T. Åberg and B. Crasemann, in *Resonant Anomalous X-ray Scattering: Theory and Applications* (Ref. 37), p. 431.
61. C. J. Sparks, Jr., *Phys. Rev. Lett.* **33**, 262 (1974).
62. P. Eisenberger, P. M. Platzman and H. Winick, *Phys. Rev. Lett.* **36**, 623 (1976).
63. J. P. Briand, D. Girard, V. O. Kostroun, P. Chevalier, K. Wohrer and J. P. Mossé, *Phys. Rev. Lett.* **46**, 1625 (1981).
64. Y. B. Bannett and I. Freund, *Phys. Rev. Lett.* **34**, 372 (1975).
65. J. Tulkki and T. Åberg, *J. Phys.* **B15**, L435 (1982).
66. K. Hämäläinen, D. P. Siddons, J. B. Hastings and L. E. Berman, *Phys. Rev. Lett.* **67**, 2850 (1991).
67. P. Skytt, P. Glans, J.-H. Guo, K. Gunnelin, N. Wassdahl and J. Nordgren, in *Raman Emission by X-ray Scattering* (Ref. 38), p. 193.
68. J. Tulkki, *Phys. Rev.* **A27**, 3375 (1983).
69. M. A. MacDonald, S. H. Southworth, J. C. Levin, A. Henins, R. D. Deslattes, T. LeBrun, Y. Azuma, P. L. Cowan and B. A. Karlin, *Phys. Rev.* **A51**, 3598 (1995).
70. M. Breinig, M. H. Chen, G. E. Ice, F. Parente, B. Crasemann and G. S. Brown, *Phys. Rev.* **A22**, 520 (1980).
71. T. Mooney, E. Lindroth, P. Indelicato, E. G. Kessler, Jr. and R. D. Deslattes, *Phys. Rev.* **A45**, 1531 (1992).
72. M. Ohno and R. E. LaVilla, *Phys. Rev.* **A45**, 4713 (1992).
73. To simplify the description of characteristic XFS, we neglect satellite features due to multi-electron excitations which have energy onsets above the primary threshold (see Ref. 39).
74. F. K. Richtmyer, S. W. Barnes and E. Ramberg, *Phys. Rev.* **46**, 843 (1934).
75. T. LeBrun, in *Raman Emission by X-ray Scattering* (Ref. 38), p. 142.
76. P. Carra, M. Fabrizio and B. T. Thole, *Phys. Rev. Lett.* **74**, 3700 (1995).
77. G. B. Armen, S. H. Southworth, J. C. Levin, U. Arp, T. LeBrun and M. A. MacDonald, *Phys. Rev.* **A56**, R1079 (1997).
78. G. B. Armen, J. C. Levin and I. A. Sellin, *Phys. Rev.* **A53**, 772 (1996).
79. T. LeBrun, in *X-ray and Inner-Shell Processes*, Eds. R. L. Johnson, H. Schmidt-Böcking and B. F. Sonntag, American Institute of Physics, Woodbury, New York, 1997, p. 647.
80. S. J. Schaphorst, A. F. Kodre, J. Ruscheinski, B. Crasemann, T. Åberg, J. Tulkki, M. H. Chen, Y. Azuma and G. S. Brown, *Phys. Rev.* **A47**, 1953 (1993).
81. W. Heitler, *The Quantum Theory of Radiation*, 3rd Edition (Dover, New York, 1984), p. 204.
82. U. Fano and J. H. Macek, *Rev. Mod. Phys.* **45**, 553 (1973).
83. C. H. Greene and R. N. Zare, *Ann. Rev. Phys. Chem.* **33**, 119 (1982).
84. F. Kh. Gel'mukhanov and L. N. Mazalov, *Opt. Spectrosc. (USSR)* **42**, 371 (1977).
85. J. A. Guest, K. H. Jackson and R. N. Zare, *Phys. Rev.*, **A28**, 2217 (1983).

86. S. T. Manson and D. Dill, in *Electron Spectroscopy: Theory, Techniques and Applications*. Vol. 2, Eds. C. R. Brundle and A. D. Baker (Academic Press, New York, 1978), p. 157.
87. P. P. Feofilov, *The Physical Basis of Polarized Emission* (Consultans Bureau, New York, 1961).
88. R. N. Zare, *Angular Momentum* (Wiley, New York, 1988).
89. J. D. Mills and P. W. Langhoff, in *Raman Emission by X-ray Scattering* (Ref. 38), p. 169.
90. Y. Luo, H. Ågren and F. Gel'mukhanov, *Phys. Rev.* **A53**, 1340 (1996).
91. R. C. C. Perera, P. L. Cowan, D. W. Lindle, R. E. laVilla, T. Jach and R. D. Deslattes, *Phys. Rev.* **A43**, 3609 (1991).
92. F. Gel'mukhanov and H. Ågren, *Phys. Lett.* **A185**, 407 (1994).
93. E. D. Poliakoff, in *Raman Emission by X-ray Scattering* (Ref. 38), p. 220.
94. Y. Ma, *Chem. Phys. Lett.* **230**, 451 (1994).
95. P. Glans, K. Gunnelin, P. Skytt, J.-H. Guo, N. Wassdahl, J. Nordgren, H. Ågren, F. Kh. Gel'mukhanov, T. Warwick and E. Rotenberg, *Phys. Rev. Lett.* **76**, 2448 (1996).
96. P. Glans, P. Skytt, K. Gunnelin, J.-H. Guo and J. Nordgren, in *X-ray and Inner-Shell Processes*, Eds. R. L. Johnson, H. Schmidt-Böcking and B. F. Sonntag, American Institute of Physics, Woodbury, New York, 1997, p. 723.
97. P. Skytt, P. Glans, J.-H. Guo, K. Gunnelin, C. Såthe, J. Nordgren, F. Kh. Gel'mukhanov, A. Cesar and H. Ågren, *Phys. Rev. Lett.* **77**, 5035 (1996).
98. L. S. Cederbaum, *J. Chem. Phys.* **103**, 562 (1995).
99. S. Bodeur, J. L. Maréchal, C. Reynaud, D. Bazin and I. Nenner, *Z. Phys.* **D17**, 291 (1990).
100. B. M. Kincaid and P. Eisenberger, *Phys. Rev. Lett.* **34**, 1361 (1975).
101. L. Frost, A. M. Grisogono, E. Weigold, C. E. Brion, A. O. Bawagan, P. Tomasello and W. von Niessen, *Chem. Phys.* **119**, 253 (1988).

INDEX